RTC Limerick
3 9002 00004099 9

The Finite Element Method
Fourth Edition

Volume 2

Solid and Fluid Mechanics

Dynamics and Non-Linearity

Plate 1 Impact dynamics of a projectile showing large plastic deformations. Three-dimensional finite element computation with an explicit code (DYNA) showing deformation and effective stresses at 6 and 20 ms. For comparison the actual deformed projectile is shown. The analysis has 6074 nodes and 4356 elements. In addition, five separate slidelines are modelled. (By permission of the University of California, Lawrence Livermore National Laboratory, Livermore and US Department of Energy.)

Plate 2 Steady state, inviscid, flow for supersonic (Mach 2) and subsonic aircraft. (NASA high-speed prototype and Boeing 747.) Surface mesh and pressure contours shown.

Details of analysis (noting symmetry):

	Supersonic aircraft	*Boeing 747*
No. of elements (tetrahedra)	76 522	388 614
No. of faces on aircraft	7262	26 060
Computing time for steady state (Cray XMP 48)	45 min	180 min

This analysis was carried out at the Institute for Numerical Methods in Engineering, University College, Swansea (J. Peraire, J. Peiro, L. Formaggia, K. Morgan, and O.C. Zienkiewicz, 'Finite element Euler computations in three dimensions', *Int. J. Meth. Eng.*, **26**, 2135–59 1988).

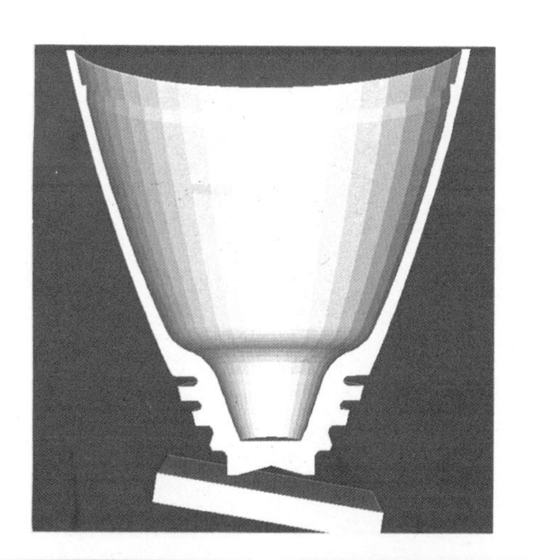

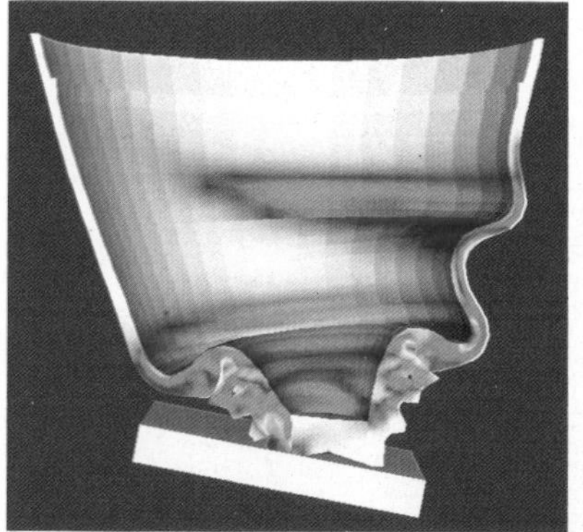

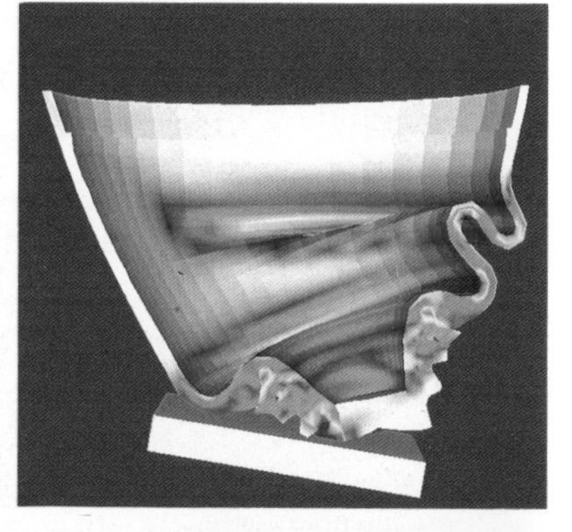

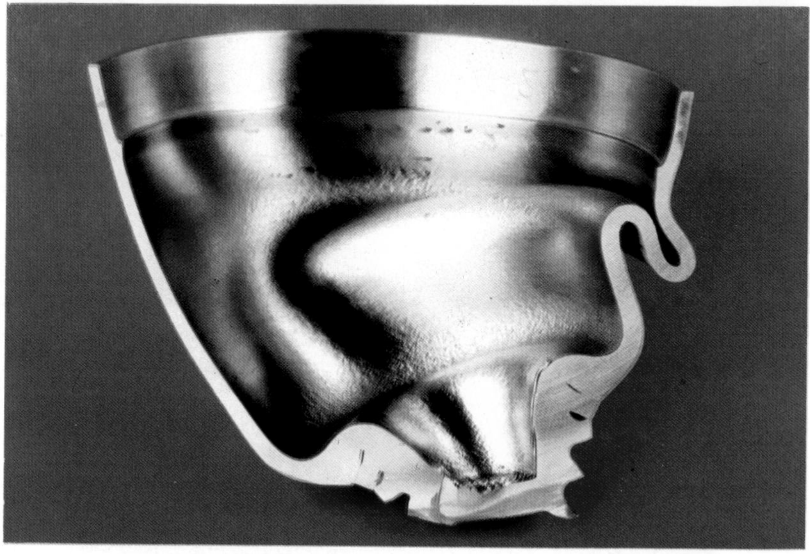

THE FINITE ELEMENT METHOD

Fourth Edition

Volume 2

Solid and Fluid Mechanics
Dynamics and Non-Linearity

O. C. Zienkiewicz, CBE, FRS

Unesco Professor of Numerical Methods in Engineering
International Centre for Numerical Methods in Engineering, Barcelona
Previously Director of the Institute of Numerical Methods in Engineering, University of Wales, Swansea

and

R. L. Taylor

Professor of Civil Engineering
University of California at Berkeley
Berkeley, California

McGRAW-HILL BOOK COMPANY

London · New York · St Louis · San Francisco · Auckland · Bogotá · Caracas
Hamburg · Lisbon · Madrid · Mexico · Milan · Montreal · New Delhi · Panama · Paris
San Juan · São Paulo · Singapore · Sydney · Tokyo · Toronto

Published by
McGRAW-HILL Book Company Europe

SHOPPENHANGERS ROAD · MAIDENHEAD · BERKSHIRE · ENGLAND
TELEPHONE: 0628 23432 FAX: 0628 770224

British Library Cataloguing in Publication Data

Zienkiewicz, O.C. (Olgierd Cecil)
The finite element method.
Vol. 2. Solid and fluid mechanics Dynamics and
non-linearity – 4th ed.
1. Engineering. Mathematics. Finite element method
I. Title II. Taylor, R.L.
680.001'515353

ISBN 0-07-084176-6

Library of Congress Cataloging-in-Publication Data

Zienkiewicz, O.C.
The finite element method.

Includes bibliographies and indexes.
Contents: v. 1. Basic formulation and linear
problems – v. 2. Solid and fluid mechanics.
1. Structural analysis (Engineering) 2. Continuum
mechanics. 3. Finite element method. I. Taylor, R.L.
II. Title.
TA640.2.Z5 1989 620'.001'515353 88-682

ISBN 0-07-084175-6 (v. 2)

2345 IP 9432

Typeset, printed and bound in Malta by Interprint Limited.

To
Our Families

Contents

Preface

The first volume of this edition covered basic aspects of finite element approximation in the context of linear, self-adjoint problems. Typical examples of two- and three-dimensional elasticity, heat conduction and electromagnetic problems in a steady state were dealt with and an 'essential' finite element computer code structure was introduced. However, many aspects of formulation and approximation had to be relegated to the second volume in which we hope the reader will find the answer to more perplexing problems, most of which are of continuing practical interest.

The 'division line' between the contents of the two volumes is by necessity not clear cut and indeed the choice of title presented some difficulty, which could only have been overcome by further classification and subdivision—for this reason the somewhat all-embracing subtitle.

In essence four general areas are covered here:

1. *Plates and shells* (Chapters 1 to 6). This section is of course of most interest to those interested in solid mechanics and deals with a specific class of problems. However, as this application was one of the first to which finite elements were directed and which still is a subject of continuing research, we considered its inclusion of importance. Those with interests in other fields may well omit this part on first reading, though by analogy the methods exposed have quite wide applications outside structural mechanics.
2. *Non-linear problems* (Chapters 7 and 8). In this the special problems of solving non-linear equations systems are addressed and we hope that the presentation is such that readers of all backgrounds will find it of interest. Indeed, such non-linear applications are today of great importance and practical interest in most areas of engineering and physics. However, problems of plasticity, viscoplasticity, etc., are covered in some detail.
3. *Time domain problems* (Chapters 9 to 11). This section, which could well be studied as a 'basic approach', focuses its attention on the time

dimension. Here the eigenvalue-vibration-type problems are treated and direct finite element approximation in the time domain considered. Obviously again application to a very wide range of problems exists and, as with the previous section, it is of interest to all. In particular we would like here to draw the readers' attention to Chapter 11 where coupled problems are dealt with. Here many of the approaches are novel and possibilities of iterative solutions of general applicability are introduced.

4. *Fluid mechanics* (*and non-self-adjoint problems*) (Chapters 12 to 15). We were much tempted to publish this section as a separate volume. This is not only because it deals with a field of application of its own wide interest but also because it extends the field of finite element applications to a difficult area in which 'variational principles' do not exist naturally. Those with interest in 'non-fluid' problems of this category (e.g. semi-conductor modelling) may find the introductory Chapter 12 of special interest.

   The whole field of computational fluid mechanics in which finite difference approximation has been the mainstay is today in a transition stage in which the advantages of finite elements are being realized.

   We hope that this presentation will show to the reader the present achievements and directions of current research.

Volume 2 concludes with a chapter on Computer Procedures, where we extend the basic program presented in Volume 1 to the time domain and non-linear problems. Clearly the variety of problems presented in the text does not permit a detailed treatment of all subjects discussed but we hope that the 'skeletal' format presented will allow readers to make their own extensions.

The reader familiar with the third edition will notice that 10 out of the 16 chapters are either new or totally rewritten. The remaining six (Chapters 3, 4, 5, 6, 8 and 9) are essentially from the third edition with some updating.

The time gap of two years between publication of the two volumes is to a large extent due to the 'ongoing' research in many of the areas discussed in Volume 2 and we trust the readers will make the necessary allowances to the authors. It is symptomatic of the subject that the matters introduced in the first edition of 1967 should still be worked on extensively as the continuing flood of published papers indicates.

We would like at this stage to thank once again our collaborators and friends for many helpful comments and suggestions. Our particular gratitude goes to Professor Peter Bettess for contributing a section of Chapter 15 summarizing some of his research work and for the work he and his wife, Jackie, have done on the subject index and to Professor Y. K. Cheung for summarizing his work on the strip methods in Chapter 6.

O. C. Z. would like to take this opportunity to thank his friends at the Texas Institute of Computation Mechanics (University of Texas, Austin) and at the International Center for Numerical Methods in Engineering (University of Catalunya, Barcelona) for providing stimulating environments where much of Volume 2 was written.

O. C. ZIENKIEWICZ and R. L. TAYLOR

1

Plate and shell bending approximation: thin (Kirchhoff) plates and C_1 continuity requirements

1.1 Introduction

The subject of bending of plates and indeed its extension to shells was one of the first to which the finite element method was applied in the early sixties. At that time the various difficulties that were to be encountered were not fully appreciated and for this reason the topic remains one in which research is active to the present day. It is therefore fitting that we should address it at the opening of this volume. Although the subject is of direct interest only to applied mechanicians and structural engineers, there is much that has more general applicability, and many of the procedures which we shall introduce can be directly translated to other fields of application.

Plates and shells are but a particular form of a three-dimensional solid, the treatment of which presents no theoretical difficulties, at least in the case of elasticity. However, the thickness of such structures (denoted throughout this and later chapters as t) is very small when compared with other dimensions, and complete three-dimensional numerical treatment would not only be costly but in addition could lead to serious equation conditioning problems. To ease the solution, even long before numerical approaches became possible, several classical assumptions regarding the behaviour of such structures were introduced. Clearly such assumptions result in a series of approximations. Thus the numerical treatment which we shall discuss will, in general, concern itself with the approximation to an already approximate theory (or mathematical model), the validity of which is restricted. On occasion we shall point out the shortcomings of the original assumptions, and indeed modify these as necessary or convenient. This can be done simply because now we are granted more freedom than that which existed in the 'precomputer' era.

The *thin plate* theory is based on assumptions formalized by Kirchhoff in 1850,[1] and indeed his name is often associated with this theory, though an early version was presented by Sophie Germain in 1811.[2,3] A relaxation of the assumptions was made by Reissner in 1945[4] and in a slightly different manner by Mindlin[5] in 1951. These modified theories extend the field of application of the theory to *thick plates* and we shall associate this name with the Reissner–Mindlin postulates.

It turns out that the *thick* plate theory is simpler to implement in the finite element method, though in the early days of analytical treatment it presented more difficulties. As it is more convenient to introduce first the thick plate theory and by imposition of additional assumptions to limit it to *thin* plate theory we shall follow this path in the present chapter. However, when discussing numerical solution we shall reverse the process and follow the historical procedure of dealing with the thin plate situations first in this chapter. The extension to thick plates and to what turns out to be a *mixed* formulation will be the subject of Chapter 2.

In the thin plate theory it is possible to represent the state of deformation by one quantity w. This is the lateral displacement of the middle plane of the plate. Clearly such a formulation is *irreducible*. The achievement of this irreducible form introduces second derivatives of w to the strain definition and continuity conditions between elements have now to be imposed not only on this quantity but on its derivatives. This is to ensure that the plate remains continuous and does not 'kink'.† Thus at nodes on element interfaces it will always be necessary to use both the values of w and of its slopes to impose continuity.

Determination of suitable shape functions is now much more complex. Indeed, as complete slope continuity is required on the interfaces between various elements, the mathematical and computational difficulties often rise disproportionately fast. It is, however, relatively simple to obtain shape functions which, while preserving continuity of w, may violate its slope continuity between elements, though naturally not at the node where such continuity is imposed. If such chosen functions satisfy the 'patch test' then convergence will still be found. The first part of this chapter will be concerned with such 'non-conforming' or 'incompatible' shape functions. In further parts new functions will be introduced by which continuity can be restored. The solution with such 'conforming' shape functions will now give bounds to the energy of the correct

† If 'kinking' occurs the second derivative or curvature becomes infinite and certain infinite terms occur in the energy expression.

solution, but, on many occasions, will yield inferior accuracy to that achieved with non-conforming elements. For practical usage the methods of the first part of the chapter are often recommended.

The simplest type of element shape is a rectangle and this will be introduced first. Triangular and quadrilateral elements are more complex and will be introduced later for solutions of plates of arbitrary shape or, for that matter, for dealing with shell problems where such elements are essential.

The problem of thin plates, where the potential energy functional contains *second derivatives* of the unknown functions, is characteristic of a large class of physical problems associated with *fourth-order differential equations.* Thus, although the chapter concentrates on the structural problem, the reader with other physical problems in mind will find that the procedures developed will be equally applicable elsewhere.

The difficulty of imposing C_1 continuity on the shape functions has resulted in many alternative approaches to the problems in which this difficulty is side-stepped. Several possibilities exist. Two of the most important are:

(*a*) independent interpolation of rotations $\boldsymbol{\theta}$ and displacements w imposing continuity as a special constraint often applied at discrete points only and

(*b*) the introduction of lagrangian variables or indeed other variables to avoid the necessity of C_1 continuity.

Both approaches fall into the class of mixed formulations and we shall discuss these briefly at the end of the chapter. However, a fuller statement of mixed approaches will be made in the second chapter where both thick and thin approximations will be dealt with simultaneously.

1.2 The plate problem: *thick* and *thin* formulations

1.2.1 *Governing equations.* The mechanics of plate and shell action is perhaps best illustrated in one dimension as shown in Fig. 1.1. Here we show a long beam of unit width subject to some stress resultants M_x, P_x and S_x.

It is intuitively obvious that at some distance from the point of application of such resultants plane sections will remain plane during the deformation process. This is particularly important for the sections that originally are normal to the middle plane of the plate.

The postulate that sections normal to the middle plane remain plane during deformation is thus the *first* and most important assumption of the theory of plates and shells. To this is added the *second* assumption. This simply observes that the direct stresses in the normal direction, z, are small, that is of the order of applied lateral load intensities, q, and

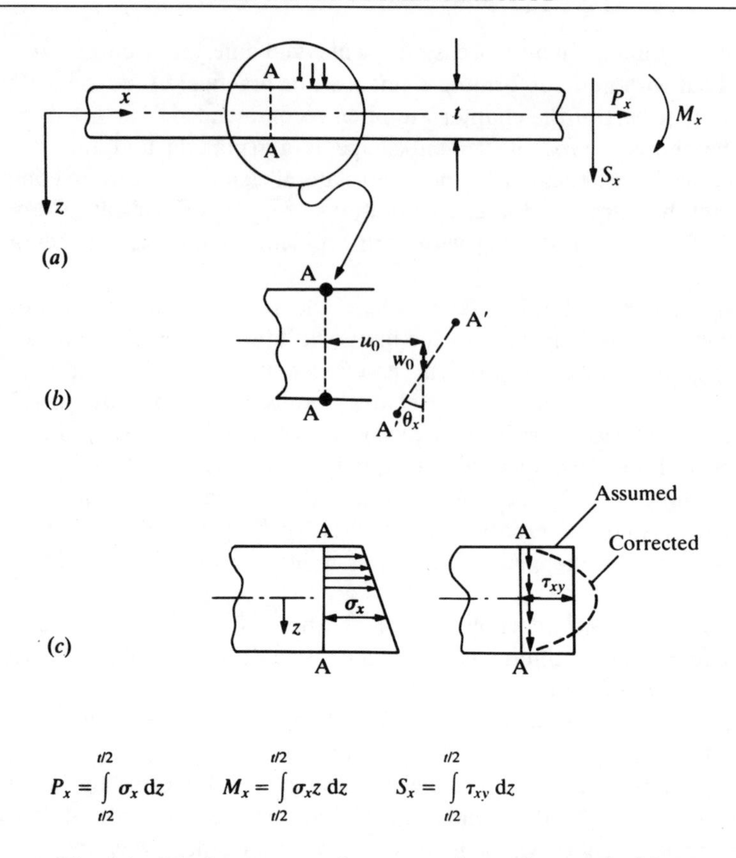

Fig. 1.1 Displacements and stress resultants for a typical beam

hence direct strains in that direction can be neglected. This 'inconsistency' in approximation is compensated for by assuming plane stress conditions in each lamina.

With these two assumptions it is easy to see that the total state of deformation can be described by displacements u_0 and w_0 of the middle surface ($z=0$) and a rotation θ_x of the normal. Thus the local displacements in the directions of the x and z axes are

$$
\begin{aligned}
u &= -\theta_x z + u_0 \qquad & \text{with } u_0 &= u_0(x) \\
w &= w_0 & w_0 &= w_0(x) \\
& & \theta_x &= \theta_x(x)
\end{aligned}
\tag{1.1}
$$

Immediately the strains in the x and z component directions are available as

$$\varepsilon_x = \frac{\partial u}{\partial x} = -\frac{\partial \theta_x}{\partial x} z + \frac{\partial u_0}{\partial x}$$

$$\varepsilon_z = 0 \qquad (1.2)$$

$$\gamma_{xz} = \frac{\partial u}{\partial z} + \frac{\partial w}{\partial x} = -\theta_x + \frac{\partial w_0}{\partial x}$$

Writing the appropriate constitutive relations the stresses σ_x and τ_{xz} can be evaluated and hence the stress resultants are obtained as

$$M_x = -\int_{-t/2}^{t/2} \sigma_x z \, dz = -\frac{Et^3}{12} \frac{\partial \theta_x}{\partial x} \qquad (1.3a)$$

$$P_x = Et \frac{\partial u_0}{\partial x} \qquad (1.3b)$$

$$S_x = \beta Gt \left(-\theta_x + \frac{\partial w_0}{\partial x} \right) \qquad (1.3c)$$

where E and G are direct and shear elastic moduli respectively.†

Three equations of equilibrium complete the formulation:

$$\frac{\partial M_x}{\partial x} + S_x = 0 \qquad (1.4a)$$

$$\frac{\partial S_x}{\partial x} + q = 0 \qquad (1.4b)$$

$$\frac{\partial P_x}{\partial x} = 0 \qquad (1.4c)$$

In the elastic case of a straight beam it is easy to see that the in-plane displacements and forces u_0 and P_x decouple and the problem of lateral deformations can be dealt with separately. We shall thus only consider bending in the present chapter, returning to the combined problem, characteristic of shell behaviour, later.

Equations (1.1) to (1.4) are typical for thick beams and the well-known thin beam theory adds an additional assumption. This simply neglects the

† A constant β has been added here to account for the fact that the shear stresses are not constant across the section. A value of $\beta = \frac{5}{6}$ is exact for a rectangular, homogeneous section and corresponds to a parabolic shear stress distribution.

shear deformation and puts $G=\infty$. Equation (1.3c) thus becomes

$$-\theta_x + \frac{\partial w}{\partial x} = 0 \tag{1.5}$$

This thin beam assumption is equivalent to stating that the normals to the middle plane remain normal to it during deformation and is the well-known Bernoulli–Euler assumption. Thin, constrained theory is very widely used in practice and proves adequate for a large number of structural problems, though of course should not be taken literally as the true behaviour near supports or where local load action is important and is three dimensional.

In Fig. 1.2 we illustrate some of the boundary conditions imposed on

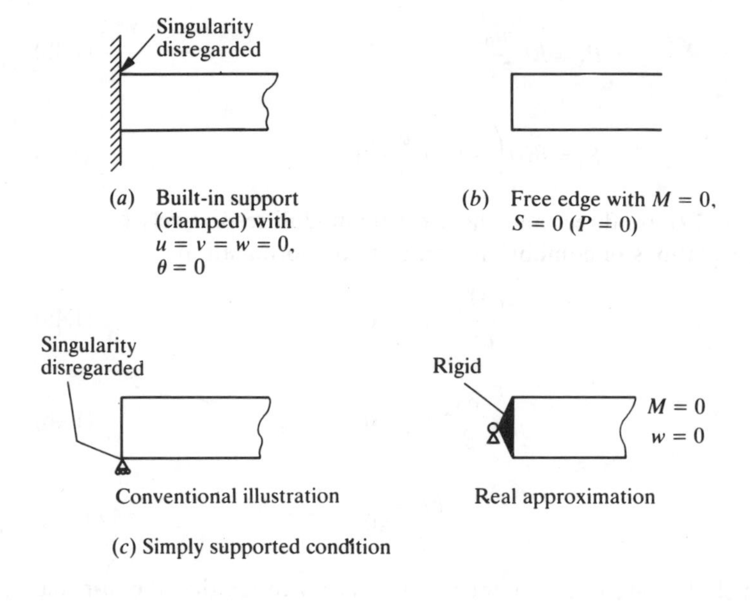

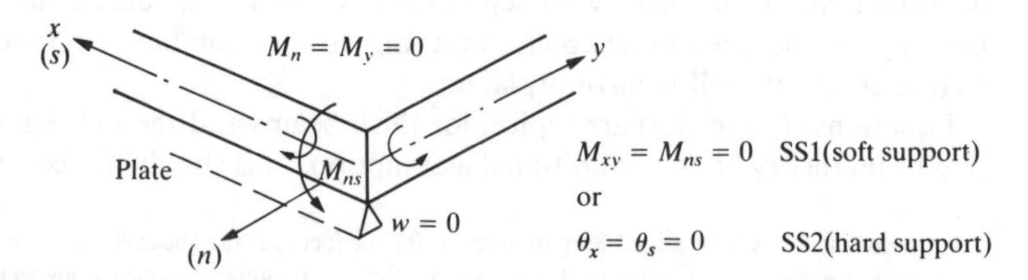

Fig. 1.2 Support (end) conditions for a beam and a plate. (*Note*: conventionally illustrated simple support leads to infinite displacement—reality is different)

beams (and plates) and immediately note that the diagrammatic representation of a simple support as a knife edge would lead to infinite displacements and stresses. Of course, if a rigid bracket is added in the manner shown this will alter the behaviour to that which we shall generally assume.

The one-dimensional problem of beams and the introduction of thick and thin assumptions translates directly to plates. In Fig. 1.3 we illustrate the extensions necessary and write, in place of Eq. (1.1) (assuming u_0 and v_0 to be zero.

$$u = -\theta_x z \qquad v = -\theta_y z \qquad w = w_0 \tag{1.6}$$

with θ_x, θ_y and w_0 being functions of x and y only.

The strains will now be separated into in plane (bending) and trans-

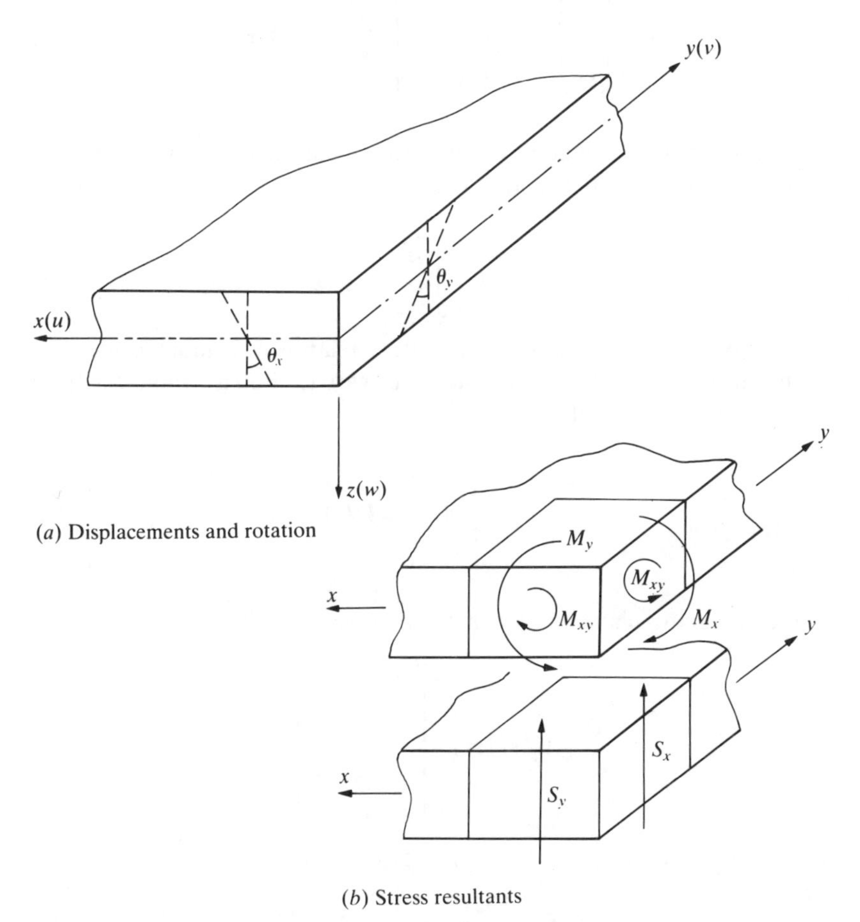

(b) Stress resultants

Fig. 1.3 Definitions of variables for plate approximation

verse shear groups and we have, in place of Eq. (1.2),

$$\boldsymbol{\varepsilon}=\begin{Bmatrix}\varepsilon_x\\ \varepsilon_y\\ \gamma_{xy}\end{Bmatrix}=-z\begin{bmatrix}\dfrac{\partial}{\partial x} & 0\\ 0 & \dfrac{\partial}{\partial y}\\ \dfrac{\partial}{\partial y} & \dfrac{\partial}{\partial x}\end{bmatrix}\begin{Bmatrix}\theta_x\\ \theta_y\end{Bmatrix}\equiv -z\mathbf{L}\boldsymbol{\theta} \tag{1.7a}$$

and

$$\boldsymbol{\gamma}=\begin{Bmatrix}\gamma_{xz}\\ \gamma_{yz}\end{Bmatrix}=-\begin{Bmatrix}\theta_x\\ \theta_y\end{Bmatrix}+\begin{Bmatrix}\dfrac{\partial w}{\partial x}\\ \dfrac{\partial w}{\partial y}\end{Bmatrix}=-\boldsymbol{\theta}+\boldsymbol{\nabla} w \tag{1.7b}$$

We note that now in addition to normal bending moments, defined by expression (1.3a) in the x and y directions, a twisting moment arises defined by

$$M_{xy}=-\int_{-t/2}^{t/2}\tau_{xy}z\,\mathrm{d}z \tag{1.8}$$

and introducing appropriate constitutive relations all moment components can be related to displacement derivatives. For isotropic elasticity we can thus write, in place of Eq. (1.3),

$$\mathbf{M}=\begin{Bmatrix}M_x\\ M_y\\ M_{xy}\end{Bmatrix}=\mathbf{DL}\boldsymbol{\theta} \tag{1.9}$$

where, assuming plane stress behaviour in each layer,

$$\mathbf{D}=\frac{Et^3}{12(1-\nu^2)}\begin{bmatrix}1 & \nu & 0\\ \nu & 1 & 0\\ 0 & 0 & \dfrac{1-\nu}{2}\end{bmatrix}$$

where ν is Poisson's ratio. Further, the shear force resultants are

$$\mathbf{S}=\begin{Bmatrix}S_x\\ S_y\end{Bmatrix}=\boldsymbol{\alpha}(-\boldsymbol{\theta}+\boldsymbol{\nabla} w) \tag{1.10}$$

For isotropic elasticity (though here we deliberately have not related G to E and ν to allow for possibly different shear rigidities)

$$\boldsymbol{\alpha}=\alpha=\beta G t \tag{1.11}$$

Of course, the constitutive relations can be simply generalized to anisotropic or inhomogeneous behaviour such as can be manifested if several layers of materials are assembled to form a *composite*. The only apparent difference is the structure of $\mathbf{D}$ and $\boldsymbol{\alpha}$ matrices, which can always be found by simple integration.

The governing equations of thick and thin plate behaviour are completed by writing the equilibrium relations. Again omitting the 'in-plane' behaviour we have, in place of Eq. (1.4a),

$$\begin{bmatrix} \dfrac{\partial}{\partial x} & 0 & \dfrac{\partial}{\partial y} \\ 0 & \dfrac{\partial}{\partial y} & \dfrac{\partial}{\partial x} \end{bmatrix} \begin{Bmatrix} M_x \\ M_y \\ M_{xy} \end{Bmatrix} + \begin{Bmatrix} S_x \\ S_y \end{Bmatrix} \equiv \mathbf{L}^{\mathrm{T}}\mathbf{M}+\mathbf{S}=0 \tag{1.12}$$

and, in place of Eq. (1.4b),

$$\left[\frac{\partial}{\partial x}, \frac{\partial}{\partial y}\right] \begin{Bmatrix} S_x \\ S_y \end{Bmatrix} + q \equiv \boldsymbol{\nabla}^{\mathrm{T}}\,\mathbf{S}+q=0 \tag{1.13}$$

The equations (1.9) to (1.13) are the basis from which the solution of both thick and thin plates can start. For thick plates any (or all) of the independent variables can be approximated independently, leading to a mixed formulation which we shall discuss in Chapter 2 and also briefly in Sec. 1.16 of this chapter.

For thin plates in which the shear deformations are suppressed Eq. (1.10) is rewritten as

$$-\boldsymbol{\theta}+\boldsymbol{\nabla}\mathbf{w}=\mathbf{0} \tag{1.14}$$

and both irreducible and mixed forms can now be written. In particular it is an easy matter to eliminate $\mathbf{M}$, $\mathbf{S}$ and $\boldsymbol{\theta}$ and leave only w as the variable.

Applying the operator $\boldsymbol{\nabla}^{\mathrm{T}}$ to (1.12), inserting (1.13) and (1.9) and finally replacing $\boldsymbol{\theta}$ by the use of (1.14) gives a scalar equation

$$(\mathbf{L}\boldsymbol{\nabla})^{\mathrm{T}}\,\mathbf{D}\mathbf{L}\boldsymbol{\nabla}w+q=0 \tag{1.15}$$

where

$$(\mathbf{L}\boldsymbol{\nabla})^{\mathrm{T}} \equiv \left[\frac{\partial^2}{\partial x^2}, \frac{\partial^2}{\partial y^2}, 2\frac{\partial^2}{\partial x \partial y}\right]$$

In the case of constant **D** this becomes the well-known biharmonic equation of plate flexure

$$\frac{\partial^4 w}{\partial x^4}+2\frac{\partial^4 w}{\partial x^4 \partial y^4}+\frac{\partial^4 w}{\partial y^4}+q\frac{12(1-\nu^2)}{Et^3}=0 \qquad (1.16)$$

In the first part of the chapter we shall be concerned with the above formulation [starting from Eq. (1.15)], and the presence of the fourth derivative indicates quite clearly that, even after integration by parts, we shall need C_1 continuity of the shape functions.

1.2.2 *The boundary conditions.* The boundary conditions which have to be imposed on the problem (viz. Figs 1.2 and 1.4) range from:

(*a*) *traction boundary*, where stress resultants M_n, M_{ns} and S_n are given prescribed values; here n and s are directions orthogonal and tangential to the boundary (a free edge being a special case with zero values assigned);

(*b*) *fixed boundary*, where the displacements conjugate to the stress resultants, i.e. θ_n, θ_s and w, are specified† (a clamped edge being a special case with zero values assigned);

to

(*c*) '*mixed*' *boundary conditions*, where both tractions and displace-

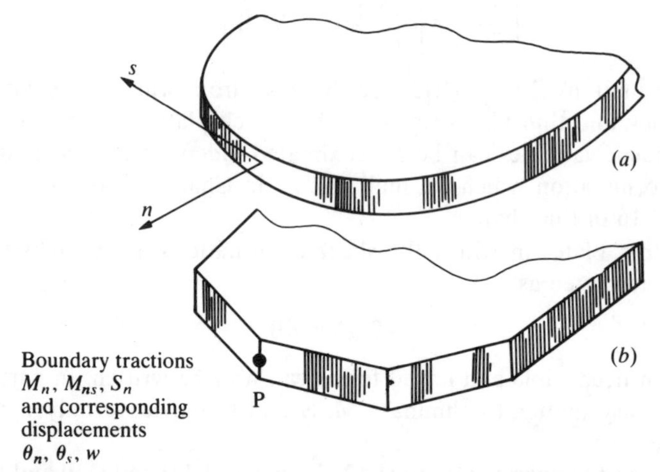

Fig. 1.4 Boundary traction and conjugate displacement. (*Note*: the simply supported condition requiring $M_n=0$, $\theta_s=0$ and $w=0$ is identical at a corner node to specifying $\theta_n=\theta_s=0$, i.e. a clamped support. This leads to a paradox if a curved boundary (*a*) is modelled as a polygon (*b*)

† Note that in thin plates the specification of w along s automatically specifies θ_s by Eq. (1.14), but this is not the case in thick plates where the quantities have to be independently prescribed.

ments can be specified. Typical here is the simply supported edge (viz. Fig. 1.2). For this clearly $M_n = 0$ and $w = 0$, but it is less clear whether M_{ns} or θ_s needs to be given. Specification of $M_{ns} = 0$ is *physically* a more acceptable condition and does not lead to difficulties. This should always be adopted for thick plates. In thin plates θ_s is automatically specified as zero and we shall find certain difficulties, and indeed anomalies, associated with this assumption.[6,7] For instance, in Fig. 1.4 we see how a specification of $\theta_s = 0$, at corner nodes implicit in thin plates, leads to the prescription of all boundary parameters which is identical to boundary conditions of a clamped plate.

1.2.3 *The irreducible, thin plate approximation.* The thin plate approximation when cast in terms of a single variable w is clearly irreducible and is in fact typical of a displacement formulation. The equations (1.12) and (1.13) can be written together as

$$(\mathbf{L}\nabla)^{\mathrm{T}}\mathbf{M} + q = 0 \tag{1.17}$$

and the constitutive relation (1.9) can be recast using (1.14) as

$$\mathbf{M} = \mathbf{D}\mathbf{L}\nabla w \tag{1.18}$$

With a discretization

$$w = \mathbf{N}\bar{\mathbf{a}} \tag{1.19}$$

where $\bar{\mathbf{a}}$ are appropriate parameters, we can obtain for a linear case standard displacement approximation equations

$$\mathbf{K}\bar{\mathbf{a}} = \mathbf{f} \tag{1.20}$$

with

$$\mathbf{K}\bar{\mathbf{a}} = \left(\int_\Omega \mathbf{B}^{\mathrm{T}}\,\mathbf{D}\mathbf{B}\,\mathrm{d}\Omega\right)\bar{\mathbf{a}} \equiv \int_\Omega \mathbf{B}^{\mathrm{T}}\,\mathbf{M}\,\mathrm{d}\Omega \tag{1.21a}$$

and

$$\mathbf{f} = \int_\Omega \mathbf{N}^{\mathrm{T}} q\,\mathrm{d}\Omega + \mathbf{f}_{\mathrm{b}} \tag{1.21b}$$

where $\mathbf{f}_{\mathrm{b}}$ is the boundary contribution to be discussed later and

$$\mathbf{M} = \mathbf{D}\mathbf{B}\bar{\mathbf{a}} \tag{1.21c}$$

with

$$\mathbf{B} = (\mathbf{L}\nabla)\mathbf{N} \tag{1.21d}$$

The derivation of the above can follow either a weak form of Eq. (1.17) obtained by weighting with $\mathbf{N}^T$ and integration by parts (done twice) or, more directly, by application of the virtual work equivalence. Whichever way is adopted the reader will recognize the well-known ingredients of a displacement formulation (see Chapter 2 of Volume 1) and the procedures are almost automatic once $\mathbf{N}$ is chosen.

It is of interest, and indeed important to note, that when tractions are prescribed to non-zero values the force term $\mathbf{f}_b$ includes all prescribed values of M_n, M_{ns} and S irrespective of whether the thick or thin formulation is used. The reader can verify that this term is

$$\mathbf{f}_b = \int_\Gamma \left(\boldsymbol{\theta}_n^* \bar{M}_n + \boldsymbol{\theta}_s^* \bar{M}_{ns} + \mathbf{w}^* S_n \right) \mathrm{d}\Gamma \tag{1.22}$$

where $\bar{M}_n$, etc., are prescribed values and the virtual rotation and displacements are given by

$$\boldsymbol{\theta}_n^* = \frac{\partial}{\partial n}\mathbf{N}^T \qquad \boldsymbol{\theta}_s^* = \frac{\partial}{\partial s}\mathbf{N}^T \qquad \mathbf{w}^* = \mathbf{N}^T \tag{1.23}$$

for thin plates [though of course relation (1.22) is valid also for thick plates].

1.2.4 *Continuity requirement for shape functions* (C_1 *continuity*). As we have already mentioned, it is necessary for the shape functions to be C_1 continuous for the irreducible, thin plate, formulation. This continuity is difficult to achieve and reasons for this are given below.

To ensure the continuity of both w and its normal slope across an interface we must have both w and $\partial w/\partial n$ uniquely defined by values of nodal parameters along such an interface. Consider Fig. 1.5 depicting the side 1–2 of a rectangular element. The normal direction n is in fact that of y and we desire w and $\partial w/\partial y$ to be uniquely determined by values of w, $\partial w/\partial x$, $\partial w/\partial y$ at the nodes lying along this line.

Following the principles expounded in Chapter 7 of Volume 1, we

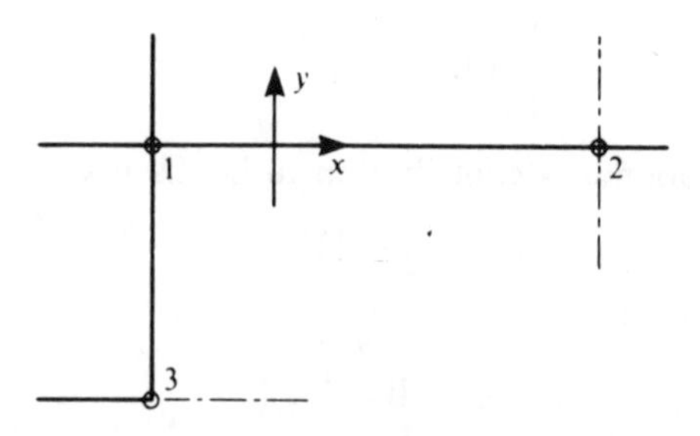

Fig. 1.5 Continuity requirement for normal slopes

would write along side 1–2,

$$w = A_1 + A_2 x + A_3 x^2 + \cdots$$

and
$$\frac{\partial w}{\partial y} = B_1 + B_2 x + B_3 x^2 + \cdots \tag{1.24}$$

with a number of constants in each expression just sufficient to determine the expressions by nodal parameters associated with the line.

Thus, for instance, if only two nodes are present a cubic variation of w would be permissible noting that $\partial w/\partial x$ and w are specified at each node. Similarly, only a linear, or two-term, variation of $\partial w/\partial y$ would be permissible.

Note, however, that a similar exercise could be performed along the side placed in the y direction preserving continuity of $\partial w/\partial x$ along this. We thus have along side 1–2,

$\dfrac{\partial w}{\partial y}$ depending on nodal parameters of line 1–2 only

and along side 1–3,

$\dfrac{\partial w}{\partial x}$ depending on nodal parameters of line 1–3 only.

Differentiating the first with respect to x we have on line 1–2,

$\dfrac{\partial^2 w}{\partial x \partial y}$ depending on nodal parameters of line 1–2 only

and on line 1–3 similarly,

$\dfrac{\partial^2 w}{\partial y \partial x}$ depending on nodal parameters of line 1–3 only.

At the common point, 1, an inconsistency arises immediately as we cannot automatically have there the necessary identity for continuous functions

$$\frac{\partial^2 w}{\partial x \partial y} \equiv \frac{\partial^2 w}{\partial y \partial x} \tag{1.25}$$

for arbitrary values of parameters at nodes 2 and 3. *It is thus impossible to specify simple polynomial expressions for shape functions ensuring full compatibility when only w and its slopes are prescribed at corner nodes.*[8]

Thus if any functions satisfying the compatibility are found with the three nodal variables, they must be such that at corner nodes these functions are not continuously differentiable and the cross-derivative is not unique.

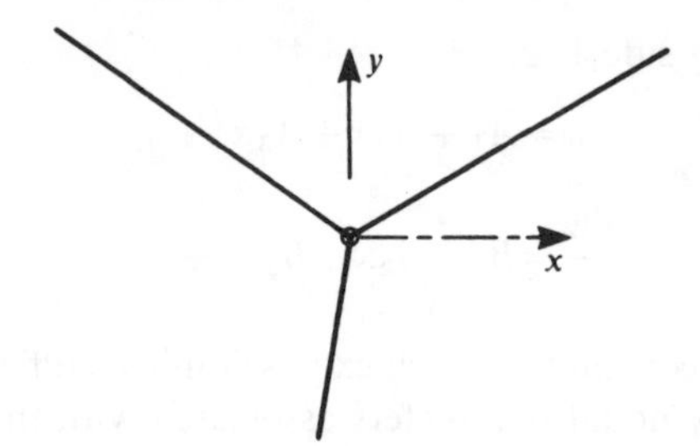

Fig. 1.6 Nodes where elements meet in arbitrary directions

Some such functions are discussed in the second part of this chapter.[9-15]

The above proof has been given for a rectangular element. Clearly the arguments can be extended for any two arbitrary directions of interfaces at the corner node 1.

A way out of this difficulty appears to be obvious. We could specify the cross-derivative as one of the nodal parameters. This, for an assembly of rectangular elements, is convenient and indeed permissible. Simple functions of that type have been suggested by Bogner *et al.*[16] and used with some success. Unfortunately the extension to nodes at which a number of element interfaces meet with different angles (Fig. 1.6) is not in general permissible. Here the continuity of cross-derivatives in several sets of orthogonal directions implies in fact a specification of *all second derivatives at a node.*

This, however, violates physical requirements if the plate stiffness varies abruptly from element to element, for then equality of moments normal to the interfaces cannot be maintained. However, this process has been used with some success in homogeneous plate situations,[17-24] although Smith and Duncan[17] comment adversely on the effect of imposing such *excessive continuities* on several orders of higher derivatives.

The difficulties of finding compatible displacement functions have led to many attempts at ignoring the complete slope continuity while still continuing with the other necessary criteria. Proceeding perhaps from a naïve but intuitive idea that the imposition of slope continuity at nodes only must, in the limit, lead to a complete slope continuity, several very successful, 'non-conforming', elements have been developed.[10,25-39]

The convergence of such elements is not obvious but can be proved either by the application of the patch test or by comparison with finite difference algorithms. We have discussed the importance of the patch test extensively in Chapter 11 of Volume 1. Full details are available in references 40 to 42.

In plate problems the importance of the patch test in both design and testing of elements is paramount and this test should never be omitted.

In the first part of this chapter dealing with non-conforming elements we shall repeatedly make use of it. Indeed, we shall show how some of the most successful elements currently used have developed via this analytical interpretation.[43-48]

NON-CONFORMING SHAPE FUNCTIONS

1.3 Rectangular element with corner nodes (12 DOF)[25,36-38]

1.3.1 *Shape functions.* Consider a rectangular element of a plate *ijkl* coinciding with the xy plane as shown in Fig. 1.7. At each node, n, displacements $\mathbf{a}_n$ are introduced. These have three components: the first a displacement in the z direction, w_n, the second a rotation about the x axis, $(\hat{\theta}_x)_n$,† and the third a rotation about the y axis $(\hat{\theta}_y)_n$.

The nodal displacement vectors are defined below as $\mathbf{a}_i$. The element displacements will, as usual, be given by a listing of the nodal displace-

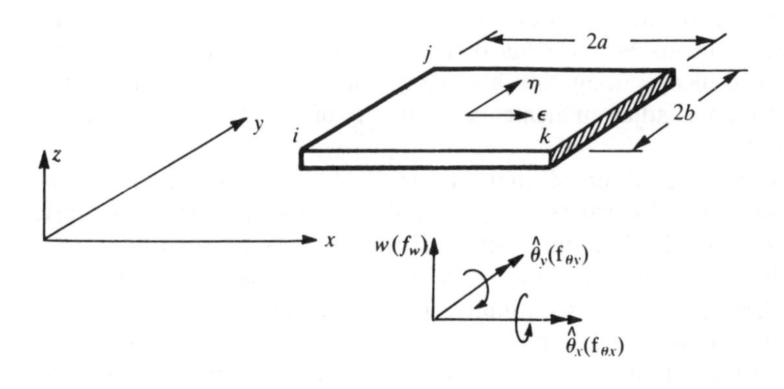

Fig. 1.7 A rectangular plate element

† Note that we have changed here the convention from that of Fig. 1.2 and we shall follow the definition of Fig. 1.7 in this chapter. This allows transformations needed for shells to be carried out in an easier manner. However, when manipulating the equations in Chapter 2 we shall return to the original definitions of Fig. 1.2. Similar difficulties are discussed by Hughes[49] and a simple transformation is shown below:

$$\hat{\boldsymbol{\theta}} = T\boldsymbol{\theta} \qquad \text{into} \qquad T = \begin{bmatrix} 0 & 1 \\ -1 & 0 \end{bmatrix}$$

ments, now totalling four:

$$\mathbf{a}^e = \begin{Bmatrix} \mathbf{a}_i \\ \mathbf{a}_j \\ \mathbf{a}_l \\ \mathbf{a}_k \end{Bmatrix} \qquad \mathbf{a}_i - \begin{Bmatrix} w \\ \hat{\theta}_x \\ \hat{\theta}_y \end{Bmatrix}_i \tag{1.26}$$

A polynomial expression is conveniently used to define the shape functions in terms of the 12 parameters. Certain terms must be omitted from a complete fourth-order polynomial. Writing

$$\begin{aligned} w &= \alpha_1 + \alpha_2 x + \alpha_3 y + \alpha_4 x^2 + \alpha_5 xy + \alpha_6 y^2 + \alpha_7 x^3 + \alpha_8 x^2 y \\ &\quad + \alpha_9 xy^2 + \alpha_{10} y^3 + \alpha_{11} x^3 y + \alpha_{12} xy^3 \\ &\equiv \mathbf{P}\boldsymbol{\alpha} \end{aligned} \tag{1.27}$$

has certain advantages. In particular, along any $x = \text{constant}$ or $y = \text{constant}$ line, the displacement w will vary as a cubic. The element boundaries or interfaces are composed of such lines. As a cubic is uniquely defined by four constants, the two end values of slopes and displacements at the ends of the boundaries will therefore define the displacements along this boundary uniquely. As such end values are common to adjacent elements, continuity of w will be imposed along any interface.

It will be observed that the gradient of w normal to any of the boundaries also varies along it in a cubic way. (Consider, for instance, $\partial w/\partial x$ along a line on which x is constant.) As on such lines only two values of the normal slope are defined, the cubic is not specified uniquely and, in general, a discontinuity of normal slope will occur. The function is thus 'non-conforming'.

The constants α_1 to α_{12} can be evaluated by writing down the 12 simultaneous equations linking the values of w and its slopes at the nodes when the coordinates take up their appropriate values. For instance,

$$w_i = \alpha_1 + \alpha_2 x_i + \alpha_3 y_i + \text{etc.}$$

$$\left(-\frac{\partial w}{\partial y}\right)_i = \hat{\theta}_{xi} = -\alpha_3 + \text{etc.}$$

$$\left(\frac{\partial w}{\partial x}\right)_i = \hat{\theta}_{yi} = \alpha_2 + \text{etc.}$$

Listing all 12 equations we can write, in matrix form,

$$\mathbf{a}^e = \mathbf{C}\boldsymbol{\alpha} \tag{1.28}$$

where $\mathbf{C}$ is a 12×12 matrix depending on nodal coordinates and $\boldsymbol{\alpha}$ is a vector of the 12 unknown constants. Inverting we have

$$\boldsymbol{\alpha} = \mathbf{C}^{-1}\mathbf{a}^e \tag{1.29}$$

This inversion can be carried out by the computer or, if an explicit expression for the stiffnesses, etc., is desired, can be performed algebraically. This was in fact done by Zienkiewicz and Cheung.[25]

It is now possible to write the expression for the displacement within the element in a standard form as

$$\mathbf{u} \equiv w = \mathbf{N}\mathbf{a}^e = \mathbf{P}\mathbf{C}^{-1}\mathbf{a}^e \tag{1.30}$$

where

$$\mathbf{P} = (1, x, y, x^2, xy, y^2, x^3, x^2y, xy^2, y^3, x^3y, xy^3)$$

An explicit form of the shape function $\mathbf{N}$ was derived by Melosh.[35]

The shape functions can be written simply in terms of normalized coordinates. Thus we can write for any node

$$\begin{aligned} \mathbf{N}_i = \tfrac{1}{8}\{(\xi_0+1)(\eta_0+1)(2+\xi_0+\eta_0-\xi^2-\eta^2),& \\ a\xi_i(\xi_0+1)^2(\xi_0-1)(\eta_0+1),& \\ b\eta_i(\xi_0+1)(\eta_0+1)^2(\eta_0-1)\}& \end{aligned} \tag{1.31}$$

with normalized coordinates defined below:

$$\xi = \frac{x - x_c}{a}$$

$$\xi_0 = \xi\xi_i$$

$$\eta = \frac{y - y_c}{b}$$

$$\eta_0 = \eta\eta_i$$

The form of $\mathbf{B}$ is obtained directly from Eqs (1.19) and (1.21d). We thus have

$$\mathbf{L}\boldsymbol{\nabla} w = \begin{Bmatrix} +2\alpha & +6\alpha_7 x & +2\alpha_8 y & +6\alpha_{11} xy & \\ -2\alpha_6 & -2\alpha_9 x & -6\alpha_{10} y & -6\alpha_{12} xy & \\ 2\alpha_5 & +4\alpha_8 x & +4\alpha_9 y & +6\alpha_{11} x^2 & +6\alpha_{12} y^2 \end{Bmatrix}$$

We can write the above as

$$\mathbf{L}\boldsymbol{\nabla} w = \mathbf{Q}\boldsymbol{\alpha} = \mathbf{Q}\mathbf{C}^{-1}\mathbf{a}^e = \mathbf{B}\mathbf{a}^e \qquad \text{and thus } \mathbf{B} = \mathbf{Q}\mathbf{C}^{-1} \tag{1.32}$$

in which

$$\mathbf{Q}=\begin{bmatrix} 0 & 0 & 0 & -2 & 0 & 0 & -6x & -2y & 0 & 0 & -6xy & 0 \\ 0 & 0 & 0 & 0 & 0 & -2 & 0 & 0 & -2x & -6y & 0 & -6xy \\ 0 & 0 & 0 & 0 & 2 & 0 & 0 & 4x & xy & 0 & 6x^2 & 6y^2 \end{bmatrix} \quad (1.33)$$

It is of interest to remark now that the displacement function chosen does in fact permit a state of constant strain (curvature) to exist† and therefore satisfies one of the criteria of convergence stated in Volume 1.

1.3.2 *Stiffness and load matrices.* Standard procedure can now be followed, and it is almost superfluous to recount the details.

The stiffness matrix relating the nodal *forces* (given by a lateral force and two moments at each node) to the corresponding nodal displacements is

$$\mathbf{K}^e = \iint_{V^e} \mathbf{B}^\mathrm{T}\mathbf{D}\mathbf{B}\,\mathrm{d}x\,\mathrm{d}y \quad (1.34)$$

or substituting Eq. (1.32),

$$\mathbf{K}^e = \mathbf{C}^{-1\mathrm{T}}\left(\int_{-b}^{b}\int_{-a}^{a} \mathbf{Q}^\mathrm{T}\mathbf{D}\mathbf{Q}\,\mathrm{d}x\,\mathrm{d}y\right)\mathbf{C}^{-1} \quad (1.35)$$

The terms not containing x and y have now been moved from the operation of integrating. The term within the integration sign can be multiplied out and integrated explicitly without difficulty, if $\mathbf{D}$ is constant.

An explicit expression for the stiffness matrix $\mathbf{K}$ has been evaluated for the case of an orthotropic material and the result is given in Table 1.1. The corresponding stress matrix for the internal moments at all the nodes is given in Table 1.2.

The external forces at nodes due to distributed loading can be assigned 'by inspection', allocating specific areas as contributing to any node. However, it is more logical and accurate to use once again the standard expression Eq. (1.21b) for such an allocation.

The contribution of these forces to each of the nodes is

$$\mathbf{f}_i = -\int_{-b}^{b}\int_{-a}^{a} \mathbf{N}^\mathrm{T} q\,\mathrm{d}x\,\mathrm{d}y \quad (1.36\mathrm{a})$$

or, by Eq. (1.30),

$$\mathbf{f}_i = -\mathbf{C}^{-1\mathrm{T}}\int_{-b}^{b}\int_{-a}^{a} \mathbf{P}^\mathrm{T} q\,\mathrm{d}x\,\mathrm{d}y \quad (1.36\mathrm{b})$$

† If α_7 to α_{12} are zero, then the 'strain' defined by second derivatives is constant. By Eq. (1.28), the corresponding $\mathbf{a}^e$ can be found. As there is a unique correspondence between $\mathbf{a}^e$ and α such a state is therefore unique. All this presumes that $\mathbf{C}^{-1}$ does in fact exist. The algebraic inversion shows that the matrix $\mathbf{C}$ is never singular.

Table 1.1
Stiffness matrix for a rectangular element
(Fig. 1.7 orthotropic material)

Stiffness matrix

$$\mathbf{K}=\frac{1}{60ab}\,\mathbf{L}\{D_x\mathbf{K}_1+D_y\mathbf{K}_2+D_1\mathbf{K}_3+D_{xy}\mathbf{K}_4\}\mathbf{L}$$

with

$$\begin{Bmatrix}\mathbf{f}_i\\ \mathbf{f}_j\\ \mathbf{f}_k\\ \mathbf{f}_l\end{Bmatrix}=\mathbf{K}\begin{Bmatrix}\mathbf{a}_i\\ \mathbf{a}_j\\ \mathbf{a}_k\\ \mathbf{a}_l\end{Bmatrix}$$

$$\mathbf{K}_1=p^{-2}\left[\begin{array}{rrrrrrrrrrrr}
60 \\
0 & 0 & & & & & & \multicolumn{5}{l}{p^{-2}=\frac{b^2}{a^2}} \\
30 & 0 & 20 \\
30 & 0 & 15 & 60 & & & & \multicolumn{5}{l}{\text{symmetric}} \\
0 & 0 & 0 & 0 & 0 \\
15 & 0 & 10 & 30 & 0 & 20 \\
-60 & 0 & -30 & -30 & 0 & -15 & 60 \\
0 & 0 & 0 & 0 & 0 & 0 & 0 & 0 \\
30 & 0 & 10 & 15 & 0 & 5 & -30 & 0 & 20 \\
-30 & 0 & -15 & -60 & 0 & -30 & 30 & 0 & -15 & 60 \\
0 & 0 & 0 & 0 & 0 & 0 & 0 & 0 & 0 & 0 & 0 \\
15 & 0 & 5 & 30 & 0 & 10 & -15 & 0 & 10 & -30 & 0 & 20
\end{array}\right]$$

$$\mathbf{K}_2=p^{2}\left[\begin{array}{rrrrrrrrrrrr}
60 \\
-30 & 20 & & & & & & \multicolumn{5}{l}{p^{-2}=\frac{a^2}{b^2}} \\
0 & 0 & 0 \\
-60 & 30 & 0 & 60 & & & & \multicolumn{5}{l}{\text{symmetric}} \\
-30 & 10 & 0 & 30 & 20 \\
0 & 0 & 0 & 0 & 0 & 0 \\
30 & -15 & 0 & -30 & -15 & 0 & 60 \\
-15 & 10 & 0 & 15 & 5 & 0 & -30 & 20 \\
0 & 0 & 0 & 0 & 0 & 0 & 0 & 0 & 0 \\
-30 & 15 & 0 & 30 & 15 & 0 & -60 & 30 & 0 & 60 \\
-15 & 5 & 0 & 15 & 10 & 0 & -30 & 10 & 0 & 30 & 20 \\
0 & 0 & 0 & 0 & 0 & 0 & 0 & 0 & 0 & 0 & 0 & 0
\end{array}\right]$$

$$\mathbf{K}_3=\left[\begin{array}{rrrrrrrrrrrr}
30 \\
-15 & 0 \\
15 & -15 & 0 \\
-30 & 0 & -15 & 30 & & & & \multicolumn{5}{l}{\text{symmetric}} \\
0 & 0 & 0 & 15 & 0 \\
-15 & 0 & 0 & 15 & 15 & 0 \\
-30 & 15 & 0 & 30 & 0 & 0 & 30 \\
15 & 0 & 0 & 0 & 0 & 0 & -15 & 0 \\
0 & 0 & 0 & 0 & 0 & 0 & -15 & 15 & 0 \\
30 & 0 & 0 & -30 & -15 & 0 & -30 & 0 & 15 & 30 \\
0 & 0 & 0 & -15 & 0 & 0 & 0 & 0 & 0 & 15 & 0 \\
0 & 0 & 0 & 0 & 0 & 0 & 15 & 0 & 0 & -15 & -15 & 0
\end{array}\right]$$

TABLE 1.1—*contd*

$$
\mathbf{K}_4 = \begin{bmatrix}
84 & & & & & & & & & & & \\
-6 & 8 & & & & & & & & & & \\
6 & 0 & 8 & & & & & & & & & \\
-84 & 6 & -6 & 84 & & & & \text{symmetric} & & & & \\
-6 & -2 & 0 & 6 & 8 & & & & & & & \\
-6 & 0 & -8 & 6 & 0 & 8 & & & & & & \\
-84 & 6 & -6 & 84 & 6 & 6 & 84 & & & & & \\
6 & -8 & 0 & -6 & 2 & 0 & -6 & 8 & & & & \\
6 & 0 & -2 & -6 & 0 & 2 & -6 & 0 & 8 & & & \\
84 & -6 & 6 & -84 & -6 & -6 & 84 & 6 & 6 & 84 & & \\
6 & 2 & 0 & -6 & -8 & 0 & -6 & -2 & 0 & 6 & 8 & \\
-6 & 0 & 2 & 6 & 0 & -2 & 6 & 0 & -8 & -6 & 0 & 8
\end{bmatrix}
$$

$$
\mathbf{L} = \begin{bmatrix} \mathbf{l} & 0 & 0 & 0 \\ 0 & \mathbf{l} & 0 & 0 \\ 0 & 0 & \mathbf{l} & 0 \\ 0 & 0 & 0 & \mathbf{l} \end{bmatrix}, \quad \text{where } \mathbf{l} = \begin{bmatrix} 1 & 0 & 0 \\ 0 & 2b & 0 \\ 0 & 0 & 2a \end{bmatrix}
$$

The integral is again evaluated simply. It will now be noted that, in general, all three components of external force at any node will have non-zero values. This is a result that the simple allocation of external loads would have missed. Table 1.3 shows the nodal load vector for a uniform loading q.

If initial strains are introduced into the plate the vector of nodal forces due to such initial strains and the initial stresses can be found in a similar way. It is necessary to remark in this connection that initial strain, such as may be due to a temperature rise, is seldom confined in its effects on curvatures. Usually, direct strains in the plate are introduced additionally, and the complete problem can be solved only by consideration of the plane stress problem as well as that of bending.

1.4 Quadrilateral and parallelogram elements

The rectangular element developed in the preceding passes the patch test[40] and is always convergent. However, it cannot be easily generalized into a quadrilateral shape. Transformation of coordinates of the type described in Chapter 8 of Volume 1 can be performed but unfortunately now it will be found that the constant curvature criterion is violated. As expected, such elements behave badly but by arguments given in Chapter 8 of Volume 1 convergence may still occur providing the patch test is passed in the curvilinear coordinates. Henshell *et al.*[39] study the performance of such an element (and also some of a higher order) and conclude

TABLE 1.2

STRESS MATRIX ($p = a/b$).

RECTANGULAR ELEMENT OF FIG. 1.7 ORTHOTROPIC MATERIAL

$$
\begin{Bmatrix} \mathbf{M}_i \\ \mathbf{M}_j \\ \mathbf{M}_k \\ \mathbf{M}_l \end{Bmatrix} = \frac{1}{4ab}
\begin{bmatrix}
\begin{matrix}6p^{-1}D_x\\+6pD_1\end{matrix} & -8aD_1 & 8bD_x & -6pD_1 & -4aD_1 & 0 & -6p^{-1}D_x & 0 & 4bD_x & 0 & 0 & 0 \\
\begin{matrix}6pD_y\\+6p^{-1}D_1\end{matrix} & -8aD_y & 8bD_1 & -6pD_y & -4aD_y & 0 & -6p^{-1}D_1 & 0 & 4bD_1 & 0 & 0 & 0 \\
-2D_{xy} & 4bD_{xy} & -4aD_{xy} & 2D_{xy} & 0 & 4aD_{xy} & 2D_{xy} & -4bD_{xy} & 0 & -2D_{xy} & 0 & 0 \\
-6pD_1 & 4aD_1 & 0 & \begin{matrix}6p^{-1}D_x\\+6pD_1\end{matrix} & 8aD_1 & 8bD_x & 0 & 0 & 0 & -6p^{-1}D_x & 0 & 4bD_x \\
-6pD_y & 4aD_y & 0 & \begin{matrix}6pD_y\\+6p^{-1}D_1\end{matrix} & 8aD_y & 8bD_1 & 0 & 0 & 0 & -6p^{-1}D_1 & 0 & 4bD_1 \\
-2D_{xy} & 0 & -4aD_{xy} & 2D_{xy} & 4bD_{xy} & 4aD_{xy} & 2D_{xy} & 0 & 0 & -2D_{xy} & -4bD_{xy} & 0 \\
-6p^{-1}D_x & 0 & -4bD_x & 0 & 0 & 0 & \begin{matrix}6p^{-1}D_x\\+6pD_1\end{matrix} & -8aD_1 & -8bD_x & -6pD_1 & -4aD_1 & 0 \\
-6p^{-1}D_1 & 0 & -4bD_1 & 0 & 0 & 0 & \begin{matrix}6pD_y\\+6p^{-1}D_1\end{matrix} & -8aD_y & -8bD_1 & -6pD_y & -4aD_y & 0 \\
-2D_{xy} & 4bD_{xy} & 0 & 2D_{xy} & 0 & 0 & 2D_{xy} & -4bD_{xy} & -4aD_{xy} & -2D_y & 0 & 4aD_{xy} \\
0 & 0 & 0 & -6p^{-1}D_x & 0 & -4bD_x & -6pD_1 & 4aD_1 & 0 & \begin{matrix}6p^{-1}D_x\\+6pD_1\end{matrix} & 8aD_1 & -8bD_x \\
0 & 0 & 0 & -6p^{-1}D_1 & 0 & -4bD_1 & -6pD_y & 4aD_y & 0 & \begin{matrix}6pD_y\\+6p^{-1}D_1\end{matrix} & 8aD_y & -8bD_1 \\
-2D_{xy} & 0 & 0 & 2D_{xy} & 4bD_{xy} & 0 & 2D_{xy} & 0 & -4aD_{xy} & -2D_{xy} & -4bD_{xy} & 4aD_{xy}
\end{bmatrix}
\begin{Bmatrix} \mathbf{a}_i \\ \mathbf{a}_j \\ \mathbf{a}_k \\ \mathbf{a}_l \end{Bmatrix}
$$

TABLE 1.3
LOAD MATRIX FOR A RECTANGULAR ELEMENT OF FIG 1.7 UNDER UNIFORM LOAD q

$$\begin{Bmatrix} \mathbf{f}_i \\ \\ \mathbf{f}_j \\ \\ \mathbf{f}_k \\ \\ \mathbf{f}_l \end{Bmatrix} = 4qab \begin{Bmatrix} 1/4 \\ -b/12 \\ a/12 \\ 1/4 \\ b/12 \\ a/12 \\ 1/4 \\ -b/12 \\ -a/12 \\ 1/4 \\ b/12 \\ -a/12 \end{Bmatrix} \qquad f_i = \begin{Bmatrix} f_{wi} \\ f_{\theta xi} \\ f_{\theta yi} \end{Bmatrix}$$

that reasonable accuracy is attainable. Their paper gives all the details of transformations required for an isoparametric mapping and the resulting need for numerical integration.

Only for the case of a parallelogram is it possible to achieve states of constant curvature exclusively using functions of ξ and η and the patch test is satisfied. Such an element is suggested in the discussion in reference 25 and the stiffness matrices have been worked out by Dawe.[27]

A somewhat different set of shape functions was suggested by Argyris.[28] For a parallelogram the local coordinates can be related to the

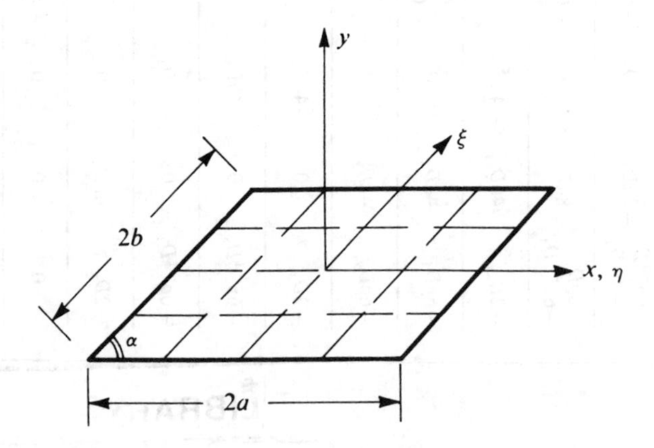

Fig. 1.8 Parallelogram element and skew coordinates

global ones by an explicit expression (Fig. 1.8)

$$\xi = \frac{x - y \cot\text{an}\,\alpha}{a}$$
$$\eta = \frac{y \operatorname{cosec} \alpha}{b} \tag{1.37}$$

and all expressions can therefore be also derived directly.

1.5 Triangular element with corner nodes (9 DOF)

1.5.1 *Shape functions.* At first sight, it would seem that once again a simple polynomial expansion could be used in a manner identical to that of the previous section. As only nine independent movements are imposed, only nine terms of the expansion are permissible. Here, an immediate difficulty arises as the full cubic expansion contains 10 terms [Eq. (1.27)] and any omission has to be made rather arbitrarily. To retain a certain symmetry of appearance all 10 terms could be retained and two coefficients made equal (for example $\alpha_8 = \alpha_9$) to limit the number of unknowns to nine. Several such possibilities have been investigated but a further, much more serious, problem arises. The matrix corresponding to **C** of Eq. (1.28) becomes singular for certain orientations of the triangle sides. This happens, for instance, when two sides of the triangle are parallel to the x and y axes.

An 'obvious' alternative is to add a central node to the formulation and eliminate this by static condensation. This would allow a complete cubic to be used, but again it was found that an element derived on this basis does not converge.

Difficulties of asymmetry can be avoided by the use of area coordinates described in Volume 1. These are indeed nearly always a natural choice for triangles, viz. Fig. 1.9.

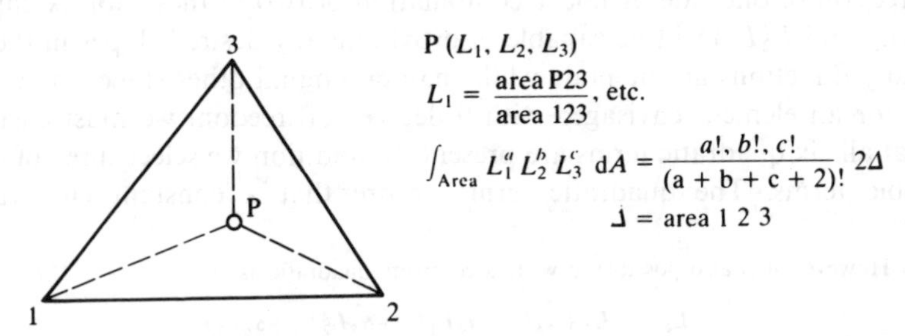

Fig. 1.9 Area coordinates

As before we shall use polynomial expansion terms, and it is worth remarking that these are given in area coordinates in an unusual form. For instance,

$$\alpha_1 L_1 + \alpha_2 L_2 + \alpha_3 L_3 \tag{1.38}$$

gives the three terms of a *complete* linear polynomial and

$$\alpha_1 L_1 L_2 + \alpha_2 L_2 L_3 + \alpha_3 L_3 L_1 + \alpha_4 L_1^2 + \alpha_5 L_2^2 + \alpha_6 L_3^2 \tag{1.39}$$

gives all six terms of a quadratic (containing within it the linear terms).† The 10 terms of a cubic expression are similarly formed by the products of all possible cubic combinations, i.e.

$$L_1^3, L_2^3, L_3^3, L_1^2 L_2, L_2^2 L_3, L_3^2 L_1, L_1 L_2^2, L_2 L_3^2, L_3 L_1^2, L_1 L_2 L_3 \tag{1.40}$$

For a 9 degree of freedom element any of the above terms can be used in a suitable combination, remembering, however, that only nine independent functions are needed and that constant curvature states have to be obtained. Figure 1.10 shows some functions that are of importance. The first [Fig. 1.10(*a*)] gives one of three functions representing a simple, unstrained, translation of the plate. Obviously these modes must be available. Further, functions of the type $L_1^2 L_2$, of which there are six in the cubic expression, will be found to take up a form similar (though not identical) to Fig. 1.10(*b*).

The cubic function $L_1 L_2 L_3$ is shown in Fig. 1.10(*c*), illustrating that this is a purely internal mode with zero values and slopes at all three corners. This function could thus be useful for a nodeless or internal variable but will not, in isolation, be used as it cannot be prescribed in terms of corner variables. It can, however, be added to any other basic shape in any proportion.

The functions of the second kind are of special interest. They have zero values of w at all corners and indeed always have a zero slope in the direction of one side. A linear combination of two of these (for example $L_1^2 L_3$ and $L_2^2 L_1$) will be capable of providing any desired slopes in the x and y directions at one node while maintaining all other slopes at zero.

For an element envisaged with 9 degrees of freedom we must ensure that all six quadratic terms are present. In addition we select three of the cubic terms. The quadratic terms ensure that a constant curvature

† However, it is also possible to write a complete quadratic as

$$\alpha_1 L_1 + \alpha_2 L_2 + \alpha_3 L_3 + \alpha_4 L_1 L_2 + \alpha_5 L_2 L_3 + \alpha_6 L_2 L_1$$

etc., for higher orders. This has the advantage of explicitly stating all retained terms of polynomials of lower order.

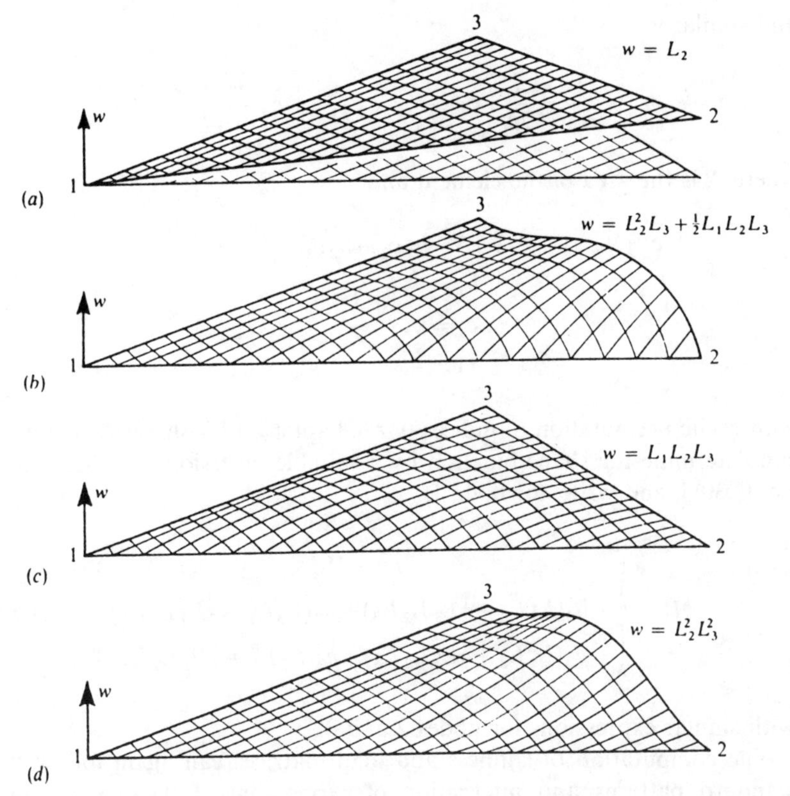

Fig. 1.10 Some basic functions in area coordinate polynomials

necessary for the patch test satisfaction is present if we write

$$
\begin{aligned}
w = & \alpha_1 L_1 + \alpha_2 L_2 + \alpha_3 L_3 + \alpha_4 L_1 L_2 + \alpha_5 L_2 L_3 + \alpha_6 L_3 L_1 \\
& + \alpha_7 L_1^2 L_2 + \alpha_8 L_2^2 L_3 + \alpha_9 L_3^2 L_1 \\
= & [L_1, L_2, L_3, \ldots]\boldsymbol{\alpha}
\end{aligned}
\tag{1.41}
$$

Identifying the nine nodal values

$$w_i, (\hat{\theta}_x)_i, (\hat{\theta}_y)_i$$

and noting that

$$
\begin{aligned}
\frac{\partial}{\partial x} = & \frac{\partial L_1}{\partial x}\frac{\partial}{\partial L_1} + \frac{\partial L_2}{\partial x}\frac{\partial}{\partial L_2} + \frac{\partial L_3}{\partial x}\frac{\partial}{\partial L_3} \\
\equiv & \frac{1}{2\Delta}\left(b_1\frac{\partial}{\partial L_1} + b_2\frac{\partial}{\partial L_2} + b_3\frac{\partial}{\partial L_3}\right)
\end{aligned}
\tag{1.42}
$$

and similarly

$$\frac{\partial}{\partial y}=\frac{1}{2\Delta}\left(c_1\frac{\partial}{\partial L_1}+c_2\frac{\partial}{\partial L_2}+c_3\frac{\partial}{\partial L_3}\right)$$

where Δ is the area of the element and

$$\begin{aligned}&2\Delta=b_1c_2-b_2c_1\\&b_1=y_2-y_3\\&c_1=x_3-x_2\\&\text{etc.}\end{aligned}$$

with cyclic permutation of indices (viz. Chapter 8 of Volume 1). We now can determine the shape function by a suitable inversion [viz. Sec. 1.3.1, Eq. (1.30)], and write, for node 1,

$$\mathbf{N}_1^{\mathrm{T}}=\begin{Bmatrix}3L_1^2-2L_1^3\\-b_3(L_1^2L_2+\tfrac{1}{2}L_1L_2L_3)+b_2(L_3L_1^2+\tfrac{1}{2}L_1L_2L_3)\\-c_3(L_1^2L_2+\tfrac{1}{2}L_1L_2L_3)+c_2(L_3L_1^2+\tfrac{1}{2}L_1L_2L_3)\end{Bmatrix}\qquad(1.43)$$

with similar expressions for nodes 2 and 3.

The computation of stiffness and load matrices can again follow the standard patterns and integration of expressions (1.21) can be done exactly using the general integrals given in Fig. 1.9. However, numerical quadrature is generally used and proves equally efficient (see Chapter 8 of Volume 1).

The element just derived is one first developed in reference 10. Although it satisfies the constant strain criterion (due to being able to render constant curvature states) it unfortunately does not pass the patch test for arbitrary mesh configurations. Indeed, this was pointed out in the original reference (which also was the one in which the patch test was mentioned for the first time). However, the patch test is fully satisfied with this element for triangular meshes created by three sets of equally spaced straight lines and its general performance, despite this shortcoming, made the element quite popular[37] in practical applications.

It is, however, possible to amend the element shape functions so that the resulting element passes the patch test in all configurations. Bergan[43-46] and Samuelsson[47] showed a way of doing this, but a simple successful modification is one derived by Specht.[48] This modification uses three fourth-order terms in place of the three cubic terms of Eq. (1.41). The particular form of these is so designed that the patch test criterion which we shall discuss later in Sec. 1.7 is identically satisfied. We now write

$$
\begin{aligned}
w = [&L_1, L_2, L_3, L_1L_2, L_2L_3, L_3L_1,\\
&L_1^2L_2+\tfrac{1}{2}L_1L_2L_3\{3(1-\mu_3)L_1-(1+3\mu_3)L_2+(1+3\mu_3)L_3\},\\
&L_2^2L_2+\tfrac{1}{2}L_1L_2L_3\{3(1-\mu_1)L_2-(1+3\mu_1)L_3+(1+3\mu_1)L_1\},\\
&L_3^2L_1+\tfrac{1}{2}L_1L_2L_3\{3(1-\mu_2)L_3-(1+3\mu_2)L_1+(1+3\mu_2)L_2\}]\boldsymbol{\alpha}\\
=\mathbf{P}&\boldsymbol{\alpha}
\end{aligned} \tag{1.44}
$$

where

$$
\mu_1=\frac{l_3^2-l_2^2}{l_1^2} \qquad \mu_2=\frac{l_1^2-l_3^2}{l_2^2} \qquad \mu_3=\frac{l_2^2-l_1^2}{l_3^2} \tag{1.45}
$$

and l_1, l_2, l_3 are lengths of the triangle sides.†

On identification of nodal values and inversion, the shape functions can be written explicitly in terms of the components of the vector $\mathbf{P}$ defined by Eq. (1.44) as

$$
\mathbf{N}_i^{\mathrm{T}}=\begin{Bmatrix} P_i-P_{i+3}+P_{k+3}+2(P_{i+6}-P_{k+6}) \\ -b_j(P_{k+6}-P_{k+3})-b_kP_{i+6} \\ -c_j(P_{k+6}-P_{k+3})-c_kP_{i+6} \end{Bmatrix} \tag{1.46}
$$

where i, j, k are the cyclic permutations of 1,2,3.

Once again stiffness and load matrices can be determined either explicitly or using numerical quadrature.

The element now derived passes all the patch tests[40] and performs excellently, as we shall show later. Indeed, if the quadrature is carried out in a 'reduced' manner using three internal quadrature points (viz. Volume 1, page 176) then the element is one of the best of that form currently available, as we shall show in the section dealing with numerical comparisons.

1.6 Triangular element of the simplest form[29,30] (6 DOF)

If conformity (C_1 continuity) is to be abandoned, it is possible to introduce even simpler elements than those already described by reducing the element interconnections. A very simple element of that type was first proposed by Morley.[29] In this element, illustrated in Fig. 1.11, the interconnections require continuity of the displacement w at the triangle vertices and of normal slopes at element sides.

† The constants μ_1, etc., are geometric parameters occurring in the expression for normal derivatives. Thus on side l_1 the normal derivative is given by

$$
\frac{\partial}{\partial n}=\frac{l_1}{4\Delta}\left[\frac{\partial}{\partial L_2}+\frac{\partial}{\partial L_3}-2\frac{\partial}{\partial L_1}+\mu_1\left(\frac{\partial}{\partial L_3}-\frac{\partial}{\partial L_2}\right)\right], \qquad \text{etc.}
$$

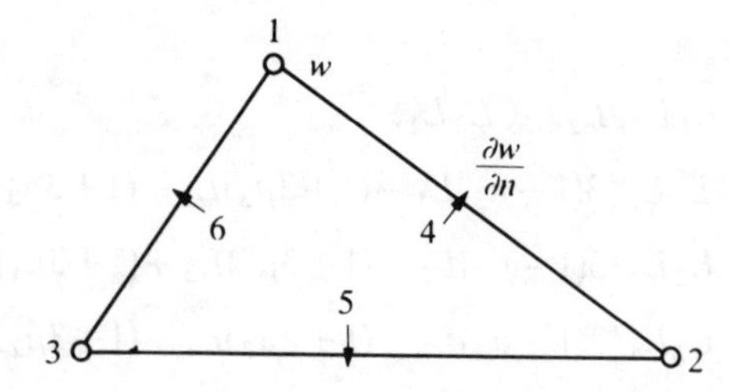

Fig. 1.11 The simplest non-conforming triangle due to Morley[29] with 6 degrees of freedom

With 6 degrees of freedom the expansion can be limited to quadratic terms alone, which one can write as

$$\mathbf{w}=[L_1, L_2, L_3, L_1L_2, L_2L_3, L_3L_1]\boldsymbol{\alpha}=\mathbf{P}\boldsymbol{\alpha} \tag{1.47}$$

Identification of nodal variables and inversion leads to the following shape functions:

For corner nodes:

$$N_1=L_1-L_1(1-L_1)-\frac{b_1b_3-c_1c_3}{b_2^2+c_2^2}(1-L_2)L_2-\frac{b_1b_2-c_1c_2}{b_3^2+c_3^2}(1-L_3)L_3 \tag{1.48a}$$

For 'normal gradient' nodes:

$$N_4=\frac{2\Delta}{\sqrt{b_1^2+c_1^2}}L_1(1-L_1) \tag{1.48b}$$

where the symbols are identical to those used in Eqs (1.41) to (1.45) and the other functions are obtained by cyclic permutation of indices.

Establishment of stiffness matrices, etc., follows the standard pattern and we find that once again the element passes fully all the patch tests required. This simple element performs reasonably, as we shall show later, though its accuracy is of course less than that of the preceding ones.

It is of interest to remark that the moment field described by the element satisfies exactly interelement equilibrium conditions, as the reader can verify. Indeed, originally the element was derived as an equilibrating one using the complementary energy principle,[29] and for this reason it always gives an upper bound on the strain energy of flexure. This is the simplest possible element as it simply represents the minimum requirements of a constant moment field. An explicit form of stiffness subroutines for this element has been derived by Wood.[30]

1.7 The patch test—an analytical requirement

The patch test in its different forms[40] (discussed also fully in Chapter 11 of Volume 1) is generally applied numerically to test the final form of an element. However, the basic requirements for its satisfaction by shape functions that violate compatibility can be forecast accurately if certain conditions are satisfied in the choice of such functions. These conditions follow from the requirement that for constant strain states the virtual work done by internal forces acting at the discontinuity must be zero. Thus if the tractions acting on an element interface of a plate are (viz. Fig. 1.4)

$$M_n,\ M_{ns} \text{ and } S_n \tag{1.49a}$$

and if the corresponding mismatch of virtual displacements is

$$\Delta\theta_n \equiv \Delta\left(\frac{\partial w}{\partial n}\right) \qquad \Delta\theta_s \equiv \Delta\left(\frac{\partial w}{\partial s}\right) \qquad \Delta w \tag{1.49b}$$

then ideally we would like the integral given below to be zero at least for constant stress states:

$$\int_{\Gamma^e} M_n \Delta\left(\frac{\partial w}{\partial n}\right) \mathrm{d}\Gamma + \int_{\Gamma^e} M_{ns} \Delta\left(\frac{\partial w}{\partial s}\right) \mathrm{d}\Gamma + \int_{\Gamma^e} S \Delta w \, \mathrm{d}\Gamma = 0 \tag{1.50}$$

The last term will always be zero identically for constant M_x, M_y, M_{xy} fields as then $S_x = S_y = 0$ [in the absence of applied couples, viz. Eq. (1.12)] and we can ensure the satisfaction of the remaining conditions if

$$\int_{\Gamma^s} \Delta\left(\frac{\partial w}{\partial n}\right) \mathrm{d}\Gamma = 0 \quad \text{and} \quad \int_{\Gamma^s} \Delta\left(\frac{\partial w}{\partial s}\right) \mathrm{d}\Gamma = 0 \tag{1.51a,b}$$

is satisfied for each straight side Γ^e of the element.

For elements joining at vertices where $\partial w/\partial n$ is prescribed, these integrals will be identically zero only if antisymmetric cubic terms arise in the departure from linearity and a quadratic variation of the normal gradients is absent, as shown in Fig. 1.12(a). This is the motivation for the rather special form of shape function basis chosen to describe the incompatible triangle in Eq. (1.44), and here Eq. (1.51a) is automatically satisfied. The satisfaction of Eq. (1.51b) is always ensured if the function w and its derivatives are prescribed at the corner nodes.

For the purely quadratic triangle of Sec. 1.6 the situation is even simpler. Here the gradients can only be linear and if their value is prescribed at the element mid-point as shown in Fig. 1.11(*b*) the integral is identically zero.

The same arguments apparently fail when the rectangular element with the function basis given in Eq. (1.30) is examined. However, the reader

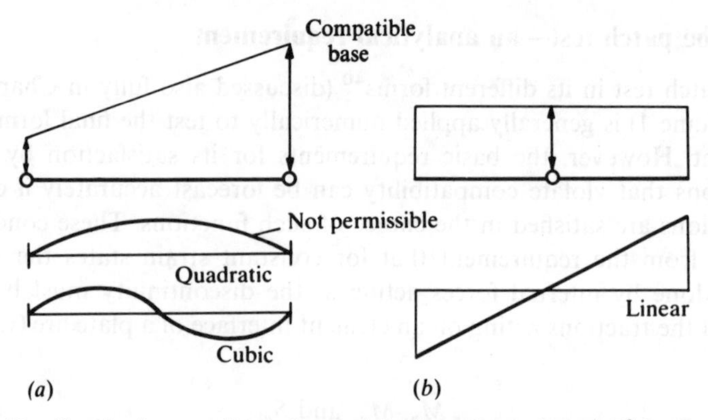

Fig. 1.12 Continuity condition for satisfaction of patch test [$\int(\partial w/\partial n)\mathrm{d}s=0$]. Variation of $\partial w/\partial n$ along side. (*a*) Definition by corner nodes (linear component compatible). (*b*) Definition by one central node (constant component compatible)

can verify by direct algebra that the integrals of Eqs (1.51) are identically satisfied. Thus, for instance,

$$\int_{-a}^{a} \frac{\partial w}{\partial y}\,\mathrm{d}x=0 \qquad \text{when } y=\pm b$$

and $\partial w/\partial y$ is taken as zero at the two nodes (i.e. departure from prescribed linear variation only is considered).

The remarks of this section are verified in numerical tests and lead to an intelligent, *a priori*, determination of conditions which make shape functions convergent for incompatible elements.

1.8 Numerical examples

The various plate bending elements already derived—and those derived in subsequent sections—have been introduced to many commercial and industrial programmes and are being used daily in the solution of engineering problems. The reader familiar with the type of approximation that can be expected probably does not require the graphic illustrations given below, but for all users it is important to realize the possibilities available. We shall thus give below two specific illustrations and follow these with a general convergence study of elements discussed.

Figure 1.13 shows the deflections and moments in a square plate clamped along its edges and solved by the use of the rectangular element derived in Sec. 1.3[25] and a uniform mesh. Table 1.4[38] gives numerical results for a set of similar examples solved with the same element and Table 1.5 presents another square plate with more complex boundary

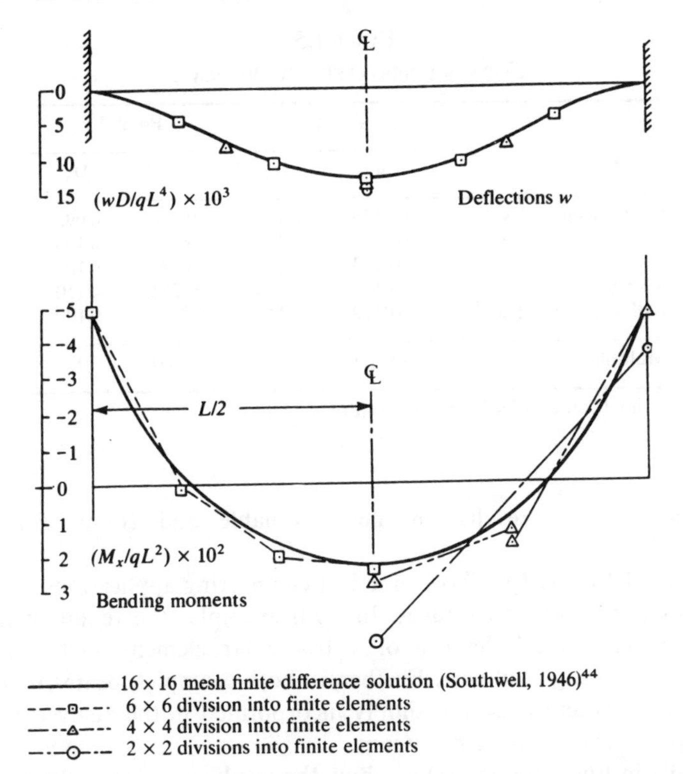

Fig. 1.13 A square plate with clamped edges. Uniform load q. Square elements

TABLE 1.4

COMPUTED CENTRAL DEFLECTION OF A SQUARE PLATE FOR SEVERAL MESHES (RECTANGULAR ELEMENTS)

Mesh	Total number of nodes	Simply supported plate		Clamped plate	
		α (uniform load)	β (concentrated load)	α (uniform load)	β (concentrated load)
2×2	9	0.003446	0.013784	0.001480	0.005919
4×4	25	0.003939	0.012327	0.001403	0.006134
8×8	81	0.004033	0.011829	0.001304	0.005803
12×12	169	0.004050	0.011715	0.001283	0.005710
16×16	289	0.004056	0.011671	0.001275	0.005672
Exact (Timoshenko)		0.004062	0.01160	0.00126	0.00560

$w_{max} = \alpha qL^4/D$ for a uniformly distributed load q;
$w_{max} = \beta PL^2/D$ for a central concentrated load P.

(Based on Tocher and Kapur.[38])

(Subdivision of whole plate given above.)

TABLE 1.5
CORNER SUPPORTED SQUARE PLATE

	Point 1		Point 2	
	w	M_x	w	M_x
Finite element 2×2	0.0126	0.139	0.0176	0.095
4×4	0.0165	0.149	0.0232	0.108
6×6	0.0173	1.150	0.0244	0.109
Marcus[50]	0.0180	0.154	0.0281	0.110
Ballesteros and Lee[51]	0.0170	0.140	0.0265	0.109
Multiplier	qL^4/D	qL^2	qL^4/D	qL^2

Point 1, centre of side; point 2, centre of plate.

conditions. Exact results are here available and comparisons are made.[50,51]

Figures 1.14 and 1.15 show practical engineering applications to more complex shapes of slab bridges. In both examples the requirements of geometry necessitate the use of a triangular element—with that of reference 10 being used here. Further, in both examples, beams reinforce the slab edges and these are simply incorporated in the analysis on the assumption of concentric behaviour.

Finally in Fig. 1.16(*a*) to (*d*) we show the results of a convergence study of the square plate with simply supported and clamped edge conditions for various triangular and rectangular elements and two load types. This type of diagram is conventionally used for assessing the behaviour of various elements, and we show on it the performance of the elements already described as well as others to which we shall refer later. Table 1.6 gives the key to the various element 'codes' which include elements yet to be described.[52-55]

This comparison singles out only one displacement and the plot uses the number of mesh divisions in a quarter plate as abscissa. It is therefore difficult to deduce the convergence rate and the performance of elements with multiple nodes. A more convenient plot gives the energy norm, $\|u\|$, versus the number of degrees of freedom N on a logarithmic scale. We show such a comparison for some elements in Fig. 1.17 for a problem of a slightly skewed, simply supported plate.[6] It is of interest to observe that, due to the singularity, both high- and low-order elements converge at almost identical rates (though of course the former give higher accuracy). Different rates of convergence would of course be obtained if no singularity existed (see Chapter 14 of Volume 1).

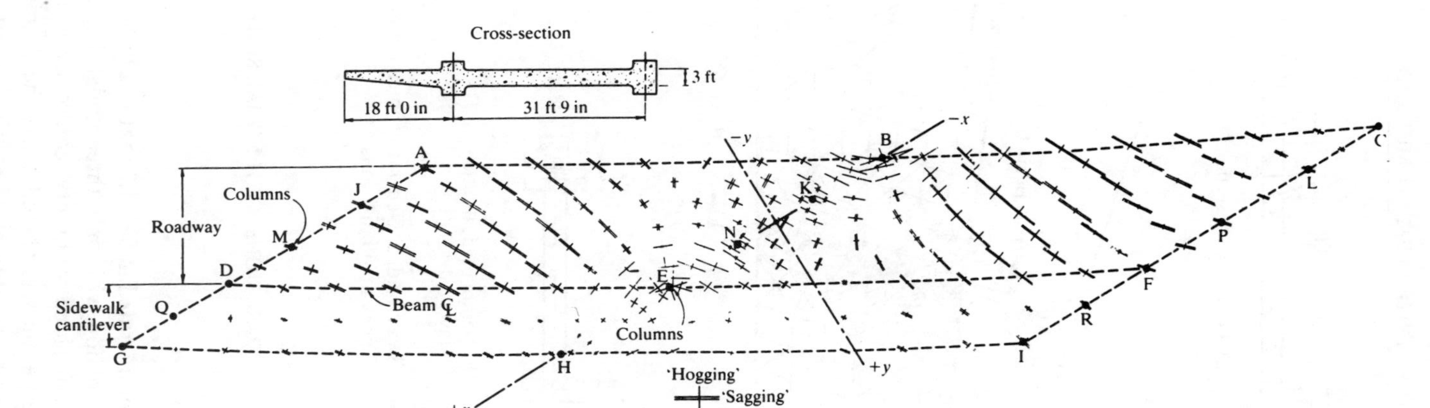

Fig. 1.14 A skew, curved, bridge with beams and non-uniform thickness. Computer plot of principal moments under dead load

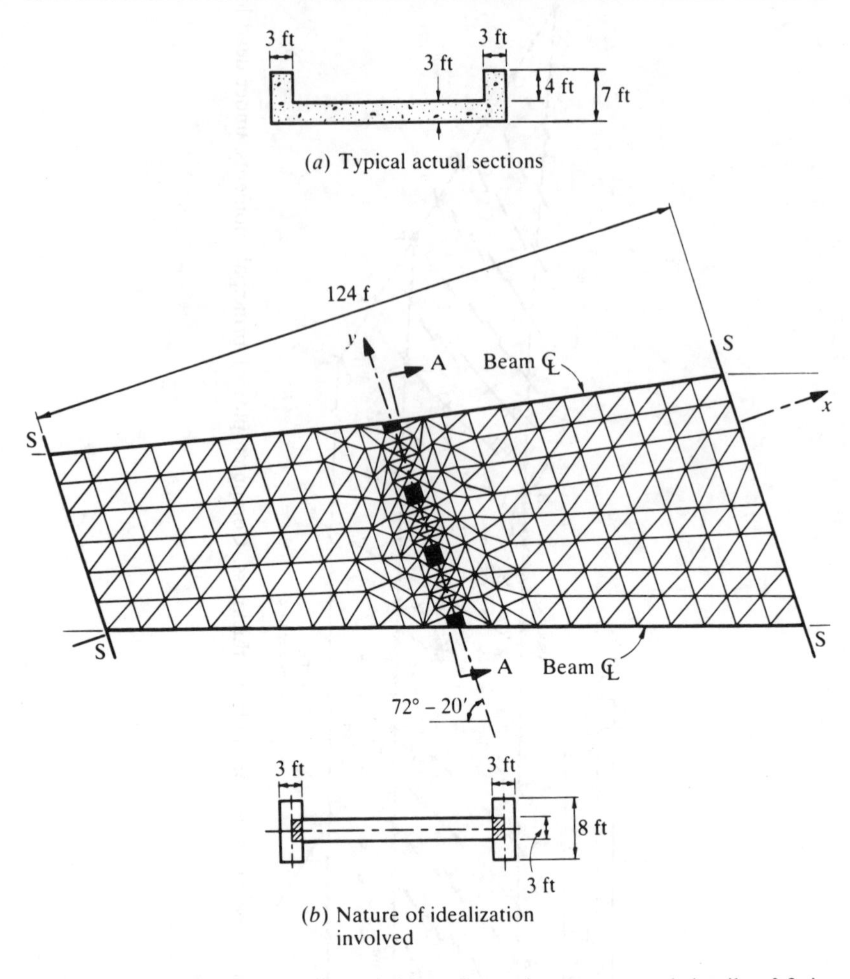

Fig. 1.15 The Castleton railway bridge. General geometry and details of finite element subdivisions

CONFORMING SHAPE FUNCTIONS WITH NODAL SINGULARITIES

1.9 General remarks

It has already been demonstrated in Sec. 1.3 that it is impossible to devise a simple polynomial function with only three nodal degrees of freedom that will be able to satisfy slope continuity requirements. The alternative of imposing curvature parameters at nodes has the disadvantage, however, of imposing excessive conditions of continuity. Furthermore, it is

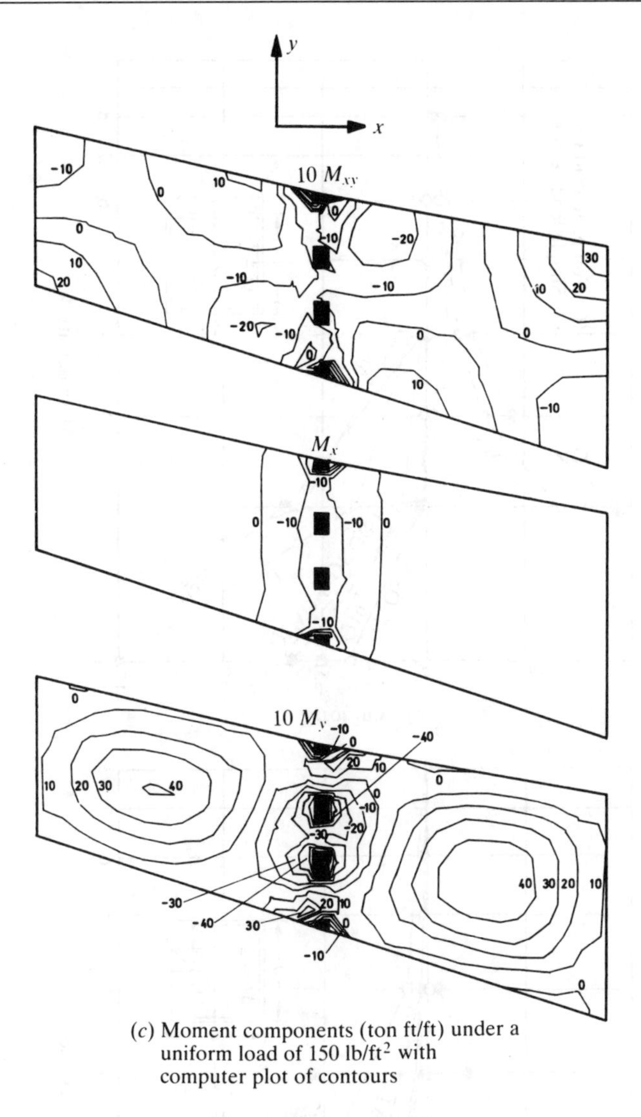

(*c*) Moment components (ton ft/ft) under a uniform load of 150 lb/ft² with computer plot of contours

Fig. 1.15 (*continued*)

desirable from many points of view to limit the nodal variables to three quantities only. These, with a simple physical interpretation, allow the generalization of plate elements to shells to be easily interpreted.

It is, however, possible to achieve C_1 continuity by provision of additional shape functions for which *second-order derivatives have non-unique values at nodes.* Providing the patch test conditions are satisfied, convergence is again assured.

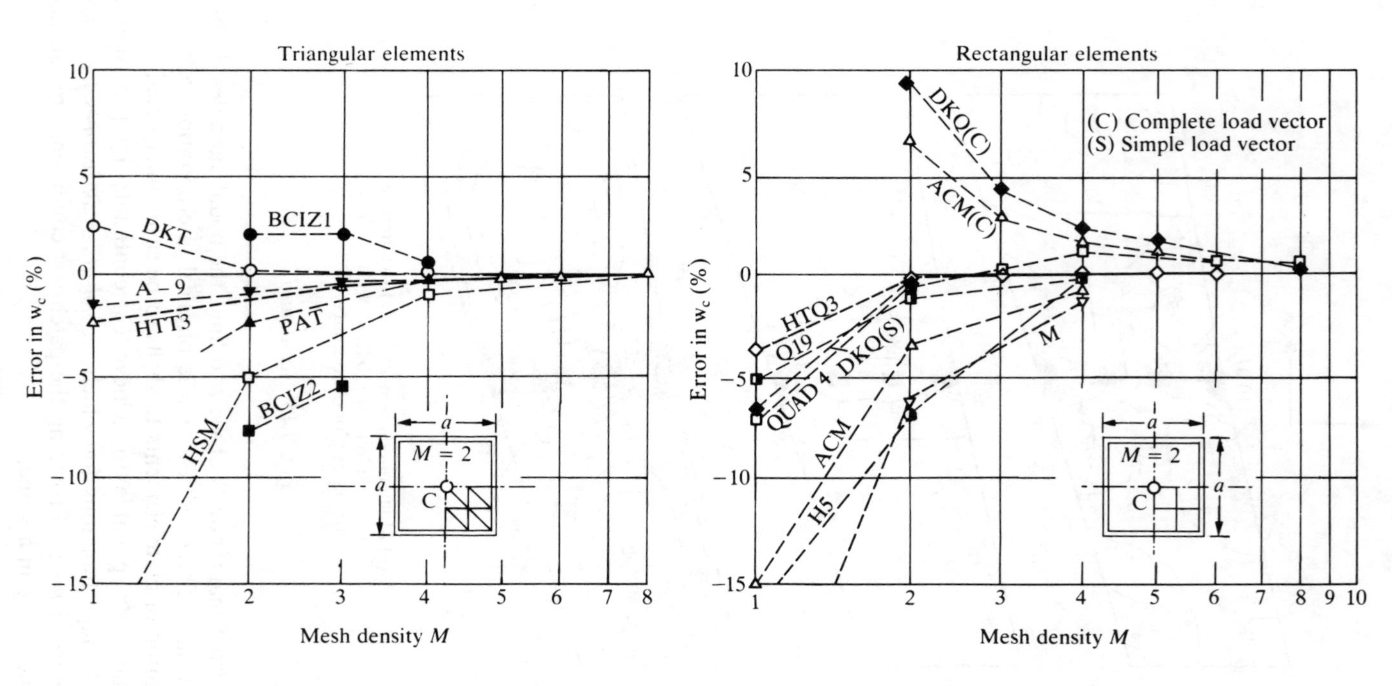

Fig. 1.16*a* Simply supported uniformly loaded square plate: percentage error in central displacement (see Table 1.6 for key)

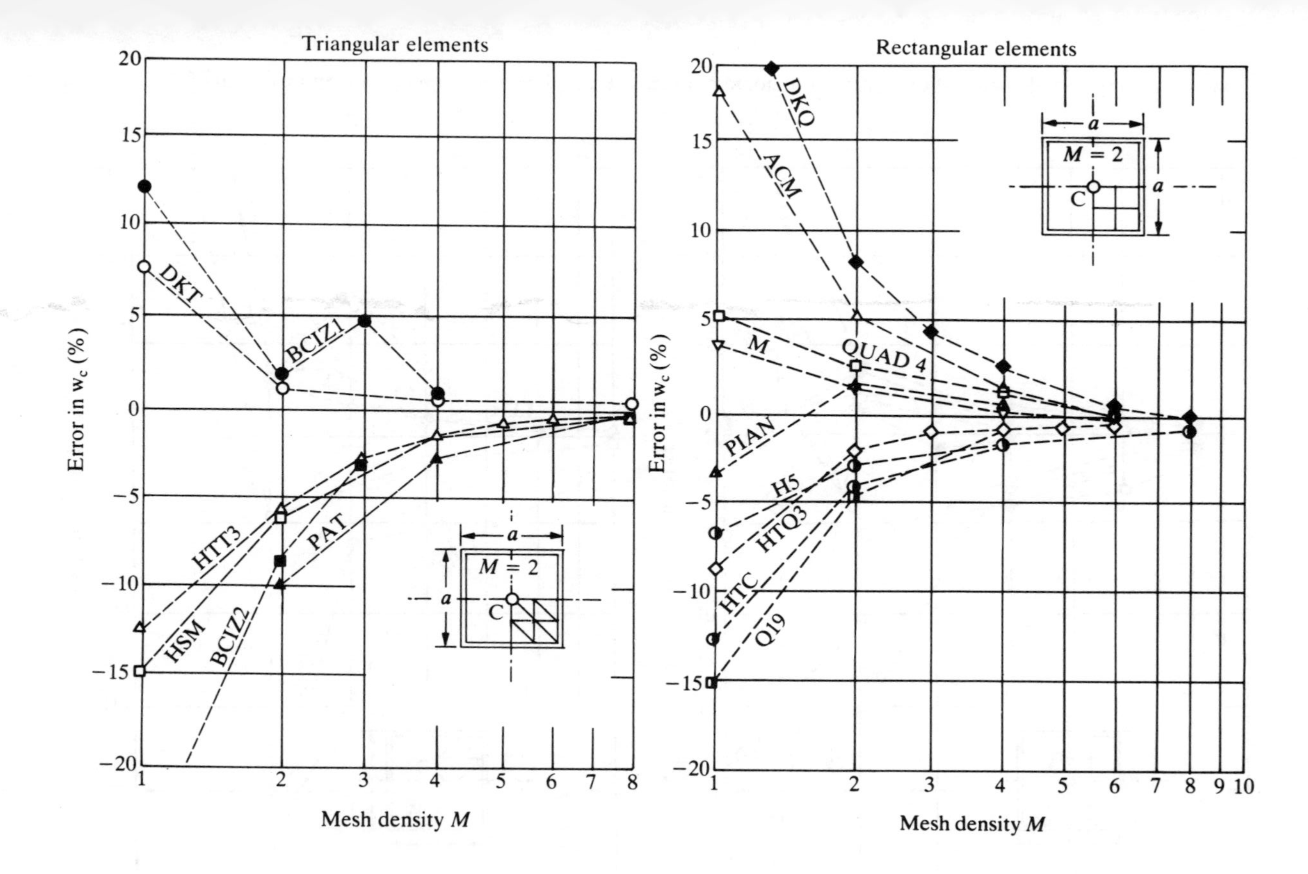

Fig. 1.16*b* Simply supported square plate with concentrated central load: percentage error in central displacement (see Table 1.6 for key)

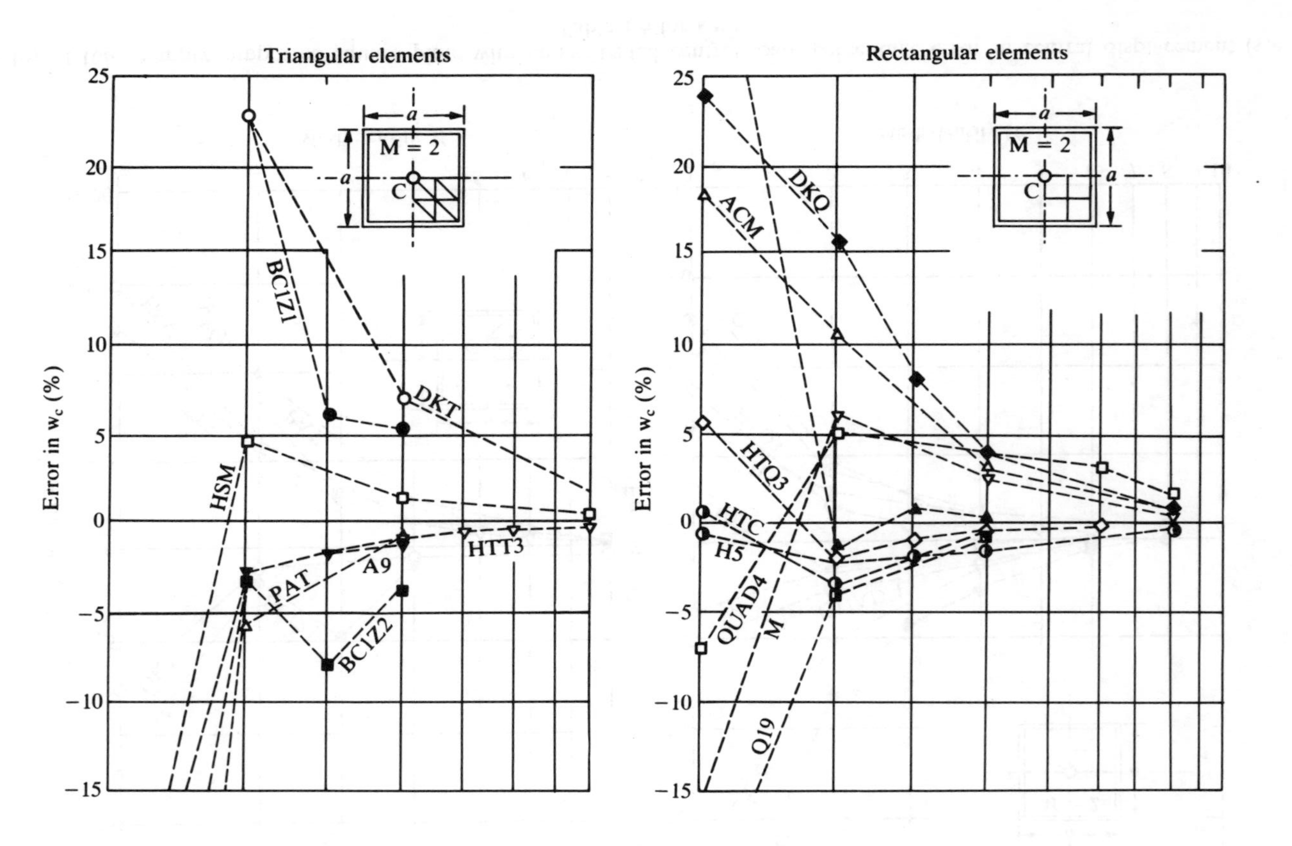

Fig. 1.16c Clamped uniformly loaded square plate: percentage error in central displacement (see Table 1.6 for key)

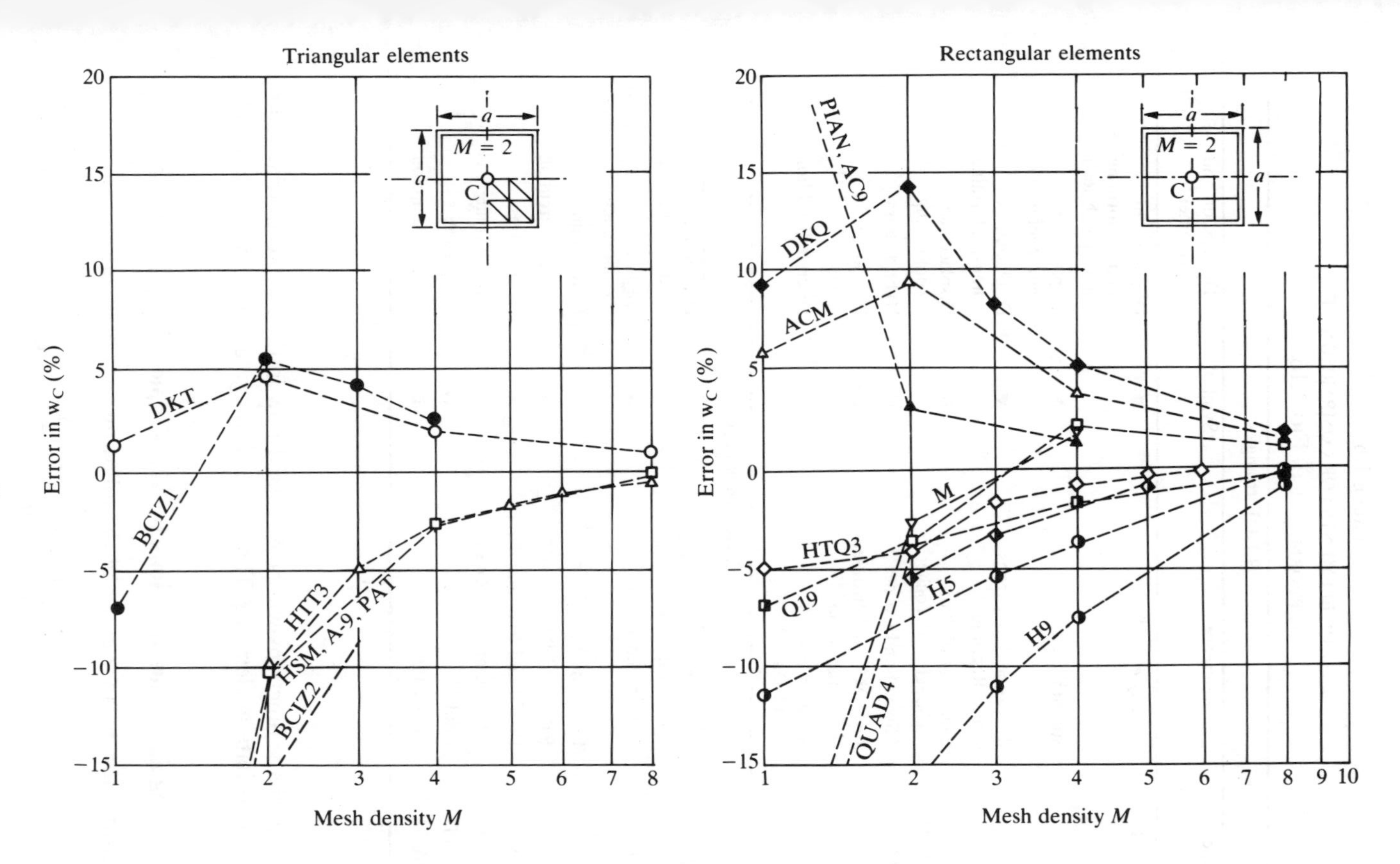

Fig. 1.16d Clamped square plate with concentrated central load: percentage error in central displacement (see Table 1.6 for key)

TABLE 1.6
LIST OF ELEMENTS FOR COMPARISON OF PERFORMANCE IN FIG. 1.16

Code	Reference	Symbol	Description and comment
	9 DOF triangles		
BCIZ 1	Bazeley *et al.*[10]	●	Displacement non-conforming (foils patch test)
PAT	Specht[48]	▲	Displacement non-conforming
A 9	Razzaque[32] and Irons and Razzaque[33]	▼	Displacement non-conforming
BCIZ 2 (HCT)	Bazeley *et al.*[10] or Clough and Tocher[9]	■	Displacement conforming
HSM	Allwood and Cornes[76]	□	Hybrid stress
HTT 3	Jirousek[85] and Jirousek and Lan Guex[54]	△	Hybrid Trefftz
DKT	Stricklin *et al.*[87] and Dhatt[88]		Discrete Kirchhoff
	12 DOF rectangles		
ACM	Zienkiewicz and Cheung[25] and Adini and Clough[36]	△	Displacement non-conforming
Q 19	Clough and Felippa[14]	◘	Displacement conforming
M	Fraeijs de Veubeke[72]	▽	Equilibrium
DKQ	Batoz and Ben Tohar[94]	◆	Discrete Kirchhoff
PIAN	Pian[74] and Pian and Tong[75]	▲	Hybrid stress
HTQ 3	Jirousek and Lan Guex[54]	◇	Hybrid Trefftz
H 5/HTC	Cook[80]	◑	Hybrid stress
QUADA	McNeal[103]	□	Direct assumption

Such shape functions will be discussed now in the context of triangular and quadrilateral elements. The simple rectangular shape will be omitted.

1.10 Singular shape functions for the simple triangular element

Consider for instance either of the following sets of functions:

$$\varepsilon_{23} = \frac{L_1 L_2^2 L_3^2 (L_3 - L_2)}{(L_1 + L_2)(L_2 + L_3)}, \qquad \text{etc.} \tag{1.52}$$

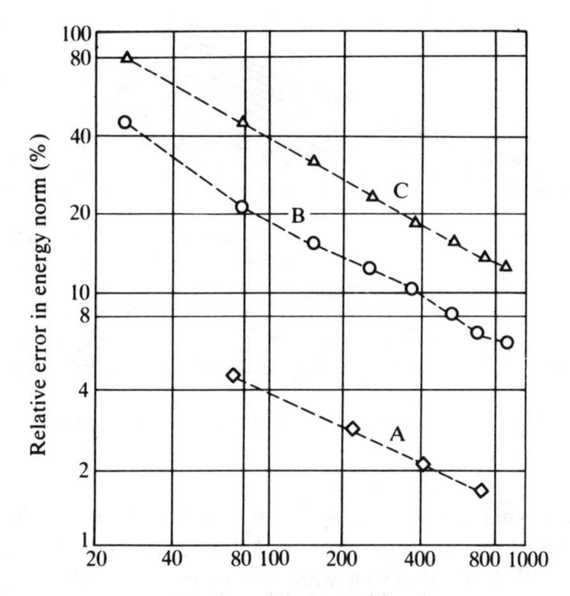

A Fifth-order conforming triangle[18–22]
B Low-order conforming element ($p = 2$)[9,10]
C Hybrid[76]

Fig. 1.17 Rate of convergence in energy norm versus degrees of freedom for three elements. Problem of a slightly skewed, simply supported plate (80°) with uniform mesh subdivision[6]

or

$$\varepsilon_{23} = \frac{L_1 L_2^2 L_3^2 (1 + L_1)}{(L_1 + L_2)(L_2 + L_3)}, \qquad \text{etc.} \tag{1.53}$$

Both have the property that along two sides (1–2 and 1–3) of a triangle (Fig. 1.18), their values and the values of their normal slope are zero. On the third side (2–3) their value is zero but a normal slope exists. In both, its variation is parabolic. Now, all the functions used to define the non-conforming triangle [see Eq. (1.41)] were cubic and hence permitted also a parabolic variation of the normal slope which is not uniquely defined by the two, end, nodal values (and hence resulted in non-conformity). However, if we specify as an additional variable the *normal slope of* w at a mid-side point of each side then, by combining the new functions ε_{23}, etc., with the other functions previously given, a *unique parabolic variation of the normal slope* along interelement faces is achieved and a compatible element will result.

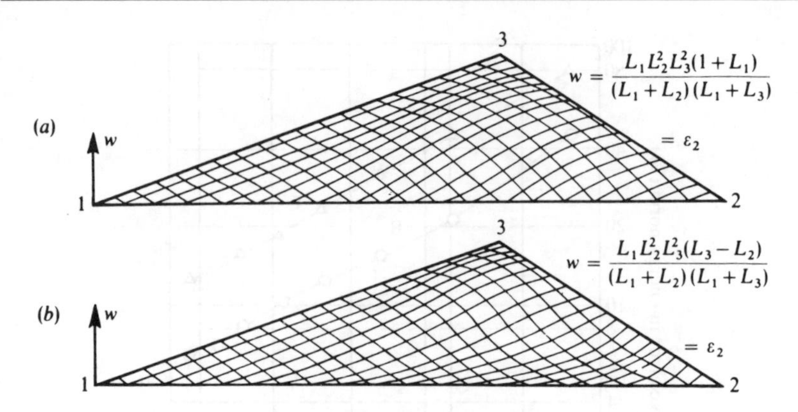

Fig. 1.18 Some singular area coordinate functions

Apparently, this can be achieved by adding three such additional degrees of freedom to expression (1.41) and proceeding as described above. This will result in an element shown in Fig. 1.19(*a*) which has six nodes, three corner ones as before and three additional ones at which only normal slope is specified. Such an element presents some assembly difficulties as different numbers of degrees of freedom are associated with the nodes, but this is commonplace. Further, a unique way of defining the normal slope is needed.

However, to avoid the above difficulties the mid-side node degree of freedom can be constrained. For instance, we can assume that the normal slope at the centre-point of a line is given as the average of the two slopes at the end of that side. This, after suitable transformation, results in a compatible element with exactly the same degrees of freedom as that described in previous sections [see Fig. 1.19(*b*)].

The algebra involved in the generation of suitable shape functions on the lines described here is tedious and so will not be given fully. It is developed most simply on the following lines.

First the normal slopes at the mid-sides are calculated from the basic element shape functions [Eq. (1.43)] as

$$\begin{Bmatrix} \left(\dfrac{\partial w}{\partial n}\right)_4 \\ \left(\dfrac{\partial w}{\partial n}\right)_5 \\ \left(\dfrac{\partial w}{\partial n}\right)_6 \end{Bmatrix} = \mathbf{Z}\mathbf{a}^e \tag{1.54}$$

Similarly the average values of the nodal slopes in directions normal to

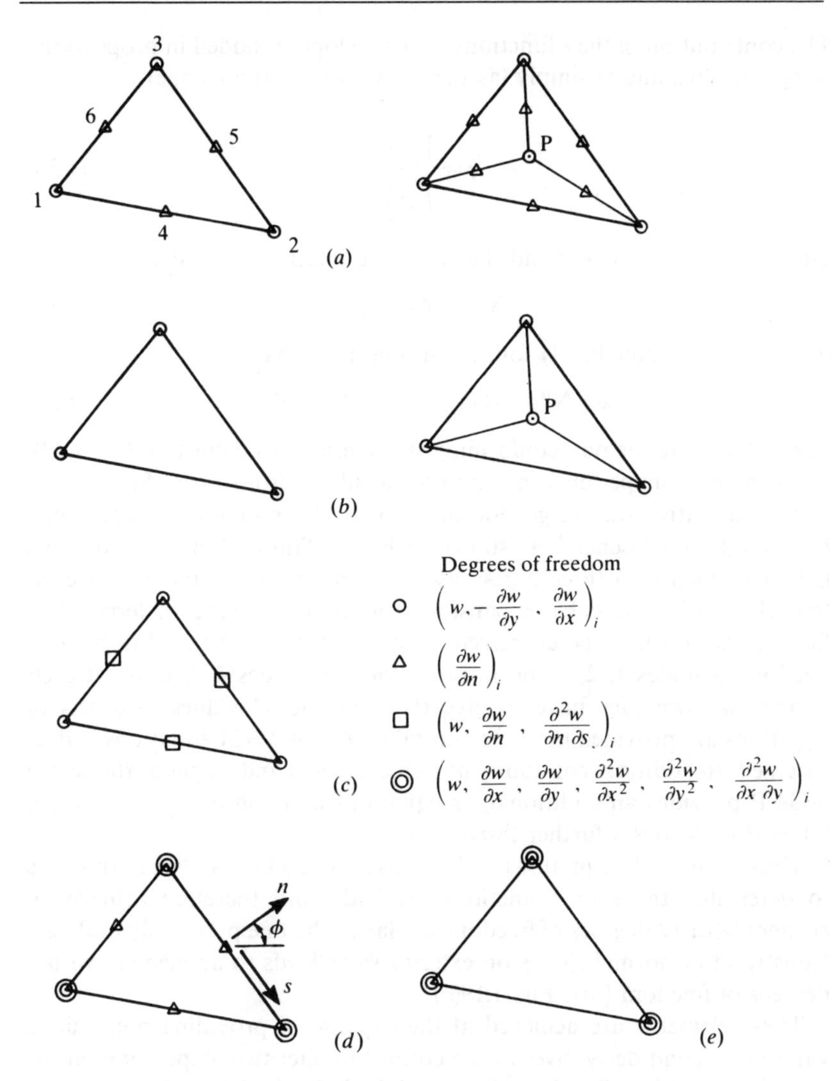

Fig. 1.19 Various conforming triangular elements

the sides are calculated for these points from these functions:

$$\begin{Bmatrix} \left(\dfrac{\partial w}{\partial n}\right)_4^a \\ \left(\dfrac{\partial w}{\partial n}\right)_5^a \\ \left(\dfrac{\partial w}{\partial n}\right)_6^a \end{Bmatrix} = \mathbf{Y}\mathbf{a}^e \tag{1.55}$$

The contribution of the ε functions to these slopes is added in proportions of $\varepsilon_{23}-\gamma_1$, etc., and is simply (as these give unit normal slope)

$$\boldsymbol{\gamma}=\begin{Bmatrix}\gamma_1\\ \gamma_2\\ \gamma_3\end{Bmatrix} \tag{1.56}$$

On combining Eq. (1.43) and the last three relations we have

$$\mathbf{Y}\mathbf{a}^e=\mathbf{Z}\mathbf{a}^e+\boldsymbol{\gamma} \tag{1.57}$$

from which it immediately follows on finding $\boldsymbol{\gamma}$ that

$$w=\mathbf{N}^0\mathbf{a}^e+[\varepsilon_{23},\varepsilon_{31},\varepsilon_{13}](\mathbf{Y}-\mathbf{Z})\mathbf{a}^e \tag{1.58}$$

in which $\mathbf{N}^0$ are the non-conforming shape functions defined in Eq. (1.43). Thus the new shape functions are now available from Eq. (1.58).

An alternative way of generating compatible triangles was developed by Clough and Tocher.[9] As shown in Fig. 1.19(*a*) each element triangle is first divided into three parts based on an internal point P. For each triangle a complete cubic expansion is written involving 10 terms. The final expansion is to be expressed in terms of 9 conventional degrees of freedom at nodes 1, 2, 3 and normal slopes at nodes 4, 5, 6. As at each corner two triangles have to give the same nodal values, two sets of equations are provided there, i.e. a total of $9\times 2+3=21$ equations is thus specified. In addition, continuity of displacements and slopes at the centre node P provides an additional six equations and continuity of slopes of internal mid-sides a further three.

Thus we have 30 equations and 30 unknowns which suffice in this case to determine the shape functions explicitly and therefore achieve an element with 12 degrees of freedom similar to the one previously outlined. Constraint of normal slopes on exterior sides leads to an element with 9 degrees of freedom [viz. Fig. 1.19(*b*)].

These elements are achieved at the expense of providing non-unique values of second derivatives at the corners (in fact two, depending on the direction in which the corner is approached). In the previously discussed set, in fact, the shape functions ε provide an infinite number of derivatives depending on the direction in which the corner is approached. Indeed, the derivation of the Clough and Tocher triangles[9] can be obtained by defining an alternative set of ε functions, as has been shown in reference 10.

As both types of elements lead to almost identical numerical results the preferable one is that leading to simplified computation. If numerical integration is used (as indeed is strongly recommended for such elements) the form of functions continuously defined over the whole triangle as given in Eqs (1.43) and (1.58) is advantageous, although a fairly high order of numerical integration is necessary due to the singular nature of the functions.

1.11 An 18 degree of freedom triangular element with conforming shape functions

An element that presents a considerable improvement over the type illustrated in Fig. 1.19(*a*) is shown in Fig. 1.19(*c*). Here the 12 degrees of freedom are increased to 18 by considering both the value of w and its cross derivative $\partial^2 w/\partial s \partial n$, in addition to the normal slope of $\partial w/\partial n$, at element sides.†

Thus an equal number of degrees of freedom is presented at each node giving a computational advantage. Imposition of the continuity of cross derivatives at *mid-sides* does not involve an additional constraint as this indeed must be continuous in physical situations.

The derivation of this element is given by Irons[13] and it will suffice here to say that in addition to the modes already discussed, fourth-order terms of the type illustrated in Fig. 1.10(*d*) and 'twist' functions of Fig. 1.18(*b*) are used. Indeed, it can be simply verified that the element contains *all* the fifteen terms of the quartic expansion in addition to the 'singularity' functions.

1.12 Compatible quadrilateral elements

Any of the previous triangles can be combined to produce compatible quadrilateral elements with or without internal degrees of freedom. Three such quadrilaterals are illustrated in Fig. 1.20 and, in all, no mid-side nodes exist on the external boundaries. This avoids the difficulties of assembly already mentioned.

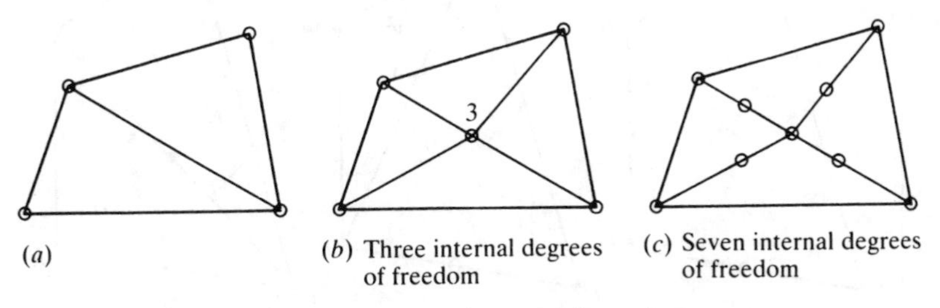

Fig. 1.20 Some composite quadrilateral elements

† This is, in fact, identical to specifying both $\partial w/\partial n$ and $\partial w/\partial s$ at the mid-side.

In the first, no internal degrees of freedom are present and indeed no improvement on the comparable triangles is expected. In the following two, 3 and 7 internal degrees of freedom exist respectively. Here, normal slope continuity imposed in the last one does not interfere with the assembly, as internal degrees of freedom are in all cases eliminated by static condensation. Much improved accuracy with these elements has been demonstrated by Clough and Felippa.[14]

An alternative direct derivation of a quadrilateral element was proposed by Sander[11] and Fraeijs de Veubeke.[12,15] This is along the following lines. Within a quadrilateral of Fig. 1.21(*a*) a complete cubic with 10 constants is taken, giving the first component of the displacement which is defined by three functions. Thus

$$w = w^a + w^b + w^c$$

$$w^a = \alpha_1 + \alpha_2 x + \cdots + \alpha_{10} y^3 \tag{1.59}$$

The second function w^b is defined in a piecewise manner. In the lower triangle of Fig. 1.21(*b*) it is taken as zero; in the upper triangle a cubic expression with three constants merges without slope discontinuity into the field of the lower triangle. Thus in *jkm*,

$$w^b = \alpha_{11} y'^2 + \alpha_{12} y'^3 + \alpha_{13} x' y'^2 \tag{1.60}$$

in terms of the locally specified coordinates x' and y'. Similarly for the third function, Fig. 1.21(*c*), $w^c = 0$ in the lower triangle and in *imj* we define

$$w^c = \alpha_{14} y''^2 + \alpha_{15} y''^3 + \alpha_{16} x'' y''^2 \tag{1.61}$$

The 16 external degrees of freedom are provided by three usual corner

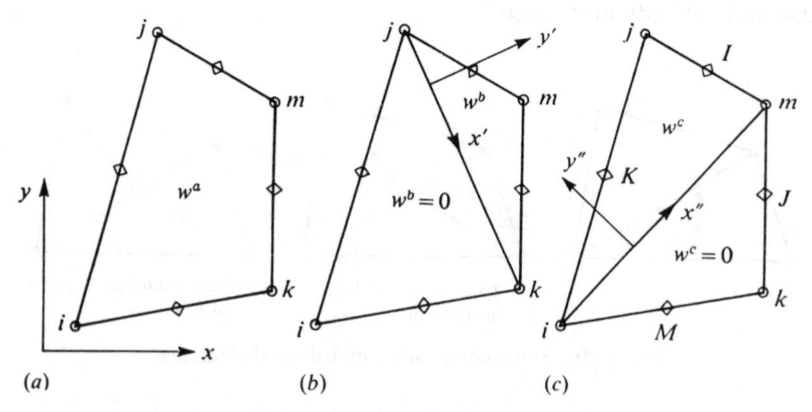

Fig. 1.21 The compatible functions of Fraeijs de Veubeke[12,15]

variables and normal mid-side slopes and allow the 16 constants α_{1-16} to be found by inversion. Compatibility is assured and once again non-unique second derivatives arise at corners.

Again it is possible to constrain the mid-side nodes if desired and thus obtain a 12 degree of freedom element. The expansion can be found explicitly as shown by Fraeijs de Veubeke[15] and a useful element generated.

The element described above cannot be formulated if a corner of the quadrilateral is reentrant. This is not a serious limitation but needs to be considered on occasion if such an element degenerates to a near triangular shape.

1.13 Quasi-conforming elements

The performance of some of the conforming elements discussed in Secs 1.10 to 1.12 is shown in the comparative graphs of Fig. 1.16. It should be noted that although monotonic convergence in the energy norm is now guaranteed the conforming triangular elements of references 9 and 10 perform almost identically but are considerably stiffer and hence less accurate than the non-conforming elements previously derived.

To overcome this inaccuracy a *quasi-conforming* or *smoothed* element was derived by Razzaque and Irons.[32,33] For the derivation of this element *substitute shape functions* are used.

The substitute functions are cubics (in area coordinates) so designed as to approximate in the least square sense the singular functions ε_i and their derivatives used to enforce continuity [viz. Eqs (1.52) to (1.58)], as shown in Fig. 1.22.

The algebra involved is complex but a full Fortran routine of stiffness computations is available in reference 32. It is noted that this element performs very similarly to the simpler, non-conforming element previously derived for the triangle. It is interesting to observe that here the non-conforming element is developed by choice and not to avoid difficulties.

CONFORMING SHAPE FUNCTION WITH ADDITIONAL DEGREES OF FREEDOM

1.14 Hermitian rectangle shape function

With the rectangular element of Fig. 1.7 the specification of $\partial^2 w/\partial x \partial y$ as a nodal parameter is always permissible as it does not involve 'excessive continuity'. It is easy to show that for such an element polynomial shape functions giving compatibility can be easily determined.

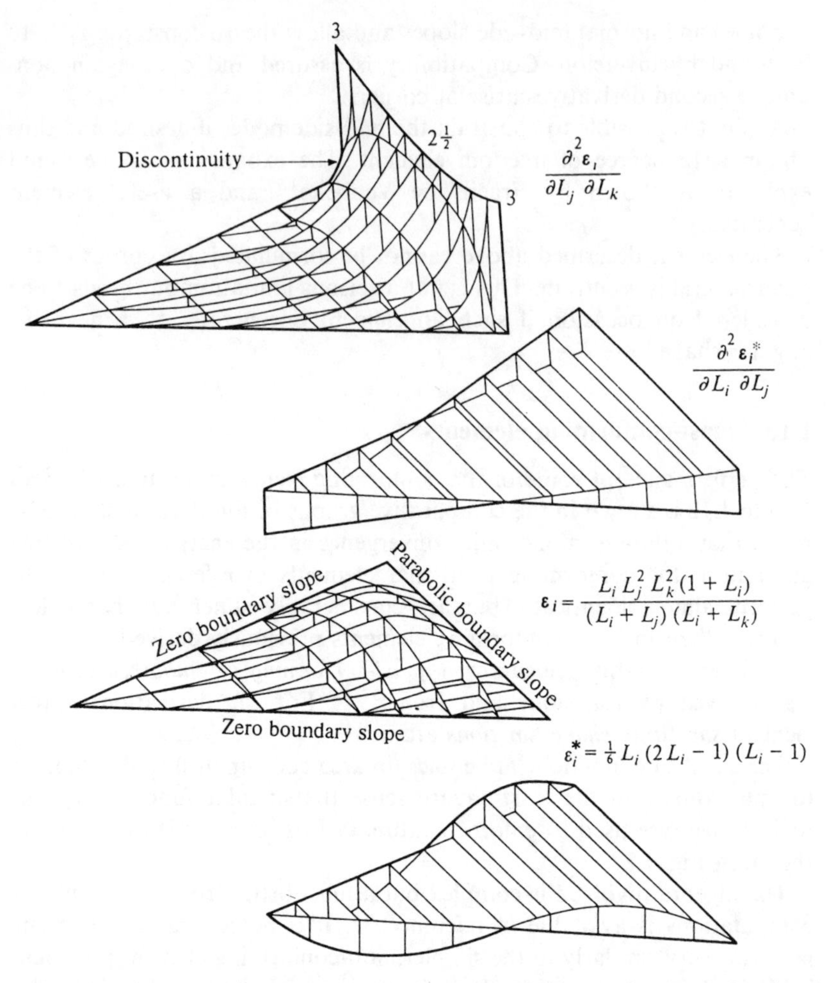

Fig. 1.22 Least square substitute cubic shape function ε_i in place of rational function ε_i for plate bending triangles

A polynomial expansion involving 16 constants (equal to the number of nodal parameters) could, for instance, be written retaining terms that do not produce a higher-order variation of w or its normal slope along the sides. Many alternatives will be present here and some may not produce invertible **C** matrices [viz. Eq. (1.29)].

An alternative derivation uses hermitian polynomials which permit the writing down of suitable functions directly. An hermitian polynomial

$$H_{mi}^{n}(x) \tag{1.62}$$

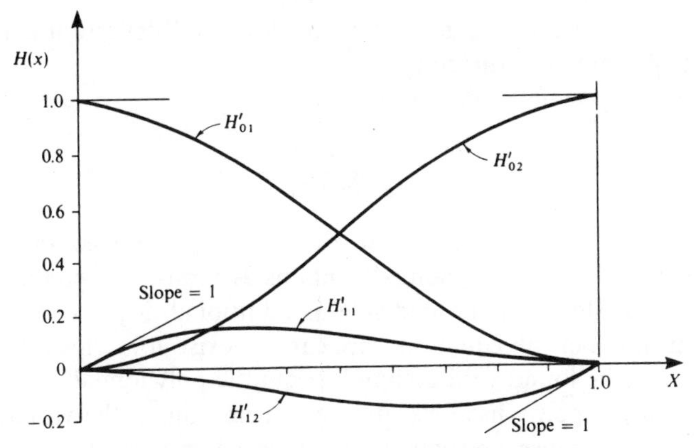

Fig. 1.23 First-order hermitian functions

is a polynomial of order $2n+1$ which gives, when $x=x_i$,

$$\frac{d^k H}{dx^k}=1, \qquad k=m \text{ for } m=0 \text{ to } n$$

and

$$\frac{d^k H}{dx^k}=0, \qquad k\neq m \text{ or when } x=x_j$$

A set of first-order hermitian polynomials is thus a set of cubics giving shape functions for a line element *ij* at the ends of which slopes and values of the function are used as variables. Figure 1.23 shows such a set of cubics.

It is easy to verify that the following shape functions

$$\mathbf{N}_i=[H_{0i}^{(1)}(x)H_{0i}^{(1)}(y),\ H_{1i}^{(1)}(x)H_{0i}^{(1)}(y),\ H_{0i}^{(1)}(x)H_{1i}^{(1)}(y),\ H_{1i}^{(1)}(x)H_{1i}^{(1)}(y)] \quad (1.63)$$

correspond to values of

$$w, \frac{\partial w}{\partial y}, \frac{\partial w}{\partial x}, \frac{\partial^2 w}{\partial x \partial y}$$

specified at the corner nodes, taking successively unit values at node i and zero elsewhere.

An element based on these shape functions has been developed by Bogner *et al.*[16] and used with some success. A development of this type of element to include continuity of higher derivatives is simple and is outlined in reference 17. In their undistorted form the above elements are, as for all rectangles, of very limited applicability.

1.15 The 21 and 18 degree of freedom triangles

If continuity of higher derivatives than the first is accepted at nodes (thus imposing a certain constraint on non-homogeneous situations as explained in Sec. 1.2.4), the generation of slope and deflection compatible elements presents less difficulty.

Considering as nodal degrees of freedom

$$w, \frac{\partial w}{\partial x}, \frac{\partial w}{\partial y}, \frac{\partial^2 w}{\partial x^2}, \frac{\partial^2 w}{\partial y^2}, \frac{\partial^2 w}{\partial x \partial y}$$

a triangular element will involve at least 18 degrees of freedom. However, a complete fifth-order polynomial contains 21 terms. If, therefore, we add three normal slopes at the mid-side as additional degrees of freedom a sufficient number of equations appear to exist for which the shape function can be found with a complete quintic polynomial.

Along any edge we have six quantities determining the variation of w (displacement, slopes and curvature at corner nodes), i.e. specifying a fifth-order variation. Thus this is uniquely defined and therefore w is continuous between elements. Similarly $\partial w/\partial n$ is prescribed by five quantities and varies as a fourth-order polynomial. Again this is as required by the slope continuity between elements.

If we write the complete quintic†

$$w = \alpha_1 + \alpha_2 x + \cdots + \alpha_{21} y^5 \tag{1.64}$$

we can proceed along the lines of the argument used to develop the rectangle in Sec. 1.3 and write

$$w_1 = \alpha_1 + \alpha_2 x_1 + \cdots + \alpha_{21} y_1^5$$

$$\left(\frac{\partial w}{\partial x}\right)_1 = \alpha_2 + \cdots + \alpha_{20} y_1^3$$

$$\left(\frac{\partial^2 w}{\partial x^2}\right)_1 = 2\alpha_4 + \cdots + 2\alpha_{19} y_1^3$$

and finally obtain an expression

$$\mathbf{a}^e = \mathbf{C}\boldsymbol{\alpha} \tag{1.65}$$

in which $\mathbf{C}$ is a 21×21 matrix.

The only apparent difficulty in the process that the reader may experience in forming this is the definition of the normal slopes at the mid-side nodes. However, if one notes that

$$\frac{\partial w}{\partial n} = \cos\phi \frac{\partial w}{\partial x} + \sin\phi \frac{\partial w}{\partial y} \tag{1.66}$$

† For this derivation use of simple cartesian coordinates is recommended in preference to area coordinates. Symmetry is assured as the polynomial is complete.

in which ϕ is the angle of a particular side to the x axis, the matter of formulation becomes simple. However, it is not easy to determine an explicit inverse of $\mathbf{C}$, and the stiffness expressions, etc., are evaluated as in Eqs (1.21) by a numerical inversion.

The existence of the mid-side nodes with their single degree of freedom is an inconvenience. It is possible, however, to constrain these by allowing only a cubic variation of the normal slope along each triangle side. Now, explicitly, the matrix $\mathbf{C}$ and the degrees of freedom can be reduced to 18, giving an element illustrated in Fig. 1.19(*e*) with three corner nodes and 18 degrees of freedom. This in fact is the more useful element in practice.

Both these elements were described in several publications appearing during 1968 and obviously quite independently arrived at. This 'simultaneous' discovery fact is one of the curiosities of scientific progress and seems to occur in many fields where the stage for a particular development is reached. Thus, the 21 degree of freedom element was described by Argyris *et al.*,[22] Bell,[18] Bosshard,[21] Irons[13] and Visser,[23] listing the authors alphabetically.

The reduced 18 degree of freedom version was developed by Argyris *et al.*,[22] Bell[18] and Cowper *et al.*[20] An essentially similar, but more complicated, formulation has been developed by Butlin and Ford,[19] and mention of the element shape functions was made earlier by Withum[56] and Felippa.[57]

It is clear that many more elements of this type could be developed and indeed some are suggested in the above references. A very full study is included in the work of Zenisek,[58] Peano[59] and others.[60-62] However, it should always be borne in mind that they involve an inconsistency when discontinuous variation of material properties occurs. Further, the existence of higher-order derivatives makes it more difficult to impose boundary conditions and indeed the simple interpretation of energy derivatives as 'nodal forces' disappears. Thus the engineer may still feel a justified preference for the more intuitive formulation previously described, despite the fact that very good accuracy has been demonstrated in the many references quoted for these elements.

AVOIDANCE OF CONTINUITY DIFFICULTIES—
MIXED AND CONSTRAINED ELEMENTS

1.16 Mixed formulations—general remarks

Equations (1.9) to (1.13) of this chapter give a plethoras of possibilities for approximation of both thick and thin plates using mixed (i.e. reducible) forms. In these, more than one set of the variables is approximated directly, and generally continuity requirements for such approximations can be relaxed. The procedures used in mixed formulations generally have been described in Chapters 12 and 13 of Volume 1, and the reader is referred to these for the general principles involved.

The options open are large and indeed so is the number of publications proposing various alternatives. We shall therefore limit the discussion to those that are most useful, but even here the presentation will perforce be short.

To avoid constant reference to the beginning of this chapter, the four governing equations (1.9) to (1.13) are rewritten below in their abbreviated form with dependent variable sets $\mathbf{M}$, $\boldsymbol{\theta}$, $\mathbf{S}$ and w:

$$\mathbf{M}-\mathbf{DL}\boldsymbol{\theta}=0 \tag{1.67a}$$

$$\mathbf{L}^{\mathrm{T}}\mathbf{M}+\mathbf{S}=0 \tag{1.67b}$$

$$\frac{\mathbf{S}}{\boldsymbol{\alpha}}+\boldsymbol{\theta}-\boldsymbol{\nabla}w=0 \tag{1.67c}$$

$$\boldsymbol{\nabla}^{\mathrm{T}}\mathbf{S}=-q \tag{1.67d}$$

To these, of course, the appropriate boundary conditions can be added. For the details of the operators, etc., the fuller form previously quoted needs to be consulted.

Mixed forms that utilize direct approximation to all the four variables are rare. The most obvious set arises from elimination of the moments $\mathbf{M}$, i.e.

$$\mathbf{L}^{\mathrm{T}}\mathbf{DL}\boldsymbol{\theta}+\mathbf{S}=0 \tag{1.68a}$$

$$\frac{\mathbf{S}}{\boldsymbol{\alpha}}+\boldsymbol{\theta}-\boldsymbol{\nabla}w=0 \tag{1.68b}$$

$$\boldsymbol{\nabla}^{\mathrm{T}}\mathbf{S}=q \tag{1.68c}$$

and is the basis of a formulation directly related to the three-dimensional elasticity consideration. This is so important that we shall devote Chapter 2 entirely to it, although of course it can be used for both thick and thin plates. However, we shall return to one of its derivatives in a later section of this chapter.

One of the earliest mixed approaches leaves the variables $\mathbf{M}$ and w to be approximated and eliminates $\mathbf{S}$ and $\boldsymbol{\theta}$. The form given is restricted to thin plates and thus $\boldsymbol{\alpha}=\infty$ is taken.

We now can write for Eqs (1.67a) and (1.67c),

$$\mathbf{D}^{-1}\mathbf{M}-\mathbf{L}\boldsymbol{\nabla}w=0 \tag{1.69a}$$

and for Eqs (1.67b) and (1.67d),

$$\boldsymbol{\nabla}^{\mathrm{T}}\mathbf{L}^{\mathrm{T}}\mathbf{M}=-q \tag{1.69b}$$

The approximation can now be made directly putting

$$\mathbf{M} = \mathbf{N}_M \bar{\mathbf{M}} \qquad \text{and} \qquad w = \mathbf{N}_w \bar{\mathbf{w}} \tag{1.70}$$

where $\bar{\mathbf{M}}$ and $\bar{\mathbf{w}}$ list the nodal (or other) parameters of the expansion and $\mathbf{N}_M$ and $\mathbf{N}_w$ are appropriate shape functions.

The approximation equations can, as is well known (viz. Chapter 9 of Volume 1), be made either via a suitable variational principle or directly in a weighted residual, Galerkin form, both leading to identical results. We choose here the latter, although the first presentations of this approximation by Herrmann[63] and later others[64–73] all use the Hellinger–Reissner principle.

Weighting the first of equations (1.69) by $\mathbf{N}_M^{\mathrm{T}}$ and the second by $\mathbf{N}_w^{\mathrm{T}}$ we have on integration by parts the following equation set:

$$\begin{bmatrix} \mathbf{A} & \mathbf{B} \\ \mathbf{B}^{\mathrm{T}} & \mathbf{0} \end{bmatrix} \begin{Bmatrix} \bar{\mathbf{M}} \\ \bar{\mathbf{w}} \end{Bmatrix} = \begin{Bmatrix} \mathbf{f}_1 \\ \mathbf{f}_2 \end{Bmatrix} \tag{1.71}$$

where

$$\begin{aligned} \mathbf{A} &= \int_\Omega \mathbf{N}_M^{\mathrm{T}} \mathbf{D}^{-1} \mathbf{N}_M \, \mathrm{d}\Omega \qquad & f_1 &= \int_{\Gamma_t} \nabla \mathbf{N}_w^{\mathrm{T}} \begin{Bmatrix} \tilde{M}_n \\ \tilde{M}_{ns} \end{Bmatrix} \mathrm{d}\Gamma \\ \mathbf{B} &= \int_\Omega (\mathbf{L}\mathbf{N}_M)^{\mathrm{T}} \nabla \mathbf{N}_w \, \mathrm{d}\Omega \qquad & f_2 &= \int_\Omega \mathbf{N}_w^{\mathrm{T}} \mathbf{q} \, \mathrm{d}\Omega + \int_\Gamma N_w^{\mathrm{T}} S_n \, \mathrm{d}\Gamma \end{aligned} \tag{1.72}$$

where $\tilde{M}_n$ and $\tilde{M}_{ns}$ are the prescribed boundary moments and S_n the prescribed shear forces.

Immediately it is evident that only C_0 continuity is required for both **M**† and w interpolation—and many forms of elements are therefore applicable. Of course, appropriate patch tests for mixed formulation must be enforced[41] and this requires a necessary condition that

$$n_{\mathrm{m}} \geqslant n_{\mathrm{w}} \tag{1.73}$$

where n_{m} stands for the number of parameters describing the moment field and n_{w} the displacement field.

Many useful elements have been developed using this type of approximation though their use is limited due to the difficulty of interconnection with other structures as well as the fact that the coefficient matrix in Eq. (1.71) is indefinite with many zero diagonal forms.

Indeed, a similar fate has overtaken numerous 'equilibrium element' forms in which the moment field is chosen *a priori* in a manner satisfying Eq. (1.69b). Here the research of Fraeijs de Veubeke[72] and others[11,29] has

† It should be observed that, if C_0 continuity to the whole **M** field is taken, excessive continuity will arise and it is usual to ensure the continuity of M_n and M_{ns} at interfaces only.

to be noted. It must, however, be noted that the second of these elements[29] is in fact identical to the mixed element developed by Herrmann[65] and Hellan[64] (see also reference 73).

1.17 Hybrid plate elements

Hybrid elements are essentially mixed elements in which the field inside the element is defined by one set of parameters and the one on the element frame by another, as shown in Fig. 1.24. The latter are generally chosen to be of a type identical to other displacement models and thus can be readily incorporated in a general program and indeed used in conjunction with the standard displacement types we have already discussed. The internal parameters can be readily eliminated (being confined to a single element) and thus the difference from displacement forms are confined to the element subroutine. The original concept is due to T. H. H. Pian[74,75] who pioneered this very successful approach, and today many variants of the procedures exist in the context of thin plate theory.[54,76–85]

In the majority of approximations, an equilibrating stress field was assumed to be given by a number of suitable shape functions and unknown parameters. In others, a mixed stress field is taken in the interior. A more refined procedure, introduced by Jirousek,[54,85] assumes in the interior a series solution exactly satisfying all the differential equations involved for a homogeneous field.

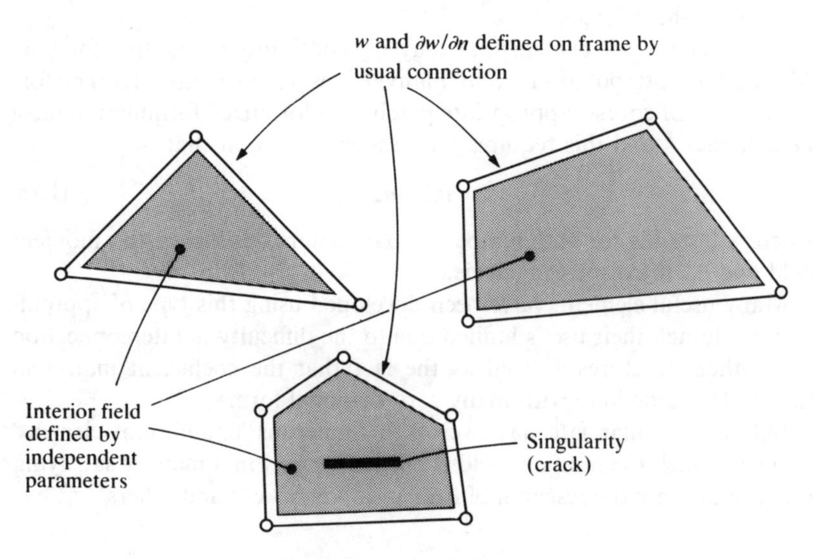

Fig. 1.24 Hybrid elements

All procedures use a suitable linking of the interior parameters with those defined on the boundary by the 'frame parameters'. The procedures for doing this are fully described in Chapter 13 of Volume 1 in the context of elasticity equations, and only a small change of variables is needed to adapt these to the present case. We leave this extension to the reader who can also consult the appropriate references for details.

Some remarks need to be made in the context of hybrid elements.

Remark 1. The first is that the number of internal parameters, n_I, must be at least as large as the number of frame parameters, n_F, which describe the displacements there, less the rigid body motions if singularity of the final (stiffness) matrices is to be avoided. Thus we require that

$$n_I \geqslant n_F - 3 \tag{1.74}$$

for plates.

Remark 2. The second remark is a simple statement that it is possible, but counterproductive, to introduce an excessive number of internal parameters that simply give a more exact solution to a 'wrong' problem in which the frames are constraining the interior of elements. Thus an additional accuracy is not achieved overall.

Remark 3. Most of the formulations are available for non-homogeneous plates (and hence non-linear problems of the type that we shall discuss later). However, this is not true for the Trefftz–hybrid elements[54,85] where an exact solution to the differential equation needs to be available for the interior. Such solutions are not known for arbitrary non-homogeneous interiors and hence the procedure fails. However, for homogeneous problems the elements can be made much more accurate than any of the others and indeed allow a general polygonal element with singularities and/or internal boundaries to be developed (viz. Fig. 1.24) by the use of special functions. Obviously this advantage needs to be borne in mind.

A number of elements matching (or duplicating) the displacement method have been developed and the performance of some of the simpler ones is shown in Fig. 1.16. Indeed, it can be shown that many hybrid-type elements duplicate precisely the various incompatible elements that pass the convergence requirement. Thus it is interesting to note that the triangle of Allman[84] gives precisely the same results as the 'smoothed', Razzaque element of references 32 and 33 or, indeed, the element of Sec. 1.5.

1.18 Discrete Kirchhoff constraints

We conclude this chapter by outlining yet another procedure for achieving excellent element performance by a constrained (mixed) method.

Here it is convenient (though by no means essential) to use a variational principle to describe Eqs (1.68). This can be written simply as the minimization of total potential energy

$$(\min)\ \Pi = \frac{1}{2}\int_{\Omega} (\mathbf{L}\boldsymbol{\theta})^{\mathrm{T}}\mathbf{D}(\mathbf{L}\boldsymbol{\theta})\,\mathrm{d}\Omega + \frac{1}{2}\int_{\Omega} \mathbf{S}^{\mathrm{T}}\frac{1}{\alpha}\mathbf{S}\,\mathrm{d}\Omega - \int_{\Omega} wq\,\mathrm{d}\Omega + \text{boundary terms} \tag{1.75a}$$

subject to the constraint that Eq. (1.68b) is satisfied, i.e. that

$$\frac{\mathbf{S}}{\boldsymbol{\alpha}} + \boldsymbol{\theta} - \boldsymbol{\nabla} w = 0 \tag{1.75b}$$

We shall use this form for general, thick, plates in Chapter 2, but in the case of thin plates with which this chapter is concerned, we can specialize by putting $\boldsymbol{\alpha} = \infty$ and rewrite the above as

$$(\min)\ \Pi = \frac{1}{2}\int_{\Omega} (\mathbf{L}\boldsymbol{\theta})^{\mathrm{T}}\mathbf{D}(\mathbf{L}\boldsymbol{\theta})\,\mathrm{d}\Omega - \int wq\,\mathrm{d}\Omega \tag{1.76a}$$

subject to

$$\boldsymbol{\theta} - \boldsymbol{\nabla} w = 0 \tag{1.76b}$$

and we note that the explicit mention of shear forces $\mathbf{S}$ is no longer necessary.

To solve the problem posed by Eqs (1.76) we can

(*a*) approximate w and $\boldsymbol{\theta}$ by independent interpolations of C_0 continuity as

$$w = \mathbf{N}_w \bar{\mathbf{w}} \qquad \text{and} \qquad \boldsymbol{\theta} = \mathbf{N}_\theta \bar{\boldsymbol{\theta}} \tag{1.77}$$

(*b*) impose a discrete approximation to the constraint of Eq. (1.76b) and solve the minimization problems resulting from substitution of (1.77) in (1.76a) by either discrete elimination, use of suitable lagrangian multipliers or penalty procedures.

In the application of the so-called *discrete Kirchhoff constraints*, Eq. (1.76b) is approximated by point (or subdomain) *collocation* and direct elimination is used to *reduce the number of nodal parameters*. Of course, the other means of imposing the constraints could be used with *identical* effect and we shall return to these in the next chapter. However, direct elimination is advantageous in reducing the final total number of variables and can be used effectively.

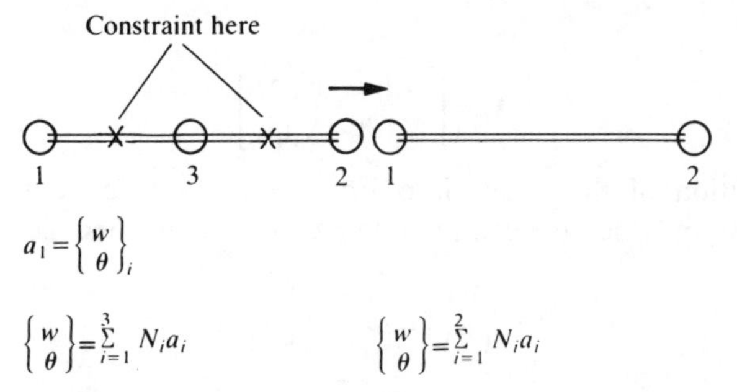

Fig. 1.25 A beam element with independent, lagrangian, interpolation of w and θ with constraint $\partial w/\partial x-\theta=0$ applied at points $\times$

We illustrate the process on a simple, one-dimensional, example of a beam shown in Fig. 1.25. In this, initially the displacements and rotations are taken as determined by a quadratic interpolation of an identical kind and we write in place of Eq. (1.77),

$$\begin{Bmatrix} w \\ \theta \end{Bmatrix} = \sum_{i=1}^{3} N_i \begin{Bmatrix} \bar{w} \\ \bar{\theta} \end{Bmatrix}_i \tag{1.78}$$

where i are the three element nodes.

The constraint is now applied by point collocation at coordinates x_α and x_β of the beam, i.e. we require that at these points

$$\theta-\frac{\partial w}{\partial x}=0 \tag{1.79}$$

This can be written using the interpolation of Eq. (1.78) as two simultaneous equations

$$\sum_{i=1}^{3} \dot{N}_i(\alpha)\bar{w}_i - \sum_{i=1}^{3} N_i(\alpha)\bar{\theta}_i = 0$$

and

$$\sum_{i=1}^{3} \dot{N}_i(\beta)\bar{w}_i - \sum_{i=1}^{3} N_i(\beta)\bar{\theta}_i = 0 \tag{1.80}$$

where

$$N(\alpha)\equiv N(x_\alpha) \qquad \text{and} \qquad \dot{N}(\alpha)\equiv\left(\frac{\mathrm{d}N}{\mathrm{d}x}\right)_{x=x_\alpha}, \text{ etc.}$$

Equations (1.80) can be used to eliminate $\bar{w}_3$ and $\bar{\theta}_3$. Writing Eqs (1.80) explicitly we have

$$\mathbf{A}_3\begin{Bmatrix} \bar{w}_3 \\ \bar{\theta}_3 \end{Bmatrix} = -\mathbf{A}_1\begin{Bmatrix} \bar{w}_1 \\ \bar{\theta}_1 \end{Bmatrix} - \mathbf{A}_2\begin{Bmatrix} \bar{w}_2 \\ \bar{\theta}_2 \end{Bmatrix} \tag{1.81}$$

where

$$\mathbf{A}_i = \begin{bmatrix} \dot{N}_i(\alpha) - N_i(\alpha) \\ \dot{N}_i(\beta) - N_i(\beta) \end{bmatrix}$$

Substitution of the above into Eq. (1.78) results directly in shape functions from which the centre node has been eliminated, i.e.

$$\begin{Bmatrix} w \\ \theta \end{Bmatrix} = \sum_{i=1}^{2} \bar{\mathbf{N}}_i \begin{Bmatrix} \bar{w} \\ \bar{\theta} \end{Bmatrix}_i \tag{1.82}$$

with

$$\bar{\mathbf{N}}_i = N_i \mathbf{I} - \mathbf{A}_3^{-1} \mathbf{A}_i$$

where **I** is a 2×2 identity matrix.

If these functions are used for a beam, we arrive at an element that is convergent. Indeed, in the particular case where x_α and x_β are chosen to coincide with the two Gauss points the element coincides precisely with that given by a displacement formulation involving a cubic w interpolation. This in fact is *exact* for a uniform beam.

For two-dimensional plate elements the situation is a little more complex, but if we imagine x to coincide with the direction tangent to an element side precisely identical elimination enforces *complete compatibility* along an element side when both gradients of w are specified at the ends. However, with discrete imposition of the constraints it is not clear *a priori* that convergence will always occur—though of course one can argue heuristically that collocation applied in numerous directions should result in an acceptable element. Indeed, patch tests turn out to be satisfied by most elements in which the w interpolation (and hence the $\partial w/\partial s$ interpolation) have C_0 continuity.

The constraints frequently applied in practice involve the use of line or subdomain collocation to increase their number (which must of course always be less than the number of remaining variables) and such additional constraint equations as

$$I_\Gamma \equiv \int_{\Gamma_e} \left(\frac{\partial w}{\partial s} - \theta_s \right) \mathrm{d}s = 0$$

$$I_{\Omega x} \equiv \int_{\Omega} \left(\frac{\partial w}{\partial x} - \theta_x \right) \mathrm{d}\Omega = 0 \tag{1.83}$$

or

$$I_{\Omega y} \equiv \int_{\Omega} \left(\frac{\partial w}{\partial y} - \theta_y \right) \mathrm{d}\Omega$$

are frequently used. The algebra involved in the elimination is not always

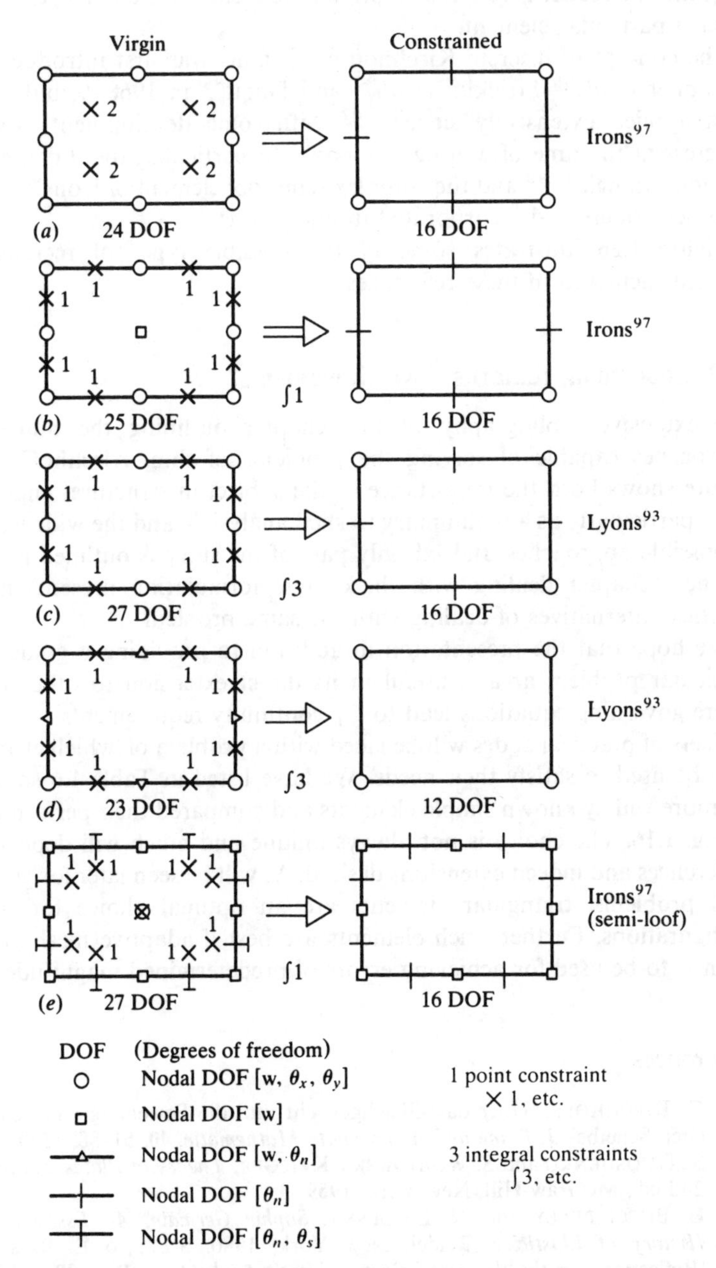

Fig. 1.26 A series of DKT-type elements of quadrilateral type

easy and the reader is referred to original references for details pertaining to each particular element.

The concept of discrete Kirchhoff constraints was first introduced by Wempner *et al.*,[86] Stricklin *et al.*[87] and Dhatt[88] in 1968–9, but it has been applied extensively since[89-102] with some developments still in progress at the time of writing this book. In particular, the 9 degrees of freedom triangle[89,90] and the complex semi-loof element of Irons[97,98] are high performers and incorporated in many codes.

Figure 1.26 illustrates some of the possible types of rectangular elements achieved in these references.

1.19 Concluding remarks—which elements?

The extensive bibliography of this chapter outlining the numerous approaches capable of solving the problems of thin, Kirchhoff, plate flexure shows both the importance of the subject in structural engineering—particularly as a preliminary to shell analysis—and the wide variety of possible approaches. Indeed only part of the story is outlined here as the next chapter dealing with thick plate formulation presents many practical alternatives of dealing with the same problem.

We hope that the presentation in addition to providing a guide to a particular problem area is useful in its direct extension to other fields where governing equations lead to C_1 continuity requirements.

Users of practical codes will be faced with a problem of 'which element' is to be used to satisfy their needs. We have listed in Table 1.6 some of the more widely known simple elements and compared their performance in Fig. 1.16. The choice is not always unique and much will depend on preferences and indeed extensions desired. As will be seen later for general shell problems, triangular elements are an optimal choice for many configurations. Further, such elements are best if adaptive mesh generation is to be used for achieving errors of predetermined magnitude.

References

1. G. KIRCHHOFF, 'Über das Gleichgewicht und die Bewegung einer elastischen Scheibe', *J. Reine und Angewandte Mathematic*, **40**, 51–88, 1850.
2. S. TIMOSHENKO and S. WOINOWSKY-KRIEGER, *Theory of Plates and Shells*, 2nd ed., McGraw-Hill, New York, 1959.
3. L. BUCCIARELLY and N. DWORSKY, *Sophie Germain, An Essay on the History of Elasticity*, Reidel, New York, 1980. See also E. REISSNER, 'Reflections on the theory of elastic plates', *Appl. Mech. Rev.*, **38**, 1453–64, 1985.
4. E. REISSNER, 'The effect of transverse shear deformation on the bending of elastic plates', *J. Appl. Mech.*, **12**, 69–76, 1945.

5. R. D. MINDLIN, 'Influence of rotatory inertia and shear in flexural motions of isotropic elastic plates', *J. Appl. Mech.*, **18**, 31–8, 1951.
6. I. BABUSKA and T. SCAPOLLA, 'Benchmark computation and performance evaluation for a rhombic plate bending problem', *Int. J. Num. Meth. Eng.*, **28**, 155–180, 1989.
7. I. BABUSKA, 'The stability of domains and the question of formulation of plate problems', *Appl. Math.*, 463–7, 1962.
8. B. M. IRONS and J. K. DRAPER, 'Inadequacy of nodal connections in a stiffness solution for plate bending', *JAIAA*, **3**, 5, 1965.
9. R. W. CLOUGH and J. L. TOCHER, 'Finite element stiffness matrices for analysis of plates in bending', in *Proc. Conf. on Matrix Methods in Structural Mechanics*, Air Force Institute of Technology, Wright-Patterson AF Base, Ohio, 1965.
10. G. P. BAZELEY, Y. K. CHEUNG, B. M. IRONS and O. C. ZIENKIEWICZ, 'Triangular elements in bending—conforming and non-conforming solutions', in *Proc. Conf. on Matrix Methods in Structural Mechanics*, Air Force Institute of Technology, Wright-Patterson AF Base, Ohio, 1965.
11. G. SANDER, 'Bornes supérieures et inférieures dans l'analyse matricielle des plaques en flexion-torsion', *Bull. Soc. Royale des Sci. de Liège*, **33**, 456–94, 1964.
12. B. FRAEIJS DE VEUBEKE, 'Bending and stretching of plates', *Proc. Conf. on Matrix Methods in Structural Mechanics*, Air Force Institute of Technology, Wright-Patterson AF Base, Ohio, 1965.
13. B. M. IRONS, 'A conforming quartic triangular element for plate bending', *Int. J. Num. Meth. Eng.*, **1**, 29–46, 1969.
14. R. W. CLOUGH and C. A. FELIPPA, 'A refined quadrilateral element for analysis of plate bending', *Proc. 2nd Conf. on Matrix Methods in Structural Mechanics*, Air Force Institute of Technology, Wright-Patterson AF Base, Ohio, 1968.
15. B. FRAEIJS DE VEUBEKE, 'A conforming finite element for plate bending', *Int. J. Solids Struct.*, **4**, 95–108, 1968.
16. F. K. BOGNER, R. L. FOX and L. A. SCHMIT, 'The generation of interelement-compatible stiffness and mass matrices by the use of interpolation formulae', *Proc. Conf. on Matrix Methods in Structural Mechanics*, Air Force Institute of Technology, Wright-Patterson AF Base, Ohio, 1965.
17. I.M. SMITH and W. DUNCAN, 'The effectiveness of nodal continuities in finite element analysis of thin rectangular and skew plates in bending', *Int. J. Num. Meth. Eng.*, **2**, 253–8, 1970.
18. K. BELL, 'A refined triangular plate bending element', *Int. J. Num. Meth. Eng.*, **1**, 101–22, 1969.
19. G. A. BUTLIN and R. FORD, 'A compatible plate bending element', University of Leicester Engineering Department report 68–15, 1968.
20. G. R. COWPER, E. KOSKO, G. M. LINDBERG and M. D. OLSON, 'Formulation of a new triangular plate bending element', *Trans. Canad. Aero-Space Inst.*, **1**, 86–90, 1968. (See also NRC Aero report LR514, 1968.)
21. W. BOSSHARD, 'Ein neues vollverträgliches endliches Element für Plattenbiegung', *Mt. Ass. Bridge Struct. Eng. Bull.*, **28**, 27–40, 1968.
22. J. H. ARGYRIS, I. FRIED and D. W. SCHARPF, 'The TUBA family of plate elements for the matrix displacement method', *The Aeronaut. J., Roy. Aeronaut. Soc.*, **72**, 701–9, 1968.
23. W. VISSER, 'The finite element method in deformation and heat conduction problems', Dr Wiss. Dissertation, Tech. Hoch., Delft, 1968.

24. B. M. IRONS, 'Comments on "Complete polynomial displacement fields for finite element method", by P. C. Dunne', *The Aeronaut. J., Roy. Aeronaut. Soc.*, **72**, 709, 1968.
25. O. C. ZIENKIEWICZ and Y. K. CHEUNG, 'The finite element method for analysis of elastic isotropic and orthotropic slabs', *Proc. Inst. Civ. Eng.*, **28**, 471–88, 1964.
26. R. W. CLOUGH, 'The finite element method in structural mechanics', in *Stress Analysis* (eds O. C. Zienkiewicz and G. S. Holister), chap. 7, Wiley, Chichester, 1965.
27. D. J. DAWE, 'Parallelogram element in the solution of rhombic cantilever plate problems', *J. Strain Anal.*, **1**, 223–30, 1966.
28. J. G. ARGYRIS, 'Continua and discontinua', *Proc. Conf. on Matrix Methods in Structural Mechanics*, Air Force Institute of Technology, Wright-Patterson AF Base, Ohio, 1965.
29. L. S. D. MORLEY, 'On the constant moment plate bending element', *J. Strain Anal.*, **6**, 20–4, 1971.
30. R. D. WOOD, 'A shape function routine for the constant moment triangular plate bending element', *Eng. Computations*, **1**, 189–98, 1984.
31. R. NARAYANASWAMI, 'New triangular plate-bending element with transverse shear flexibility', *JAIAA*, **12**, 1761–3, 1974.
32. A. RAZZAQUE, 'Program for triangular bending element with derivative smoothing', *Int. J. Num. Meth. Eng.*, **5**, 588–9, 1973.
33. B. M. IRONS and A. RAZZAQUE, 'Shape function formulation for elements other than displacement models', in *Proc. 2nd Conf. on Variational Methods in Engineering*, Southampton University, 1972.
34. J. E. WALZ, R. E. FULTON and N. J. CYRUS, 'Accuracy and convergence of finite element approximation', in *Proc. 2nd Conf. on Matrix Methods in Structural Mechanics*, Air Force Institute of Technology, Wright-Patterson AF Base, Ohio, 1968.
35. R. J. MELOSH, 'Basis of derivation of matrices for the direct stiffness method', *JAIAA*, **1**, 1631–7, 1963.
36. A. ADINI and R. W. CLOUGH, 'Analysis of plate bending by the finite element method', Report to National Science Foundation USA G.7337, 1961.
37. Y. K. CHEUNG, I. P. KING and O. C. ZIENKIEWICZ, 'Slab bridges with arbitrary shape and support conditions—a general method of analysis based on finite elements', *Proc. Inst. Civ. Eng.*, **40**, 9–36, 1968.
38. J. L. TOCHER and K. K. KAPUR, 'Comment on basis of derivation of matrices for direct stiffness method by R. J. Melosh', *JAIAA*, **3**, 1215–16, 1965.
39. R. D. HENSHELL, D. WALTERS and G. B. WARBURTON, 'A new family of curvilinear plate bending elements for vibration and stability', *J. Sound Vibr.*, **20**, 327–43, 1972.
40. R. L. TAYLOR, O. C. ZIENKIEWICZ, J. C. SIMO and A. H. C. CHAN, 'The patch test—a condition for assessing FEM convergence', *Int. J. Num. Meth. Eng.*, **22**, 39–62, 1986.
41. O. C. ZIENKIEWICZ, S. QU, R. L. TAYLOR and S. NAKAZAWA, 'The patch test for mixed formulations', *Int. J. Num. Meth. Eng.*, **23**, 1873–83, 1986.
42. O. C. ZIENKIEWICZ and D. LEFEBVRE, 'Three field mixed approximation and the plate bending problem', *Comm. Appl. Num. Meth.*, **3**, 301–9, 1987.
43. P. G. BERGAN and L. HANSSEN, 'A new approach for deriving "good" element stiffness matrices', in *The Mathematics of Finite Elements and*

Applications (ed. J. R. Whiteman), pp. 483–97, Academic Press, London, 1977.
44. R. V. SOUTHWELL, 'Relaxation methods in theoretical physics', Clarendon Press, Oxford, 1946.
45. P. G. BERGAN and M. K. NYGARD, 'Finite elements with increased freedom in choosing shape functions', *Int. J. Num. Meth. Eng.*, **20**, 643–63, 1984.
46. C. A. FELIPPA and P. G. BERGAN, 'A triangular plate bending element based on energy orthogonal free formulation', *Comp. Meth. Appl. Mech. Eng.*, **61**, 129–60, 1987.
47. A. SAMUELSSON, 'The global constant strain condition and the patch test', in *Energy Methods in Finite Element Analysis* (eds R. Glowinski, E. Y. Rodin and O. C. Zienkiewicz), chap. 3, pp. 49–68, Wiley, Chichester, 1979.
48. B. SPECHT, 'Modified shape functions for the three node plate bending element passing the patch test', *Int. J. Num. Meth. Eng.*, **26**, 705–15, 1988.
49. T. J. R. HUGHES, *The Finite Element Method*, pp. 311–12, Prentice-Hall, New Jersey, 1987.
50. H. MARCUS, *Die Theorie elastischer Gewebe und ihre Anwendung auf die Berechnung biegsamer Platten*, Springer, Berlin, 1932.
51. P. BALLESTEROS and S. L. LEE, 'Uniformly loaded rectangular plate supported at the corners', *Int. J. Mech. Sci.*, **2**, 206–11, 1960.
52. J. L. BATOZ, K. J. BATHE and L. W. HO, 'A study of three-node triangular plate bending elements', *Int. J. Num. Meth. Eng.*, **15**, 1771–812, 1980.
53. M. M. HRABOK and T. M. HRUDEY, 'A review and catalogue of plate bending finite elements', *Comp. Struct.*, **19**, 479–95, 1984.
54. J. JIROUSEK and LAN GUEX, 'The hybrid–Trefftz finite element model and its application to plate bending', *Int. J. Num. Meth. Eng.*, **23**, 651–93, 1986.
55. A. RAZZAQUE, 'Finite element analysis of plates and shells', Ph.D. thesis, Civil Engineering Department, University of Wales, Swansea, 1972.
56. D. WITHUM, 'Berechnung von Platten nach dem Ritzschen Verfahren mit Hilfe dreieckförmiger Meshnetze', Mittl. Inst. Statik Tech. Hochschule, Hanover, 1966.
57. C. A. FELIPPA, 'Refined finite element analysis of linear and non-linear two-dimensional structures', Ph.D. thesis, Structural Engineering Department, University of California, Berkeley, 1966.
58. A. ZENISEK, 'Interpolation polynomials on the triangle', *Int. J. Num. Meth. Eng.*, **10**, 283–96, 1976.
59. A. PEANO, 'Conforming approximation for Kirchhoff plates and shells', *Int. J. Num. Meth. Eng.*, **14**, 1273–91, 1979.
60. J. J. GÖEL, Construction of basic functions for numerical utilization of Ritz's method', *Numerische Math.*, **12**, 435–47, 1968.
61. G. BIRKHOFF and L. MANSFIELD, 'Compatible triangular finite elements', *J. Math. Anal. Appl.*, **47**, 531–53, 1974.
62. C. L. LAWSON, 'C^1-compatible interpolation over a triangle', NASA Jet Propulsion Laboratory, TM 33–770, 1976.
63. L. R. HERRMANN, 'A bending analysis for plates', in *Proc. Conf. on Matrix Methods in Structural Mechanics*, Air Force Institute of Technology, Wright-Patterson AF Base, Ohio, pp. 577–602, 1965.
64. K. HELLAN, 'Analysis of elastic plates in flexure by a simplified finite element method', Acta Polytechnica Scandinavica report C146, 28 pp., Trondheim, 1967.
65. L. R. HERRMANN, 'Finite element bending analysis of plates', *Proc. Am. Soc. Civ. Eng.*, **93**, EM 5, 13–26, 1967.

66. J. BRON and G. DHATT, 'Mixed quadrilateral elements for plate bending', *JAIAA*, **10**, 1359–61, 1972.
67. W. VISSER, 'A refined mixed type plate bending element', *JAIAA*, **7**, 1801–3, 1969.
68. J. C. BOOT, 'On a problem arising from the derivation of finite element matrices using Reissner's principle', *Int. J. Num. Meth. Eng.*, **12**, 1879–82, 1978.
69. A. CHATERJEE and A. V. SETLUR, 'A mixed finite element formulation for plate problems', *Int. J. Num. Meth. Eng.*, **4**, 67–84, 1972.
70. J. W. HARVEY and S. KELSEY, 'Triangular plate bending elements with enforced compatibility', *JAIAA*, **9**, 1023–6, 1971.
71. B. FRAEIJS DE VEUBEKE and O. C. ZIENKIEWICZ, 'Strain energy bounds in finite element analysis by slab analogy', *J. Strain Anal.*, **2**, 265–71, 1967.
72. B. FRAEIJS DE VEUBEKE, 'An equilibrium model for plate bending', *Int. J. Solids Struct.*, **4**, 447–68, 1968.
73. L. S. D. MORLEY, 'The triangular equilibrium element in the solution of plate bending problems', *Aero. Q.*, **19**, 149–69, 1968.
74. T. H. H. PIAN, 'Derivation of element stiffness matrices by assumed stress distribution', *JAIAA*, **2**, 1332–6, 1964.
75. T. H. H. PIAN and P. TONG, 'Basis of finite element methods for solid continua', *Int. J. Num. Meth. Eng.*, **1**, 3–28, 1969.
76. R. J. ALLWOOD and G. M. M. CORNES, 'A polygonal finite element for plate bending problems using the assumed stress approach', *Int. J. Num. Meth. Eng.*, **1**, 135–60, 1969.
77. P. TONG, 'New displacement hybrid models for solid continua', *Int. J. Num. Meth. Eng.*, **2**, 73–83, 1970.
78. I. TORBE and K. CHURCH, 'A general quadrilateral plate element', *Int. J. Num. Meth. Eng.*, **9**, 855–68, 1975.
79. B. E. GREENE, R. E. JONES, R. W. MCLAY and D. R. STROME, 'Generalized variational principles in the finite element method', *JAIAA*, **7**, 1254–60, 1969.
80. R. D. COOK, 'Two hybrid elements for analysis of thick, thin and sandwich plates', *Int. J. Num. Meth. Eng.*, **5**, 277–99, 1972.
81. R. D. COOK and S. G. LADKANY, 'Observations regarding assumed-stress hybrid plate elements', *Int. J. Num. Meth. Eng.*, **8**, 513–20, 1974.
82. B. K. NEALE, R. D. HENSHELL and G. EDWARDS, 'Hybrid plate bending elments', *J. Sound Vibr.*, **22**, 101–12, 1972.
83. C. JOHNSON, 'On the convergence of a mixed finite-element method for plate bending problems', *Num. Math.*, **21**, 43–62, 1973.
84. D. J. ALLMAN, 'A simple cubic displacement model for plate bending', *Int. J. Num. Meth. Eng.*, **10**, 263–81, 1976.
85. J. JIROUSEK, 'Improvement of computational efficiency of the 9 DOF triangular hybrid–Trefftz plate bending element' (Letter to Editor), *Int. J. Num. Meth. Eng.*, **23**, 2167–8, 1986.
86. G. A. WEMPNER, J. T. ODEN and D. K. CROSS, 'Finite element analysis of thin shells', *Proc. Am. Soc. Civ. Eng.*, **EM6**, 1273–94, 1968.
87. J. H. STRICKLIN, W. HAISLER, P. TISDALE and K. GUNDERSON, 'A rapidly converging triangle plate element', *JAIAA*, **7**, 180–1, 1969.
88. G. S. DHATT, 'Numerical analysis of thin shells by curved triangular elements based on discrete Kirchhoff hypothesis', in *Proc. Symp. on Applications of FEM in Civil Engineering* (eds W. R. Rowan and R. M. Hackett), Vanderbilt University, Nashville, Tennessee, 1969.

89. G. DHATT, 'An efficient triangular shell element', *JAIAA*, **8**, 2100–2, 1970.
90. J. L. BATOZ and G. DHATT, 'Development of two simple shell elements', *JAIAA*, **10**, 237–8, 1972.
91. J. T. BALDWIN, A. RAZZAQUE and B. M. IRONS, 'Shape function subroutine for an isoparametric thin plate element', *Int. J. Num. Meth. Eng.*, **7**, 431–40, 1973.
92. M. A. CRISFIELD, 'A qualitative mindlin element using shear constraints', *Comp. Struct.*, **18**, 833–52, 1984.
93. L. P. R. LYONS, 'A general finite element system with special analysis of cellular structures', Ph.D. thesis, Imperial College of Science and Technology, London, 1977.
94. J. L. BATOZ and M. BEN TOHAR, 'Evaluation of a new quadrilateral thin plate bending element', *Int. J. Num. Meth. Eng.*, **18**, 1655–77, 1982.
95. J. L. BATOZ, 'An explicit formulation for an efficient triangular plate bending element', *Int. J. Num. Meth. Eng.*, **18**, 1077–89, 1982.
96. G. DHATT, L. MARCOTTE and Y. MATTE, 'A new triangular discrete Kirchhoff plate-shell element', *Int. J. Num. Meth. Eng.*, **23**, 453–70, 1986.
97. B. M. IRONS, 'The semi loof shell element', in *Finite Elements for Thin Shells and Curved Members* (eds D. G. Ashwell and R. H. Gallagher), chap. 11, pp. 197–222, Wiley, Chichester, 1976.
98. R. A. F. MARTINS and D. R. J. OWEN, 'Thin plate semi-loof element for structural analysis including stability and structural vibration', *Int. J. Num. Meth. Eng.*, **12**, 1667–76, 1978.
99. J. L. BATOZ and G. DHATT, 'Development of two simple shell elements', *JAIAA*, **10**, 237–8, 1972.
100. J. L. BATOZ and M. B. TEHAR, 'Evaluation of a new quadrilateral thin plate bending element', *Int. J. Num. Meth. Eng.*, **18**, 1655–77, 1982.
101. M. A. CRISFIELD, 'A new model thin plate bending element using shear constraints; a modified version of Lyons' element', *Comp. Meth. Appl. Mech. Eng.*, **38**, 93–120, 1983.
102. M. A. CRISFIELD, *Finite Elements and Solution Procedures for Structural Analysis*, Vol. 1, *Linear Analysis*, Pineridge Press, Swansea, 1986.
103. R. H. MACNEAL, 'A simple quadrilateral shell element', *Comp. Struct.*, **8**, 175–83, 1978.

2

'Thick' Reissner–Mindlin plates — irreducible and mixed formulations

2.1 Introduction

We have already introduced in Chapter 1 the full theory of thick plates from which the thin plate, Kirchhoff, theory arises as the limiting case. In this chapter we shall show how the numerical solution of thick plates can easily be achieved and how, in the limit, an alternative procedure for solving all the problems of Chapter 1 arises.

To ensure continuity we repeat below the governing equations [viz. Eqs (1.9) to (1.13) or (1.67a) to (1.67d)]. Referring to Fig. 1.3 of Chapter 1 and the text for definitions, we remark that all the equations could equally well be derived from full three-dimensional analysis of a flat and relatively thin portion of an elastic continuum illustrated in Fig. 2.1. All that it is now necessary to do is to assume that, whatever form of the approximating shape functions in the xy plane, those in the z direction are only linear. Further, it is assumed that σ_{zz} stresses are zero valued,† thus eliminating the effect of vertical strain. The first approximations of that type were introduced quite early[1,2] and the elements then derived are exactly of the Reissner–Mindlin type discussed in Chapter 1.

The equations from which we shall start and on which we shall base all the subsequent discussion are thus

$$\mathbf{M}-\mathbf{DL\theta}=\mathbf{0} \tag{2.1a}$$

[viz. Eqs (1.9) and (1.67a)],

$$\mathbf{L}^{\mathrm{T}}\mathbf{M}+\mathbf{S}=\mathbf{0} \tag{2.1b}$$

† Reissner includes the effect of σ_{zz} in bending but for simplicity this is disregarded here.

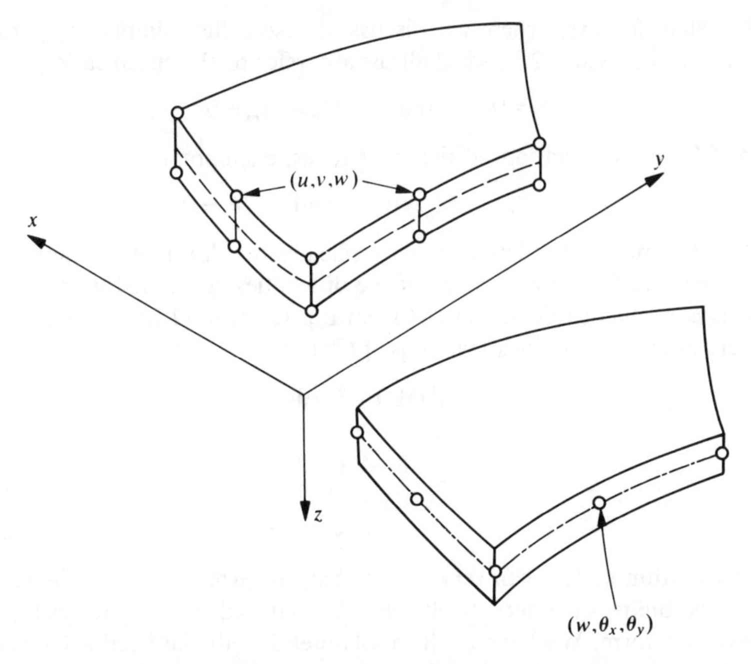

Fig. 2.1 An isoparametric three-dimensional element with linear interpolation in the transverse direction and the 'thick' plate element

[viz. Eqs (1.12) and (1.67b)],

$$\frac{\mathbf{S}}{\alpha}+\boldsymbol{\theta}-\boldsymbol{\nabla}w=\mathbf{0} \tag{2.1c}$$

where $\alpha=\beta Gt$ is the shear rigidity [viz. Eqs (1.10) and (1.67c)] and

$$\boldsymbol{\nabla}^{\mathrm{T}}\mathbf{S}=-q \tag{2.1d}$$

[viz. Eqs (1.13) and (1.67d)]. In the above the moments **M**, the shear forces **S** and the elastic matrices **D** are as defined in Chapter 1 and

$$\mathbf{L}^{\mathrm{T}}=\begin{bmatrix}\dfrac{\partial}{\partial x} & 0 & \dfrac{\partial}{\partial y}\\[2ex] 0 & \dfrac{\partial}{\partial y} & \dfrac{\partial}{\partial x}\end{bmatrix} \tag{2.1e}$$

Boundary conditions are of course imposed on w and $\boldsymbol{\theta}$ or the corresponding tractions S_n, M_n, M_{ns} in the manner discussed in Sec. 1.2.2.

In what follows, when we discuss the so-called simply supported condition (viz. Sec. 1.2.2), we shall usually refer to the specification

$$w=0 \qquad \text{and} \qquad M_n=M_{ns}=0$$

as 'soft' support (and indeed the most realistic support) and to

$$w=0, \qquad M_n=0 \qquad \text{and} \qquad \theta_s=0$$

as 'hard' support. The latter in fact replicates the thin plate assumptions and, incidentally, leads to some of the difficulties associated with it.

It is convenient to eliminate **M** from Eqs (2.1a) to (2.1d) and write the system of three equations [viz. Eqs (1.68)] as

$$\mathbf{L}^{\mathrm{T}}\mathbf{D}\mathbf{L}\boldsymbol{\theta}+\mathbf{S}=\mathbf{0} \tag{2.2a}$$

$$\frac{\mathbf{S}}{\alpha}+\boldsymbol{\theta}-\boldsymbol{\nabla}w=\mathbf{0} \tag{2.2b}$$

$$\boldsymbol{\nabla}^{\mathrm{T}}\mathbf{S}=-q \tag{2.2c}$$

This equation system can serve as the base on which a mixed discretization is built—or alternatively can be reduced further to yield an irreducible form. We have dealt in Chapter 1 with the irreducible form which was given by a fourth-order equation in terms of w alone and which could only serve for solution of *thin plate* problems, i.e. when $\alpha=\infty$ [Eq. (1.15)]. On the other hand, it is easy to derive an alternative irreducible form which is valid only if $\alpha\neq\infty$. Now the shear forces can be eliminated yielding two equations:

$$\mathbf{L}^{\mathrm{T}}\mathbf{D}\mathbf{L}\boldsymbol{\theta}+\alpha(\boldsymbol{\nabla}w-\boldsymbol{\theta})=\mathbf{0} \tag{2.3a}$$

$$-\boldsymbol{\nabla}^{\mathrm{T}}(\alpha\boldsymbol{\theta})+\boldsymbol{\nabla}^{\mathrm{T}}(\alpha\boldsymbol{\nabla}w)=-q \tag{2.3b}$$

This is an *irreducible* system corresponding to minimization of the total potential energy

$$\Pi=\frac{1}{2}\int_{\Omega}(\mathbf{L}^{\mathrm{T}}\boldsymbol{\theta})\mathbf{D}(\mathbf{L}\boldsymbol{\theta})\,\mathrm{d}\Omega+\frac{1}{2}\int_{\Omega}(\boldsymbol{\nabla}w-\boldsymbol{\theta})^{\mathrm{T}}\alpha(\boldsymbol{\nabla}w-\boldsymbol{\theta})\,\mathrm{d}\Omega$$
$$-\int_{\Omega}w^{\mathrm{T}}q\,\mathrm{d}\Omega+\text{boundary terms} \tag{2.4}$$

as can easily be verified.

In the above the first term is simply the bending energy and the second the shear distortion energy [viz. Eq. (1.75a)].

Clearly this irreducible system is only possible when $\alpha\neq\infty$, but it can, obviously, be interpreted as a solution of the potential energy given by Eq. (1.75a) for 'thin' plates with the constraint of Eq. (1.75b) being imposed in a *penalty manner* with α being now a pure penalty parameter. Thus, as indeed is physically evident, the thin plate formulation is simply a limiting case of such analysis.

We shall see that the penalty form can only yield a satisfactory solution when discretization of the corresponding mixed formulation satisfies the necessary convergence criteria.

2.2 The irreducible formulation—reduced integration

The procedures for discretizing Eq. (2.3) are straightforward. Firstly, the two displacement variables are approximated by appropriate shape functions and parameters as

$$\boldsymbol{\theta} = \mathbf{N}_\theta \bar{\boldsymbol{\theta}} \qquad \text{and} \qquad w = \mathbf{N}_w \bar{\mathbf{w}} \tag{2.5}$$

Then, the approximation equations are obtained directly by the use of the Galerkin process and appropriate integration by parts, or, equivalently, by the use of virtual work expressions. Here we note that the appropriate, generalized strain components, corresponding to the moments $\mathbf{M}$ and shear forces $\mathbf{S}$ are

$$\boldsymbol{\varepsilon}_{\mathrm{m}} = \mathbf{L}\boldsymbol{\theta} = (\mathbf{L}\mathbf{N}_\theta)\bar{\boldsymbol{\theta}} \tag{2.6a}$$

and

$$\boldsymbol{\varepsilon} = \boldsymbol{\nabla} w - \boldsymbol{\theta} = \boldsymbol{\nabla}\mathbf{N}_w \bar{\mathbf{w}} - \mathbf{N}_\theta \bar{\boldsymbol{\theta}} \tag{2.6b}$$

We thus obtain

$$\left(\int_\Omega (\mathbf{L}\mathbf{N}_\theta)^{\mathrm{T}} \mathbf{D} \mathbf{L}\mathbf{N}_\theta \,\mathrm{d}\Omega + \int_\Omega \mathbf{N}_\theta^{\mathrm{T}} \alpha \mathbf{N}_\theta \,\mathrm{d}\Omega\right)\bar{\boldsymbol{\theta}} - \left(\int_\Omega \mathbf{N}_\theta^{\mathrm{T}} \alpha \boldsymbol{\nabla}\mathbf{N}_w \,\mathrm{d}\Omega\right)\bar{\mathbf{w}} = \mathbf{f}_\theta \tag{2.7a}$$

and

$$-\left(\int_\Omega (\boldsymbol{\nabla}\mathbf{N}_w)^{\mathrm{T}} \alpha \mathbf{N}_\theta \,\mathrm{d}\Omega\right)\bar{\boldsymbol{\theta}} + \left(\int_\Omega (\boldsymbol{\nabla}\mathbf{N}_w)^{\mathrm{T}} \alpha \boldsymbol{\nabla}\mathbf{N}_w \,\mathrm{d}\Omega\right)\bar{\mathbf{w}} = \mathbf{f}_w \tag{2.7b}$$

or simply

$$\mathbf{K}\begin{Bmatrix} \bar{\mathbf{w}} \\ \bar{\boldsymbol{\theta}} \end{Bmatrix} = \mathbf{K}\mathbf{a} = (\mathbf{K}_b + \mathbf{K}_s)\mathbf{a} = \begin{Bmatrix} \mathbf{f}_w \\ \mathbf{f}_\theta \end{Bmatrix} = \mathbf{f} \tag{2.8}$$

with

$$\mathbf{a}^{\mathrm{T}} = [\bar{\mathbf{w}}, \bar{\boldsymbol{\theta}}_x, \bar{\boldsymbol{\theta}}_y]$$

$$\mathbf{K}_b = \begin{bmatrix} 0 & 0 \\ 0 & \tilde{\mathbf{K}}_b \end{bmatrix}$$

$$\mathbf{K}_s = \begin{bmatrix} \tilde{\mathbf{K}}_s & \tilde{\mathbf{K}}_{bs} \\ \mathbf{K}_{bs} & 0 \end{bmatrix}$$

where $\tilde{\mathbf{K}}_b, \tilde{\mathbf{K}}_{bs}$ and $\tilde{\mathbf{K}}_s$ are as defined in Eq. (2.7) and

$$\mathbf{f}_\theta = \int_{\Gamma_t} \mathbf{N}_\theta^{\mathrm{T}} \tilde{\mathbf{M}} \, \mathrm{d}\Omega \tag{2.9a}$$

where $\tilde{\mathbf{M}}$ are prescribed moments on boundary Γ_t, and

$$\mathbf{f}_w = \int_\Omega \mathbf{N}_w^{\mathrm{T}} q \, \mathrm{d}\Omega + \int_{\Gamma_t} \mathbf{N}_w^{\mathrm{T}} \tilde{S}_n \, \mathrm{d}\Gamma \tag{2.9b}$$

where $\tilde{S}_n$ is the prescribed shear on boundary Γ_t.

The formulation is straightforward and there is little to be said about it *a priori.* Apparently any shape functions of a two-dimensional kind interpolating the two rotations and the lateral displacement could be used. Figure 2.2 shows some rectangular (or with isoparametric distortion, quadrilateral) elements used in the early work.[1-3] All should, in principle, be convergent as C_0 continuity exists and constant strain states are available. In Fig. 2.3 we show what in fact happens with a fairly fine subdivision of quadratic serendipity and lagrangian rectangles as the ratio of thickness to span, t/L, varies.

We note that the magnitude of the coefficient α is best measured by the ratio of the shear to bending rigidities and we could assess its value in a non-dimensional form. Thus for an isotropic material with $\alpha = Gt$ this ratio becomes

$$\frac{Gt}{Et^3} 12(1-\nu^2)L^2 \propto \left(\frac{L}{t}\right)^2 \tag{2.10}$$

Obviously 'thick' and 'thin' behaviour therefore depends on the t/L ratio.

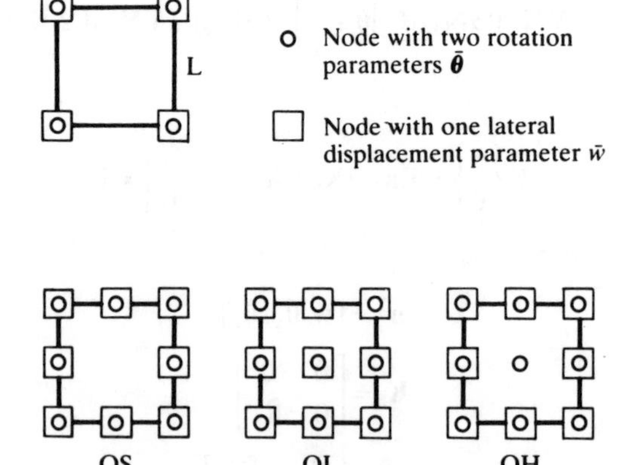

Fig. 2.2 Some early thick plate elements

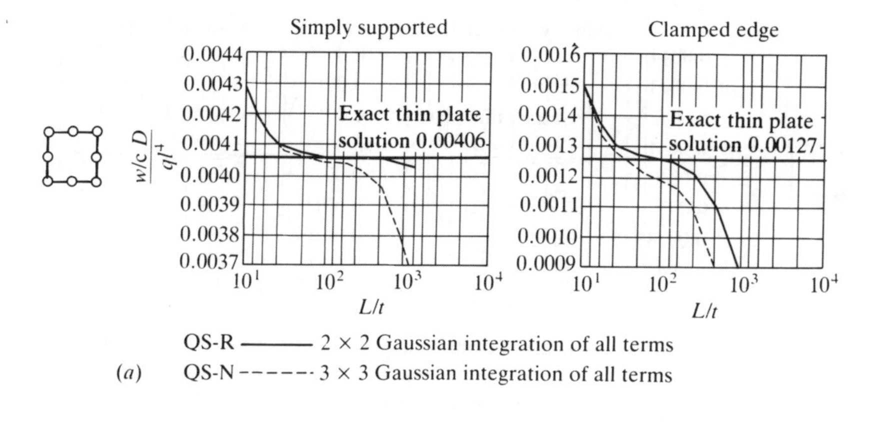

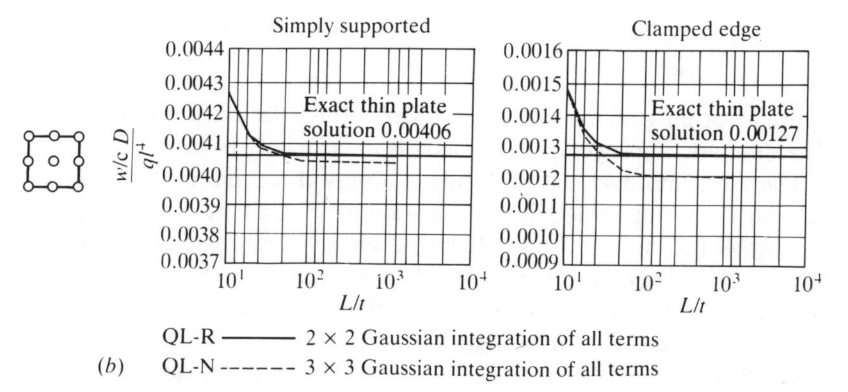

Fig. 2.3 Performance of (*a*) quadratic serendipity (QS) and (*b*) Lagrangian (QL) elements with varying L/t ratios, uniform load on a square plate with 4×4 normal subdivisions in a quarter. R is reduced 2×2 integration and N is normal 3×3 integration

It is immediately evident from Fig. 2.3 that while the answers are quite good for larger t/L ratios the serendipity, quadratic, fully integrated elements (QSN) rapidly depart from the thin plate solution, and in fact tend to zero results (locking) when this ratio becomes small. For lagrangian quadratics (QLN) the answers are better, but again as the plate tends to be thin they err on the small side.

What is the reason for this? As both elements contain full quadratic expansions their performance should be similar. Why is there locking behaviour? Various physical reasons were advanced quite early, and the one most valid is that the shear constraint implied by Eq. (2.2b) and used to eliminate the shear resultant is too strong if the terms in which this is involved are fully integrated. The problem was thus eliminated by using a *reduced* numerical integration, either on all terms, which we label R in the

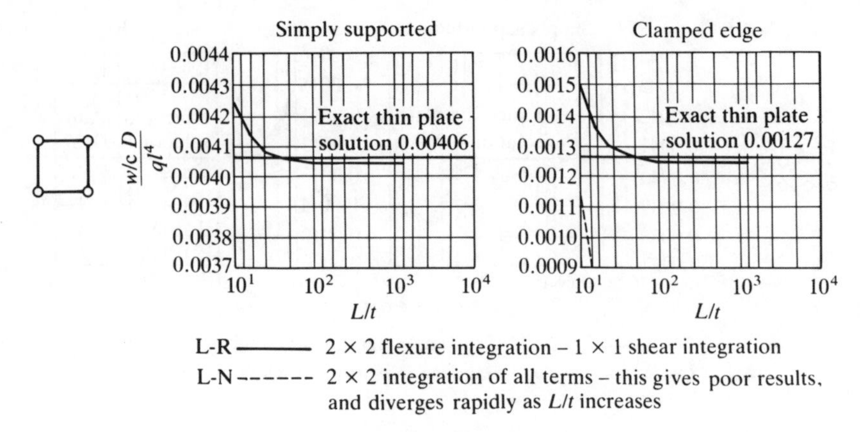

Fig. 2.4 Performance of bilinear elements with varying L/t ratio

figure,[4,5] or only on the offending, shear, terms selectively[6,7] (labelled S). The dramatic improvement in results is immediately noted.

The same improvement in results is observed for linear quadrilaterals in which the full (exact) integration gives results that are totally unacceptable (as shown in Fig. 2.4), but where a reduced integration on the shear terms (single point) gives excellent performance. (Reduced integration on all terms gives, of course, a matrix singularity.)

A remedy thus has been suggested; however, it is not universal. We note in Fig. 2.3 that even without reduction of integration order, lagrangian elements perform better in the quadratic expansion. In cubic elements (Fig. 2.5), however, we note that (*a*) almost no change occurs when integration is 'reduced' and (*b*), again, lagrangian-type elements perform very much better.

Many heuristic arguments have been advanced for devising better elements,[8–12] all making use of reduced integration concepts. Some of these perform quite well, for example the so-called 'heterosis' element of Hughes and Cohen[9] illustrated in Fig. 2.3 (in which the serendipity type of interpolation is used on w and a lagrangian one on θ), but all of the elements suggested in that era fail on some occasions, either locking or exhibiting singular behaviour. Thus such elements are not 'robust' and should not be used universally.

Surely a better explanation of their failure is needed and hence an understanding of how such elements could be designed. In the next section we shall address this problem by consideration of the mixed formulation.

The reader will recognize here arguments used in Chapter 12 of Volume 1, which led to a better understanding of the failure of some

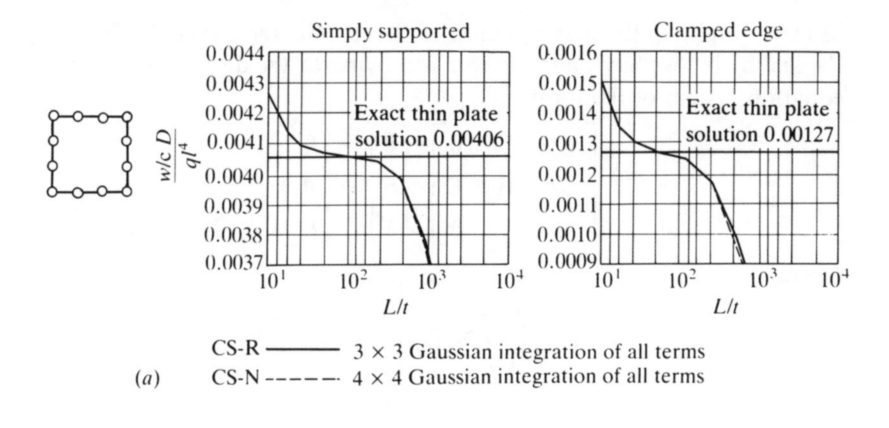

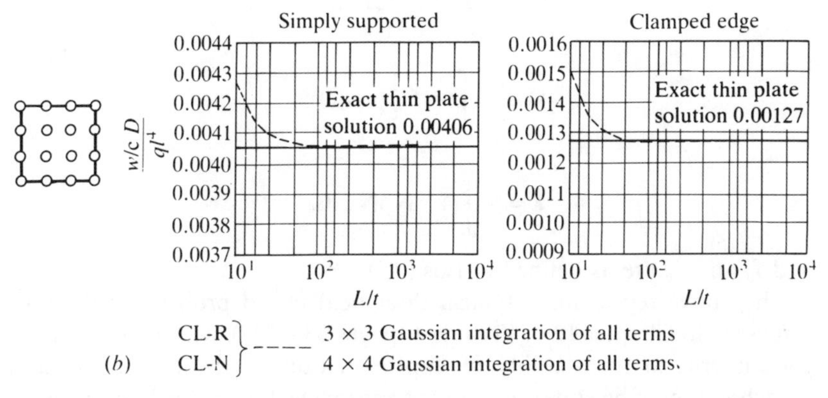

Fig. 2.5 Performance of cubic quadrilaterals (*a*) serendipity (CS) and (*b*) Lagrangian (CL) with varying L/t ratios

straightforward elasticity elements as incompressible behaviour was approached. The situation is completely parallel here.

2.3 Mixed formulation for thick plates and numerical integration equivalence

2.3.1 *The approximation.* The problem of thick plates can, of course, be solved as a mixed one starting from Eqs (2.2) and approximating directly each of the variables $\boldsymbol{\theta}$, $\mathbf{S}$ and w independently. We thus write

$$\boldsymbol{\theta}=\mathbf{N}_\theta\bar{\boldsymbol{\theta}}, \qquad \mathbf{S}=\mathbf{N}_s\bar{\mathbf{S}} \qquad \text{and} \qquad w=\mathbf{N}_w\bar{\mathbf{w}} \tag{2.11}$$

and obtain the approximating equation by the standard Galerkin procedure applied to each of the equations (though of course other weighting functions could be used, as we shall note later).

Thus weighting the first equation of (2.2) by $\mathbf{N}_\theta^{\mathrm{T}}$, the second by $\mathbf{N}_s^{\mathrm{T}}$ and the last by $\mathbf{N}_w^{\mathrm{T}}$ we have, after integration by parts, the discrete, symmetric equation system of the form

$$\begin{bmatrix} \mathbf{K}_b & \mathbf{C} & \mathbf{0} \\ \mathbf{C}^{\mathrm{T}} & \mathbf{H} & \mathbf{E} \\ \mathbf{0} & \mathbf{E}^{\mathrm{T}} & \mathbf{0} \end{bmatrix} \begin{Bmatrix} \bar{\boldsymbol{\theta}} \\ \bar{\mathbf{S}} \\ \bar{\mathbf{w}} \end{Bmatrix} = \begin{Bmatrix} \mathbf{f}_\theta \\ \mathbf{0} \\ \mathbf{f}_w \end{Bmatrix} \tag{2.12}$$

where

$$\begin{aligned} \mathbf{K}_b &= \int_\Omega (\mathbf{L}\mathbf{N}_\theta)^{\mathrm{T}}\, \mathbf{D}(\mathbf{L}\mathbf{N}_\theta)\, \mathrm{d}\Omega \\ \mathbf{C} &= -\int_\Omega \mathbf{N}_\theta^{\mathrm{T}}\, \mathbf{N}_s\, \mathrm{d}\Omega \\ \mathbf{H} &= -\int_\Omega \mathbf{N}_s^{\mathrm{T}}\, \frac{1}{\alpha} \mathbf{N}_s\, \mathrm{d}\Omega \\ \mathbf{E} &= -\int (\boldsymbol{\nabla}^{\mathrm{T}}\, \mathbf{N}_s)\mathbf{N}_w\, \mathrm{d}\Omega \end{aligned} \tag{2.13}$$

and $\mathbf{f}_\theta$ and $\mathbf{f}_w$ are as defined in Eqs (2.9).

The above represents a typical three-field mixed problem of the type discussed in Chapter 12 of Volume 1 (pages 333–334), which has to satisfy certain criteria for stability of approximation as the thin plate limit (which can now be solved exactly) is approached. For this limit we have

$$\alpha = \infty$$

and

$$\mathbf{H} = \mathbf{0} \tag{2.14}$$

In this limiting case it can readily be shown that one of the most important criteria of stability for any element assembly and boundary conditions is that

$$n_\theta + n_w \geqslant n_s \qquad \text{or} \qquad \alpha_{\mathrm{P}} \equiv \frac{n_\theta + n_w}{n_s} \geqslant 1 \tag{2.15a}$$

and

$$n_s \geqslant n_w \qquad \text{or} \qquad \beta_{\mathrm{P}} \equiv \frac{n_s}{n_w} \geqslant 1 \tag{2.15b}$$

where n_θ, n_s and n_w are the numbers of parameters defining the approximations of $\boldsymbol{\theta}$, $\mathbf{S}$ and w in Eqs (2.11).

If this necessary condition is not satisfied then the *equation system will always be singular*. Of course, this must be satisfied for the whole system but in addition it needs to be satisfied for element patches[13-15] if local instabilities and oscillations are to be avoided.

The above criteria will, as we shall see later, help us to design suitable thick plate elements which show convergence to thin plate solutions.

2.3.2 *Continuity requirements.* The approximation of the form given in Eqs (2.12) and (2.13) implies certain continuities. It is immediately evident that C_0 continuity is needed for the rotation shape functions $\mathbf{N}_\theta$ (as first derivatives are present in approximation), but that either $\mathbf{N}_s$ or $\mathbf{N}_w$ can be discontinuous (though of course some interelement connection of w variables is necessary physically).

In all the early approximations discussed in the previous section, C_0 continuity was assumed for both $\boldsymbol{\theta}$ and w variables, this being very easy to impose. We note that such continuity can not be described as *excessive* (as no physical conditions are violated), but we shall show later that very successful elements can be generated with discontinuous w interpolation.

For $\mathbf{S}$ it is obviously more convenient to use a completely discontinuous interpolation as then (for $1/\alpha$ not zero) the shear can be eliminated at the element level and the final stiffness matrices written simply in standard $\bar{\boldsymbol{\theta}}$, $\bar{\mathbf{w}}$ terms for element boundary nodes.

The continuous interpolation of the normal component of $\mathbf{S}$ is of course physically correct in the absence of line or point loads. However, with such interpolation elimination of $\bar{\mathbf{S}}$ is not possible and the retention of such additional system variables appears too costly to be used in practice and so far has not been adopted. However, we should note that an iterative solution process applicable to mixed forms described in Chapter 12 of Volume 1, page 357, can reduce substantially the cost of such additional variables[16] and that this procedure could possibly be applied here. Research is in fact in progress on such iterative formulation at the time of writing this edition, but to date no practical elements with C_0 interpolation of shear forces have been developed.

2.3.3 *Equivalence of forms with discontinuous* **S** *interpolation and reduced (selective) integration.* The equivalence of penalized mixed forms with a discontinuous interpolation of the constraint variable and of the corresponding irreducible forms with the same penalty variable was demonstrated in Chapter 12 of Volume 1, page 351, following the work of Malkus and Hughes[17] for incompressible problems. Indeed, an exactly analogous proof can be used for the present case and we leave the details to the reader.

Thus, for instance, if we consider a serendipity quadrilateral, shown in Fig. 2.6(a), in which integration of the shear terms (involving α) is made

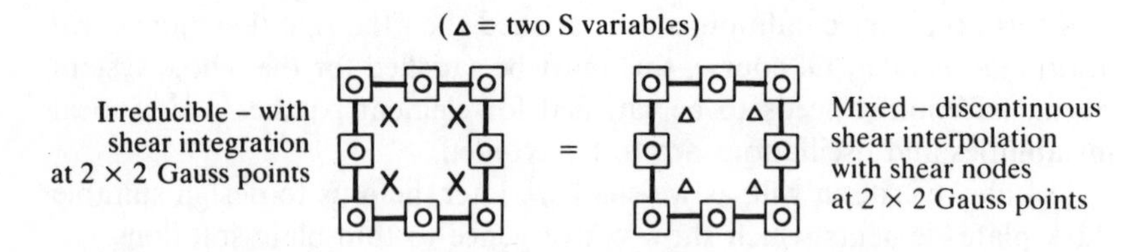

Fig. 2.6 Equivalence of mixed form and reduced shear integration in quadratic serendipity rectangle

at four Gauss points in an irreducible formulation [viz. Eqs (2.7) to (2.9)], we find that the answers are *identical* to a mixed form in which the **S** variables are given by a bilinear interpolation from nodes placed at the same Gauss points.

This result can also be argued from the limitation principle first given by Fraeijs de Veubeke.[18] This states that if the mixed form in which the stress is independently interpolated is precisely capable of reproducing the stress variation which is given in a corresponding irreducible form then the analysis results will be identical. It is clear that the four Gauss points at which the shear stress is sampled can only define a bilinear variation and the identity applies here.

The equivalence of reduced integration with the mixed, discontinuous, interpolation of **S** will be useful in our discussions for pointing out the reasons why many elements discussed in the previous section failed. However, in practice, it will be found equally convenient (and often more effective) to use the mixed interpolation explicitly and eliminate the **S** variables by element level condensation rather than to use special integration rules.

It must be pointed out that the equivalence fails if α varies within an element or indeed if the isoparametric mapping implies different interpolations. In such cases the mixed procedures are generally more accurate.

2.4 The patch test for plate bending elements

2.4.1 *Why elements fail.* The nature and application of the patch test have changed considerably since its early introduction. As shown in references 13 to 15 (and indeed as discussed in Chapter 11 of Volume 1 in detail), this test can prove, in addition to consistency requirements (which were initially the only item tested), the stability of the approximation by requiring that for a patch consisting of an assembly of one or more elements the stiffness matrices are non-singular whatever the boundary conditions imposed.

To be absolutely sure of such non-singularity the test must, at the final stage, be performed numerically. However, we find that the 'count' conditions necessary for such non-singularity, and given in Eqs (2.15), frequently also prove sufficient and make the numerical test only a final confirmation.[14,15] We shall demonstrate how the simple application of such counts immediately indicates *which elements fail* and which have a chance of survival. Indeed, it is easy to show why the original quadratic serendipity element with reduced integration (QS–R) is not robust.

In Fig. 2.7 we consider this element in a single- and four-element patch subject to so-called *constrained* boundary conditions, in which all displacements on the external boundary of the patch are prescribed and a *relaxed* boundary condition in which only three displacements (conveniently two θ's and one w) eliminate the rigid body modes. To ease the presentation in this figure, as well as in subsequent tests, we shall simply quote the values of α_P and β_P parameters as defined in Eqs (2.15) with a suffix C or R denoting the constrained and relaxed tests. The symbol (F) will be given to any failure to satisfy the *necessary* condition. In the tests of Fig. 2.7 we note that both patch tests fail with the parameter α_C being less than 1, and hence the elements will lock under certain circumstances (or show a singularity in the evaluation of **S**). A failure in the relaxed tests generally predicts a singularity in the final stiffness matrix of the assembly, and this is also where frequently computational failures have been observed.

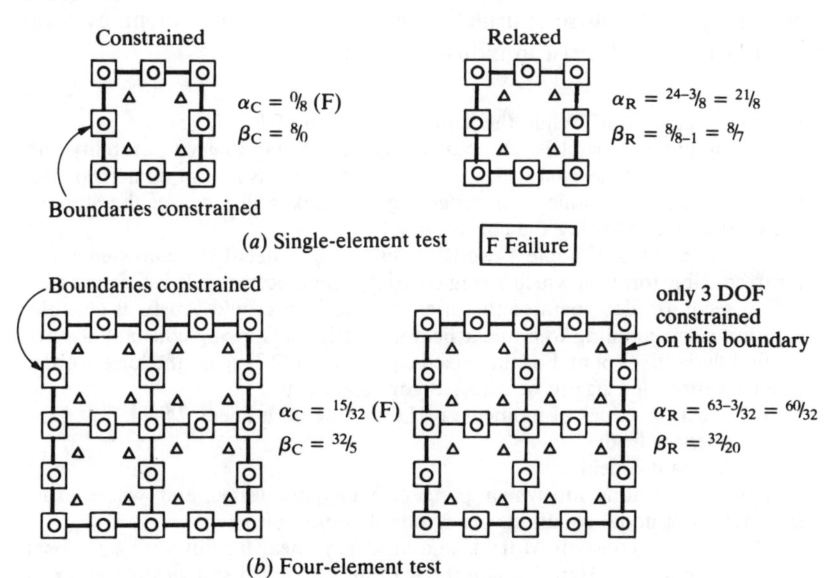

Fig. 2.7 'Constrained' and 'relaxed' patch test/count for serendipity quadrilateral. (In the C test all boundary displacements are fixed. In the R test only three boundary displacements are fixed, eliminating rigid body movement)

As the mixed and reduced integration elements are identical in this case we see immediately why the element fails in the problem of Fig. 2.3 (more spectacularly under clamped conditions). Indeed it is clear why in general the performance of lagrangian-type elements is better as it adds further degrees of freedom to increase n_θ (and also n_w).

In Table 2.1 we show a list of the α_P and β_P values for single- and four-element patches of various rectangles, and again we note that none of these satisfies completely the *necessary* requirements, and therefore none can be considered robust. However, it is interesting to note that the elements closest to satisfaction of the count perform best, and this explains why the heterosis element[9] is quite popular and used in many codes, and indeed why the lagrangian cubic is nearly robust and is used with success.[19]

Of course, similar approximations and counts can be made for various triangular elements. We list some typical and obvious ones, together with patch tests, in the first part of Table 2.2. Again, none perform adequately and all will result in locking and spurious modes in finite element programs.

We should note again that the failure of the patch test (with regard to stability) means that under some circumstances the element will fail. However, in many problems a reasonable performance can still be obtained and non-singularity observed in its performance, providing consistency is of course satisfied. It is for this reason that many non-robust elements still exist in industrial codes.

Numerical patch test. While the 'count' condition of Eqs (2.15) is a necessary one for stability of patches, on occasion singularity (and hence instability and locking) can still arise even with its satisfaction. For this reason numerical tests should always be conducted ascertaining the rank sufficiency of the stiffness matrices and also testing the consistency.

In Chapter 11 of Volume 1, we have discussed in detail the consistency test for irreducible forms in which a single variable set $\mathbf{u}$ occurred. It was found that with a second-order operator the discrete equations should satisfy *at least* the solution corresponding to a linear field $\mathbf{u}$ exactly, thus giving constant stresses (or first derivatives of $\mathbf{u}$). For the mixed equation set (2.2) again the lowest-order exact solution that has to be satisfied corresponds to:

(*a*) constant values of moments or $\mathbf{L\theta}$ and hence a linear θ field;
(*b*) linear w field;
(*c*) constant $\mathbf{S}$ field.

The exact solutions for which plate elements are tested and where exact satisfaction of nodal equations are required consist of:

(*a*) arbitrary constant $\mathbf{M}$ fields and arbitrary linear $\boldsymbol{\theta}$ fields with zero shear forces ($\mathbf{S}=0$). Here a quadratic w form is assumed still yielding an exact finite element solution.
(*b*) constant $\mathbf{S}$ and linear w fields yielding a constant $\boldsymbol{\theta}$ field. The solution requires a distributed couple on the right-hand side of Eq. (2.2a) and this was not included in the original formulation. A simple procedure is

TABLE 2.1

QUADRILATERAL MIXED ELEMENTS—PATCH COUNT

Element	Reference	Single element patch				Four element patch			
		α_C	β_C	α_R	β_R	α_C	β_C	α_R	β_R
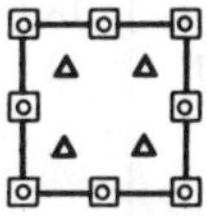	4 Q8S Q8R	$\frac{0}{8}$ (F)	$\frac{8}{0}$	$\frac{21}{8}$	$\frac{8}{7}$	$\frac{15}{32}$ (F)	$\frac{32}{5}$	$\frac{60}{32}$	$\frac{32}{20}$
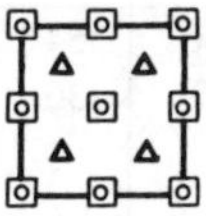	6,7 Q9S	$\frac{3}{8}$ (F)	$\frac{8}{1}$	$\frac{16}{8}$	$\frac{8}{8}$	$\frac{27}{32}$ (F)	$\frac{32}{9}$	$\frac{72}{32}$	$\frac{32}{24}$
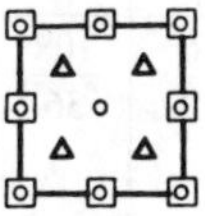	9 Q9H	$\frac{2}{8}$ (F)	$\frac{8}{0}$	$\frac{15}{8}$	$\frac{8}{7}$	$\frac{23}{32}$ (F)	$\frac{32}{5}$	$\frac{68}{32}$	$\frac{32}{20}$
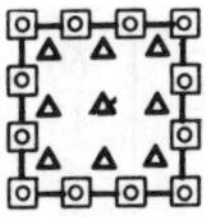	4 Q12R	$\frac{0}{18}$ (F)	$\frac{18}{0}$	$\frac{23}{18}$	$\frac{18}{11}$	$\frac{27}{72}$ (F)	$\frac{72}{9}$	$\frac{96}{72}$	$\frac{72}{32}$
	19 Q16R	$\frac{12}{18}$ (F)	$\frac{18}{4}$	$\frac{45}{18}$	$\frac{18}{15}$	$\frac{75}{72}$	$\frac{72}{25}$	$\frac{150}{72}$	$\frac{72}{50}$
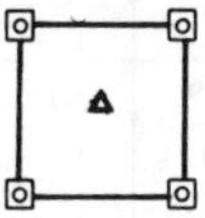	4,7 Q4S Q4R	$\frac{0}{2}$ (F)	$\frac{2}{0}$	$\frac{9}{2}$	$\frac{2}{3}$ (F)	$\frac{3}{8}$ (F)	$\frac{8}{1}$	$\frac{24}{8}$	$\frac{8}{8}$
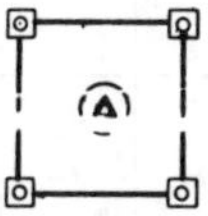	Q4BS	$\frac{2}{2}$	$\frac{2}{0}$	$\frac{11}{2}$	$\frac{2}{3}$ (F)	$\frac{11}{8}$	$\frac{8}{1}$	$\frac{32}{8}$	$\frac{8}{8}$

□ w – 1DOF
○ $\boldsymbol{\theta}$ – 2DOF
△ S – 2DOF

to disregard the satisfaction of the moment equilibrium in this test. This is done simply by inserting a very large value of the bending rigidity **D**.

2.4.2 *Design of some useful elements.* The simple patch count test indicates how elements could be designed to pass it, and thus avoid the

TABLE 2.2

TRIANGULAR MIXED ELEMENTS—PATCH COUNT

Element	Reference	Single element patch				Six element patch			
		α_C	β_C	α_R	β_R	α_C	β_C	α_R	β_R
I 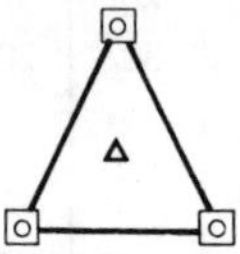		$\frac{0}{2}$ (F)	$\frac{2}{0}$	$\frac{6}{2}$	$\frac{2}{2}$	$\frac{3}{12}$ (F)	$\frac{12}{1}$	$\frac{18}{12}$	$\frac{12}{6}$
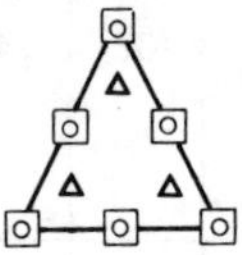		$\frac{0}{6}$ (F)	$\frac{6}{0}$	$\frac{15}{6}$	$\frac{6}{5}$	$\frac{21}{36}$ (F)	$\frac{36}{7}$	$\frac{54}{36}$	$\frac{36}{18}$
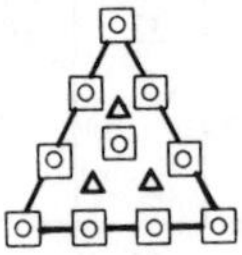		$\frac{3}{6}$ (F)	$\frac{6}{6}$	$\frac{27}{6}$	$\frac{6}{9}$ (F)	$\frac{57}{36}$	$\frac{36}{19}$	$\frac{108}{36}$	$\frac{36}{36}$
II 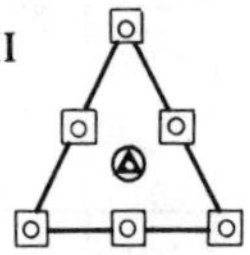		$\frac{2}{2}$	$\frac{2}{0}$	$\frac{17}{2}$	$\frac{2}{5}$ (F)	$\frac{33}{12}$	$\frac{12}{7}$	$\frac{66}{12}$	$\frac{12}{18}$ (F)
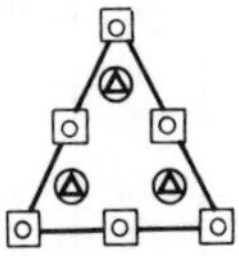	20 T6B3	$\frac{6}{6}$	$\frac{6}{0}$	$\frac{21}{6}$	$\frac{6}{5}$	$\frac{75}{36}$	$\frac{36}{7}$	$\frac{108}{36}$	$\frac{36}{18}$
III 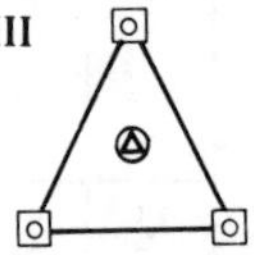	22 T3B1*	$\frac{2}{2}$	$\frac{2}{0}$	$\frac{8}{2}$	$\frac{2}{2}$	$\frac{15}{12}$	$\frac{12}{1}$	$\frac{30}{12}$	$\frac{12}{6}$
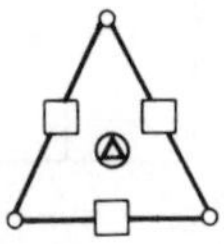	21 T3B1	$\frac{2}{2}$	$\frac{2}{0}$	$\frac{8}{2}$	$\frac{2}{2}$	$\frac{20}{12}$	$\frac{12}{6}$	$\frac{35}{12}$	$\frac{12}{11}$

singularity (instability). Two triangular elements introduced very recently on this basis are found to be robust, and at the same time excellent performers. Neither of these elements is 'obvious', and in both the interpolation of rotations is of a higher or equal order than that of w.

This is a clear violation of 'common sense', but in this case is justified by patch counts and performance.

Figure 2.8 shows both triangular elements and the second part of Table 2.2 shows again their performance in patches. The higher-order, quadratic, element was devised by Zienkiewicz and Lefebvre[20] and the lower-order one by Arnold and Falk.[21] Table 2.2 shows that both pass

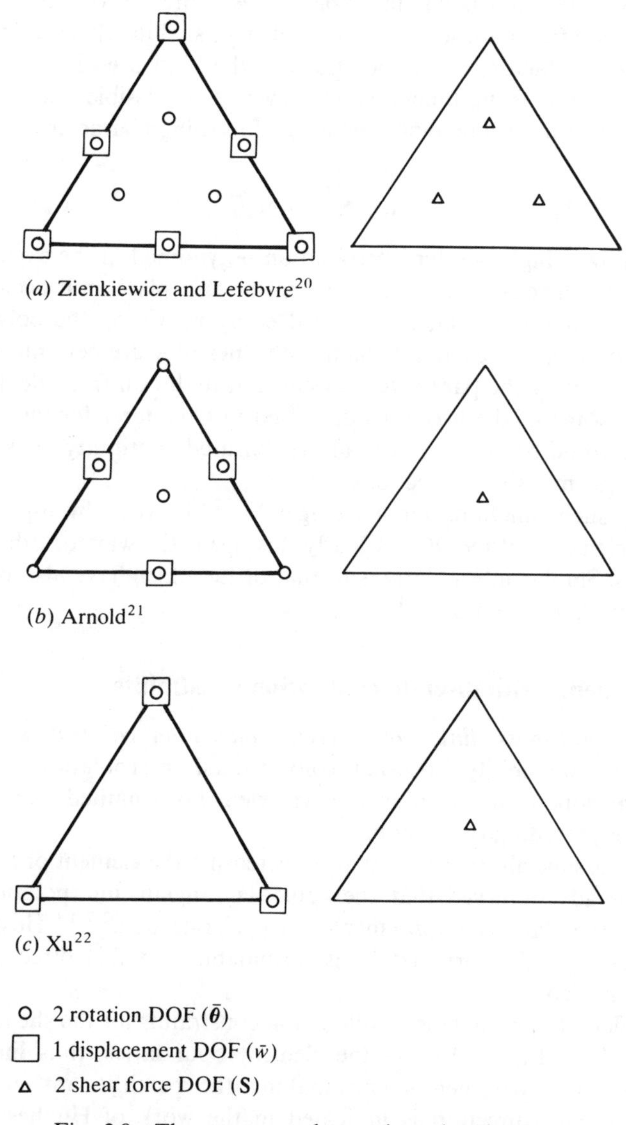

Fig. 2.8 Three recent, robust, triangular elements

the test, at least as far as the count condition goes, with flying colours, and indeed the numerical test is also found to be satisfied. It is of interest to note that the second element uses a discontinuous w interpolation, and is, in some ways, a direct opposite of the triangular element of Morley discussed in Chapter 1. We shall defer consideration of the numerical results obtainable for these elements to a later section, but remark here that their performance is excellent.

The last element shown in Table 2.2 is more conventional and as shown passes the count test. This element suggested in reference 15 proved not to be satisfactory if interpolations of the kind used in Eq. (2.5) are adopted for the shape functions. However, it is possible and frequently convenient to use shape functions of the following character:

$$\begin{aligned} \boldsymbol{\theta} &= \mathbf{N}_{\theta}\bar{\boldsymbol{\theta}} \\ w &= \mathbf{N}_{w}\bar{\mathbf{w}} + \bar{\mathbf{N}}_{w}\bar{\boldsymbol{\theta}} \end{aligned} \tag{2.16}$$

which allow a higher-order interpolation polynomial in the expansion of w. This of course is advantageous when the thin plate limit occurs. Now indeed C_0 continuity can be achieved easily providing the polynomials are of a correct degree and if again the values of w are determined along any side only by the parameters specified at nodes on this side. Here, for instance, some of the functions discussed in Chapter 1 for incompatible elements could be used, as these always ensured continuity of w and only failed in normal slope directions.

Using such functions for a triangle Xu[22] derives a well-performing, robust, element with 9 DOF. Clearly this opens the way for other similar elements. Similar interpolations to that of Eq. (2.16) have also been used by Tessler and Hughes.[23,24]

2.5 Elements with discrete collocation constraints

2.5.1 *General possibilities of discrete collocation constraints–quadrilaterals.* The possibility of using conventional interpolation to achieve satisfactory performance of mixed-type elements is limited, as is apparent from the preceding discussion.

One possible alternative is that of increasing the element order, and we have already observed that the cubic, lagrangian, interpolation nearly satisfies the stability requirement and performs well.[2,7,19] However, the complexity of the formulation is formidable and this direction is not recommended.

A different approach uses collocation constraints for the shear approximation [viz. Eq. (2.2b)] on the element boundaries, thus limiting the number of $\mathbf{S}$ parameters and making the patch count more easily satisfied. This direction is indicated in the work of Hughes and Tezduyar,[25] Bathe and Dvorkin[26,27] and Hinton and Huang,[28,29] as well as more recently by the generalization of Zienkiewicz *et al.*[30] and others.[31,32] The procedure bears much relation to the so-called DKT (discrete Kirchhoff theory) developed in Chapter 1 (viz. Sec. 1.18) and indeed explains why these, essentially thin plate, approximations are successful.

The key to the discrete formulation is evident if we consider Fig. 2.9, where a simple bilinear element is illustrated. We observe that with a C_0 interpolation of $\boldsymbol{\theta}$ and w, the shear strain

$$\gamma_x = \left[\frac{\partial w}{\partial x} - \theta_x\right] \tag{2.17}$$

is uniquely determined at any point of the side 1–2 (such as point I, for instance) and that hence [by Eq. (2.1c)]

$$S_x = \alpha \gamma_x \tag{2.18}$$

is also uniquely determined there.

Thus, if a node specifying the shear resultant distribution were placed at that point and if the constraints [or satisfaction of Eq. (2.1c)] were only imposed there, then

(*a*) the nodal value of S_x would be shared by adjacent elements (assuming continuity of α) and

(*b*) the nodal values of S_x would be prescribed if the $\boldsymbol{\theta}$ and $\mathbf{w}$ values were constrained as they are in the constrained patch test.

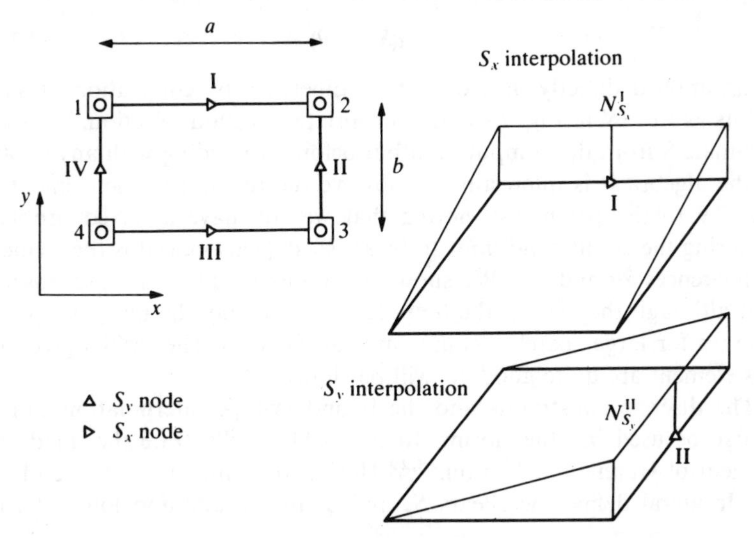

Fig. 2.9 Collocation constraints on a bilinear element. Independent interpolation of S_x and S_y

Indeed if α, the shear rigidity, varied between adjacent elements the values of S_x would only differ by a multiplying constant and arguments remain essentially the same.

The prescription of the shear field in terms of such boundary values is simple. In the case illustrated in Fig. 2.9 we interpolate independently

$$S_x = \mathbf{N}_{sx}\bar{\mathbf{S}}_x \quad \text{and} \quad S_y = \mathbf{N}_{sy}\bar{\mathbf{S}}_y \tag{2.19}$$

using the shape functions illustrated. Such an interpolation of course defines $\mathbf{N}_s$ of Eq. (2.11).

The introduction of the discrete constraint into the analysis is a little more involved. We can proceed by using different (Petrov–Galerkin) weighting functions, and in particular applying the Dirac delta weighting or point collocation to Eq. (2.1c) in the approximate form. However, it is advantageous here to return to the constrained variational principle and seek stationarity of

$$\Pi = \frac{1}{2}\int (L\theta)^{\mathrm{T}} D(L\theta)\,\mathrm{d}\Omega + \frac{1}{2}\int \mathbf{S}^{\mathrm{T}}\frac{1}{\alpha}\mathbf{S}\,\mathrm{d}\Omega - \int wq\,\mathrm{d}\Omega \tag{2.20a}$$

where the first term denotes the *bending* and the second *transverse shear* energies. In the above

$$\begin{aligned} &\theta = \mathbf{N}_\theta\bar{\theta} && w = \mathbf{N}_w\bar{w} \\ \text{and} \quad &\mathbf{S} = \mathbf{N}_s\bar{\mathbf{S}} && \mathbf{N}_s = [\mathbf{N}_{sx}, \mathbf{N}_{sy}] \end{aligned} \tag{2.20b}$$

subject to the constraint Eq. (2.1c):

$$\mathbf{S} = \alpha(\nabla w - \theta) \tag{2.20c}$$

being applied directly in a discrete manner, i.e. by collocation at such points as I to IV in Fig. 2.9 and appropriate direction selection. We shall eliminate **S** from the computation but before proceeding with any details of the algebra it is interesting to observe the relation of the element of Fig. 2.9 to the patch test, noting that we still have a mixed problem requiring the count conditions to be satisfied. (This indeed is the element of references 26 and 27.) We show the counts on Fig. 2.10 and observe that although they fail in the four-element assembly the margin is small here (as for larger patches counts are satisfactory). The results given by this element are quite good, as will be shown later.†

The discrete constraints and the boundary-type interpolation can of course be used in other forms. In Fig. 2.11 we illustrate the quadratic element of Huang and Hinton.[28,29] Here two points on each side of the quadrilateral define the shears S_x and S_y but in addition four internal

† Reference 31 reports a mathematical study of stability for this element.

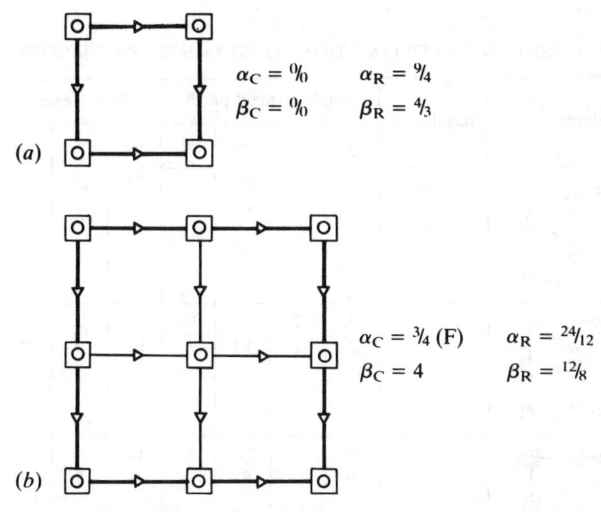

Fig. 2.10 Patch test on (*a*) one and (*b*) four elements of the type given in Fig. 2.9. (Observe that in a constrained test boundary values of **S** are prescribed)

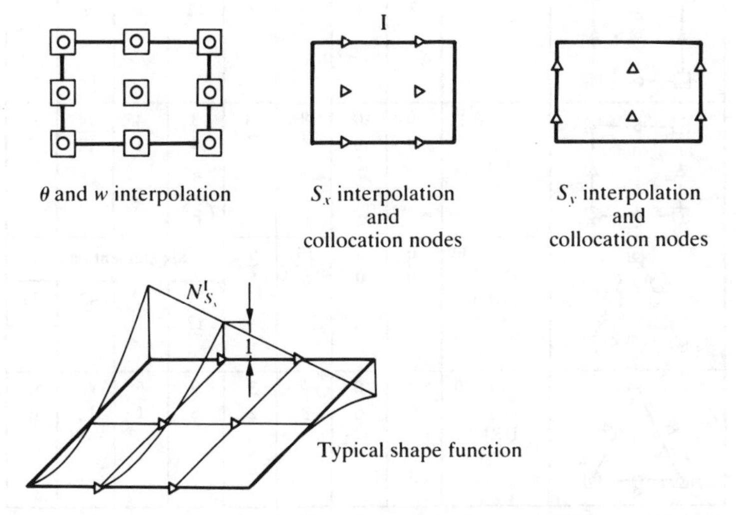

Fig. 2.11 The quadratic lagrangian with collocation constraints on boundaries and in the internal domain[28,29]

parameters are introduced as shown. Now both the boundary and internal 'nodes' are again used as collocation points for imposing the constraints.

The count for single- and four-element patches is given in Table 2.3. This element only fails in a single-element patch under constrained

TABLE 2.3

ELEMENTS WITH COLLOCATION CONSTRAINTS—PATCH COUNT

Element	Reference	Single element patch				Four element patch			
		α_C	β_C	α_R	β_R	α_C	β_C	α_R	β_R
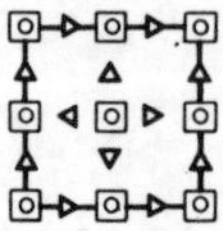	28,29 Q9*	$\frac{3}{4}$ (F)	$\frac{4}{1}$	$\frac{24}{12}$	$\frac{12}{8}$	$\frac{27}{24}$	$\frac{24}{9}$	$\frac{72}{40}$	$\frac{40}{24}$
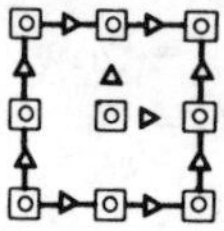		$\frac{3}{2}$	$\frac{2}{1}$	$\frac{24}{10}$	$\frac{10}{8}$	$\frac{27}{16}$	$\frac{16}{9}$	$\frac{72}{32}$	$\frac{32}{24}$
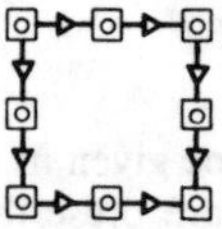		$\frac{0}{0}$	$\frac{0}{0}$	$\frac{21}{8}$	$\frac{8}{7}$	$\frac{15}{8}$	$\frac{8}{5}$	$\frac{60}{24}$	$\frac{24}{21}$
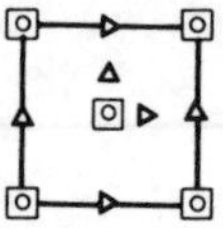		$\frac{3}{2}$	$\frac{2}{1}$	$\frac{12}{6}$	$\frac{6}{4}$	$\frac{15}{12}$	$\frac{12}{5}$	$\frac{36}{20}$	$\frac{20}{12}$
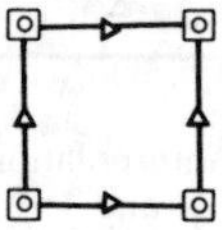	26,27 Q4*	$\frac{0}{0}$	$\frac{0}{0}$	$\frac{8}{4}$	$\frac{4}{3}$	$\frac{3}{4}$ (F)	$\frac{4}{1}$	$\frac{24}{12}$	$\frac{12}{8}$
						Six element patch			
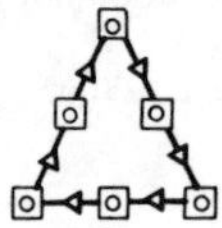	30 TRI-6	$\frac{0}{0}$	$\frac{0}{0}$	$\frac{15}{6}$	$\frac{6}{5}$	$\frac{21}{12}$	$\frac{12}{7}$	$\frac{43}{24}$	$\frac{24}{23}$
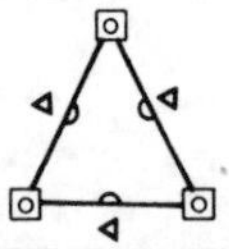	30 DRM	$\frac{0}{0}$	$\frac{0}{0}$	$\frac{9}{3}$	$\frac{3}{2}$	$\frac{9}{6}$	$\frac{6}{1}$	$\frac{45}{12}$	$\frac{12}{6}$

□ w – 1DOF
○ $\boldsymbol{\theta}$ – 2DOF
△ S – 1DOF
⌓ θ – 1DOF (normal)

conditions, and again numerical verification shows generally excellent performance. Details of numerical examples will be given later.

It is clear that with discrete constraints many more alternatives for design of satisfactory elements that pass the patch test are present. In

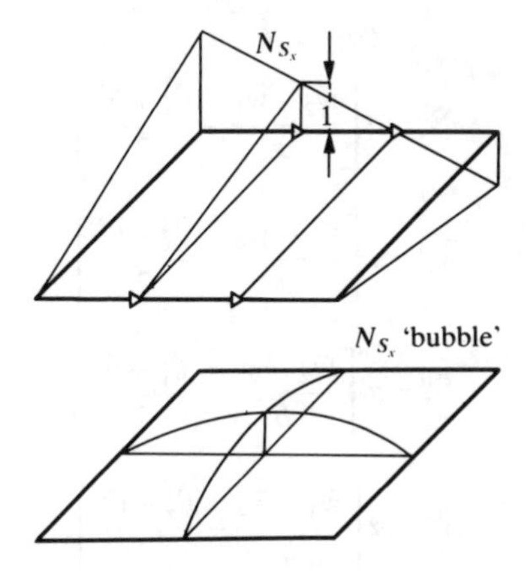

Fig. 2.12 A biquadratic hierarchical bubble for S_x

Table 2.3 several quadrilaterals that satisfy the count conditions are illustrated. In the first a modification of the Hinton–Huang element with reduced internal shear constraints is shown. Here biquadratic 'bubble functions' are used in the interior shear component interpolation, as shown in Fig. 2.12. Similar improvements in the count can be achieved by using a serendipity-type interpolation, but now of course the distorted performance of the element may be impaired (for reasons we discussed in Volume 1, Sec. 8.7, pages 166–170). Addition of bubble functions on all the w and θ parameters can, as shown, make the Dvorkin–Bathe element fully robust.

All quadrilateral elements can of course be mapped isoparametrically, remembering of course that now components of shear S_ξ and S_η parallel to the $\xi\eta$ coordinates have to be used to ensure the preservation of the desirable constraint properties previously discussed. Such 'directional' shear interpolation is essential when considering triangular elements, to which the next section is devoted. Before doing this, however, we shall complete the algebraic derivation of element properties.

2.5.2 *Element matrices for discrete, collocation, constraints.* The starting point here will be to use the variational principle given by Eq. (2.20a) with the shear variables eliminated directly.

The application of the discrete constraints of Eq. (2.20c) allows the 'nodal' parameters $\bar{\mathbf{S}}$ defining the shear force distribution to be determined explicitly in terms of the $\bar{\mathbf{w}}$ and $\bar{\boldsymbol{\theta}}$ parameters. This gives in general terms

$$\bar{\mathbf{S}} = \alpha\,[\mathbf{Q}_w \bar{\mathbf{w}} + \mathbf{Q}_\theta \bar{\boldsymbol{\theta}}] \tag{2.21}$$

in each element. For instance, for the rectangular element of Fig. 2.9

we can write

$$\bar{\mathbf{S}}_x^{\mathrm{I}} = \alpha \left[\frac{\bar{w}_2 - \bar{w}_1}{a} - \frac{\bar{\theta}_x^1 + \bar{\theta}_x^2}{2} \right]$$

$$\bar{S}_x^{\mathrm{III}} = \alpha \left[\frac{\bar{w}_3 - \bar{w}_4}{a} - \frac{\bar{\theta}_x^3 + \bar{\theta}_x^2}{2} \right]$$

$$\bar{\mathbf{S}}_y^{\mathrm{II}} = \alpha \left[\frac{\bar{w}_2 - \bar{w}_3}{b} - \frac{\bar{\theta}_y^2 + \bar{\theta}_y^1}{2} \right] \tag{2.22}$$

$$\bar{\mathbf{S}}_y^{\mathrm{IV}} = \alpha \left[\frac{\bar{w}_1 - \bar{w}_4}{b} - \frac{\bar{\theta}_x^1 + \bar{\theta}_x^2}{2} \right]$$

which can readily be rearranged into the form of Eq. (2.21).

In the general case, returning to the variational principle of Eq. (2.20a) we write this in discrete form as

$$\Pi = \frac{1}{2} \int_\Omega (\mathbf{L}\mathbf{N}_\theta \bar{\boldsymbol{\theta}})^{\mathrm{T}} \, \mathbf{D}(\mathbf{L}\mathbf{N}_\theta \bar{\boldsymbol{\theta}}) \, \mathrm{d}\Omega$$

$$+ \frac{1}{2} \int_\Omega (\mathbf{Q}_\theta \bar{\boldsymbol{\theta}} + \mathbf{Q}_w \bar{\mathbf{w}})^{\mathrm{T}} \, \mathbf{N}_s^{\mathrm{T}} \, \alpha \mathbf{N}_s \, \mathrm{d}\Omega (\mathbf{Q}_\theta \bar{\boldsymbol{\theta}} + \mathbf{Q}_w \bar{w})$$

$$- \int_\Omega (\mathbf{N}_w \bar{\mathbf{w}})^{\mathrm{T}} \, \bar{q} \, \mathrm{d}\Omega \tag{2.23}$$

This is a constrained potential energy principle from which on minimization we obtain the system of equations

$$\begin{bmatrix} \mathbf{K}_{\theta\theta} & \mathbf{K}_{\theta w} \\ \mathbf{K}_{\theta w}^{\mathrm{T}} & \mathbf{K}_{ww} \end{bmatrix} \begin{Bmatrix} \bar{\boldsymbol{\theta}} \\ \bar{\mathbf{w}} \end{Bmatrix} = \begin{Bmatrix} \mathbf{f}_\theta \\ \mathbf{f}_w \end{Bmatrix} \tag{2.24}$$

The element contributions are simply

$$\mathbf{K}_{\theta\theta}^e = \int_{\Omega^e} [(\mathbf{L}\mathbf{N}_\theta)^{\mathrm{T}} \, \mathbf{D}\mathbf{L}\mathbf{N}_\theta + (\mathbf{N}_s \mathbf{Q}_\theta)^{\mathrm{T}} \, \alpha (\mathbf{N}_s \mathbf{Q}_\theta)] \, \mathrm{d}\Omega$$

$$\mathbf{K}_{\theta w}^e = \int_{\Omega^e} (\mathbf{N}_s \mathbf{Q}_\theta)^{\mathrm{T}} \, \alpha (\mathbf{N}_s \mathbf{Q}_\theta) \, \mathrm{d}\Omega \tag{2.25}$$

$$\mathbf{K}_{ww}^e = \int_{\Omega^e} (\mathbf{N}_s \mathbf{Q}_w)^{\mathrm{T}} \, \alpha (\mathbf{N}_s \mathbf{Q}_w) \, \mathrm{d}\Omega$$

with the force terms identical to those defined in Eqs. (2.9).

These general expressions derived above can be used for any form of discrete constraint elements described and presents no computational difficulties.

In the preceding we have imposed the constraints by point collocation of nodes placed on external boundaries or indeed the interior of the element. Other integrals could be used without introducing any difficulties in the final construction of the stiffness matrix. One could, for instance, require integrals such as $\int_\Gamma W[S_s - \alpha(\partial w/\partial s - \theta_s)]\, d\Gamma$ to vanish on segments of the boundary or $\int_\Omega W[S_s - \alpha(\partial w/\partial s - \theta_s)]\, d\Omega$ to vanish in the interior. All would achieve the same objective providing elimination of the S_s parameters is still possible.

The use of discrete constraints can easily be shown to be equivalent to use of *substitute shear strain matrices* in the irreducible formulation of Eq. (2.8). This makes introduction of such forms easy in standard programs. Details of such an approach are given by Oñate *et al.*[32,33]

2.5.3 *Relation to the discrete Kirchhoff (DKT) formulation.* In Chapter 1, Sec. 1.18, we have discussed in some detail the so-called DKT formulation in which the Kirchhoff constraints [i.e. Eq. (2.20c) with $\alpha = \infty$] were applied in a discrete manner. The reason for the success of such discrete constraints was not obvious previously, but we believe that the formulation here presented in terms of the mixed form fully explains its basis. It is well known that the study of mixed forms frequently reveals the robustness or otherwise of irreducible approaches.

In Chapter 12 of Volume 1 we have explained why certain elements of irreducible form perform well as limits of incompressibility are approached and why others fail. Here an analogous situation is illustrated.

It is clear that every one of the elements so far discussed has its analogue in the DKT form. Indeed, the thick plate approach we have adopted here with $\alpha \neq \infty$ is simply a penalty approach to the DKT constraints in which direct elimination of variables was used. Many opportunities for development of interesting and perhaps effective plate elements are thus available for both the thick and thin range.

We shall show in the next section some particularly useful triangular elements with their DKT counterparts. Perhaps the whole range of the present elements should be termed DRM (discrete Reissner–Mindlin) elements in order to ease the classification.

2.5.4 *Collocation constraints for triangular elements.* Figure 2.13 illustrates a triangle in which a straightforward quadratic interpolation of $\boldsymbol{\theta}$ and w is used. In this we shall take the shear forces to be given as a complete linear field defined by six shear force values on the element boundaries in directions parallel to these. The shear nodes are located at Gauss points and the constraint collocation is made at the same position.

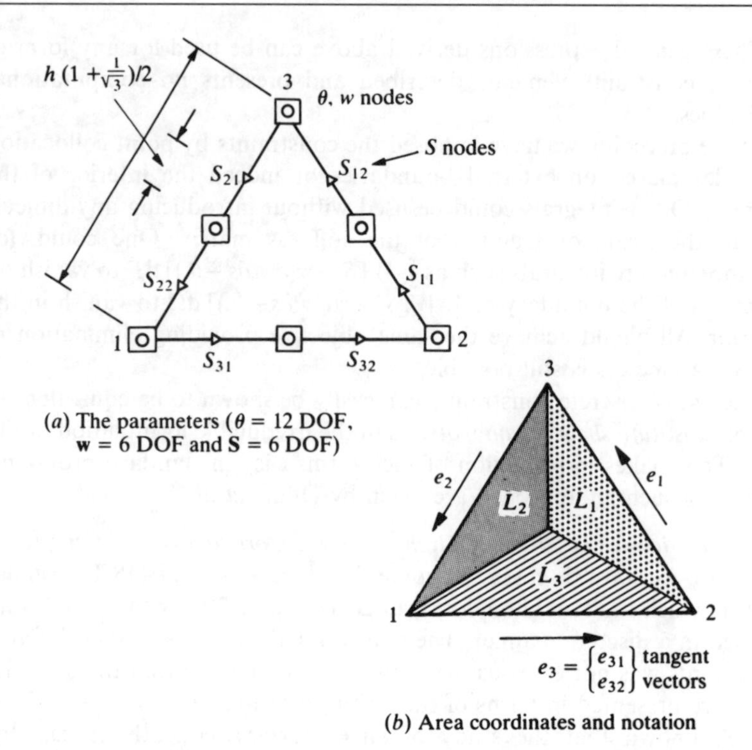

Fig. 2.13 The new quadratic triangular plate element

Writing the interpolation in the area coordinates we have

$$\mathbf{S} = \sum_{i=1}^{3} L_i \mathbf{a}_i \tag{2.26}$$

where $\mathbf{a}_i$ are six as yet undetermined parameters. These can be determined by writing the expressions for tangential shear at the six constraint nodes, obtaining finally

$$\mathbf{S} = \sum_{i=1}^{3} \frac{L_i}{\Delta_i} \begin{bmatrix} e_{ky}, & -e_{jy} \\ -e_{kx}, & e_{jx} \end{bmatrix} \begin{Bmatrix} g_1 \bar{S}_{j1} + g_2 \bar{S}_{j2} \\ g_2 \bar{S}_{k1} + g_1 \bar{S}_{k2} \end{Bmatrix} \tag{2.27}$$

This defines uniquely the shape functions of Eqs (2.11) and, on application of constraints, expresses finally the shear field of nodal displacements $\bar{\mathbf{w}}$ and rotations $\bar{\boldsymbol{\theta}}$ in the manner of Eq. (2.21).

In Eq. (2.27) $\bar{S}_{j1}$ and $\bar{S}_{j2}$ are the tangential shear stress resultants on the two Gauss points of the jth edge (on which $L_j = 0$) and

$$g_1 = \tfrac{1}{2}(1 - \sqrt{3}) \qquad g_2 = \tfrac{1}{2}(1 + \sqrt{3}) \tag{2.28a}$$

$$\Delta_i = e_{jx} e_{ky} - e_{jy} e_{kx} \tag{2.28b}$$

with
$$\mathbf{e}_j = \begin{Bmatrix} e_{jx} \\ e_{jy} \end{Bmatrix} \tag{2.28c}$$

defining the direction cosines on the jth side. The full derivation of the above expression is given in reference 27 and the final derivation of element matrices follows the procedures of Eqs (2.23) to (2.25).

The element derived satisfies fully the patch test count conditions as shown in Table 2.3 as the TRI-6 element. This element performs quite satisfactorily in all configurations, but is somewhat overflexible, as will be shown later. An alternative triangular element which shows considerable improvement in performance is outlined in Fig. 2.14. Here the w displacement variable is interpolated linearly and θ is *nearly* quadratic, resulting from a fully quadratic interpolation with the rotation normal to the sides constrained to be varying linearly, and only a single shear variable and one constrained point are introduced on the element sides.

The 'count' conditions are again fully satisfied for single- and multiple-element patches, as shown in Table 2.3.

The element is of particular interest as it turns out to be an exact equivalent of the DKT triangle with 9 degrees of freedom which gave a very satisfactory solution for thin plates.[34–36] Indeed, in the limit the two elements have an identical performance, though of course the DRM element is applicable also to plates with shear deformation.†

Perhaps the only point of detail worth mentioning is interpolation of the **S** variable. In the TRI-6 quadratic element a full linear expansion of **S** was available. With only three shear parameters a shear field differs

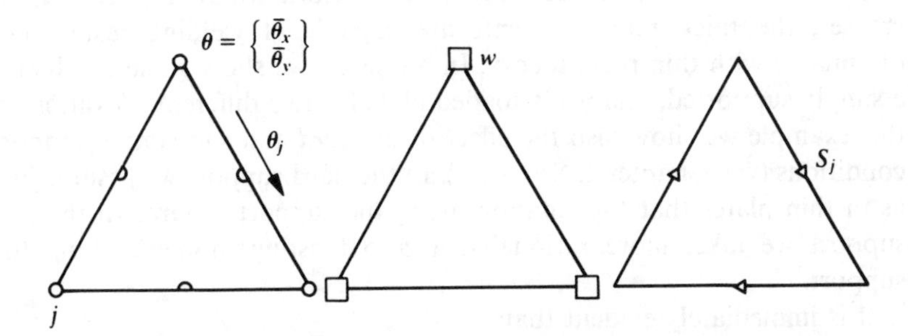

Fig. 2.14 The DRM (discrete Reissner–Mindlin) triangle of reference 30 with θ interpolated by values at corners and rotation parallel to sides $\bar{\theta}_j$, w linear, and $\bar{S}$ stress parallel to sides $\bar{S}_j$ constant

† It is of interest to note that the original DKT element can be modified in a different manner to achieve shear deformability[37] and obtain similar results. However, the element introduced in that reference is not fully convergent.

only slightly from constant values defined by two parameters and the simplest definition uses the expression Eq. (2.27) with

$$\bar{S}_{j1} = \bar{S}_{j2} = \bar{S}_j \tag{2.29}$$

2.6 Performance of various 'thick' plate elements—limitations of thin plate theory

The performance of both 'thick' and 'thin' elements is frequently compared for the examples of clamped and simply supported square plates, though of course more stringent tests can and should be devised. Figure 2.15(*a*) to (*d*) illustrates the behaviour of various elements here discussed in the case of a span to thickness (a/t) ratio of 100, which generally is considered well within the scope of thin plate theory. The results are indeed directly comparable to those of Fig. 1.16 of Chapter 1, and it is evident that here the thick plate elements perform as well as the best of the thin plate forms.

It is of interest to note that in Fig. 2.15 we have included some elements that do not fully pass the patch test and hence are not robust. Many such elements are still used as their failure occurs only occasionally.

All 'robust' elements of the thick plate kind can be easily mapped isoparametrically and their performance remains excellent and convergent. Figure 2.16 shows isoparametric mapping used on a curved sided mesh in the solution of a circular plate for two of the elements previously discussed. Obviously such a lack of sensitivity to distortion will be of considerable advantage when shells are considered, as we shall show in Chapter 5.

Of course when thickness and (shear deformation) importance increases, the thick plate elements are capable of yielding results not obtainable with thin plate theory. In Table 2.4 we show some results for a simply supported, uniformly loaded plate for two different a/t ratios. In this example we show also the effect of the *hard* and *soft* simple support conditions (viz. Chapter 1, Sec. 1.2.2). In the hard support we assume just as in thin plates that the rotation along the support is zero. In the soft support we take, more rationally, a zero twisting moment along the support.

It is immediately evident that:

1. The thick plate ($a/t = 10$) shows deflections converging to very different values depending on the support conditions, both being considerably larger than those given by thin plate theory.
2. For the thin plate ($a/t = 100$) the deflections converge uniformly to the thin (Kirchhoff) plate results for *hard* support conditions, but for *soft* support conditions give answers some 0.3 per cent higher in deflection.

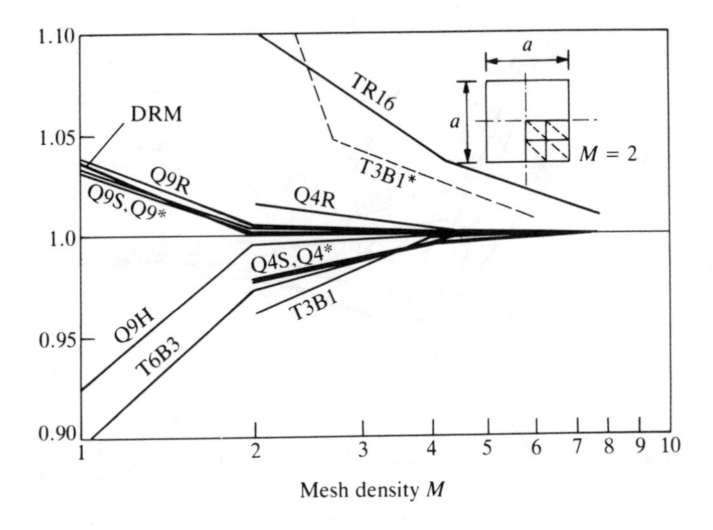

(*a*) Centre displacement normalized with respect to thin plate theory for simply supported, uniformly loaded square plate

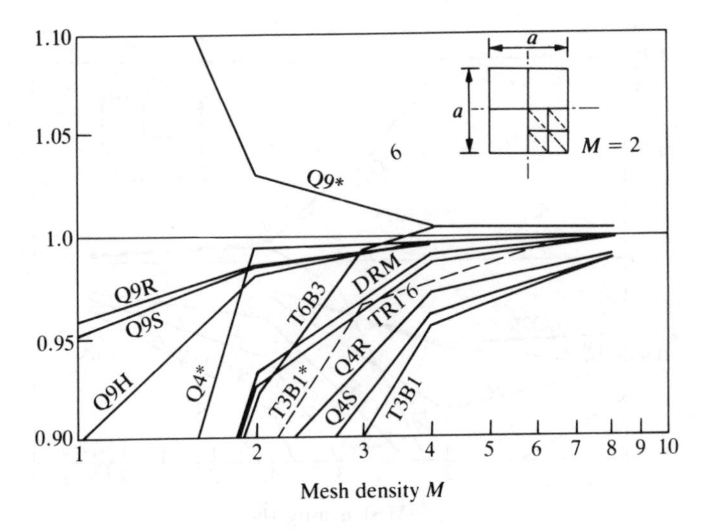

(*b*) Moment at Gauss point nearest centre (or central point) normalized by centre moment of thin plate theory for simply supported, uniformly loaded square plate

Fig. 2.15 Convergence studies for a relatively thin plate ($a/t = 100$). Tables 2.1 to 2.3 give keys to elements used

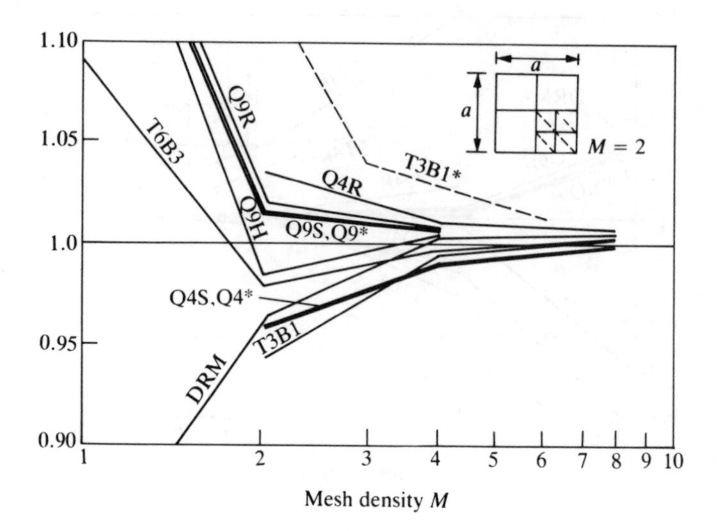

(*c*) **Centre displacement normalized with respect to thin plate theory for clamped, uniformly loaded square plate**

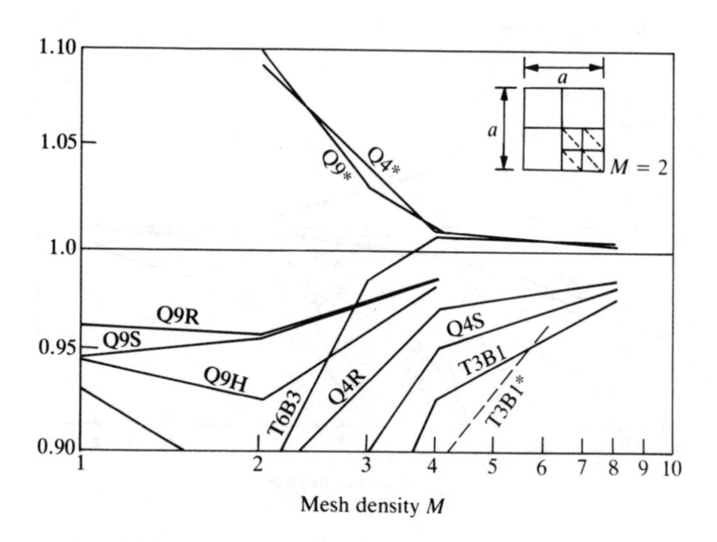

(*d*) **Moment at Gauss point nearest centre (or central point) normalized by centre moments of thin plate theory for clamped, uniformly loaded square plate**

Fig. 2.15 (*continued*)

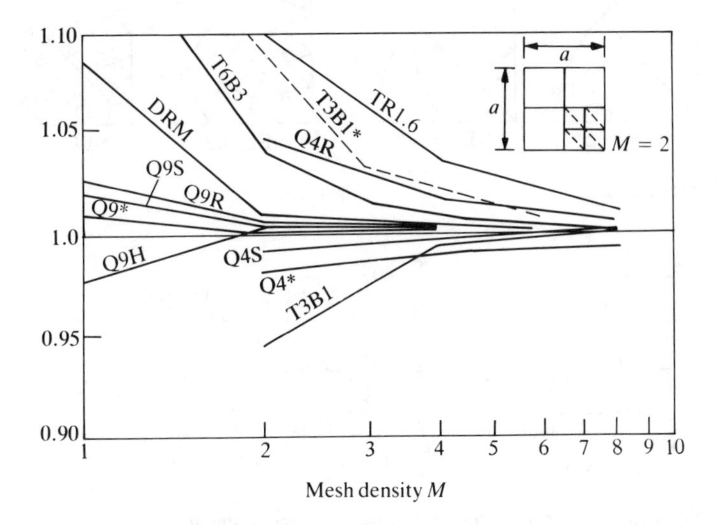

(e) Centre displacement normalized with respect to thin plate theory for thin, simply supported, square plate under concentrated load (note that the solution by thick plate theory is infinite under concentrated load)

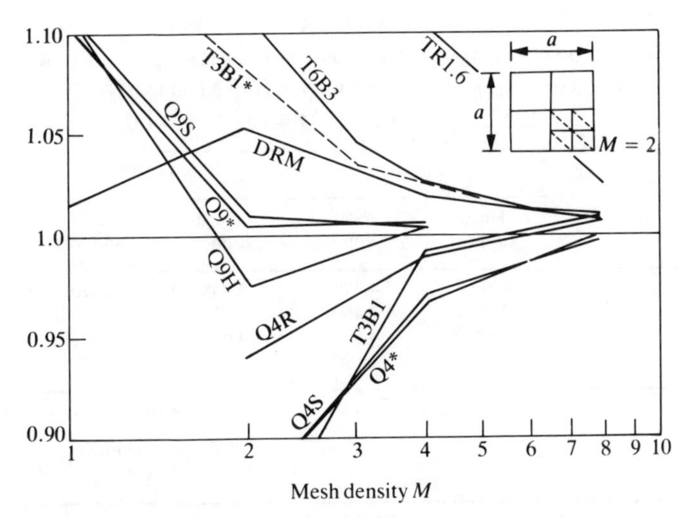

(f) Centre displacement normalized with respect to thin plate theory for thin, clamped, square plate under concentrated load

Fig. 2.15 (*continued*)

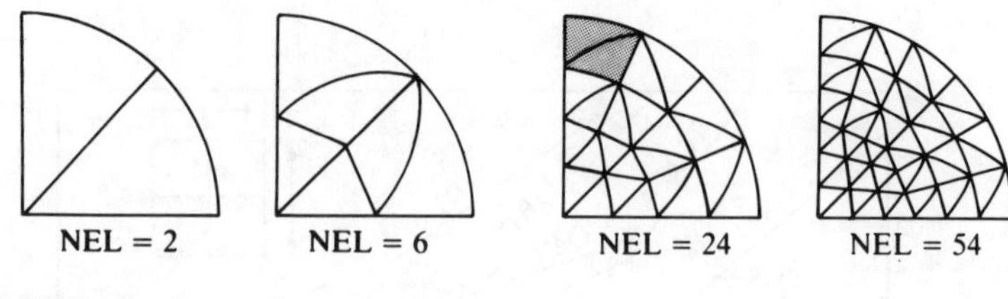

(*a*) Meshes used

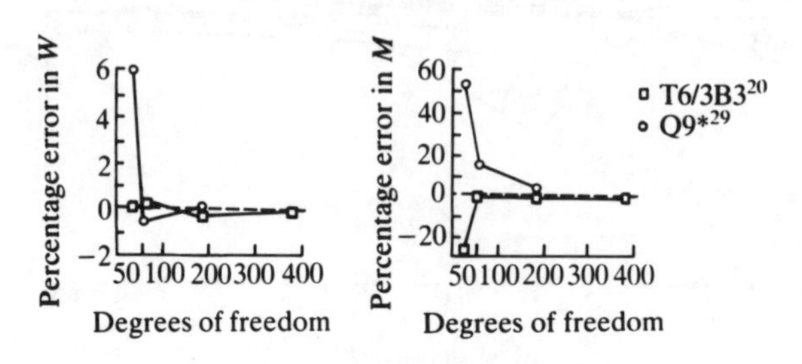

(*b*) Percentage of errors in centre displacement and moment

Fig. 2.16 Mapped curvilinear elements in solution of a circular clamped plate under uniform load

TABLE 2.4

CENTRE DISPLACEMENT OF A SIMPLY SUPPORTED PLATE UNDER UNIFORM LOAD FOR TWO a/t RATIOS (DRM ELEMENT[30])

$E = 10.92 \quad \nu = 0.3 \quad a = 10 \quad q = 1.0$

	$a/t = 10$		$a/t = 100$	
Mesh M	Hard support	Soft support	Hard support	Soft support
2	4.8665	4.3992	4.0582	4.0903
4	4.2829	4.4600	4.0671	4.0737
8	4.2739	4.5393	4.0659	4.0719
16	4.2728	4.5906	4.0649	4.0756
	$\times 10^{-1}$		$\times 10^{-4}$	
			Thin plate solution 4.0623	

It is perhaps an insignificant difference that occurs in this example between the support conditions, but this can be more pronounced in different plate configurations.

In Fig. 2.17 we show the results of a study of a simply supported

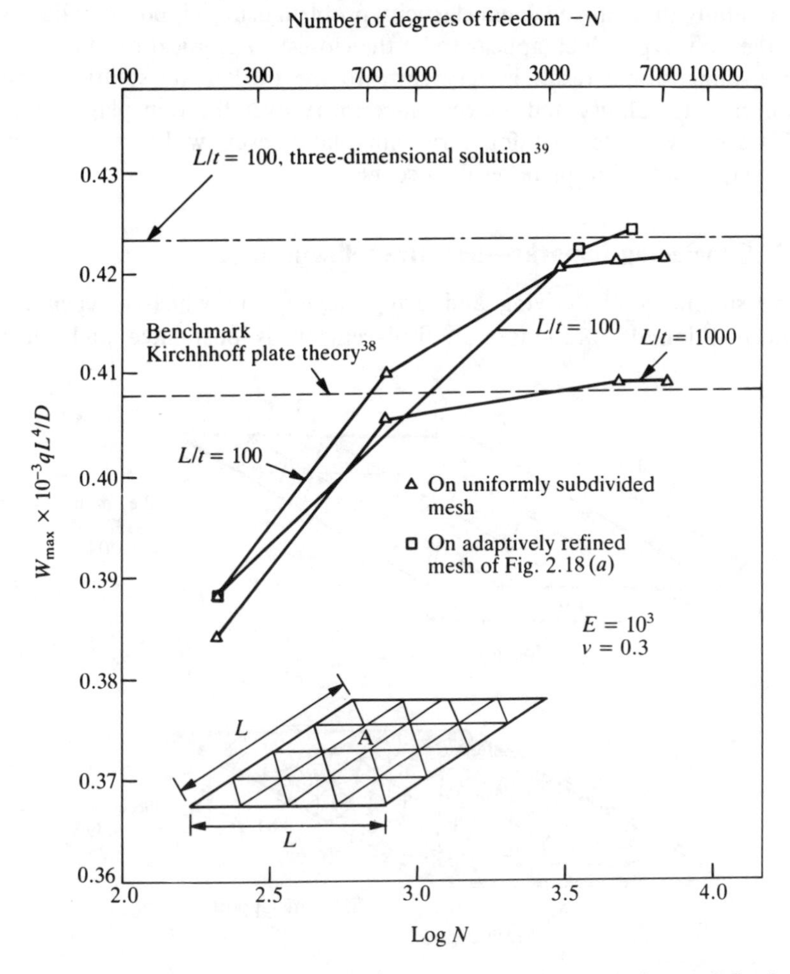

Fig. 2.17 Skew 30° simply supported plate (soft support). Maximum deflection at A, the centre for various degrees of freedom N. The triangular element of reference 20 is used

rhombic plate with $a/t = 100$ and 1000. For this problem an exact Kirchhoff plate theory solution is available,[38] but as will be noticed the thick plate results converge uniformly to a displacement nearly 4% per cent in excess of the thin plate solutions for the $a/t = 100$ cases.

This problem is illustrative of the substantial difference that can on occasion arise in situations that fall well within the limits assumed by conventional thin plate theory ($a/t = 100$), and for this reason the problem has been thoroughly investigated by Babuska and Scapolla,[39] who solve it as a fully three-dimensional elasticity problem using support conditions of the 'soft' type which appear to be the closest to physical reality. Their three-dimensional result is very close to the thick plate solution, and confirms its validity and, indeed, superiority over the thin plate forms. However, we note that for very thin plates, even with soft support, convergence to thin plate results occurs.

2.7 Concluding remarks—adaptive refinement

The simplicity of deriving and using elements in which independent interpolation of rotations and displacements is postulated and shear

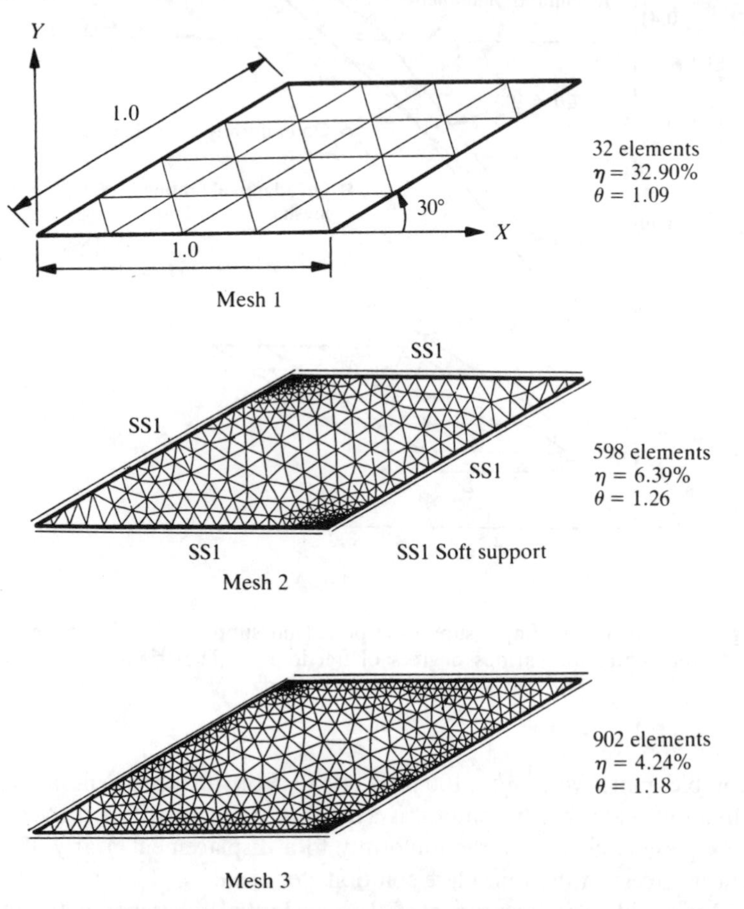

Fig. 2.18*a* Simply supported 30° skew plate with uniform load problem of Fig. 2.17. Adaptive analysis to achieve 5 per cent accuracy. $a/t = 100$, $\nu = 0.3$, six-node element of reference 20; θ = effectivity index, η = percentage error in energy norm of estimator

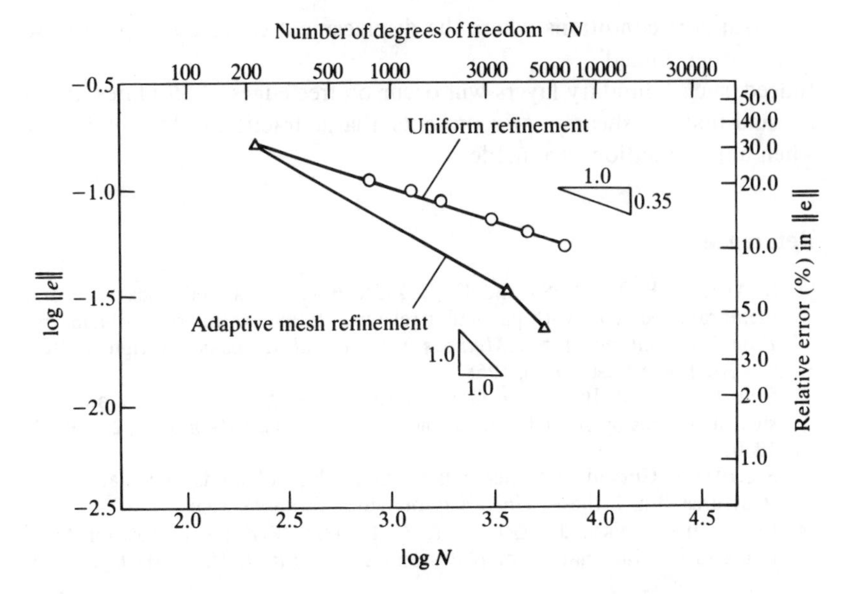

Fig. 2.18*b* Experimental energy norm rate of convergence for the 30° skew plate of Fig. 2.17 for uniform and adaptive refinement. Adaptive analysis to achieve 5 per cent accuracy

deformations are included assures popularity of the approach. The final degrees of freedom used are exactly the same type as those used in the direct approach to thin plate forms in Chapter 1, and at no additional cost shear deformability is included in the analysis.

Providing care is used in ensuring robustness, elements of the type discussed in this chapter are generally applicable and indeed could be used with similar restrictions to other finite element approximations requiring C_1 continuity in the limit.

The ease of element distortion will make elements of the type discussed here the first choice for curved shell solutions and they can easily be adapted to non-linearity due to material or geometric behaviour.

In Volume 1 we discussed the need for an adaptive approach in which error estimation is used in conjunction with mesh generation to obtain answers of specified accuracy. Such adaptive procedures are easily used in plate bending problems with an almost identical form of error estimation.[40]

In Fig. 2.18 we show a sequence of automatically generated meshes for the problem of the skew plate. It is of particular interest to note:

(*a*) the initial refinement in the vicinity of corner singularity and

(*b*) the final refinement near the boundary layer caused by the simple support conditions where the displacement stress assumptions are rather 'forced'.

Indeed, such boundary layers will occur on free edges of all plates and it is usual that the shear error represents a large fraction of the total error when approximations are made.

References

1. S. AHMAD, B. M. IRONS and O. C. ZIENKIEWICZ, 'Curved thick shell and membrane elements with particular reference to axi-symmetric problems', in *Proc. 2nd Conf. on Matrix Methods in Structural Mechanics*, Wright-Patterson Air Force Base, Ohio, 1968.
2. S. AHMAD, B. M. IRONS and O. C. ZIENKIEWICZ, 'Analysis of thick and thin shell structures by curved finite elements', *Int. J. Num. Meth. Eng.*, **2**, 419–51, 1970.
3. S. AHMAD, 'Curved finite elements in the analysis of solids, shells and plate structures', Ph.D. thesis, University of Wales, Swansea, 1968.
4. O. C. ZIENKIEWICZ, J. TOO and R. L. TAYLOR, 'Reduced integration technique in general analysis of plates and shells', *Int. J. Num. Meth. Eng.*, **3**, 275–90, 1971.
5. S. F. PAWSEY and R. W. CLOUGH, 'Improved numerical integration of thick slab finite elements', *Int. J. Num. Meth. Eng.*, **3**, 575–86, 1971.
6. O. C. ZIENKIEWICZ and E. HINTON, 'Reduced integration, function smoothing and non-conformity in finite element analysis', *J. Franklin Inst.*, **302**, 443–61, 1976.
7. E. D. L. PUGH, E. HINTON and O. C. ZIENKIEWICZ, 'A study of quadrilateral plate bending elements with reduced integration', *Int. J. Num. Meth. Eng.*, **12**, 1059–79, 1978.
8. T. J. R. HUGHES, R. L. TAYLOR and W. KANOKNUKULCHAI, 'A simple and efficient finite element for plate bending', *Int. J. Num. Meth. Eng.*, **11**, 1529–43, 1977.
9. T. J. R. HUGHES and M. COHEN, 'The "heterosis" finite element for plate bending', *Comp. Struct.*, **9**, 445–50, 1978.
10. T. J. R. HUGHES, M. COHEN and M. HAROU, 'Reduced and selective integration techniques in the finite element analysis of plates', *Nuclear Eng. Des.*, **46**, 203–22, 1978.
11. E. HINTON and N. BICANIC, 'A comparison of Lagrangian and serendipity Mindlin plate elements for free vibration analysis', *Comp. Struct.*, **10**, 483–93, 1979.
12. R. D. COOK, *Concepts and Applications of Finite Element Analysis*', Wiley, Chichester, 1982.
13. R. L. TAYLOR, O. C. ZIENKIEWICZ, J. C. SIMO and A. H. C. CHAN, 'The patch test—a condition for assessing FEM convergence', *Int. J. Num. Meth. Eng.*, **22**, 39–62, 1986.
14. O. C. ZIENKIEWICZ, S. QU, R. L. TAYLOR and S. NAKAZAWA, 'The patch test for mixed formulations', *Int. J. Num. Meth. Eng.*, **23**, 1873–83, 1986.
15. O. C. ZIENKIEWICZ and D. LEFEBVRE, 'Three field mixed approximation and the plate bending problem', *Comm. Appl. Num. Meth.*, **3**, 301–9, 1987.
16. O. C. ZIENKIEWICZ, J. P. VILOTTE, S. TOYOSHIMA and S. NAKAZAWA,

'Iterative method for constrained and mixed approximation. An inexpensive improvement of FEM performance', *Comp. Meth. Appl. Mech. Eng.*, **51**, 3–29, 1985.
17. D. S. MALKUS and T. J. R. HUGHES, 'Mixed finite element methods—reduced and selective integration techniques: a unification of concepts', *Comp. Meth. Appl. Mech. Eng.*, **15**, 63–81, 1978.
18. B. FRAEIJS DE VEUBEKE, 'Displacement and equilibrium models in finite element method', in *Stress Analysis* (eds O. C. Zienkiewicz and G. S. Holister). chap. 9, pp. 1–20, Wiley, Chichester, 1982.
19. K. J. BATHE and L. W. HO, 'Some results in the analysis of thin shell structures', in *Nonlinear Finite Elements Analysis in Structural Mechanics* (eds W. Wunderlich *et al.*), pp. 122–56, Springer-Verlag, Berlin, 1981.
20. O. C. ZIENKIEWICZ and D. LEFEBVRE, 'A robust triangular plate bending element of the Reissner–Mindlin type', *Int. J. Num. Meth. Eng.*, **26**, 1169–84, 1988.
21. D. N. ARNOLD and R. S. FALK, 'A uniformly accurate finite element method for Mindlin–Reissner plate', IMA Preprint Series No. 307, Institute for Mathematics and its Applications, University of Minnesota, April 1987.
22. Z. XU, 'A simple and efficient triangular finite element for plate bending', *Acta Mechanica Sinica*, **2**, 185–92, 1986.
23. A. TESSLER and T. J. R. HUGHES, 'A three node Mindlin plate element with improved transverse shear', *Comp. Math. Appl. Mech. Eng.*, **50**, 71–101, 1985.
24. A. TESSLER, 'A C^0 anisoparametric three node shallow shell element', *Comp. Meth. Appl. Mech. Eng.*, **78**, 89–103, 1990.
25. T. J. R. HUGHES and T. E. TEZDUYAR, 'Finite elements based upon Mindlin plate theory with particular reference to the four node bilinear isoparametric element', *J. Appl. Mech.*, **46**, 587–96, 1981.
26. E. N. DVORKIN and K. J. BATHE, 'A continuum mechanics based four node shell element for general non-linear analysis', *Eng. Comp.*, **1**, 77–88, 1984.
27. K. J. BATHE and E. N. DVORKIN, 'A four node plate bending element based on Mindlin/Reissner plate theory and mixed interpolation', *Int. J. Num. Meth. Eng.*, **21**, 367–383, 1985.
28. H. C. HUANG and E. HINTON, 'A nine node Lagrangian Mindlin element with enhanced shear interpolation', *Eng. Comp.*, **1**, 369–80, 1984.
29. E. HINTON and H. C. HUANG, 'A family of quadrilateral Mindlin plate elements with substitute shear strain fields', *Comp. Struct.*, **23**, 409–31, 1986.
30. O. C. ZIENKIEWICZ, R. L. TAYLOR, P. PAPADOPOULOS and E. OÑATE, 'Plate bending elements with discrete constraints; new triangular elements', *Comp. Struct.*, **35**, 505–22, 1990.
31. K. J. BATHE and F. BREZZI, 'On the convergence of a four node plate bending element based on Mindlin–Reissner plate theory and a mixed interpolation', in *The Mathematics of Finite Elements. Applications* (ed. J. Whiteman), Vol. V, pp. 491–503, Academic Press, London, 1985.
32. E. OÑATE, R. L. TAYLOR and O. C. ZIENKIEWICZ, 'Consistent formulation of shear constrained Reissner–Mindlin plate elements', in *Discretization Methods in Structural Mechanics* (eds C. Kuhn and H. Mang), pp. 169–80, Springer-Verlag, Berlin, 1990.
33. E. OÑATE, O. C. ZIENKIEWICZ, B. SUÁREZ and R. L. TAYLOR, 'A general methodology for deriving shear constrained Reissner–Mindlin plate elements', *Int. J. Num. Meth. Eng.* (to be published).
34. G. S. DHATT, 'Numerical analysis of thin shells by curved triangular elements

based on discrete Kirchhoff hypothesis', in *Proc. Symp. on Applications of FEM in Civil Engineering* (eds W. R. Rowan and R. M. Hackett), Vanderbilt University, Nashville, Tennessee, 1969.
35. J. L. BATOZ, 'An explicit formulation for an efficient triangular plate bending element', *Int. J. Num. Meth. Eng.*, **18**, 1077–89, 1982.
36. J. L. BATOZ, K. J. BATHE and L. W. HO, 'A study of three node triangular plate bending elements', *Int. J. Num. Meth. Eng.*, **15**, 1771–812, 1980.
37. J. L. BATOZ and P. LARDEUR, 'A discrete shear triangular nine d.o.f. element for the analysis of thick to very thin plates', *Int. J. Num. Meth. Eng.*, **28**, 533–60, 1989.
38. L. S. D. MORLEY, *Skew Plates and Structures*, International Series of Monographs in Aeronautics and Astronautics, Macmillan, New York, 1963.
39. I. BABUSKA and T. SCAPOLLA, 'Benchmark computation and performance evaluation for a rhombic plate bending problem', *Int. J. Num. Meth. Eng.*, **28**, 155–79, 1989.
40. O. C. ZIENKIEWICZ and J. Z. ZHU, 'Error estimation and adaptive refinement for plate bending problems', *Int. J. Num. Meth. Eng.*, **28**, 2839–53, 1989.

3

Shells as an assembly of flat elements

3.1 Introduction

A shell is, in essence, a structure that can be derived from a thin plate by initially forming the middle plane to a singly (or doubly) curved surface. Although the same assumptions regarding the transverse distribution of strains and stresses are again valid, the way in which the shell supports external loads is quite different from that of a flat plate. The stress resultants acting parallel to the middle plane of the shell now have components normal to the surface and carry a major part of the load, a fact that explains the economy of shells as load-carrying structures and their well-deserved popularity.

The derivation of detailed governing equations for a curved shell problem presents many difficulties and, in fact, leads to many alternative formulations, each depending on the approximations introduced. For details of classical shell treatment the reader is referred to standard texts on the subject, e.g. the well-known treatise by Flügge.[1]

In the finite element treatment of shell problems to be described in this chapter the difficulties referred to above are eliminated, at the expense of introducing a further approximation. This approximation is of a physical, rather than mathematical, nature. In this it is assumed that the behaviour of a continuously curved surface can be adequately represented by the behaviour of a surface built up of small, flat, elements. Intuitively, as the size of the subdivision decreases it would seem that convergence must occur, and indeed experience indicates such a convergence.

It will be argued by many shell experts that when we compare the *exact* solution of a shell approximated by flat facets to the exact solution of a truly curved shell considerable differences in the distribution of bending moments, etc., occur. This is undoubtedly true, *but for simple elements* the discretization error is approximately of the same order and excellent results can be obtained with the flat shell element approximation. The mathematics of this problem is discussed in detail by Ciarlet.[2]

In a shell, the element will be subject, generally, both to bending and 'in-plane' forces. For a flat element these cause independent deformations, provided the local deformations are small, and therefore the ingredients for obtaining the necessary stiffness matrices are available in the material already covered in both volumes of this book.

In the division of an arbitrary shell into flat elements only triangular elements can be used. Although the concept of the use of such elements in the analysis has been suggested as early as 1961 by Greene *et al.*,[3] the success of such analysis was hampered by the lack of a good stiffness matrix for triangular plate elements in bending.[4-7] The developments described in Chapters 1 and 2 open the way to adequate models for representing the behaviour of shells with such a division.

Some shells, for example those with general cylindrical shapes, can be well represented by flat elements of rectangular or quadrilateral shape. With good stiffness matrices available for such elements the progress here has been more satisfactory. Practical problems of arch dam design, and others for cylindrical shape roofs, have been solved quite early with such subdivisions.[8,9]

Clearly, the possibilities of analysis of shell structures by the finite element method are enormous. Problems presented by openings, variation of thickness or anisotropy are no longer of consequence once general programs are written.

A special case is presented by axisymmetrical shells. Although it is obviously possible to deal with these in the way described in this chapter, a simpler approach can be used. This will be presented in Chapter 4.

As an alternative to the type of analysis described here, curved shell elements could be used. Here curvilinear coordinates are essential and general procedures in Chapter 8 of Volume 1 can be extended to define these. The physical approximation involved in flat elements is now avoided at the expense of reintroducing an arbitrariness of various shell theories. Several approaches using a direct displacement approach are given in references 10 to 30 and 'mixed' variation principles in references 31 to 34.

A very simple and effective way of deriving curved shell elements is to use the so-called 'shallow' shell theory approach.[19,20,35,36] Here the displacement components, w, u, v, define the *normal and tangential* components of displacement to the curved surface. If all the elements are assumed tangent to each other, no need arises to transfer those from local to global values.

The element is assumed to be 'shallow' with respect to a local coordinate system representing its projection on to a plane defined by nodal points and its strain energy is defined by appropriate equations that include derivatives with respect to *coordinates in the plane of projection*. Thus, precisely the same shape functions can be used as in flat elements discussed in this chapter and all integrations are in fact carried out in the plane as before.

Such shallow shell elements, by coupling the effects of membrane and bending strain in the energy expression, are slightly more efficient than flat ones where such coupling occurs on the interelement boundary only. For simple, small elements the gains are marginal but with few complex large elements advantages show up. A particularly good discussion of such a formulation is given in reference 21.

However, for many practical purposes the flat element approximation gives very adequate answers and indeed permits an easy coupling with edge beam and rib members, a facility sometimes not present in curved element formulation. Indeed, in many practical problems the structure is in fact composed of flat surfaces at least in part and these can be simply reproduced. For these reasons curved general thin shell forms will not be discussed here and instead a general formulation of thick curved shells (based directly on three-dimensional behaviour and avoiding the shell equation ambiguities) will be presented in Chapter 5.

The development of curved elements for general shell theories can be effected in a direct manner; however, additional transformations over those discussed in this chapter are involved. The interested reader is referred to references 37 and 38 for an additional discussion on this approach.

In the context of axisymmetric shells given in the next chapter both straight and curved elements will be considered.

In most arbitrary shaped, curved shell elements so far derived the coordinates used are such that complete smoothness of the surface between elements is not guaranteed. The shape discontinuity occurring there, and, indeed, on any shell where 'branching' occurs, is precisely of the same type as that encountered in this chapter and therefore the methodology of assembly discussed here is perfectly general.

3.2 Stiffness of a plane element in local coordinates

Consider a typical polygonal flat element subject simultaneously to 'in plane' and bending actions (Fig. 3.1).

Taking first the *in-plane* (plane stress) action, we know from Chapter 3 of Volume 1 that the state of strain is uniquely described in terms of the u and v displacement of each typical node i. The minimization of the total potential energy led to the stiffness matrices described there and gives

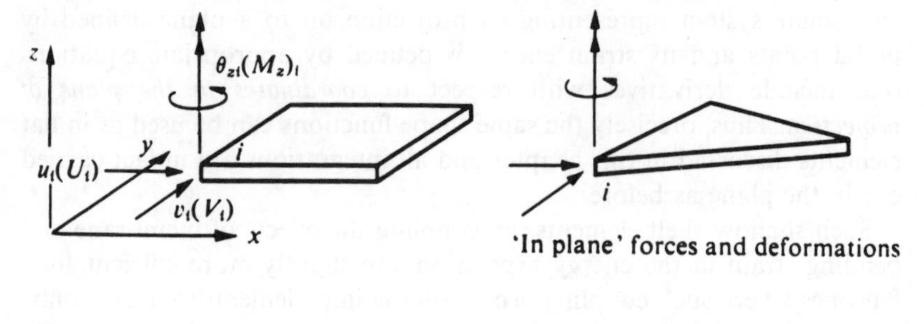

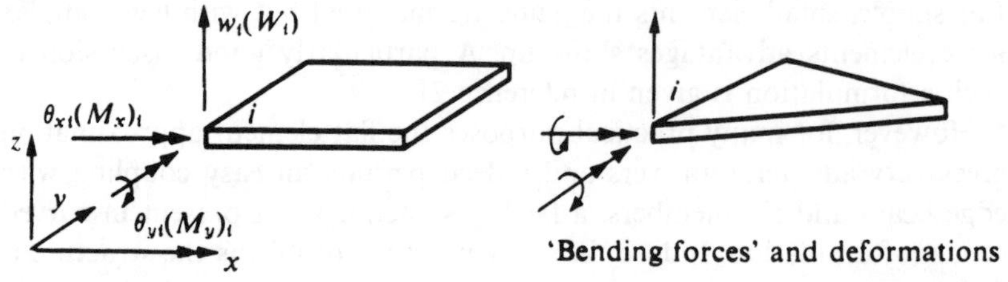

Fig. 3.1 A flat element subject to 'in-plane' and 'bending' actions

'nodal' forces due to displacement parameters $\mathbf{a}^p$ as

$$\mathbf{f}^{ep}=\mathbf{K}^{ep}\mathbf{a}^p \qquad \text{with} \qquad \mathbf{a}_i^p=\begin{Bmatrix} u_i \\ v_i \end{Bmatrix}$$

$$\mathbf{f}_i^p=\begin{Bmatrix} U_i \\ V_i \end{Bmatrix} \tag{3.1}$$

Similarly, when bending was considered, the state of strain was given uniquely by the nodal displacement in the z direction (w) and the two rotations θ_x and θ_y. This resulted in stiffness matrices of the type

$$\mathbf{f}^{eb}=\mathbf{K}^{eb}\mathbf{a}^b \qquad \text{with} \qquad \mathbf{a}_i^b=\begin{Bmatrix} w_i \\ \theta_{xi} \\ \theta_{yi} \end{Bmatrix}$$

$$\mathbf{f}_i^b=\begin{Bmatrix} W_i \\ M_{xi} \\ M_{yi} \end{Bmatrix} \tag{3.2}$$

Before combining these stiffnesses it is important to note two facts. The first is that the displacements prescribed for 'in-plane' forces do not affect the bending deformations and vice versa. The second is that the rotation θ_z does not enter as a parameter into the definition of deformations in either mode. While one could neglect this entirely at the present stage it is convenient, for reasons which will be apparent later when assembly is considered, to take this rotation into account now, and associate with it a fictitious couple M_z. The fact that it does not enter into the minimization procedure can be accounted for simply by inserting an appropriate number of zeros into the stiffness matrix.

Redefining now the combined nodal displacements as

$$\mathbf{a}_i = \begin{Bmatrix} u_i \\ v_i \\ w_i \\ \theta_{xi} \\ \theta_{yi} \\ \theta_{zi} \end{Bmatrix} \tag{3.3}$$

and the appropriate 'forces' as

$$\mathbf{f}_i^e = \begin{Bmatrix} U_i \\ V_i \\ W_i \\ M_{xi} \\ M_{yi} \\ M_{zi} \end{Bmatrix} \tag{3.4}$$

we can write

$$\mathbf{f}^e = \mathbf{K}^e \mathbf{a} \tag{3.5}$$

The stiffness matrix is now made up from the following submatrices:

$$\mathbf{K}_{rs} = \left[\begin{array}{cc|ccc|c} \multicolumn{2}{c|}{\multirow{2}{*}{$\mathbf{K}_{rs}^p$}} & 0 & 0 & 0 & 0 \\ & & 0 & 0 & 0 & 0 \\ \hline 0 & 0 & & & & 0 \\ 0 & 0 & & \mathbf{K}_{rs}^b & & 0 \\ 0 & 0 & & & & 0 \\ \hline 0 & 0 & 0 & 0 & 0 & 0 \end{array}\right] \tag{3.6}$$

if we note that

$$\mathbf{a}_i = \begin{Bmatrix} \mathbf{a}_i^p \\ \mathbf{a}_i^b \\ \theta_{zi} \end{Bmatrix} \tag{3.7}$$

The above formulation is valid for any shape of polygonal element and, in particular, for the two important types illustrated in Fig. 3.1.

3.3 Transformation to global coordinates and assembly of the elements

The stiffness matrix derived in the previous section used a system of local coordinates as the 'in plane', and bending components are originally derived for this system.

Transformation of coordinates to a common global system (which now will be denoted by xyz, and the local system by $x'y'z'$) will be necessary to assemble the elements and to write the appropriate equilibrium equations.

In addition it will be initially more convenient to specify the element nodes by their global coordinates and to establish from these the local coordinates, thus requiring an inverse transformation. Fortunately, all the transformations are accomplished by a simple process.

The two systems of coordinates are shown in Fig. 3.2. The forces and displacements of a node transform from the global to the local system by a matrix $\mathbf{L}$ giving

$$\mathbf{a}'_i = \mathbf{L}\mathbf{a}_i \qquad \mathbf{f}'_i = \mathbf{L}\mathbf{f}_i \tag{3.8}$$

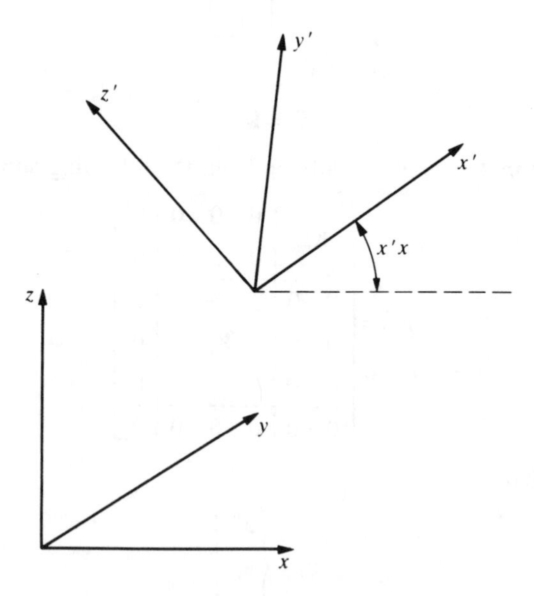

Fig. 3.2 Local and global coordinates

in which

$$\mathbf{L}=\begin{bmatrix} \boldsymbol{\lambda} & 0 \\ 0 & \boldsymbol{\lambda} \end{bmatrix} \tag{3.9}$$

with $\boldsymbol{\lambda}$ being a 3×3 matrix of direction cosines of angles formed between the two sets of axes, i.e.

$$\boldsymbol{\lambda}=\begin{bmatrix} \lambda_{x'x} & \lambda_{x'y} & \lambda_{x'z} \\ \lambda_{y'x} & \lambda_{y'y} & \lambda_{y'z} \\ \lambda_{z'x} & \lambda_{z'y} & \lambda_{z'z} \end{bmatrix} \tag{3.10}$$

in which $\lambda_{x'x}$ = cosine of angle between the x' and x axes, etc.

For the whole set of forces acting on the nodes of an element we can therefore write

$$\mathbf{a}'^{e}=\mathbf{T}\mathbf{a}^{e}, \qquad \text{etc.} \tag{3.11}$$

By the rules of orthogonal transformation (see Sec. 1.8 in Volume 1, pages 17–19) the stiffness matrix of an element in the global coordinates becomes

$$\mathbf{K}^{e}=\mathbf{T}^{\mathrm{T}}\mathbf{K}'^{e}\mathbf{T} \tag{3.12}$$

In both of the above equations $\mathbf{T}$ is given by

$$\mathbf{T}=\begin{bmatrix} \mathbf{L} & 0 & 0 & \cdots \\ 0 & \mathbf{L} & 0 & \\ 0 & 0 & \mathbf{L} & \\ \vdots & & & \end{bmatrix} \tag{3.13}$$

a diagonal matrix built up of $\mathbf{L}$ matrices in a number equal to that of the nodes in the element.

It is simple to show that the typical stiffness submatrix now becomes

$$\mathbf{K}_{rs}^{e}=\mathbf{L}^{\mathrm{T}}\mathbf{K}_{rs}'^{e}\mathbf{L} \tag{3.14}$$

in which $\mathbf{K}_{rs}'^{e}$ is determined by Eq. (3.6) in the local coordinates.

The determination of local coordinates follows a similar pattern. If the origins of both local and global systems are identical then

$$\begin{Bmatrix} x' \\ y' \\ z' \end{Bmatrix}=\boldsymbol{\lambda}\begin{Bmatrix} x \\ y \\ z \end{Bmatrix} \tag{3.15}$$

As in the computation of stiffness matrices for flat plane and bending elements the position of the origin is immaterial, this transformation will always suffice for determination of the local coordinates in the plane (or a plane parallel to the element).

Once the stiffness matrices of all the elements have been determined in the common, global coordinate system the assembly of the elements and the final solution follow the standard pattern. The resulting displacements calculated are referred to the global system, and before the stresses can be computed it is necessary to change these for each element of the local system. The usual stress matrices for 'in-plane' bending components can then be used.

3.4 Local direction cosines

Once the direction cosine matrix $\boldsymbol{\lambda}$ has been determined for each element the problem presents no difficulties, and the solution follows the usual lines. The determination of the direction cosine matrix gives rise to some algebraic difficulties and, indeed, is not unique since the direction of one of the axes is arbitrary, provided it lies in the plane of the element.

We shall first deal with the assembly of rectangular elements in which this problem is particularly simple.

3.4.1 *Rectangular elements.* Such elements are limited in use to representing a cylindrical or box type of surface and it is convenient to take one side of the elements and the corresponding coordinate x' parallel to the global, x, axis. For a typical element $ijkm$, illustrated in Fig. 3.3, it is now easy to calculate all the relevant direction cosines. Direction cosines of x' are, obviously,

$$
\begin{aligned}
\lambda_{x'x} &= 1 \\
\lambda_{x'y} &= 0 \\
\lambda_{x'x} &= 0
\end{aligned}
\tag{3.16}
$$

The direction cosines of the y' axis have to be obtained by consideration of the coordinates of the various nodal points. Thus

$$
\begin{aligned}
\lambda_{y'x} &= 0 \\
\lambda_{y'y} &= +\frac{y_j - y_i}{\sqrt{[(z_j - z_i)^2 + (y_j - y_i)^2]}} \\
\lambda_{y'z} &= +\frac{z_j - z_i}{\sqrt{[(z_j - z_i)^2 + (y_j - y_i)^2]}}
\end{aligned}
\tag{3.17}
$$

simple geometrical relations which can be obtained by consideration of the sectional plane passing vertically through ij.

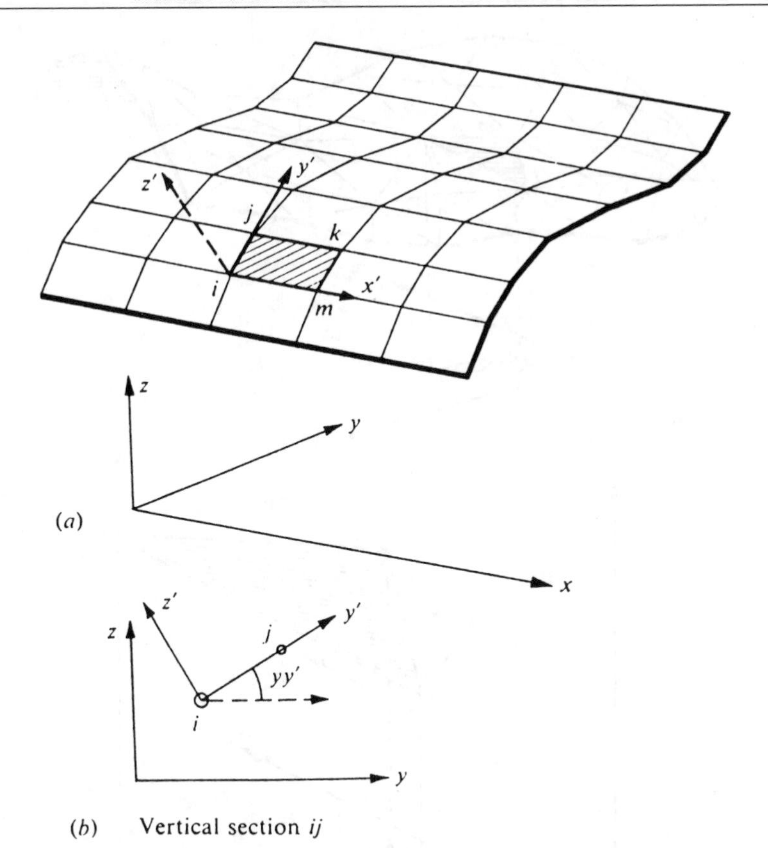

Fig. 3.3 A cylindrical shell as an assembly of rectangular elements. Local and global coordinates

Similarly, from the same section, we have for the z' axis

$$
\begin{aligned}
\lambda_{z'x} &= 0 \\
\lambda_{z'y} &= -\frac{z_j - z_i}{\sqrt{[(z_j - z_i)^2 + (y_j - y_i)^2]}} \\
\lambda_{z'z} &= +\frac{y_j - y_i}{\sqrt{[(z_j - z_i)^2 + (y_j - y_i)^2]}}
\end{aligned}
\tag{3.18}
$$

Clearly, the numbering of points in a consistent fashion is important to preserve the correct signs of the expression.

3.4.2 *Triangular elements arbitrarily oriented in space.* An arbitrary shell divided into triangular elements is shown in Fig. 3.4(*a*). Each element is in an orientation in which the angles with the coordinate planes are

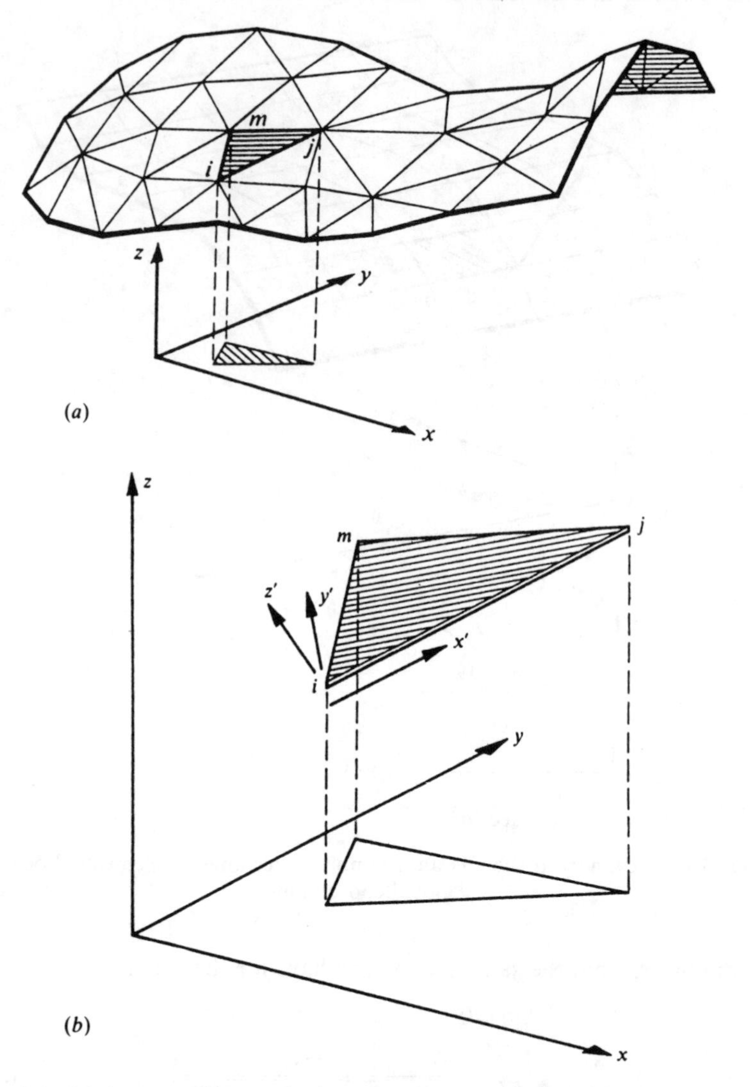

Fig. 3.4 (*a*) An assemblage of triangular elements representing an arbitrary shell. (*b*) Local and gobal coordinates for a triangular element

arbitrary. The problems of defining local axes and their direction cosines are therefore considerably more complex than in the previous simple example. The most convenient way of dealing with the problem is to use some features of geometrical vector algebra and for readers who may have forgotten some of this background a brief resumé of its essentials is included in Appendix 5 of Volume 1.

One arbitrary but convenient choice of local axis direction is given here. We shall specify in this that the x' axis is to be directed along the side ij of the triangle, as shown in Fig. 3.4(b).

The vector $\mathbf{V}_{ij}$ defines this side and in terms of global coordinates we have

$$\mathbf{V}_{ij}=\begin{Bmatrix} x_j-x_i \\ y_j-y_i \\ z_j-z_i \end{Bmatrix} \tag{3.19}$$

The direction cosines are given by dividing the components of this vector by its length, i.e. defining a vector of unit length

$$\mathbf{v}_{x'}=\begin{Bmatrix} \lambda_{x'x} \\ \lambda_{x'y} \\ \lambda_{x'z} \end{Bmatrix}=\frac{1}{l_{ij}}\begin{Bmatrix} x_{ji} \\ y_{ji} \\ z_{ji} \end{Bmatrix} \tag{3.20}$$

with

$$l_{ij}=\sqrt{x_{ji}^2+y_{ji}^2+z_{ji}^2}$$

in which $x_{ji}=x_j-x_i$, etc., for brevity.

Now the z' direction, which must be normal to the plane of the triangle, needs to be established. By properties of the cross-product of two vectors we can obtain this direction from a 'vector' (cross) product of two sides of the triangle. Thus

$$\mathbf{V}_{z'}=\mathbf{V}_{ij}\times\mathbf{V}_{im}=\begin{Bmatrix} y_{ji}z_{mi}-z_{ji}y_{mi} \\ \cdot\ \cdot\ \cdot\ \cdot\ \cdot \\ \cdot\ \cdot\ \cdot\ \cdot\ \cdot \end{Bmatrix} \tag{3.21}$$

represents a vector normal to the plane of the triangle whose length, by definition (see Appendix 5 of Volume 1), is equal to twice the area of the triangle. Thus

$$l_{z'}=\sqrt{(y_{ji}z_{mi}-z_{ji}y_{mi})^2+(\cdots)^2+(\cdots)^2}=2\Delta$$

The direction cosines of the z' axis are available simply as the direction cosines of $\mathbf{V}_{z'}$ and we have a unit vector

$$\mathbf{v}_{z'}=\begin{Bmatrix} \lambda_{z'x} \\ \lambda_{z'y} \\ \lambda_{z'z} \end{Bmatrix}=\frac{1}{2\Delta}\begin{Bmatrix} y_{ji}z_{mi}-z_{ji}y_{mi} \\ \cdot\ \cdot\ \cdot\ \cdot \\ \cdot\ \cdot\ \cdot\ \cdot \end{Bmatrix} \tag{3.22}$$

Finally, the direction cosines of the y' axis are established in a similar manner as the direction cosines of a vector normal to both the x' and z'

directions. If vectors of unit length are taken in each of these directions as in fact defined by Eqs (3.20) and (3.22) we have simply

$$\mathbf{v}_{y'} = \begin{Bmatrix} \lambda_{y'x} \\ \lambda_{y'y} \\ \lambda_{y'z} \end{Bmatrix} = \mathbf{v}_{z'} \times \mathbf{v}_{x'} = \begin{Bmatrix} \lambda_{z'y}\lambda_{x'z} - \lambda_{z'z}\lambda_{x'y} \\ \cdot \quad \cdot \quad \cdot \quad \cdot \quad \cdot \\ \cdot \quad \cdot \quad \cdot \quad \cdot \quad \cdot \end{Bmatrix} \tag{3.23}$$

without having to divide by the length of the vector which is now simply unity.

Indeed, the vector operations involved can be written as a special computer routine in which vector products, normalizing (i.e. division by length), etc., are automatically carried out[39] and there is no need to specify in detail the various operations given above.

In the preceding outline the direction of the x' axis was taken as lying along one side of the element. A useful alternative is to specify this by the section of the triangle plane with a plane parallel to one of the coordinate planes. Thus, for instance, if we should desire to erect the x' axis along a horizontal contour of the triangle (i.e. a section parallel to the xy plane) we can proceed as follows.

Firstly the normal direction cosines $\mathbf{v}_{z'}$ are defined as in Eq. (3.22). Now, the matrix of direction cosines of x' has to have a zero component in the z direction. Thus we have

$$\mathbf{v}_{x'} = \begin{Bmatrix} \lambda_{x'x} \\ \lambda_{x'y} \\ 0 \end{Bmatrix} \tag{3.24}$$

As the length of the vector is unity

$$\lambda_{x'x}^2 + \lambda_{x'y}^2 = 1 \tag{3.25}$$

and as further the *scalar* product of the $\mathbf{v}_{x'}$ and $\mathbf{v}_{z'}$ must be zero, we can write

$$\lambda_{x'x}\lambda_{z'x} + \lambda_{x'y}\lambda_{z'y} = 0 \tag{3.26}$$

and from these two equations $\mathbf{v}_{x'}$ can be uniquely determined. Finally, as before

$$\mathbf{v}_{y'} = -\mathbf{v}_{x'} \times \mathbf{v}_{z'} \tag{3.27}$$

Yet another alternative of a unique specification of the x' axis is given in Chapter 5.

3.5 'Drilling' rotational stiffness—6 degree of freedom assembly

In the formulation described above a difficulty arises if all the elements meeting at a node are co-planar. This situation will arise for flat (folded) shell segments and at straight boundaries of cylindrical shaped shells. The difficulty is due to the assignment of a zero stiffness in the θ_{zi} direction of Fig. 3.1 and the fact that classical shell equations do not produce equations associated with this rotational parameter. Inclusion of the third rotation and of 'forces' associated with it has obvious benefits for a finite element model in that both rotations and displacements at nodes may be treated in a very simple manner using the transformations just presented.

If the set of assembled equilibrium equations *in local coordinates* is considered at such a point we have six equations of which the last (corresponding to the θ_z direction) is simply

$$0=0 \tag{3.28}$$

As such, an equation of this type presents no special difficulties (although in the programs included in this book a warning is issued). However, if the global coordinate directions differ from the local ones and a transformation is accomplished, the six equations mask the fact that the equations are singular. Detection of this singularity is somewhat more difficult.

A number of alternatives have been presented that avoid the presence of the singular behaviour. Two simple ones are

(*a*) to assemble the equations at points where elements are co-planar in local coordinates (and to delete the $0=0$ equation) or/and

(*b*) to insert an arbitrary stiffness coefficient $k'_{\theta z}$ at such points only.

This leads in the local coordinates to replacing Eq. (3.28) by

$$k'_{\theta z}\theta_{zi}=0 \tag{3.29}$$

which, on transformation, leads to a perfectly well-behaved set of equations from which, by usual processes, all displacements, now including θ_{zi}, are obtained. As θ_{zi} does not affect the stresses and indeed is uncoupled from all equilibrium equations any value of $k'_{\theta z}$ similar to values already in Eq. (3.6) can be inserted as an external stiffness without affecting the result.

These two approaches lead to programming complexity (as a decision on the coplanar nature is necessary) and an alternative is to modify the formulation such that the rotational parameters arise naturally and have a real physical significance. This has been a topic of much recent study[40-52] and the θ_z parameter introduced in this way is commonly called a *drilling* degree of freedom on account of its action to the surface of the shell. An early application considering the rotation as an additional degree of freedom in plane analysis is contained in reference 20. In reference 7 for a general shell program a set of rotational stiffness coefficients was used in all elements whether co-planar or not. These were defined such that in local coordinates overall equilibrium is not disturbed. This may be accomplished by adding to the formulation for each element the term

$$\Pi^* = \Pi + \int_\Omega \alpha_n E t^n (\theta_z - \overline{\theta}_z)^2 D\Omega \tag{3.30}$$

in which the parameter α_n is a fictitious elastic parameter and $\overline{\theta}_z$ is a mean rotation of each element which permits the element to satisfy local equilibrium in a weak sense. The above is a generalization of that proposed in reference 7 in the scaling value, where the value of n is unity. Since the term will lead to a stiffness that will be in terms of rotation parameters the scaling indicated above permits values proportional to those generated by the bending rotations—namely, proportional to t cubed. In numerical experiments this scaling leads to less sensitivity in the choice of α_n. For a triangular element minimization with respect to $\overline{\theta}_z$ and θ_z leads to the form (after elimination of $\overline{\theta}_z$)

$$\begin{Bmatrix} M_{zi} \\ M_{zj} \\ M_{zm} \end{Bmatrix} = \alpha_n E t^n \Delta \begin{bmatrix} 1 & -0.5 & -0.5 \\ -0.5 & 1 & -0.5 \\ -0.5 & -0.5 & 1 \end{bmatrix} \begin{Bmatrix} \theta_{zi} \\ \theta_{zj} \\ \theta_{zm} \end{Bmatrix} \tag{3.31}$$

where α_n is yet to be specified. This additional stiffness does in fact affect the results where nodes are not co-planar and indeed represents an approximation; however, effects of varying α_n over fairly wide limits are quite small in many applications. For instance, in Table 3.1 a set of displacements of an arch dam analysed in reference 3 is given for various values of α_1. For practical purposes extremely small values of α_n are possible, providing a large computer word length (e.g. REAL*8 in Fortran) is used.[53]

The analysis of the spherical test problem proposed by MacNeal and Harter as a standard test[54] is indicated in Fig. 3.5. For this test problem a constant strain triangular membrane together with the discrete Kirchhoff triangular plate bending element is combined with the rotational treatment. The results for regular meshes is shown in Table 3.2 for several values of α_3.

The above development, while quite easy to implement, retains the original form of the membrane interpolations. For triangular elements with corner nodes only, the membrane form utilizes linear displacement

TABLE 3.1

NODAL ROTATION COEFFICIENT IN DAM ANALYSIS[3]

α_1	1.00	0.50	0.10	0.03	0.00
Radial displacement (mm)	61.13	63.55	64.52	64.78	65.28

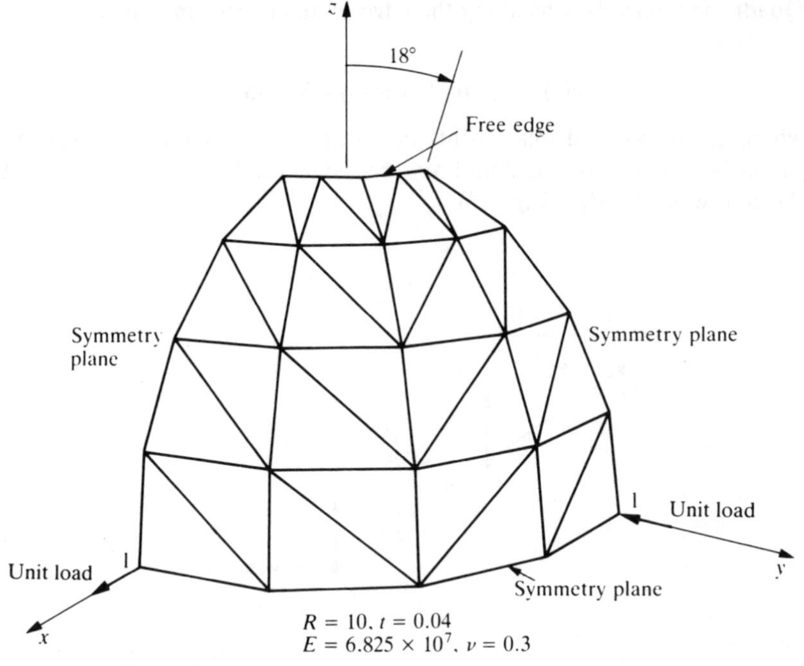

Fig. 3.5 Spherical shell test problem[54]

TABLE 3.2

SPHERE PROBLEM—RADIAL DISPLACEMENT AT LOAD

Mesh	α_3 value					
	10.0	1.00	0.10	0.01	0.0001	0.00
4 × 4	0.0639	0.0919	0.0972	0.0979	0.0980	0.0980
8 × 8	0.0897	0.0940	0.0945	0.0946	0.0946	0.0946
16 × 16	0.0926	0.0929	0.0929	0.0929	0.0930	0.0930

fields that yield complete constant strain terms. Most bending elements discussed in Chapters 1 and 2 have bending strains with higher than constant terms. Consequently, the membrane error terms will dominate the behaviour of many shell problem solutions. In order to improve the situation it is desirable to increase the order of interpolation. Using conventional interpolations this implies the introduction of additional nodes on each element (e.g. see Chapter 7 of Volume 1); however, by utilizing a drill parameter these interpolations can be transformed to a form that permits a 6 degree of freedom assembly at each vertex node.

Quadratic interpolations along the edge of an element may be expressed as

$$\mathbf{u}(\xi) = N_i(\xi)\mathbf{u}_i + N_j(\xi)\mathbf{u}_j + N_k(\xi)\Delta\mathbf{u}_k \tag{3.32}$$

where $\mathbf{u}_i$ are nodal displacements (u_i, v_i) at an end of the edge (vertex), similarly $\mathbf{u}_j$ is the other end and $\Delta\mathbf{u}_k$ are the hierarchical displacements at the centre of the edge (Fig. 3.6).

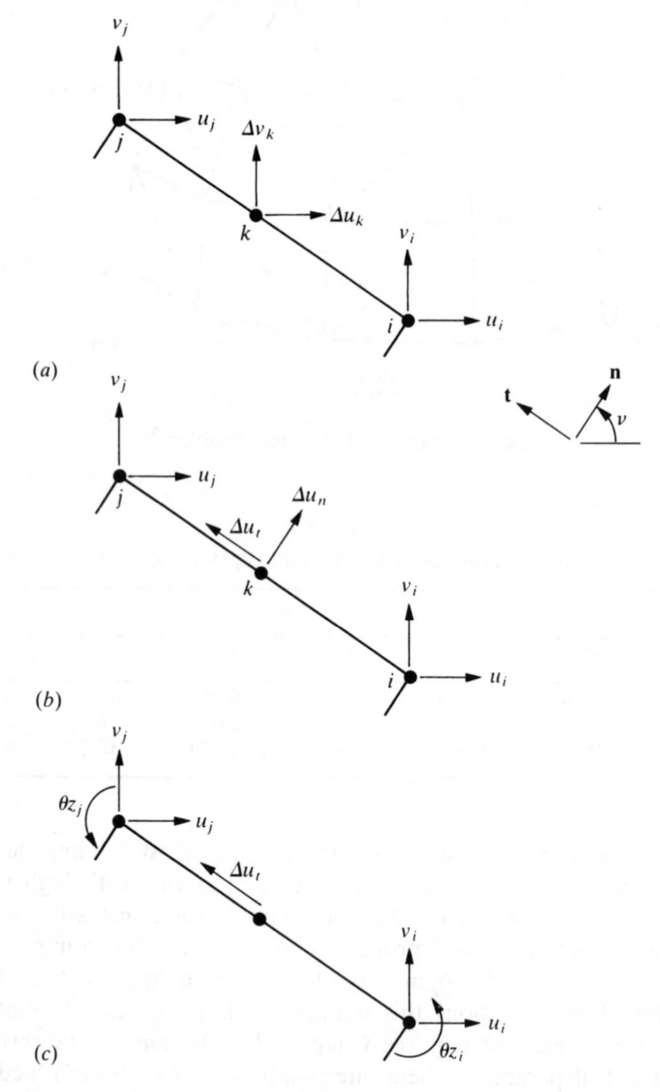

Fig. 3.6 Construction of in-plane interpolations with drilling' parameters

The centre displacement parameters may be expressed in terms of a *normal* (Δu_n) and a *tangential* (Δu_t) component as

$$\Delta u_k = \Delta u_n \mathbf{n} + \Delta u_t \mathbf{t} \tag{3.33}$$

where $\mathbf{n}$ is a unit outward normal and $\mathbf{t}$ is a unit tangential vector to the edge:

$$\mathbf{n} = \begin{Bmatrix} \cos \nu \\ \sin \nu \end{Bmatrix} \quad \text{and} \quad \mathbf{t} = \begin{Bmatrix} -\sin \nu \\ \cos \nu \end{Bmatrix} \tag{3.34}$$

where ν is the angle that the normal to the edge makes with the x axis. The normal displacement component may be expressed in terms of drilling parameters at each end of the edge[40,49] (assuming a cubic expansion). Accordingly,

$$\Delta u_n = \tfrac{1}{8} l_{ij} (\theta_{zj} - \theta_{zi}) \tag{3.35}$$

in which l_{ij} is the length of the i–j side. This construction produces an interpolation on each edge given by

$$\mathbf{u}(\xi) = N_i(\xi)\mathbf{u}_i + N_j(\xi)\mathbf{u}_j + N_k(\xi)[\tfrac{1}{8} l_{ij}(\theta_{zj} - \theta_{zi})\mathbf{n} + \Delta u_t \mathbf{t}] \tag{3.36}$$

The interpolation may be simplified by constraining to zero the Δu_t parameters. We note, however, that these terms are beneficial in a three-node triangular element. If a common sign convention is used for the hierarchical tangential displacement at each edge, this tangential component maintains compatibility of displacement even in the presence of a kink between adjacent elements. For example, an appropriate sign convention can be accomplished by directing a positive component in the direction in which the end (vertex) node numbers increase. The above structure for the in-plane displacement interpolations may be used for either an irreducible or mixed element model and generates stiffness coefficients that include terms for the θ_z parameters as well as those for u and v. It is apparent, however, that the element generated in this manner must be singular (i.e. has zero energy modes) since for equal values of the end rotation the interpolation is independent of the θ_z parameters. Moreover, when used in non-flat shell applications the element is not free of local equilibrium errors. This latter defect may be removed using the procedure identified above in Eq. (3.30) and results for a quadrilateral element generated according to this scheme are given by Jetteur[50] and Taylor.[51]

A structure of the plane stress problem which includes the effects of a drill rotation field is given by Reissner[55] and extended to finite element applications by Hughes and Brezzi.[47] A variational problem for the in-plane problem may be stated as [see Eq. (2.29) in Volume 1]

$$\Pi_d(\mathbf{u}, \theta_z, \tau) = \int_\Omega \tfrac{1}{2}\boldsymbol{\varepsilon}^T \mathbf{D}\, \boldsymbol{\varepsilon}\, d\Omega + \int_\Omega \tau^T(\omega_{xy} - \theta_z)\, d\Omega \tag{3.37}$$

where τ is a *skew-symmetric* stress component and ω_{xy} is the rotational part of the displacement gradient, which for the xy plane is given by

$$\omega_{xy} = \frac{\partial v}{\partial x} - \frac{\partial u}{\partial y} \tag{3.38}$$

In addition to the terms shown in Eq. (3.37), terms associated with initial stress and strain as well as boundary and body loads must be appended for a general case as discussed in Chapter 2 of Volume 1.

A variation of Eq. (3.37) with respect to τ gives the constraint that the skew-symmetric part of the displacement gradients is the rotation θ_z. Conversely, variation with respect to θ_z gives the result that τ must vanish. Thus the equations generated from Eq. (3.37) are those of the conventional membrane but include the rotation field. A penalty form of the above equations suitable for finite element applications may be constructed by modifying Eq. (3.37) to

$$\bar{\Pi}_d = \Pi_d - \int_\Omega \frac{1}{\gamma E t} \tau^2\, d\Omega \tag{3.39}$$

where γ is a penalty number.

It is important to use this mixed representation of the problem with the mixed patch test to construct viable finite element models. Use of constant τ and isoparametric interpolation of θ_z in each element together with the interpolations for the displacement approximation given by Eq. (3.36) lead to good triangular and quadrilateral membrane elements. Applications to shell solutions using this form are given by Ibrahimbegovic *et al.*[56] Also the solution for a standard barrel vault problem is contained in Sec. 3.8.

3.6 Elements with mid-side slope connections only

Many of the difficulties encountered with the nodal assembly in global coordinates disappear if the element is so constructed as to require only the continuity of displacements u, v and w at the corner nodes with continuity of the normal slope being imposed along the element sides. Clearly, the corner assembly is now simple and the introduction of the sixth nodal variable is unnecessary. As the normal slope rotation along the sides is the same both in local and global coordinates its transformation there is unnecessary.

Elements of this type arise naturally in hybrid forms (see Chapter 13 of Volume 1) and we have already referred to a plate bending element of a suitable type in Sec. 1.6. This element of the simplest possible kind has been used in shell problems by Dawe[25] with some success. A considerably more sophisticated and complex element of such a type is derived by Irons[26] with a suggested curious name of 'semi-loof'. This element is briefly mentioned in Chapter 1 (page 90) and although its derivation is far from simple it performs well in many situations.

3.7 Choice of element

Numerous 'in-plane' and bending element formulations are now available, and, in both, conformity was achievable in flat assemblies. Clearly, if the elements are not co-planar conformity will, in general, be violated (except in the limit as smooth shell conditions are reached).

It would appear consistent to use expansions of similar accuracy in both the membrane and bending approximations, but much depends on which action is predominant. For thin shells, the simplest triangular element would thus appear to be one with a linear in-plane displacement field and a quadratic bending displacement—thus approximating the stresses as constants in plane and in bending. Such an element is used by Dawe[25] but gives rather poor (though convergent) results.

In the examples shown we use the following elements which give adequate performance:

Element A. Mixed in-plane rectangle with four corner nodes (Chapter 13 of Volume 1, page 382) combined with the non-conforming bending rectangle with four corner nodes (Chapter 1, page 15). This was first used in references 8 and 9.

Element B. Constant strain triangle with three nodes (basic element of Chapter 3 of Volume 1) combined with the incompatible bending triangle with 9 degrees of freedom (Chapter 1, page 23). Use of this in the shell context is given in references 7 and 57.

Element C. In this a more consistent linear strain triangle with six nodes is combined with a 12 degree of freedom bending triangle using shape function smoothing. This element has been introduced by Razzaque.[58]

Element D. Four-node rectangle with drilling degrees of freedom [Eq. (3.36) with Δu_t constrained to zero] combined with a discrete Kirchhoff quadrilateral.[59]

3.8 Some practical examples

The first example given here is that of the solution of an arch dam shell. A simple geometrical configuration, shown in Fig. 3.7, was taken for this particular problem as results of model experiments and alternative numerical approaches were available.

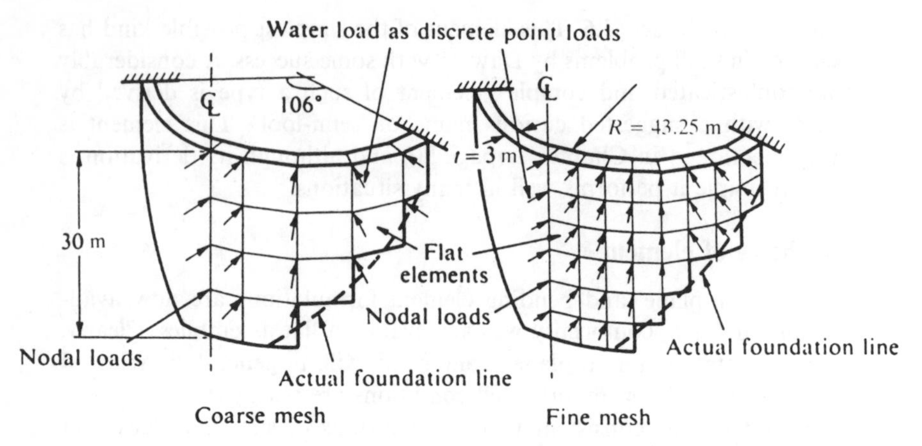

Fig. 3.7 An arch dam as an assembly of rectangular elements

A division based on rectangular elements (type A) was used as the simple cylindrical shape permitted this, although a rather crude approximation to the fixed foundation line had to be used.

Two sizes of division into elements were used, and the results given in Figs 3.8 and 3.9 for both deflections and stresses on the centre-line section show that little refinement was achieved by the use of the finer mesh. This indicates that the convergence of both the physical approximation to the true shape by flat elements and of the mathematical approximation involved in the finite element formulation is excellent. For comparison, stresses and deflection obtained by another, approximate, method of calculation are shown.

A doubly curved arch dam was similarly analysed using the triangular flat element (type B) representation. The results show an even better approximation.[7]

A large number of examples have been computed by Parekh[57] using the triangular, non-conforming element (B), and indeed show for equal division its general superiority over the conforming triangular version presented by Clough and Johnson.[6] Some examples of such analyses are now shown.

3.8.1 *Cooling tower.* This problem of a general axisymmetric shape could, obviously, be more efficiently dealt with by the processes of Chapters 4 or 6. However, here this example is used as a general illustration of the accuracy attainable. The answers against which the numerical solution is compared have been derived by Albasiny and Martin.[60] Figures 3.10 to 3.12 show the geometry of the mesh used and some results. Unsymmetric wind loading is used here.

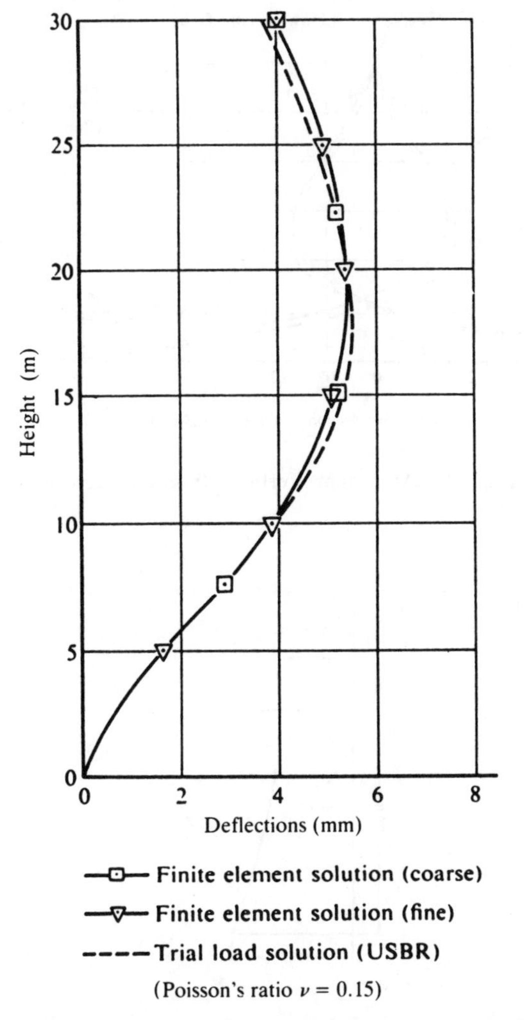

Fig. 3.8 Arch dam. Horizontal deflections on centre-line

3.8.2 *Barrel vault.* This typical shell used in civil engineering is analysed using analytical methods by Scordelis and Lo[61] and Scordelis.[62] The barrel is supported on rigid diaphragms and is loaded by its own weight. Figures 3.13 and 3.14 show some comparative answers, obtained by elements of type B, C and D of the previous section. The latter are obviously more accurate involving more degrees of freedom, and with a mesh of 6 × 6 elements of element C the results are almost indistinguishable from exact ones. This problem has become a classic one on which

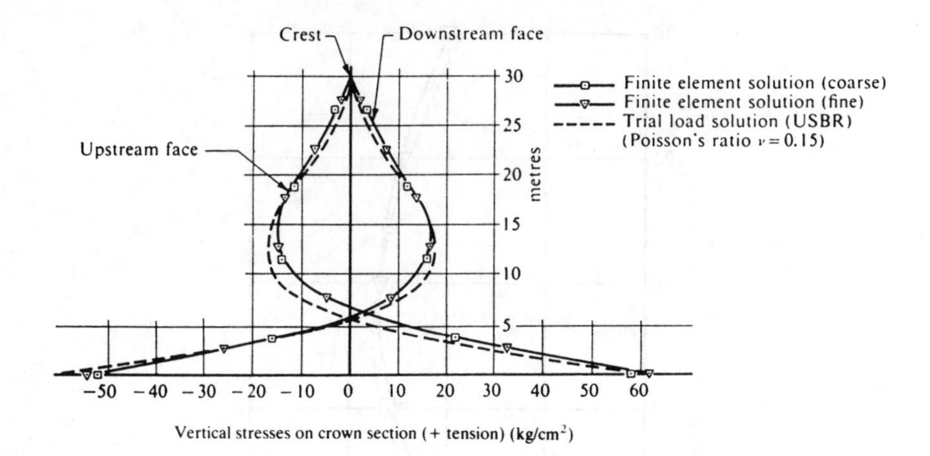

Fig. 3.9 Arch dam. Vertical stresses on centre-line

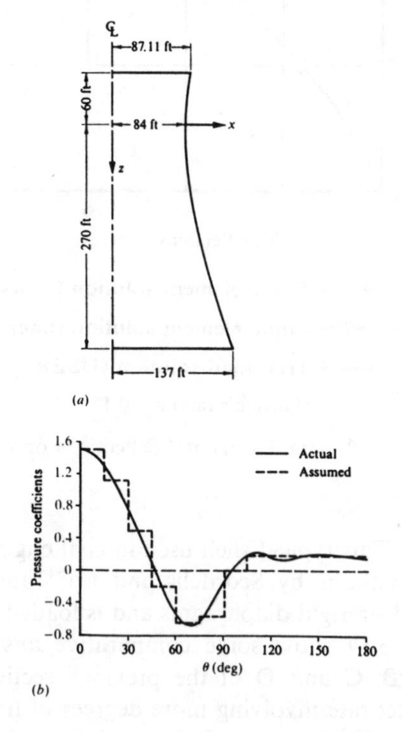

Fig. 3.10 Cooling tower. Geometry and pressure load variation about circumference

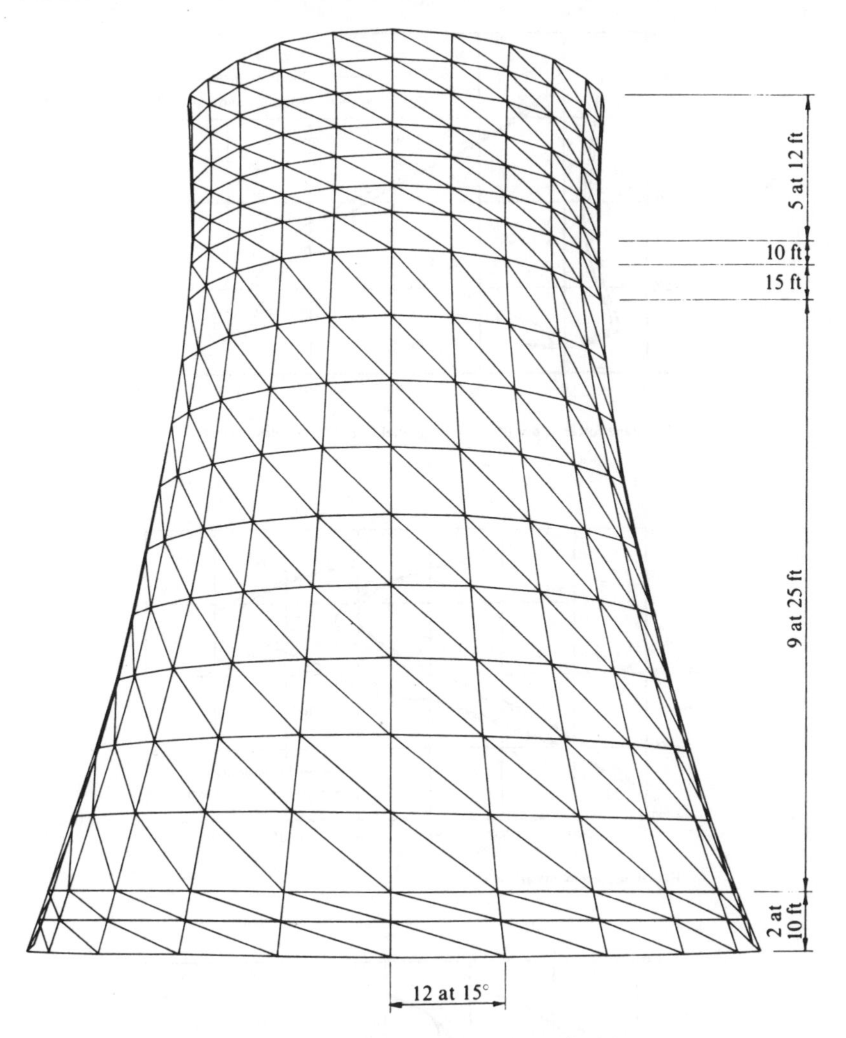

Fig. 3.11 Cooling tower. Mesh subdivisions

various shell elements are compared and we shall return to it in Chapter 5. It is worth while remarking that only a few, second-order, curved elements give superior results to those presented here with a flat element approximation.

3.8.3 *Folded plate structure.* As no exact solution of this problem is known comparison is made with a set of experimental results obtained by Mark and Riesa.[63]

This example demonstrates a problem in which actual flat finite

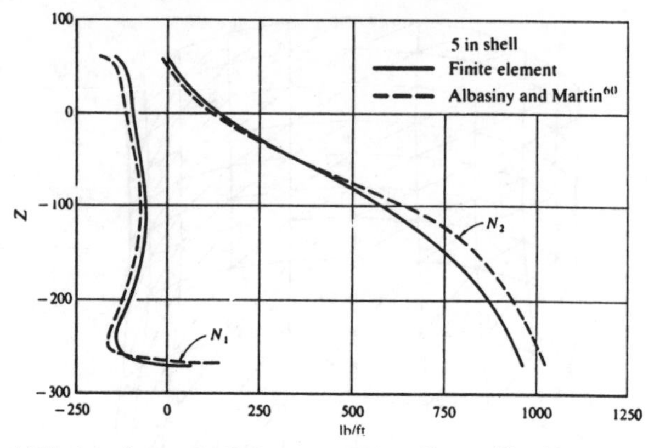

(*a*) Membrane forces at $\theta = 0°$, N_1 = tangential force, N_2 = meridional force

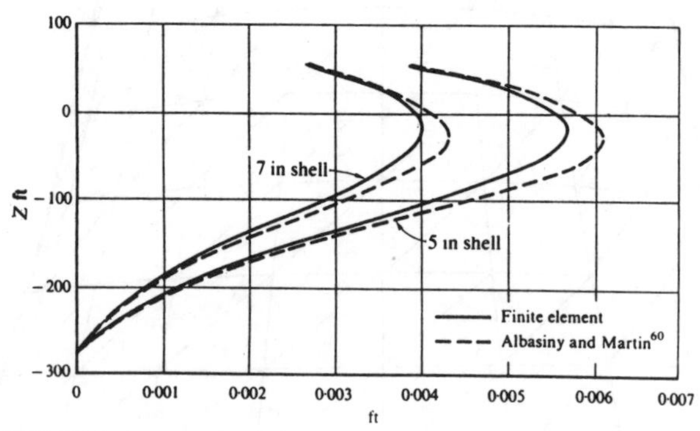

(*b*) Radial displacements at $\theta = 0°$

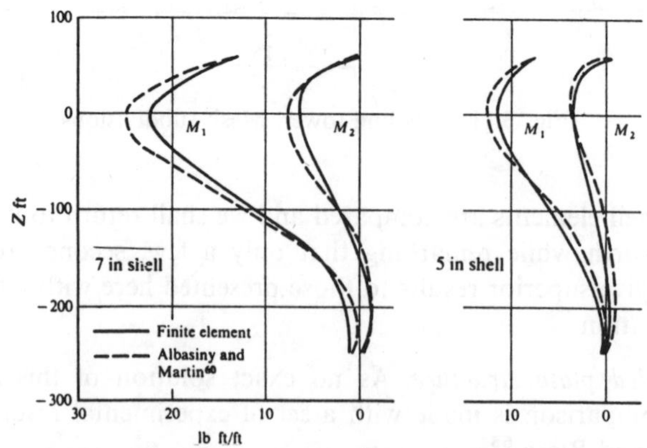

(*c*) Bending moments at $\theta = 0°$, M_1 = tangential moment, M_2 = meridional moment

Fig. 3.12 Cooling tower of Fig. 3.10

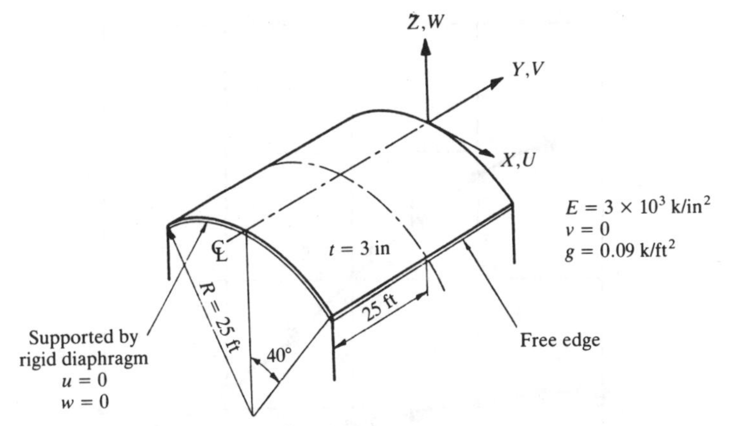

(*a*) Finite element and exact[62] solutions under dead loads

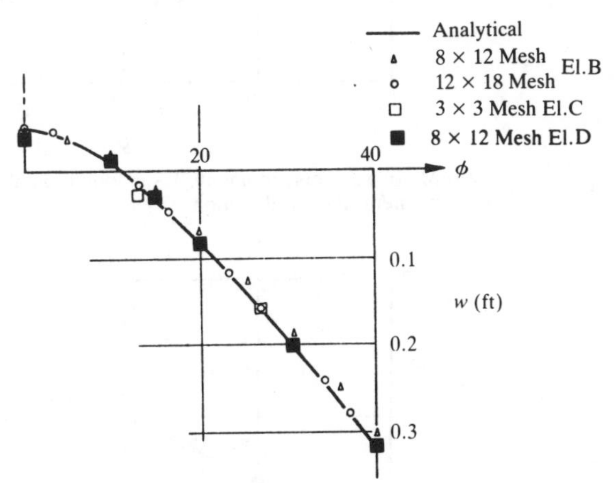

(*b*) Vertical displacement of central sections

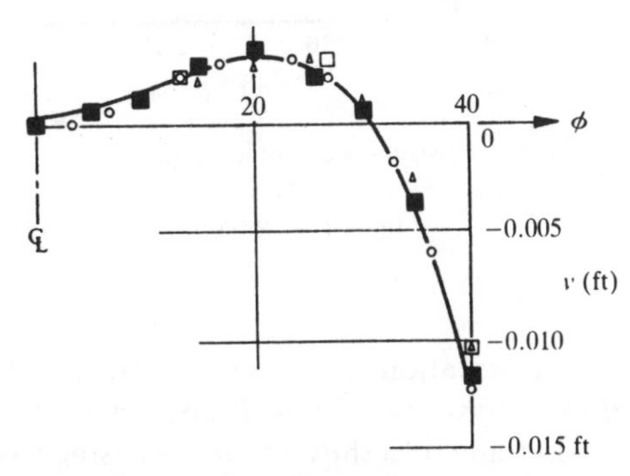

(*c*) Longitudinal displacement of support

Fig. 3.13 A barrel (cylindrical) vault. $E = 3 \times 10^6$ lb/in^2, $\nu = 0$, weight of shell = 90 lb/ft^2

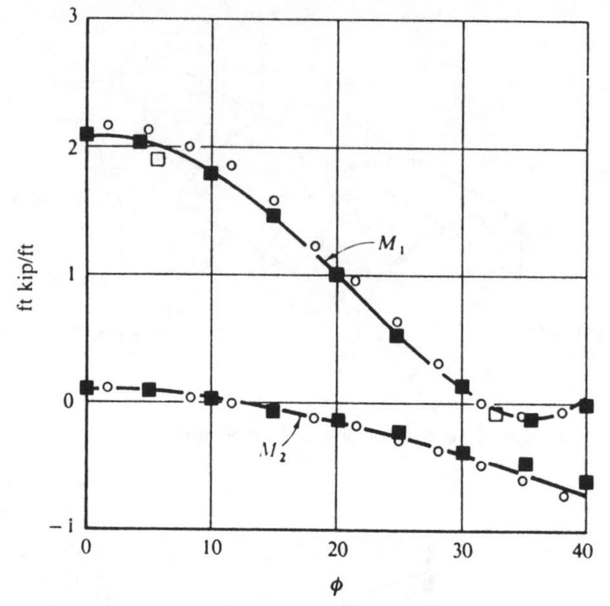

(a) M_1 = transverse moment, M_2 = longitudinal moment at central section

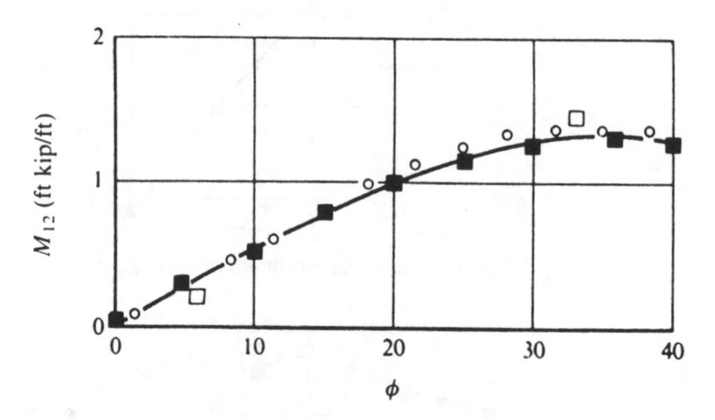

(b) M_{12} = twisting moment at support

Fig. 3.14 Barrel vault of Fig. 3.13

element representation is physically exact. Also a frame stiffness is included in analysis by suitable beam elements.

Figures 3.15 and 3.16 show the results using elements of type B. Similar applications are of considerable importance in the analysis of box-type structures, etc.

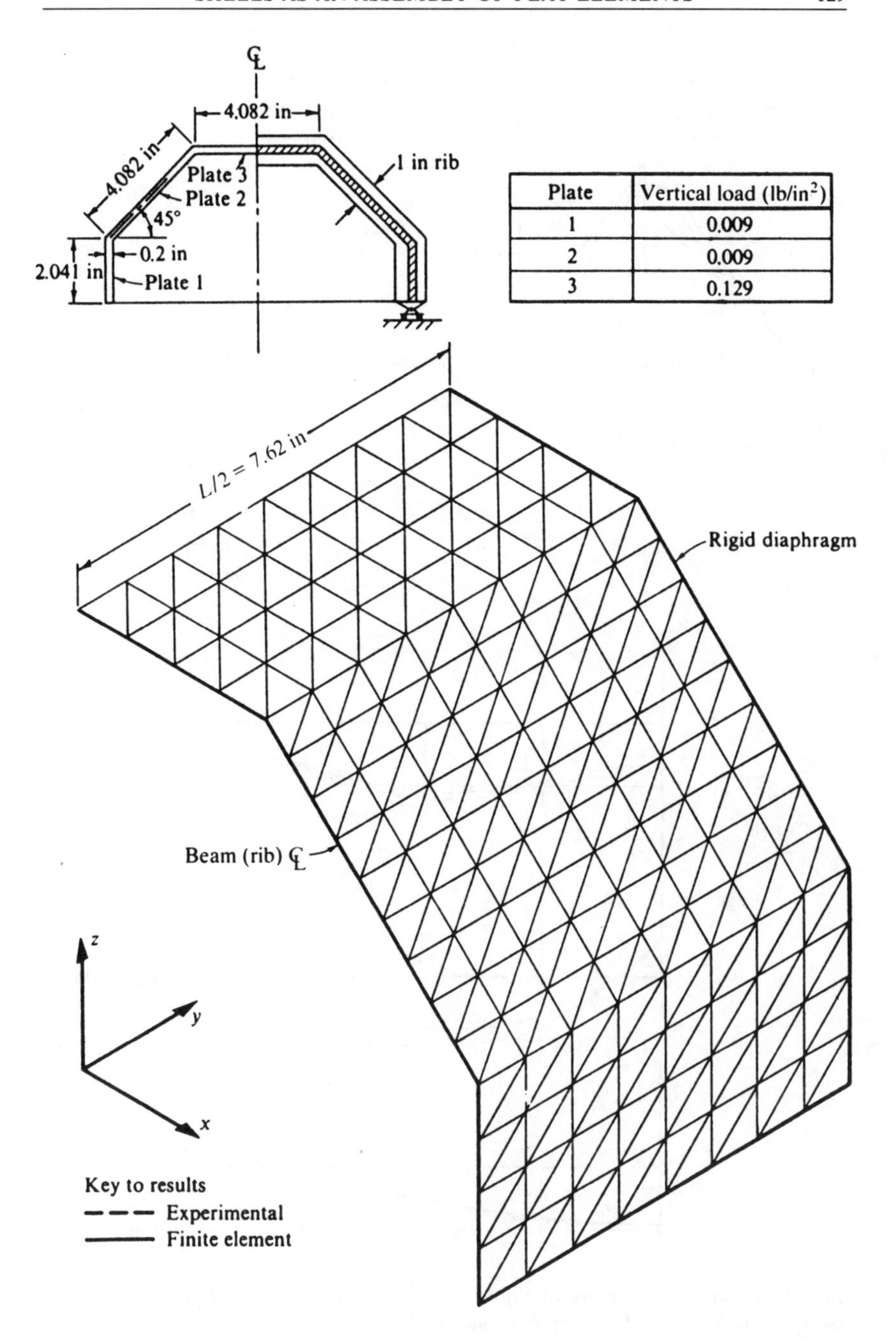

Plate	Vertical load (lb/in^2)
1	0.009
2	0.009
3	0.129

Fig. 3.15 A folded plate structure.[63] Model geometry, loading and mesh, $E = 3560\ \text{lb/in}^2$, $\nu = 0.43$

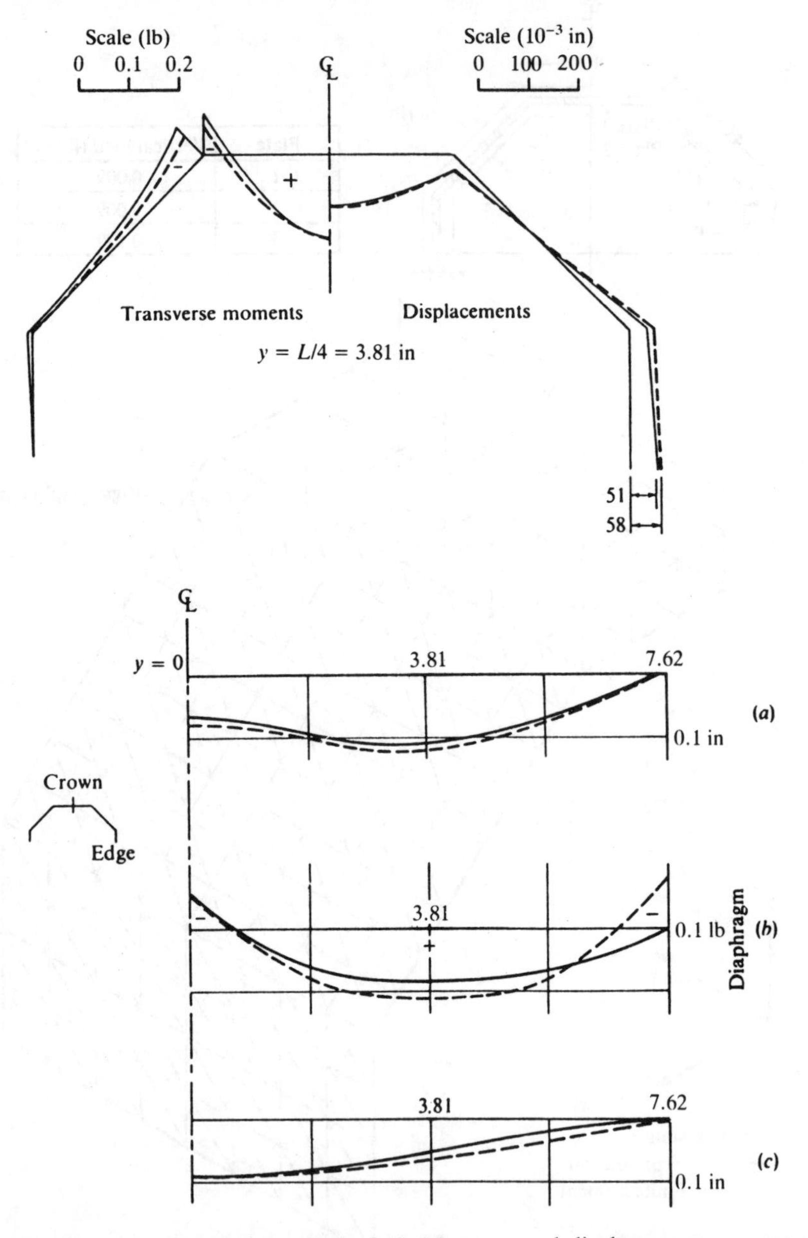

Fig. 3.16 The folded plate of Fig. 3.15. Moments and displacements on centre section. (*a*) Vertical displacements along the crown; (*b*) longitudinal moments along the crown; (*c*) horizontal displacements along edge

References

1. W. FLÜGGE, *Stresses in Shells*, Springer-Verlag, Berlin, 1960.
2. P. G. CIARLET, 'Conforming finite element method for shell problem', in *The Mathematics of Finite Elements and Applications* (ed. J. Whiteman), Vol. II, Academic Press, London, 1977.
3. B. E. GREEN, D. R. STROME and R. C. WEIKEL, 'Application of the stiffness method to the analysis of shell structures', in *Proc. Aviation Conf. of American Society of Mechanical Engineers*, Los Angeles, March 1961.
4. R. W. CLOUGH and J. L. TOCHER, 'Analysis of thin arch dams by the finite element method', in *Proc. Symp. on Theory of Arch Dams, Southampton University, 1964*, Pergamon Press, Oxford, 1965.
5. J. H. ARGYRIS, 'Matrix displacement analysis of anisotropic shells by triangular elements', *J. Roy. Aero. Soc.*, **69**, 801–5, 1965.
6. R. W. CLOUGH and C. P. JOHNSON, 'A finite element approximation for the analysis of thin shells', *J. Solids Struct.*, **4**, 43–60, 1968.
7. O. C. ZIENKIEWICZ, C. J. PAREKH and I. P. KING, 'Arch dams analysed by a linear finite element shell solution program', in *Proc. Symp. on Arch Dams*, Institution of Civil Engineers, London, 1968.
8. O. C. ZIENKIEWICZ and Y. K. CHEUNG, 'Finite element procedures in the solution of plate and shell problems', in *Stress Analysis* (eds O. C. Zienkiewicz and G. S. Holister), chap. 8, Wiley, Chichester, 1965.
9. O. C. ZIENKIEWICZ and Y. K. CHEUNG, 'Finite element methods of analysis for arch dam shells and comparison with finite difference procedures', in *Proc. Symp. on Theory of Arch Dams, Southampton University, 1964*, Pergamon Press, Oxford, 1965.
10. R. H. GALLAGHER, 'Shell elements', in *World Conf. on Finite Element Methods in Structural Mechanics*, Bournemouth, 1975.
11. D. J. DAWE, 'Rigid-body motions and strain–displacement equations of curved shell finite elements', *Int. J. Mech. Sci.*, **14**, 569–78, 1972.
12. G. CANTIN, 'Strain–displacement relationships for cylindrical shells', *JAIAA*, **6**, 1787–8, 1968.
13. D. G. ASHWELL, 'Strain elements with applications to arches, rings, and cylindrical shells', in *Finite Elements for Thin Shells and Curved Members* (eds. D. G. Ashwell and R. H. Gallagher), Wiley, Chichester, 1976.
14. F. K. BOGNER, R. L. FOX and L. A. SCHMIT, 'A cylindrical shell element', *JAIAA*, **5**, 745–50, 1967.
15. G. CANTIN and R. W. CLOUGH, 'A refined, curved cylindrical shell element', AIAA Conf. paper 68–176, New York, 1968.
16. G. BONNEW, G. DHATT, Y. M. GIROUX and L. P. A. ROBICHAUD, 'Curved triangular elements for analysis of shells, in *Proc. 2nd Conf. on Matrix Methods in Structural Mechanics*, Air Force Institute of Technology, Wright-Patterson AF Base, Ohio, 1968.
17. G. E. STRICKLAND and W. A. LODEN, 'A doubly curved triangular shell element', in *Proc. 2nd Conference Matrix Methods in Structural Mechanics*, Air Force Institute of Technology, Wright-Patterson AF Base, Ohio, 1968.
18. B. E. GREENE, R. E. JONES and D. R. STROME, 'Dynamic analysis of shells using doubly curved finite elements', in *Proc. 2nd Conf. on Matrix Methods in Structural Mechanics*, Air Force Institute of Technology, Wright-Patterson AF Base, Ohio, 1968.

19. J. CONNOR and C. BREBBIA, 'Stiffness matrix for shallow rectangular shell element', *Proc. Am. Soc. Civ. Eng.*, **93**, EM1, 43–65, 1967.
20. A. J. CARR, 'A refined element analysis of thin shell structures including dynamic loading', SEL report 67–9, University of California, Berkeley, 1967.
21. G. R. COWPER, G. M. LINDBERG and M. D. OLSON, 'A shallow shell finite element of triangular shape', *Int. J. Solids Struct.*, **6**, 1133–56, 1970.
22. S. UTKU, 'Stiffness matrices for thin triangular elements of non-zero Gaussian curvature', *JAIAA*, **5**, 1659–67, 1967.
23. S. AHMAD, 'Curved finite elements in the analysis of solids, shell and plate structures', Ph.D. thesis, University of Wales, Swansea, 1969.
24. S. W. KEY and Z. E. BEISINGER, 'The analysis of thin shells by the finite element method', in *High Speed Computing of Elastic Structures*, Vol. 1, pp. 209–52, University of Liége Press, 1971.
25. D. J. DAWE, 'The analysis of thin shells using a facet element', CEGB report RD/B/N2038, Berkeley Nuclear Laboratory, England, 1971.
26. B. M. IRONS, 'The semi loof shell element', in *Finite Elements for Thin Shells and Curved Members* (eds D. G. Ashwell and R. H. Gallagher), chap. 11, pp. 197–222, Wiley, Chichester, 1976.
27. D. G. ASHWELL and A. SABIR, 'A new cylindrical shell finite element based on simple independent strain functions', *Int. J. Mech. Sci.*, **4**, 37–47, 1973.
28. G. R. THOMAS and R. H. GALLAGHER, 'A triangular thin shell finite element: linear analysis', NASA report CR-2582, 1975.
29. G. DUPUIS and J. J. GOËL, 'A curved finite element for thin elastic shells', *Int. J. Solids Struct.*, **6**, 987–96, 1970.
30. N. CARPENTER, H. STOLARSKI and T. BELYTSCHKO, 'A flat triangular shell element with improved membrane interpolation', *Comm. Appl. Num Meth.*, **1**, 161–8, 1985.
31. C. PRATT, 'Shell finite element via Reissner's principle', *Int. J. Solids Struct.*, **5**, 1119–33, 1969.
32. J. CONNOR and G. WILL, 'A mixed finite element shallow shell formulation', in *Advances in Matrix Methods of Structural Analysis and Design* (eds R. H. Gallagher *et al.*), pp. 105–37, University of Alabama Press, 1969.
33. L. R. HERRMANN and W. E. MASON, 'Mixed formulations for finite element shell analysis', in *Conf. on Computer-Oriented Analysis of Shell Structures*, paper AFFDL-TR-71-79, June 1971.
34. G. EDWARDS and J. J. WEBSTER, 'Hybrid cylindrical shell elements', in *Finite Elements for Thin Shells and Curved Members* (eds D. G. Ashwell and R. H. Gallagher), Wiley, Chichester, 1976.
35. H. STOLARSKI and T. BELYTSCHKO, 'Membrane locking and reduced integration for curved elements,' *J. Appl. Mech.*, **49**, 172–6, 1982.
36. Ph. JETTEUR and F. FREY, 'A four node Marguerre element for non-linear shell analysis,' *Eng. Comp.*, **3**, 276–82, 1986.
37. J. C. SIMO and D. D. FOX, 'On a stress resultant geometrically exact shell model. Part I: Formulation and optimal parametrization', *Comp. Meth. Appl. Mech. Eng.*, **72**, 267–304, 1989.
38. J. C. SIMO, D. D. FOX and M. S. RIFAI, 'On a stress resultant geometrically exact shell model. Part II: The linear theory; computational aspects', *Comp. Meth. Appl. Mech. Eng.*, **73**, 53–92, 1989.
39. S. AHMAD, B. M. IRONS and O. C. ZIENKIEWICZ, 'A simple matrix-vector handling scheme for three-dimensional and shell analysis', *Int. J. Num. Meth. Eng.*, **2**, 509–22, 1970.

40. D. J. ALLMAN, 'A compatible triangular element including vertex rotations for plane elasticity analysis', *Comp. Struct.*, **19**, 1–8, 1984.
41. D. J. ALLMAN, 'A quadrilateral finite element including vertex rotations for plane elasticity analysis', *Int. J. Num. Meth. Eng.*, **26**, 717–30, 1988.
42. D. J. ALLMAN, 'Evaluation of the constant strain triangle with drilling rotations', *Int. J. Num. Meth. Eng.*, **26**, 2645–55, 1988.
43. P. G. BERGAN and C. A. FELIPPA, 'A triangular membrane element with rotational degrees of freedom', *Comp. Meth. Appl. Mech. Eng.*, **50**, 25–69, 1985.
44. P. G. BERGAN and C. A. FELIPPA, 'Efficient implementation of a triangular membrane element with drilling freedoms', in *Finite Element Methods for Plate and Shell Structures* (eds T. J. R. Hughes and E. Hinton), Vol. 1, pp. 128–52, Pineridge Press, Swansea, 1986.
45. R. D. COOK, 'On the Allman triangle and a related quadrilateral element', *Comp. Struct.*, **2**, 1065–7, 1986.
46. R. D. COOK, 'A plane hybrid element with rotational d.o.f. and adjustable stiffness', *Int. J. Num. Meth. Eng.*, **24**, 1499–508, 1987.
47. T. J. R. HUGHES and F. BREZZI, 'On drilling degrees-of-freedom', *Comp. Meth. Appl. Mech. Eng.*, **72**, 105–21, 1989.
48. T. J. R. HUGHES, F. BREZZI, A. MASUD and I. HARARI, 'Finite elements with drilling degrees-of-freedom: 'theory and numerical evaluations', 1989 (preprint).
49. R. L. TAYLOR and J. C. SIMO, 'Bending and membrane elements for analysis of thick and thin shells', in *Proc. NUMETA 85 Conf.* (eds G. N. Pande and J. Middleton), Vol. 1, pp. 587–91, A. A. Balkema, Rotterdam, 1985.
50. Ph. JETTEUR, 'Improvement of the quadrilateral JET shell element for a particular class of shell problems', IREM internal report 87/1, Ecole Polytechnique Federale de Lausanne, February 1987.
51. R. L. TAYLOR, 'Finite element analysis of linear shell problems', in *The Mathematics of Finite Elements and Applications* (ed. J. R. Whiteman), Vol. VI, pp. 191–205, Academic Press, London, 1988.
52. R. H. MACNEAL and R. L. HARTER, 'A refined four-noded membrane element with rotational degrees of freedom', *Comp. Struct.*, **28**, 75–88, 1988.
53. R. W. CLOUGH and E. L. WILSON, 'Dynamic finite element analysis of arbitrary thin shells', *Comp. Struct.*, **1**, 33–56, 1971.
54. R. H. MACNEAL and R. L. HARTER, 'A proposed standard set of problems to test finite element accuracy', *J. Finite Elements in Anal. Des.*, **1**, 3–20, 1985.
55. E. REISSNER, 'A note on variational theorems in elasticity', *Int. J. Solids Struct.*, **1**, 93–5, 1965.
56. A. IBRAHIMBEGOVIC, R. L. TAYLOR and E. L. WILSON, 'A robust quadrilateral membrane finite element with drilling degrees of freedom,' *Int. J. Num. Meth. Eng.*, **30**, 445–57, 1990.
57. C. J. PAREKH, 'Finite element solution system', Ph.D. thesis, University of Wales, Swansea, 1969.
58. A. RAZZAQUE, 'Finite element analysis of plates and shells', Ph.D. thesis, University of Wales, Swansea, 1972.
59. J-L. BATOZ and M. B. TAHAR, 'Evaluation of a new quadrilateral thin plate bending element', *Int. J. Num. Meth. Eng.*, **18**, 1655–77, 1982.
60. E. L. ALBASINY and D.W. MARTIN, 'Bending and mebrane equilibrium in cooling towers', *Proc. Am. Soc. Civ. Eng.*, **93**, EM3, 1–17, 1967.

61. A. C. Scordelis and K. S. Lo, 'Computer analysis of cylindrical shells', *J. Am. Concr. Inst.*, **61**, 539–61, 1964.
62. A. C. Scordelis, 'Analysis of cylindrical shells and folded plates', in *Concrete Thin Shells*, American Concrete Institute report SP 28-N, 1971.
63. R. Mark and J. D. Riesa, 'Photoelastic analysis of folded plate structures', *Proc. Am. Soc. Civ. Eng.*, **93**, EM4, 79–83, 1967.

4

Axisymmetric shells

4.1 Introduction

The problem of axisymmetric shells is of sufficient practical importance to include in this chapter special methods of dealing with their solution.

While the general method described in the previous chapter is obviously applicable here, it will be found that considerable simplification can be achieved if account is taken of axial symmetry of the structure. In particular, if both the shell and the loading are axisymmetric it will be found that the elements become 'one dimensional'. This is the simplest type of element, to which little attention was given in earlier chapters.

The first approach to the finite element solution of axisymmetric shells was presented by Grafton and Strome.[1] In this, the elements are simple conical frustra and a direct approach via displacement functions is used. Refinements in the derivation of element stiffnesses are presented in Popov *et al.*[2] and Jones and Strome,[3] and an extension to the case of unsymmetrical loads, which was suggested in Grafton and Strome,[1] is elaborated in Percy *et al.*,[4] Klein[5] and others.[6,7]

Later much work was accomplished to extend the processes to curved elements and indeed to refine the approximations involved. The literature on the subject is considerable, no doubt promoted by the interest in space flight, and a complete bibliography is here impracticable. References 8 to 16 show how curvilinear coordinates of various kinds can be introduced to the analysis while 11 and 13 discuss the use of additional nodeless degrees of freedom in improving the accuracy. 'Mixed' formulations (Chapter 12 of Volume 1) have found here some use.[17] The subject is reviewed comprehensively by Gallagher[18,19] and others,[20] where very full bibliographies can be found.

In axisymmetric shells, in common with all other shells, both bending and 'in-plane' or 'membrane' forces will occur. These will be specified uniquely in terms of the generalized 'strains', which now involve extensions and curvatures of the middle surface. If the displacement of each point of the middle surface is specified, such 'strains' and the internal stress resultants, or simply 'stresses', can be determined by formulae available in standard texts dealing with shell theory.

For example, in an axisymmetric shell under axisymmetric loading, such as is shown in Fig. 4.1, the displacement of a point on the middle surface is uniquely determined by two components u and w in the tangential and normal directions respectively.

The four strain components are given by the following expression, using the Kirchhoff–Love assumption, provided the angle ϕ does not vary (i.e. elements are straight):[21-23]

$$\{\varepsilon\} = \begin{Bmatrix} \varepsilon_s \\ \varepsilon_\theta \\ \chi_s \\ \chi_\theta \end{Bmatrix} = \begin{Bmatrix} \mathrm{d}u/\mathrm{d}s \\ (w \cos\phi + u \sin\phi)/r \\ -\mathrm{d}^2 w/\mathrm{d}s^2 \\ -(\sin\phi/r)(\mathrm{d}w/\mathrm{d}s) \end{Bmatrix} \tag{4.1}$$

This results in four internal stress resultants, shown in Fig. 4.1, and related to the strains by an elasticity matrix $\mathbf{D}$:

$$\boldsymbol{\sigma} = \begin{Bmatrix} N_s \\ N_\theta \\ M_s \\ M_\theta \end{Bmatrix} = \mathbf{D}\boldsymbol{\varepsilon} \tag{4.2}$$

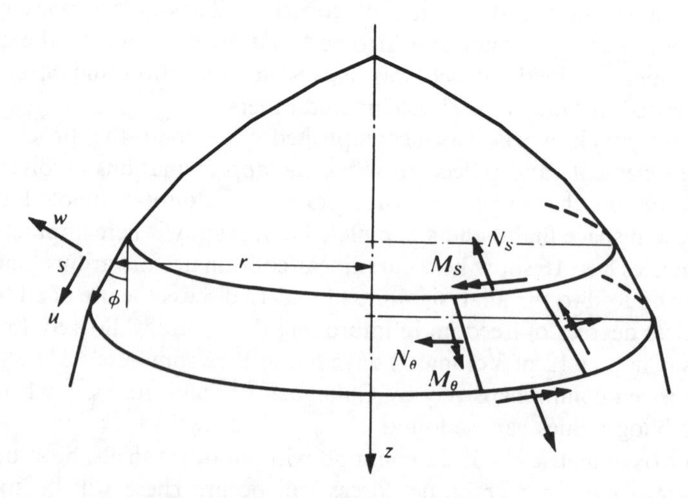

Fig. 4.1 Axisymmetric shell, loading, displacements and stress resultants. Shell represented as a stack of conical frustra

For an isotropic shell the matrix **D** becomes

$$\mathbf{D}=\frac{Et}{1-\nu^2}\begin{bmatrix}1 & \nu & 0 & 0\\ \nu & 1 & 0 & 0\\ 0 & 0 & t^2/12 & \nu t^2/12\\ 0 & 0 & \nu t^2/12 & t^2/12\end{bmatrix} \tag{4.3}$$

the upper part being a plane stress and the lower a bending stiffness matrix, with the shear terms omitted in both as 'thin' conditions are assumed.

4.2 Element characteristics—axisymmetrical loads—straight elements

Let the shell be divided by nodal surfaces into a series of conical frustra, as shown in Fig. 4.2. The nodal displacements at points such as i and j will have to define uniquely the deformations of the element via prescribed shape functions.

At each node the axial and radial movements and a rotation will be used as the parameters. All three components are necessary as the shell can carry bending moments. The displacements of a node i can thus be defined by three components, the first two being in global directions,

$$\mathbf{a}_i=\begin{Bmatrix}\bar{u}_i\\ \bar{w}_i\\ \beta_i\end{Bmatrix} \tag{4.4}$$

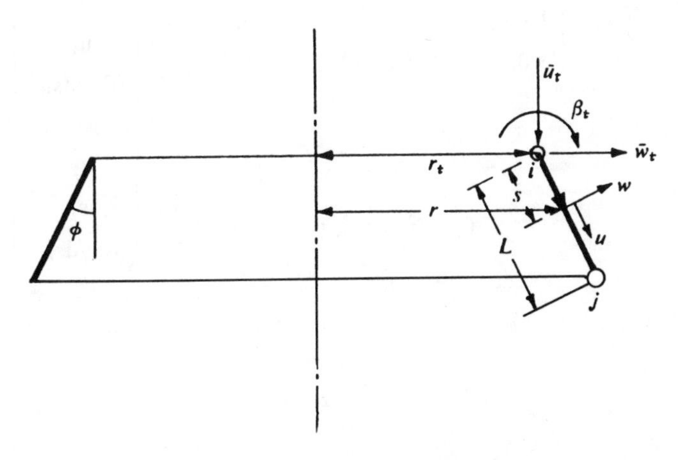

Fig. 4.2 An element of an axisymmetric shell

The simplest elements with two nodes, i and j, thus possess 6 degrees of freedom, determined by the element displacements

$$\mathbf{a}^e = \begin{Bmatrix} \mathbf{a}_i \\ \mathbf{a}_j \end{Bmatrix} \tag{4.5}$$

The displacements within the element have to be uniquely determined by the nodal displacements $\mathbf{a}^e$ and the position s, and maintain slope and displacement continuity.

Thus in local coordinates we have

$$\mathbf{u} = \begin{Bmatrix} u \\ w \end{Bmatrix} = \mathbf{N}\mathbf{a}^e \tag{4.6}$$

If u is taken as varying linearly with s and w as a cubic in s we shall have six undetermined constants, which can be determined from the nodal values of $\bar{u}$, $\bar{w}$ and β.

At the node i,

$$\begin{Bmatrix} u_1 \\ w_i \\ (\mathrm{d}w/\mathrm{d}s)_i \end{Bmatrix} = \begin{bmatrix} \cos\phi, & +\sin\phi, & 0 \\ -\sin\phi, & \cos\phi, & 0 \\ 0, & 0, & 1 \end{bmatrix} \begin{Bmatrix} \bar{u}_i \\ \bar{w}_i \\ \beta_i \end{Bmatrix} = \boldsymbol{\lambda}\mathbf{a}_i \tag{4.7}$$

Writing

$$\begin{aligned} u &= \alpha_1 + \alpha_2 s \\ w &= \alpha_3 + \alpha_4 s + \alpha_5 s^2 + \alpha_6 s^3 \end{aligned} \tag{4.8}$$

it is an easy matter to state the six end conditions and arrive at†

$$\begin{Bmatrix} u \\ w \end{Bmatrix} = \left[\begin{matrix} 1-s', & 0, & 0, \\ 0, & 1-3s'^2+2s'^3, & L(s'-2s'^2+s'^3), \end{matrix} \quad \begin{matrix} s', & 0, & 0 \\ 0, & 3s'^2-2s'^3, & (-s'^2+s'^3)L \end{matrix}\right] \begin{Bmatrix} u_i \\ w_i \\ (\mathrm{d}w/\mathrm{d}s)_i \\ u_j \\ w_j \\ (\mathrm{d}w/\mathrm{d}s)_j \end{Bmatrix} \tag{4.9}$$

in which

$$s' = \frac{s}{L}$$

† The functions that occur there are, in fact, hermitian polynomials of order 0 and 1 (see Chapter 1, Sec. 1.14).

Calling the above 2×6 matrix $\mathbf{N}'$ we can now write

$$\begin{Bmatrix} u \\ w \end{Bmatrix} = \mathbf{N}' \begin{bmatrix} \boldsymbol{\lambda} & 0 \\ 0 & \boldsymbol{\lambda} \end{bmatrix} \mathbf{a}^e = [\mathbf{N}_i'\boldsymbol{\lambda}, \ \mathbf{N}_j'\boldsymbol{\lambda}]\mathbf{a}^e = \mathbf{N}\mathbf{a}^e \tag{4.10}$$

From Eq. (4.10) it is a simple matter to obtain the strain matrix $\mathbf{B}$ by the use of the definition Eq. (4.1). This gives

$$\boldsymbol{\varepsilon} \equiv \mathbf{B}\mathbf{a}^e = [\mathbf{B}_i'\boldsymbol{\lambda}, \ \mathbf{B}_j'\boldsymbol{\lambda}]\mathbf{a}^e \tag{4.11}$$

in which

$$\mathbf{B}_i' = \begin{bmatrix} -1/L, & 0, & 0 \\ (1-s')\sin\phi/r, & (1-3s'^2+2s'^3)\cos\phi/r, & L(s'-2s'^2+s'^3)\cos\phi/r \\ 0, & (6-12s')/L^2, & (4-6s')/L \\ 0, & (6s'-6s'^2)\sin\phi/rL, & (-1+4s'-3s'^2)\sin\phi/r \end{bmatrix} \tag{4.12}$$

$$\mathbf{B}_j' = \begin{bmatrix} 1/L, & 0, & 0 \\ s'\sin\phi/r, & (3s'^2-2s'^3)\cos\phi/r, & L(-s'^2+s'^3)\cos\phi/r \\ 0, & (-6+12s')/L^2, & (2-6s')/L \\ 0, & (-6s'+6s'^2)\sin\phi/rL, & (2s'-3s'^2)\sin\phi/r \end{bmatrix}$$

Now all the 'ingredients' required for computing the stiffness matrix (or load, stress and initial stress matrices) by standard formulae are known. The integrations required are carried out over the area, A, of the element, i.e. with

$$dA = 2\pi \, ds = 2\pi r L \, ds' \tag{4.13}$$

with s' varying from 0 to 1.

Thus, the stiffness matrix $\mathbf{K}$ becomes in local coordinates

$$\mathbf{K} = \int_0^1 \mathbf{B}'^T\mathbf{D}\mathbf{B}' \, 2\pi r L \, ds' \tag{4.14}$$

On transformation, the element $\mathbf{K}_{rs}$ of the global matrix is given by

$$\mathbf{K}_{rs} = \boldsymbol{\lambda}^T\left(\int_0^1 \mathbf{B}_r'^T\mathbf{D}\mathbf{B}_s' r \, ds'\right)\boldsymbol{\lambda} \, 2\pi L \tag{4.15}$$

The radius r has to be expressed as a function of s before such integrations are carried out.

Once again it is convenient to use numerical integration. Grafton and Strome[1] give an explicit formula for the stiffness matrix based on a single average value of the integrand and using a $\mathbf{D}$ matrix corresponding to an orthotropic material. Even with this crude approximation extremely good results can be obtained, provided small elements are used.

Percy *et al.*[4] and Klein[5] carry out a seven-point numerical integration and a slightly improved matrix is obtained.

It should be remembered that if any external line loads or moments are present, their full circumferential value must be used in the analysis, just as was the case with axisymmetric solids discussed in Chapter 4 of Volume 1.

4.3 Examples and accuracy

In the treatment of axisymmetric shells described here continuity is satisfied at all times. For a polygonal shape of shell, therefore, convergence will always occur.

The problem of the physical approximation to a curved shell by a polygonal shape is similar to the one discussed in Chapter 3. Intuitively, convergence can be expected, and indeed numerous examples indicate this.

When the loading is such as to cause predominantly membrane stresses, discrepancies in bending moment values have been found to exist even with reasonably fine subdivision. Again, however, these disappear as the size of the subdivision decreases, particularly if correct (consistent) sampling is used (viz. Chapter 12 of Volume 1). This is necessary to eliminate the physical approximation involved in representing the shell as a series of conical frustra.

Figures 4.3 and 4.4 illustrate some typical examples taken from Grafton and Strome,[1] which show quite remarkable accuracy.

4.4 Curved elements and their shape functions

Use of curved elements has already been described in Chapter 8 of Volume 1, in the context of analysis which involved, in the definition of strain, only first derivatives. Here second derivatives exist [see Eq. (4.1)] and some of the theorems of Chapter 8 are no longer applicable.

It was previously mentioned that many possible definitions of curved elements have been proposed and used in the context of axisymmetric shells.[9-12] The derivation used here is one due to Delpak[11] and, to use the nomenclature of Chapter 8, is of the subparametric type.

The basis of curved element definition is one that gives a common tangent between adjacent elements (or alternatively a specified tangent direction). This is physically necessary to avoid 'kinks' in the description of what in practice is possibly a smooth shell.

If a general curved form of a shell of revolution is considered, as shown in Fig. 4.5, the expressions for strain quoted in Eq. (4.1) have to be

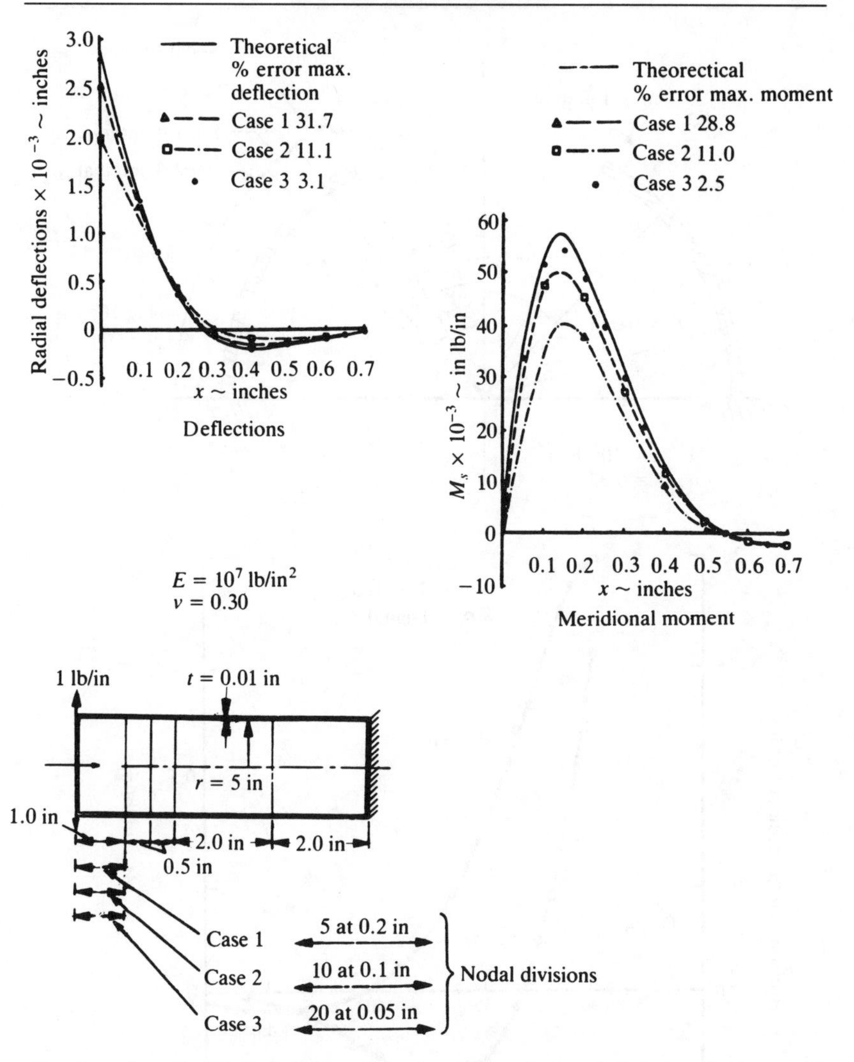

Fig. 4.3 A cylindrical shell solution by finite elements. (From Grafton and Strome[1])

modified to take into account the curvature of the shell in the meridional plane.[21-23] These now become

$$\boldsymbol{\varepsilon} = \begin{Bmatrix} \varepsilon_s \\ \varepsilon_\theta \\ \chi_s \\ \chi_\theta \end{Bmatrix} = \begin{Bmatrix} \mathrm{d}u/\mathrm{d}s + w/R_s \\ (w \cos\phi + u \sin\phi)/r \\ -\mathrm{d}^2 w/\mathrm{d}s^2 + \mathrm{d}(u/R_s)/\mathrm{d}s \\ -(\sin\phi/r)(\mathrm{d}w/\mathrm{d}s - u/R_s) \end{Bmatrix} \tag{4.16}$$

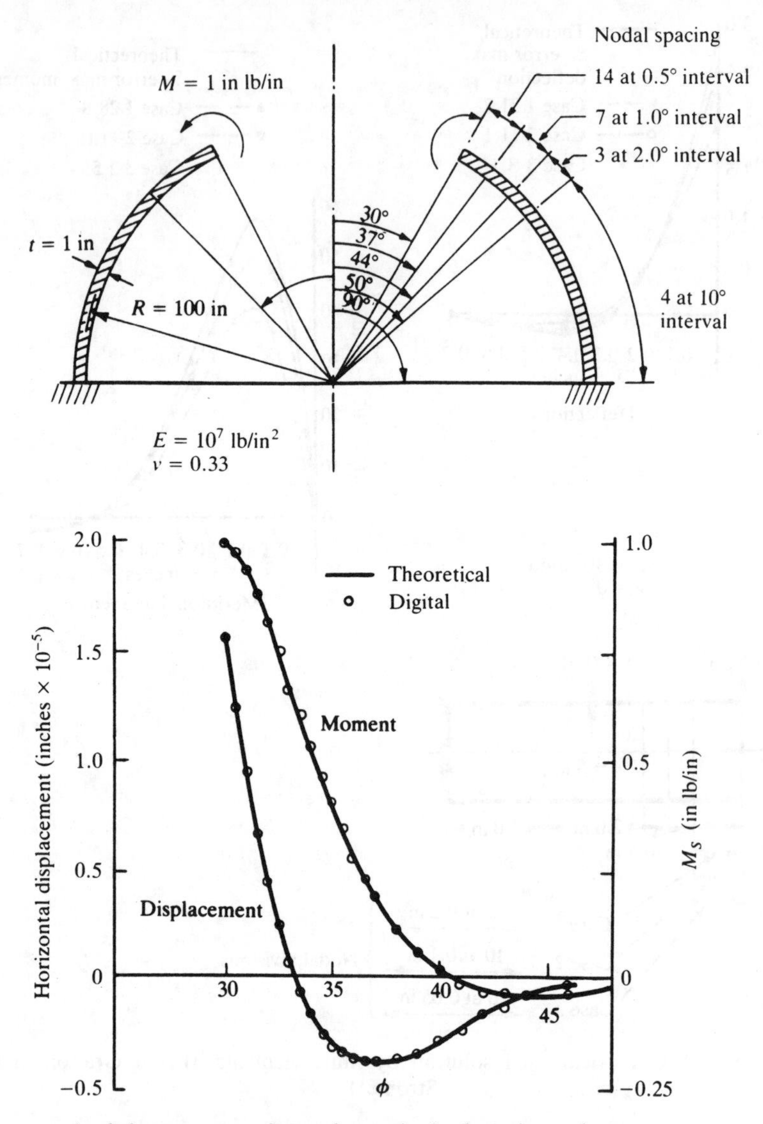

Fig. 4.4 A hemispherical shell solution by finite elements. (From Grafton and Strome[1])

In the above the angle ϕ is a function of s, i.e.

$$\frac{\mathrm{d}r}{\mathrm{d}s} = \sin\phi$$

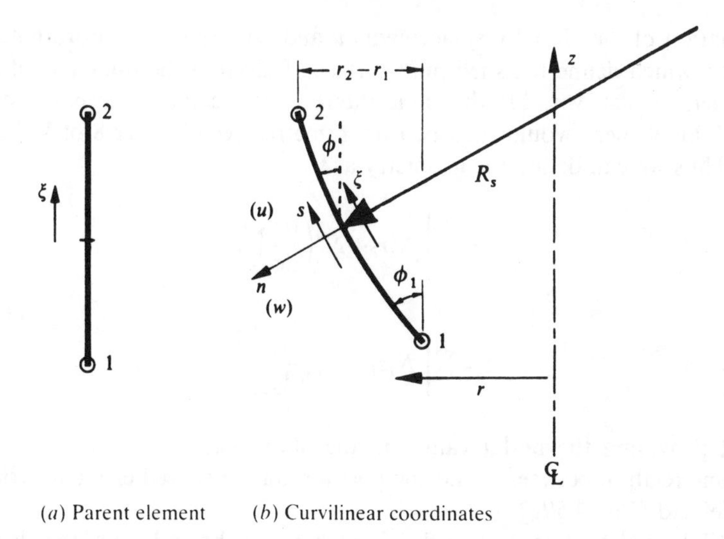

(a) Parent element (b) Curvilinear coordinates

Fig. 4.5 Curved, isoparametric, shell element for axisymmetric problems

R_s is the principal radius in the meridianal plane and the second principal curvature radius R_θ is given by

$$R_\theta = r \cos \phi$$

The reader can verify that for $R_s = \infty$ Eq. (4.16) coincides with Eq. (4.1).

We shall now consider a curved element 1–2 shown in Fig. 4.5(*b*), where the coordinate in its 'parent' form $-1 \leqslant \xi \leqslant 1$ is shown in Fig. 4.5(*a*). The coordinates and the unknowns are 'mapped' in the manner of Chapter 8 of Volume 1. As we wish to interpolate a quantity ψ with slope continuity we can write

$$\psi = \sum_{i=1}^{2} \left[N'_i \psi_i + N''_i \left(\frac{d\psi}{d\xi} \right)_i \right] = \mathbf{N\Psi}^e \tag{4.17}$$

In this N' and N'' are scalar shape functions and for simplest representation will be cubics [similar to those used in Eq. (4.9) for the variation of w].

Explicitly we can write these cubic functions as

$$\begin{aligned} N'_i &= \tfrac{1}{4}\{\xi_0 \xi^2 - 3\xi_0 + 2\} \\ N''_i &= \tfrac{1}{4}(1-\xi_0)^2(1+\xi_0) \qquad \text{with } \xi_0 = \xi_i \xi \end{aligned} \tag{4.18}$$

Now we can simultaneously use the above functions to describe the variation of the global displacements $\bar{u}$ and $\bar{w}$† *and* of the coordinates r and z which define the shell (mid-surface). Indeed, if the thickness of the element is also variable the same interpolation could be applied to it. Such an element would then be isoparametric (see Chapter 8 of Volume 1). Thus we can define the geometry as

$$r = \sum_1^2 \left[N_i' r_i + N_i'' \left(\frac{\mathrm{d}r}{\mathrm{d}\xi} \right)_i \right]$$

and

$$z = \sum_1^2 \left[N_i' z_i + N_i'' \left(\frac{\mathrm{d}z}{\mathrm{d}\xi} \right)_i \right] \tag{4.19}$$

and providing the nodal values in the above can be specified, a one-to-one relation between ξ and the position on the curved element surface is defined [Fig. 4.5(*b*)].

While specification of r_i and z_i is obvious, at the ends only the slope

$$(\tan \phi)_i = \left(\frac{\mathrm{d}r}{\mathrm{d}z} \right) \tag{4.20}$$

is defined. What specification is to be adopted with regard to the derivatives occurring in Eq. (4.19) depends on the *scaling* of ξ along the tangent length s.

Only the ratio

$$\left(\frac{\mathrm{d}r}{\mathrm{d}z} \right)_i = \frac{(\mathrm{d}r/\mathrm{d}\xi)_i}{(\mathrm{d}z/\mathrm{d}\xi)_i} \tag{4.21}$$

is unambiguously specified. Thus $(\mathrm{d}r/\mathrm{d}\xi)_i$ or $(\mathrm{d}z/\mathrm{d}\xi)_i$ can be given an arbitrary value. Here, however, practical considerations intervene as with the wrong choice of value a very uneven relationship between s and ξ will be achieved. Indeed, with an unsuitable choice the shape of the curve can depart from the smooth one illustrated and loop between the end values.

To achieve a reasonably uniform spacing it suffices for well-behaved

† One immediate difference will be observed from that of the previous formulation. Now both displacement components vary in at least a cubic manner along an element while previously a linear variation of the tangental displacement was permitted. This additional degree of freedom does not, however, introduce any excessive continuities in this case providing the shell is itself continuous in thickness.

surfaces to approximate

$$\frac{dr}{d\xi}=\frac{\Delta r}{\Delta \xi}=\frac{r_2-r_1}{2} \tag{4.22}$$

noting that the whole range of ξ is 2 between the nodal points.

4.5 Strain expressions and properties of curved elements

The variation of global displacements has been specified while, by Eq. (4.16), the strains are determinate in terms of the derivatives of locally directed displacements with respect to the tangent, s. Some transformations are therefore necessary before the strains can be determined.

If thus we take the global displacement variation to be defined by the shape function, Eq. (4.17), as

$$\begin{aligned} \bar{u} &= \sum_{i=1}^{2}\left[N_i'\bar{u}_i + N_i''\left(\frac{d\bar{u}}{d\xi}\right)_i\right] \\ \bar{w} &= \sum_{i=1}^{2}\left[N_i'\bar{w}_i + N_i''\left(\frac{d\bar{w}}{d\xi}\right)_i\right] \end{aligned} \tag{4.23}$$

we can find the locally directed displacements u, w from the transformation implied in Eq. (4.7), i.e.

$$\begin{Bmatrix} u \\ w \end{Bmatrix} = \begin{bmatrix} \cos\theta & \sin\theta \\ -\sin\theta & \cos\theta \end{bmatrix} \begin{Bmatrix} \bar{u} \\ \bar{w} \end{Bmatrix} = \mathbf{L}\begin{Bmatrix} \bar{u} \\ \bar{w} \end{Bmatrix} \tag{4.24}$$

where ϕ is the angle of the tangent to the curve and z axis (Fig. 4.5). However, before we can proceed further it is necessary to express this transformation in terms of the ξ coordinate. We have

$$\tan\phi = \frac{dr/d\xi}{dz/d\xi} \tag{4.25}$$

and hence this can now be accomplished by (4.19).

Before proceeding further we must consider whether continuity can be imposed at nodes on the parameters of Eq. (4.23). Clearly the global displacements must be continuous. However, on previous occasions we have specified a continuity of *rotation* of the tangent only. Here we shall allow usually the continuity of both the s derivatives in displacements.

Thus the parameters

$$\frac{\mathrm{d}\bar{u}}{\mathrm{d}s} \quad \text{and} \quad \frac{\mathrm{d}\bar{w}}{\mathrm{d}s}$$

will be given common values at nodes. As

$$\frac{\mathrm{d}\bar{u}}{\mathrm{d}s}=\frac{\dfrac{\mathrm{d}\bar{u}}{\mathrm{d}\xi}}{\dfrac{\mathrm{d}s}{\mathrm{d}\xi}}$$

$$\frac{\mathrm{d}\bar{w}}{\mathrm{d}s}=\frac{\dfrac{\mathrm{d}\bar{w}}{\mathrm{d}\xi}}{\dfrac{\mathrm{d}s}{\mathrm{d}\xi}} \tag{4.26}$$

and

$$\frac{\mathrm{d}s}{\mathrm{d}\xi}=\sqrt{\left(\frac{\mathrm{d}r}{\mathrm{d}\xi}\right)^2+\left(\frac{\mathrm{d}z}{\mathrm{d}\xi}\right)^2}$$

no difficulty exists in substituting these new variables in Eqs (4.23) and (4.24) which now take the form

$$\begin{Bmatrix} u \\ w \end{Bmatrix}=[N(\xi)]\mathbf{a}^e \qquad \text{with } \mathbf{a}_i \begin{Bmatrix} \bar{u}_i \\ \bar{w}_i \\ (\mathrm{d}\bar{u}/\mathrm{d}s)_i \\ (\mathrm{d}\bar{w}/\mathrm{d}s)_i \end{Bmatrix} \tag{4.27}$$

The form of the 2×4 submatrices is complicated but can be explicitly determined.[11] We note that the curvature radius R_s can be explicitly calculated from the mapped, parametric, form of the element. We thus write

$$R_s=\frac{\left[\left(\dfrac{\mathrm{d}r}{\mathrm{d}\xi}\right)^2+\left(\dfrac{\mathrm{d}z}{\mathrm{d}\xi}\right)^2\right]^{3/2}}{\dfrac{\mathrm{d}r}{\mathrm{d}\xi}\dfrac{\mathrm{d}^2z}{\mathrm{d}\xi^2}-\dfrac{\mathrm{d}z}{\mathrm{d}\xi}\dfrac{\mathrm{d}^2r}{\mathrm{d}\xi^2}} \tag{4.28}$$

in which all the derivatives are directly evolved from expression (4.19).

If shells that branch or in which abrupt thickness changes occur are to be treated, the nodal parameters specified in Eq. (4.27) are not satisfactory. It is better then to rewrite these as

$$\mathbf{a}_i = \begin{Bmatrix} \bar{u}_i \\ \bar{w} \\ \beta_i \\ (\mathrm{d}\bar{u}/\mathrm{d}s)_i \end{Bmatrix} \tag{4.29}$$

where $\beta_i = \mathrm{d}w/\mathrm{d}s$ is the nodal rotation, and to connect only the first three parameters. The fourth is now an unconnected element parameter with respect to which, however, the usual minimization is still carried out. Transformations needed in the above are implied in Eq. (4.24).

In the derivation of the **B** matrix expressions which define the strains, both first and second derivatives with respect to s occur, as seen in the definition of Eq. (4.16). If we observe that the derivatives can be obtained by the simple rules already implied in Eq. (4.26) for any function F we can write

$$\frac{\mathrm{d}F}{\mathrm{d}s} = \frac{\dfrac{\mathrm{d}F}{\mathrm{d}\xi}}{\dfrac{\mathrm{d}s}{\mathrm{d}\xi}}$$

(4.30)

and

$$\frac{\mathrm{d}^2 F}{\mathrm{d}s^2} = \frac{\dfrac{\mathrm{d}^2 F}{\mathrm{d}\xi^2}}{\left(\dfrac{\mathrm{d}s}{\mathrm{d}\xi}\right)^2} - \frac{\mathrm{d}F}{\mathrm{d}\xi}\frac{\dfrac{\mathrm{d}^2 s}{\mathrm{d}\xi^2}}{\left(\dfrac{\mathrm{d}s}{\mathrm{d}\xi}\right)^3}$$

and all the expressions of **B** can be found.

Finally the stiffness matrix is obtained in a similar way as in Eq. (4.14) changing the variable

$$\mathrm{d}s = \frac{\mathrm{d}s}{\mathrm{d}\xi}\mathrm{d}\xi \tag{4.31}$$

and integrating ξ within the limits -1 and $+1$.

Once again the quantities contained in the integral expressions prohibit explicit integration and numerical integration must be used. As this is carried out in one coordinate only it is not very time-consuming and an adequate number of Gauss points can be used to determine the stiffness very accurately.

Stress and other matrices are similarly obtained.

The particular isoparametric formulation presented in outline here differs somewhat from the alternatives of references 8, 9, 10 and 12 and has the advantage that due to its *isoparametric* form rigid body displacement modes and indeed the states of constant first derivatives are available. Proof of this is similar to that contained in Sec. 8.5 in Chapter 8 of Volume 1. The fact that the forms given in the alternative formulations strain under rigid body displacements may not be serious in some applications, as discussed by Haisler and Stricklin.[24] However, in some modes of non-axisymmetric loads (see Chapter 6) this incompleteness may be a serious drawback and may indeed lead to very wrong results.

Constant states of curvature cannot be obtained for a *finite* element of any kind described here and indeed are not physically possible. When the size of the element decreases it will be found that such arbitrary constant curvature states are available in the limit.

4.6 Additional nodeless variables

Addition of nodeless variables in the analysis of axisymmetric shells is particularly valuable as large curved elements are capable of reproducing with good accuracy the geometric shapes. Thus an addition of a set of internal, hierarchical, element variables

$$\sum_{j=1}^{n} N_j''' a_j \tag{4.32}$$

to the definition of the normal displacement defined in Eq. (4.6) or (4.23), in which a_j is a set of internal element parameters and N_j''' is a set of functions having zero values and zero first derivatives at the nodal points, allows considerable improvement in representation of the displacements to be achieved without violating any of the convergence requirements (see Chapter 2 of Volume 1). For tangental displacements the requirements of zero first derivatives of nodes can be omitted.

Webster[13] uses such additional functions in the context of straight elements.

Whether the element is in fact straight or curved does not matter and indeed we can supplement the definitions of displacements contained in Eq. (4.23) by Eq. (4.32) for each of the components. If this is done only in the displacement definition and *not* in the coordinate definition [Eq. (4.19)] the element becomes now of the category of subparametric.† As

† While it would obviously be possible to include the new shape function in the element shape definition, little practical advantage would be gained as a cubic represents the realistic shapes adequately.

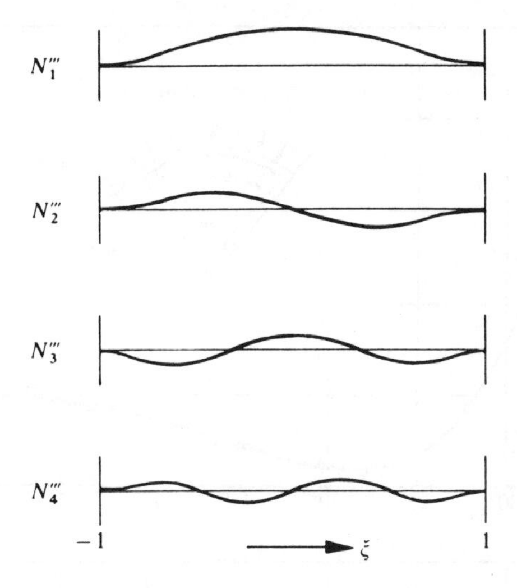

Fig. 4.6 Internal shape functions for a linear element

proved in Chapter 8 of Volume 1, the same advantages are retained as in isoparametric forms.

The question as to the expression to be used for the additional, internal shape functions is of some importance though the choice is wide. While it is no longer necessary to use polynomial representation, Delpak[11] does so and uses a special form of Légendre polynomials (hierarchical functions). The general shapes are shown in Fig. 4.6.

A series of examples shown in Figs 4.7, 4.8 and 4.9 illustrate the applications of the isoparametric curvilinear element of the previous section with additional internal parameters.

In Fig. 4.7 a spherical dome with clamped edges is analysed and compared with analytical results of reference 22. Figures 4.8 and 4.9 show, respectively, more complex examples. In the first a torus analysis is made and compared with alternative finite element results.[12,15,25,26] The second case is one where branching occurs, and here alternative analytical results are given by Kraus.[27]

4.7 Independent slope–displacement interpolation with penalty functions (thick or thin shell formulations)

In Chapter 2 we discussed, in the context of beams and plates, the use of independent slope and displacement interpolation. Continuity was assured by the introduction of the shear force as an independent *mixed* variable which was defined within each element. The elimination of the

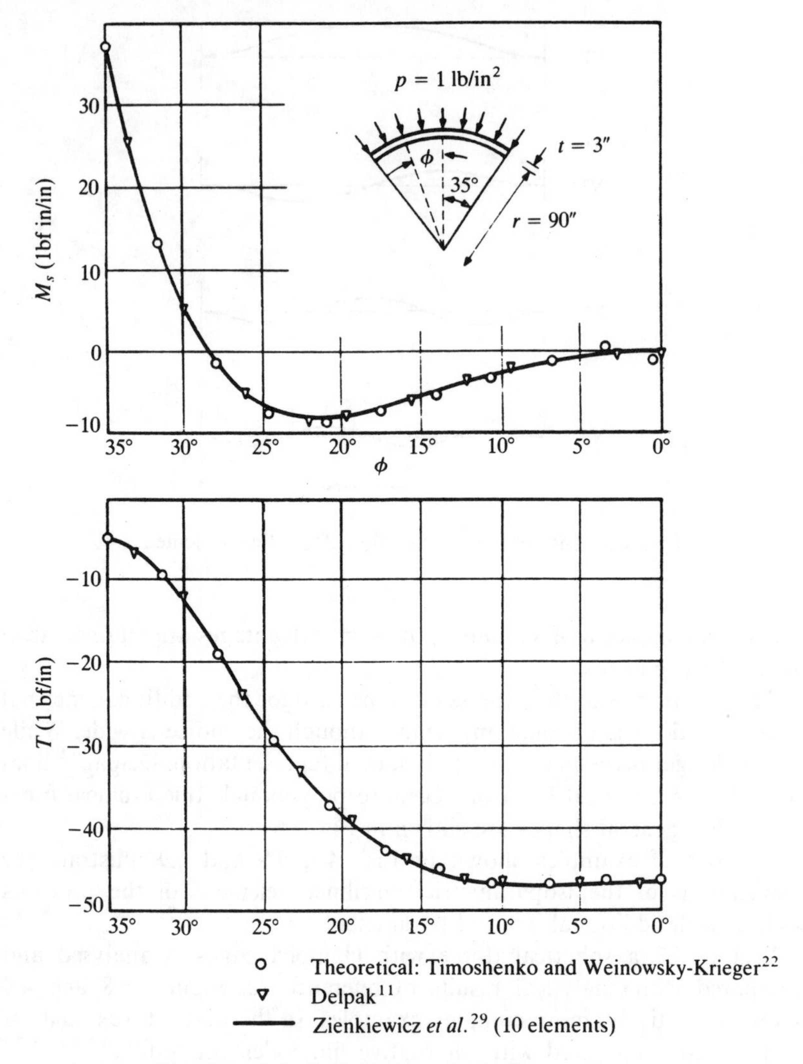

Fig. 4.7 Spherical dome under uniform pressure

shear variable led to a penalty-type formulation in which the shear rigidity played the role of the penalty parameter. The equivalence of the number of parameters used in defining the shear variation and the number of integration points used in evaluating the penalty terms was demonstrated there (and also in Chapter 12 of Volume 1) in special cases, and this justified the success of *reduced* integration methods. This equivalence is not exact in the case of the axisymmetric problem in which the radius, r, enters the integrals, and hence slightly different results can

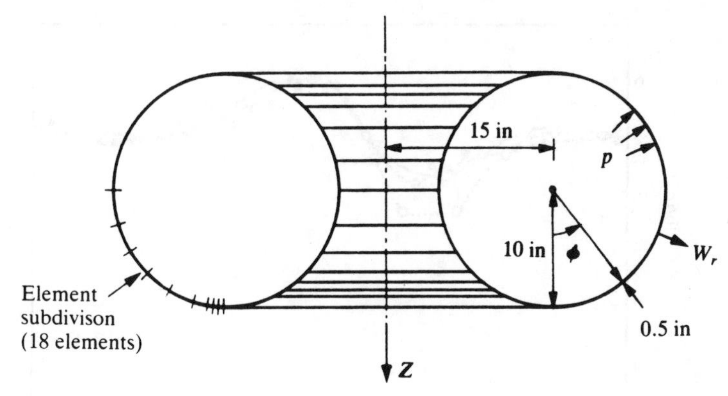

(*a*) Element subdivision

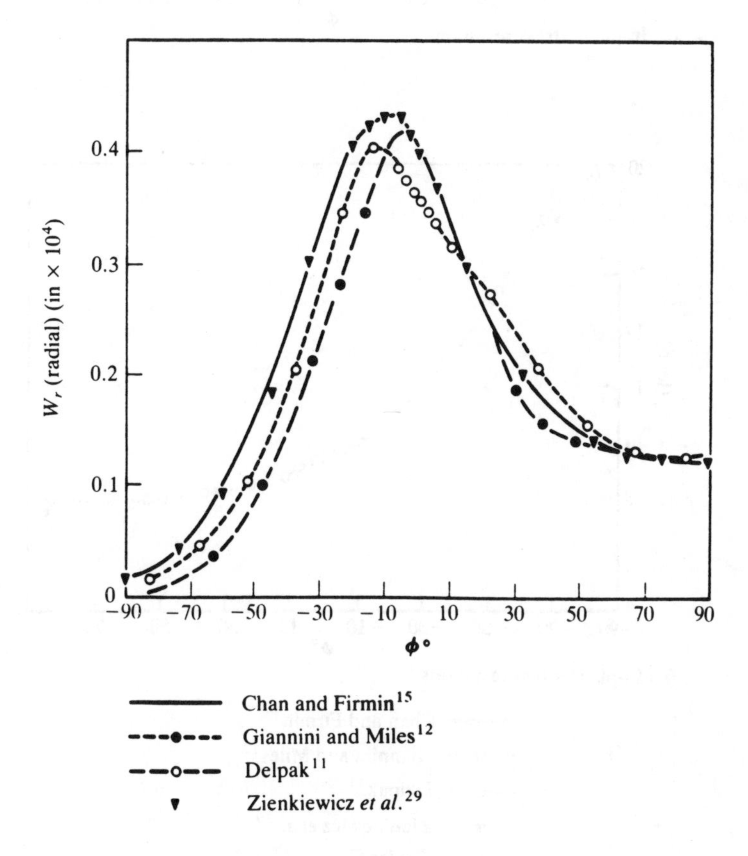

(*b*) Radial displacements

Fig. 4.8 Toroidal shell under internal pressure

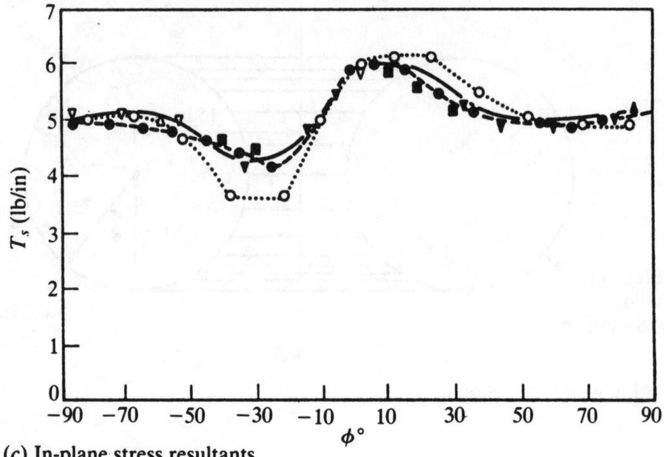

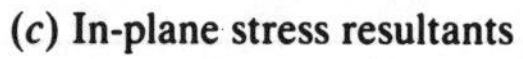
(*c*) In-plane stress resultants

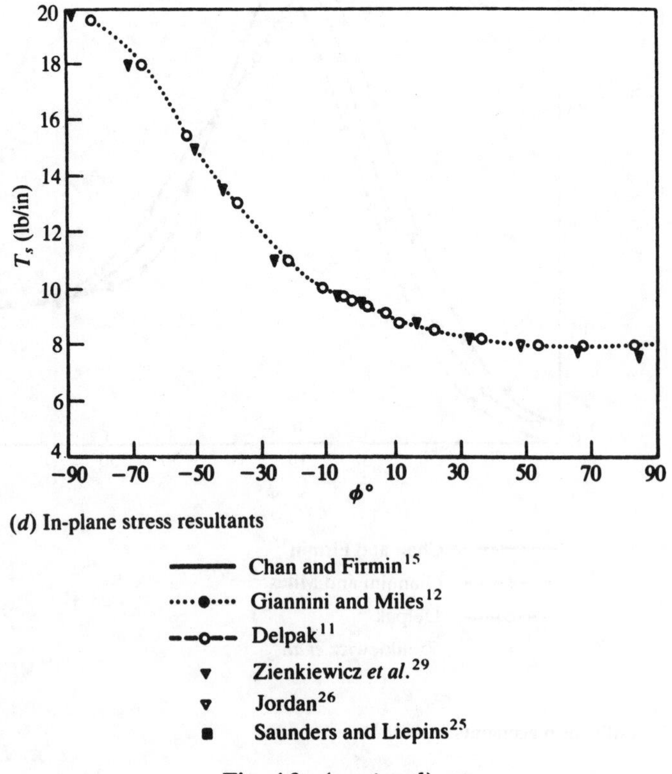

Fig. 4.8 (*continued*)

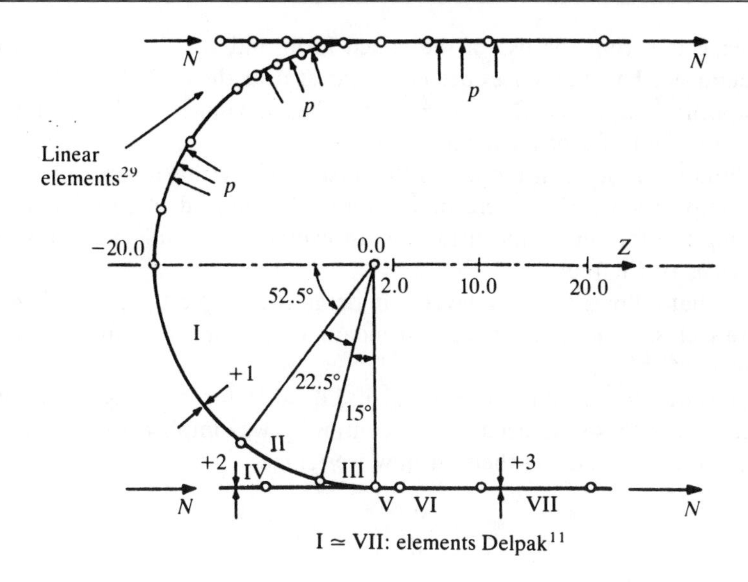

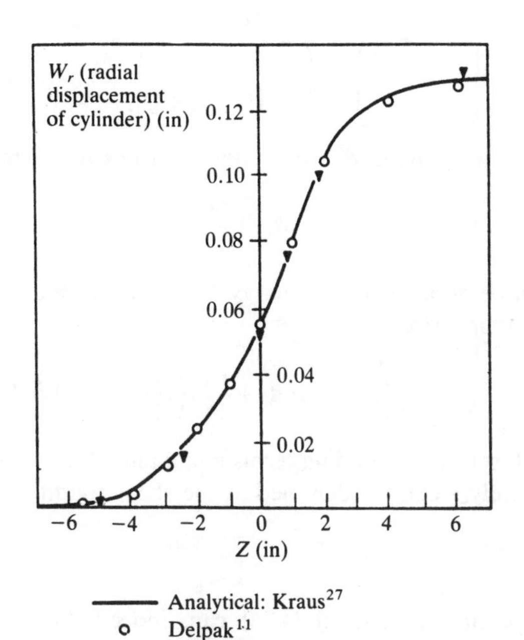

Fig. 4.9 Branching shell

be expected from the use of the mixed form and simple use of reduced integration. The differences become greatest near the axis of rotation and disappear completely when $r \to \infty$ where the axisymmetric plate results in an equivalent of a beam element.

Although in general the use of the mixed form yields a superior result, for simplicity we shall here derive only the reduced integration form, leaving the former to the reader as an exercise accomplished following the rules of Chapter 2.

In what follows we shall develop in detail the simplest possible element of this class. This is a direct descendant of the linear beam and plate elements.[28,29]

Consider the strain expressions of Eq. (4.1) for a straight element. When using these the need for C_1 continuity was implied by the second derivative of w existing there. If now we replace

$$\frac{\mathrm{d}w}{\mathrm{d}s} = -\beta \tag{4.33}$$

the strain expression becomes

$$\boldsymbol{\varepsilon} = \begin{Bmatrix} \varepsilon_s \\ \varepsilon_\theta \\ \chi_s \\ \chi_\theta \end{Bmatrix} = \begin{Bmatrix} \mathrm{d}u/\mathrm{d}s \\ (u \sin\phi + w \cos\phi)/r \\ \mathrm{d}\beta/\mathrm{d}s \\ (\beta \sin\phi)/r \end{Bmatrix} \tag{4.34}$$

As β can vary independently a constraint has to be imposed:

$$C(w, \beta) \equiv \frac{\mathrm{d}w}{\mathrm{d}s} + \beta = 0 \tag{4.35}$$

This can be done using the energy functional with a penalty multiplier α. We can thus write

$$\Pi = \frac{1}{2}\int \boldsymbol{\varepsilon}^{\mathrm{T}} \mathbf{D} \boldsymbol{\varepsilon} 2\pi r \,\mathrm{d}s + \frac{1}{2}\int \alpha \left(\frac{\mathrm{d}w}{\mathrm{d}s} + \beta\right)^2 2\pi r \,\mathrm{d}s + \text{l.t.} \tag{4.36}$$

where l.t. stands for loading terms and $\boldsymbol{\varepsilon}$ and $\mathbf{D}$ are defined as previously. Immediately α can be identified as the shear rigidity

$$\begin{aligned} \alpha &= \kappa G t \\ \kappa &= \tfrac{5}{6} \end{aligned} \tag{4.37}$$

The penalty functional (4.36) can, indeed, be identified on purely physical grounds. Washizu[23] quotes this on pp. 199–201, and the general theory indeed follows that earlier suggested by Naghdi[30] for shells with shear deformation.

With first derivatives only occurring in the energy expression C_0, continuity is now required only in the interpolation for u, w, β, and in place of Eqs (4.6) to (4.10) we can write directly

$$\mathbf{u} = \begin{Bmatrix} u \\ w \\ \beta \end{Bmatrix} = N\boldsymbol{\lambda}\mathbf{a}^e \qquad \text{where } N = N(\xi)$$

$$\mathbf{a}_i^{\mathrm{T}} = [\bar{u}, \bar{w}, \beta]_i \tag{4.38}$$

Here for $N(\xi)$ we can use any of the one-dimensional C_0 interpolations in Chapter 7 of Volume 1. Once again isoparametric transformation could be used for curvilinear elements with strains defined now by Eq. (4.16), and indeed a formulation that we shall discuss in Chapter 5 is but an alternative to this process. If linear elements are used, we can write the expression without consequent use of isoparametric transformation. With the symbols used in Eq. (4.8) we can now simply use

$$\begin{aligned} \bar{u} &= \bar{u}_i(1-s') + \bar{u}_j s' \\ \bar{w} &= \bar{w}_i(1-s') + \bar{w}_j s' \\ \beta &= \beta_i(1-s') + \beta_j s' \end{aligned} \tag{4.39}$$

and evaluate the integrals arising from expression (4.36) at one Gauss point, which is sufficient to maintain convergence and yet here does not give a singularity.

This extremely simple form will of course give very poor results with exact integration, even for thick shells, but now with reduced integration shows an excellent performance.

In Figs 4.7 to 4.9 we superpose results obtained with this simple, straight element and the results speak for themselves.

For other examples the reader can consult reference 29, but in Fig. 4.10 we show a very simple example of a bending of a circular plate using different numbers of equal elements. This purely bending problem shows the type of results and convergence attainable.

Interpreting the single integrating point as a single shear variable and applying the patch test count of Chapter 2, the reader can verify that the simple element passes this in assemblies of two or more. In a similar way it can be verified that a quadratic interpolation of displacements and the use of two quadrature points (or a linear shear force) will result in a robust element of excellent performance.

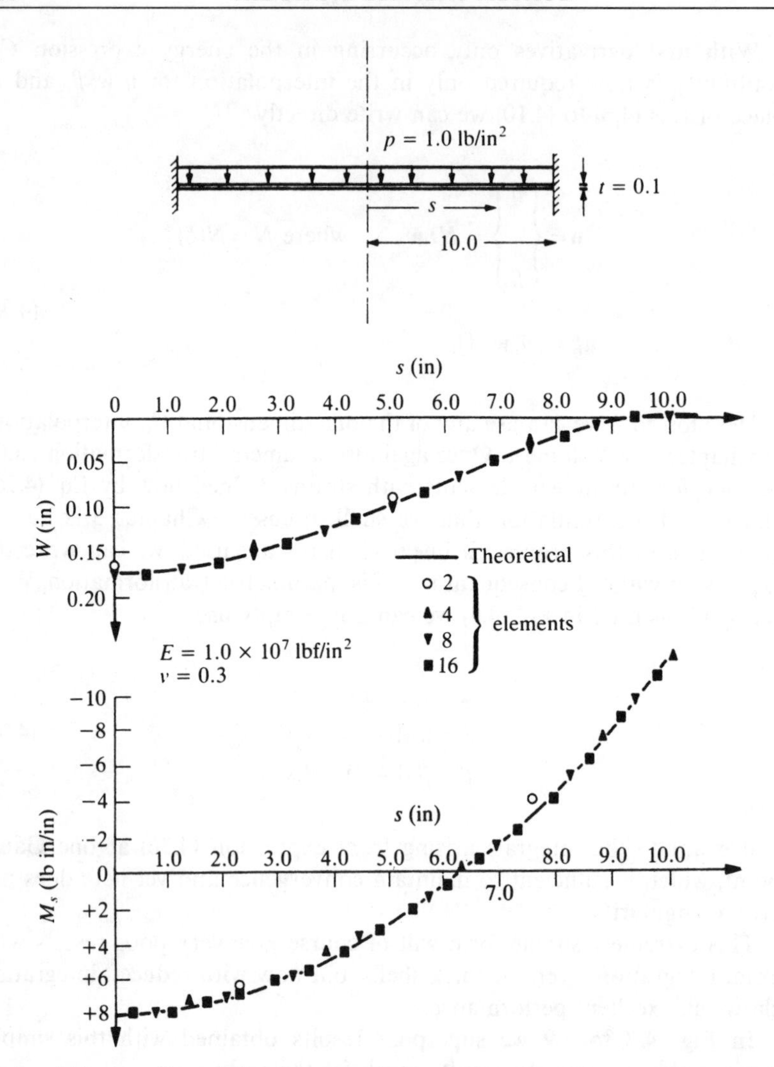

Fig. 4.10 Bending of circular plate under uniform load. Convergence study

References

1. P. E. GRAFTON and D. R. STROME, 'Analysis of axi-symmetric shells by the direct stiffness method', *JAIAA*, **1**, 2342–7, 1963.
2. E. P. POPOV, J. PENZIEN and Z. A. LU, 'Finite element solution for axisymmetric shells', *Proc. Am. Soc. Civ. Eng.*, **EM5**, 119–45, 1964.
3. R. E. JONES and D. R. STROME, 'Direct stiffness method of analysis of shells of revolution utilising curved elements', *JAIAA*, **4**, 1519–25, 1966.
4. J. H. PERCY, T. H. H. PIAN, S. KLEIN and D. R. NAVARATNA, 'Application of

matrix displacement method to linear elastic analysis of shells of revolution', *JAIAA*, **3**, 2138–45, 1965.
5. S. KLEIN, 'A study of the matrix displacement method as applied to shells of revolution', in *Proc. Conf. on Matrix Methods in Structural Mechanics*, Air Force Institute of Technology, Wright-Patterson AF Base, Ohio, October 1965.
6. R. E. JONES and D. R. STROME, 'A survey of analysis of shells by the displacement method', in *Proc. Conf. on Matrix Methods in Structural Mechanics*, Air Force Institute of Technology, Wright-Patterson AF Base, Ohio, October 1965.
7. O. E. HANSTEEN, 'A conical element for displacement analysis of axisymmetric shells', in *Finite Element Methods*, TAPIR, Trondheim, 1969.
8. P. L. GOULD and S. K. SEN, 'Refined mixed method finite elements for shells of revolution', in *3rd Conf. on Matrix Methods in Structural Mechanics*, Wright-Patterson AF Base, Ohio, 1971.
9. J. A. STRICKLIN, D. R. NAVARATNA and T. H. H. PIAN, 'Improvements in the analysis of shells of revolution by matrix displacement method (curved elements)', *JAIAA*, **4**, 2069–72, 1966.
10. M. KHOJASTEH-BAKHT, 'Analysis of elastic–plastic shells of revolution under axi-symmetric loading by the finite element method', report SESA 67–8, Department of Civil Engineering, University of California, Berkeley, 1967.
11. R. DELPAK, 'Role of the curved parametric element in linear analysis of thin rotational shells', Ph.D. thesis, Department of Civil Engineering and Building, The Polytechnic of Wales, 1975.
12. M. GIANNINI and G. A. MILES, 'A curved element approximation in the analysis of axi-symmetric thin shells', *Int. J. Num. Meth. Eng.*, **2**, 459–76, 1970.
13. J. J. WEBSTER, 'Free vibration of shells of revolution using ring elements', *Int. J. Mech. Sci.*, **9**, 559–70, 1967.
14. S. AHMAD, B. M. IRONS and O. C. ZIENKIEWICZ, 'Curved thick shell and membrane elements with particular reference to axi-symmetric problems', in *Proc. 2nd Conf. on Matrix Methods in Structural Mechanics*, Air Force Institute of Technology, Wright-Patterson AF Base, Ohio, 1968.
15. A. S. L. CHAN and A. FIRMIN, 'The analysis of cooling towers by the matrix finite element method', *Aeronaut. J.*, **74**, 826–35, 1970.
16. E. A. WITMER and J. J. KOTANCHIK, 'Progress report on discrete element elastic and elastic-plastic analysis of shells of revolution subjected to axisymmetric and asymmetric loading', in *Proc. 2nd Conf. on Matrix in Structural Mechanics*, Air Force Institute of Technology, Wright-Patterson AF Base, Ohio, 1968.
17. Z. M. ELIAS, 'Mixed finite element method for axisymmetric shells', *Int. J. Num. Meth. Eng.*, **4**, 261–77, 1972.
18. R. H. GALLAGHER, 'Analysis of plate and shell structures', in *Applications of Finite Element Method in Engineering*, pp. 155–205, Vanderbilt University, ASCE, 1969.
19. R. H. GALLAGHER, 'Shell elements', in *World Conf. on Finite Element Methods in Structural Mechanics*, Bournemouth, Dorset, England, October 1975.
20. J. A. STRICKLIN, 'Geometrically nonlinear static and dynamic analysis of shells of revolution', in *High Speed Computing of Elastic Structures*, pp. 383–411, University of Liége, 1976.
21. V. V. NOVOZHILOV, *Theory of Thin Shells* (translation), Noordhoff, Dordrecht, 1959.

22. S. TIMOSHENKO and S. WOINOWSKY-KRIEGER, *Theory of Plates and Shells*, 2nd ed., pp. 533–5, McGraw-Hill, New York, 1959.
23. K. WASHIZU, *Variational Methods in Elasticity and Plasticity*, 2nd ed., pp. 189–99, Pergamon Press, Oxford, 1975.
24. W. E. HAISLER and J. A. STRICKLIN, 'Rigid body displacements of curved elements in the analysis of shells by the matrix displacement method', *JAIAA*, **5**, 1525–7, 1967.
25. J. L. SANDERS, Jr and A. LIEPINS, 'Toroidal membrane under internal pressure', *JAIAA*, **1**, 2105–10, 1963.
26. F. F. JORDAN, 'Stresses and deformations of the thin-walled pressurized torus', *J. Aero. Sci.*, **29**, 213–25, 1962.
27. H. KRAUS, *Thin Elastic Shells*, pp. 168–78, Wiley, New York, 1967.
28. T. J. R. HUGHES, R. L. TAYLOR and W. KANOKNUKULCHAI, 'A simple and efficient finite element for plate bending', *Int. J. Num. Meth. Eng.*, **11**, 1529–43, 1977.
29. O. C. ZIENKIEWICZ, J. BAUER, K. MORGAN and E. OÑATE, 'A simple element for axi-symmetric shells with shear deformation', *Int. J. Num. Meth. Eng.*, **11**, 1545–58, 1977.
30. P. M. NAGHDI, 'Foundations of elastic shell theory', in *Progress in Solid Mechanics* (eds I. N. Sneddon and R. Hill), Vol. IV, chap. 1, North-Holland, Amsterdam, 1963.

5

Shells as a special case of three-dimensional analysis—Reissner–Mindlin assumptions

5.1 Introduction

In Chapters 8 and 9 of Volume 1 the formulation and use of complex, curved, two- and three-dimensional elements was illustrated. It seems obvious that use of such elements could be made directly in the analysis of curved shells simply by reducing their dimension in the shell thickness direction as shown in Fig. 5.1. Indeed, in an axisymmetric situation such an application has been illustrated in the example of Fig. 8.25 in Chapter 8 of Volume 1.

With a straightforward use of the three-dimensional concept, however, certain difficulties will be encountered.

In the first place the retention of 3 degrees of freedom at each note leads to large stiffness coefficients for relative displacements along an edge corresponding to the shell thickness. This presents numerical problems and may lead to ill-conditioned equations when shell thicknesses become small compared with the other dimensions in the element.

The second factor is that of economy. The use of several nodes across the shell thickness ignores the well-known fact that even for thick shells the 'normals' to the middle surface remain practically straight after deformation. Thus an unnecessarily high number of degrees of freedom has to be carried, involving penalties of computer time.

Here, specialized formulation is presented overcoming both these difficulties.[1-3] The constraint of straight 'normals' is introduced to improve economy and the strain energy corresponding to stresses perpendicular to the middle surface is ignored to improve numerical conditioning. With these modifications an efficient tool for analysing curved thick

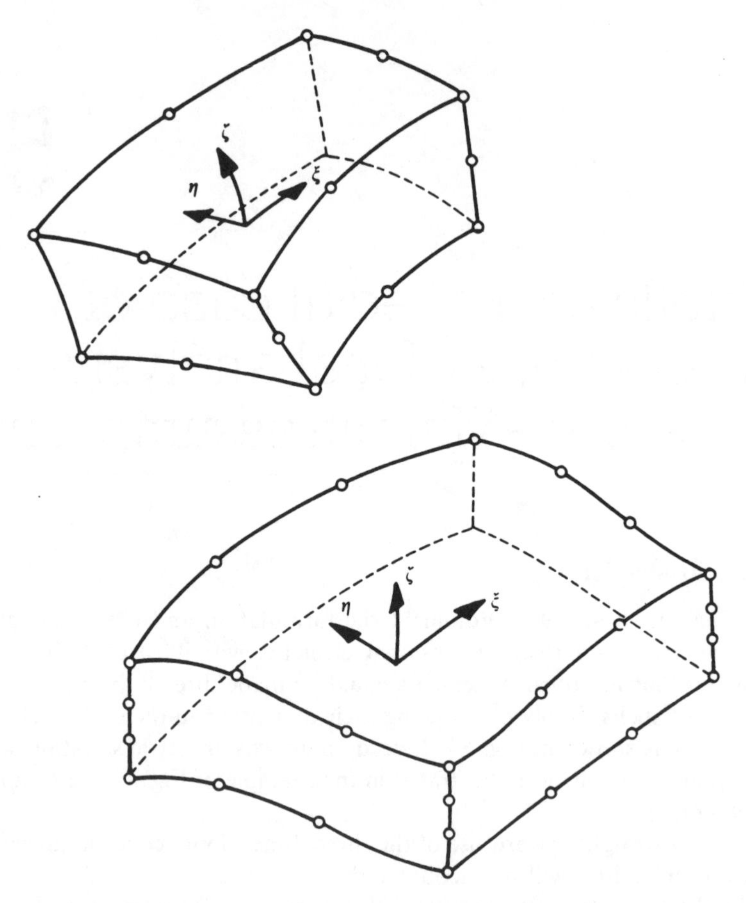

Fig. 5.1 Curved, isoparametric hexahedra in a direct approximation to a curved shell

shells becomes available. Its accuracy and wide range of applicability is demonstrated in several examples.

The reader will note that the two constraints introduced correspond precisely to those introduced in Chapter 2 to describe the behaviour of thick plates and are the so-called Reissner–Mindlin assumptions. The omission of the third constraint associated with the thin plate theory (normals remaining normal to the middle plane after deformation) permits the shell to experience shear deformations—an important feature in thick shell situations.

The formulation presented here leads to some more complication than that of straightforward use of a three-dimensional element—and indeed the reader may be tempted to a direct use of three-dimensional formulation, especially as it appears that the use of an element with only a linear variation of displacements across the thickness would be permissible. Only 6 degrees of freedom corresponding to a specified mid-plane point would be necessary *vis-à-vis* five which will arise with the formulation of this chapter, and this would appear to be a small penalty if the ill-conditioning due to large stiffness ratios could be overcome. This is indeed feasible, as shown by Wood[4,5] and Wilson [6] who use as variables the *differences of displacements* at the two surfaces and a high precision computer. However, another difficulty now becomes apparent if linear interpolation in the direction of the shell normal is used. When Poisson's ratio is not zero the results converge to a solution which is in error by a factor $(1-\nu)^2/(1-2\nu)$. The reason for this is easily explained. With pure bending a zero strain is obtained in the direction normal to the middle plane and, consequently, stresses develop in that direction if $\nu \neq 0$ restraining the *in-plane* strains. To overcome this effect *either* artificial anisotropic properties of the material have to be assumed or a full parabolic displacement has to be used which will make the computation uneconomic.

The elements developed here are in essence an alternative formulation of the processes discussed in Chapter 2, for which an independent interpolation of slopes and displacement was used with a penalty function imposition of the continuity requirements. The use of reduced integration is thus once again imperative if thin shells are to be dealt with—and it was, indeed, in this context that this procedure was first discovered.[7-10]

Again the same restrictions for robust behaviour as those discussed in Chapter 2 become applicable and generally elements that perform well in plate situations will do well in shells.

5.2 Geometric definition of the element

Consider a typical shell element of Fig. 5.2. The external faces of the element are curved, while the sections across the thickness are generated by straight lines. Pairs of points, i_{top} and i_{bottom}, each with given cartesian coordinates, prescribe the shape of the element.

Let ξ, η be the two curvilinear coordinates in the middle plane of the shell and ζ a linear coordinate in the thickness direction. If, further, we assume that ξ, η, ζ vary between -1 and 1 on the respective faces of the element we can write a relationship between the cartesian coordinates of any point of the shell and the curvilinear coordinates in the form

$$\begin{Bmatrix} x \\ y \\ z \end{Bmatrix} = \sum N_i(\xi, \eta) \frac{1+\xi}{2} \begin{Bmatrix} x_i \\ y_i \\ z_i \end{Bmatrix}_{\text{top}} + \sum N_i(\xi, \eta) \frac{1-\xi}{2} \begin{Bmatrix} x_i \\ y_i \\ z_i \end{Bmatrix}_{\text{bottom}} \tag{5.1}$$

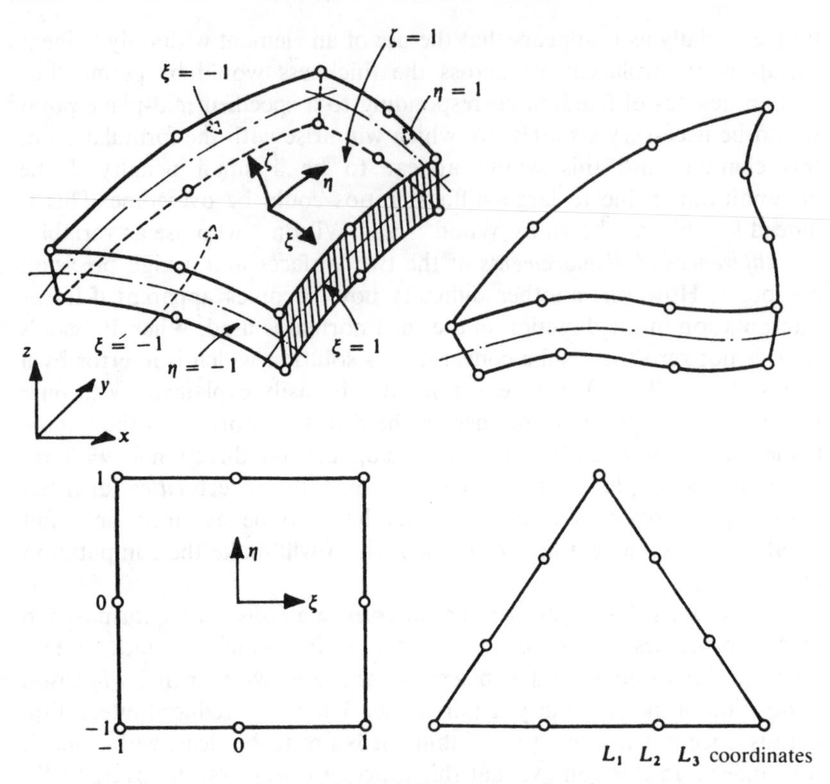

Fig. 5.2 Curved thick shell elements of various types

Here $N_i'(\xi, \eta)$ is a shape function taking a value of unity at the nodes i and zero of all other nodes (Chapter 8 of Volume 1). If the basic functions N_i are derived as 'shape functions' of a 'parent', two-dimensional element, square or triangular† in plan, and are so 'designed' that compatibility is achieved at interfaces, then the curved space elements will fit into each other. Arbitrary curved shapes of the element can be achieved by using shape functions of higher orders. Only parabolic and cubic types are shown in Fig. 5.2. By placing a larger number of nodes on the surfaces of the element more elaborate shapes can be achieved if so desired. Indeed, any of the two-dimensional shape functions in Chapter 7 of Volume 1 can be used here.

The relation between the cartesian and curvilinear coordinates is now established and it will be found desirable to operate with the curvilinear coordinates as the basis. It should be noted that the coordinate direction ζ is *only approximately normal* to the middle surface.

† Area coordinates would be used in this case in place of ξ and η as in Chapter 7 of Volume 1.

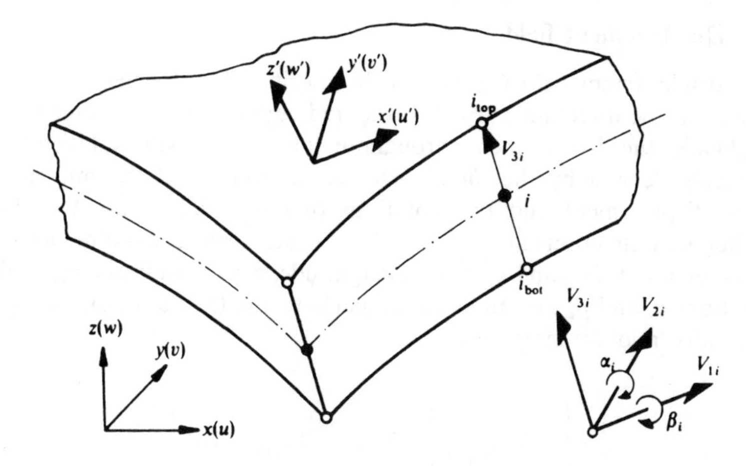

Fig. 5.3 Local and global coordinates

It is convenient to rewrite the relationship, Eq. (5.1), in a form specified by the 'vector' connecting the upper and lower points (i.e. a vector of length equal to the shell thickness t) and the mid-surface coordinates. Thus† we can rewrite Eq. (5.1) as (Fig. 5.3)

$$\begin{Bmatrix} x \\ y \\ z \end{Bmatrix} = \sum N_i \begin{Bmatrix} x_i \\ y_i \\ z_i \end{Bmatrix}_{\text{mid}} + \sum N_i \frac{\xi}{2} \mathbf{V}_{3i}$$

with (5.2)

$$\mathbf{V}_{3i} = \begin{Bmatrix} x_i \\ y_i \\ z_i \end{Bmatrix}_{\text{top}} - \begin{Bmatrix} x_i \\ y_i \\ z_i \end{Bmatrix}_{\text{bottom}}$$

defining a vector whose length is the shell thickness.

For relatively thin shells, it is convenient to replace the vector $\mathbf{V}_{3i}$ by a unit vector in the direction normal to the middle plane $\mathbf{v}_{3i}$. Now the last term is then written simply as

$$\sum N_i \xi \frac{t_i}{2} \mathbf{v}_{3i}$$

where t_i is the shell thickness at the node i. Construction of a vector normal to the middle surface is a simple process (viz. Appendix 5 of Volume 1).

† For details of vector algebra see Appendix 5 of Volume 1.

5.3 Displacement field

The displacement field has now to be specified for the element. As the strains in the direction normal to the mid-surface will be assumed to be negligible, the displacement throughout the element will be taken to be uniquely defined by the *three cartesian components* of the mid-surface node displacement and two rotations of the nodal vector $\mathbf{V}_{3i}$ about orthogonal directions normal to it. If two such orthogonal directions are given by vectors $\mathbf{v}_{2i}$ and $\mathbf{v}_{1i}$ of unit magnitude, with corresponding (scalar) rotations α_i and β_i we can write, similarly to Eq. (5.2) but now dropping the suffix 'mid' for simplicity,

$$\begin{Bmatrix} u \\ v \\ w \end{Bmatrix} = \sum N_i \begin{Bmatrix} u_i \\ v_i \\ w_i \end{Bmatrix} + \sum N_i \zeta \frac{t_i}{2} [\mathbf{v}_{1i}, \ -\mathbf{v}_{2i}] \begin{Bmatrix} \alpha_i \\ \beta_i \end{Bmatrix} \tag{5.3}$$

from which the usual form is readily obtained as

$$\begin{Bmatrix} u \\ v \\ w \end{Bmatrix} = \mathbf{N} \begin{Bmatrix} \mathbf{a}_i^e \\ \vdots \\ \mathbf{a}_j^e \end{Bmatrix} \qquad \text{with } \mathbf{a}_i^e = \begin{Bmatrix} u_i \\ v_i \\ w_i \\ \alpha_i \\ \beta_i \end{Bmatrix}$$

where u, v and w are displacements in the directions of the global x, y and z axes.

As an infinity of vector directions normal to a given direction can be generated, a particular scheme has to be devised to ensure a *unique* definition. Some such schemes were discussed in Chapter 3. Here a simpler unique alternative will be given,[2] but other possibilities are open.[10]

Thus if $\mathbf{V}_{3i}$ is the vector to which a normal direction is to be constructed we form the first normal axis in a direction perpendicular to the plane defined by this vector and the x axis.†

A vector $\mathbf{V}_{1i}$ of the description is given by the cross-product

$$\mathbf{V}_{1l} = \mathbf{i} \times \mathbf{V}_{3i} \tag{5.4}$$

† This process fails if $\mathbf{V}_{3i}$ corresponds in direction with the x axis. A program checking this possibility is easily written and in such a case the local directions are obtained using the y axis.

In this

$$\mathbf{i}=\begin{Bmatrix}1\\0\\0\end{Bmatrix}$$

is a unit vector in the direction of the x axis. Dividing this by its length we can write the unit vector $\mathbf{v}_{1i}$.

The last vector normal to the other two is simply

$$\mathbf{V}_{2i}=\mathbf{V}_{1i}\times\mathbf{V}_{3i} \tag{5.5}$$

and all the direction cosines of the local axes can be determined by normalizing this to $\mathbf{v}_{2i}$. We have thus three local, orthogonal axes defined by unit vectors

$$\mathbf{v}_{1i}, \mathbf{v}_{2i} \text{ and } \mathbf{v}_{3i} \tag{5.6}$$

Once again if N_i are compatible functions then displacement compatibility is maintained between adjacent elements.

The element coordinate definition is now given by the relation Eq. (5.1), which has more degrees of freedom than the definition of the displacements. The element is therefore of the superparametric kind (see Chapter 8 of Volume 1) and the constant strain criteria are not automatically satisfied. Nevertheless, it will be seen from the definition of strain components involved that both rigid body motions and constant strain conditions are available.

Physically, it has been assumed in the definition of Eq. (5.3) that no strains occur in the 'thickness' direction ζ. While this direction is not exactly normal to the middle surface it still represents to a good approximation one of the usual shell assumptions.

At each mid-surface node i of Fig. 5.3 we now have the 5 basic degrees of freedom and the connection of elements will follow precisely the patterns described in Chapter 3 (Secs 3.3 and 3.4).

5.4 Definition of strains and stresses

To derive the properties of a finite element the essential strains and stresses have to be defined. The components in directions of *orthogonal axes* related to the surface $\zeta=\text{constant}$ are essential if account is to be taken of the basic shell assumptions. Thus, if at any point in this surface we erect a normal z' with two other orthogonal axes x' and y' tangent to it (Fig. 5.3), the strain components of interest are given simply by the three-dimensional relationships in Chapter 6 of Volume 1:

$$\boldsymbol{\varepsilon}' = \begin{Bmatrix} \varepsilon_{x'} \\ \varepsilon_{y'} \\ \gamma_{x'y'} \\ \gamma_{x'z'} \\ \gamma_{y'z'} \end{Bmatrix} = \begin{Bmatrix} \dfrac{\partial u'}{\partial x'} \\ \dfrac{\partial v'}{\partial y'} \\ \dfrac{\partial u'}{\partial y'} + \dfrac{\partial v'}{\partial x'} \\ \dfrac{\partial w'}{\partial x'} + \dfrac{\partial u'}{\partial z'} \\ \dfrac{\partial w'}{\partial y'} + \dfrac{\partial v'}{\partial z'} \end{Bmatrix} \tag{5.7}$$

with the strain in direction z' neglected so as to be consistent with the usual shell assumptions. It must be noted that in general none of these directions coincide with those of the curvilinear coordinates ξ, η, ζ, although x', y' are in the $\xi\eta$ plane (ζ = constant).†

The stresses corresponding to these strains are defined by a matrix $\boldsymbol{\sigma}'$ and are related by the usual elasticity matrix $\mathbf{D}'$. Thus

$$\boldsymbol{\sigma}' = \begin{Bmatrix} \sigma_{x'} \\ \sigma_{y'} \\ \tau_{x'y'} \\ \tau_{x'z'} \\ \tau_{y'z'} \end{Bmatrix} = \mathbf{D}'(\boldsymbol{\varepsilon}' - \boldsymbol{\varepsilon}'_0) + \boldsymbol{\sigma}'_0 \tag{5.8}$$

where $\boldsymbol{\varepsilon}'_0$ in $\boldsymbol{\sigma}'_0$ may represent any 'initial' strains or stresses.

The 5×5 matrix $\mathbf{D}'$ can now include any anisotropic properties and indeed may be prescribed as a function of ζ if sandwich (layered) construction is used. For the present moment we shall define it only for an isotropic material. Here

$$\mathbf{D}' = \frac{E}{1-\nu^2} \begin{bmatrix} 1 & \nu & 0 & 0 & 0 \\ & 1 & 0 & 0 & 0 \\ & & \dfrac{1-\nu}{2} & 0 & 0 \\ & & & \dfrac{1-\nu}{2k} & 0 \\ & \text{symmetric} & & & \dfrac{1-\nu}{2k} \end{bmatrix} \tag{5.9}$$

† Indeed, these directions will only approximately agree with the nodal direction $\mathbf{v}_{1i}$, etc., previously derived, as in general the vector $\mathbf{v}_{3i}$ is only approximately normal to the mid-surfaces.

in which E and ν are Young's modulus and Poisson's ratio respectively. The factor k included in the last two shear terms is taken as 6/5 and its purpose is to improve the shear displacement approximation. From the displacement definition it will be seen that the shear distribution is approximately constant through the thickness, whereas in reality the shear distribution is approximately parabolic. The value $k=6/5$ is the ratio of relevant strain energies.

It is important to note that this matrix is *not* derived simply by deleting appropriate terms from the equivalent three-dimensional stress matrix. It must be derived by substituting $\sigma'_z=0$ into Eq. (5.13) in Volume 1 and a suitable elimination so that this important shell assumption is satisfied.

5.5 Element properties and necessary transformations

The stiffness matrix—and indeed all other 'element' property matrices—involve integrals over the volume of the element, which are quite generally of the form

$$\int_{V^e} \mathbf{S}\, \mathrm{d}x\, \mathrm{d}y\, \mathrm{d}z \tag{5.10}$$

where the matrix $\mathbf{S}$ is a function of the coordinates.

In the stiffness matrix

$$\mathbf{S}=\mathbf{B}^{\mathrm{T}}\mathbf{D}\mathbf{B} \tag{5.11}$$

for instance and with the usual definition of Chapter 2 of Volume 1,

$$\boldsymbol{\varepsilon}'=\mathbf{B}\mathbf{a}^e \tag{5.12}$$

we have $\mathbf{B}$ defined in terms of the displacement derivatives with respect to the local cartesian coordinates $x'y'z'$ by Eq. (5.7). Now, therefore, *two sets of transformations* are necessary before the element can be integrated with respect to the curvilinear coordinates ξ, η, ζ.

Firstly, by identically the same process as we used in Chapter 8 of Volume 1, the derivatives with respect to the x, y, z directions are obtained. As Eq. (5.3) relates the global displacements u, v, w to the curvilinear coordinates, the derivatives of these displacements with respect to the global x, y, z coordinates are given by a matrix relation:

$$\begin{bmatrix} \dfrac{\partial u}{\partial x} & \dfrac{\partial v}{\partial x} & \dfrac{\partial w}{\partial x} \\ \dfrac{\partial u}{\partial y} & \dfrac{\partial v}{\partial y} & \dfrac{\partial w}{\partial y} \\ \dfrac{\partial u}{\partial z} & \dfrac{\partial v}{\partial z} & \dfrac{\partial w}{\partial z} \end{bmatrix} = \mathbf{J}^{-1} \begin{bmatrix} \dfrac{\partial u}{\partial \xi} & \dfrac{\partial v}{\partial \xi} & \dfrac{\partial w}{\partial \xi} \\ \dfrac{\partial u}{\partial \eta} & \dfrac{\partial v}{\partial \eta} & \dfrac{\partial w}{\partial \eta} \\ \dfrac{\partial u}{\partial \zeta} & \dfrac{\partial v}{\partial \zeta} & \dfrac{\partial w}{\partial \zeta} \end{bmatrix} \tag{5.13}$$

In this, the jacobian matrix is defined as before:

$$\mathbf{J}=\begin{bmatrix} \dfrac{\partial x}{\partial \xi} & \dfrac{\partial y}{\partial \xi} & \dfrac{\partial z}{\partial \xi} \\ \dfrac{\partial x}{\partial \eta} & \dfrac{\partial y}{\partial \eta} & \dfrac{\partial z}{\partial \eta} \\ \dfrac{\partial x}{\partial \zeta} & \dfrac{\partial y}{\partial \zeta} & \dfrac{\partial z}{\partial \zeta} \end{bmatrix} \tag{5.14}$$

and is calculated from the coordinate definitions of Eq. (5.2).

Now, for every set of curvilinear coordinates the global displacement derivatives can be obtained numerically. A further transformation to the local displacement directions x', y', z' will allow the strains, and hence the **B** matrix, to be evaluated.

Secondly, the directions of the local axes have to be established. A vector normal to the surface ζ constant can be found as a vector product of any two vectors tangent to the surface. Thus

$$\mathbf{V}_3=\begin{bmatrix} \dfrac{\partial x}{\partial \xi} \\ \dfrac{\partial y}{\partial \xi} \\ \dfrac{\partial z}{\partial \xi} \end{bmatrix} \mathbf{x} \begin{bmatrix} \dfrac{\partial x}{\partial \eta} \\ \dfrac{\partial y}{\partial \eta} \\ \dfrac{\partial z}{\partial \eta} \end{bmatrix} = \begin{bmatrix} \dfrac{\partial y}{\partial \xi}\dfrac{\partial z}{\partial \eta} - \dfrac{\partial y}{\partial \eta}\dfrac{\partial z}{\partial \xi} \\ \dfrac{\partial x}{\partial \eta}\dfrac{\partial z}{\partial \xi} - \dfrac{\partial x}{\partial \xi}\dfrac{\partial z}{\partial \eta} \\ \dfrac{\partial x}{\partial \xi}\dfrac{\partial y}{\partial \eta} - \dfrac{\partial x}{\partial \eta}\dfrac{\partial y}{\partial \zeta} \end{bmatrix} \tag{5.15}$$

Following the process that defines uniquely two perpendicular vectors, given previously, and reducing these to unit magnitudes, we construct a matrix of unit vectors in the x', y', z' directions (which is in fact the direction cosine matrix):

$$\boldsymbol{\theta}=[\mathbf{v}_1, \mathbf{v}_2, \mathbf{v}_3] \tag{5.16}$$

The global derivatives of displacements u, v and w are now transformed to the local derivatives of the local orthogonal displacements by a standard operation

$$\begin{bmatrix} \dfrac{\partial u'}{\partial x'} & \dfrac{\partial v'}{\partial x'} & \dfrac{\partial w'}{\partial x'} \\ \dfrac{\partial u'}{\partial y'} & \dfrac{\partial v'}{\partial y'} & \dfrac{\partial w'}{\partial y'} \\ \dfrac{\partial u'}{\partial z'} & \dfrac{\partial v'}{\partial z'} & \dfrac{\partial w'}{\partial z'} \end{bmatrix} = \boldsymbol{\theta}^{\mathrm{T}} \begin{bmatrix} \dfrac{\partial u}{\partial x} & \dfrac{\partial v}{\partial x} & \dfrac{\partial w}{\partial x} \\ \dfrac{\partial u}{\partial y} & \dfrac{\partial v}{\partial y} & \dfrac{\partial w}{\partial y} \\ \dfrac{\partial u}{\partial z} & \dfrac{\partial v}{\partial z} & \dfrac{\partial w}{\partial z} \end{bmatrix} \boldsymbol{\theta} \tag{5.17}$$

From this the components of the $\mathbf{B}'$ matrix can now be found explicitly, noting that 5 degrees of freedom exist at each node:

$$\boldsymbol{\varepsilon}' = \mathbf{B}' \begin{Bmatrix} \mathbf{a}_i^e \\ \vdots \\ \mathbf{a}_j^e \end{Bmatrix} \qquad \mathbf{a}_i^e = \begin{Bmatrix} u_i \\ v_i \\ w_i \\ \alpha_i \\ \beta_i \end{Bmatrix} \tag{5.18}$$

The infinitesimal volume is given in terms of the curvilinear coordinates as

$$\mathrm{d}x\,\mathrm{d}y\,\mathrm{d}z = \det|\mathbf{J}|\,\mathrm{d}\xi\,\mathrm{d}\eta\,\mathrm{d}\zeta \tag{5.19}$$

and this standard expression completes the basic formulation.

Numerical integration within the appropriate $-1, +1$ limits is carried out in exactly the same way as for three-dimensional elements discussed in Chapter 8 of Volume 1. Identical processes serve to define all the other relevant element matrices.

As the variation of the strain quantities in the thickness, or ζ direction, is linear, only two Gauss points in that direction are required for isotropic elasticity, while three or four in the ξ, η directions are used for parabolic and cubic shape functions respectively.

It should be remarked here that, in fact, the integration with respect to ζ can be performed exactly if desired, thus saving computation time.[1]

5.6 Some remarks on stress representation

The element properties are now defined, and the assembly and solution are standard processes.

It remains to discuss the presentation of the stresses, and this problem is of some consequence. The strains being defined in local directions, $\boldsymbol{\sigma}'$ is readily available. Such components are indeed directly of interest but as the directions of local axes are not easily visualized it is sometimes convenient to transfer the components to the global system using the following expression:

$$\begin{bmatrix} \sigma_x & \tau_{xy} & \tau_{xz} \\ \tau_{xy} & \sigma_y & \tau_{yz} \\ \tau_{xz} & \tau_{yz} & \sigma_z \end{bmatrix} = \boldsymbol{\theta} \begin{bmatrix} \sigma_{x'} & \tau_{x'y'} & \tau_{x'z'} \\ \tau_{x'y'} & \sigma_{y'} & \tau_{y'z'} \\ \tau_{x'z'} & \tau_{y'z'} & 0 \end{bmatrix} \boldsymbol{\theta}^{\mathrm{T}} \tag{5.20}$$

If the stresses are calculated at a nodal point where several elements meet then they are averaged.

In a general shell structure, the stresses in a global system do not, however, give a clear picture of shell surface stresses. It is thus convenient always to compute the principal stresses by a suitable transformation.

However, regarding the shell surface stresses more rationally, one may note that the shear components $\tau_{x'z'}$ and $\tau_{y'z'}$ are in fact zero there, and can indeed be made zero at the stage before converting to global components. The values directly obtained for these shear components are the average values across the section. The maximum transverse shear value occurs on the neutral axis and is equal to 1.5 times the average value.

5.7 Special case of axisymmetric, curved, thick shells

For axisymmetric shells the formulation is, obviously, simplified.[1] Now the element mid-surface is defined by only two coordinates ξ, η and a considerable saving in computer effort is obtained.

The element now is derived in a similar manner but starting from a two-dimensional definition of Fig. 5.4.

Equations (5.1) and (5.2) are now replaced by their two-dimensional equivalents defining the relation between the coordinates as

$$\begin{Bmatrix} r \\ z \end{Bmatrix} = \sum N_i(\xi)\,\frac{1+\eta}{2}\begin{Bmatrix} r_i \\ z_i \end{Bmatrix}_{\text{top}} + \sum N_i(\xi)\,\frac{1-\eta}{2}\begin{Bmatrix} r_i \\ z_i \end{Bmatrix}_{\text{bottom}}$$

$$= \sum N_i(\xi)\begin{Bmatrix} r_i \\ z_i \end{Bmatrix}_{\text{mid}} + \sum N_i(\xi)\,\frac{\eta}{2}\,\mathbf{V}_{3i} \tag{5.21}$$

with
$$\mathbf{V}_{3i} = t_i \begin{Bmatrix} \cos\phi_i \\ \sin\phi_i \end{Bmatrix}$$

in which ϕ_i is the angle defined in Fig. 5.4(*b*) and t_i is the shell thickness. Similarly the displacement definition is specified by following the lines of Eq. (5.3).

For generality we shall consider the case of non-symmetric loading only noting the terms that can be eliminated *a priori* for the simple case of symmetry. Indeed, the decomposition into trigonometric components will be tacitly assumed as it follows precisely the lines to be described in Chapter 6.

Thus generally we specify the three displacement components of the

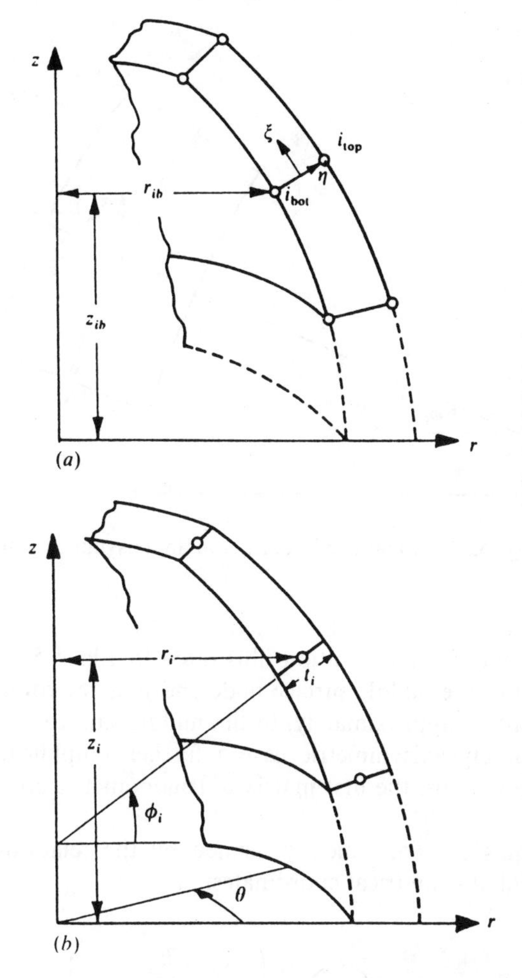

Fig. 5.4 Coordinates for an axisymmetric shell problem

nth harmonic as

$$\begin{Bmatrix} u^n \\ v^n \\ w^n \end{Bmatrix} = \begin{bmatrix} \cos n\theta & 0 & 0 \\ 0 & \cos n\theta & 0 \\ 0 & 0 & \sin n\theta \end{bmatrix} \times \left(\sum N_i \begin{Bmatrix} u_i^n \\ v_i^n \\ w_i^n \end{Bmatrix} + \sum N_i \eta \, \frac{t_i}{2} \begin{bmatrix} -\sin \phi_i & 0 \\ \cos \phi_i & 0 \\ 0 & 1 \end{bmatrix} \begin{Bmatrix} \alpha_i^n \\ \beta_i^n \end{Bmatrix} \right) \qquad (5.22)$$

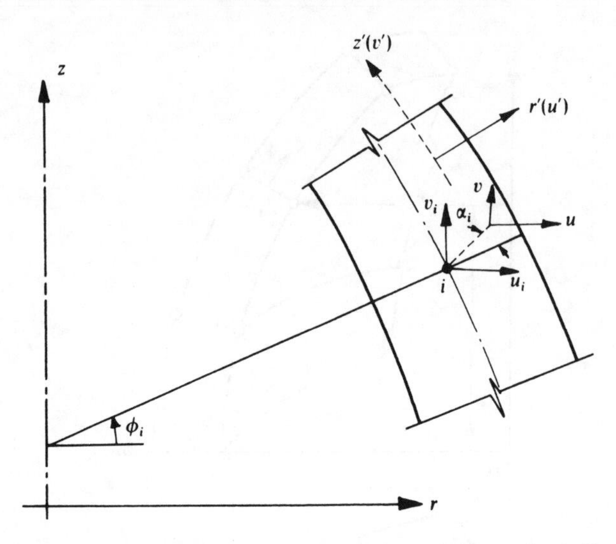

Fig. 5.5 Global displacements in an axisymmetric shell

In this α_i stands for the rotation illustrated in Fig. 5.5, u_i, etc., for the displacement of the middle surface node and β_i is the rotation about the vector tangential (approximately) to the middle surface.

For the purely axisymmetric case a further simplification arises by omitting the w terms, the first matrix of trigonometric constants and the rotation β_i.

Local strains are conveniently defined by the relationship Eq. (5.7) written in global cylindrical coordinates:

$$\boldsymbol{\varepsilon} = \begin{Bmatrix} \varepsilon_r \\ \varepsilon_z \\ \varepsilon_\theta \\ \gamma_{rz} \\ \gamma_{r\theta} \\ \gamma_{z\theta} \end{Bmatrix} = \begin{Bmatrix} \dfrac{\partial u}{\partial r} \\ \dfrac{\partial v}{\partial z} \\ \dfrac{u}{r} + \dfrac{1}{r}\dfrac{\partial w}{\partial \theta} \\ \dfrac{\partial u}{\partial z} + \dfrac{\partial v}{\partial r} \\ \dfrac{1}{r}\dfrac{\partial u}{\partial \theta} + \dfrac{\partial w}{\partial r} - \dfrac{w}{r} \\ \dfrac{1}{r}\dfrac{\partial u}{\partial \theta} + \dfrac{\partial w}{\partial z} \end{Bmatrix} \tag{5.23}$$

These strains are transformed to the local coordinates and the component normal to η=constant is neglected.

The $\mathbf{D}'$ matrix takes, however, a form identical to that defined by Eq. (5.9). For the axisymmetric case once again the appropriate terms are simply deleted.

All the transformations follow the pattern described in previous sections and need not be further commented upon except perhaps to remark that they are now only carried out between sets of directions ξ, η; r, z; and r', z' involving only two variables.

Similarly the integration of element properties is carried out numerically with respect to ξ and η only, noting, however, that the volume element is

$$dx\, dy\, dz = \det|\mathbf{J}|\, d\xi\, d\eta r\, d\theta \tag{5.24}$$

By suitable choice of shape functions $N_i(\xi)$, straight parabolic or cubic shapes of variable thickness elements can be used as shown in Fig. 5.6.

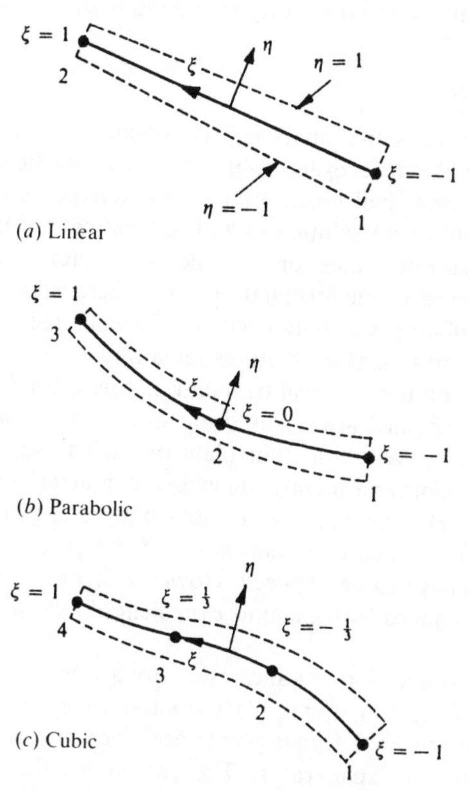

Fig. 5.6 Axisymmetric shell elements

5.8 Special case of thick plates

The transformations necessary in this chapter are somewhat involved and indeed the programming needed is sophisticated. However, the application of the principle involved is available for thick plates and readers are advised to test their comprehension on such a simple problem.

Here the following obvious simplifications arise:

1. $\zeta = z$ and unit vectors $\mathbf{v}_{1i}, \mathbf{v}_{2i}, \mathbf{v}_{3i}$ can be taken in the directions of the x, y and z axes respectively.
2. α_i and β_i are simply the rotations θ_y and θ_x (see Chapter 2).
3. It is no longer necessary to transform stress and strain components to a local system of axes $x' y' z'$ and global definitions can be used throughout. For elements of this type, numerical integration can be avoided and as an exercise readers are encouraged to derive stiffnesses, etc., for, say, linear, rectangular, elements. They will then find forms identical to those derived in Chapter 2 with an independent displacement and rotation interpolation and using shear constraints. This shows the essential identity of the alternative procedures.

5.9 Convergence

While in three-dimensional analysis it is possible to talk about absolute convergence to the true exact solution of the elasticity problem, in equivalent plate and shell problems such a convergence cannot happen. The so-called convergent solution of a plate bending problem approaches, as the element size decreases, only to the exact solution of the approximate model implied in the formulation. Thus, here again convergence of the above formulation will only occur to the exact solution constrained by the requirement that plane sections remain plane during deformation.

In elements of finite size it will be found that pure bending deformation modes are accompanied always by some shear stresses which in fact do not exist in the conventional thin plate or shell bending theory. Thus large elements deforming mainly under bending action (as would be the case of the shell element degenerated to a flat plate) tend to be appreciably too stiff. In such cases certain limits of the ratio of side of element to its thickness have to be imposed. However, it will be found that such restrictions are relaxed by the simple expedient of *reducing the integration order*.[7]

Figure 5.7 shows, for instance, the application of the quadratic, eight-node, element to a square plate situation. Here results for integration with 3×3 and 2×2 Gauss points are given and results plotted for different thickness to span ratios. For reasonably thick situations, the results are similar and both give the additional shear deformation not

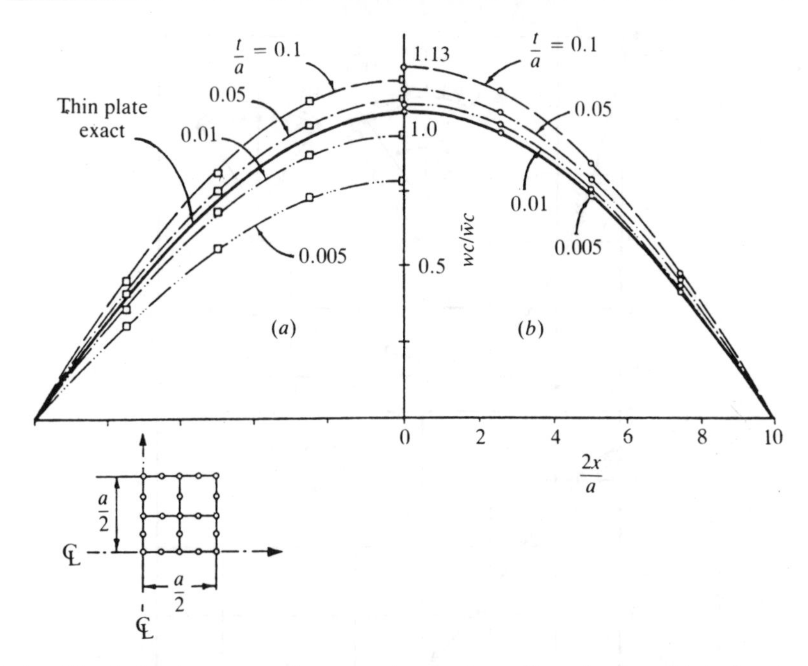

Fig. 5.7 A simply supported square plate under uniform load q_0. Plot of central deflection w_c for eight-node elements (*a*) with 3×3 Gauss point integration and (*b*) with 2×2 (reduced) Gauss point integration. Central deflection is $\bar{w}_c$ for thin plate theory

available by thin plate theory, but for thin plates the results with the more exact integration tend to diverge rapidly from the now correct thin plate results whereas the reduced integration still gives excellent results. The reasons for this improved performance are fully discussed in Chapter 2.

The reader is referred to Chapter 2 for further plate examples using different types of shape functions.

5.10 Some shell examples

A limited number of examples which show the accuracy and range of application of the shell formulation just described will be given. For a fuller selection the reader is referred to references 1, 2, 3, 7 to 10.

5.10.1 *Spherical dome under uniform pressure.* The 'exact' solution of shell theory is known for this axisymmetrical problem illustrated in Fig. 5.8. Twenty-four cubic-type elements were used here. These were of graded size more closely spaced towards the abutments.

The solution appears to be more accurate than the 'exact', shell theory,

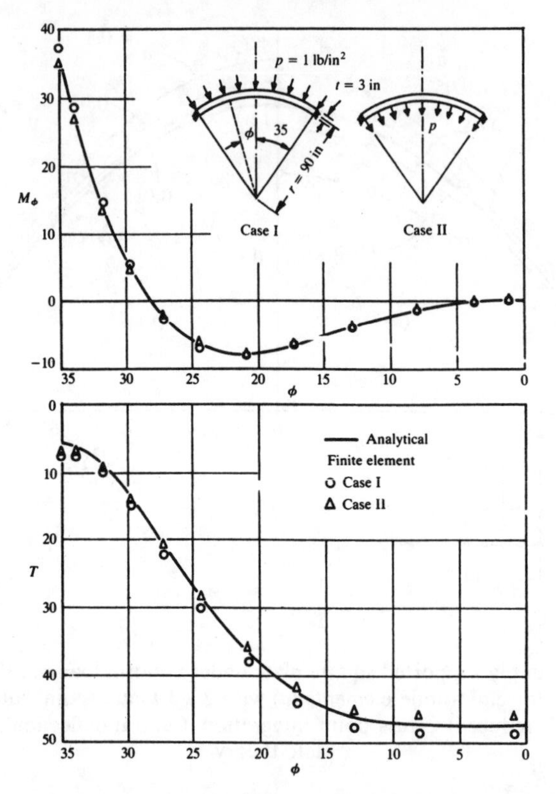

Fig. 5.8. Spherical dome under uniform pressure analysed with 24 cubic elements (first element subtends an angle of 0.1° from fixed end, others in arithmetic progression)

one distinguishing between the application of pressure on the inner and outer surfaces.

5.10.2 *Edge loaded cylinder.* A further axisymmetric example is shown in Fig. 5.9. to study the effect of subdivision. Two, six or fourteen elements of unequal length are used and the results for both the latter subdivisions are almost coincident with the exact solution. Even the two-element solution gives reasonable results and departs only in the vicinity of the loaded edge.

Once again the solutions are basically identical to those derived with independent slope and displacement interpolation in the manner presented in Chapter 2.

5.10.3 *Cylindrical vault.* This is a test example of application of the full process to a shell in which bending action is severe, due to supports restraining deflection at the ends (see also Sec. 3.8.2).

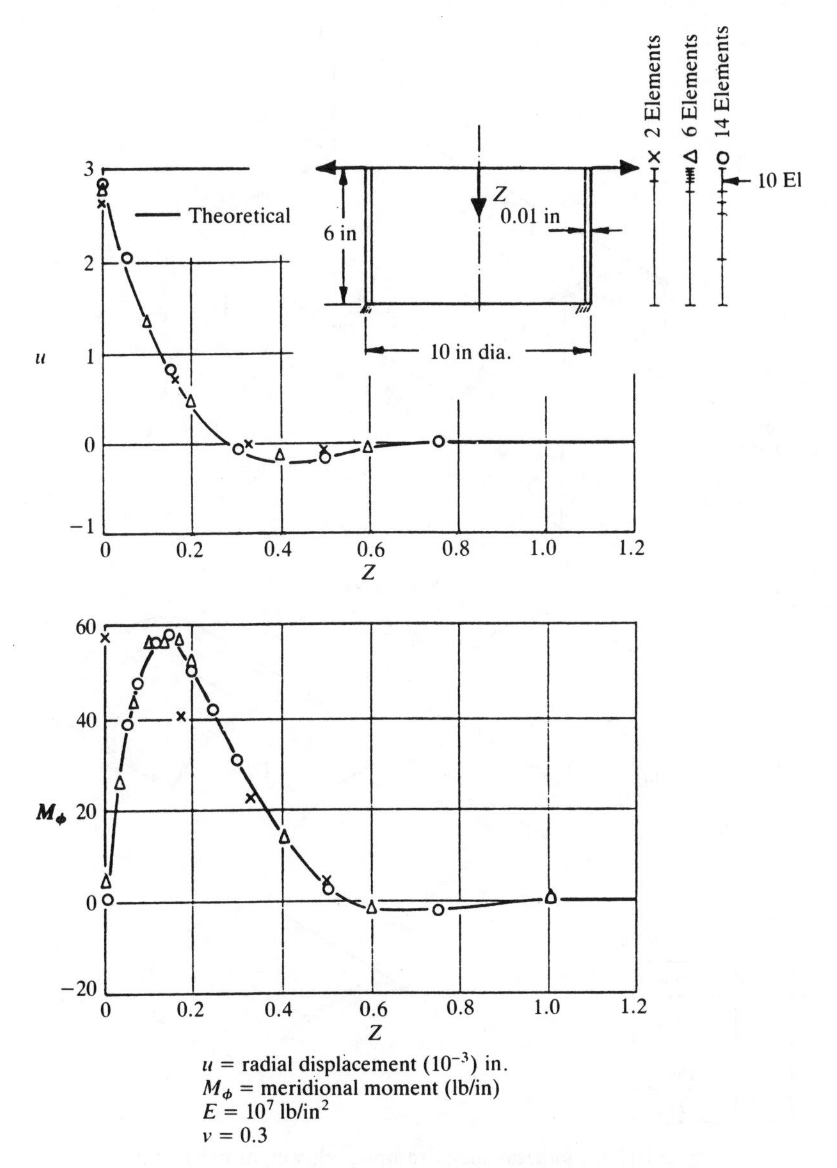

u = radial displacement (10^{-3}) in.
M_ϕ = meridional moment (lb/in)
$E = 10^7$ lb/in^2
$\nu = 0.3$

Fig. 5.9 Thin cylinder under a unit radial edge load

In Fig. 5.10 the geometry, physical details of the problem and subdivision are given while in Fig. 5.11 the comparision of the effects of 3×3 and 2×2 integration using parabolic elements is shown on the displacements calculated. Both integrations result as expected in convergence. For the more exact integration, this is rather slow while, with

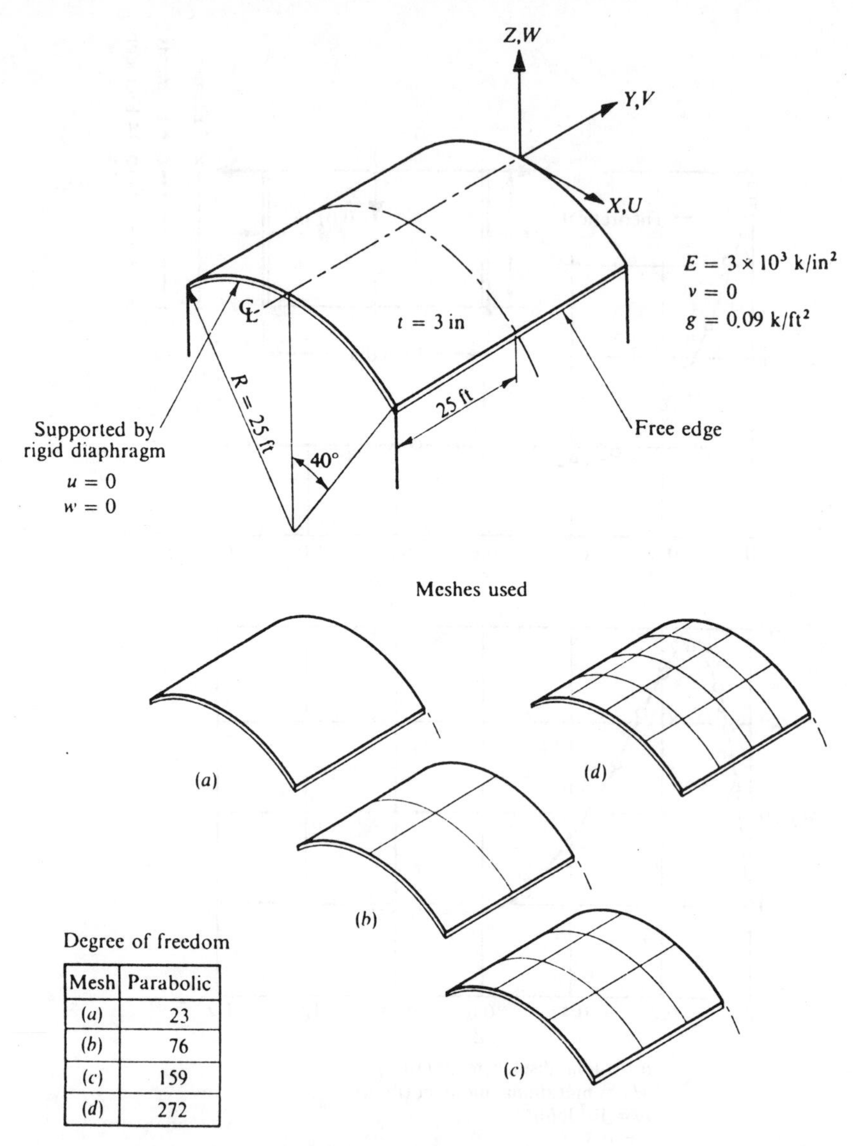

Mesh	Parabolic
(a)	23
(b)	76
(c)	159
(d)	272

Fig. 5.10 Cylindrical shell example, self-weight behaviour

reduced integration order, very accurate results are obtained, even with one element. This example illustrates most dramatically the advantages of this simple expedient and is described more fully in references 7 and 9. The 'exact' solution for this problem is one derived on more conventional lines by Scordelis and Lo.[11]

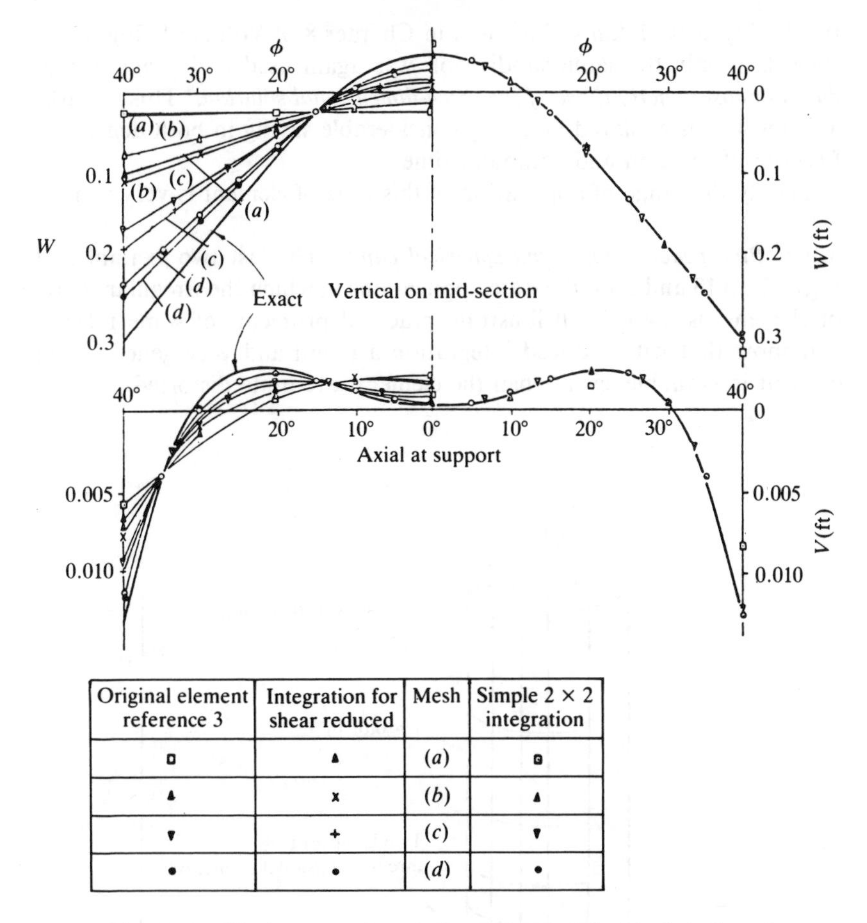

Original element reference 3	Integration for shear reduced	Mesh	Simple 2 × 2 integration
□	▲	(a)	□
▲	×	(b)	▲
▼	+	(c)	▼
•	•	(d)	•

Fig. 5.11 Displacement (parabolic element), cylindrical shell roof

The improved convergence of displacements is matched by rapid convergence of stress components.

5.10.4 *Cooling tower.* The cooling tower already referred to in Chapter 3 (Figs 3.11 and 3.12) has been again analysed dividing the axisymmetric shell into 15 elements of cubic type. Using 10 harmonics the unsymmetric (wind) loading is adequately represented and the results coincide with those of the test analysis against which the results of Chapter 3 are compared so that additional plots are not necessary.

5.10.5 *Curved dams.* All the previous examples were rather thin shells and indeed demonstrated the applicability of the process to these situations. At the other end of the scale this formulation has been applied to the doubly curved dams illustrated in Chapter 8 of Volume 1 (Fig. 8.28). Indeed, exactly the same subdivision was again used and *results reproduced almost exactly those of the three-dimensional solution.*[3] This remarkable result was achieved at a very considerable saving in both degrees of freedom of solution and computer time.

Clearly the range of application of this type of element is very wide.

5.10.6 *Pipe penetration*[12] *and spherical cap.*[10] The last two examples of Figs 5.12/5.13 and 5.14 illustrate applications in which the irregular shape of elements is used. Both illustrate practical problems of some interest and show that with reduced integration a useful and very general shell element is available, even when the elements are quite distorted.

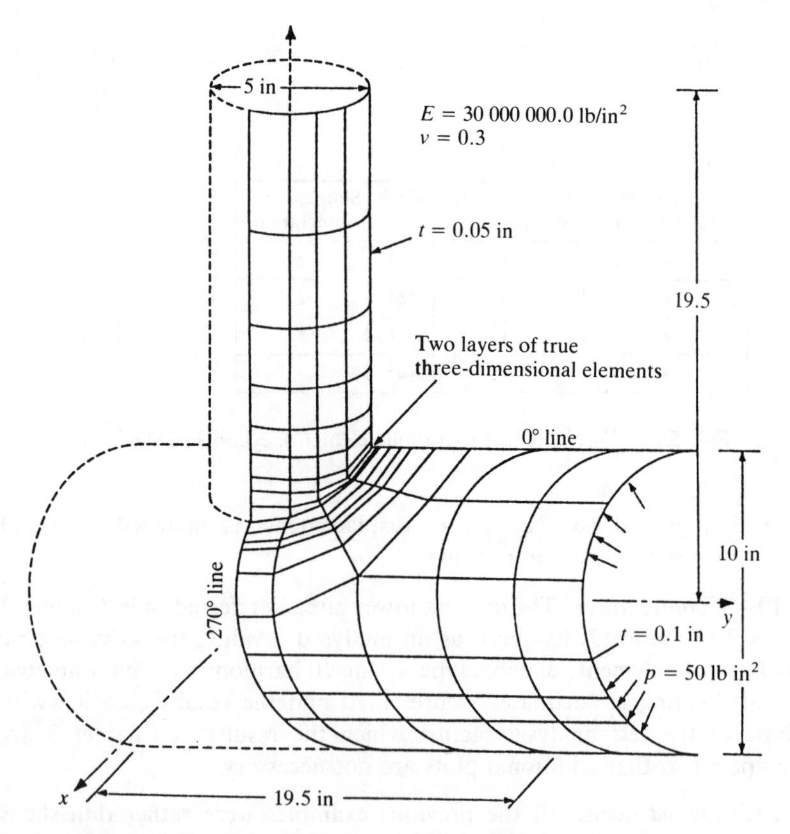

Fig. 5.12 An analysis of cylinder intersection using reduced integration shell-type elements[12]

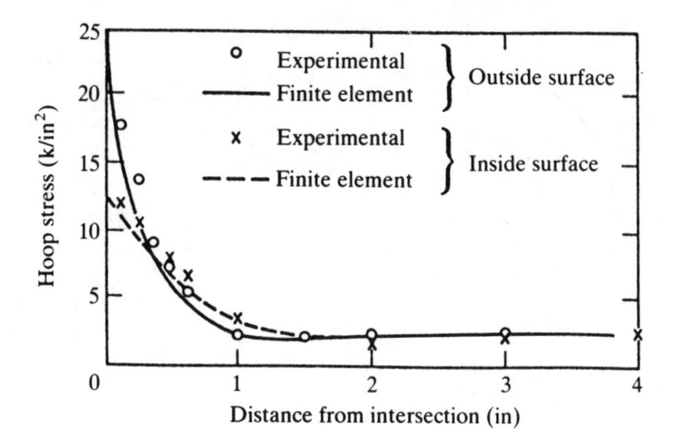

(a) Hoop stresses near 0° line

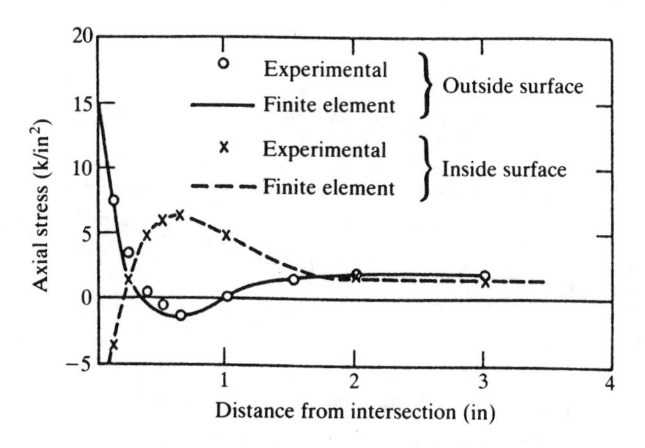

(b) Axial stresses near 0° line

Fig. 5.13 Cylinder-to-cylinder intersections of Fig. 5.12

5.11 Concluding remarks

The elements described in this chapter using a degeneration of solid elements have been noted in plate and axisymmetric problems to be identical with those in which an independent slope and displacement interpolation was directly used in the middle plane in Chapters 2 and 4. For the general curved shell the analogy is less obvious but clearly still exists. We should therefore expect that the conditions established in Chapter 2 for robustness of plate elements to be still valid. Further, it appears possible that other, additional, conditions on the various inter-

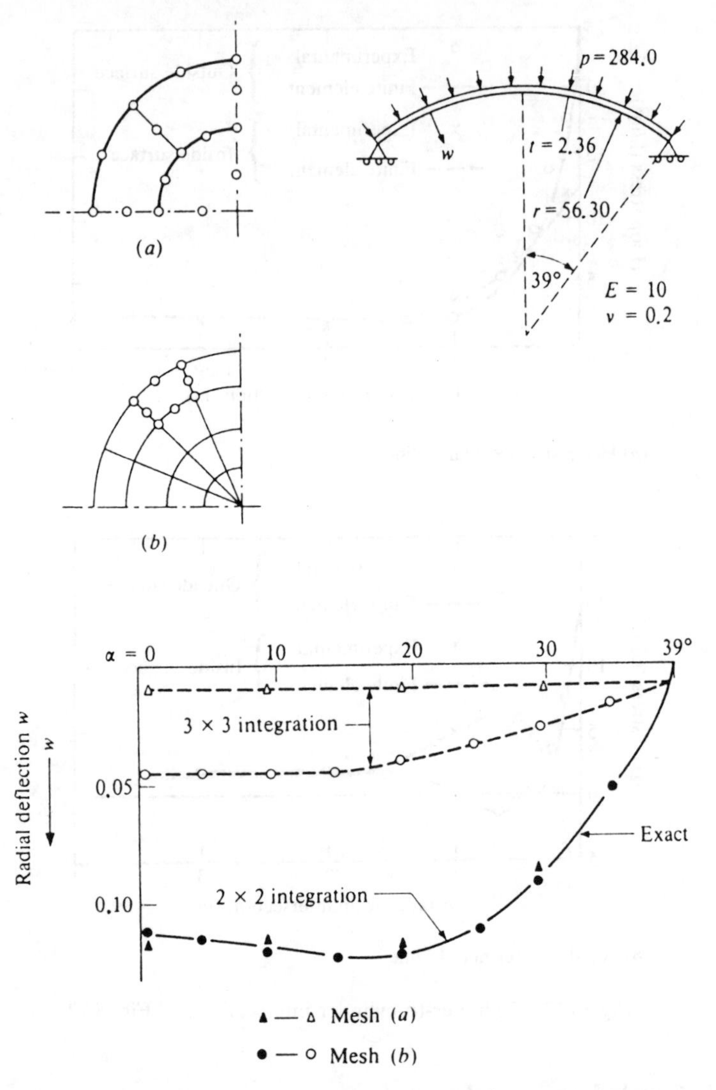

Fig. 5.14 A spherical cap analysis with irregular isoparametric shell elements using full 3×3 and reduced 2×2 integration

polations may have to be imposed in curved element forms. Both statements are true. The eight- and nine-node elements which we have shown in the previous section to perform admirably will fail under certain circumstances and for this reason many of the more successful plate elements have been adapted to the shell problem.

The introduction of additional degrees of freedom in the interior of the eight-node serendipity element was first suggested by Cook[13,14] and later by Hughes[15,16] without, however, achieving complete robustness. The full lagrangian cubic interpolation as shown in Chapter 2 is quite effective and has been shown to perform well. However, the best results achieved to date appear to be those in which 'local constraints' are applied (viz. page 82) and such elements as those due to Dvorkin and Bathe[17] or Huang and Hinton[18] fall into this category.

While the importance of transverse shear strain constraints is now fully understood, the constraints introduced by the 'in-plane' stress resultants are less amenable to analysis (although the elastic parameters Et associated with these are of the same order as those of shear Gt). It is well known that *membrane locking* can occur in situations that do not permit inextensional bending. Such locking has been extensively discussed[19,20] but to date the problem has not been rigorously solved and further developments are in progress.

Much effort is continuing to improve the formulation of the processes described in this chapter as they offer probably the optimal solution of the curved shell problem.[20-23]

References

1. S. AHMAD, B. M. IRONS and O. C. ZIENKIEWICZ, 'Curved thick shell and membrane elements with particular reference to axi-symmetric problems', in *Proc. 2nd Conf. on Matrix Methods in Structural Mechanics*, Wright-Patterson AF Base, Ohio, 1968.
2. S. AHMAD, 'Curved finite elements in the analysis of solid, shell and plate structures', Ph.D. thesis, University of Wales, Swansea, 1969.
3. S. AHMAD, B. M. IRONS and O. C. ZIENKIEWICZ, 'Analysis of thick and thin shell structures by curved elements', *Int. J. Num. Meth. Eng.*, **2**, 419–51, 1970.
4. R. D. WOOD, 'The application of finite element methods to geometrically non-linear analysis', Ph.D. thesis, University of Wales, Swansea, 1973.
5. R. D. WOOD and O. C. ZIENKIEWICZ, 'Geometrically non-linear finite element analysis of beams, frames, arches and axisymmetric shells', *Comp. Struct.*, **7**, 725–35, 1977.
6. E. L. WILSON, 'Finite elements for foundations, joints and fluids', in *Finite Elements in Geomechanics* (ed. G. Gudehus), chap. 10, pp. 319–50, Wiley, Chichester, 1977.
7. O. C. ZIENKIEWICZ, J. TOO and R. L. TAYLOR, 'Reduced integration technique in general analysis of plates and shells', *Int. J. Num. Meth. Eng.*, **3**, 275–90, 1971.
8. S. F. PAWSEY and R. W. CLOUGH, 'Improved numerical integration of thick slab finite elements', *Int. J. Num. Meth. Eng.*, **3**, 575–86, 1971.
9. S. F. PAWSEY, Ph.D. thesis, Department of Structural Mechanics, University of California, Berkeley, 1970.
10. J. J. M. TOO, 'Two dimensional, plate, shell and finite prism isoparametric elements and their applications', Ph.D. thesis, University of Wales, Swansea, 1971.

11. A. C. SCORDELIS and K. S. LO, 'Computer analysis of cylindrical shells', *J. Am. Concr. Inst.*, **61**, 539–61, 1969.
12. S. A. BAKHREBAH and W. C. SCHNOBRICH, 'Finite element analysis of interesting cylinders', UILU-ENG-73-2018, University of Illinois, Civil Engineering Studies, 1973.
13. R. D. COOK, 'More on reduced integration and isoparametric elements', *Int. J. Num. Meth. Eng.*, **5**, 141–2, 1972.
14. R. D. COOK, *Concepts and Applications of Finite Element Analysis*, Wiley, New York, 1974.
15. T. J. R. HUGHES and M. COHEN, 'The "heterosis" finite element for plate bending', *Comp. Struct.*, **9**, 445–50, 1978.
16. T. J. R. HUGHES and W. K. LIU, 'Non linear finite element analysis of shells', *Comp. Meth. Appl. Mech. Eng.*, **26**, 331–62; **27**, 167–81, 1981.
17. E. N. DVORKIN and K. J. BATHE, 'A continuum mechanics based four noded shell element for non linear analysis', *Eng. Comput.*, **1**, 77–88, 1984.
18. E. C. HUANG and E. HINTON, 'Elastic, plastic and geometrically non linear analysis of plates and shells using a new, nine noded element', in *Finite Elements for Non Linear Problems* (eds P. Bergan *et al.*), pp. 283–97, Springer-Verlag, Berlin, 1986.
19. H. STOLARSKI and T. BELYTCHKO, 'Shear and membrane locking in curved C^0 elements', *Comp. Mech. Appl. Mech. Eng.*, **41**, 279–96, 1983.
20. R. V. MILFORD and W. C. SCHNOBRICH, 'Degenerated isoparametric finite elements using explicit integration', *Int. J. Num. Mech. Eng.*, **23**, 133–54, 1986.
21. D. BUSHNELL, 'Computerized analysis of shells—governing equation', *Comp. Struct.*, **18**, 471–536, 1984.
22 M. A. CELIA and N. G. GRAY, 'Improved coordinate transformations for finite elements. The Lagrange cubic case', *Int. J. Num. Mech. Eng.*, **23**, 1529–45, 1986.
23. S. VLACHOUTSIS, 'Explicit integration for three dimensional degenerated shell finite elements', *Int. J. Num. Mech. Eng.*, **29**, 861–80, 1990.

6

Semi-analytical finite element processes—use of orthogonal functions and 'finite strip' methods

6.1 Introduction

The standard finite element methods have been shown to be capable, in principle, of dealing with any two- or three- (or even four-)† dimensional situations. Nevertheless, the cost of solutions increases greatly with each dimension added and indeed, on occasion, overtaxes the capabilities of available machines. It is therefore always desirable to search for alternatives that may reduce the computational labour. One such class of processes of quite a wide applicability will be illustrated here.

In many physical problems the situation is such that the *geometry* and *material properties* do not vary along one coordinate direction. However, the 'load' terms may still exhibit a variation in that direction preventing the use of such simplifying assumptions as those that, for instance, permitted a two-dimensional, plane strain, analysis to be substituted for a full three-dimensional treatment. In such cases it is possible still to consider a 'substitute' problem, not involving the particular coordinate (along which the properties do not vary), and to synthesize the true answer from a series of such simplified solutions.

The method to be described is of quite general use and, obviously, is not limited to structural situations. It will be convenient, however, to use the nomenclature of structural mechanics and to use the potential energy minimization as an example.

† See finite elements in the time domain in Chapter 10.

We shall confine our attention to problems of minimizing a quadratic functional such as described in Chapters 2 and 9 of Volume 1. The interpretation of the processes involved as the application of partial discretization in Chapter 9 of Volume 1 (page 225) followed by the use of a Fourier series expansion should be noted.

Let (x, y, z) be the coordinates describing the domain (in this context these do not necessarily have to be the cartesian coordinates). The last one of these, z, is the coordinate along which the geometry and material properties do not change and which is limited to lie between two values

$$0 \leqslant z \leqslant a$$

The boundary values are thus specified at $z=0$ and $z=a$.

We shall assume that the shape functions defining the variation of displacements $\mathbf{u}$ can be written in a product form as

$$\begin{aligned} \mathbf{u} &= \mathbf{N}(x, y, z)\mathbf{a}^e \\ &= \sum_{l=1}^{L} \left\{ \bar{\mathbf{N}}(x, y) \cos \frac{l\pi z}{a} + \bar{\bar{\mathbf{N}}}(x, y) \sin \frac{l\pi z}{a} \right\} \mathbf{a}_l^e \end{aligned} \tag{6.1}$$

In this type of representation completeness is preserved in view of the capability of the Fourier series to represent any continuous function within a given region (naturally assuming that the shape functions $\bar{\mathbf{N}}$ and $\bar{\bar{\mathbf{N}}}$ in the domain x, y satisfy the same requirements).

The loading terms will similarly be given a form

$$\mathbf{b} = \sum_{l=1}^{L} \left(\bar{\mathbf{b}}_l \cos \frac{l\pi z}{a} + \bar{\bar{\mathbf{b}}}_l \sin \frac{l\pi z}{a} \right) \tag{6.2}$$

with similar form for concentrated loads and boundary tractions (see Chapter 2 of Volume 1). Indeed, initial strains and stresses, if present, would be expanded again in the above form.

Applying the standard processes of Chapter 2 of Volume 1 to the determination of the element contribution to the equation minimizing the potential energy, and limiting our attention to the contribution of forces $\mathbf{b}$ only, we can write

$$\frac{\partial \Pi}{\partial \mathbf{a}^e} = \mathbf{K}^e \begin{Bmatrix} \mathbf{a}_1^e \\ \vdots \\ \mathbf{a}_L^e \end{Bmatrix} + \begin{Bmatrix} \mathbf{f}_1^e \\ \vdots \\ \mathbf{f}_1^e \end{Bmatrix} \tag{6.3}$$

In the above, to avoid summation signs, the vectors $\mathbf{a}^e$, etc., are expanded, listing the contribution of each value of l separately.

Now a typical submatrix of $\mathbf{K}^e$ is

$$(\mathbf{K}^{lm})^e = \iiint_V \mathbf{B}^{l\mathrm{T}} \mathbf{D} \mathbf{B}^m \, dx \, dy \, dz \tag{6.4}$$

and a typical term of the 'force' vector becomes

$$(\mathbf{f}^l)^e = \iiint_V \mathbf{N}^{lT}\mathbf{b}^l \, \mathrm{d}x \, \mathrm{d}y \, \mathrm{d}z \tag{6.5}$$

Without going into details it is obvious that the matrix given by Eq. (6.4) will contain the following integrals as products of various sub-matrices:

$$\begin{aligned} I_1 &= \int_0^a \sin\frac{l\pi z}{a}\cos\frac{m\pi z}{a}\,\mathrm{d}z \\ I_2 &= \int_0^a \sin\frac{l\pi z}{a}\sin\frac{m\pi z}{a}\,\mathrm{d}z \\ I_3 &= \int_0^a \cos\frac{l\pi z}{a}\cos\frac{m\pi z}{a}\,\mathrm{d}z \end{aligned} \tag{6.6}$$

These integrals arise from products of the derivatives contained in the definition of $\mathbf{B}$ and, due to the well-known orthogonality property, give

$$I_2 = I_3 = 0 \qquad \text{for } l \neq m \tag{6.7}$$

when $l = 1, 2, \ldots$ and $m = 1, 2, \ldots$. The first integral I_1 is only zero when l and m are both even or odd numbers. The term involving I_1, however, vanishes in most applications.

This means that the matrix $\mathbf{K}^e$ becomes a diagonal one and that the assembled final equations of the system have the form

$$\begin{bmatrix} \mathbf{K}^{11} & & & \\ & \mathbf{K}^{22} & & \\ & & \ddots & \\ & & & \mathbf{K}^{LL} \end{bmatrix} \begin{Bmatrix} \mathbf{a}_1 \\ \vdots \\ \mathbf{a}_L \end{Bmatrix} + \begin{Bmatrix} \mathbf{f}_1 \\ \vdots \\ \mathbf{f}_L \end{Bmatrix} \tag{6.8}$$

and the large system of equations splits into L separate problems:

$$\mathbf{K}^{ll}\mathbf{a}_l + \mathbf{f}^l = 0 \tag{6.9}$$

in which

$$\mathbf{K}_{ij}^{ll} = \iiint_V \mathbf{B}_i^{lT}\mathbf{D}\mathbf{B}_j^l \, \mathrm{d}x \, \mathrm{d}y \, \mathrm{d}z, \qquad \text{etc.} \tag{6.10}$$

Further, from Eqs (6.5) and (6.2) we observe that due to the orthogonal property of the integrals given by Eqs (6.6), the typical load term becomes simply

$$\mathbf{f}_i^l = \iiint_V \mathbf{N}_i^{lT}\mathbf{b}^l \, \mathrm{d}x \, \mathrm{d}y \, \mathrm{d}z \tag{6.11}$$

This means that the force term of the lth harmonic only affects the lth system of Eq. (6.9) and contributes nothing to the other equations. This extremely important property is of considerable practical significance for, *if the expansion of the loading factors involves only one term, only one set of equations need be solved.* The solution of this will tend to the exact one with increasing subdivision in the xy domain only. Thus, what was originally a three-dimensional problem has now been reduced to a two-dimensional one with the consequent reduction of computational effort.

The preceding derivation was illustrated on a three-dimensional, elastic situation. Clearly the arguments could be equally well applied for reduction of two-dimensional problems to one-dimensional ones, etc., and the arguments are not restricted to problems of elasticity. Any physical problem governed by a minimization of a quadratic functional (Chapter 9 of Volume 1) or by linear differential equations is amenable to the same treatment, which under various guises has been used since time immemorial in applied mechanics.

A word of warning should be added regarding the boundary conditions imposed on $\mathbf{u}$. For a complete decoupling to be possible these must be satisfied separately by each and every term of the expansion given by Eq. (6.1). Insertion of a zero displacement in the final reduced problem implies in fact a zero displacement fixed throughout all terms in the z direction by definition. Care must be taken not to treat the final matrix therefore as a simple reduced problem. Indeed this is one of the limitations of the process described.

When the loading is complex and many Fourier components need to be considered the advantages of the approach outlined here reduce and the full solution sometimes becomes superior in economy.

Other permutations of the basic definitions of the type given by Eq. (6.1) are obviously possible. For instance, two independent sets of parameters $\mathbf{a}^e$ may be specified with each of the trigonometric terms. Indeed, on occasion use of other orthogonal functions may be possible.

As trigonometric functions will arise frequently it is convenient to remind readers of the following integrals:

$$
\begin{aligned}
&\int_0^a \sin\frac{l\pi z}{a}\cos\frac{l\pi z}{a}\,\mathrm{d}z = 0 && \text{when } l=0, 1, \ldots \\
&\int_0^a \sin^2\frac{l\pi z}{a}\,\mathrm{d}z = \int_0^a \cos^2\frac{l\pi z}{a}\,\mathrm{d}z = \frac{a}{2} && \text{when } l=1, 2, \ldots
\end{aligned}
\tag{6.12}
$$

6.2 Prismatic bar

Consider a prismatic bar illustrated in Fig. 6.1 which is assumed to be held at $z=0$ and $z=a$ in a manner preventing all displacements in the xy

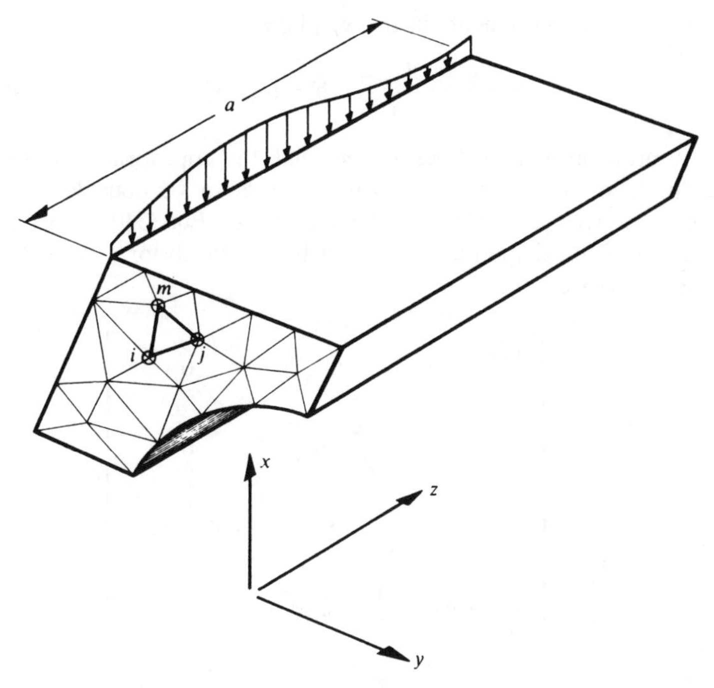

Fig. 6.1 A prismatic bar reduced to a series of two-dimensional finite element solutions

plane but permitting unrestricted motion in the z direction (traction $t_z = 0$). The problem is fully three dimensional and three components of displacement u, v and w have to be considered.

Subdividing into finite elements in the xy plane we can prescribe the lth displacement components in the x direction as

$$u^l = [N_1, N_2, \dots] \sin \frac{l\pi z}{a} \mathbf{u}^l \tag{6.13}$$

with similar expressions for the v^l and w^l but with a cosine term in the last.

In this N, etc., are simply the (scalar) shape functions appropriate to the elements used in the xy plane. If, as shown in Fig. 6.1, simple triangles are used then the shape functions are given by Eq. (4.8) in Chapter 4 of Volume 1, but any of the more elaborate elements described in Chapter 7 of Volume 1 (with or without the transformation of Chapter 8 of Volume 1) would be equally suitable.

The expansion ensures zero u and w displacements and zero axial stresses at the ends and hence satisfies the boundary conditions.

The load terms can still be expressed in terms of a similar Fourier series, giving, for components in the xy plane,

$$\mathbf{b}^l = \bar{\mathbf{b}}^l \sin \frac{l\pi z}{a} \qquad \bar{\mathbf{b}}^e = \bar{\mathbf{b}}^e(x, y) \tag{6.14}$$

As the problem is fully three dimensional, the appropriate expression for strain involving all six components needs to be considered. This expression is given in Chapter 5 of Volume 1 by Eqs (5.9) to (5.11). On substitution of the shape function given by Eq. (6.13) for a typical term of the $\mathbf{B}$ matrix we have

$$\mathbf{B}_i^l = \begin{bmatrix} \frac{\partial N_i}{\partial x}\sin\gamma & 0 & 0 \\ 0 & \frac{\partial N_i}{\partial y}\sin\gamma & 0 \\ 0 & 0 & -N_i \frac{l\pi}{a}\sin\gamma \\ \frac{\partial N_i}{\partial y}\sin\gamma & \frac{\partial N_i}{\partial x}\sin\gamma & 0 \\ 0 & N_i \frac{l\pi}{a}\cos\gamma & \frac{\partial N_i}{\partial y}\cos\gamma \\ N_i \frac{l\pi}{a}\cos\gamma & 0 & \frac{\partial N_i}{\partial x}\cos\gamma \end{bmatrix} \tag{6.15}$$

with $\gamma = l\pi z/a$. It is convenient to separate the above as

$$\mathbf{B}_i^l = \bar{\mathbf{B}}_i^l \sin \frac{\pi l z}{a} + \bar{\bar{\mathbf{B}}}_i^l \cos \frac{\pi l z}{a} \tag{6.16}$$

In all of the above it is assumed that the parameters are listed in the usual order:

$$\mathbf{a}_i^l = \begin{Bmatrix} u_i^l \\ v_i^l \\ w_i^l \end{Bmatrix} \tag{6.17}$$

and that the axes are as shown in Fig. 6.1.

The stiffness matrix can be computed in the usual manner noting that

$$(\mathbf{K}_{ij}^{ll})^e = \iiint_{V^e} \mathbf{B}_i^{lT} \mathbf{D} \mathbf{B}_j^l \, dx \, dy \, dz \tag{6.18}$$

On substitution of Eq. (6.16), multiplying out and noting the value of the integrals from Eq. (6.12), this reduces to

$$(\mathbf{K}_{ij}^{ll})^e = \frac{a}{2}\iint_{A^e} \{\mathbf{B}_i^{l\mathrm{T}}\mathbf{D}\bar{\mathbf{B}}_j^l + \bar{\mathbf{B}}_i^{l\mathrm{T}}\mathbf{D}\bar{\mathbf{B}}_j^l\}\,\mathrm{d}x\,\mathrm{d}y \tag{6.19}$$

when $l=1, 2, \ldots$. The integration is now simply carried out over the element *area*.†

Similarly, the contributions due to distributed loads, initial stresses, etc., are found as the loading terms. Concentrated line loads, for instance, would be expressed directly as nodal forces

$$\mathbf{f}_i^l = \int_0^a \sin\frac{\pi l z}{a}\begin{Bmatrix}\bar{\mathbf{f}}_{xi}^l\\ \bar{\mathbf{f}}_{yi}^l\\ \bar{\mathbf{f}}_{zi}^l\end{Bmatrix}\sin\frac{\pi l z}{a}\,\mathrm{d}z = \bar{\mathbf{f}}_i^l\,\frac{a}{2} \tag{6.20}$$

in which $\bar{\mathbf{f}}_i^l$ are intensities per unit length.

The boundary conditions used here have been of a type ensuring *simply supported* conditions for the prism. Other conditions can be inserted by suitable expansions.

The method of analysis outlined here can be applied to a range of practical problems—one of these being a popular type of concrete bridge illustrated in Fig. 6.2. Here a particularly convenient type of element is the distorted, 'serendipity', quadratic or cubic of Chapters 7 and 8 of Volume 1.[1] Finally it should be mentioned that some restrictions placed on the general shapes defined by Eqs (6.1) or (6.13) can be raised by doubling the number of parameters and writing expansions in the form of two sums:

$$\mathbf{u} = \sum_{l=1}^{L}\bar{\mathbf{N}}(x,y)\cos\frac{l\pi z}{a}\mathbf{a}^{Al} + \sum_{l=1}^{L}\bar{\mathbf{N}}(x,y)\sin\frac{l\pi z}{a}\mathbf{a}^{Bl} \tag{6.21}$$

Parameters $\mathbf{a}^{Al}$ and $\mathbf{a}^{Bl}$ are independent and for every component of displacement two values have to be found and two equations formed.

An alternative to the above process is to write the expansion as

$$\mathbf{u} = \sum[\mathbf{N}(x,y)\,\mathrm{e}^{i(l\pi z/a)}]\,\mathbf{a}^e$$

and to observe that both $\mathbf{N}$ and $\mathbf{a}$ are complex quantities.

Complex algebra is available on standard computers and the identity of the above expression with Eq. (6.21) will be observed noting that

$$\mathrm{e}^{i\theta} = \cos\theta + \mathrm{i}\sin\theta$$

† It should be noted that now, even for a simple triangle, the integration is not trivial as some linear terms will remain in $\bar{\mathbf{B}}$.

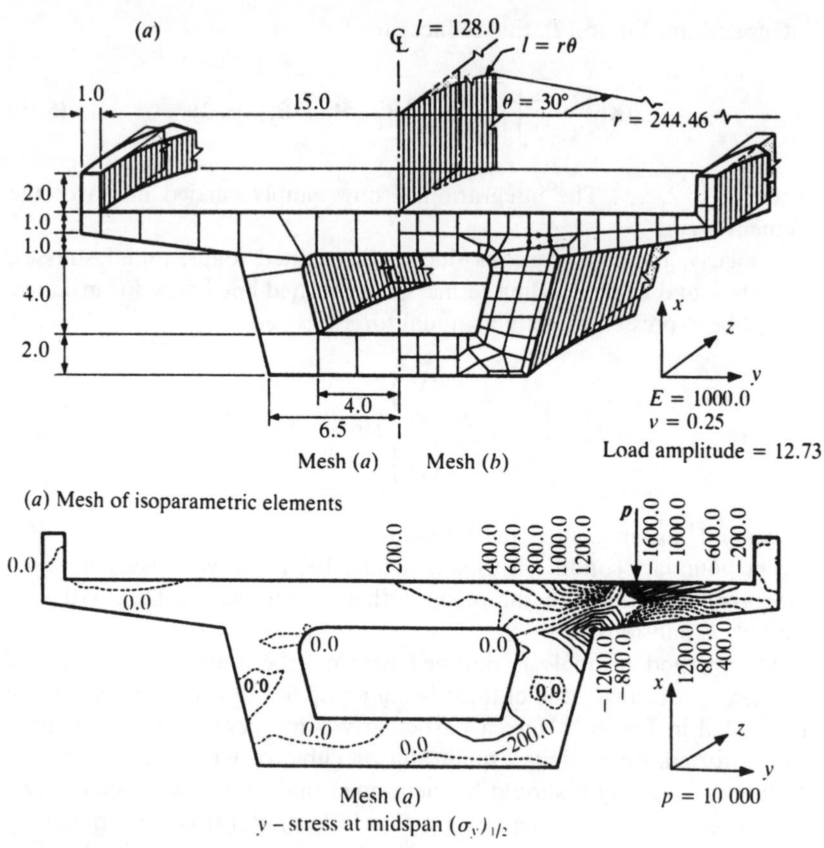

(*a*) Mesh of isoparametric elements

(*b*) Distribution of σ_y stress on mid-span: computer stress plot. Point load on cantilevered span

Fig. 6.2 A thick box bridge prism of straight or curved platform

6.3 Thin membrane box structures

In the previous section a three-dimensional problem was reduced to that of two dimensions. Here we shall see how a somewhat similar problem can be reduced to one-dimensional elements (Fig. 6.3).

A box-type structure is made up of thin sheet components capable of sustaining stresses only in its own plane. Now, just as in the previous case, three displacements have to be considered at every point and indeed similar variation can be prescribed for these. However, a typical element *ij* is 'one dimensional' in the sense that integrations have to be carried out only along the line *ij* and stresses in that direction only considered. Indeed, it will be found that the situation and the solution are similar to that of a pin-jointed framework.

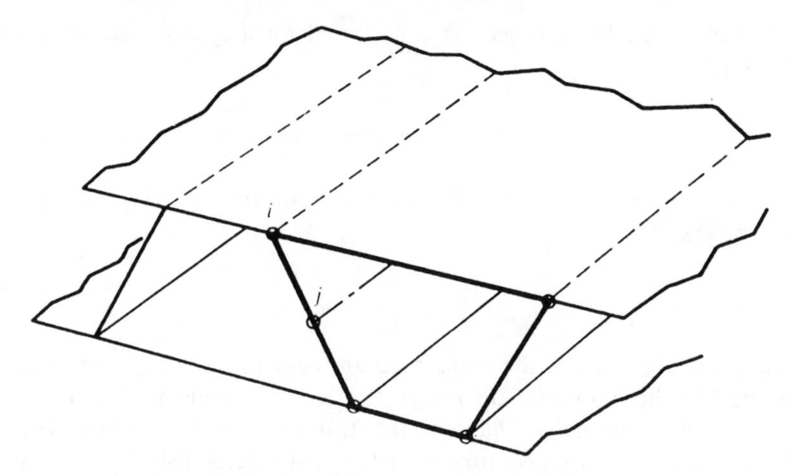

Fig. 6.3 A 'membrane' box with one-dimensional elements

6.4 Plates and boxes with flexure

Consider now a rectangular plate simply supported at the ends and in which all strain energy is contained in flexure. Only one displacement, w, is needed to specify fully the state of strain (see Chapter 1).

For consistency of notation, the direction in which geometry and material properties do not change has been taken as y (see Fig. 6.4). To preserve slope continuity the functions need to include now the 'rotation' parameter θ_i.

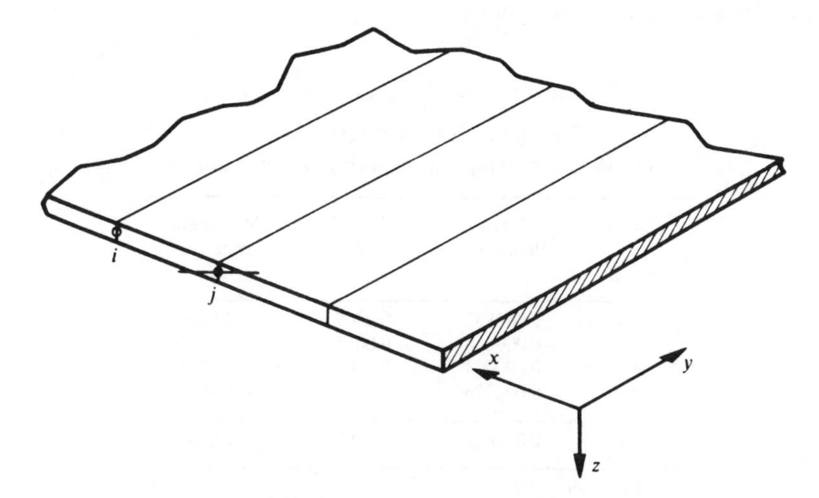

Fig. 6.4 The 'strip' method in slabs

Use of simple beam functions is easy and for a typical element ij we can write

$$w^l = \bar{\mathbf{N}}(x) \sin \frac{l\pi y}{a} (\mathbf{a}^l)^e \tag{6.22}$$

ensuring *simply supported* end conditions. In this, the typical nodal parameters are

$$\mathbf{a}_i^l = \begin{Bmatrix} w_i \\ \theta_i \end{Bmatrix} \tag{6.23}$$

The shape functions of the cubic type are easy to write and are in fact identical to those used for the axisymmetric shell problem (Chapter 4).

Using all definitions of Chapter 1 the strains (curvatures) are found and the $\mathbf{B}$ matrices determined; now with C_1 continuity satisfied in a trivial manner, the problem of a two-dimensional kind has here been reduced to that of one dimension.

This application has been developed by Cheung and others[2-17], named by him the 'finite strip' method, and used to solve many rectangular plate problems, box girders, shells and various folded plates.

It is illuminating to quote an example from the above papers here. This refers to a square, uniformly loaded plate with three sides simply supported and one free. Ten strips or elements in the x direction were used in the solution and Table 6.1 gives the results corresponding to the first three harmonics.

Not only is an accurate solution of each l term a simple one involving only some nine unknowns but the importance of higher terms in the series is seen to decrease rapidly.

TABLE 6.1
SQUARE PLATE, UNIFORM LOAD q
THREE SIDES SIMPLY SUPPORTED, ONE CLAMPED

$\nu = 0.3$	Central deflection	Central M_x	Maximum negative M
$l = 1$	0.002832	0.0409	−0.0858
$= 2$	−0.000050	−0.0016	0.0041
$= 3$	0.000004	0.0003	−0.0007
Σ	0.002786	0.0396	−0.0824
Exact	0.0028	0.039	−0.084
Multiplier qa^4/D		qa^2	

Extension of the process to box structures in which both *membrane and bending effects* are present is almost obvious when this example is considered together with the ones of the previous section.

In another paper Cheung[5] shows how functions other than trigonometric ones can be used to advantage, although only partial decoupling then occurs.

In the examples just quoted a thin plate theory using the single displacement variable w and enforcing C_1 compatibility in the x direction was employed. Obviously any of the independently interpolated slope and displacement elements of Chapter 2 could be used here again employing reduced integration. Parabolic-type elements are thus employed in references 6 and 7, and the linear interpolation with a simple integration point is shown to be effective in reference 8.

Other applications for plate- and box-type structures abound and additional information is given in the text of reference 17.

6.5 Axisymmetric solids with non-symmetrical load

One of the most natural and indeed earliest applications of the one-way Fourier expansion occurs in axisymmetric bodies subject to non-axisymmetric loads.

Now, not only the radial (u) and axial (v) displacements (as in Chapter 4 of Volume 1) will have to be considered but also a tangential component (w) associated with angular direction θ (Fig. 6.5). It is in this direction that the geometric and material properties do not vary and hence here that the elimination will be applied.

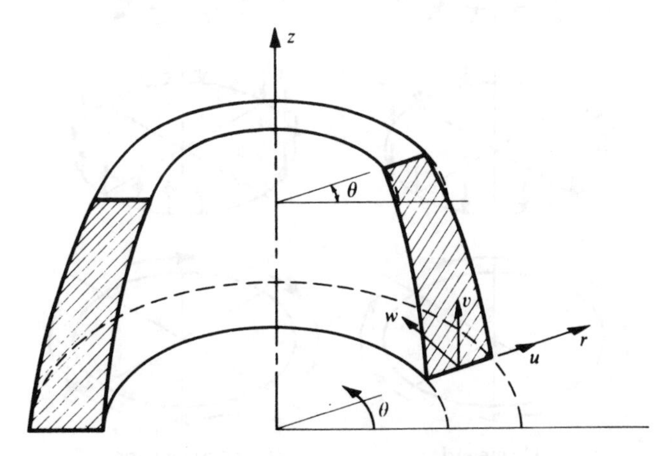

Fig. 6.5 An axisymmetric solid. Coordinates and displacement components in an axisymmetric body

To simplify matters we shall consider first components of load which are symmetric about the $\theta=0$ axis and later those which are antisymmetric. Describing now only the nodal loads (with similar expansions holding for body forces, boundary conditions, initial strains, etc.) we specify forces per unit length of circumference as

$$R=\sum_{1}^{L}\bar{R}^{l}\cos l\theta$$

$$Z=\sum_{1}^{L}\bar{Z}^{l}\cos l\theta \tag{6.24}$$

$$T=\sum_{1}^{L}\bar{T}^{l}\sin l\theta$$

in the direction of the various coordinates for symmetric loads [Fig. 6.6(*a*)]. The apparently non-symmetric sine expansion is used for T, as to achieve symmetry the direction of T has to change for $\theta>\pi$.

The displacement components are described again in terms of the two-dimensional (r, z) shape functions appropriate to the element subdivision,

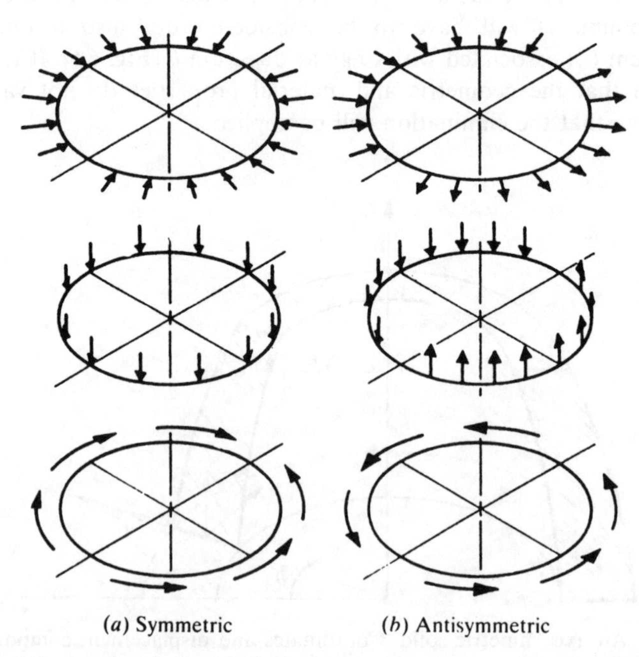

(*a*) Symmetric (*b*) Antisymmetric

Fig. 6.6 Load and displacement components in an axisymmetric body

and observing symmetry we write, as in Eq. (6.13),

$$\begin{aligned} u^l &= [N_1, N_2, \dots] \cos l\theta \mathbf{u}^{le} \\ v^l &= [N_1, N_2, \dots] \cos l\theta \mathbf{v}^{le} \\ w^l &= [N_1, N_2, \dots] \sin l\theta \mathbf{w}^{le} \end{aligned} \tag{6.25}$$

To proceed further it is necessary to specify the general, three-dimensional expression for strains in cylindrical coordinates. This is (see Love[18])

$$\boldsymbol{\varepsilon} = \begin{Bmatrix} \varepsilon_r \\ \varepsilon_z \\ \varepsilon_\theta \\ \gamma_{rz} \\ \gamma_{r\theta} \\ \gamma_{z\theta} \end{Bmatrix} = \begin{Bmatrix} \dfrac{\partial u}{\partial r} \\ \dfrac{\partial v}{\partial z} \\ \dfrac{u}{r} + \dfrac{1}{r}\dfrac{\partial w}{\partial \theta} \\ \dfrac{\partial u}{\partial z} + \dfrac{\partial v}{\partial r} \\ \dfrac{1}{r}\dfrac{\partial u}{\partial \theta} + \dfrac{\partial w}{\partial r} - \dfrac{w}{r} \\ \dfrac{1}{r}\dfrac{\partial v}{\partial \theta} + \dfrac{\partial w}{\partial z} \end{Bmatrix} \tag{6.26}$$

As before, uncoupling will occur between the modes and we can proceed to evaluation of the stiffness matrices, etc., in each harmonic. Typically, we have on substitution of Eq. (6.25) into Eq. (6.26), and grouping the variables as in Eq. (6.17):

$$\mathbf{B}_i = \begin{bmatrix} \dfrac{\partial N_i}{\partial r} \cos l\theta & 0 & 0 \\ 0 & \dfrac{\partial N_i}{\partial z} \cos l\theta & 0 \\ \dfrac{N_i}{r} \cos l\theta & 0 & \dfrac{lN_i}{r} \cos l\theta \\ \dfrac{\partial N_i}{\partial z} \cos l\theta & \dfrac{\partial N_i}{\partial r} \cos l\theta & 0 \\ -\dfrac{lN_i}{r} \sin l\theta & 0 & \left(\dfrac{\partial N_i}{\partial r} - \dfrac{N_i}{r}\right) \sin l\theta \\ 0 & -\dfrac{lN_i}{r} \sin l\theta & \dfrac{\partial N_i}{\partial z} \sin l\theta \end{bmatrix} \tag{6.27}$$

The remaining steps of the formulation follow precisely the previous derivations and could be repeated by the reader as an exercise.

For the antisymmetric loading of Fig. 6.6(*b*) we shall simply replace the sine by cosine and vice versa in Eqs (6.24) and (6.25).

The load terms in each harmonic will be obtained by virtual work as

$$\mathbf{f}_i^l = \int_0^{2\pi} \begin{Bmatrix} \bar{R}^l \cos^2 l\theta \\ \bar{Z}^l \cos^2 l\theta \\ \bar{T}^l \sin^2 l\theta \end{Bmatrix} \mathrm{d}\theta = \pi \begin{Bmatrix} \bar{R}^l \\ \bar{Z}^l \\ \bar{T}^l \end{Bmatrix} \quad \text{when } l = 1, 2, \ldots$$

$$= 2\pi \begin{Bmatrix} \bar{R}^l \\ \bar{Z}^l \\ 0 \end{Bmatrix} \quad \text{when } l = 0 \tag{6.28}$$

for the symmetric case. Similarly for the antisymmetric case

$$\mathbf{f}_i^l = \pi \begin{Bmatrix} \bar{R}^l \\ \bar{Z}^l \\ \bar{T}^l \end{Bmatrix} \quad \text{when } l = 1, 2, \ldots$$

$$= \begin{Bmatrix} 0 \\ 0 \\ \bar{T}^l \end{Bmatrix} \quad \text{when } l = 0 \tag{6.29}$$

We see from this and from the expansion of $\mathbf{K}^e$ that, as expected, for $l = 0$ the problem reduces to only two variables and the axisymmetric case is retrieved when symmetric terms only are involved.

Similarly, when $l = 0$ only one set of equations remains in the variable w for the antisymmetric case. This corresponds to constant tangential traction and solves simply the torsion problems of shafts subject to known torques (Fig. 6.7). This problem is classically treated by the use of a stress function[19] and indeed has been solved using a finite element formulation.[20] Here an alternative, more physical, approach is available.

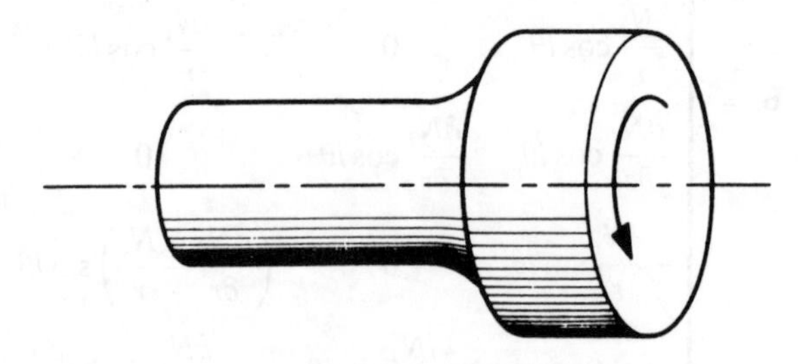

Fig. 6.7 Torsion of a variable section bar

The first application of the above concepts to the analysis of axisymmetric solids was made by Wilson.[21]

A simple example illustrating the effects of various harmonics is shown in Fig. 6.8(*a*) and (*b*).

6.6 Axisymmetric shells with non-symmetric loading

The extension of analysis of axisymmetric shells as described in Chapter 4 to the case of non-axisymmetric loads is simple and will again follow the standard pattern.

It is, however, necessary to extend the definition of strains and to include now all three displacements and force components (Fig. 6.9). Three membrane and three bending effects are now present and extending Eq. (6.1) involving straight generators we now define strains as[22,23]†

$$\boldsymbol{\varepsilon}=\begin{Bmatrix}\varepsilon_s\\ \varepsilon_\theta\\ \gamma_{s\theta}\\ \chi_s\\ \chi_\theta\\ \chi_{s\theta}\end{Bmatrix}=\begin{Bmatrix}\dfrac{\partial u}{\partial s}\\ \dfrac{1}{r}\dfrac{\partial v}{\partial\theta}+(w\cos\phi+u\sin\phi)\dfrac{1}{r}\\ \dfrac{1}{r}\dfrac{\partial u}{\partial\theta}+\dfrac{\partial v}{\partial s}-v\sin\phi\dfrac{1}{r}\\ -\dfrac{\partial^2 w}{\partial s^2}\\ -\dfrac{1}{r^2}\dfrac{\partial^2 w}{\partial\theta^2}+\dfrac{\partial v}{\partial\theta}\dfrac{\cos\phi}{r^2}-\dfrac{\sin\phi}{r}\dfrac{\partial w}{\partial s}\\ 2\left(-\dfrac{1}{r}\dfrac{\partial^2 w}{\partial s\,\partial\theta}+\dfrac{\sin\phi}{r^2}\dfrac{\partial w}{\partial\theta}+\dfrac{\cos\phi}{r}\dfrac{\partial v}{\partial s}-\dfrac{\sin\phi\cos\phi}{r^2}v\right)\end{Bmatrix} \tag{6.30}$$

The corresponding stress matrix is

$$\boldsymbol{\sigma}=\begin{Bmatrix}N_s\\ N_\theta\\ N_{s\theta}\\ M_s\\ M_\theta\\ M_{s\theta}\end{Bmatrix} \tag{6.31}$$

† Various alternatives are here present due to the multiplicity of shell theories. This one is fairly generally accepted.

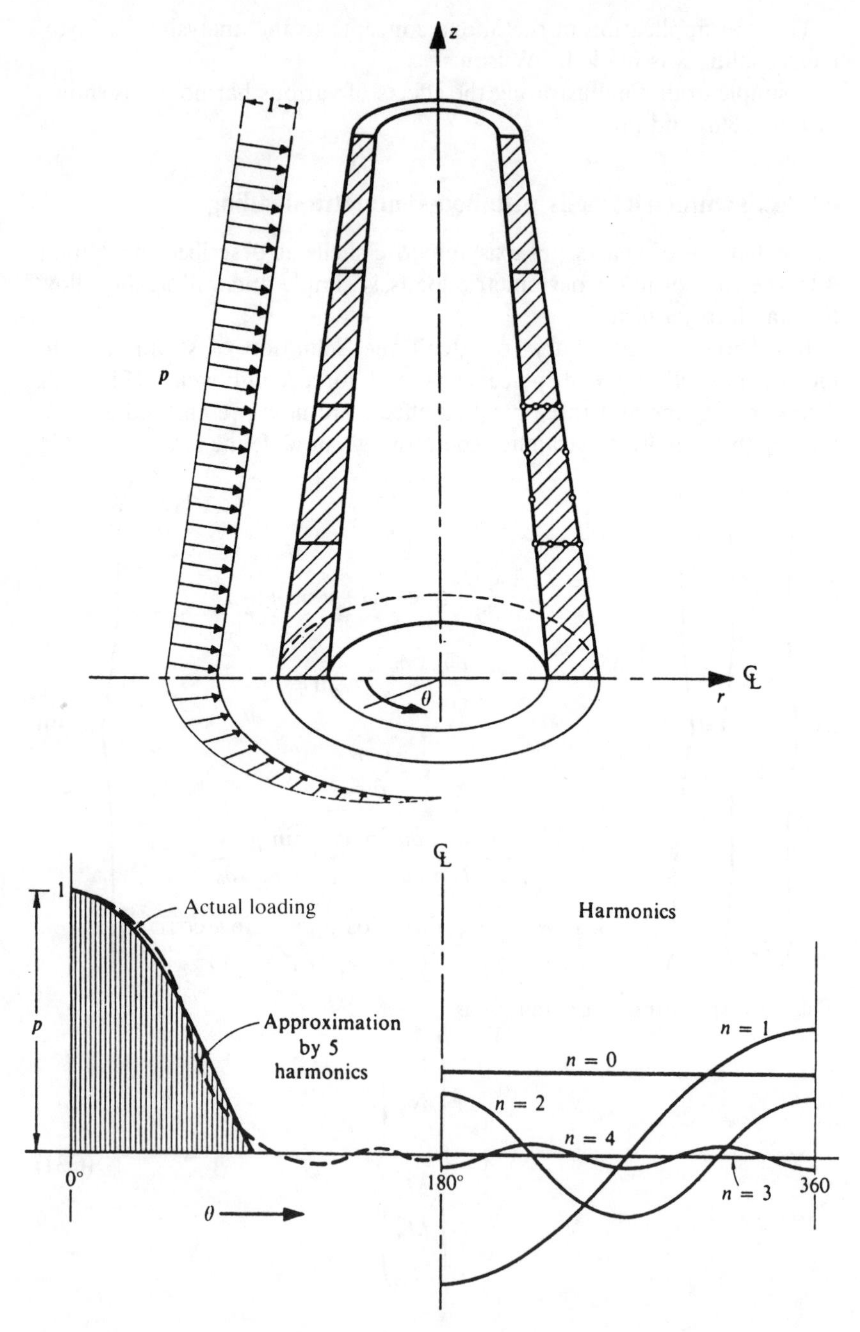

Fig. 6.8*a* An axisymmetric tower under non-symmetric load. Four cubic elements used in solution. Harmonics of load expansion used in analysis are shown

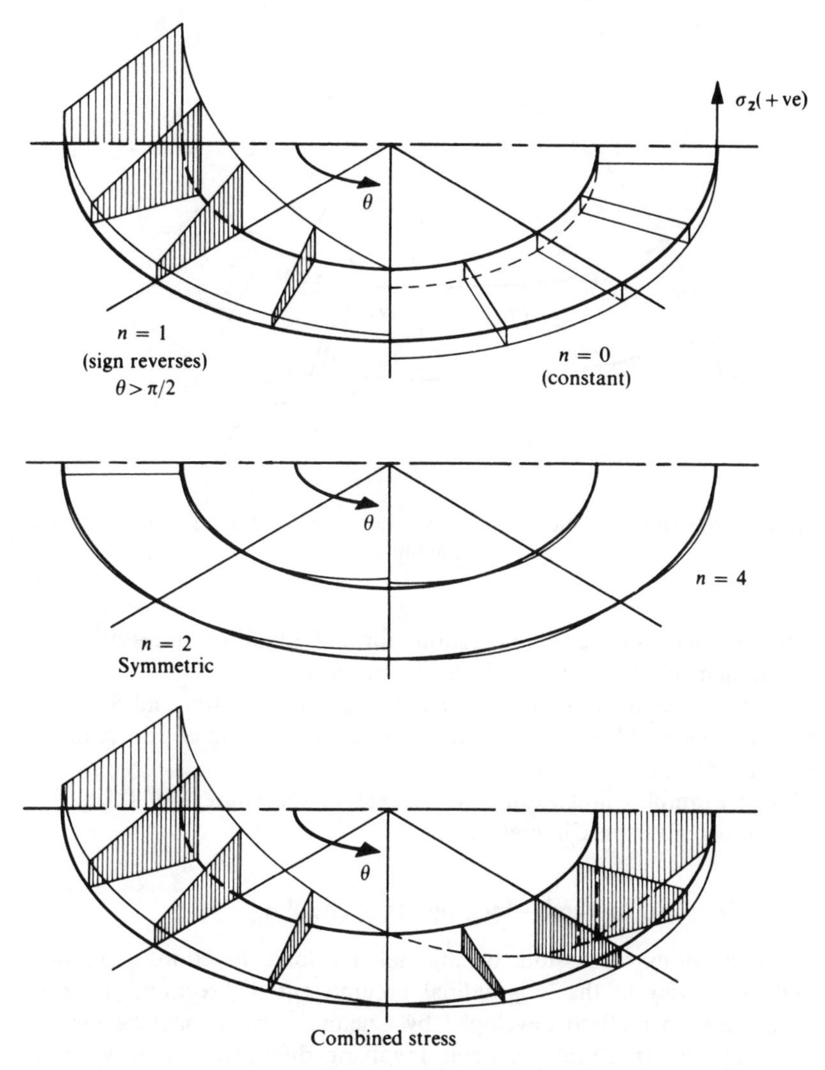

Combined stress

Fig. 6.8*b* Distribution of σ_z—the vertical stress on base due to various harmonics and their combination (third harmonic identically zero). First two harmonics give practically complete answer

with the three membrane and bending 'stresses' defined as in Fig. 6.9.

Once again symmetric and antisymmetric variation of loads and displacements can be assumed as in the previous section.

As the processes involved in executing this extension of the application are now obvious no further elaboration is needed here, but note again

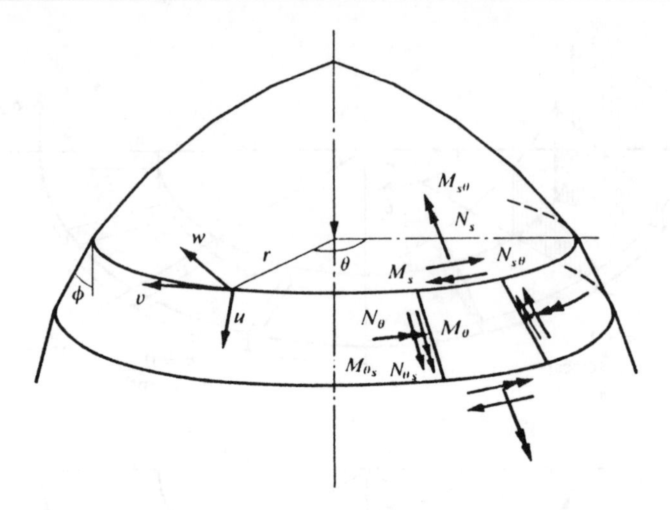

Fig. 6.9 Axisymmetric shell with non-symmetric load. Displacements and stress resultants

should be made of the more elaborate form of equations necessary when curved elements are involved [see Chapter 4, Eq. (4.16)].

The reader is referred to the original paper by Grafton and Strome[23] in which this problem is first treated and to the many later papers on the subject listed in Chapter 4.

Some examples illustrating the process in the context of thick shell analysis are given in Chapter 5.

6.7 Finite strip method—incomplete decoupling

In the previous discussion, orthogonal harmonic functions have been used exclusively in the longitudinal/circumferential direction. However, the finite strip method developed by Cheung[17] can in fact be used to solve various structural problems involving different boundary conditions and arbitrary geometrical shapes at the expense of introducing a limited amount of coupling.

As already stated, the finite strip method calls for the use of displacement functions of the multiplicative type (similar to the use of separation of variables in differential equations), in which simple, finite element, polynomials are used in one direction, and continuously differentiable smooth series or spline functions in the other. The first type, similar to that previously discussed, is called the semi-analytical finite strip, and the series must be chosen in such a way that they satisfy *a priori* the boundary conditions at the ends of the strip. The second type is called the spline finite strip, where usually cubic (B_3) spline functions are used and the boundary conditions are incorporated *a posteriori*. Here for a strip, in which a two-dimensional problem is to be reduced to a one-dimensional one, the displacement previously defined by Eq. (6.22) is assumed to be of the form

$$w^e = \sum_{n=1}^{r} \bar{\mathbf{N}}(x) Y_n(y) (\mathbf{a}'')^e \tag{6.32}$$

where $Y_n(y)$ are suitable continuous functions.

Semi-analytical finite strips with orthogonal series Y_n have been developed for plates, and shells with regular shapes. The method is definitely one of the best techniques for solving single-span plates and prismatic thin-walled structures under arbitrary loading due to the uncoupling of the terms of the series. The method is also highly efficient for dynamic and stability analysis and for static analysis of multispan structures under uniformly distributed loads because only a few coupled terms are required to yield a fairly accurate solution. Spline finite strips are better suited for plates with arbitrary shapes (parallelogram quadrilateral, S-shaped, etc.), for plates and shells with multispans and for concentrated loading and point support conditions.

Displacement functions are of two types, the polynomial part made up of the shape functions $\bar{N}(x)$ of standard type and the series or spline function part.

The most commonly used series are the basic functions[24] (or eigenfunctions) which are derived from the solution of the beam vibration differential equation for a single span

$$Y'''' = \frac{\mu^4 Y}{a^4} \tag{6.33}$$

where a is length of the beam (strip) and μ is a parameter.

The general form of such basic functions is

$$Y_n(y) = C_1 \sin\left(\frac{\mu_n y}{a}\right) + C_2 \cos\left(\frac{\mu_n y}{a}\right) + C_3 \sinh\left(\frac{\mu_n y}{a}\right) + C_4 \cosh\left(\frac{\mu_n y}{a}\right) \tag{6.34}$$

To a much more limited extent the buckling modes of a beam are used for stability analysis,[24] and the series takes up the form

$$Y_n(y) = C_1 \sin\left(\frac{\mu_n y}{a}\right) + C_2 \cos\left(\frac{\mu_n y}{a}\right) + C_3 y + C_4 \tag{6.35}$$

The constants C_i are determined by the end conditions.

Another form of series solution commonly used for shear walls[25] is of the form

$$Y_1(y)=\frac{y}{a}$$

$$Y_n(y)=\sin\left(\frac{\mu_n y}{a}\right)-\sinh\left(\frac{\mu_n y}{a}\right)-\left[\cos\left(\frac{\mu_n y}{a}\right)-\cosh\left(\frac{\mu_n y}{a}\right)\right] \tag{6.36}$$

$$\times\frac{\sin(\mu_n)+\sinh\mu_n}{\cos(\mu_n)+\cosh\mu_n} \qquad \text{for } n=2,\ldots,r$$

where

$$\mu_n=1.875,\ 4.694,\ldots,(2n-1)\pi/2$$

For multiple spans such as illustrated in Fig. 6.10 similar series can be used in each span with the constant appropriately adjusted to ensure continuity. However, spline functions are useful here.

Spline, which is originally the name of a small flexible wooden strip employed by draftsman as a tool for drawing a continuous smooth curve segment by segment, became a mathematical tool after the seminal work of Schoenberg.[26] A variety of spline functions are available. The spline function chosen here (Fig. 6.11) to represent the displacement is the B_3 spline of equal section length (B_3 splines of unequal section length have been discussed in a paper by Li *et al.*[27]) and is given as

$$Y=\sum_{i=-1}^{m+1}\alpha_i\psi_i \tag{6.37}$$

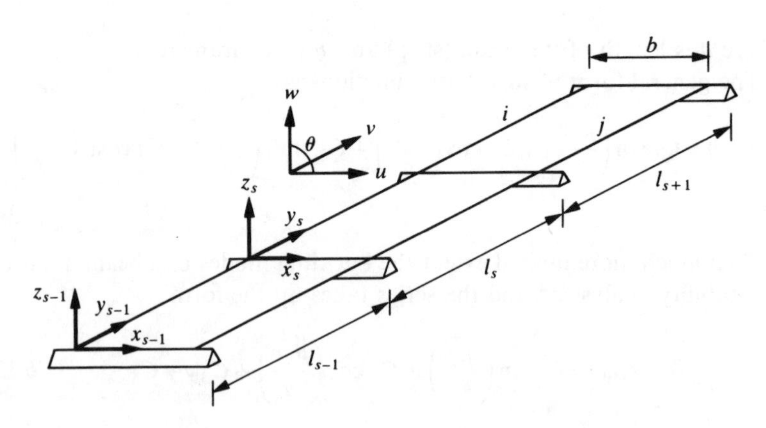

Fig. 6.10 A typical continuous finite strip

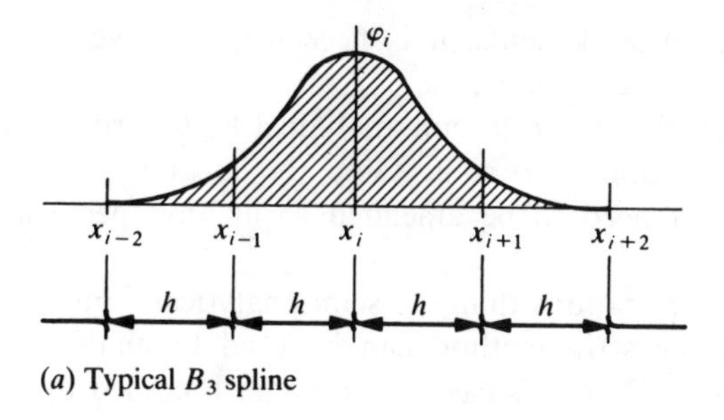

(*a*) Typical B_3 spline

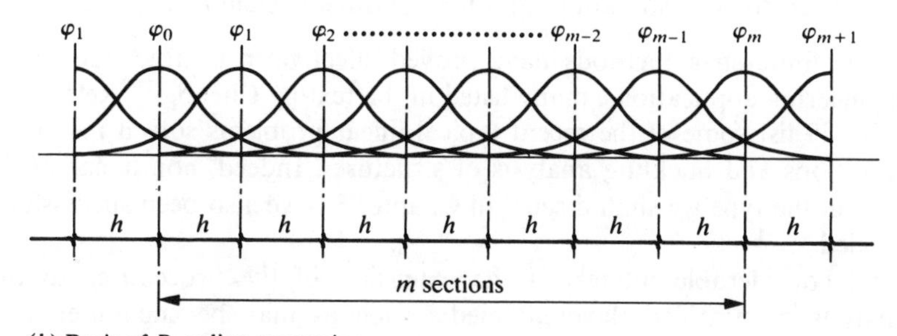

(*b*) Basis of B_3 spline expression

Fig. 6.11 Typical spline approximations

in which each local B_3 spline ψ_i has non-zero values over four consecutive sections with the section-knot $x = x_i$ as the centre and is defined by

$$\psi_i = \frac{1}{6H^3}\begin{cases} 0, & x < x_{i-2} \\ (x - x_{i-2})^3, & x_{i-2} \leqslant x \leqslant x_{i-1} \\ h^3 + 3h^2(x - x_{i-1}) + 3h(x - x_{i-1})^2 - 3(x - x_{i-1})^3, & x_{i-1} \leqslant x \leqslant x_i \\ h^3 + 3h^2(x_{i+1} - x) + 3h(x_{i+1} - x)^2 - 3(x_{i+1} - x)^3, & x_i \leqslant x \leqslant x_{i+1} \\ (x_{i+2} - x)^3, & x_{i+1} \leqslant x \leqslant x_{i+2} \\ 0, & x_{i+2} < x \end{cases} \tag{6.38}$$

The use of B_3 splines offers certain distinct advantages when compared with the conventional finite element method and the semi-analytical finite strip method:

1. It is computationally efficient. When using B_3 splines as displacement functions, continuity is ensured up to the second order (C_2 continuity). However, to achieve the same continuity conditions for the

conventional finite elements, it is necessary to have three times as many unknowns at the nodes.
2. It is more flexible than the semi-analytical finite strip method in the boundary condition treatment. Only the local splines around the boundary point need to be amended to fit any specified boundary conditions.
3. It has wider applications than the semi-analytical finite strip method. The spline finite strip method can be used to analyse plates with arbitrary shapes.[28] In this case any domain bounded by four curved (straight) sides can be mapped into a rectangular one (viz. Chapter 8 of Volume 1) and all operations for one system (x, y) can be transformed to corresponding ones for the other system (ξ, η).

The finite strip methods have proved effective in a large number of engineering applications, many listed in the text by Cheung.[17] References 29 to 39 list some of the recent typical linear problems solved in statics, vibrations and buckling analysis of structures. Indeed, non-linear problems of the type we shall discuss in Chapter 8 have also been successfully tackled.[40,41]

Of considerable interest is the extension of the procedures to the analysis of stratified (layered) media such as may be encountered in laminar structures or foundations.[42 44]

6.8 Concluding remarks

A fairly general process combining some of the advantages of finite element analysis with the economy of expansion in terms of generally orthogonal functions has been illustrated in several applications. Certainly these only touch on the possibilities offered, but it should be borne in mind that the economy is only achieved in certain geometrically constrained situations and those to which the number of terms requiring solution is limited.

Similarly other 'prismatic' situations can be dealt with in which only a segment of a body of revolution is developed (Fig. 6.12.) Clearly, the expansion must now be taken in terms of the angle $l\pi\theta/\alpha$ but otherwise the approach is identical to that described previously.[1]

In the methods of this chapter it was assumed that material properties remain invariant with one coordinate direction. This restriction can on occasion be lifted with the same general process maintained. An interesting example of this type is outlined by Stricklin and De Andrade.[45] Indeed, in non-linear problems this situation is encountered and can be successfully dealt with.[46]

In Chapter 9 of Volume 1 dealing with the general formulation of the finite element discretization we have referred to semi-discretization (see

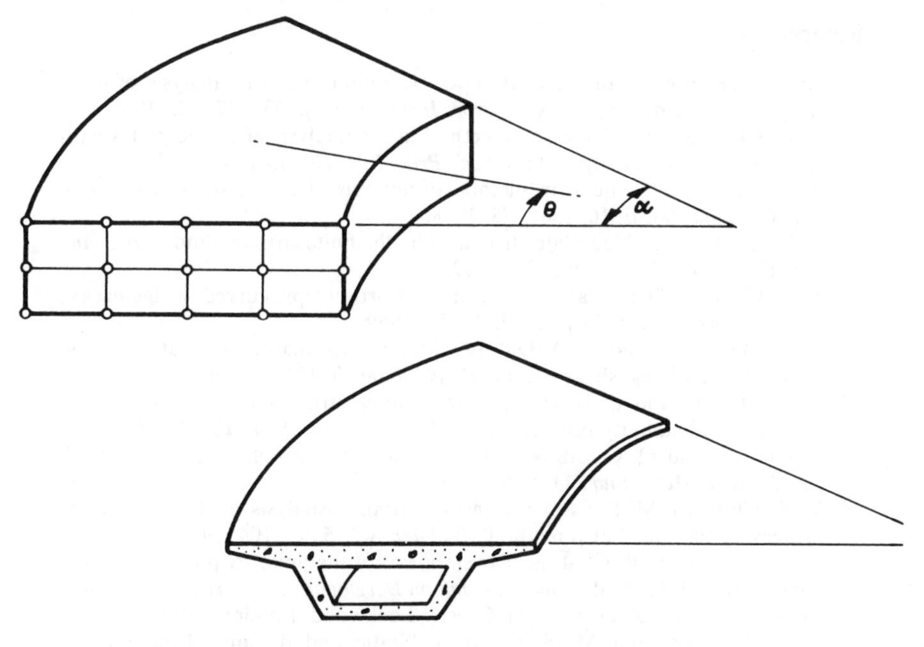

Fig. 6.12 Other segmental, prismatic situations

Sec. 9.7). In this, one of the problem variables (say z) is retained and the problem is reduced to an ordinary differential equation in terms of the nodal parameters $\mathbf{a}$ and its derivatives with respect to z.

We shall have occasion to make use of such partial discretization in Chapters 10 and 11 where the variable z is the time domain in which the problem is 'prismatic'. However, all the problems we have described in this chapter could be derived in terms of such semi-discretization. We would thus *firstly* semi-discretize, describing the problem in terms of a differential equation of the form

$$\mathbf{K}_1 \frac{\mathrm{d}^2\mathbf{a}}{\mathrm{d}z^2} + \mathbf{K}_2 \frac{\mathrm{d}\mathbf{a}}{\mathrm{d}z} + \mathbf{K}_3 \mathbf{a} + \mathbf{f} = 0$$

Secondly, the above equation system would be solved in the domain $0 < z < a$ using orthogonal functions that *naturally* enter the problem as solutions at ordinary differential equations *with constant coefficients*. This second solution is most easily found using a diagonalization process described in dynamic applications (Chapter 9).

Clearly the final result of such computations would turn out to be identical with the procedures here described—but on occasion the above formulation is more self-evident.

References

1. O. C. ZIENKIEWICZ and J. J. M. TOO, 'The finite prism in analysis of thick simply supported bridge boxes', *Proc. Inst. Civ. Eng.*, **53**, 147–72, 1972.
2. Y. K. CHEUNG, 'The finite strip method in the analysis of elastic plates with two opposite simply supported ends', *Proc. Inst. Civ. Eng.*, **40**, 1–7, 1968.
3. Y. K. CHEUNG, 'Finite strip method of analysis of elastic slabs', *Proc. Am. Soc. Civ. Eng.*, **94**, EM6, 1365–78, 1968.
4. Y. K. CHEUNG, 'Folded plate structures by the finite strip method', *Proc. Am. Soc. Civ. Eng.*, **95**, ST2, 963–79, 1969.
5. Y. K. CHEUNG, 'The analysis of cylindrical orthotropic curved bridge decks', *Publ. Int. Ass. Struct. Eng.*, **29**-II, 41–52, 1969.
6. A. S. MAWENYA and J. D. DAVIES, 'Finite strip analysis of plate bending including transverse shear effects', *Building Sci.*, **9**, 175–80, 1974.
7. P. R. BENSON and E. HINTON, 'A thick finite strip solution for static, free vibration and stability problems', *Int. J. Num. Meth. Eng.*, **10**, 665–78, 1976.
8. E. HINTON and O. C. ZIENKIEWICZ, 'A note on a simple thick finite strip', *Int. J. Num. Meth. Eng.*, **11**, 905–9, 1977.
9. Y. K. CHEUNG, M. S. CHEUNG and A. GHALI, 'Analysis of slab and girder bridges by the finite strip method', *Building Sci.*, **5**, 95–104, 1970.
10. Y. C. LOO and A. R. CUSENS, 'Development of the finite strip method in the analysis of cellular bridge decks', *Conf. on Developments in Bridge Design and Construction* (eds Rockey *et al.*), Crosby Lockwood, London, 1971.
11. Y. K. CHEUNG and M. S. CHEUNG, 'Static and dynamic behaviour of rectangular plates using higher order finite strips', *Building Sci.*, **7**, 151–8, 1972.
12. T. G. BROWN and A. GHALI, 'Semi-analytic solution of skew plates in bending', *Proc. Inst. Civ. Eng.*, **57**-II, 165–75, 1974.
13. G. S. TADROS and A. GHALI, 'Convergence of semi-analytical solution of plates', *Proc. Am. Soc. Civ. Eng.*, **99**, EM5, 1023–35, 1973.
14. A. R. CUSENS and Y. C. LOO, 'Application of the finite strip method in the analysis of concrete box bridges', *Proc. Inst. Civ. Eng.*, **57**-II, 251–73, 1974.
15. Y. K. CHEUNG, 'Folded plate structures by the finite strip method', *Proc. Am. Soc. Civ. Eng.*, ST12, **95**, 63–79, 1969.
16. H. C. CHAN and O. FOO, 'Buckling of multilayer plates by the finite strip method', *Int. J. Mech. Sci.*, **19**, 447–56, 1977.
17. Y. K. CHEUNG, *Finite Strip Method in Structural Analysis*, Pergamon Press, Oxford, 1976.
18. A. E. H. LOVE, *The Mathematical Theory of Elasticity*, 4th ed., p. 56, Cambridge University Press, 1927.
19. S. TIMOSHENKO and J. N. GOODIER, *Theory of Elasticity*, 2nd ed., McGraw-Hill, New York, 1951.
20. O. C. ZIENKIEWICZ and Y. K. CHEUNG, 'Stresses in shafts', *The Engineer*, 24 Nov. 1967.
21. E. L. WILSON, 'Structural analysis of axi-symmetric solids', *JAIAA*, **3**, 2269–74, 1965.
22. V. V. NOWOZHILOV, *Theory of Thin Shells* (translation), Noordhoff, Dordrecht, 1959.
23. P. E. GRAFTON and D. R. STROME, 'Analysis of axi-symmetric shells by the direct stiffness method', *JAIAA*, **1**, 2342–7, 1963.
24. O. FOO, 'Application of finite strip method in structural analysis with

particular reference to sandwich plate structure', Ph.D. thesis, The Queen's University of Belfast, 1977.
25. Y. K. CHEUNG, 'Computer analysis of tall buildings', in *Proc. of 3rd Int. Conf. on Tall Buildings*, Hong Kong and Guangzhou, pp. 8–15, December 1984.
26. I. J. SCHOENBERG, 'Contributions to the problem of approximation of equidistant data by analytic functions', *Q. Appl. Math.*, **4**, 45–99 and 112–14, 1946.
27. W. Y. LI, Y. K. CHEUNG and L. G. THAM, 'Spline finite strip analysis of general plates', *J. Eng. Mech.*, **112**, 43–54, 1986.
28. Y. K. CHEUNG, L. G. THAM and W. Y. LI, 'Free vibration and static analysis of general plates by spline finite strip', *Comput. Mech.*, **3**, 187–97, 1988.
29. D. BUCCO, J. MAZUMDAR and G. SVED, 'Application of the finite strip method combined with the deflection contour method to plate bending problems', *J. Comp. Struct.*, **10**, 827–30, 1979.
30. Y. K. CHEUNG, 'Orthotropic right bridges by the finite strip method', in *Concrete Bridge Design*, American Concrete Institute report SP-26, pp. 812–905, 1971.
31. Y. K. CHEUNG, L. G. THAM and W. Y. LI, 'Application of spline-finite-strip method in the analysis of curved slab bridge', *Proc. Inst. Civ. Engrs*, **81**-II, 111–24, 1986.
32. H. C. CHAN and Y. K. CHEUNG, ' Static and dynamic analysis of multilayered sandwich plates', *Int. J. Mech. Sci.*, **14**, 399–406, 1972.
33. Y. K. CHEUNG and S. SWADDIWUDHIPONG, 'Analysis of frame shear wall structures using finite strip elements', *Proc. Inst. Civ. Engrs*, **65**-II, 517–35, 1978.
34. C. MEYER and A. C. SCORDELIS, 'Analysis of curved folded plate structures', *Proc. Am. Soc. Civ. Eng.*, **97**, ST10, 2459–80, 1979.
35. W. Y. LI, L. G. THAM and Y. K. CHEUNG, 'Curved box-girder bridges', *Proc. Am. Soc. Civ. Eng.*, **114**, ST6, 1324–38, 1988.
36. D. J. DAWE, 'Finite strip models for vibration of Mindlin plates', *J. Sound Vibr.*, **59**, 441–52, 1978.
37. Y. K. CHEUNG, W. Y. LI and L. G. THAM, 'Free vibration analysis of singly curved shell by spline finite strip method', *J. Sound Vibr.*, **128**, 411–22, 1989.
38. D. J. DAWE, 'Finite strip buckling of curved plate assemblies under biaxial loading', *Int. J. Solids Struct.*, **13**, 1141–55, 1977.
39. Y. K. CHEUNG and C. DELCOURT, 'Buckling and vibration of thin, flat-walled structures continuous over several spans', *Proc. Inst. Civ. Engrs*, **64**-II, 93–103, 1977.
40. Y. K. CHEUNG and DASHAN ZHU, 'Large deflection analysis of arbitrary shaped thin plates', *Comp. Struct.*, **26**, 811–14, 1987.
41. D. S. ZHU and Y. K. CHEUNG, 'Postbuckling analysis of shells by spline finite strip method', *Comp. Struct.*, **31**, 357–64, 1989.
42. D. J. GUO, L. G. THAM and Y. K. CHEUNG, 'Infinite layer for the analysis of a single pile', *J. Comp. Geotechnics*, **3**, 229–49, 1987.
43. Y. K. CHEUNG, L. G. THAM and D. J. GUO, 'Analysis of pile group by infinite layer method', *Geotechnique*, **38**, 415–31, 1988.
44. S. B. DONG and R. B. NELSON, 'On natural vibrations and waves in laminated orthotropic plates', *J. Appl. Mech.*, **30**, 739, 1972.
45. J. A. STRICKLIN and J. C. DE ANDRADE, 'Linear and non linear analysis of shells of revolution with asymmetrical stiffness properties', in *Proc. 2nd Conf.*

on Matrix Methods in Structural Mechanics, Air Force Institute of Technology, Wright-Patterson AF Base, Ohio, 1968.

46. L. A. WINNICKI and O. C. ZIENKIEWICZ, 'Plastic or visco-plastic behaviour of axisymmetric bodies subject to non-symmetric loading; semi-analytical finite element solution, *Int. J. Num. Meth. Eng.*, **14**, 1399–412, 1979.

7

Non-linear problems—plasticity, creep (viscoplasticity), non-linear field problems, etc.

7.1 Introduction

In all the problems discussed so far the governing differential equations were linear and self-adjoint, leading to the standard quadratic form of the functional. In elastic solid mechanics this was implied in

(*a*) a linear form of strain–displacement relationships [see Eq. (2.2), Chapter 2 of Volume 1] and

(*b*) a linear form of stress–strain relationships [see Eq. (2.5), Chapter 2 of Volume 1].

In various field problems similar linearity was imposed by such 'constants' as the permeability k remaining independent of the variation of the unknown potential ϕ [see Eq. (10.6), Chapter 10 of Volume 1].

Many problems of practical consequence exist in which linearity is not preserved and it is of interest to extend the numerical processes described to cover these. In this context we have a whole range of *solid mechanics* situations in which such phenomena as plasticity, creep or other *complex constitutive relations* supersede the simple linear elasticity assumptions.

Similarly, in flow-type situations the dependence of viscosity on velocity distribution, the inapplicability of Darcy's seepage laws in a porous medium due to onset of turbulence or dependence of magnetic permeability on flux densities give non-linearity with respect to material properties.

These classes of problems can often be simply dealt with without reformulation of the discretization process (i.e. without recourse to rewriting of the basic variational statements). Indeed, if a solution to the 'linear' problem can be arrived at by some 'trial and error' process in which, at the final stage, the material constants are so adjusted that the appropriate new constitutive law is satisfied, then a solution is achieved.

However, if the strain–displacement relationship is non-linear, then a more fundamental reorganization of the formulation is necessary. For this reason alone such problems have been removed from this chapter and will be dealt with separately in Chapter 8. It will nevertheless be found that the basic iteration processes involved in the solution remain unchanged and indeed combination of both types of non-linearities may easily be dealt with.

One important point needs, however, to be mentioned. While in linear problems the solution was always unique this no longer is the case in many non-linear situations. Thus, if *a solution* is achieved it may not necessarily be *the solution* sought. Physical insight into the nature of the problem and, usually, small-step incremental approaches are essential to obtain physically significant answers. Such increments are indeed always required if the constitutive law relating, say, stress and strain changes is path dependent.

The general problem is therefore always formulated (in terms of the discretization parameter $\mathbf{a}$) as the solution of

$$\mathbf{\Psi}_{n+1} \equiv \mathbf{\Psi}(\mathbf{a}_{n+1}) = \mathbf{P}(\mathbf{a}_{n+1}) - \mathbf{f} = 0 \tag{7.1}$$

which starts from a (near) equilibrium solution at

$$\mathbf{a} = \mathbf{a}_n \qquad \mathbf{\Psi}_n = 0 \qquad \mathbf{f} = \mathbf{f}_n \tag{7.2}$$

and arises generally due to the changes in the forcing functions $\mathbf{f}$ from $\mathbf{f}_n$ to

$$\mathbf{f}_{n+1} = \mathbf{f}_n + \Delta\mathbf{f}_n \tag{7.3}$$

The determination of the change $\Delta\mathbf{a}_n$ such that

$$\mathbf{a}_{n+1} = \mathbf{a}_n + \Delta\mathbf{a}_n \tag{7.4}$$

will be the objective and generally the increments of $\Delta\mathbf{f}_n$ will be kept reasonably small so that path dependence can be followed. Further, such incremental procedures will be useful in avoiding excessive numbers of iterations and indeed in following the physically correct path. In Fig. 7.1 we show a typical non-uniqueness which may occur if the function $\mathbf{\Psi}$ decreases and subsequently increases as the parameter $\mathbf{a}$ uniformly increases. It is clear that to follow the path $\Delta\mathbf{f}_n$ with a negative sign will be required at some stages of the computation.

Only in the case of mild non-linearity (and no path dependence) is it

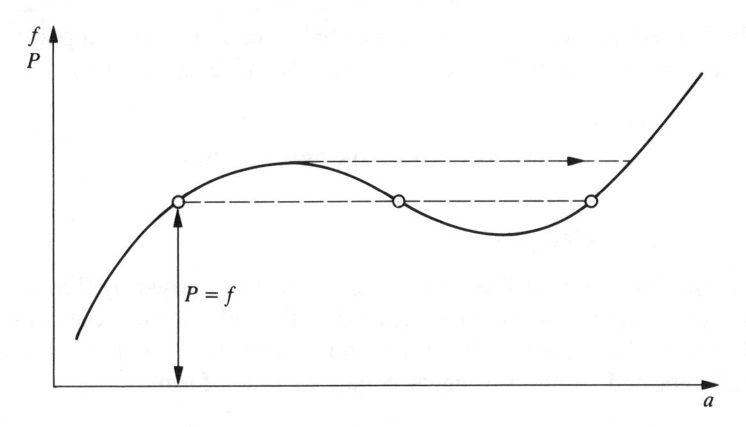

Fig. 7.1 Possibility of multiple solutions

possible to obtain solutions in a single increment of **f**, i.e. with

$$\Delta \mathbf{f}_n = \mathbf{f}_{n+1} \tag{7.5}$$

The general equations arise naturally in materials exhibiting inelastic constitutive laws in which the stress $\boldsymbol{\sigma}$ depends in some complex fashion on the state of strain and its history. Following the standard discretization with displacement as the unknown, i.e.

$$\mathbf{u} \approx \hat{\mathbf{u}} = \mathbf{N}\mathbf{a} \tag{7.6}$$

the term **P** of Eq. (7.1) is simply the vector of *internal forces*

$$\mathbf{P} = \int_{\Omega} \mathbf{B}^{\mathrm{T}} \boldsymbol{\sigma} \, \mathrm{d}\Omega \tag{7.7}$$

while **f** gives the external forces in the manner described in Chapter 2 of Volume 1.

Now the parameters **a** and their history together with the constitutive relations prescribed will define $\boldsymbol{\sigma}$ leading to the non-linear problem. The situation is completely analogous in many field problems, as we shall illustrate later.

The literature on general solution approaches and on particular applications is expanding very rapidly so that in the orbit of a single chapter it is not possible to encompass fully all the variants. However, we shall attempt to give a comprehensive picture by outlining first the *general solution procedures* and then focusing on *rate-independent material non-linearity in solid mechanics* (plasticity) and *rate-dependent material non-linearity of solid mechanics* (creep). Later *some non-linear field problems* and other special examples will be discussed.

In further chapters of this book we shall encounter other applications where again the general solution procedures will be applicable.

GENERAL SOLUTION PROCEDURES

7.2 Iterative techniques

7.2.1 *General remarks.* The solution of the problem posed by Eqs (7.1) to (7.4) cannot, of course, be approached directly and some form of iteration will always be required. We shall concentrate here on procedures in which repeated solution of linear equations of the form

$$\mathbf{K}\boldsymbol{\delta}=\mathbf{r} \tag{7.8}$$

is used. Here gaussian elimination techniques of the type discussed in Volume 1 will be applicable. The application of direct iteration processes may of course prove more economical and in later chapters we shall frequently refer to such possibilities which as yet have not been fully explored.

Many of the iterative techniques currently used originated by intuitive application of physical reasoning. However, each of such techniques has a good pedigree in numerical analysis and in what follows we shall use the nomenclature generally accepted in texts on this subject.[1-5]

We shall first illustrate the procedures used in the context of a single, scalar, equation. This, though useful from the pedagogical viewpoint, is dangerous as convergence of problems with numerous degrees of freedom may depart from the simple pattern.

7.2.2 *The Newton–Raphson method.* The Newton–Raphson method is probably the most rapidly convergent process for solution of non-linear problems (providing of course that the initial solution is within the 'zone of attraction' and divergence does not occur). Indeed, it is the only process in which convergence is quadratic. The method is sometimes simply called the Newton process but it appears to have been simultaneously derived by Raphson and the interesting history of its origins is given in reference 6.

In this iteration we note that, to the first order, Eq. (7.1) can be approximated as

$$\boldsymbol{\Psi}(\mathbf{a}_{n+1}^{i+1})\approx\boldsymbol{\Psi}(\mathbf{a}_{n+1}^{i})+\left(\frac{\partial\boldsymbol{\Psi}}{\partial\mathbf{a}}\right)_{n+1}^{i}\delta\mathbf{a}_{n}^{i}=0 \tag{7.9}$$

Here i is the iteration counter starting from

$$\mathbf{a}_{n+1}^{1}=\mathbf{a}_{n} \tag{7.10}$$

and

$$\frac{\partial \boldsymbol{\Psi}}{\partial \mathbf{a}} = \frac{\partial \mathbf{P}}{\partial \mathbf{a}} = \mathbf{K}_T \tag{7.11}$$

is the jacobean matrix (or in structural terms the stiffness matrix) corresponding to the tangent direction. Equation (7.9) gives immediately the iterative correction as

$$\mathbf{K}_T^i \, \delta\mathbf{a}_n^i = -\boldsymbol{\Psi}_{n+1}^i \tag{7.12}$$

or

$$\delta\mathbf{a}_n^i = -(\mathbf{K}_T^i)^{-1} \boldsymbol{\Psi}_{n+1}^i \tag{7.13}$$

A series of successive approximations gives

$$\mathbf{a}_{n+1}^{i+1} = \mathbf{a}_n + \Delta\mathbf{a}_n^i = \mathbf{a}_{n+1}^i + \delta\mathbf{a}_n^i \tag{7.14}$$

with

$$\Delta\mathbf{a}_n^i = \sum_{k=1}^{i} \delta\mathbf{a}_n^k \tag{7.15}$$

The process is illustrated in Fig. 7.2 and shows the very rapid convergence that can be achieved.

The need for the introduction of the total increment $\Delta\mathbf{a}_n^i$ and its storage is perhaps not obvious here but in fact it is essential if the constitutive non-linearity is path dependent, as we shall see later.

The Newton–Raphson process, despite its rapid convergence, can be expensive and inconvenient. The main reasons for this are:

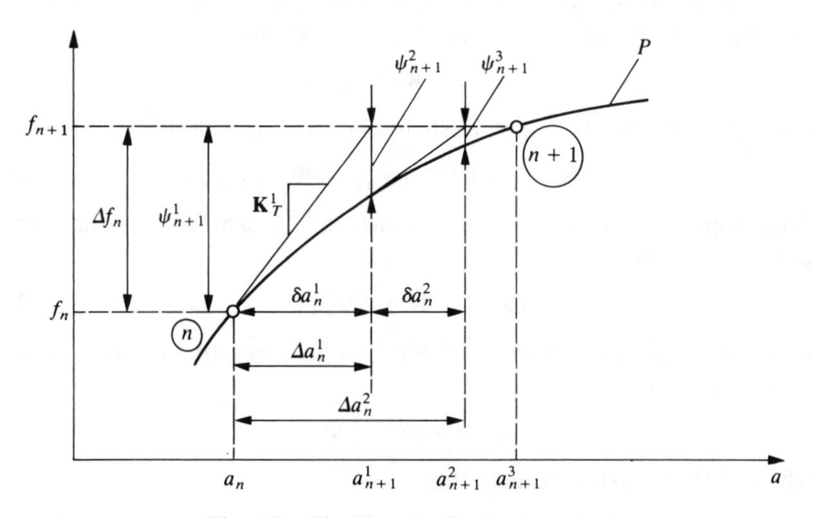

Fig. 7.2 The Newton–Raphson method

1. A new $\mathbf{K}_T$ matrix has to be formed and refactorized (solved) for each integration step.
2. On some occasions, as with non-associative elastoplasticity, for instance, the $\mathbf{K}_T$ matrix though initially symmetric becomes non-symmetric and consequently non-symmetric solvers are required.

Some of these drawbacks may be avoided by alternative procedures.

7.2.3 *Modified Newton–Raphson methods.* This method uses essentially the same algorithm as the Newton–Raphson process but replaces the variable jacobian stiffness $\mathbf{K}_T^i$ by a constant approximation:

$$\mathbf{K}_T^i \approx \bar{\mathbf{K}}_T \tag{7.16}$$

giving, in place of Eq. (7.13),

$$\delta \mathbf{a}_n^i = -\bar{\mathbf{K}}_T^{-1} \boldsymbol{\Psi}_{n+1}^i \tag{7.17}$$

Many possible choices exist here. For instance, $\bar{\mathbf{K}}_T$ can be chosen as the matrix corresponding to the first iteration $\bar{\mathbf{K}}_T^1$ [as shown in Fig. 7.3(*a*)] or may even be one corresponding to some previous (initial) step of load incrementation $\mathbf{K}^0$ [as shown in Fig. 7.3(*b*)].

Obviously the procedure will converge generally at a slower rate but difficulties mentioned above for the Newton–Raphson process disappear. Indeed, frequently the 'zone of attraction' for this process is increased and previously divergent approaches can be made to converge, albeit slowly. Many variants of this process are used in practice and symmetric solvers can generally be used providing a symmetric form of $\bar{\mathbf{K}}_T$ is chosen.

7.2.4 *Incremental-secant of quasi-Newton methods.* Once the first iterate in the preceding section has been established giving

$$\delta \mathbf{a}_n^1 = -\mathbf{K}_T^{-1} \boldsymbol{\Psi}_{n+1}^1 \tag{7.18}$$

a secant 'slope' can be found, as shown in Fig. 7.4, such that

$$\delta \mathbf{a}_n^1 = -(\mathbf{K}_s^2)^{-1} (\boldsymbol{\Psi}_{n+1}^1 - \boldsymbol{\Psi}_{n+1}^2) \tag{7.19}$$

This 'slope' can now be used to establish $\delta \mathbf{a}_n^2$ by expression of the form of Eq. (7.13), giving

$$\delta \mathbf{a}_n^2 = -(\mathbf{K}_s^2)^{-1} \boldsymbol{\Psi}_{n+1}^2 \tag{7.20}$$

Quite generally one could write for $i > 1$ in place of Eq. (7.20), now dropping subscripts,

$$\delta \mathbf{a}^i = -(\mathbf{K}_s^i)^{-1} \boldsymbol{\Psi}^i \tag{7.21}$$

where $(\mathbf{K}_s^i)_i^{-1}$ is determined so that

$$\delta \mathbf{a}^{i-1} = -(\mathbf{K}_s^i)^{-1} (\boldsymbol{\Psi}^{i-1} - \boldsymbol{\Psi}^i) = -(\mathbf{K}_s^i)^{-1} \boldsymbol{\gamma}^{i-1} \tag{7.22}$$

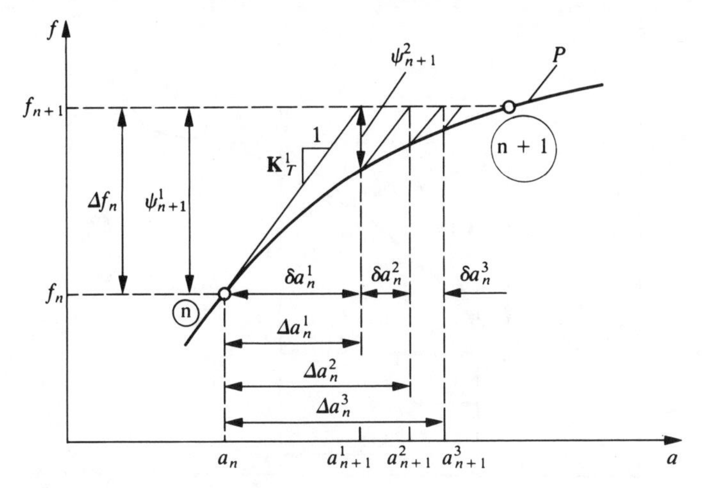

(a) With initial tangent in increment

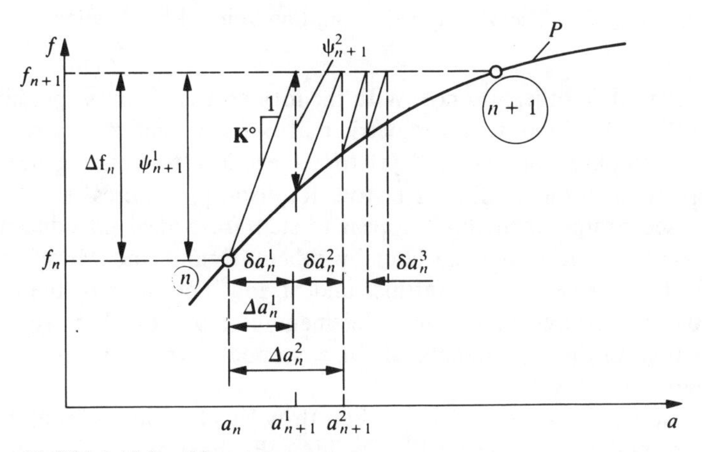

(b) With initial problem tangent

Fig. 7.3 The modified Newton–Raphson method

For the scalar system illustrated in Fig. 7.4 the determination of $\mathbf{K}_s^i$ is trivial and, as shown, the convergence is almost as rapid as with the Newton–Raphson process.

For systems with more than 1 degree of freedom the determination of $\mathbf{K}_s^i$ or its inverse is more difficult and indeed is not unique. Many different forms of the matrix $\mathbf{K}_s^i$ can satisfy relation (7.22) and as expected many alternatives can be used in practice. All of these use some form of updating of a previously determined matrix or of its inverse in a manner that satisfies identically Eq. (7.22). Some such updates preserve the matrix

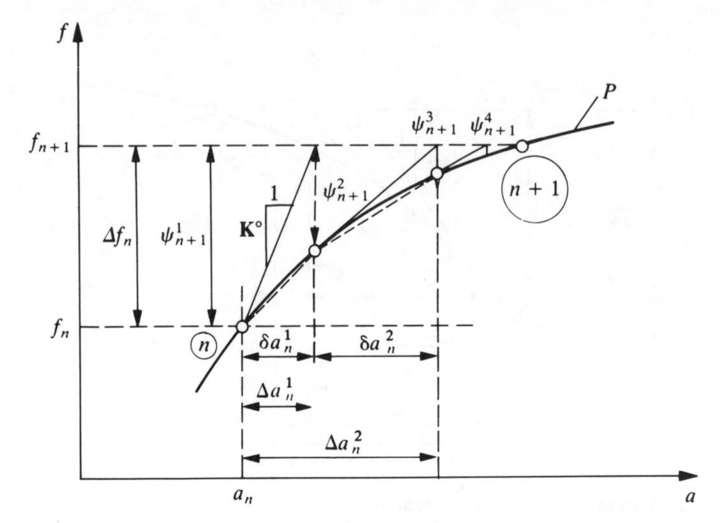

Fig. 7.4 The Secant method starting from a $\mathbf{K}^0$ prediction

symmetry while others do not. With the former it is of course possible to avoid the difficulty of non-symmetric matrix forms that may arise in the Newton–Raphson process and yet to achieve a faster convergence than is possible with the modified Newton–Raphson procedures.

The secant update methods appear to stem from ideas introduced first by Davidon[7] and developed later by others. Dennis and More[8] survey the field extensively while Matthies and Strang[9] appear to be the first to use the procedures in the finite element context. Further work and assessment of the performance of various update procedures is available in references 10 to 17.

The so-called BFGS[8] (named after Broyden, Fletcher, Goldfarb and Shanno) update and the DFP[8] (Davidon, Fletcher and Powell) update preserve matrix symmetry and positive definiteness and both are widely used. We shall quote below the former which can be written as

$$(\mathbf{K}^i)^{-1} = (\mathbf{I} + \mathbf{w}_i\mathbf{v}_i^{\mathrm{T}})(\mathbf{K}^{i-1})^{-1}(\mathbf{I} + \mathbf{v}_i\mathbf{w}_i^{\mathrm{T}}) \tag{7.23}$$

where $\mathbf{I}$ is an identity matrix and

$$\mathbf{v}_i = \boldsymbol{\Psi}^{i-1}\left[1 - \frac{(\delta\mathbf{a}^{i-1})^{\mathrm{T}}\boldsymbol{\gamma}^{i-1}}{(\delta\mathbf{a}^i)^{\mathrm{T}}\boldsymbol{\Psi}^{i-1}}\right] - \boldsymbol{\Psi}^i$$
$$\mathbf{w}_i = \frac{\delta\mathbf{a}^{i-1}}{\delta\mathbf{a}^{(i-1)\mathrm{T}}\boldsymbol{\gamma}^{i-1}} \tag{7.24}$$

Here γ is as defined by Eq. (7.22). Some algebra will readily verify that substitution of (7.23) and (7.24) into Eq. (7.22) results in an identity.

Further, the form of Eq. (7.23) guarantees the preservation of the symmetry of the original matrix.

The nature of the update does not guarantee the preservation of the matrix sparsity. For this reason it is convenient at every iteration to return to the original (sparse) matrix $\mathbf{K}_s^1$, used in the first iteration, and to reapply the multiplication of Eq. (7.23) through all previous iterations. This necessitates the storage of the vectors $\mathbf{v}_i$ and $\mathbf{w}_i$ for all previous iterations and their successive multiplications. Details of this operation are well described in references 9 and 17.

When the number of iterations is large ($i > 15$) the efficiency of the update decreases due to incipient instability. Various procedures are now open, usually involving the omission of some previous updates. An obvious possibility is to disregard *all* the previous updates and always return to the original matrix $\mathbf{K}_s^1$.

Such a procedure was first suggested by Crisfield[18,19] in the finite element context and is illustrated in Fig. 7.5. It is seen to be convergent at a slightly slower rate but avoids totally the stability difficulties previously encountered and reduces the storage and number of operations needed.

The procedure of Fig. 7.5 is identical to that generally known as direct (or Picard) iteration and is particularly useful in the solution of non-linear problems which can be written as

$$\boldsymbol{\Psi}(\mathbf{a}) \equiv \mathbf{K}(\mathbf{a})\mathbf{a} - \mathbf{f} = 0 \tag{7.25}$$

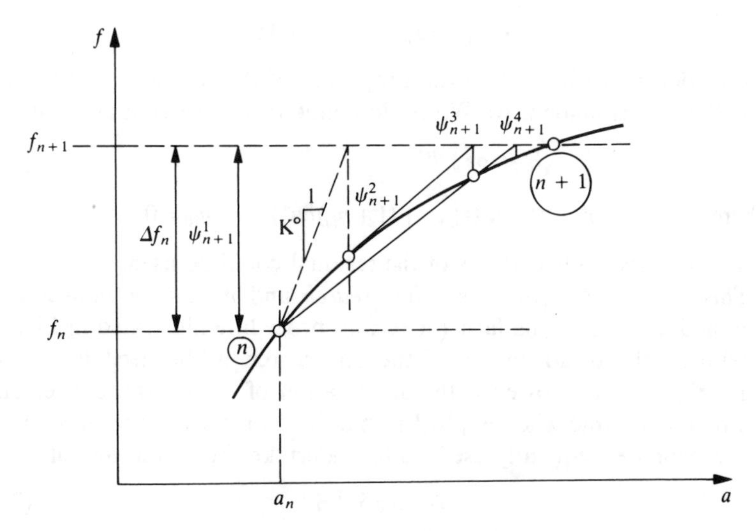

Fig. 7.5 Direct (or Picard) iteration

In such a case $\mathbf{a}_n=0$ is taken and the iteration proceeds without increments, writing

$$\mathbf{a}_{n+1}^{i+1}=[\mathbf{K}(\mathbf{a})_{n+1}^{i}]^{-1}\mathbf{f}_{n+1} \tag{7.26}$$

7.3 Acceleration of convergence and line search procedures

All the iterative methods of the preceding section have an identical structure described by Eqs (7.12) to (7.15) in which various approximations to the Newton matrix $\mathbf{K}_T^i$ are used. For all of these an iterative vector $\delta\mathbf{a}_n^i$ is determined and the new value of the unknowns found as

$$\mathbf{a}_{n+1}^{i+1}=\mathbf{a}_{n+1}^{i}+\delta\mathbf{a}_n^i \tag{7.27}$$

starting from

$$\mathbf{a}_{n+1}^{1}=\mathbf{a}_n$$

The objective, of course, is to achive the reduction of $\mathbf{\Psi}_{n+1}^{i+1}$ to zero, although this is not easily achieved by any of the procedures described even in the scalar example illustrated. To get here a solution approximately satisfying such a non-linear problem would have been in fact easier by simply evaluating the scalar Ψ_{n+1} for various values of a_{n+1} and by suitable interpolation arriving at the required answer. For multi degrees of freedom systems such an approach is obviously not possible unless some scalar norm of the residual is considered. One possible approach is to write

$$\mathbf{a}_{n+1}^{i+1,j}=\mathbf{a}_{n+1}^{i}+(1+\eta_{ij})\delta\mathbf{a}_n^i \tag{7.28}$$

and to determine η_{ij} so that the projection of the residual on the *search direction* $\delta\mathbf{a}_n^i$ is made zero. We could define this projection as below:

$$G_{i,j}\equiv\delta\mathbf{a}_n^{i\,\mathrm{T}}\mathbf{\Psi}^{i+1,j} \tag{7.29}$$

where $$\mathbf{\Psi}^{i+1,j}\equiv\mathbf{\Psi}(\mathbf{a}_{n+1}^{i}+(1+\eta_{ij})\delta\mathbf{a}_n^i),\qquad \eta_{i0}=0$$

Here, of course, other norms of the residual could be used.

This process is known as a *line search* and η_{ij} can be conveniently obtained using a 'regula falsi' (or secant) procedure illustrated in Fig. 7.6. Obviously the disadvantage of the line search is the need for several evaluations of $\mathbf{\Psi}$. However, the acceleration of the overall convergence can be remarkable when applied to modified or quasi Newton methods. A compromise frequently used[9] is to undertake the search only if

$$G_{i0}>\varepsilon\,\delta\mathbf{a}_n^{i\,\mathrm{T}}\mathbf{\Psi}^{i} \tag{7.30}$$

where the tolerance ε is set close to 0.5. This means that if the iteration

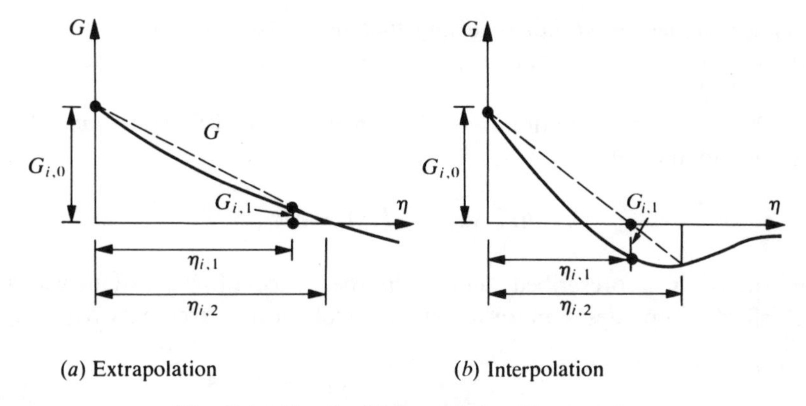

(*a*) Extrapolation (*b*) Interpolation

Fig. 7.6 'Regula falsi' applied to line search

process has directly resulted in reduction of the residual projection to one half or less of its original value then a line search is not used.

Other acceleration procedures have from time to time been suggested, e.g. the use of the Aitken extrapolation[20, 21] or the so-called α method.[22] None are, however, as efficient as the line search process described.

7.4 'Softening' behaviour and displacement control

In all of the preceding we have assumed that the iteration is associated with positive increments of the forcing vector, **f** in Eq. (7.1). In structural problems this is usually a set of loads that can be assumed to be proportional to each other so that one can write

$$\Delta \mathbf{f}_n = \Delta \lambda_n \mathbf{f}_0 \tag{7.31}$$

In many problems the situation will arise that no solution exists above a certain maximum value of **f** and that the real solution is a 'softening' branch, as shown in Fig. 7.1. In such cases $\Delta\lambda_n$ will need to be negative unless the problem can be recast as one in which the forcing can be applied by displacement control. In a simple case of a single load or a single displacement forcing it is easy to recast the general formulation to increments of a single prescribed displacement and much effort has gone into such solutions.[23–29]

In all the successful approaches the original problem of Eq. (7.1) is rewritten as the solution of

$$\boldsymbol{\Psi}_{n+1} \equiv \mathbf{P}(\mathbf{a}_{n+1}) - \lambda_{n+1}\mathbf{f}_0 = 0 \tag{7.32}$$

with

$$\begin{aligned} \mathbf{a}_{n+1} &= \mathbf{a}_n + \Delta \mathbf{a}_n \\ \lambda_{n+1} &= \lambda_n + \Delta \lambda_n \end{aligned} \tag{7.33}$$

and

being included as variables in any increment. Now of course an additional equation (constraint) needs to be provided to solve for the extra variable $\Delta\lambda_n$.

This additional equation can take various forms. Riks[25] assumes that in each increment

$$\Delta \mathbf{a}_n^{\mathrm{T}} \Delta \mathbf{a}_n + \Delta\lambda_n^2 \mathbf{f}_0^{\mathrm{T}} \mathbf{f}_0 = \Delta l^2 \tag{7.34}$$

where Δl is a prescribed 'length' in the space of $N+1$ dimensions. Crisfield[10] provides a more natural control on displacements requiring that

$$\Delta \mathbf{a}_n^{\mathrm{T}} \Delta \mathbf{a}_n = \Delta l^2 \tag{7.35}$$

This so-called arc length and spherical path controls are but some of the possible constraints.

Direct addition of the constraint equation [(7.34) or (7.35)] to the system of equations (7.32) is now possible and the previously described iterative methods could again be used. However, the 'tangent' equation system would lose its symmetry and banded structure so an alternative procedure is generally used.

We note that for a given iteration i we can write quite generally the solution as

$$\begin{aligned}
\delta \mathbf{a}_n^i &= -\mathbf{K}_i^{-1} \mathbf{\Psi}_{n+1}(\Delta\lambda_n^i + \delta\lambda_n^i) \\
&= -\mathbf{K}_i^{-1}(\mathbf{\Psi}_{n+1}^i(\Delta\lambda_n^i) - \delta\lambda_n^i \mathbf{f}_0) \\
&= \delta\tilde{\mathbf{a}}_n^i(\Delta\lambda_n^i) + \delta\lambda_n^i \delta\hat{\mathbf{a}}_n^i, \qquad \delta\hat{\mathbf{a}}_n^i \equiv \mathbf{K}_i^{-1} \mathbf{f}_0
\end{aligned} \tag{7.36}$$

Now an additional equation is cast using the constraint. Thus, for instance, with Eq. (7.35) we have

$$(\Delta \mathbf{a}_n^{i-1} + \delta \mathbf{a}_n^i)^{\mathrm{T}} (\Delta \mathbf{a}_n^{i-1} + \delta \mathbf{a}_n^i) = \Delta l^2 \tag{7.37}$$

On substitution of (7.36) in which $\delta\lambda_i$ is still undetermined into (7.37) a quadratic equation is available for the solution of $\delta\lambda_i$ (which may well turn out to be negative). Details of the algebra are given in reference 19.

The procedure suggested by Bergan[26,29] is somewhat different from those just described. Here a fixed load increment $\Delta\lambda_n$ is first assumed and any of the previously introduced iterative procedures are used for calculating the increment $\delta \mathbf{a}_n^i$. Now a new increment $\Delta\lambda_n^*$ is calculated so that it minimizes a norm of the residual

$$[(\mathbf{P}_{n+1}^{i+1} + \Delta\lambda_n^* \mathbf{f}_0)^{\mathrm{T}} (\mathbf{P}_{n+1}^{i+1} + \Delta\lambda_n^* \mathbf{f}_0)]^{1/2} = d \tag{7.38}$$

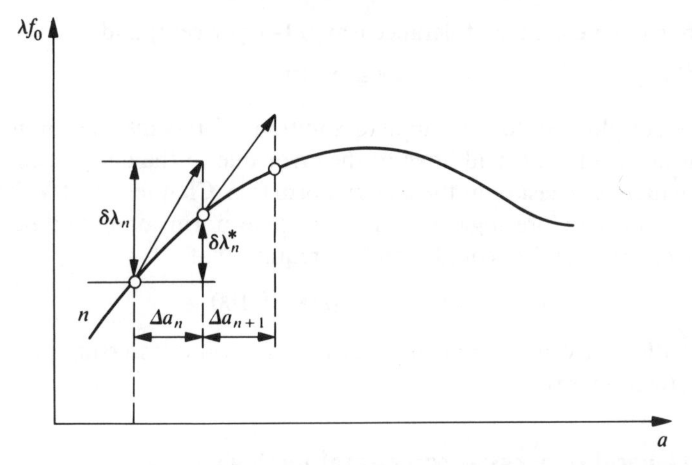

Fig. 7.7 One-dimensional interpretation of the Bergan procedure

This results in

$$\frac{\mathrm{d}(d)}{\mathrm{d}(\Delta\lambda_n^*)}=0$$

and

$$\Delta\lambda_n^* = \frac{\mathbf{f}_0^{\mathrm{T}}\mathbf{P}_{n+1}^{i+1}}{\mathbf{f}_0^{\mathrm{T}}\mathbf{f}_0} \tag{7.39}$$

This quantity may again well be negative requiring a load decrease, and indeed results in a rapid residual reduction in all cases, but precise control of displacement magnitudes becomes more difficult. The interpretation of the Bergan method in a one-dimensional example, shown in Fig. 7.7, is illuminating. Here it gives the exact answers—with a displacement control the magnitude of which is determined by the initial $\Delta\lambda_n$ assumed and the slope $\mathbf{K}$ used in the first iteration.

7.5 Convergence criteria

In all the iterative processes described the numerical solution is only approximately achieved and some tolerance limits have to be set to terminate the iteration.

Frequently the criteria used involve a norm of the displacement parameter changes $\|\delta\mathbf{a}_n^i\|$ or, more logically, that of the residuals $\|\boldsymbol{\Psi}_{n+1}\|$. In the latter case the limit is expressed as a percentage of the norm of the external forces $\|\mathbf{f}_{n+1}\|$. Thus we may require that

$$\|\boldsymbol{\Psi}_{n+1}\| \leqslant \varepsilon \|\mathbf{f}_{n+1}\|/100 \tag{7.40}$$

where ε is a percentage tolerance (say 0.1–1 per cent) and

$$\|\mathbf{\Psi}\| \equiv (\mathbf{\Psi}^{\mathrm{T}}\mathbf{\Psi})^{1/2} \tag{7.41}$$

The error due to the incomplete solution of the discrete, non-linear, equations is of course additive to the error due to the discretization that we frequently measure in the energy norm (viz. Chapter 14 of Volume 1). It appears therefore logical to use the same norm for bounding of the iteration process. We could therefore require that

$$(\mathbf{\Psi}^{\mathrm{T}}\mathbf{a})^{1/2} \leqslant \varepsilon(\mathbf{f}^{\mathrm{T}}\mathbf{a})^{1/2}/100 \tag{7.42}$$

and set for ε a value of similar order to that used in assessing permissible discretization errors.

7.6 General remarks—incremental methods

The various iterative methods described provide the essential tool-kit for the solution of *all non-linear problems* in which finite element discretization has been used. The precise choice of the optimal methodology is problem dependent and although many comparative solution cost studies have been published[9,15,18] the differences are often marginal. There is, however, little doubt that

(*a*) exact Newton–Raphson processes have to be used when convergence is difficult to achieve and

(*b*) the advantage of symmetric update matrices in the quasi Newton procedures frequently make these the only economically qualifying candidate (when non-symmetric tangent moduli exist).

We have not discussed in the preceding *direct iterative methods* such as the various forms of the conjugate gradient method[30-34] or *dynamic relaxation methods* in which an explicit dynamic transient analysis (viz. Chapter 10) is carried out to achieve a static solution.[35,36] All such processes are characterized by

(*a*) a diagonal form of the matrix used in computing trial increments $\delta\mathbf{a}$ (and hence very low cost of an iteration) and

(*b*) a large number of total iterations and hence of evaluations of the residual $\mathbf{\Psi}$.

These opposing trends have not yet resulted in dramatic cost reductions for medium size problems but there is little doubt that in large problems such methods are competitive. We shall illustrate their use later in this book in the context of fluid mechanics (Chapters 14 and 15) where currently they provide the favoured approach.

One final remark concerns the size of increments $\Delta\mathbf{f}$ or $\Delta\lambda$ to be adopted. Firstly, it is clear that small increments reduce the total number of iterations required per increment and in many applications automatic guidance on the size of an increment to preserve a (nearly) constant number of iterations is needed. Here such processes as the use of the 'current stiffness parameter' used by Bergan[26] can be commended.

Secondly, if the constitutive relation is *path dependent* the use of small increments is desirable to preserve the accuracy of evaluation of stress changes (or of similar parameters in other field problems). We have already emphasized in this context the need for calculating such changes always using the accumulated $\Delta \mathbf{a}_n^i$ change and not (as sometimes is practised) in adding changes due to each iterative $\delta \mathbf{a}_n^i$ step in an increment.

Thirdly, if only a single Newton–Raphson step is used in each increment of $\Delta\lambda$ then the procedure is equivalent to the solution of a standard incremental problem by direct forward integration. Here we note that if Eq. (7.1) is rewritten as

$$\mathbf{P}(\mathbf{a}) - \lambda \mathbf{f}_0 = 0 \tag{7.43}$$

we can, on differentiation, obtain

$$\frac{\mathrm{d}\mathbf{P}}{\mathrm{d}\mathbf{a}} \frac{\mathrm{d}\mathbf{a}}{\mathrm{d}\lambda} = \mathbf{f}_0 \tag{7.44}$$

and write this as

$$\frac{\mathrm{d}\mathbf{a}}{\mathrm{d}\lambda} = \mathbf{K}_T^{-1} \mathbf{f}_0 \tag{7.45}$$

or explicitly

$$\Delta \mathbf{a}_n = \mathbf{K}_{Tn}^{-1} \mathbf{f}_0 \Delta\lambda_n \tag{7.46}$$

This direct integration is illustrated in Fig. 7.8 and can frequently be divergent. However, use of Runge–Kutta procedures which we shall discuss later can provide improved accuracy.

RATE-INDEPENDENT PROBLEMS OF SOLID MECHANICS

7.7 General remarks—non-linear elasticity

With standard displacement formulation all problems of solid mechanics can be written as a set of equations

$$\boldsymbol{\Psi}(\mathbf{a}) = \int \mathbf{B}^{\mathrm{T}} \boldsymbol{\sigma} \, \mathrm{d}\Omega - \mathbf{f} \equiv \mathbf{P}(\mathbf{a}) - \mathbf{f} = 0 \tag{7.47}$$

where the displacements and strains are approximated as

$$\mathbf{u} = \mathbf{N}\mathbf{a} \qquad \boldsymbol{\varepsilon} = \mathbf{B}\mathbf{a} \tag{7.48}$$

in the usual manner.

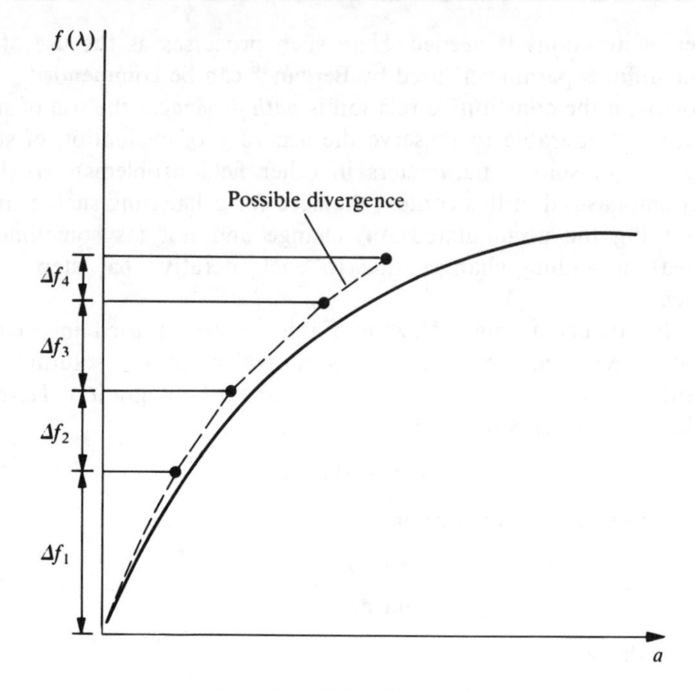

Fig. 7.8 Direct integration

Quite generally $\boldsymbol{\sigma}$ will be dependent on the strain $\boldsymbol{\varepsilon}$ in some path-dependent manner and the calculation of changes $\Delta\boldsymbol{\sigma}$ corresponding to changes of $\Delta\boldsymbol{\varepsilon}$ (or $\Delta\mathbf{a}$) will require special treatment. However, in all cases any of the iterative procedures previously discussed can be used. We note that

$$\frac{\partial \boldsymbol{\Psi}}{\partial \mathbf{a}} \equiv \frac{\partial \mathbf{P}}{\partial \mathbf{a}} = \mathbf{K}_T = \int_\Omega \mathbf{B}^{\mathrm{T}} \frac{\partial \boldsymbol{\sigma}}{\partial \boldsymbol{\varepsilon}} \frac{\partial \boldsymbol{\varepsilon}}{\partial \mathbf{a}} \, \mathrm{d}\Omega$$

$$= \int_\Omega \mathbf{B}^{\mathrm{T}} \mathbf{D}_{\mathrm{T}} \mathbf{B} \, \mathrm{d}\Omega \tag{7.49}$$

where

$$\boldsymbol{\sigma} = \boldsymbol{\sigma}(\boldsymbol{\varepsilon}) \tag{7.50}$$

and

$$\mathbf{D}_T = \frac{\partial \boldsymbol{\sigma}}{\partial \boldsymbol{\varepsilon}} \tag{7.51}$$

is known as the *tangential elasticity matrix*. The above form is particularly convenient as it is precisely the same as that pertaining to elasticity, and if $\mathbf{D}_T$ is a symmetric matrix, the same computational routines as those for the solution of elasticity problems can once again be used.

It is useful to interpret the residual vector $\boldsymbol{\Psi}(\mathbf{a}_n)$ as a *residual* or *unbalanced* force vector. This provides a convenient physical measure of the error by which the equations are not satisfied.

As, frequently, departures from linearity occur only at higher values of stress or strain, it is convenient to compare relationship (7.50) with the linear one, i.e.

$$\boldsymbol{\sigma} = \mathbf{D}(\boldsymbol{\varepsilon} - \boldsymbol{\varepsilon}_0) + \boldsymbol{\sigma}_0 \tag{7.52}$$

Clearly it is possible to make both expressions equal by expressing $\boldsymbol{\varepsilon}_0$ or $\boldsymbol{\sigma}_0$ as functions of the strain level where these functions have zero values at small strain levels and, indeed, are applied as a *correction* to the linear process. If all the non-linearity is expressed in the initial stress term $\boldsymbol{\sigma}_0 = \boldsymbol{\sigma}_0(\varepsilon)$, a purely linear elastic solution with a constant matrix $\mathbf{D}$ will result in error of the force term given by

$$\int_\Omega \mathbf{B}^{\mathrm{T}} \boldsymbol{\sigma}_0 \, \mathrm{d}\Omega \tag{7.53}$$

and this *unbalanced force vector* has to be corrected.

The correction can be carried out by a subsequent elastic solution using either the tangent or the original elastic modulus, and the reader will recognize that this is precisely the application of the Newton–Raphson or modified Newton–Raphson processes discussed in the previous section in which we test the error by comparing the stress that has been computed as a linear elastic one with that determined from the non-linear relationship. Such techniques are, therefore, known as the *stress transfer* or *initial stress methods* and are essentially identical to those described as the modified Newton–Raphson algorithms.

The case of non-linear elasticity (hyperelasticity) in which the relation (7.50) is unique and not path dependent is one in which all of the techniques described previously can be simply used. Here the size of the increment $\Delta\mathbf{a}$ is completely arbitrary and can indeed correspond to the total value of $\mathbf{a}$ providing the non-linearity is sufficiently mild to allow a reasonable number of iterations to be used.[37] With the stress $\boldsymbol{\sigma}$ depending uniquely on $\boldsymbol{\varepsilon}$ the strain energy W (potential) must be such that it is unique, i.e. that

$$W = W(\boldsymbol{\varepsilon}) \qquad \text{and} \qquad \boldsymbol{\sigma} = \frac{\partial W}{\partial \boldsymbol{\varepsilon}} \tag{7.54}$$

as otherwise the first law of thermodynamics would be violated. The consequence of this is that $\mathbf{D}_T$ must always be a symmetric matrix as

$$\frac{\partial W}{\partial \varepsilon_i \, \partial \varepsilon_j} \equiv \frac{\partial W}{\partial \varepsilon_j \, \partial \varepsilon_i} \tag{7.55}$$

7.8 Plasticity

7.8.1 *Classical plasticity theory.* 'Plastic' behaviour of solids is characterized by a non-unique stress–strain relationship—as opposed to that of non-linear elasticity discussed previously. Indeed, one definition of plasticity may be the presence of irrecoverable strains on load removal.

If uniaxial behaviour of a material is considered, as shown in Fig. 7.9(*a*), a non-linear relationship on loading alone does not determine whether non-linear elastic or plastic behaviour is exhibited. Unloading will immediately discover the difference, with the elastic material following the same path and the plastic material showing a *history-dependent*, different, path.

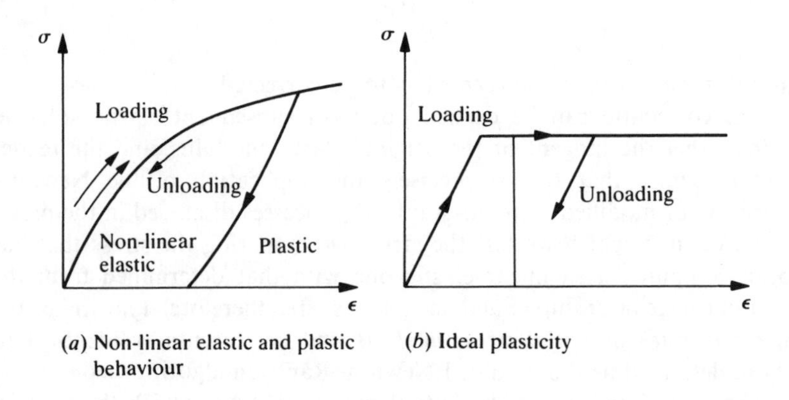

(*a*) Non-linear elastic and plastic behaviour (*b*) Ideal plasticity

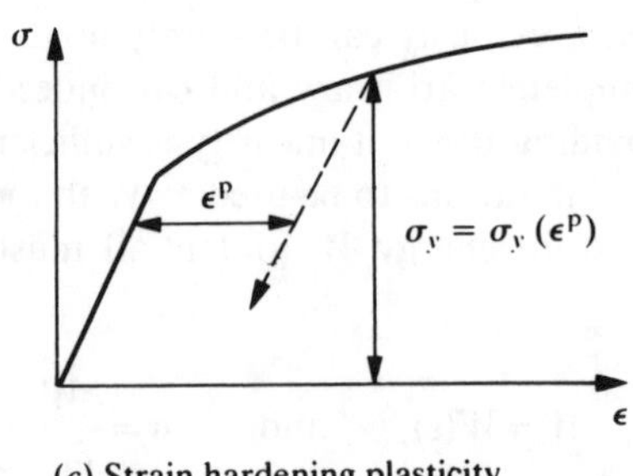

(*c*) Strain hardening plasticity

Fig. 7.9 Uniaxial behaviour of real materials

Many materials show an *ideal plastic* behaviour in which a limiting yield stress, σ_y, exists at which the strains are indeterminate. For all stresses below such yield a linear (or non-linear) elasticity relationship is assumed. Figure 7.9(*b*) illustrates this. A further refinement of this model is one of *hardening/softening plastic material* in which the yield stress depends on some parameter κ (such as plastic strain ε^p) [Fig. 7.9(*c*)]. It is with such kinds of plasticity that this section is concerned and for which much theory has been developed.[38]

In a multiaxial rather than uniaxial state of stress $\boldsymbol{\sigma}$ the concepts of yield need to be generalized.

Yield surface. It is quite generally postulated, as an experimental fact, that yielding can occur only if the stresses σ satisfy the general yield criterion

$$F(\boldsymbol{\sigma}, \kappa) = 0 \tag{7.56}$$

where κ is a 'hardening' parameter. This yield condition can be visualized as a surface in *n*-dimensional space of stress with the position of the surface dependent on the instantaneous value of the state parameter κ (Fig. 7.10).

Flow rule (*normality principle*). Von Mises[38] first suggested that basic behaviour defining the plastic strain increments is related to the yield surface. Heuristic arguments for the validity of the relationship proposed have been given by various workers in the field[39–46] and at the present time the following hypothesis appears to be generally accepted for many

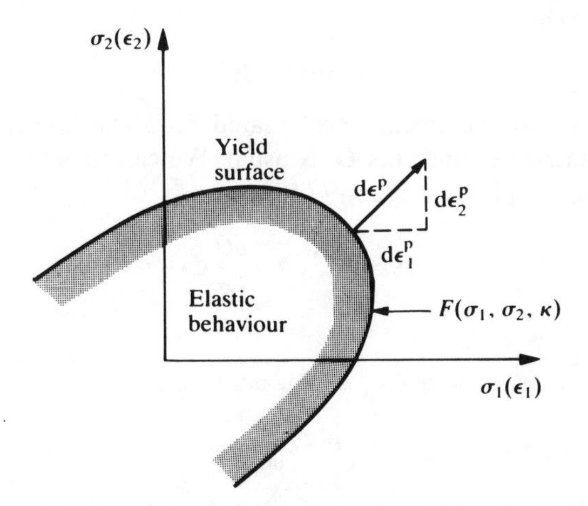

Fig. 7.10 Yield surface and normality criterion in two-dimensional stress space

materials; if $d\boldsymbol{\varepsilon}^{p}$ denotes the increment of plastic strain then†

$$d\boldsymbol{\varepsilon}^{p} = d\lambda \frac{\partial F}{\partial \boldsymbol{\sigma}} \tag{7.57}$$

or, for any component n,

$$d\varepsilon_n^{p} = d\lambda \frac{\partial F}{\partial \sigma_n}$$

In this $d\lambda$ is a proportionality constant, as yet undetermined. The rule is known as the *normality* principle because relation (7.57) can be interpreted as requiring the normality of the plastic strain increment 'vector' to the yield surface in the space of n stress and strain dimensions.

Restrictions of the above rule can be removed by specifying separately a *plastic potential*

$$Q = Q(\boldsymbol{\sigma}, \kappa) \tag{7.58}$$

which defines the plastic strain increment similarly to Eq. (7.57), i.e. giving this as

$$d\boldsymbol{\varepsilon}^{p} = d\lambda \frac{\partial Q}{\partial \boldsymbol{\sigma}} \tag{7.59}$$

The particular case of $Q = F$ is known as *associated plasticity.* When this relation is not satisfied the plasticity is *non-associated.* In what follows this more general form will be considered.

Incremental stress–strain relations. During an infinitesimal increment of stress, changes of strain are assumed to be divisible into elastic and plastic parts. Thus

$$d\boldsymbol{\varepsilon} = d\boldsymbol{\varepsilon}^{e} + d\boldsymbol{\varepsilon}^{p} \tag{7.60}$$

The elastic strain increments are related to stress increments by a symmetric matrix of constants $\mathbf{D}$ as usual. We can thus write Eq. (7.60) incorporating the plastic relation (7.59) as

$$d\boldsymbol{\varepsilon} = \mathbf{D}^{-1} d\boldsymbol{\sigma} + \frac{\partial Q}{\partial \boldsymbol{\sigma}} d\lambda \tag{7.61}$$

† Some authors prefer to write Eq. (7.57) in a rate form, i.e.

$$\dot{\boldsymbol{\varepsilon}}^{p} = \lambda \frac{dF}{d\boldsymbol{\sigma}}$$

where $\dot{\boldsymbol{\varepsilon}}^{p} \equiv d\boldsymbol{\varepsilon}^{p}/dt$ and t is some pseudo-time variable. To avoid confusion with real time (or rate) dependence we prefer to operate with infinitesimal increments.

The plastic increment of strain will occur only if the 'elastic' stress increment

$$\mathrm{d}\boldsymbol{\sigma}^e \equiv \mathbf{D}\,\mathrm{d}\boldsymbol{\varepsilon} \tag{7.62}$$

tends to put the stress outside the yield surface, i.e. is in the *plastic loading* direction. If, on the other hand, this stress change is such that *unloading* occurs then of course no plastic straining will be present, as illustrated for the one-dimensional case in Fig. 7.9. The test of the above relation is therefore crucial in differentiating between loading and unloading operations and underlines the importance of the straining path in computing stress changes.

When plastic loading is occurring the stresses are on the yield surface given by Eq. (7.56). Differentiating this we can therefore write

$$\mathrm{d}F = \frac{\partial F}{\partial \sigma_1}\mathrm{d}\sigma_1 + \frac{\partial F}{\partial \sigma_2}\mathrm{d}\sigma_2 + \cdots + \frac{\partial F}{\partial \kappa}\mathrm{d}\kappa = 0$$

or

$$\left\{\frac{\partial F}{\partial \boldsymbol{\sigma}}\right\}^{\mathrm{T}} \mathrm{d}\boldsymbol{\sigma} - A\,\mathrm{d}\lambda = 0 \tag{7.63}$$

in which we make the substitution

$$A = -\frac{\partial F}{\partial \kappa}\frac{\mathrm{d}\kappa}{\mathrm{d}\lambda} \tag{7.64}$$

Equations (7.61) and (7.63) can now be written in a single matrix form as

$$\begin{Bmatrix} \mathrm{d}\boldsymbol{\varepsilon} \\ 0 \end{Bmatrix} = \begin{bmatrix} \mathbf{D}^{-1} & \left\{\dfrac{\partial Q}{\partial \boldsymbol{\sigma}}\right\} \\ \left\{\dfrac{\partial F}{\partial \boldsymbol{\sigma}}\right\}^{\mathrm{T}} & -A \end{bmatrix} \begin{Bmatrix} \mathrm{d}\boldsymbol{\sigma} \\ \mathrm{d}\lambda \end{Bmatrix} \tag{7.65}$$

The indeterminate constant $\mathrm{d}\lambda$ can now be eliminated (taking care not to multiply or divide by A which may be zero in ideal plasticity).† This results in an explicit expansion that determines the *stress changes* in terms of imposed *strain changes* with

† To accomplish the elimination multiply the first set of Eq. (7.65) by $(\partial F/\partial \boldsymbol{\sigma})^{\mathrm{T}}\mathbf{D}$, giving

$$\left\{\frac{\partial F}{\partial \boldsymbol{\sigma}}\right\}^{\mathrm{T}} \mathrm{d}\boldsymbol{\sigma} = \left\{\frac{\partial F}{\partial \boldsymbol{\sigma}}\right\}^{\mathrm{T}} \mathbf{D}\,\mathrm{d}\boldsymbol{\varepsilon} - \left\{\frac{\partial F}{\partial \boldsymbol{\sigma}}\right\}^{\mathrm{T}} \mathbf{D} \left\{\frac{\partial Q}{\partial \boldsymbol{\sigma}}\right\} \mathrm{d}\lambda$$

Substituting above into the second set yields

$$\left\{\frac{\partial F}{\partial \boldsymbol{\sigma}}\right\}^{\mathrm{T}} \mathbf{D}\,\mathrm{d}\boldsymbol{\varepsilon} - \left[\left\{\frac{\partial F}{\partial \boldsymbol{\sigma}}\right\}^{\mathrm{T}} \mathbf{D} \left\{\frac{\partial Q}{\partial \boldsymbol{\sigma}}\right\} + A\right] \mathrm{d}\lambda = 0$$

Elimination of $\mathrm{d}\lambda$ from the first equation now gives Eqs. (7.66) and (7.67).

imposed *strain changes* with

$$d\boldsymbol{\sigma} = \mathbf{D}_{ep}^{*} d\boldsymbol{\varepsilon} \tag{7.66}$$

and
$$\mathbf{D}_{ep}^{*} = \mathbf{D} - \mathbf{D}\left\{\frac{\partial Q}{\partial \boldsymbol{\sigma}}\right\}\left\{\frac{\partial F}{\partial \boldsymbol{\sigma}}\right\}^{\mathrm{T}} \mathbf{D}\left[A + \left\{\frac{\partial F}{\partial \boldsymbol{\sigma}}\right\}^{\mathrm{T}} \mathbf{D}\left\{\frac{\partial Q}{\partial \boldsymbol{\sigma}}\right\}\right]^{-1} \tag{7.67}$$

The elastoplastic matrix $\mathbf{D}_{ep}^{*}$ takes the place of the elasticity matrix $\mathbf{D}_T$ in incremental analysis.

This matrix is symmetric only when the plasticity is associated. The non-associated material will present special difficulties if tangent modulus procedures other than the modified Newton–Raphson method are used.

The matrix is defined even for ideal plasticity when $A=0$. Explicit formulation of plasticity in this form was first introduced by Yamada *et al.*[47] and Zienkiewicz *et al.*[48]

Significance of parameter 'A'. Clearly for ideal plasticity with no hardening, A is simply zero. If hardening is considered, attention must be given to the nature of the parameter (or parameters) κ on which the shifts of the yield surface depend.

With a 'work hardening' material κ is sometimes taken to be the amount of plastic work done during plastic deformation. Thus

$$d\kappa = \sigma_1\, d\varepsilon_1^{p} + \sigma_2\, d\varepsilon_2^{p} + \cdots = \sigma^{\mathrm{T}}\, d\varepsilon^{p} \tag{7.68}$$

Using the flow rule [Eq. (7.59)] we have alternatively

$$d\kappa = d\lambda \boldsymbol{\sigma}^{\mathrm{T}} \frac{\partial Q}{\partial \boldsymbol{\sigma}} \tag{7.69}$$

Substituting Eq. (7.69) into Eq. (7.64) we see that $d\lambda$ disappears and we can write

$$A = \frac{\partial F}{\partial \kappa} \boldsymbol{\sigma}^{\mathrm{T}} \frac{\partial Q}{\partial \boldsymbol{\sigma}} \tag{7.70}$$

This assumes a determinate form if the explicit relationship between F and κ is known. A similar interpretation occurs for different hardening assumptions. For a generalization of the concepts to a yield surface possessing 'corners' where $\partial Q/\partial \boldsymbol{\sigma}$ is indeterminate, the reader is referred to the work of Koiter.[40]

7.8.2 *Some typical examples of classical plasticity*

Prandtl–Reuss relations. To illustrate some of the concepts consider the particular case of the well-known Huber–von Mises yield surface with an associated flow rule. This is given by

$$F=[\tfrac{1}{2}(\sigma_1-\sigma_2)^2+\tfrac{1}{2}(\sigma_2-\sigma_3)^2+\tfrac{1}{2}(\sigma_3-\sigma_1)^2+3\sigma_4^2+3\sigma_5^2+3\sigma_6^2]^{1/2}-\sigma_y$$
$$\equiv\bar{\sigma}-Y \tag{7.71}$$

in which the suffixes 1, 2, 3 refer to the normal stress components and 4, 5, 6 to shear stress components in a general three-dimensional stress state. In the above, $\bar{\sigma}$ is the second stress invariant.

On differentiation it will be found that

$$\frac{\partial F}{\partial \sigma_1}=\frac{3s_1}{2\bar{\sigma}},\qquad \frac{\partial F}{\partial \sigma_2}=\frac{3s_2}{2\bar{\sigma}},\qquad \frac{\partial F}{\partial \sigma_3}=\frac{3s_3}{2\bar{\sigma}}$$
$$\frac{\partial F}{\partial \sigma_4}=\frac{3s_4}{\bar{\sigma}},\qquad \frac{\partial F}{\partial \sigma_5}=\frac{3s_5}{\bar{\sigma}},\qquad \frac{\partial F}{\partial \sigma_6}=\frac{3s_6}{\bar{\sigma}} \tag{7.72}$$

in which the dashes stand for deviatoric stresses, i.e.

$$s_1=\sigma_1-\frac{\sigma_1+\sigma_2+\sigma_3}{3},\qquad \text{etc.} \tag{7.73}$$

The quantity $Y(\kappa)$ is the uniaxial stress at yield. If a plot of the uniaxial test giving $\bar{\sigma}$ versus the *plastic* uniaxial strain ε_u' is available and if simple work hardening is assumed, then

$$d\kappa = Y\, d\varepsilon_u^p$$

and

$$-\frac{\partial F}{\partial \kappa}=\frac{\partial Y}{\partial \kappa}=\frac{\partial Y}{\partial \varepsilon_u^p}\frac{1}{Y}=\frac{H}{Y} \tag{7.74}$$

in which H is the slope of the plot at the particular value of $\bar{\sigma}$.

On substituting into Eq. (7.70) we obtain, after some transformation, simply

$$A=H \tag{7.75}$$

where H is known as the *plastic modulus.* This reestablishes the well-known Prandtl–Reuss stress–strain relations.

Other yield surfaces. Clearly the general procedures outlined allow determination of the tangent matrices for almost any yield surface applicable in practice. If the yield surface (and the material) is isotopic it is convenient to express it in terms of the three stress invariants. A particularly useful form of these is given below, introducing here also the indicial notation:[49]

$$\sigma_m = \frac{J_1}{3} = \frac{\sigma_x + \sigma_y + \sigma_z}{3} \equiv \frac{\sigma_{ii}}{3}$$

$$\bar{\sigma} = J_2^{1/2} = [\tfrac{1}{2}(s_x^2 + s_y^2 + s_z^2) + \tau_{xy}^2 + \tau_{yz}^2 + \tau_{zx}^2]^{1/2} \equiv \sqrt{\tfrac{1}{2} s_{ij} s_{ij}} \tag{7.76}$$

$$\theta = \tfrac{1}{3} \sin^{-1}\left(-\frac{3\sqrt{3}}{2} \frac{J_3}{\bar{\sigma}} \right) \qquad \text{with } -\frac{\pi}{6} < \theta < \frac{\pi}{6}$$

where

$$J_3 = s_x s_y s_z + 2\tau_{xy}\tau_{yz}\tau_{zx} - s_x \tau_{yz}^2 - s_y \tau_{xz}^2 - s_z \tau_{xy}^2 \equiv \det \mathbf{s}$$

and

$$s_x = \sigma_x - \sigma_m, \quad s_y = \sigma_y - \sigma_m, \quad s_2 = \sigma_z - \sigma_m \quad \text{or}$$

$$s_{ij} = \sigma_{ij} - \frac{\delta_{ij}\sigma_{ii}}{3}$$

It is shown in reference 49 that the yield surface for several classical yield conditions can be given as:

1. Tresca:

$$F = 2\bar{\sigma} \cos\theta - Y(\kappa) = 0 \tag{7.77}$$

where $Y(\kappa)$ is the yield stress from uniaxial tests.

2. Huber–von Mises:

$$F = \sqrt{3}\bar{\sigma} - Y(\kappa) = 0 \tag{7.78}$$

Both conditions 1 and 2 are well verified in metal plasticity. For soils, concrete and other 'frictional' materials the Mohr–Coulomb law and its approximation given by Drucker and Prager are frequently used.[50]

3. Mohr–Coulomb:

$$F = \sigma_m \sin\phi + \bar{\sigma}\cos\theta - \frac{\bar{\sigma}}{\sqrt{3}} \sin\phi \sin\theta - c\cos\phi = 0 \tag{7.79}$$

where $c(\kappa)$ and $\phi(\kappa)$ are the cohesion and angle of friction respectively, which can depend on some strain hardening parameter κ.

4. Drucker–Prager:[50]

$$F = 3\alpha' \sigma_m + \bar{\sigma} - K = 0 \tag{7.80}$$

where

$$\alpha' = \frac{2\ \sin\phi}{\sqrt{3}(3-\ \sin\phi)} \qquad K = \frac{6c\ \cos\phi}{\sqrt{3}(3-\ \sin\phi)}$$

and again c and ϕ can depend on a strain hardening parameter.

These forms lead to a very convenient definition of the gradient vectors $\partial F/\partial \boldsymbol{\sigma}$ or $\partial Q/\partial \boldsymbol{\sigma}$, irrespective of whether the surface is used as a yield condition or potential. Thus we can always write:

$$\frac{\partial F}{\partial \boldsymbol{\sigma}} = \frac{\partial F}{\partial \sigma_m}\frac{\partial \sigma_m}{\partial \boldsymbol{\sigma}} + \frac{\partial F}{\partial J_2}\frac{\partial J_2}{\partial \boldsymbol{\sigma}} + \frac{\partial F}{\partial J_3}\frac{\partial J_3}{\partial \boldsymbol{\sigma}} \tag{7.81}$$

Noting

$$\frac{\partial F}{\partial J_3} = \frac{\partial F}{\partial \theta}\frac{\partial \theta}{\partial J_3} \tag{7.82}$$

and using Eq. (7.76) one can write the gradient vector as

$$\frac{\partial F}{\partial \sigma} = \left(\frac{\partial F}{\partial \sigma_m}\mathbf{M}^0 + \frac{\partial F}{\partial J_2}\mathbf{M}^{\mathrm{I}} + \frac{\partial F}{\partial J_3}\mathbf{M}^{\mathrm{II}}\right)\boldsymbol{\sigma} \tag{7.83}$$

where the form of the square matrices $\mathbf{M}^0, \mathbf{M}^{\mathrm{I}}$ and $\mathbf{M}^{\mathrm{II}}$ is given in Table 7.1.

The values of the three derivatives with respect to the invariants are

TABLE 7.1
MATRICES **M** OF EQ. (7.83)

$$\mathbf{M}^0 = \frac{1}{9\sigma_m}\begin{bmatrix} 1 & 1 & 1 & 0 & 0 & 0 \\ & 1 & 1 & 0 & 0 & 0 \\ & & 1 & 0 & 0 & 0 \\ & & & 0 & 0 & 0 \\ \text{symmetric} & & & & 0 & 0 \\ & & & & & 0 \end{bmatrix} \qquad \mathbf{M}^{\mathrm{I}} = \begin{bmatrix} \frac{2}{3} & -\frac{1}{3} & -\frac{1}{3} & 0 & 0 & 0 \\ & \frac{2}{3} & -\frac{1}{3} & 0 & 0 & 0 \\ & & \frac{2}{3} & 0 & 0 & 0 \\ & & & 2 & 0 & 0 \\ \text{symmetric} & & & & 2 & 0 \\ & & & & & 2 \end{bmatrix}$$

$$\mathbf{M}^{\mathrm{II}} = \begin{bmatrix} \frac{1}{3}\sigma_x & \frac{1}{3}\sigma_z & \frac{1}{3}\sigma_y & -\frac{2}{3}\tau_{yz} & \frac{1}{3}\tau_{zx} & \frac{1}{3}\tau_{xy} \\ & \frac{1}{3}\sigma_y & \frac{1}{3}\sigma_x & +\frac{1}{3}\tau_{yz} & -\frac{2}{3}\tau_{zx} & \frac{1}{3}\tau_{xy} \\ & & \frac{1}{3}\sigma_z & +\frac{1}{3}\tau_{yz} & \frac{1}{3}\tau_{zx} & -\frac{2}{3}\tau_{xy} \\ & & & -\sigma_x & \tau_{xy} & \tau_{zx} \\ \text{symmetric} & & & & -\sigma_y & \tau_{yz} \\ & & & & & -\sigma_z \end{bmatrix} + \sigma_m \begin{bmatrix} -\frac{1}{3} & -\frac{1}{3} & -\frac{1}{3} & 0 & 0 & 0 \\ & -\frac{1}{3} & -\frac{1}{3} & 0 & 0 & 0 \\ & & -\frac{1}{3} & 0 & 0 & 0 \\ & & & 1 & 0 & 0 \\ \text{symmetric} & & & & 1 & 0 \\ & & & & & 1 \end{bmatrix}$$

TABLE 7.2
INVARIANT DERIVATIVES FOR VARIOUS YIELD CONDITIONS

Yield conditions	$\frac{\partial F}{\partial \sigma_m}$	$\sqrt{J_2}\frac{\partial F}{\partial J_2}$	$J_2\frac{\partial F}{\partial J_3}$
Tresca	0	$2\cos\theta(1+\tan\theta\tan 3\theta)$	$\frac{\sqrt{3}\sin\theta}{\cos 3\theta}$
Huber-von Mises	0	$\sqrt{3}$	0
Mohr–Coulomb	$\sin\phi$	$\frac{\cos\theta}{2}\left[(1+\tan\theta\sin 3\theta) + \sin\phi(\tan 3\theta-\tan\theta)/\sqrt{3}\right]$	$\frac{\sqrt{3}\sin\theta+\sin\phi\cos\theta}{2\cos 3\theta}$
Drucker–Prager	$3\alpha'$	1.0	0

shown in Table 7.2 for the various yield surfaces mentioned. The reader can verify that the Prandtl–Reuss relations (7.72) are herein contained. The form of the various yield surfaces given above is shown in principal stress space in Fig. 7.11, though many more elaborate ones have been developed, particularly for soil (geomechanics) problems.[51-53]

Generalized hardening/softening rules. We have so far assumed that the parameter κ is associated with the plastic work done and this, being a scalar quantity, will obviously change the yield surface by a simple expansion or contraction (isotropic hardening). Such models are found to be deficient in reproducing the true behaviour of some materials and kinematic hardening theories have been introduced in which dependence on the direction of plastic straining is noted.[42-44, 54-56]

An alternative, utilizing the finite element procedure directly, is that of modelling the material by an 'overlay' technique.[57,58] In this, two or

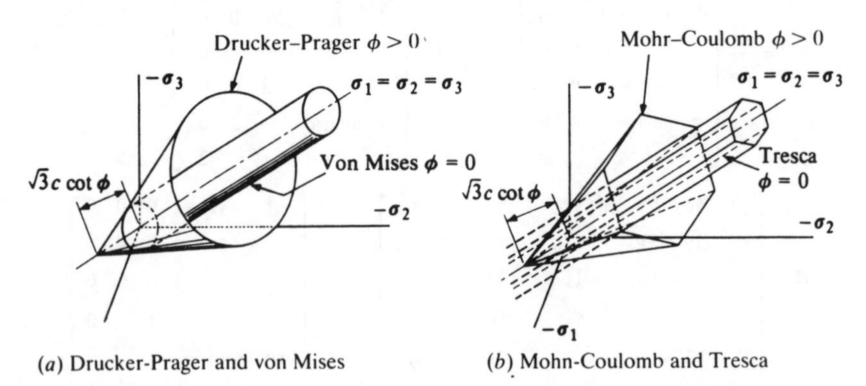

(*a*) Drucker-Prager and von Mises (*b*) Mohn-Coulomb and Tresca

Fig. 7.11 Some isotropic yield surfaces in principal stress space

more simple, ideally plastic materials are assumed to act in parallel subject to the same strain increments. This of course produces an effect of a hardening material with almost the same characteristics as those achievable by kinematic hardening models and indeed is physically more realistic. Here the finite element model describes effectively separate material components demonstrating once again its versatility.

7.8.3 *Generalized plasticity.* Plastic behaviour characterized by irreversibility of stress paths and the development of permanent strain changes after a stress cycle can be described in a variety of ways. One form of such a description has been given in Sec. 7.8.1. Another more general and indeed simpler form is presented here. This assumes *a priori* the existence of the incremental relationship

$$\mathrm{d}\boldsymbol{\sigma} = \mathbf{D}^* \, \mathrm{d}\boldsymbol{\varepsilon} \tag{7.84}$$

in which the matrix $\mathbf{D}^*$ depends not only on the stress $\boldsymbol{\sigma}$, the state parameters $\boldsymbol{\kappa}$ but also on the direction of the applied stress (or strain) increment $\mathrm{d}\boldsymbol{\sigma}$ (or $\mathrm{d}\boldsymbol{\varepsilon}$).[59] A slightly less ambitious description arises if we accept the dependence of $\mathbf{D}^*$ only on two directions—those of loading and unloading. If in the general stress space we specify a 'loading' direction by a unit vector $\mathbf{n}$ given at every point (and also depending on the state parameters $\boldsymbol{\kappa}$), as shown in Fig. 7.12, we can describe plastic loading and unloading by the sign of the projection $\mathbf{n}^{\mathrm{T}} \, \mathrm{d}\boldsymbol{\sigma}$. Thus

$$\begin{aligned} \mathbf{n}^{\mathrm{T}} \, \mathrm{d}\boldsymbol{\sigma} > 0 \qquad & \text{for loading} \\ \mathbf{n}^{\mathrm{T}} \, \mathrm{d}\boldsymbol{\sigma} < 0 \qquad & \text{for unloading} \end{aligned} \tag{7.85}$$

while $\mathbf{n}^{\mathrm{T}} \, \mathrm{d}\boldsymbol{\sigma} = 0$ is a neutral direction in which only elastic straining occurs.

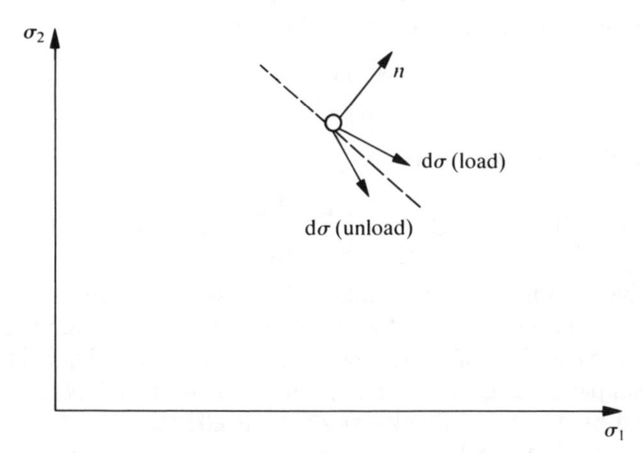

Fig. 7.12 Loading and unloading directions in stress space

One can now write quite generally that

$$\begin{aligned} d\boldsymbol{\sigma} &= \mathbf{D}_L^* \, d\boldsymbol{\varepsilon} \quad \text{for loading} \\ d\boldsymbol{\sigma} &= \mathbf{D}_U^* \, d\boldsymbol{\varepsilon} \quad \text{for unloading} \end{aligned} \tag{7.86}$$

where the matrices $\mathbf{D}_L^*$ and $\mathbf{D}_U^*$ depend only on the state described by $\boldsymbol{\sigma}$ and $\boldsymbol{\kappa}$.

The specification of $\mathbf{D}_L^*$ and $\mathbf{D}_U^*$ must be such that in the neutral direction of the stress increment $d\boldsymbol{\sigma}$, the strain increments corresponding to this are equal. Thus we require

$$d\boldsymbol{\varepsilon} = \mathbf{D}_L^{*-1} \, d\boldsymbol{\sigma} = \mathbf{D}_U^{*-1} d\boldsymbol{\sigma} \qquad \text{when } \mathbf{n}^T \, d\boldsymbol{\sigma} = 0 \tag{7.87}$$

The most general way of achieving this end is to write

$$\mathbf{D}_L^{*-1} \equiv \mathbf{D}^{-1} + \frac{\mathbf{n}_{gL} \mathbf{n}^T}{H_L}$$

and

$$\mathbf{D}_U^{*-1} = \mathbf{D}^{-1} + \frac{\mathbf{n}_{gU} \mathbf{n}^T}{H_U} \tag{7.88}$$

where $\mathbf{D}$ is the elastic matrix, $\mathbf{n}_{gL}$ and $\mathbf{n}_{gU}$ are arbitrary unit stress vectors for loading and unloading directions and H_L and H_U are appropriate plastic moduli which in general depend on $\boldsymbol{\sigma}$ and $\boldsymbol{\kappa}$.

The value of the tangent matrices $\mathbf{D}_L^*$ and $\mathbf{D}_U^*$ can be obtained by direct inversion if $H_{L/U} \neq 0$, but more generally can be written as

$$\mathbf{D}_L^* = \mathbf{D} - \mathbf{D}\mathbf{n}_{gL}\mathbf{n}^T(H_L + \mathbf{n}^T \mathbf{D}\mathbf{n}_{gL})^{-1} \tag{7.89}$$

with a similar form for $\mathbf{D}_U$. This form resembles Eq. (7.67) and indeed its derivation is almost identical. Thus if we put

$$\frac{\mathbf{n}^T \, d\boldsymbol{\sigma}}{H_L} \equiv d\lambda \tag{7.90}$$

and write Eqs (7.87) and (7.88) as

$$d\boldsymbol{\varepsilon} = \mathbf{D}_L^{*-1} \, d\boldsymbol{\sigma} = \mathbf{D}^{-1} \, d\boldsymbol{\sigma} + \frac{\mathbf{n}_{gL} \mathbf{n}^T \, d\boldsymbol{\sigma}}{H_L} = \mathbf{D}^{-1} d\boldsymbol{\sigma} + \mathbf{n}_{gL} \, d\lambda \tag{7.91}$$

we find that a form identical to that of Eq. (7.65) is obtained and indeed the same elimination would suffice to obtain $\mathbf{D}_L$ of expression (7.89) (viz. footnote to page 231). Of course the process of obtaining $\mathbf{D}_U$ is identical.

This simple and general description of *generalized plasticity* was introduced by Mróz and Zienkiewicz.[60,61] It allows:

(*a*) the full model to be specified by a direct prescription of $\mathbf{n}$, $\mathbf{n}_g$ and H for loading and unloading at any point of the stress space,

(*b*) existence of plasticity in both loading and unloading directions,
(*c*) relative simplicity for description of experimental results when these are complex and the existence of a yield surface of the kind encountered in ideal plasticity is uncertain.

For the above reasons the generalized plasticity forms have proved extremely useful in describing the complex behaviour of soils.[62-64] Here other descriptions using various interpolations of $\mathbf{n}$ and moduli from a unique yield surface, known as *bounding surface plasticity* models, are indeed particular forms of the above generalization and have proved useful.[65]

It is clear that classical plasticity is indeed a special case of the generalized models. Here the yield surface $F(\boldsymbol{\sigma}, \boldsymbol{\kappa})$ defines of course a unit vector normal to it as

$$\mathbf{n} = \frac{\partial F/\partial \boldsymbol{\sigma}}{[(\partial F/\partial \boldsymbol{\sigma})^{\mathrm{T}}(\partial F/\partial \boldsymbol{\sigma})]^{1/2}} \tag{7.92}$$

Similarly the plastic potential defines the unit vector $\mathbf{n}_g$:

$$\mathbf{n}_g = \frac{\partial Q/\partial \boldsymbol{\sigma}}{[(\partial Q/\partial \boldsymbol{\sigma})^{\mathrm{T}}(\partial Q/\partial \boldsymbol{\sigma})]^{1/2}} \tag{7.93}$$

Substitution of such values for the unit vectors into Eq. (7.89) will of course retrieve the original form of Eq. (7.67). However, interpretation of generalized plasticity in classical terms is more difficult.

The success of generalized plasticity in practical applications has allowed many complex phenomena of soil dynamics to be solved.[66] We shall refer to such applications later but in Fig. 7.13 we show how complex cyclic response with plastic loading and unloading can be followed.

While we have specified initially the loading and unloading directions in terms of the total stress change $d\boldsymbol{\sigma}$ this definition ceases to apply when strain softening occurs and the plastic modulus H becomes negative. It is therefore more convenient to check the loading or unloading direction by the elastic strain increment $d\boldsymbol{\sigma}^e$ of Eq. (7.62) and to specify

$$\mathbf{n}^{\mathrm{T}}\, d\boldsymbol{\sigma}^e \geqslant 0 \quad \text{for loading}$$

and

$$\mathbf{n}^{\mathrm{T}}\, d\boldsymbol{\sigma}^e < 0 \quad \text{for unloading} \tag{7.94}$$

This of course becomes identical to the previous definition of loading and unloading in the case of hardening.

7.9 Computation of stress increments

We have emphasized repeatedly that with the use of iterative procedures within a particular load increment $\Delta \mathbf{f}_n$ it is important to compute always

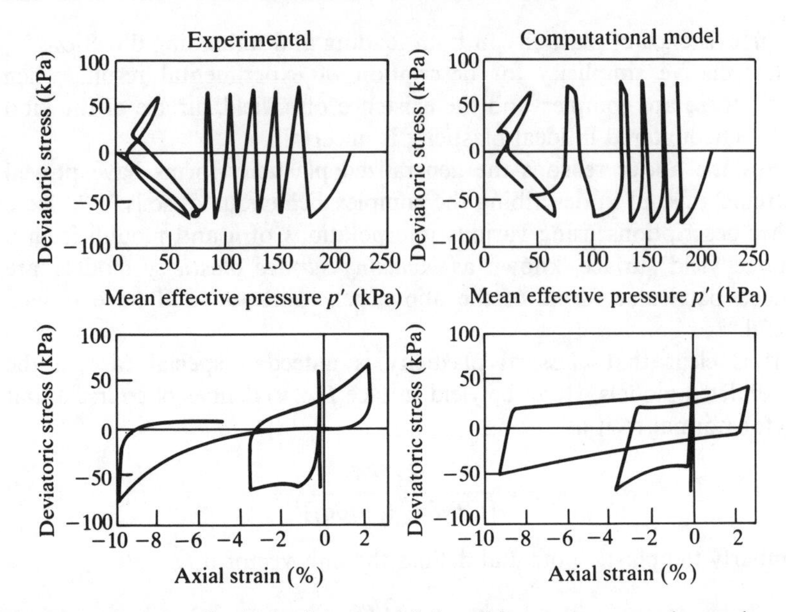

Fig. 7.13 A generalized plasticity model describing a very complex path and comparison with experimental data. Undrained two-way cyclic loading of Niigata sand (T. Tatsueka and K. Ishihera, 'Yielding of sand in triaxial compression', *Soil Found.*, **14**, 63–76, 1974). (Note that in an undrained soil test th fluid restrains all volumetric strains and pore pressures develop; see Chapter 11, pages 420–425)

the stress changes $\Delta\boldsymbol{\sigma}_n^i$ corresponding to the total change of the displacement parameters $\Delta\mathbf{a}_n^i$ and hence to the total strain change

$$\Delta\boldsymbol{\varepsilon}_n^i = \mathbf{B}\Delta\mathbf{a}_n^i \qquad \Delta\mathbf{a}_n^i = \sum_0^i \delta\mathbf{a}_n^k \tag{7.95}$$

which has accumulated in all the previous iterations. This point is of importance as during the iterative displacement changes $\delta\mathbf{a}_n^i$ different strain paths are followed, occasionally leading to stress reversal and purely elastic unloading.

In terms of the elastoplastic stiffness matrix given by Eqs (7.67) and (7.89) this means that the stresses have to be integrated as

$$\Delta\boldsymbol{\sigma}_n^i = \int_0^{\Delta\boldsymbol{\varepsilon}_n^i} \mathbf{D}_{ep}^* \, d\boldsymbol{\varepsilon} \tag{7.96}$$

incorporating into $\mathbf{D}_{ep}^*$ the dependence on $\boldsymbol{\sigma}$, $\mathbf{n}$ and $\boldsymbol{\kappa}$ in a manner corresponding to a linear increase of $\Delta\boldsymbol{\varepsilon}_n^i$ (or $\Delta\mathbf{a}_n^i$). Here of course the change of the state parameters $\boldsymbol{\kappa}$ (viz., for instance, Eq. (7.68)] has to be suitably integrated though this generally presents little problem as usually

$$\boldsymbol{\kappa} = \boldsymbol{\kappa}(\boldsymbol{\varepsilon})$$

is taken.

Since the early days of elastoplastic computation various procedures of integration of Eq. (7.96) have been adopted. These can be classified into explicit and implicit categories.

7.9.1 *Explicit methods.* In explicit procedures either a direct integration process is used or some form of the Runge–Kutta process is adopted. In the former the known increment $\Delta\boldsymbol{\varepsilon}_n^i$ is subdivided into k intervals and the integral of Eq. (7.96) is replaced by summation, writing

$$\Delta\boldsymbol{\sigma} = \sum_{l=0}^{k} \frac{{}^{l-1}\mathbf{D}^* \Delta\boldsymbol{\varepsilon}}{k} \tag{7.97}$$

where $\mathbf{D}^*$ is the tangent matrix. In the above the suffixes n and i have been omitted for clarity—as will be done in the remainder of this section. ${}^{l-1}\mathbf{D}^*$ implies the evaluation of $\mathbf{D}^*$ for $\boldsymbol{\sigma}$ and $\boldsymbol{\kappa}$ at $l-1$.

This procedure, originally introduced in reference 46 and described in detail in references 67 and 68, is known as *subincrementation.* Its accuracy increases, of course, with the number of subincrements, k, used. In general it is difficult *a priori* to decide on this number and accuracy of prediction is not easy to determine.

Such integration will generally result in the stress change departing from the yield surface by some margin. In problems such as those of ideal plasticity where the yield surface forms a meaningful limit, a proportional scaling of stresses (or radial return procedures) have been practised frequently[69,70] to obtain stresses which are on the yield surfaces at all times. This, however, makes little sense in strain hardening situations where the yield surface is really a fiction.

A more precise explicit procedure is provided by use of the Runge–Kutta method. Here first an increment of $\Delta\boldsymbol{\varepsilon}/2$ is applied in a single-step explicit manner to obtain

$$\Delta\boldsymbol{\sigma}^{1/2} = \mathbf{D}_0^* \frac{\Delta\boldsymbol{\varepsilon}}{2} \tag{7.98}$$

using the initial elastoplastic matrix. This increment of stress (and corresponding increment of $\boldsymbol{\kappa}^{1/2}$) is evaluated to compute ${}^{1/2}\mathbf{D}^*$ and finally we evaluate

$$\Delta\boldsymbol{\sigma} = {}^{1/2}\mathbf{D}^* \Delta\boldsymbol{\varepsilon} \tag{7.99}$$

This process has a second-order accuracy and in addition can give an estimate of errors incurred as

$$\Delta\boldsymbol{\sigma} - 2\Delta\boldsymbol{\sigma}_{1/2} \tag{7.100}$$

If such stress errors exceed a certain norm the size of the increment $\Delta\mathbf{f}_n$ can be modified. We highly recommend this procedure for problems of generalized plasticity where the tangent matrices are simply evaluated.

In some codes a purely explicit, single-step computation has on occasion been used. This will lead to considerable errors and indeed linearize the problem to the extent that a single Newton iteration suffices to obtain a load increment.

7.9.2 *Implicit methods.* The integration of Eq. (7.96) can, of course, be written in an implicit form. For instance, we could write in place of Eq. (7.96), during each iteration i, that

$$\Delta\boldsymbol{\sigma}_n^i = [(1-\theta)\mathbf{D}_n^{*i} + \theta\mathbf{D}_{n+1}^{*i}]\Delta\boldsymbol{\varepsilon}_n^i \tag{7.101}$$

denoting here by $\mathbf{D}_n^*$ the value of the tangential matrix at the start of the increment and by $\mathbf{D}_{n+1}^*$ that at its end.

This non-linear equation set could be solved by any of the procedures previously described; however, the derivatives of $\mathbf{D}^*$ are quite complex and in any case a serious error is committed in the approximate form of Eq. (7.101). Further, there is no guarantee that the stresses do not depart from the yield surface.

For materials with a well-defined yield surface it is desirable to return to the original plasticity equations (7.59) to (7.61) and to write alternatively

$$\begin{aligned}
\Delta\boldsymbol{\sigma}_n &= \mathbf{D}(\Delta\boldsymbol{\varepsilon}_n - \Delta\boldsymbol{\varepsilon}_n^p) \\
\Delta\boldsymbol{\varepsilon}_n^p &= \Delta\lambda\left[(1-\theta)\left.\frac{\partial Q}{\partial\boldsymbol{\sigma}}\right|_n + \theta\left.\frac{\partial Q}{\partial\boldsymbol{\sigma}}\right|_{n+1}\right] \\
F_{n+1} &= 0
\end{aligned} \tag{7.102}$$

where only the equation of strain increment (7.58) has been approximated.

The approximation is particularly simple and stable for $\theta = 1$ (backward difference) and now, eliminating $\Delta\boldsymbol{\varepsilon}_n^p$, we can write the above non-linear system simply as

$$\begin{aligned}
\Delta\boldsymbol{\varepsilon}_n - \mathbf{D}^{-1}\Delta\boldsymbol{\sigma}_n - \left.\frac{\partial Q}{\partial\boldsymbol{\sigma}}\right|_{n+1}\Delta\lambda &= 0 \\
F_{n+1} &= 0
\end{aligned} \tag{7.103}$$

Denoting by $\mathbf{R}^i$ and r^i the residuals of these equations any of the general iterative schemes described in Sec. 7.2 can be used. In particular, the full Newton–Raphson process is convenient here. Noting that $\Delta\boldsymbol{\varepsilon}_n$ is a specified constant, we can write, on differentiation,

$$\begin{bmatrix} \tilde{\mathbf{D}}^{-1} & \dfrac{\partial Q}{\partial \boldsymbol{\sigma}} \\ \dfrac{\partial F^{\mathrm{T}}}{\partial \boldsymbol{\sigma}} & -A \end{bmatrix}_{n+1}^{i} \begin{Bmatrix} \delta \boldsymbol{\sigma}^i \\ \delta \lambda^i \end{Bmatrix} = \begin{Bmatrix} \mathbf{R}^i \\ r^i \end{Bmatrix} \tag{7.104}$$

In the above,

$$\tilde{\mathbf{D}}^{-1} = \mathbf{D}^{-1} + \Delta\lambda^i \frac{\partial^2 Q}{\partial \boldsymbol{\sigma}^2} \tag{7.105}$$

and A is the same hardening parameter as that obtained in Eq. (7.64). Some complexity is introduced by the presence of the second derivatives of Q in Eq. (7.105) and frequently that term is omitted for simplicity, though analytical forms of such second derivatives are available for frequently used potential surfaces.[71–74]

It is, however, important to note that the return to the full requirement that $F_{n+1}=0$ [viz. Eq. (7.103)] ensures that the r^i residual measures precisely the departure from the yield surface. This measure is not available for the tangential form if $\mathbf{D}^{\mathrm{ep}}$ is adopted.

Of course for the solution for $\delta\lambda^i$ it is only necessary to update

$$\Delta\lambda^i = \sum_0^i \delta\lambda^k \tag{7.106}$$

and that variable can be eliminated in general. This elimination can be done in precisely the same way as it was in establishing Eq. (7.67).

We can now write

$$\delta\boldsymbol{\sigma}^i = \mathbf{D}^{**}\mathbf{R}^i - \frac{\tilde{\mathbf{D}}(\partial F/\partial\boldsymbol{\sigma})r^i}{A + (\partial F^{\mathrm{T}}/\partial\boldsymbol{\sigma})} D(\partial Q/\partial\boldsymbol{\sigma}) \tag{7.107}$$

where $\mathbf{D}^{**}$ is obtained by substituting $\tilde{\mathbf{D}}$ in place of $\mathbf{D}$ in Eq. (7.67).†

The type of algorithm here described usually starts in the first iteration with a purely elastic increment, i.e.

$$\begin{aligned} \Delta\boldsymbol{\sigma}^0 &= \mathbf{D}\Delta\boldsymbol{\varepsilon} \\ \Delta\lambda^0 &= 0 \end{aligned} \tag{7.108}$$

and iteratively reduces the stress to the yield surface if plastic defor-

† It is somewhat puzzling that the 'consistent' tangent $\mathbf{D}^{**}$ is different from that of the tangent matrix $\mathbf{D}^{*}$ in Eq. (7.67). This is entirely due to the approximation involving the finite $\Delta\boldsymbol{\varepsilon}^{\mathrm{p}}$ increment and of course disappears as this becomes small. It is debatable which of these tangents should be used in the main iteration.

mations occur. For this reason it is sometimes call a *return mapping algorithm.*

It must be remarked that the iterative stress calculation is generally combined with equilibrium iterations on a one-to-one basis. Now the stress increments in individual steps can be added as Eq. (7.103) concerns only the total increment.

In all early applications of plasticity, modified Newton procedures and simple return methods were used.[46-49, 75-81] Today the other iterative procedures described earlier such as the secant methods are becoming more popular.

7.10 Some examples of plastic computation

The finite element discretization technique in plasticity problems follows precisely the same procedures as that of corresponding elasticity problems. Any of the elements already discussed can be used and again we find that generally higher-order elements with reduced integration show an improved performance with the *best* sampling point of the same kind as those described in Chapter 12 of Volume 1.

The use of such reduced integration is important in metal plasticity as the von Mises flow rules do not permit any volume changes. As the extent of plasticity spreads at the collapse load the deformation becomes nearly incompressible, and with conventional, exactly integrated, elements the system *locks* and a true collapse load cannot be obtained.[82,83] Here of course mixed formulations provide a viable alternative.

While the elastoplastic matrix deduced previously is valid for a general three-dimensional continuum, in two-dimensional plasticity it has to be reduced to special forms. In *plane stress*, for instance, this reduction is obvious by a simple deletion of the appropriate columns in Eq. (7.65) to which zero stress components are assigned; in a *plane stress* situation, all stresses exist but appropriate strain components have to be made zero. Appropriate elimination has now to be carried out and explicit expressions will be found in reference 49. It is of interest to note that in such cases the diagonal term corresponding to A is no longer zero, even in the case of ideal plasticity.

Finally, we should remark that the possibility of solving plastic problems is not limited to a displacement formulation alone. Equilibrium fields and, indeed, most of the formulations described in Chapter 12 of Volume 1 form a suitable vehicle,[84-86] but owing to their convenient and easy interpretation displacement forms are most commonly used.

7.10.1 *Perforated plate with and without strain hardening.*[48,79] Figure 7.14 shows the configuration and the division into simple linear triangu-

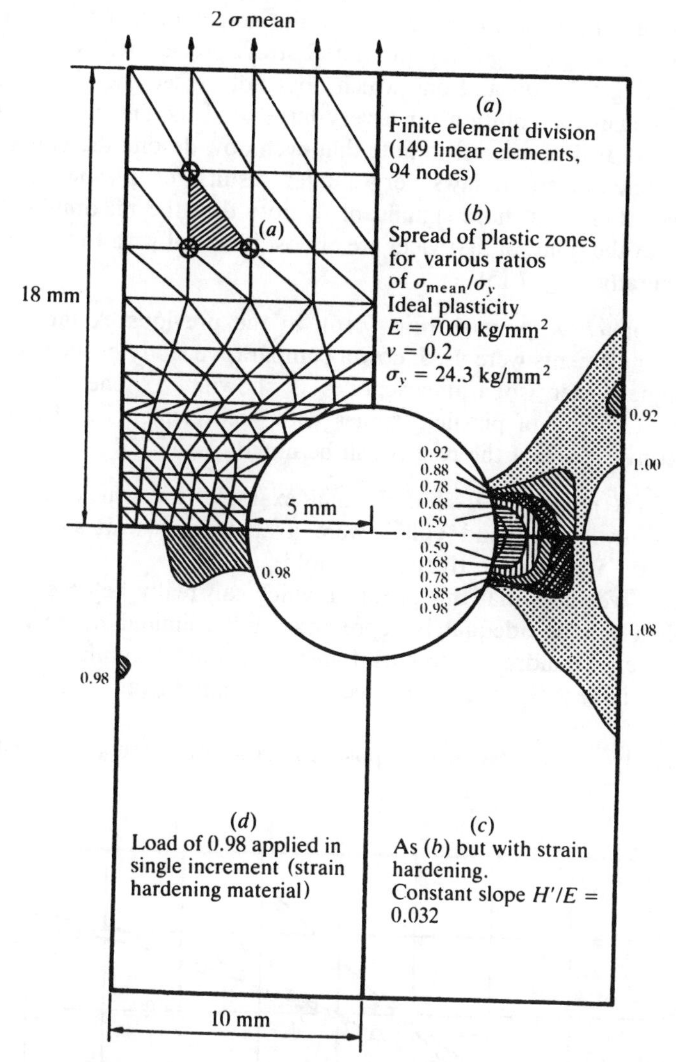

Fig. 7.14 Perforated tension strip (plane stress). Mesh used and development of plastic zones

lar elements. In this example plane stress conditions are assumed and solution is obtained for both ideal plasticity and strain hardening. The von Mises criterion is used and, in the case of strain hardening, a constant slope of the uniaxial hardening curve, H [Eq. (7.75)], was taken. The spread of plastic zones at various load levels is shown in Fig. 7.14(b) and (c).

Although the plasticity relation is only incremental, if the loads are applied in a single large step the initial stress process will still yield an equilibrating solution and one which does not exceed the yield stresses. Such a single-step solution for a very large load increment is shown in Fig. 7.14(*d*). It is of interest to note that even now despite the violation of the incremental strain laws very similar results for plastic zones are achieved. It is even more significant to note that the maximum strains reached at the point of first yield are almost identical with those achieved incrementally (Fig. 7.15).

7.10.2 *A notched specimen* (Fig. 7.16). In the previous example simple triangular elements were used; now a comparative study of such elements and higher-order isoparametric ones is shown.[67,87] The much more consistent spread of plasticity zones with such elements and the more rapid convergence of the results will be noted.

7.10.3 *Steel pressure vessel.* This final example, for which test results obtained by Dinno and Gill[88] were available, illustrates a practical application and its objectives are twofold.

Firstly, we show that this problem which can really be described as a thin shell can be adequately represented by a limited number (53) of isoparametric quadratic elements. Indeed, this model simulates well both the overall behaviour and the local stress concentration effects [Fig. 7.17(*a*)].

Secondly, it was decided to push the solution almost to the failure

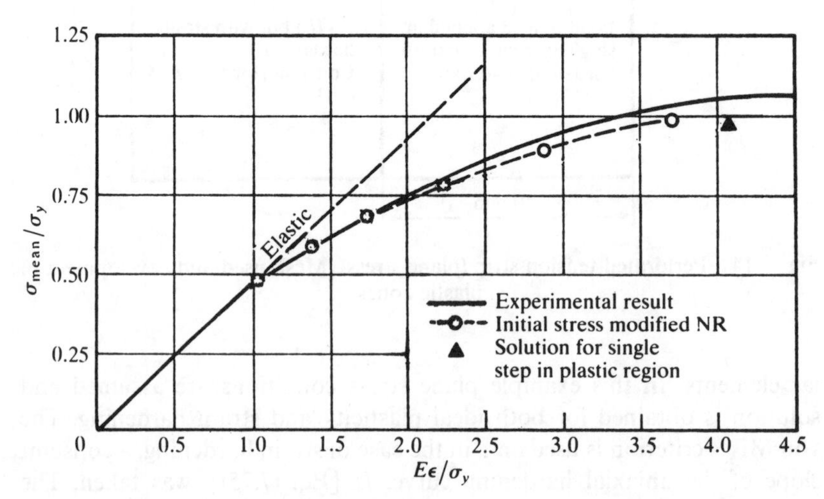

Fig. 7.15 Perforated tension strip (plane stress). Overall strain versus mean stress for $H^1/E=0.032$. Load increments of $0.2 \times$ first yield load

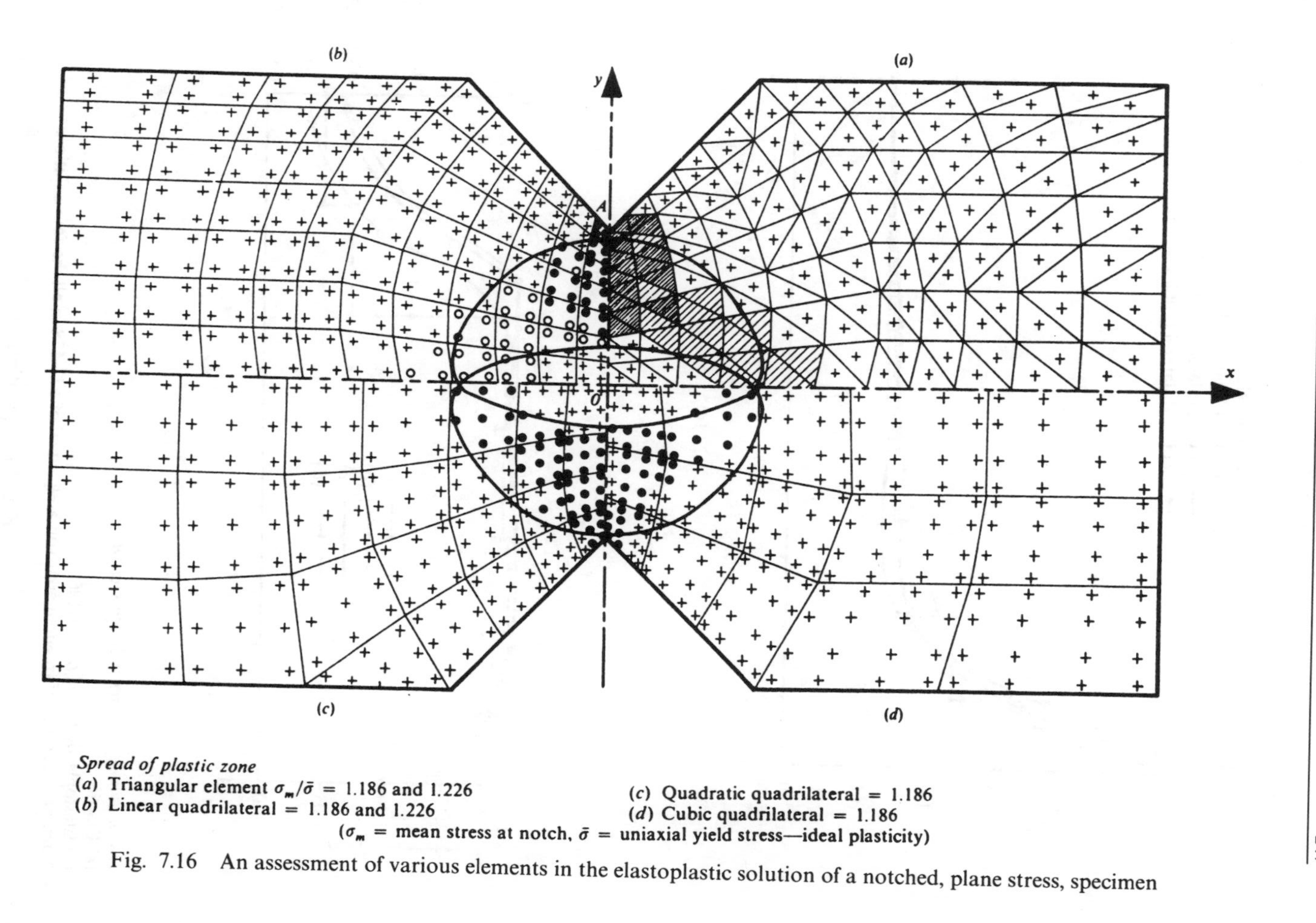

Spread of plastic zone
(*a*) Triangular element $\sigma_m/\bar{\sigma}$ = 1.186 and 1.226
(*b*) Linear quadrilateral = 1.186 and 1.226
(*c*) Quadratic quadrilateral = 1.186
(*d*) Cubic quadrilateral = 1.186
(σ_m = mean stress at notch, $\bar{\sigma}$ = uniaxial yield stress—ideal plasticity)

Fig. 7.16 An assessment of various elements in the elastoplastic solution of a notched, plane stress, specimen

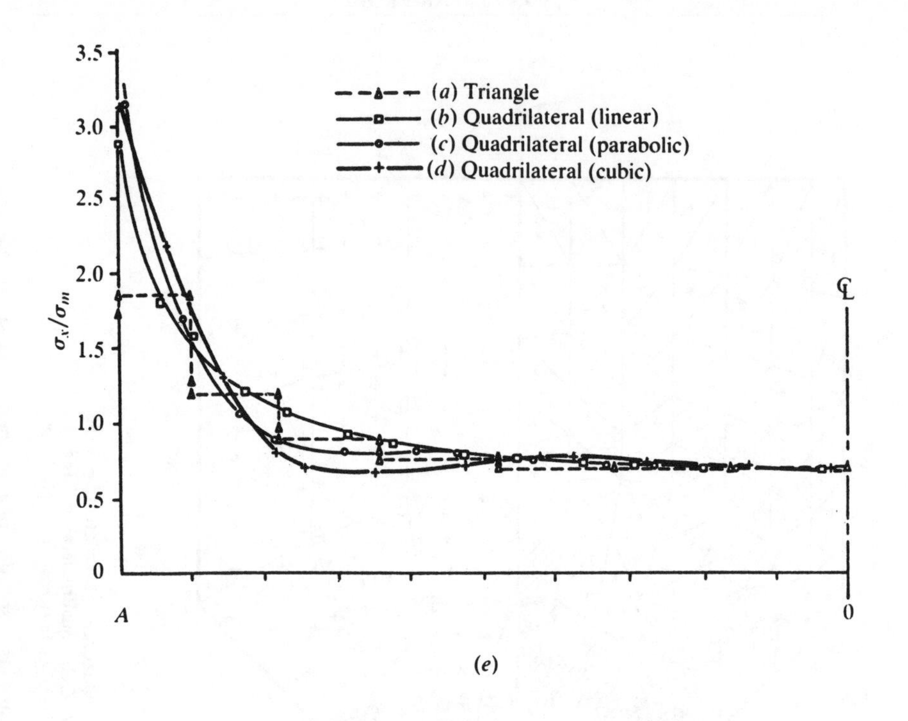

(e)

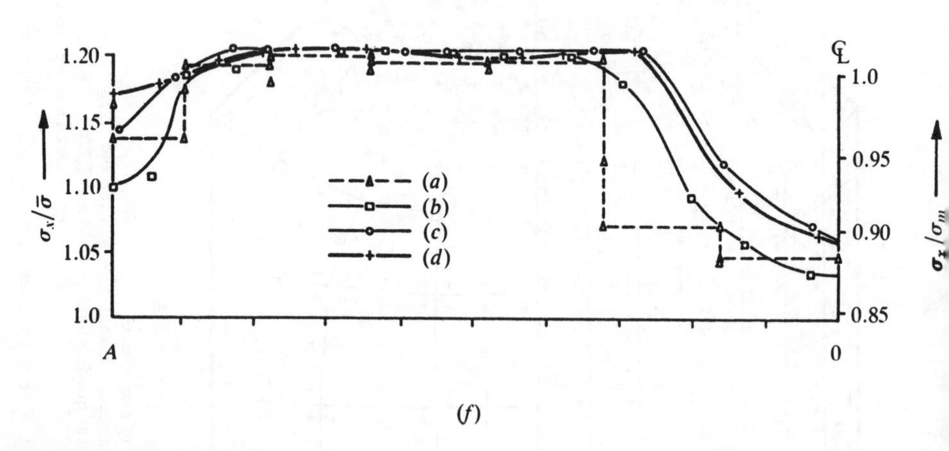

(f)

Distribution of stresses in notch section
(*e*) Elastic
(*f*) Elastoplastic for $\sigma_m/\bar{\sigma} = 1.186$
Number of degrees of freedom approximately equal of 172–178 in all the four solutions

Fig. 7.16 (continued)

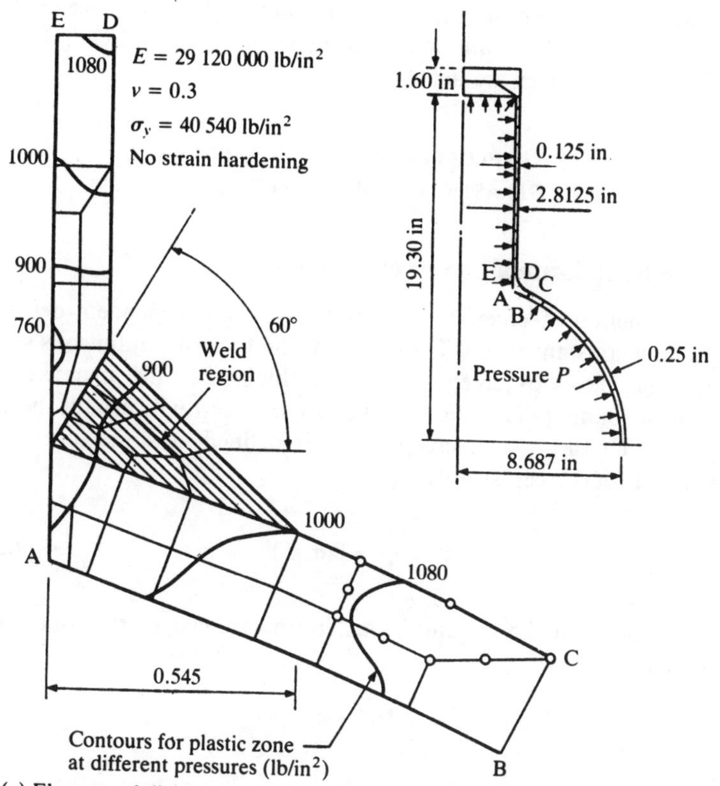

(*a*) Element subdivision and spread of plastic zones (von Mises yield and ideal plasticity)

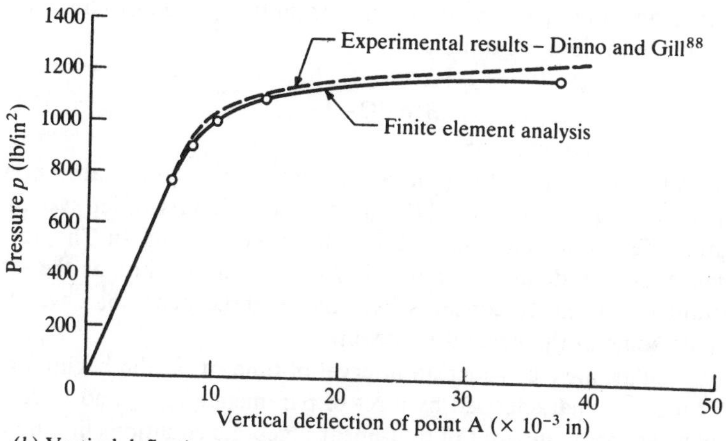

(*b*) Vertical deflection at point A with increasing pressure

Fig. 7.17 Steel pressure vessel

point while incrementing the pressure rather than displacement. A comparison of calculated and measured deflections in Fig. 17.17(*b*) shows how well the objectives are achieved.

RATE (TIME)-DEPENDENT PROBLEMS. CREEP, VISCOPLASTICITY AND VISCOELASTICITY

7.11 The basic formulation of creep problems

The phenomenon of 'creep' is manifested by a time-dependent deformation under a constant strain. Thus, in addition to an instantaneous strain, the material develops creep strains ε_c which generally increase with duration of loading. The constitutive law of creep will usually be of a form in which the *rate of creep strain* is defined as some function of stresses and total creep strains, i.e.

$$\dot{\boldsymbol{\varepsilon}}^c \equiv \frac{d\boldsymbol{\varepsilon}^c}{dt} = \boldsymbol{\beta}(\boldsymbol{\sigma}, \boldsymbol{\varepsilon}^c) \tag{7.109}$$

If we consider that the instantaneous strains are elastic, the total strain can be written as

$$\boldsymbol{\varepsilon} = \boldsymbol{\varepsilon}^e + \boldsymbol{\varepsilon}^c \tag{7.110}$$

with

$$\boldsymbol{\varepsilon}^e = \mathbf{D}^{-1}\boldsymbol{\sigma} \tag{7.111}$$

neglecting any initial (thermal) strains or initial (residual) stresses. As usual the equilibrium conditions

$$\int_\Omega \mathbf{B}^T \boldsymbol{\sigma}\, d\Omega - \mathbf{f} = 0 \tag{7.112}$$

hold at all times and, if the initial conditions of the system are known, the system of Eqs (7.109) to (7.112) gives a solvable first-order system of ordinary differential equations with non-linear coefficients. In Chapter 10 we shall discuss in detail the solution of such equations but, as the most important non-linearity concerns here the material behaviour, we shall anticipate some of the general procedures.

In particular, if we consider an interval of time Δt_n at the beginning of which the state, characterized by a set of parameters $\mathbf{a}_n$, $\boldsymbol{\sigma}_n$ and forces $\mathbf{f}_n$, is known, we can write a set of non-linear algebraic equations linking the initial conditions with the final ones at time $t_n + \Delta t = t_{n+1}$. Thus we have to solve a non-linear set of equilibrium equations of the usual type [viz.

Eq. (7.47)]

$$\boldsymbol{\Psi}_{n+1} \equiv \int_\Omega \mathbf{B}^\mathrm{T}\boldsymbol{\sigma}_{n+1}\,\mathrm{d}\Omega + \mathbf{f}_{n+1} = 0 \tag{7.113}$$

with an appropriate constitutive relation linking the stress and strain changes. Now this can be rewritten in an approximate form similar to that used in plasticity [viz. Eq. (7.102)] as

$$\begin{aligned} \Delta\boldsymbol{\sigma}_n &= \mathbf{D}(\Delta\boldsymbol{\varepsilon}_n - \Delta\boldsymbol{\varepsilon}_n^\mathrm{c}) \\ \Delta\boldsymbol{\varepsilon}_n^\mathrm{c} &= \Delta t\boldsymbol{\beta}_{n+\theta} \end{aligned} \tag{7.114}$$

where $\boldsymbol{\beta}_{n+\theta}$ is calculated as

$$\boldsymbol{\beta}_{n+\theta} = (1-\theta)\boldsymbol{\beta}_n + \theta\boldsymbol{\beta}_{n+1}$$

On eliminating $\Delta\boldsymbol{\varepsilon}_n^\mathrm{c}$ we have simply a non-linear equation

$$\mathbf{R}_{n+1} \equiv \Delta\boldsymbol{\sigma}_n - \mathbf{D}\mathbf{B}\Delta\mathbf{a}_n + \mathbf{D}\Delta t\boldsymbol{\beta}_{n+\theta} = 0 \tag{7.115}$$

The system of equations (7.113) and (7.115) can be solved iteratively using, say, the Newton–Raphson procedure.

Starting from some initial guess—say $\Delta\mathbf{a}_n^0 = 0$; $\Delta\boldsymbol{\sigma}_n^0 = 0$—the general iterative/incremental process can be written (dropping the subscripts n and $n+\theta$) as

$$\boldsymbol{\Psi}^{i+1} = 0 = \boldsymbol{\Psi}^i + \int_\Omega \mathbf{B}^\mathrm{T}\delta\boldsymbol{\sigma}^i\,\mathrm{d}\Omega \tag{7.116}$$

From Eq. (7.113) using (7.109) and (7.114) and assuming for this example that $\boldsymbol{\beta}$ does not depend on $\boldsymbol{\varepsilon}^\mathrm{c}$, we have, similarly,

$$\mathbf{R}^{i+1} = 0 = \mathbf{R}^i + \delta\boldsymbol{\sigma}^i - \mathbf{D}\mathbf{B}\delta\mathbf{a}^i + \mathbf{D}\Delta t\mathbf{C}\delta\boldsymbol{\sigma}^i \tag{7.117}$$

where

$$\mathbf{C} = \left\{\frac{\partial\boldsymbol{\beta}}{\partial\boldsymbol{\sigma}}\right\}_{n+\theta} = \theta\left\{\frac{\partial\boldsymbol{\beta}}{\partial\boldsymbol{\sigma}}\right\}_{n+1} \tag{7.118}$$

With the residuals $\boldsymbol{\Psi}^i$ and $\mathbf{R}^i$ evaluated (the latter at the element level for all integration points) the computation can proceed by eliminating $\delta\boldsymbol{\sigma}^i$ from (7.116) using (7.117) and now yielding

$$\left(\int_\Omega \mathbf{B}^T\mathbf{D}^{**}\mathbf{B}\,\mathrm{d}\Omega\right)\delta\mathbf{a}^i = -\boldsymbol{\Psi}^i - \int_\Omega \mathbf{B}^\mathrm{T}(\mathbf{I}+\mathbf{D}\Delta t\mathbf{C})^{-1}\mathbf{R}^i\,\mathrm{d}\Omega - \Delta\mathbf{f}_n \tag{7.119}$$

where

$$\mathbf{D}^{**} = (\mathbf{I}+\mathbf{D}\Delta t\mathbf{C})^{-1}\mathbf{D} \equiv [\mathbf{D}^{-1} + \Delta t\mathbf{C}]^{-1} \tag{7.120}$$

The iterative computation that follows is very similar to that used in plasticity but here Δt is equivalent to the load increment and, from the nature of time, is always positive. Indeed, the tangent matrix $\mathbf{D}^{**}$ plays here a role very similar to that used in Eq. (7.107).

While in plasticity we have generally used implicit (backward Euler) procedures, here many simple alternatives are possible and practised. In particular two schemes with a single iterative step are popular.

7.11.1 *Fully explicit–'initial strain' procedure:* $\theta=0$, $i=0$. Here, from Eqs (7.118) and (7.120) we see that

$$\mathbf{C}=\mathbf{0} \qquad \text{and} \qquad \mathbf{D}^{**}=\mathbf{D} \tag{7.121}$$

From Eq. (7.115), putting $\Delta\boldsymbol{\sigma}=0$ and $\Delta\mathbf{a}^0=0$, we have

$$\mathbf{R}^0=\mathbf{D}\Delta t\boldsymbol{\beta}_n \tag{7.122}$$

and by Eq. (7.119),

$$\delta\mathbf{a}^0=\Delta\mathbf{a}=\mathbf{K}^{-1}\left(\int_\Omega \mathbf{B}^{\mathrm{T}}\mathbf{D}\Delta t\boldsymbol{\beta}_n\,\mathrm{d}\Omega+\Delta\mathbf{f}_n\right) \tag{7.123}$$

where

$$\mathbf{K}=\int_\Omega \mathbf{B}^{\mathrm{T}}\mathbf{D}\mathbf{B}\,\mathrm{d}\Omega \tag{7.124}$$

is the standard elastic stiffness matrix. This of course is equivalent to evaluating the increment of creep strain from the initial stress values at time n and is exceedingly simple to calculate.

This process is deservedly popular[89-92] as the computation at each time step is simply one of resolution, but is obviously less accurate than other alternatives. Further, if the time interval is too large, unstable results may be obtained (see Chapter 10). Thus is it necessary for

$$\Delta t \leqslant \Delta t_{\mathrm{crit}} \tag{7.125}$$

where Δt_{crit} is determined in a suitable manner.

A rule of thumb that proves quite effective in practice is that the increment of creep strain should not exceed one half of the total elastic strain, i.e.[93]

$$\Delta t\boldsymbol{\beta}_n \leqslant \tfrac{1}{2}\mathbf{D}\boldsymbol{\sigma}_n \tag{7.126}$$

7.11.2 *Fully explicit process with modified stiffness:* $\frac{1}{2}\leqslant\theta\leqslant 1$, $i=0$. Here the main difference from the first explicit process is that the matrix $\mathbf{C}$ is not equal to zero but within a single step it is taken as a constant, i.e.

$$\mathbf{C}=\theta\left\{\frac{\partial\boldsymbol{\beta}}{\partial\boldsymbol{\sigma}}\right\}_n \tag{7.127}$$

Now

$$\mathbf{D}^{**} = (\mathbf{I} + \mathbf{D}\Delta t\mathbf{C})_n^{-1}\mathbf{D} \equiv (\mathbf{D}^{-1} + \Delta t\mathbf{C})_n^{-1}$$

Using the same substitutions as before we have

$$\delta\mathbf{a}^0 = \Delta\mathbf{a} = \tilde{\mathbf{K}}\left(\int_\Omega \mathbf{B}^T\mathbf{D}\Delta t\boldsymbol{\beta}_n \, d\Omega + \Delta\mathbf{f}_n\right) \tag{7.128}$$

where $\tilde{\mathbf{K}}$ is obtained by substitution of $\mathbf{D}^{**}$ in place of $\mathbf{D}$ in the standard stiffness matrix of Eq. (7.124).

This process is more expensive than the simple explicit one previously mentioned, as the matrix $\tilde{\mathbf{K}}$ has to be formed and solved for every time step. Further, such matrices can be non-symmetric, adding to computational difficulties.

Neither of the simplified iterations procedures described above give any attention to the errors introduced in the estimates of the creep strain. However, for accuracy the iterative process with $\theta \geqslant \frac{1}{2}$ is recommended. Such fully iterative procedures have been introduced first by Cyr and Teter,[94] Zienkiewicz and coworkers[95,96] and later by others.[97]

We shall note that the process has much similarity with the iterative solutions of plastic problems of Sec. 7.9 in the case of viscoplasticity, which we shall discuss in the next section.

7.12 Viscoplasticity

7.12.1 *General.* The purely plastic behaviour of solids postulated in Sec. 7.8 is probably a fiction as the maximum stress that can be carried is invariably associated with the rate at which this is applied. A purely elastoplastic behaviour in a uniaxial loading is described in a model of Fig. 7.18(*a*) in which the plastic strain rate is zero for stresses below yield,

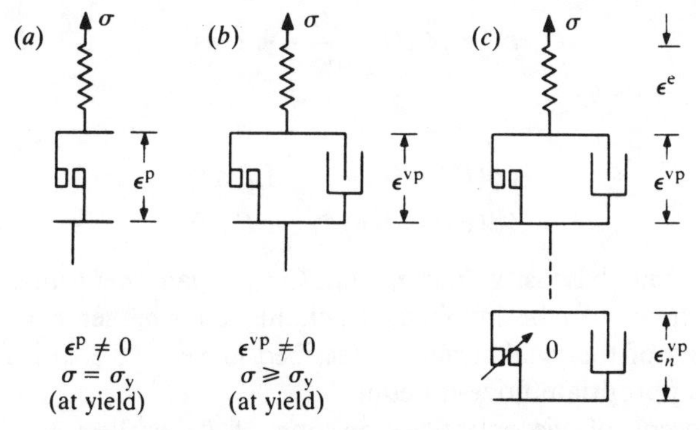

Fig. 7.18 (*a*) Elastoplastic. (*b*) Elastoviscoplastic. (*c*) A series of elastoviscoplastic models

i.e.

$$\dot{\varepsilon}^{\mathrm{p}}=0 \qquad \text{if } \sigma-\sigma_{\mathrm{y}}<0 \quad \text{and} \quad \sigma>0$$

and $\dot{\varepsilon}^{\mathrm{p}}$ is indeterminate when $\sigma-\sigma_{\mathrm{y}}=0$.

The elastoviscoplastic material, on the other hand, can be modelled as shown in Fig. 7.18(*b*), where a dashpot is placed in parallel with the plastic element. Now stresses can exceed σ_{y} for strain rates other than zero.

The viscoplastic (or creep) strain rate is now given by a general expression

$$\dot{\varepsilon}^{\mathrm{vp}}=\gamma\langle\phi(\sigma-\sigma_{\mathrm{y}})\rangle \tag{7.129a}$$

where the arbitrary function ϕ is such that

$$\begin{aligned} &\langle\phi(\sigma-\sigma_{\mathrm{y}})\rangle=0 && \text{if } \sigma-\sigma_{\mathrm{y}}\leqslant 0\\ \text{and} \quad &\langle\phi(\sigma-\sigma_{\mathrm{y}})\rangle=\phi(\sigma-\sigma_{\mathrm{y}}) && \text{if } \sigma-\sigma_{\mathrm{y}}>0 \end{aligned} \tag{7.129b}$$

The model suggested is, in fact, of a creep-type category described in the previous sections and appears to be more realistic than that of pure plasticity.

A generalization of a viscoplastic model to a general stress state follows precisely the arguments of the plasticity section.

Firstly, we note that the strain rate will be a function of the yield condition

$$F(\boldsymbol{\sigma}, \kappa)$$

defined in Eq. (7.56). If this is less than zero, no plastic flow will occur.

Secondly, we shall postulate a viscoplastic potential $Q(\boldsymbol{\sigma}, \kappa)$, the normality to this governing the ratio of the various strain rate components [see Eq. (7.59)]. Thus, quite generally we can write

$$\dot{\boldsymbol{\varepsilon}}^{\mathrm{vp}}=\gamma\langle\phi(F)\rangle\frac{\partial Q}{\partial\boldsymbol{\sigma}}\equiv\boldsymbol{\beta}(\boldsymbol{\sigma}, \kappa) \tag{7.130}$$

where

$$\begin{aligned} &\langle\phi(F)\rangle=0 && \text{if } F\leqslant 0\\ &\langle\phi(F)\rangle=\phi(F) && \text{if } F>0 \end{aligned} \tag{7.131}$$

and γ is some 'viscosity' parameter. Once again *associated* or *non-associated* flows can be invoked, depending on whether $Q=F$ or not. Further, any of the yield surfaces described in Sec. 7.8.2 can be used to define the appropriate flow in detail.

The concept of viscoplasticity in one of its earliest versions was introduced by Bingham in 1922[98] and a complete survey of such modelling is given in reference.[99] The computational procedure of using the viscoplastic model can obviously follow any of the general methods described in the previous section. Most commonly the straightforward *Euler* method of the previous section is used.[100-104] The detailed stability requirements for this have been considered for several types of yield condition by Cormeau.[105] The *tangential* process can again be used, but unless the viscoplastic flow is associated (that is $Q=F$), non-symmetric systems of equations have to be solved at each step.

7.12.2 *Iterative solution.* The complete iterative solution scheme of Eqs (7.113) to (7.119) for viscoplasticity becomes very similar to that used in plasticity and described in Eqs (7.102) to (7.107). To underline this similarity Eq. (7.114) is rewritten with $\theta=1$ in the form of Eq. (7.103) using the viscoplastic Eq. (7.130) as

$$\mathbf{R}=\Delta\boldsymbol{\varepsilon}_n-\mathbf{D}^{-1}\Delta\boldsymbol{\sigma}_n-\left(\frac{\partial Q}{\partial \boldsymbol{\sigma}}\right)_{n+1}\Delta\lambda=0 \tag{7.132a}$$

with

$$r=\frac{1}{\gamma\Delta t}\Delta\lambda-\langle\phi(F)\rangle_{n+1}=0 \tag{7.132b}$$

when $F_{n+1}\geqslant 0$; otherwise $\Delta\lambda=0$.

The Newton–Raphson iterative increments $\delta\boldsymbol{\sigma}^i$ and $\delta\lambda$ can now be determined by differentiation in the usual manner. Thus we have (again dropping the subscripts n)

$$\begin{aligned}&\mathbf{B}\delta\mathbf{a}^i-\mathbf{D}^{-1}\delta\boldsymbol{\sigma}^i-\Delta\lambda\frac{\partial^2 Q}{\partial\boldsymbol{\sigma}^2}\delta\boldsymbol{\sigma}^i-\frac{\partial Q}{\partial\boldsymbol{\sigma}}\delta\lambda^i+\mathbf{R}^i=0\\ &\frac{1}{\gamma\Delta t}\delta\lambda-\left(A\delta\lambda+\phi'\frac{\partial F}{\partial\boldsymbol{\sigma}}\delta\boldsymbol{\sigma}\right)+r^i=0\quad\left(\phi'=\frac{\mathrm{d}\phi}{\mathrm{d}F}\right)\end{aligned} \tag{7.133}$$

or

$$\begin{bmatrix}\mathbf{D}^{-1}, & \dfrac{\partial Q}{\partial\boldsymbol{\sigma}}\\ \phi'\dfrac{\partial F^{\mathrm{T}}}{\partial\boldsymbol{\sigma}}, & -\left(A+\dfrac{1}{\gamma\Delta t}\right)\end{bmatrix}\begin{Bmatrix}\delta\boldsymbol{\sigma}^i\\ \delta\lambda^i\end{Bmatrix}=\begin{Bmatrix}\mathbf{R}^i-\mathbf{B}\delta\mathbf{a}^i\\ r^i\end{Bmatrix} \tag{7.134}$$

Now the equations are almost identical to those of plasticity [viz. Eq. (7.104)] with $\tilde{\mathbf{D}}^{-1}$ being of the form given by Eq. (7.105). Further, $\phi'=1$ if $\phi(F)=F$ and now the only difference is in the time-dependent term $1/\gamma\Delta t$.

Again the consistent tangent $\mathbf{D}^{**}$ can be obtained by elimination of $\delta\lambda^i$ and a general iterative scheme is available.

Indeed, as expected, $\gamma = \infty$ or $\Delta t = \infty$ will now correspond to the exact plasticity solution. This will always be reached by any solution tending to steady state. The viscoplastic model thus provides a convenient form for solution of purely plastic, rate-independent, problems.

In this context, as we have illustrated above, the implicit iterative scheme in a single time step (corresponding to a single load increment) can be used directly. However, the explicit iteration schemes previously described are also applicable, with time incrementation being simply a computational device necessary to reach the sate at which the *strain rates are zero* and hence the plastic yield has been reached. The simple explicit, initial strain, process is equivalent to a modified Newton–Raphson method using the elastic stiffness. In such procedures a critical time step occurs[105] and this limit is roughly equivalent to the use of modified Newton–Raphson iteration with a twofold increase of $\delta \mathbf{a}$ (an accelerator process).

The viscoplastic laws can easily be generalized to include a series of components, as shown in Fig. 7.18(*c*). Now we write

$$\dot{\boldsymbol{\varepsilon}}^{vp} = \dot{\boldsymbol{\varepsilon}}_1^{vp} + \dot{\boldsymbol{\varepsilon}}_2^{vp} + \cdots = \boldsymbol{\beta}(\boldsymbol{\sigma}) \tag{7.135}$$

and again the standard formulation suffices. If, as shown in the last element of Fig. 7.18(*c*) the plastic yield is set to zero, a 'pure' creep situation arises in which flow occurs at all stress levels.

7.12.3 *Creep of metals.* If an associated form of viscoplasticity using the von Mises yield criterion of Eq. (7.78) is considered, the viscoplastic strain rate can be written as

$$\dot{\boldsymbol{\varepsilon}}^{vp} = \gamma \langle \phi(\bar{\sigma} - \sigma_y) \rangle \frac{\partial \bar{\sigma}}{\partial \boldsymbol{\sigma}} \tag{7.136}$$

If σ_y, the yield stress, is put to zero, ϕ is made an exponential form and we make use of the expressions of Table 7.1, the above can be written as

$$\dot{\boldsymbol{\varepsilon}}^{vp} = \gamma \bar{\sigma}^m \mathbf{M}^{\mathrm{I}} \boldsymbol{\sigma} \tag{7.137}$$

and we obtain the well-known Norton–Soderberg creep law. In this, generally, the parameter γ is a function of time, temperature and the total creep strain. For a survey of such laws the reader should consult specialized references.[106,107]

An example initially solved using a large number of triangular elements[92] is presented in Fig. 7.19, where a much smaller number of isoparametric quadrilaterals are used in a general viscoplastic program.[104]

7.12.4 *Plasticity solutions by viscoplastic algorithm—soil mechanics.* As we have already mentioned, the viscoplastic model provides a simple and

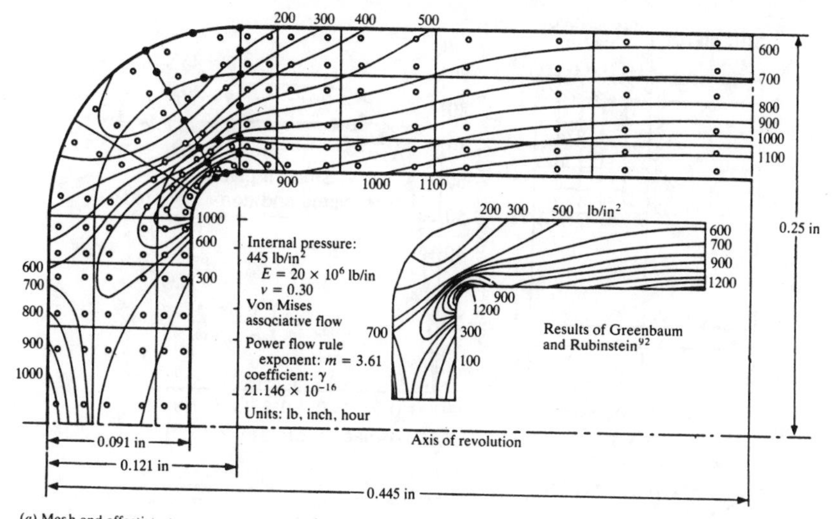

(a) Mesh end effective stress contours at start of pressurization

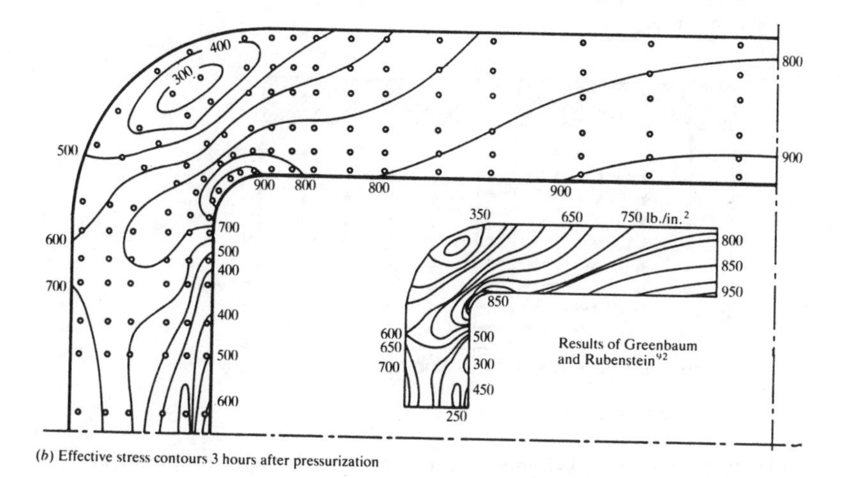

(b) Effective stress contours 3 hours after pressurization

Fig. 7.19 Creep in a pressure vessel

powerful tool for the solution of all plasticity problems. In reference 104 many classical problems are thus resolved and the reader is directed there for details. In this section some problems of soil mechanics are discussed in which the facility of the process for solving non-associated behaviour is demonstrated.[108] The whole subject of the behaviour of soils and similar porous media is one in which much has yet to be done to formulate good constitutive models. For a fuller discussion the reader is referred to recent texts, conferences and papers on the subject.[109,110]

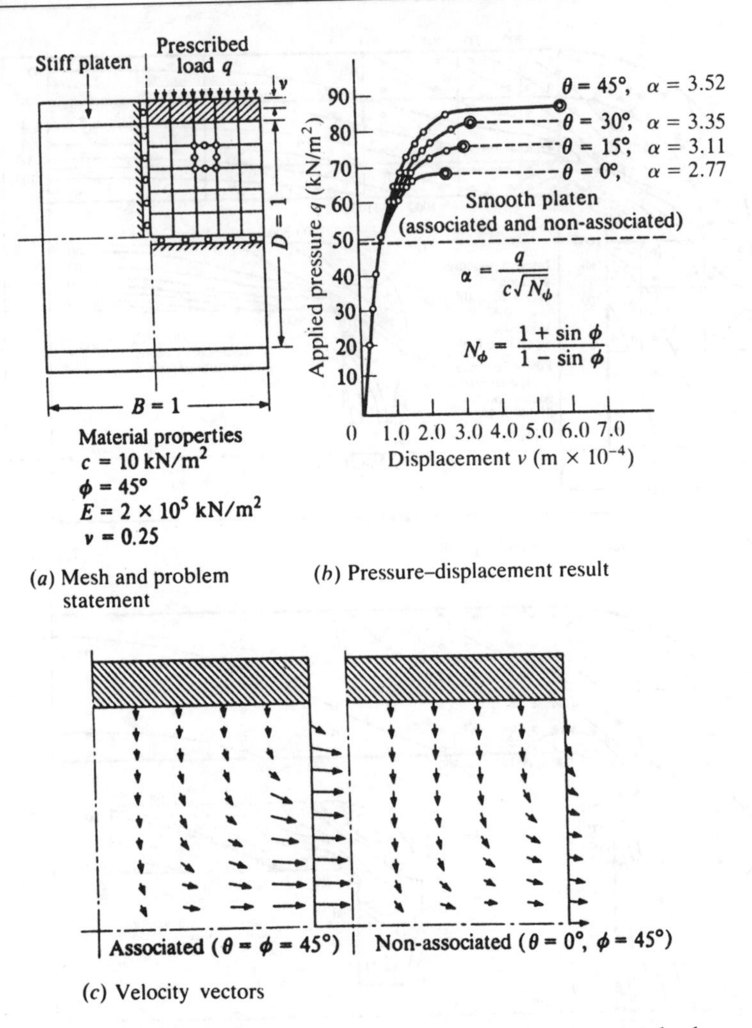

(*a*) Mesh and problem statement

(*b*) Pressure–displacement result

(*c*) Velocity vectors

Fig. 7.20 Uniaxial, axisymmetric, compression between rough plates

One particular controversy centres on the 'associated' versus 'non-associated' nature of soil behaviour. In the example of Fig. 7.20 dealing with an axisymmetric sample, the effect of these different assumptions is investigated.[108] Here a Mohr–Coulomb law is used to describe the yield surface and a similar form, but with a different friction angle, $\bar{\phi}$, is used in the plastic potential, thus reducing the plastic potential to the Tresca form of Fig. 7.11 when $\bar{\phi}=0$ and suppressing volumetric strain changes. As can be seen from the results, only moderate changes in the collapse

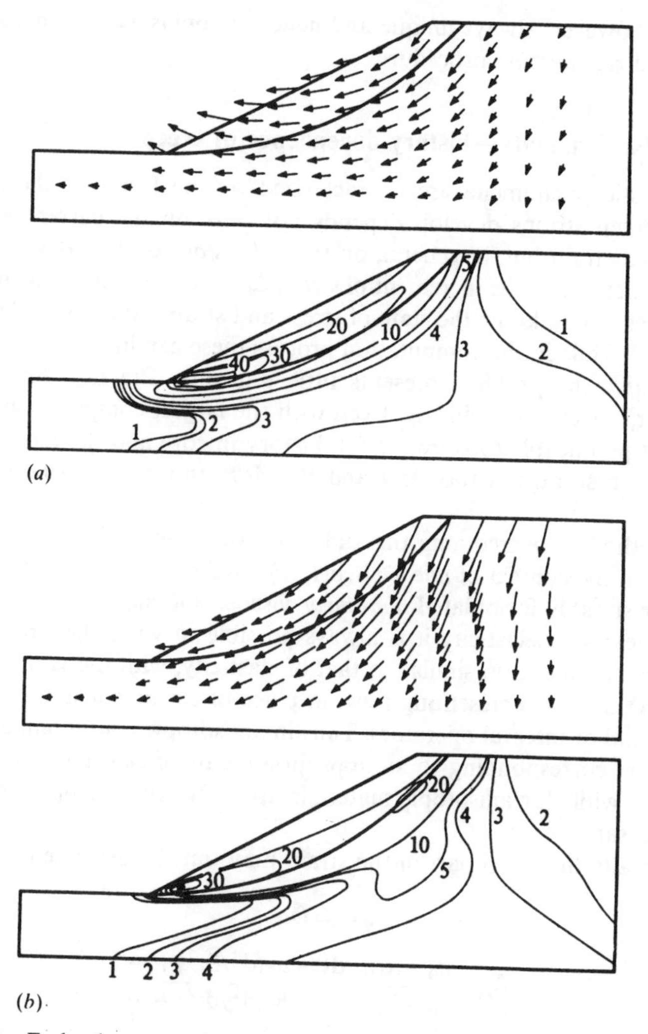

Fig. 7.21 Embankment under action of gravity, relative plastic velocities at collapse and effective shear strain rate contours at collapse; (*a*) and (*b*) show associative and non-associative (zero volume change) behaviour

load occur, although very appreciable differences in plastic flow patterns exist.

Figure 7.21 shows a similar study carried out for an embankment. Here, despite quite different flow patterns, a prediction of collapse load was almost unaffected by the flow rate law assumed.

The non-associative plasticity, in essence caused by frictional behaviour, may lead to non-uniqueness of solution. The equivalent viscoplastic form is, however, always unique and hence viscoplasticity is on occasion used as a *regularizing* procedure.

7.13 Viscoelasticity—history dependence of creep

Viscoelastic phenomena are characterized by the fact that the rate at which creep strains develop depends not only on the current state of stress and strain but, in general, on the *full history* of their development. Thus to determine the increment of strain $\Delta\varepsilon^c$ at a particular time interval it is necessary to know the state of stress and strains at all *preceding time intervals*. While in the computation process these can in fact be obtained, *in principle* the problem presents little difficulty. Practical limitations, however, appear immediately. Even with the largest computers available it is not practicable to store the full history in core and the repeated use of backing storage is too slow and therefore too costly to be contemplated.

A method of overcoming this difficulty was described by Zienkiewicz *et al.*[90] in the context of *linear viscoelastic* analysis and presents possibilities for suitably formulated *non-linear* viscoelastic materials.

In linear viscoelasticity it is always possible to write the stress–strain relationship in a form similar to that of elasticity, with the various terms of the **D** matrix representing now, in place of elastic constants, suitable differential or integral operators. Thus in an isotopic continuum a pair of operators corresponding to an appropriate pair of elastic constants will appear—while for anisotopic materials up to 21 separate operators may be necessary.[111]

Typically the creep part of the strain may thus be described by

$$\boldsymbol{\varepsilon}^c = \bar{\mathbf{D}}\boldsymbol{\sigma}$$

with

$$\bar{D}_{rs} = \frac{a_0 + a_1(\mathrm{d}/\mathrm{d}t) + a_2(\mathrm{d}^2/\mathrm{d}t^2) + \cdots}{b_0 + b_1(\mathrm{d}/\mathrm{d}t) + b_2(\mathrm{d}^2/\mathrm{d}t^2) + \cdots} \tag{7.138}$$

Here the operators are written in a differential form. If this expansion is finite, then separating any instantaneous elastic effects one can usually rewrite Eq. (7.138) in terms of partial fractions as

$$\bar{D}_{rs} = \frac{A_1}{(\mathrm{d}/\mathrm{d}t) + B_1} + \frac{A_2}{(\mathrm{d}/\mathrm{d}t) + B_2} + \cdots \tag{7.139}$$

This, as is well known, can be interpreted as a response of a series of 'Kelvin' elements illustrated in Fig. 7.22 (even though physically no significance need be attached to such models) where each term represents one Kelvin unit. A typical contribution to a strain component is thus an

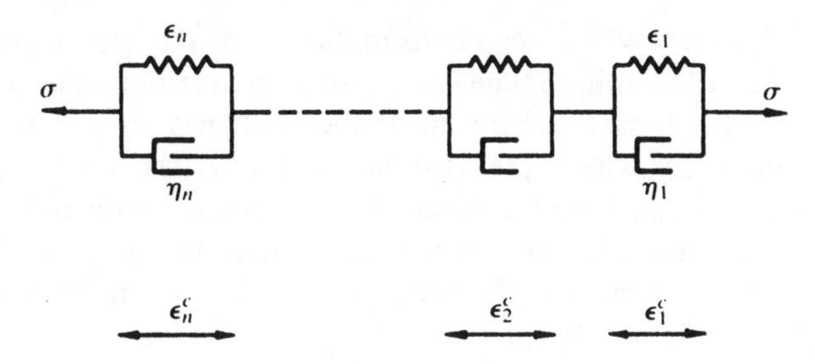

Fig. 7.22 A series of Kelvin elements

addition of terms of the form

$$\boldsymbol{\varepsilon}_k^{ve} = \frac{A_k}{(\mathrm{d}/\mathrm{d}t) + B_k} \boldsymbol{\sigma} \tag{7.140a}$$

or

$$\frac{\mathrm{d}\boldsymbol{\varepsilon}_k^{ve}}{\mathrm{d}t} = A_k \boldsymbol{\sigma} - B_k \boldsymbol{\varepsilon}_k^{ve} \tag{7.140b}$$

and the full expression for $\dot{\boldsymbol{\varepsilon}}^{ve}$ is obtained once again in the form of Eq. (7.109) for which the standard methods described in Sec. 7.12 are applicable. The Euler procedure is in fact used in reference 90, but the tangential procedures lead to a more rapid and accurate solution if we observe that the matrix **C** is constant in linear viscoelastic forms.

In practice only a limited number of Kelvin elements are needed to represent the material behaviour and, additionally, only a few 'viscoelastic' operators exist. For instance, in an isotropic incompressible material only one operator serves to define the $\bar{\mathbf{D}}^{-1}$ matrix. With two elements of expansion [Eq. (7.138)] defining the operator, only two quantities will have to be stored during the computation process, etc.

The values of A_n and B_n for each Kelvin model can be age and temperature dependent without introduction of any complexity of calculation—thus dealing with problems of thermoviscoelasticity, such as occur in the creep of concrete or plastics.[112,113]

The viscoelastic computations can, on occasion, be simplified by the use of special devices for expanding the history-dependent creep strain. A particularly efficient method is suggested by Taylor *et al.*[112]

The labour of step-by-step solutions for linear viscoelastic media can, on occasion, be substantially reduced. In the case of a homogeneous structure with linear isotopic viscoelasticity and constant Poisson ratio operator, the Alfrey–McHenry analogies allow single-step elastic solutions to be used to obtain stresses and displacements at a given time by the use of *equivalent loads, displacements and temperatures.*[114,115]

Some extensions of these analogies have been proposed by Hilton and

Russell.[116] Further, when creep deformation is of the type tending to a constant value at an infinite time it is possible to determine the final stress distribution even in cases where the above analogies are not applicable. Thus, for instance, where the viscoelastic properties are temperature dependent and the structure is subject to a system of loads and temperatures which remain constant with time, long-term 'equivalent' elastic constants can be found and the problem solved as a single, non-homogeneous elastic one of a linear kind.[117]

7.14 Some special problems of rock, concrete, etc.

7.14.1 *The no-tension material.* A hypothetical material capable of sustaining only compressive stresses and straining without resistance in tension is in many respects similar to an ideal plastic material. While in practice such an ideal material does not exist, it gives a close approximation to the behaviour of randomly jointed rock and other granular materials. While an explicit stress–strain relation cannot be generally written it suffices to carry out the analysis elastically and wherever tensile stresses develop to reduce these to zero. The initial stress process is here natural and indeed was developed in this context.[118] The steps of calculation are obvious but it is important to remember that the *principal tensile stresses* have to be eliminated.

The 'constitutive' law as stated above can at best approximate to the true situation, no account being taken of the closure of fissures on reapplication of compressive stresses. However, these results certainly give a clear insight into the behaviour of real rock structures.

An underground power station.[118] Figure 7.23(*a*) and (*b*) shows an application of this model to a practical problem. In Fig. 7.23(*a*) an elastic solution is shown for stresses in the vicinity of an underground power station with cable prestressing applied in the vicinity of the opening. The zones in which tension exists are indicated. In Fig. 7.23(*b*) a *no-tension* solution is given for the same problem, indicating the rather small general stress redistribution and the zones where 'cracking' has occurred.

Reinforced concrete. A variant on this type of material may be one in which a finite tensile strength exists but when this is once exceeded the strength drops to zero (on fissuring). Such an analysis was used by Valliappan and Nath[119] in the study of the behaviour of reinforced concrete beams. Very good correlation with experimental results for over-reinforced beams (in which development of compressive yield is not important) have been obtained. The beam is one for which test results were obtained by Krahl *et al.*[120] Figure 7.24 shows some relevant results.

Much development work on the behaviour of reinforced concrete has

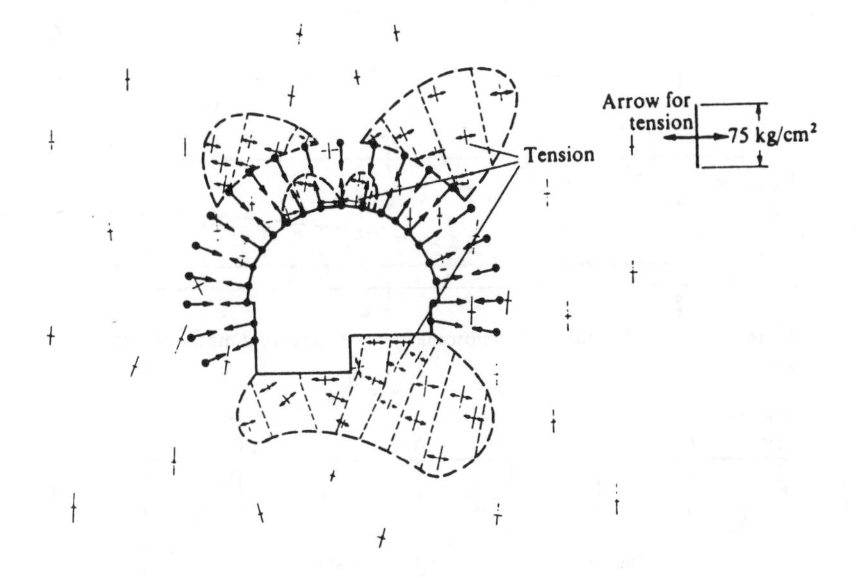

(*a*) Elastic stresses

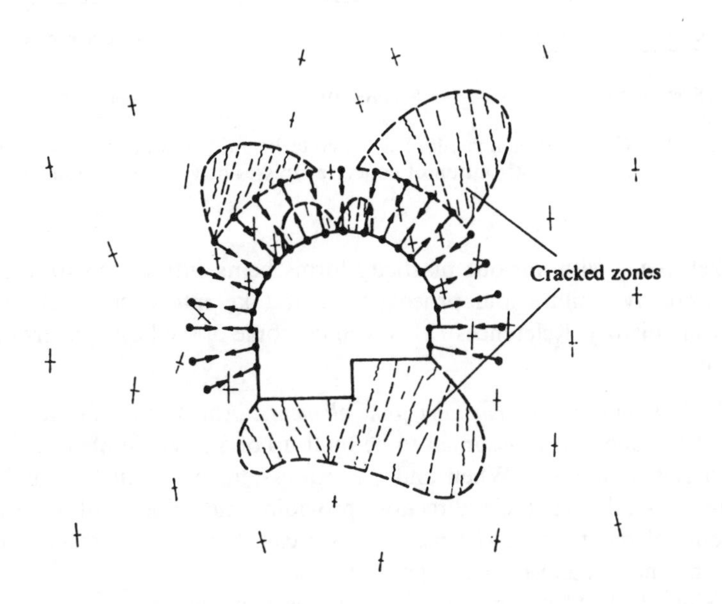

(*b*) 'No-tension' stresses

Fig. 7.23 Underground power station. Gravity and prestressing loads

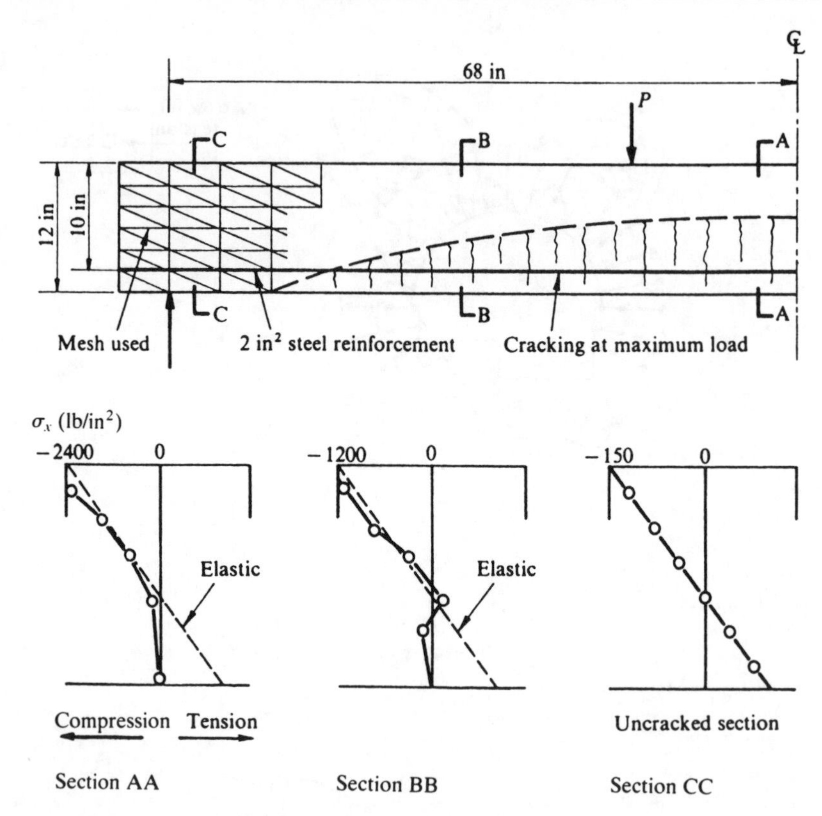

Fig. 7.24 Cracking of a reinforced concrete beam. (Maximum tensile strength $200\,\text{lb/in}^2$.) Distribution of concrete stresses at various sections[119]

taken place with various plasticity forms being introduced to allow for compressive failure and procedures that take into account the crack-closing history. References 121 to 126 list some of the basic papers on this subject.

7.14.2 *'Laminar' material and joint elements.* Another idealized material model is one that is assumed to be built up of a large number of isotopic and elastic laminae. When under compression these can transmit shear stress parallel to their direction providing this does not exceed the frictional resistance. No tensile stresses can, however, be transmitted in the normal direction to the laminae.

This idealized material has obvious uses in the study of rock masses with parallel joints but as will be seen later has a much wider applicability. Figure 7.25 shows a two-dimensional situation involving such a material. With a local coordinate axis x' oriented in the direction of the

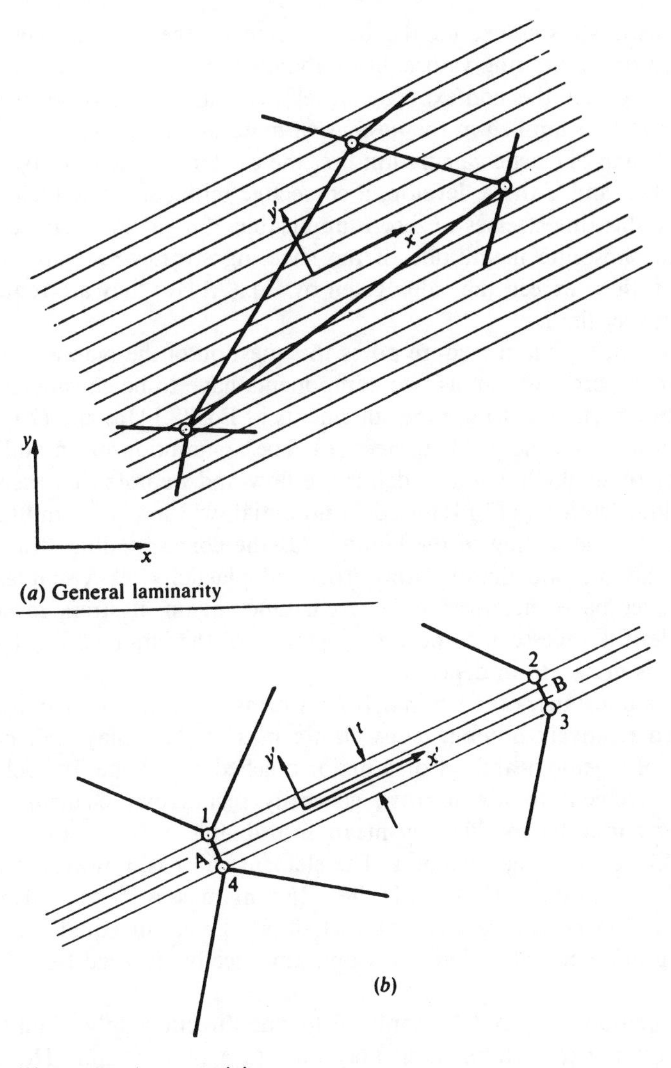

Fig. 7.25 'Laminar' material

laminae we can write for a simple friction contact joint

$$|\tau_{x'y'}| < \mu\sigma_{y'} \qquad \text{if } \sigma'_{y'} \leqslant 0 \tag{7.141a}$$

and

$$\sigma_{y'} = 0 \qquad \text{if } \varepsilon_{y'} > 0 \tag{7.141b}$$

for stresses at which purely elastic behaviour occurs. In the above, μ is the friction coefficient applicable between the laminae.

If elastic stresses exceed the limits imposed the stresses have to be reduced to the limiting values given above. The application of the initial stress process in this context is again self-evident, and the problem is very similar to that implied in the *no-tension* material of the previous section. At each step of elastic calculation first the existence of tensile stresses $\sigma_{y'}$ is checked and if these develop, a corrective initial stress reducing these and the shearing stresses to zero is applicable. If $\sigma_{y'}$ stresses are compressive, the absolute magnitude of the shearing stresses $\tau_{x'y'}$ are checked again; if these exceed the value given by Eq. (7.141a) they are reduced to their proper limit.

However, such a procedure poses the question of the manner in which the stresses are reduced, as two components have to be considered. It is, therefore, preferable to use the statements of Eqs (7.141a) and (7.141b) as definitions of plastic yield surfaces (F). The assumption of an additional plastic potential (Q) will now define the flow and we note that associated behaviour, with Eqs (7.141) used as potential, will imply a simultaneous separation and sliding of the laminae (as the corresponding strain rates $d\gamma_{x'y'}$ and $d\varepsilon_{y'}$ are finite). Non-associated plasticity (or viscoplasticity) techniques have therefore to be used. Once again, if stress reversal is possible it is necessary to note the opening of the laminae, i.e. the yield surface is made strain dependent.

In some instances the laminar behaviour is confined to a narrow joint between relatively homogeneous elastic masses. This may well be of a nature of a geological fault or a major crushed rock zone. In such cases it is convenient to use narrow, generally rectangular elements whose geometry may be specified by mean coordinates of two ends A and B [Fig. 7.25(b)] and the thickness. The element still has, however, separate points of continuity (1–4) with the adjacent masses.[127,128] Such joint elements can be simple rectangles as shown here, but equally well can take up more complex shapes if isoparametrically specified (see Chapter 8).

Laminations may not be confined to one direction only—and indeed the interlaminar material itself may possess a plastic limit. The use of such multilaminate models in the context of rock mechanics has proved very effective;[129] with a random distribution of laminations we return of course to a typical soil-like material and the possibilities of extending such models to obtain new and interesting constitutive relations have been highlighted by Pande and Sharma.[130]

7.15 Non-uniqueness and localization in solid mechanics—some outstanding problems

7.15.1 *General remarks—non-uniqueness.* In the preceding sections the general processes of dealing with complex, non-linear constitutive relations have been examined and some particular applications were discussed. Clearly, the subject is so large and of so great a practical importance that presentation in a single chapter is impracticable. For different materials different forms of constitutive relations can be proposed and experimentally verified. *Once such constitutive relations are available the standard processes of this chapter can be adopted*; indeed, it is now possible to build standard computing systems applicable to a wide variety of material properties in which new specifications of behaviour are simply inserted as an appropriate 'black box'.

What must be once more restated is that, in non-linear problems,

(*a*) non-uniqueness of solution may arise,

(*b*) convergence can never be, *a priori*, guaranteed,

(*c*) the cost of solution is invariably greater than in linear solutions.

Here of course the item of most serious concern is the first one, i.e. that of non-uniqueness, which could mean a physically irrelevant solution even if numerical convergence occurred and possibly large computation expense was incurred. Such non-uniqueness may be due to several reasons in elastoplastic computations:

(*a*) the existence of corners in the yield (or potential) surfaces at which the gradients are not uniquely defined,

(*b*) the introduction of *non-associativeness* into the formulation (to which we have already referred on pages 230 and 258),[131-134]

(*c*) the development of strain softening.[135]

The first problem is the least serious and can readily be avoided by modifying the yield (or potential) surface forms to avoid corners. A simple modification of the Mohr–Coulomb (or Tresca) surface expressions [Eq. (7.79)] is easily achieved by writing[51]

$$F = \sigma_m \sin\phi - c\cos\phi + \frac{\bar{\sigma}}{g(\theta)} \tag{7.142}$$

where

$$g(\theta) = \frac{2K}{(1+K) - (1-K)\sin 3\theta}$$

and

$$K = \frac{3 - \sin\phi}{3 + \sin\phi}$$

Figure 7.26 shows how the angular section of the Mohr-Coulomb surface in the Π plane (constant σ_m) now becomes rounded. Similar procedures have been suggested by others.[136]

Much more serious are the second and third possible causes of non-uniqueness mentioned above. Here theoretical non-uniqueness can be avoided by considering plastic deformation to be a limit state of

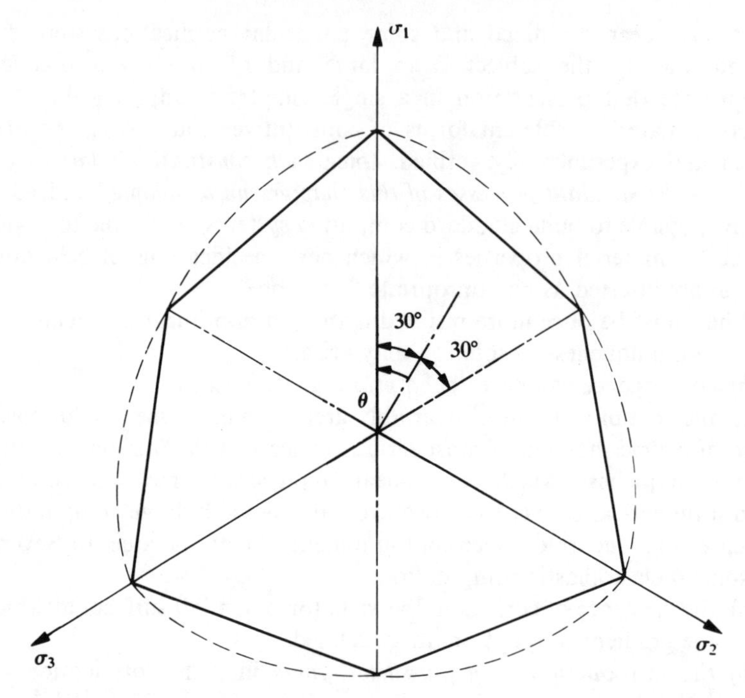

Fig. 7.26 Π plane section of Mohr–Coulomb yield surface in principal stress space with $\phi = 20°$ (solid line). Smooth approximation of Eq. (7.142) (dotted line)

viscoplastic behaviour in a manner to which we have already referred in Sec. 7.12. Such a process, mathematically known as *regularization*, has allowed us to obtain many realistic solutions for both non-associative and strain softening behaviour, as already shown.

However, on occasion (though not invariably), both forms of behaviour can lead to *localization* phenomena where strain (and displacement) discontinuities develop.[133-145] The non-uniqueness can be particularly evident in strain softening plasticity. We illustrate this in an example of Fig. 7.27 where a bar of length L, divided into elements of length h, is subject to a uniformly increasing extension u. The material is initially elastic with a modulus E and after exceeding a stress of σ_y, the yield stress softens (plastically) with a negative modulus H.

The strain–stress relation is thus [Fig. 7.27(*a*)]

$$\sigma = E\varepsilon \qquad \text{if } \varepsilon < \sigma_y/E = \varepsilon_y \tag{7.143a}$$

and for increasing ε only,

$$\sigma = \sigma_y - H(\varepsilon - \varepsilon_y) \qquad \text{if } \varepsilon > \varepsilon_y \tag{7.143b}$$

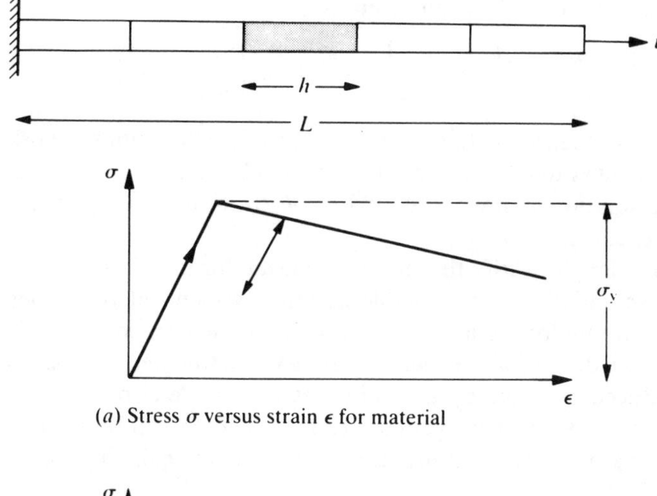

(a) Stress σ versus strain ϵ for material

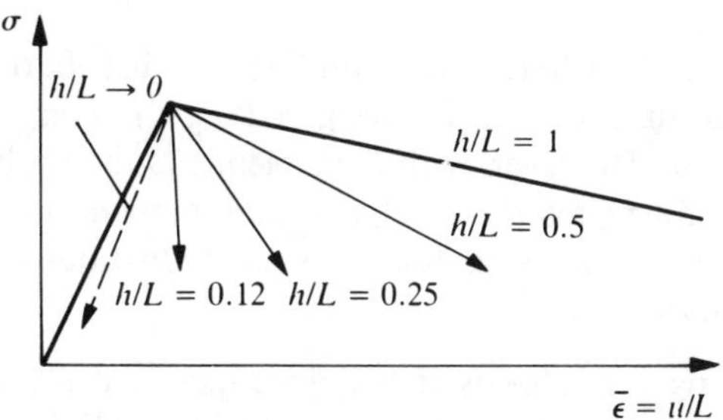

(b) Stress $\bar{\sigma}$ versus average strain $\bar{\epsilon} = u/L$ assuming yielding in a single element of length h

Fig. 7.27 Non-uniqueness—mesh size dependence in extension of a homogeneous bar with a strain softening material. (Peak value of yield stress σ_y perturbed in a single element)

For unloading from any plastic point the material behaves elastically as shown.

One possible solution is, of course, that in which all elements yield identically. Plotting the applied stress versus the elongation strain $\bar{\varepsilon} = u/L$ the material behaviour curve is obtained identically as shown in Fig. 7.27(*b*) ($h/L = 1$). However, it is equally possible that after reaching the maximum stress σ_y only one element (probably with infinitesimally lower maximum yield stress) continues into the plastic range while all others unload plastically. The total elongation is now given by

$$U = (L-h)\frac{\sigma}{E} + h\left(\frac{\sigma_y - \sigma}{H} + \frac{\sigma_y}{E}\right) \qquad (7.144a)$$

resulting in an overall strain given as

$$\bar{\varepsilon}=\frac{U}{L}=\sigma\left(\frac{L-h}{LE}-\frac{h}{LH}\right)+\sigma_y\left(\frac{h}{LH}+\frac{h}{LE}\right) \tag{7.144b}$$

and as $h \to 0$ then $\bar{\varepsilon} \to \sigma/E$. Clearly a multitude of solutions is possible for any element subdivision and in this trivial example a unique finite element solution is impossible (with localization to a singular element always occurring).

The reader can verify that if a viscoplastic form of relation is assumed the above situation is impossible as long as identical properties of the material are uniformly assumed and here uniqueness corresponding to a uniform yielding of all elements is achieved. However, if a perturbation is introduced by making the yield stress in a single element smaller than that in others then once again a *localized* yielding will occur in the limit with the nature of the limiting stress–strain curve again depending on the ratio h/L.[137]

In real problems where strain and stress distributions are seldom constant the localization will be 'triggered' by the non-uniformity of the stress distribution. However, with *finite* elements the problem of the local element size influencing the final solution remains and much current work in this area is in progress. Two main directions of avoiding the problem are advocated:

1. The softening modulus is made dependent on element size, as indicated in the work of Pietruszczak and Mróz.[135] Physical arguments for such an approach are given by Bazant and coworkers,[138,139] Biĉaniĉ *et al.*[140] and many others.[141]
2. Use of viscoplastic regularization coupled with a physical scale dimension limiting the local defect size.

The interested reader can consult many references on the subject.[142]

7.15.2 *Localization—numerical treatment.* Localization such as may be caused by strain softening (in the manner discussed in the preceding section) or indeed by non-associative behaviour[134,143] clearly requires special numerical treatment for in the local zone strains may become infinite and displacement discontinuities occur. Such discontinuities can indeed occur in ideally plastic behaviour on occasion and are only crudely indicated by simple finite element subdivisions.

To treat the problem locally discontinuous shape functions have been used with some success[144,145] but local refinement maintaining continuous functions provides a simpler alternative.[146–147] In Fig. 7.28 we illustrate the effect of h-type adaptive refinement using elongated elements of triangular shape in capturing the localized shear strain distribution at a slip surface.

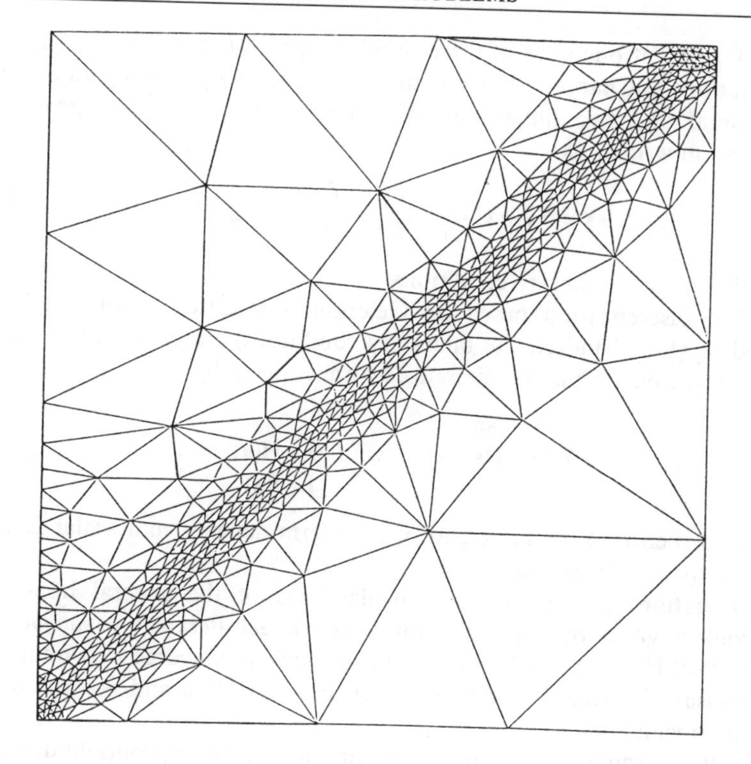

Fig. 7.28 Use of adaptive mesh refinement to capture a slip discontinuity in a homogeneous stress field

Localization due to non-linear behaviour is receiving much attention in other problems of mechanics. In particular the reader will find many similarities of the above discussion with problems posed by the formations of shocks in compressible fluid flow to be discussed in Chapter 13.

OTHER NON-LINEAR PROBLEMS

7.16 Non-linear quasi-harmonic field problems

Non-linearity may arise in many problems beyond those of solid mechanics, but the techniques described at the start of this chapter are still universally applicable. In subsequent chapters we shall touch upon such non-linearities in the context of transient problems (Chapters 10 and 11) and various fluid mechanics situations (Chapters 12 to 15). Here we shall look again at one class of these problems which is governed by the field equations of Chapter 10 of Volume 1. For simplicity treating an isotopic material and two dimensions of space, we have now the governing equation written as

$$\nabla^{\mathrm{T}} k \nabla \phi + Q \equiv \frac{\partial}{\partial x} k \frac{\partial \phi}{\partial x} + \frac{\partial}{\partial y} k \frac{\partial \phi}{\partial y} + Q = 0 \tag{7.145}$$

with suitable boundary conditions.

The discretization based on Galerkin procedures is still valid if k and/or Q (and indeed the boundary conditions) are dependent on ϕ or its derivatives. Thus the discretized form is given by

$$\begin{aligned} \phi &= \mathbf{N}\mathbf{a} \\ \boldsymbol{\Psi}(\mathbf{a}) &\equiv \mathbf{H}\mathbf{a} + \mathbf{f} \qquad \text{with } \mathbf{H} = \mathbf{H}(\mathbf{a}) \\ & \qquad\qquad\qquad\quad\;\; \mathbf{f} = \mathbf{f}(\mathbf{a}) \end{aligned} \tag{7.146}$$

and is such that the integrand of each term depends in a scalar fashion on ϕ (or its derivative).

Equations (7.146) are a particular case of the general non-linear problem given by Eq. (7.25) for which *direct iteration techniques are possible.* However, as these occasionally fail to converge it is frequently necessary to determine the tangential matrix $\mathrm{d}\boldsymbol{\Psi}/\mathrm{d}\mathbf{a}$ and use one of the other techniques.

Let us examine this in detail, making use of the appropriate discretization matrices. Using these we can write

$$\frac{\partial \boldsymbol{\Psi}}{\partial \mathbf{a}} \mathrm{d}\mathbf{a} \equiv \mathbf{H}\, \mathrm{d}\mathbf{a} + \mathrm{d}\mathbf{H}\mathbf{a} + \mathrm{d}\mathbf{f} \tag{7.147}$$

If k and Q are direct functions of ϕ, we can write†

$$(\mathrm{d}\mathbf{H})\mathbf{a} = \mathbf{A}\, \mathrm{d}\mathbf{a} \qquad A_{ij} = \int_{\Omega} \nabla N_i^{\mathrm{T}} (\nabla \mathbf{N}\mathbf{a}) N_j k'\, \mathrm{d}\Omega \tag{7.148}$$

and

$$\mathrm{d}\mathbf{f} = \left(\int_{\Omega} \mathbf{N}^{\mathrm{T}} Q' \mathbf{N}\, \mathrm{d}\Omega \right) \mathrm{d}\mathbf{a} = \mathbf{C}\, \mathrm{d}\mathbf{a} \qquad \text{with } C_{ij} = \int_{\Omega} N_i^{\mathrm{T}} Q' N_j\, \mathrm{d}\Omega \tag{7.149}$$

† To derive the matrix terms of $\mathbf{dHa}$ it is convenient to consider the jth row of $\mathrm{d}\boldsymbol{\Psi}_i$ only, i.e.

$$\mathrm{d}\left(\int (\nabla N_i^{\mathrm{T}}) k \nabla \mathbf{N}\, \mathrm{d}\Omega \right) \mathbf{a} = \int \nabla N_i^{\mathrm{T}} [\nabla N_1 k'(N_1\, \mathrm{d}a_1 + N_2\, \mathrm{d}a_2 + \cdots) a_1 + \nabla N_2 k'(N_1\, \mathrm{d}a_1 + \cdots) + \cdots]$$

The coefficient of $\mathrm{d}a_j$ immediately gives A_{ij}.

where

$$k' = \frac{dk}{d\phi} \qquad Q' = \frac{dQ}{d\phi} \tag{7.150}$$

The tangential matrix thus becomes

$$\frac{d\boldsymbol{\Psi}}{d\mathbf{a}} = \mathbf{H} + \mathbf{A} + \mathbf{B} \tag{7.151}$$

in which the second term is non-symmetric. Newton–Raphson procedures in such a case are inconvenient and modifications of the formulation are sometimes made.

Indeed, it is easy to show that a variational principle in such cases is not the same as that given in Chapter 10 of Volume 1. Special forms of a variational principle can be devised and will lead to symmetry.[149] In many physical problems, however, the values of k depend on the absolute value of the gradient of $\nabla\phi$, i.e.

$$V = \sqrt{(\nabla\phi)^{\mathrm{T}}(\nabla\phi)} = \sqrt{\left(\frac{\partial\phi}{\partial x}\right)^2 + \left(\frac{\partial\phi}{\partial y}\right)^2}$$

$$\bar{k}' = \frac{dk}{dV} \tag{7.152}$$

In such cases we can fortunately write

$$d\mathbf{Ha} = \bar{\mathbf{A}}\, d\mathbf{a} \tag{7.153}$$

where

$$\bar{A}_{ij} = \int_{\Omega} (\nabla N_i)^{\mathrm{T}} (\nabla \mathbf{Na})^{\mathrm{T}} \bar{k}' (\nabla \mathbf{Na}) \nabla N_j \, d\Omega \tag{7.154}$$

and symmetry is preserved.

Situations of this kind arise in seepage flow where the permeability is dependent on the absolute value of the flow velocity,[150–151] in magnetic fields[152–155] where magnetic formulations is a function of the absolute field strength, in slightly compressible fluid flow and indeed in many other physical situations.[156] Figure 7.29 from reference 152 illustrates a typical non-linear magnetic field solution.

Whilst many more interesting problems could be quoted we conclude with one in which the only non-linearity is that due to the heat generation term Q in Eq. (7.145). This particular problem of spontaneous ignition, in which Q depends exponentially on the temperature, serves to illustrate the point about the possibility of multiple solutions and indeed the non-existence of any solution in certain non-linear cases.[157]

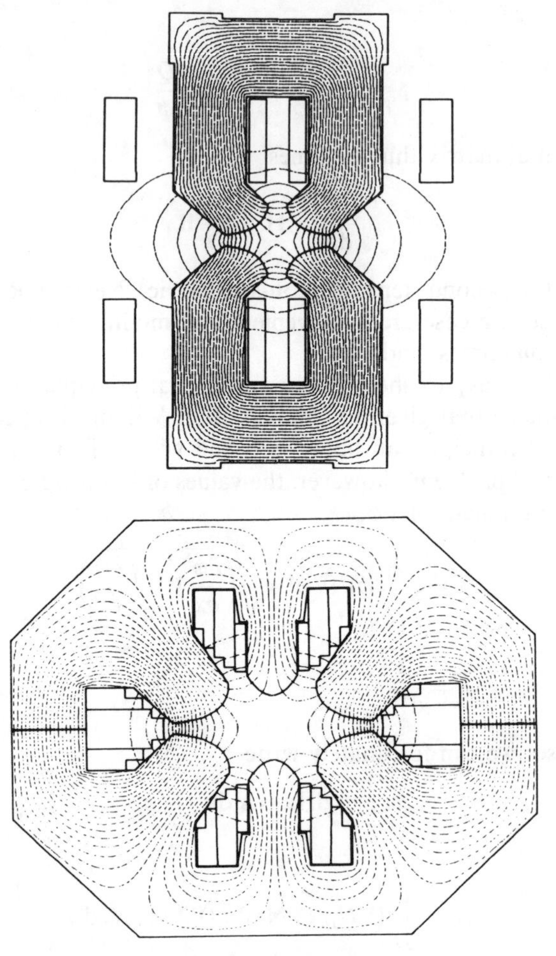

Fig. 7.29 Magnetic field in a six-pole magnet with non-linearity due to saturation[152]

Taking $k=1$ and $Q=\delta\, e^{\phi}$, an elliptic domain is examined in Fig. 7.30. Using various values of δ, a Newton–Raphson iteration is used to obtain a solution and we find that no convergence (and indeed *no solution*) exists when $\delta > \delta_{crit}$ exists. Above the critical value of δ_{crit} the temperature rises indefinitely and *spontaneous ignition* of the material occurs. For values below this *two* solutions are possible and the starting point of the iteration determines which one is in fact obtained.

This last point illustrates that an insight into the problem is, in non-linear solutions, even more important that elsewhere.

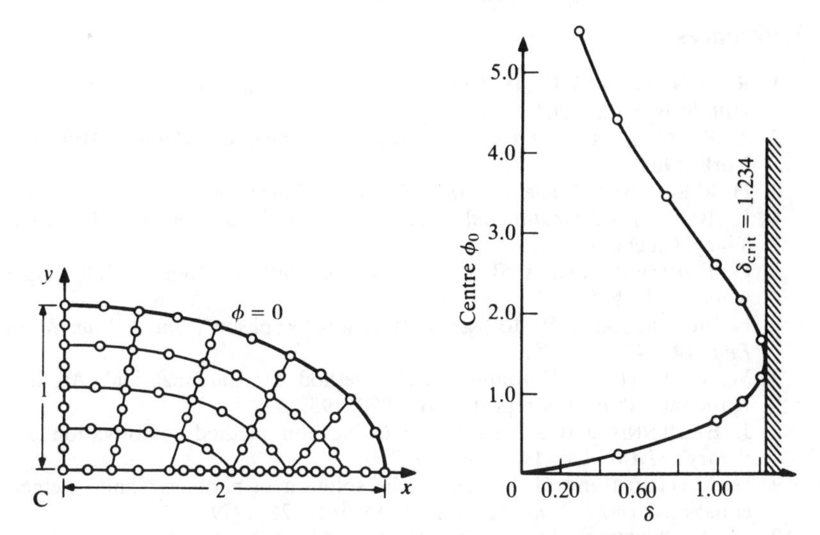

(a) Solution mesh and variation of temperature at point C with parameter δ

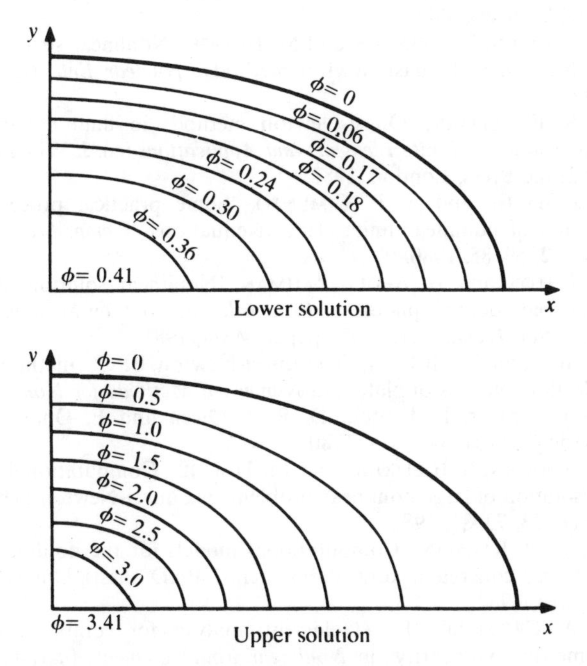

(b) Two possible temperature distributions for $\delta = 0.75$

Fig. 7.30 A non-linear heat-generation problem[157] illustrating the possibility of multiple or no solutions depending on the heat generation parameter δ ($Q = \delta\, e^{\phi}$). Spontaneous combustion

References

1. R. BECKETT and J. HURT, *Numerical Calculations and Algorithms*, McGraw-Hill, New York, 1967.
2. A. RALSTON, *A First Course in Numerical Analysis*, McGraw-Hill, New York, 1965.
3. H. R. SCHWARZ, *Numerical Analysis*, Wiley, Chichester, 1989.
4. L. B. RALL, *Computational Solution of Non-linear Operator Equations*, Wiley, Chichester, 1969.
5. G. DAHLQUIST and A. BJÖRK, *Numerical Methods*, Prentice-Hall, Englewood Cliffs, N.J., 1974.
6. N. BIĈANIĈ and K. W. JOHNSON, 'Who was "Raphson"', *Int. J. Num. Meth. Eng.*, **14**, 148–52, 1979.
7. W. C. DAVIDON, 'Variable metric method for minimization', Argonne National Laboratory report ANL-5990, 1959.
8. J. E. DENNIS and J. MORE, 'Quasi-Newton methods—motivation and theory', *SIAM Rev.*, **19**, 46–89, 1977.
9. H. MATTHIES and G. STRANG, 'The solution of nonlinear finite element equations', *Int. J. Num. Meth. Eng.*, **14**, 1613–26, 1979.
10. M. A. CRISFIELD, 'Incremental/iterative solution procedures for nonlinear structural analysis', in *Numerical Methods for Nonlinear Problems* (eds C. Taylor, E. Hinton, D. R. J. Owen and E. Oñate), pp. 261–90, Pineridge Press, Swansea, 1980.
11. M. GERADIN, S. IDELSOHN and M. HOGGE, 'Nonlinear structural dynamics via Newton and quasi-Newton methods', *Nuclear Eng. Des.*, **58**, 339–48, 1980.
12. M. S. ENGELMAN, 'Quasi-Newton methods in fluid dynamics', in *The Mathematics of Finite Elements and Applications* (ed. J. Whiteman), Vol. IV, Academic Press, London, 1982.
13. K. J. BATHE and A. P. CIMENTO, 'Some practical procedures for the solution of nonlinear finite element equations', *Comp. Meth. Appl. Mech. Eng.*, **22**, 59–85, 1980.
14. E. HINTON and H. ABDEL RAHMAN, 'Nonlinear solution algorithms for reinforced concrete plates', in *Proc. 6th Int. Conf. on Structural Mechanics in Reactor Technology*, Paris, paper M3/4, 1981.
15. A. PICA and E. HINTON, 'The quasi-Newton BFGS method in the large deflection analysis of plates', in *Numerical Methods for Nonlinear Problems* (eds C. Taylor, E. Hinton, D. R. J. Owen and E. Oñate), pp. 355–66, Pineridge Press, Swansea, 1980.
16. M. GERADIN, S. IDELSOHN and M. HOGGE, 'Computational strategies for the solution of large nonlinear problems via quasi-Newton methods', *Comp. Struct.*, **13**, 73–81, 1981.
17. H. ABDEL RAHMAN, 'Computational models for the nonlinear analysis of reinforced concrete flexural slab systems', Ph.D. thesis, University of Wales, Swansea, 1982.
18. M. A. CRISFIELD, 'Finite element analysis for combined material and geometric nonlinearity', in *Nonlinear Finite Element Analysis in Structural Mechanics* (eds W. Wunderlich *et al.*) Springer-Verlag, Berlin, 1981.
19. M. A. CRISFIELD, 'A fast incremental/iterative solution procedure that handles "snap through"', *Comp. Struct.*, **13**, 55–62, 1981.
20. A. C. AITKEN, 'The evaluation of the latent roots and latent vectors of a matrix', *Proc. Roy. Soc. Edinburgh*, **57**, 269–304, 1937.

21. B. M. IRONS and R. TUCK, 'A version of the Aitken acceleration for computer iteration', *Int. J. Num. Meth. Eng.*, **1**, 275–77, 1969.
22. G. C. NAYAK and O. C. ZIENKIEWICZ, 'Note on the "alpha"-constant stiffness method for the analysis of nonlinear problems', *Int. J. Num. Meth. Eng.*, **4**, 579–82, 1972.
23. T. H. H. PIAN and PIN TONG, 'Variational formulations of finite displacement analysis', in *Symp. on High Speed Electronic Computation of Structures*, Liège, 1970.
24. O. C. ZIENKIEWICZ, 'Incremental displacement in non-linear analysis', *Int. J. Num. Meth. Eng.*, **3**, 587–92, 1971.
25. E. RIKS, 'An incremental approach to the solution of snapping and buckling problems', *Int. J. Solids Struct.*, **15**, 529–51, 1979.
26. P. G. BERGAN, 'Solution algorithms for nonlinear structural problems', in *Int. Conf. on Engineering Applications of the Finite Element Method, Computas*, pp. 13.1–13.39, 1979.
27. J. L. BATOZ and G. DHATT, 'Incremental displacement algorithms for nonlinear problems', *Int. J. Num. Meth. Eng.*, **14**, 1261–6, 1979.
28. E. RAMM, 'Strategies for tracing nonlinear response near limit points', in *Nonlinear Finite Element Analysis in Structural Mechanics* (eds W. Wunderlich, E. Stein and K. J. Bathe), pp. 63–89, Springer-Verlag, New York, 1981.
29. P. G. BERGAN, 'Solution by iteration in displacement and load spaces', in *Nonlinear Finite Element Analysis in Structural Mechanics* (eds W. Wunderlich *et al.*), Springer-Verlag, New York, 1981.
30. M. HESTENES and E. STIEFEL, 'Method of conjugate gradients for solving linear systems', *J. Res. Nat. Bur. Stand.*, **49**, 409–36, 1952.
31. R. FLETCHER and C. M. REEVES, 'Function minimization by conjugate gradients', *The Computer Journal*, **7**, 149–54, 1964.
32. E. POLAK, *Computational Methods in Optimization: A Unified Approach*, Academic Press, London, 1971.
33. B. M. IRONS, 'The conjugate-Newton method', in *Int. Res. Semi. on Theory and Applications of the Finite Element Method*, University of Calgary, Canada, 1973.
34. B. M. IRONS and A. F. ELSAWAF, 'The conjugate Newton algorithm for solving finite element equations', in *Proc. U.S.–German Symp. on Formulations and Algorithms in Finite Element Analysis* (eds K. J. Bathe, J. T. Oden and W. Wunderlich), pp. 656–72, MIT Press, Cambridge, Mass., 1977.
35. J. R. H. OTTER, E. CASSEL and R. E. HOBBS, 'Dynamic relaxation', *Proc. Inst. Civ. Eng.*, **35**, 633–56, 1966.
36. O. C. ZIENKIEWICZ and R. LÖHNER, 'Accelerated relaxation or direct solution? Future prospects for FEM', *Int. J. Num. Meth. Eng.*, **21**, 3–11, 1986.
37. J. T. ODEN, 'Numerical formulation of non-linear elasticity problems', *Proc. Am. Soc. Civ. Eng.*, **93**, ST3, 235–55, 1967.
38. R. VON MISES, 'Mechanik der Plastischen Formänderung der Kristallen', *Z. angew. Math. Mech.*, **8**, 161–85, 1928.
39. D. C. DRUCKER, 'A more fundamental approach to plastic stress-strain solutions', *Proc. 1st U.S. Nat. Cong. Appl. Mech.*, pp. 487–91, 1951.
40. W. T. KOITER, 'Stress–strain relations, uniqueness and variational theorems for elastic plastic materials with a singular yield surface', *Q.J. Appl. Math.*, **11**, 350–4, 1953.

41. R. HILL, *The Mathematical Theory of Plasticity*, Clarendon Press, Oxford, 1950.
42. W. JOHNSON and P. W. MELLOR, *Plasticity for Mechanical Engineers*, Van Nostrand, New York, 1962.
43. W. PRAGER, *An Introduction to Plasticity*, Addison-Wesley, Reading, Mass., 1959.
44. W. F. CHEN, *Plasticity in Reinforced Concrete*, McGraw-Hill, New York, 1982.
45. D. C. DRUCKER, 'Conventional and unconventional plastic response and representation', *Appl. Mech. Rev.*, **41**, 151–67, 1988.
46. G. C. NAYAK and O. C. ZIENKIEWICZ, 'Elasto-plastic stress analysis. Generalization for various constitutive relations including strain softening', *Int. J. Num. Meth. Eng.*, **5**, 113–35, 1972.
47. Y. YAMADA, N. YISHIMURA and T. SAKURAI, 'Plastic stress–strain matrix and its application for the solution of elastic–plastic problems by the finite element method', *Int. J. Mech. Sci.*, **10**, 343–54, 1968.
48. O. C. ZIENKIEWICZ, S. VALLIAPPAN and I. P. KING, 'Elasto-plastic solutions of engineering problems. Initial-stress, finite element approach', *Int. J. Num. Meth. Eng.*, **1**, 75–100, 1969.
49. G. C. NAYAK and O. C. ZIENKIEWICZ, 'Convenient forms of stress invariants for plasticity', *Proc. Am. Soc. Civ. Eng.*, **98**, ST4, 949–53, 1972.
50. D. C. DRUCKER and W. PRAGER, 'Soil mechanics and plastic analysis or limit design', *Q. J. Appl. Math.*, **10**, 157–65, 1952.
51. O. C. ZIENKIEWICZ and G. N. PANDE, 'Some useful forms of isotropic yield surfaces for soil, and rock mechanics', in *Finite Elements in Geomechanics* (ed. G. Gudehus), chap. 5, pp. 171–90, Wiley, Chichester, 1977.
52. O. C. ZIENKIEWICZ, V. A. NORRIS, L. A. WINNICKI, D. J. NAYLOR and R. W. LEWIS, 'A unified approach to the soil mechanics problems of offshore foundations', in *Numerical Methods in Offshore Engineering* (eds O. C. Zienkiewicz, R. W. Lewis and K. G. Stagg), chap. 12, Wiley, Chichester, 1978.
53. O. C. ZIENKIEWICZ, V. A. NORRIS and D. J. NAYLOR, 'Plasticity and viscoplasticity in soil mechanics with special reference to cyclic loading problems', in *Proc. Int. Conf. on Finite Elements in Nonlinear Solid and Structural Mechanics*, Geilo, Norway, Vol. 2, pp. 455–85, Tapir Press, Trondheim, 1977.
54. O. C. ZIENKIEWICZ, C. T. CHANG, N. BIĈANIĈ and E. HINTON, 'Earthquake response of earth and concrete in the partial damage range', in *Proc. 13th Int. Cong. on Large Dams*, New Delhi, R. 14, pp. 1033–47, 1979.
55. J. F. BESSELING, 'A theory of elastic, plastic and creep deformations of an initially isotropic material', *J. Appl. Mech.*, **25**, 529–36, 1958.
56. Z. MRÓZ, 'An attempt to describe the behaviour of metals under cyclic loads using more general work hardening model', *Acta Mech.*, **7**, 199, 1969.
57. O. C. ZIENKIEWICZ, G. C. NAYAK and D. R. J. OWEN, 'Composite and "Overlay" models in numerical analysis of elasto-plastic continua', in *Foundations of Plasticity* (ed. A. Sawczuk), pp. 107–22, Noordhoff, Dordrecht, 1972.
58. D. R. J. OWEN, A. PRAKASH and O. C. ZIENKIEWICZ, 'Finite element analysis of non-linear composite materials by use of overlay systems', *Comp. Struct.*, **4**, 1251–67, 1974.
59. F. DARVE, 'An incrementally nonlinear constitutive law of second order and

its application to localisation', in *Mechanics of Engineering Materials* (eds C. S. Desai and R. H. Gallagher), chap. 9, pp. 179–96, Wiley, Chichester, 1984.

60. O. C. ZIENKIEWICZ and Z. MRÓZ, 'Generalized plasticity formulation and applications to geomechanics', in *Mechanics of Engineering Materials* (eds C. S. Desai and R. H. Gallagher), chap. 33, pp. 655–80, Wiley, Chichester, 1984.
61. Z. MRÓZ and O. C. ZIENKIEWICZ, 'Uniform formulation of constitutive equations for clays and sand', in *Mechanics of Engineering Materials* (eds C. S. Desai and R. H. Gallagher), chap. 22, pp. 415–50, Wiley, Chichester, 1984.

62a. O. C. ZIENKIEWICZ, K. H. LEUNG and M. PASTOR, 'Simple model for transient soil loading in earthquake analysis, Part I—Basic model and its application', *Int. J. Num. Anal. Meth. Geomechanics*, **9**, 453–76, 1985.

62b. M. PASTOR, O. C. ZIENKIEWICZ and K. H. LEUNG, 'Simple model for transient soil loading in earthquake analysis. Part II—Non-associative models for sands', *Int. J. Num. Anal. Meth. Geomechanics*, **9**, 477–98, 1985.

63. M. PASTOR and O. C. ZIENKIEWICZ, 'A generalized plasticity, hierarchical model for sand under monotonic and cyclic loading', in *Proc. Int. Symp. on Numerical Methods in Geomechanics* (eds G. N. Pande and W. F. Van Impe), Ghent, pp. 131–50, Jackson, England, 1986.
64. M. PASTOR and O. C. ZIENKIEWICZ, 'Generalized plasticity and modelling of soil behaviour', *Int. J. Num. Anal. Meth. Geomech.*, **14**, 151–90, 1990.
65. Y. F. DAFALIAS and E. P. POPOV, 'A model of nonlinear hardening material for complex loading', *Acta Mechanica*, **21**, 173–92, 1975.

66a. O. C. ZIENKIEWICZ, A. CHAN, M. PASTOR, D. K. PAUL and T. SHIOMI, 'Statics and dynamics of soils. A rational approach to quantitative solution. Part I. Fully saturated problems', *Proc. Roy. Soc. Lond.*, **A429**, 285–309, 1990.

66b. O. C. ZIENKIEWICZ, Y. M. XIE, B. A. SCHREFLER, A. LADESMA and N. BIĈANIĈ, Part II. Semi-saturated problems', *Proc. Roy. Soc. Lond.*, **A429**, 311–21, 1990.

67. G. C. NAYAK, 'Plasticity and large deformation problems by the finite element method', Ph.D. thesis, University of Wales, Swansea, 1971.
68. K. J. BATHE, *Finite Element Procedures in Engineering Analysis*, Prentice-Hall, Englewood Cliffs, N.J., 1982.
69. R. D. KRIEG and D. N. KRIEG, 'Accuracy of numerical solution methods for the elastic, perfectly plastic model', *Trans. ASME, J. Pressure Vessel Tech.*, **99**, 510–15, 1977.
70. M. L. WILKINS, 'Calculation of elastic-plastic flow', *Methods of Computational Physics* (eds B. Alder *et al.*), Vol. 3, Academic Press, London, 1964.
71. R. L. TAYLOR, 'Analysis of non-linear problems in inelasticity', Report of Institute of Numerical Methods in Engineering, Swansea, 1989.
72. R. L. TAYLOR and J. C. SIMO, 'A return mapping algorithm for plane stress elasto-plasticity', *Int. J. Num. Mech. Eng.*, **22**, 649–70, 1986.
73. J. C. SIMO and R. L. TAYLOR, 'Consistent tangent operators for rate dependent elasto-plasticity', *Comp. Meth. Appl. Mech. Eng.*, **48**, 101–18, 1985.
74. M. ORTIZ and J. C. SIMO, 'An analysis of a new class of integration algorithms for elasto-plastic constitutive relations', *Int. J. Num. Meth. Eng.*, **23**, 353–66, 1986.
75. R. H. GALLAGHER, J. PADLOG and P. P. BIJLAARD, 'Stress analysis of heated complex shapes', *J. Am. Rocket Soc.*, **32**, 700–7, 1962.

76. J. H. ARGYRIS, 'Elasto-plastic matrix displacement analysis of three-dimensional continua', *J. Roy. Aero. Soc.*, **69**, 633–5, 1965.
77. G. G. POPE, 'A discrete element method for analysis of plane elasto-plastic strain problems', report T.R. 65028, RAE Farnborough, 1965.
78. J. L. SWEDLOW, M. L. WILLIAMS and W. M. YANG, 'Elastic–plastic stresses in cracked plates', Calcit report SM 65-19, California Institute of Technology, 1965.
79. P. V. MARCAL and I. P. KING, 'Elastic–plastic analysis of two-dimensional stress systems by the finite element method', *Int. J. Mech. Sci.*, **9**, 143–55, 1967.
80. S. F. REYES and D. U. DEERE, 'Elasto–plastic analysis of underground openings by the finite element method', in *Proc. 1st Int. Congr. Rock Mechanics*, Lisbon, Vol. II, pp. 477–86, 1966.
81. E. P. POPOV, M. KHOJASTEH-BAKHT and S. YAGHMAI, 'Bending of circular plates of hardening material', *Int. J. Solids Struct.*, **3**, 975–88, 1967.
82. R. M. MCMEEKING and J. R. RICE, 'Finite element formulations for problems of large elastic plastic deformation', *Int. J. Solids Struct.*, **11**, 601–16, 1975.
83. J. C. NAGTEGAAL, D. M. PARKS and J. R. RICE, 'On numerically accurate finite element solutions in the fully plastic range', *Comp. Meth. Appl. Mech. Eng.*, **4**, 153–78, 1974.
84. R. H. GALLAGHER and A. K. DHALLA, 'Direct flexibility finite element elasto-plastic analysis', in *First SMIRT Conf.*, Berlin, pt 6, 1971.
85. E. F. RYBICKI and L. A. SCHMIT, 'An incremental complementary energy method of non-linear stress analysis', *JAIAA*, **8**, 1105–12, 1970.
86. J. A. STRICKLIN, W. E. HEISLER and W. VON RUSMAN, 'Evaluation of solution procedures for material and/or geometrically non-linear structural analysis', *JAIAA*, **11**, 292–9, 1973.
87. D. R. J. OWEN, G. C. NAYAK, A. P. KFOURI and J. R. GRIFFITHS, 'Stresses in a partly yielded notched bar', *Int. J. Num. Meth. Eng.*, **6**, 63–72, 1973.
88. K. S. DINNO and S. S. GILL, 'An experimental investigation into the plastic behaviour of flush nozzles in spherical pressure vessels', *Int. J. Mech. Sci.*, **7**, 817, 1965.
89. A. MENDELSON, M. H. HIRSCHBERG and S. S. MANSON, 'A general approach to the practical solution of creep problems', *Trans. ASME, J. Basic Eng.*, **81**, 85–98, 1959.
90. O. C. ZIENKIEWICZ, M. WATSON and I. P. KING, 'A numerical method of visco-elastic stress analysis', *Int. J. Mech. Sci.*, **10**, 807–27, 1968.
91. O. C. ZIENKIEWICZ, *The Finite Element Method in Structural and Continuum Mechanics*, 1st ed., McGraw-Hill, New York, 1967.
92. G. A. GREENBAUM and M. F. RUBINSTEIN, 'Creep analysis of axi-symmetric bodies using finite elements', *Nucl. Eng. Des.*, **7**, 379–97, 1968.
93. P. V. MARCAL, private communication, 1972.
94. N. A. CYR and R. D. TETER, 'Finite element elastic plastic creep analysis of two-dimensional continuum with temperature dependent material properties', *Comp. Struct.*, **3**, 849–63, 1973.
95. O. C. ZIENKIEWICZ, 'Visco-plasticity, plasticity, creep and visco-plastic flow (problems of small, large and continuing deformation)', in *Computational Mechanics*, TICOM Lecture Notes on Mathematics 461, Springer-Verlag, Berlin, 1975.
96. M. B. KANCHI, D. R. J. OWEN and O. C. ZIENKIEWICZ, 'The visco-plastic

approach to problems of plasticity and creep involving geometrically non-linear effects', *Int. J. Num. Meth. Eng.*, **12**, 169–81, 1978.
97. T. J. R. HUGHES and R. L. TAYLOR, 'Unconditionally stable algorithms for quasi-static elasto/visco-plastic finite element analysis', *Comp. Struct.*, **8**, 169–73, 1978.
98. E. C. BINGHAM, *Fluidity and Plasticity*, chap. VIII, pp. 215–18, McGraw-Hill, New York, 1922.
99. P. PERZYNA, 'Fundamental problems in viscoplasticity', *Adv. Appl. Mech.*, **9**, 243–377, 1966.
100. O. C. ZIENKIEWICZ and I. C. CORMEAU, 'Visco-plasticity solution by finite element process', *Arch. Mech.*, **24**, 873–88, 1972.
101. O. C. ZIENKIEWICZ and I. C. CORMEAU, 'Visco-plasticity and plasticity. An alternative for finite element solution of material non-linearities', in *Proc. Colloque Methodes Calcul. Sci. Tech.*, pp. 171–99, IRIA, Paris, 1973.
102. J. ZARKA, 'Généralisation de la théorie du potential multiple en visco-plasticité', *J. Mech. Phys. Solids*, **20**, 179–95, 1972.
103. Q. A. NGUYEN and J. ZARKA, 'Quelques méthodes de resolution numérique en elastoplasticité classique et en elasto-viscoplasticité', in *Seminaire Plasticité et Viscoplasticité*, Ecole Polytechnique, Paris, 1972; also *Sciences et Technique de l'Armement*, **47**, 407–36, 1973.
104. O. C. ZIENKIEWICZ and I. C. CORMEAU, 'Visco-plasticity, plasticity and creep in elastic solids—a unified numerical solution approach', *Int. J. Num. Meth. Eng.*, **8**, 821–45, 1974.
105. I. C. CORMEAU, 'Numerical stability in quasi-static elasto-visco-plasticity', *Int. J. Num. Meth. Eng.*, **9**, 109–28, 1975.
106. F. A. LECKIE and J. B. MARTIN, 'Deformation bounds for bodies in a state of creep', *Trans. ASME, J. Appl. Mech.*, **34**, 411–17, 1967.
107. I. FINNIE and W. R. HELLER, *Creep of Engineering Materials*, McGraw-Hill, New York, 1959.
108. O. C. ZIENKIEWICZ, C. HUMPHESON and R. W. LEWIS, 'Associated and non-associated visco-plasticity and plasticity in soil mechanics', *Geotechnique*, **25**, 671–89, 1975.
109. G. N. PANDE and O. C. ZIENKIEWICZ (eds), *Soil Mechanics–Transient and Cyclic Loads*, Wiley, Chichester, 1982.
110. C. S. DESAI and R. H. GALLAGHER (eds), *Mechanics of Engineering Materials*, Wiley, Chichester, 1984.
111. E. H. LEE, 'Visco-elasticity', in *Handbook of Engineering Mechanics* (ed. W. Flügge), McGraw-Hill, New York, 1962.
112. R. L. TAYLOR, K. PISTER and G. GOUDREAU, 'Thermo-mechanical analysis of visco-elastic solids', *Int. J. Num. Meth. Eng.*, **2**, 45–60, 1970.
113. O. C. ZIENKIEWICZ, 'Analysis of visco-elastic behaviour of concrete structures with particular reference to thermal stresses', *Proc. Am. Concr. Inst.*, **58**, 383–94, 1961.
114. T. ALFREY, *Mechanical Behaviour of High Polymers*, Interscience, New York, 1948.
115. D. MCHENRY, 'A new aspect of creep in concrete and its application to design', *Proc. Am. Soc. Test. Mat.*, **43**, 1064, 1943.
116. H. H. HILTON and H. G. RUSSEL, 'An extension of Alfrey's analogy to thermal stress problems in temperature dependent linear visco-elastic media', *J. Mech. Phys. Solids*, **9**, 152–64, 1961.
117. O. C. ZIENKIEWICZ, M. WATSON and Y. K. CHEUNG, 'Stress analysis by the

finite element method—thermal effects', in *Proc. Conf. on Prestressed Concrete Pressure Vessels*, Institute of Civil Engineers, London, 1967.

118. O. C. Zienkiewicz, S. Valliappan and I. P. King, 'Stress analysis of rock as a "no-tension" material', *Geotechnique*, **18**, 56–66, 1968.
119. S. Valliappan and P. Nath, 'Tensile crack propagation in reinforced concrete beams by finite element techniques', *Int. Conf. on Shear Torsion and Bond in Reinforced Concrete*, Coimbatore, India, January 1969.
120. N. W. Krahl, W. Khachaturian and C. P. Seiss, 'Stability of tensile cracks in concrete beams', *Proc. Am. Soc. Civ. Eng.*, **93**, ST1, 235–54, 1967.
121. O. Buyukozturk and P. V. Marcal, 'Strength of reinforced concrete chambers under external pressure', *2nd Nat. Congr. on Pressure Vessel and Piping*, San Francisco, California, 23–27 June 1975.
122. B. Saugy, T. Zimmermann and M. Hussain, 'Three-dimensional rupture analysis of a prestressed concrete vessel including creep effects', *Nucl. Eng. Des.*, **28**, 97–120, 1974.
123. M. Ueda, M. Kawahara, Y. Yoshioka and M. Kikuchi, 'Non-linear viscoelastic and elasto-plastic finite elements for concrete structures', in *Discrete Methods in Engineering*, CISE, 1974.
124. D. V. Phillips and O. C. Zienkiewicz, 'Finite element non-linear analysis of concrete structures', *Proc. Inst. Civ. Eng.*, pt 2, **61**, 59–88, 1976.
125. O. C. Zienkiewicz, D. V. Phillips and D. R. J. Owen, 'Finite element analysis of some concrete non-linearities. Theories and examples', in *IABSE Symp. on Concrete Structures Subjected to Triaxial Stresses*, Bergamo, 17–19 May 1974.
126. M. Suidan and W. C. Schnobrich, 'Finite element analysis of reinforced concrete', *Proc. Am. Soc. Civ. Eng.*, **99**, ST10, 2109–22, 1973.
127. O. C. Zienkiewicz and B. Best, 'Some non-linear problems in soil and rock mechanics—finite element solution', *Conf. on Rock Mechanics*, University of Queensland, Townsville, June 1969.
128. R. E. Goodman, R. L. Taylor and T. Brekke, 'A model for the mechanics of jointed rock', *Proc. Am. Soc. Civ. Eng.*, **94**, SM3, 637–59, 1968.
129. O. C. Zienkiewicz and G. N. Pande, 'Time dependent multilaminate models for rock—a numerical study of deformation and failure of rock masses', *Num. Anal. Meth. Geomech.*, **1**, 219–47, 1977.
130. G. N. Pande and K. G. Sharma, 'Multi-laminate model of clays—a numerical evaluation of the influence of rotation of the principal stress axes', *Int. J. Num. Meth. Eng.*, **7**, 397–418, 1983.
131. R. Hill, 'A general theory of uniqueness and stability in elasto-plastic solids', *J. Mech. Phys. Solids*, **6**, 236–49, 1958.
132. J. Mandel, 'Conditions de stabilité et postulat de Drucker', *Proc. IUTAM Symp. on Rheology and Soil Mechanics*, pp. 58–68, Springer-Verlag, Berlin, 1964.
133. J. W. Rudnicki and J. R. Rice, 'Conditions for the localisation of deformations in pressure sensitive dilatant materials', *J. Mech. Phys. Solids*, **23**, 371–94, 1975.
134. J. Rice, 'The localisation of plastic deformation', in *Theoretical and Applied Mechanics* (ed. W. T. Koiter), pp. 207–20, North Holland, 1977.
135. S. T. Pietruszczak and Z. Mróz, 'Finite element analysis of strain softening materials', *Int. J. Num. Meth. Eng.*, **10**, 327–34, 1981.
136. S. W. Sloan and J. R. Booker, 'Removal of singularities in Tresca and Mohr Coulomb yield functions', *Comm. Appl. Num. Meth.*, **2**, 173–9, 1986.

137. A. NEEDLEMAN, 'Material rate dependence and mesh sensitivity in localisation problems', *Comp. Meth. Appl. Mech. Eng.*, **67**, 69–85, 1988.
138. Z. P. BAZANT and F. B. LIN, 'Non-local yield limit degradation', *Int. J. Num. Meth. Eng.*, **26**, 1805–23, 1988.
139. Z. P. BAZANT and G. PIJAUDIER-CABOT, 'Non linear continuous damage, localisation instability and convergence', *J. Appl. Mech.*, **55**, 287–93, 1988.
140. N. BIĈANIĈ, E. PRAMONO, S. STURE and K. J. WILLEM, 'On numerical prediction of concrete fracture localisations', in *Proc. NUMETA Conf.* 385–92, Balkema, 1985.
141. J. MAZARS and Z. P. BAZANT (eds), *Cracking and Damage*, Elsevier Press, Dordrecht, 1989.
142. J. C. SIMO, 'Strain softening and dissipation. A unification of approaches', in *Proc. NUMETA Conf.*, Swansea, pp. 440–61, 1985.
143. Y. LEROY and M. ORTIZ, 'Finite element analysis of strain localisation in frictional materials', *Int. J. Num. Anal. Meth. Geomech.*, **13**, 53–74, 1989.
144. M. ORTIZ, Y. LEROY and A. NEEDLEMAN, 'A finite element method for localised failure analysis', *Comp. Meth. Appl. Mech. Eng.*, **61**, 189–214, 1987.
145. T. BELYTCHKO, J. FISH and A. BAYLISS, 'The spectral overlay on finite elements for problems with high gradients', *Comp. Meth. Appl. Mech. Eng.*, **6**, 71–76, 1990.
146. M. PASTOR, J. PERAIRE and O. C. ZIENKIEWICZ, 'Adaptive remeshing for shear band localisation problems', to be published.
147. O. C. ZIENKIEWICZ and G. C. HUANG, 'A note on localisation phenomena and adaptive FE analysis in forming processes', *Comm. Appl. Num. Meth.*, **6**, 71–76, 1990.
148. E. TONTI, 'Variational formulation of non-linear differential equations', *Bull. de l'Acad. Roy. Belg. (Sci.)*, **55**, 137–278, 1969.
149. M. MUSCAT, *The Flow of Homogeneous Fluids through Porous Media*, Edwards, 1964.
150. R. E. VOLKER, 'Non-linear flow in porous media by finite elements', *Proc. Am. Soc. Civ. Eng.*, **95**, H76, 1969.
151. H. AHMED and D. K. SUNEDA, 'Non-linear flow in porous media', *Proc. Am. Soc. Civ. Eng.*, **95**, H76, 1847–59, 1969.
152. A. M. WINSLOW, 'Numerical solution of the quasi-linear Poisson's equation in a non-uniform triangle mesh', *J. Comp. Phys.*, **1**, 149–72, 1967.
153. M. V. K. CHARI and P. SILVESTER, 'Finite element analysis of magnetically saturated d.c. motors', in *IEEB Winter Meeting on Power*, New York, 1971.
154. O. C. ZIENKIEWICZ, J. F. LYNESS and D. R. J. OWEN, 'Three-dimensional magnetic field determination using a scalar potential. A finite element solution', *Magn. Trans. IEEE*, **13**, 1649–56, 1977.
155. J. F. LYNESS, D. R. J. OWEN and O. C. ZIENKIEWICZ, 'The finite element analysis of engineering systems governed by a non-linear quasi-harmonic equation', *Comp. Struct.*, **5**, 65–79, 1975.
156. D. GELDER, 'Solution of the compressible flow equations', *Int. J. Num. Meth. Eng.*, **3**, 35–43, 1971.
157. C. A. ANDERSON and O. C. ZIENKIEWICZ, 'Spontaneous ignition: finite element solutions for steady and transient conditions', *Trans. ASME, J. Heat Transfer*, 398–404, August 1974.

8

Geometrically non-linear problems—large displacement and structural instability

8.1 Introduction

In the previous chapter the question of non-linearities arising from material properties was discussed and methods were developed to allow the standard linear forms to be used in an iterative way to obtain solutions. In this chapter a similar path will be followed in the treatment of geometric non-linearity of structures.

In all problems discussed so far it has been implicitly assumed that both displacements and strains developed in the structure are small. In practical terms this means that geometry of the elements remains basically unchanged during the loading process and that first-order, infinitesimal, linear strain approximations can be used.

In practice such assumptions fail frequently even though actual strains may be small and elastic limits of ordinary structural materials not exceeded. If accurate determination of the displacements is needed, geometric non-linearity may have to be considered in some structures. Here, for instance, stresses due to membrane action, usually neglected in plate flexure, may cause a considerable decrease of displacements as compared with the linear solution, even though displacements are still quite small. Conversely, it may be found that a load is reached where deflections increase more rapidly than predicted by a linear solution and indeed a state may be attained where load-carrying capacity *decreases* with continuing deformation. This classic problem is that of structural stability and obviously has many practical implications. The applications of such an analysis are clearly of considerable importance in aerospace engineering, design of radio telescopes, cooling towers, box girder bridges and other relatively slender structures.

In many cases *very large displacements* may occur without causing large strains. Typical in this context is the classical problem of the 'elastica' of which an example is a watch spring.

In this chapter an attempt is made to unify the treatment of all the above problems and to present generally applicable procedures. This is achieved by examining the basic non-linear equilibrium equations together with their solution. Such considerations also lead to the formulation of classical initial stability problems. These concepts are then illustrated by formulating the large deflection and initial stability problems for flat plates. This naturally leads to the general continuum formulation of the large displacement problem. A lagrangian approach is adopted throughout in which displacements are referred to the *original configuration.* Alternative, updated coordinate approaches are mentioned briefly at the end of the chapter.

One aspect of geometric non-linearity is not discussed in detail. This is the case of large strain such as may occur, even elastically, with such materials as rubber, etc. Here specialized relations between stress and strain have to be introduced[1] and the length of this book prohibits a full discussion. Nevertheless, the general processes of the next section are still applicable providing suitable stress and strain laws are introduced.

Geometric non-linearity may often be combined with material non-linearity of the type discussed in the previous chapter, such as small strain plasticity, etc. In principle this does not introduce additional complexities and the methods of this chapter can easily be extended to deal with this situation.[2]

8.2 General considerations

8.2.1 *The basic problem.* Whether the displacements (or strains) are large or small, equilibrium conditions between internal and external 'forces' have to be satisfied. Thus, if the displacements are prescribed in the usual manner by a finite number of (nodal) parameters $\mathbf{a}$, we can obtain the necessary equilibrium equations using the virtual work principle given in Chapter 2 of Volume 1. Now, however, different definitions of 'stresses' and 'strains' which are conjugate to each other must be used. We shall discuss some such conjugate quantities in the context of plates, shells and general elasticity later, but in all cases we find that we can write

$$\boldsymbol{\Psi}(\mathbf{a}) = \int_V \bar{\mathbf{B}}^{\mathrm{T}} \boldsymbol{\sigma}\, \mathrm{d}V - \mathbf{f} = 0 \tag{8.1}$$

where $\boldsymbol{\Psi}$ once again represents the sum of external and internal generalized forces, and in which $\bar{\mathbf{B}}$ is defined from the strain definition as

$$\mathrm{d}\boldsymbol{\varepsilon} = \bar{\mathbf{B}}\, \mathrm{d}\mathbf{a} \tag{8.2}$$

The bar suffix has now been added for, if displacements are large, the strains depend non-linearly on displacement, and the matrix $\bar{\mathbf{B}}$ is now dependent on $\mathbf{a}$. We shall see later that we can conveniently write

$$\bar{\mathbf{B}} = \mathbf{B}_0 + \mathbf{B}_L(\mathbf{a}) \tag{8.3}$$

in which $\mathbf{B}_0$ is the same matrix as in linear infinitesimal strain analysis and only $\mathbf{B}_L$ depends on the displacement. In general $\mathbf{B}_L$ will be found to be a *linear function* of such displacements.

If strains are reasonably small we can still write the general elastic relation

$$\boldsymbol{\sigma} = \mathbf{D}(\boldsymbol{\varepsilon} - \boldsymbol{\varepsilon}_0) + \boldsymbol{\sigma}_0 \tag{8.4}$$

in which $\mathbf{D}$ is the usual set of elastic constants.†

However, any non-linear stress–strain relationship could equally well be written, as the whole process of solution once again reduces to the solution of a non-linear set of equations (8.1). *It is perhaps too obvious to restate that in Eq.* (8.1) *integrals are in fact carried out element by element and contributions to 'nodal equilibrium' summed in the usual manner.*

8.2.2 *Solution processes.* Clearly the solution of Eq. (8.1) will have to be approached iteratively and the general methods described in the previous chapter (in Sec. 7.2) are applicable.

If, for instance, the Newton–Raphson process is to be adopted we have to find the relation between $\mathrm{d}\mathbf{a}$ and $\mathrm{d}\boldsymbol{\Psi}$, as explained there. Thus taking appropriate variations of Eq. (8.1) with respect to $\mathrm{d}\mathbf{a}$ we have

$$\mathrm{d}\boldsymbol{\Psi} = \int_V \mathrm{d}\bar{\mathbf{B}}^{\mathrm{T}} \boldsymbol{\sigma} \, \mathrm{d}V + \int_V \bar{\mathbf{B}}^{\mathrm{T}} \mathrm{d}\boldsymbol{\sigma} \, \mathrm{d}V = \mathbf{K}_T \, \mathrm{d}\mathbf{a} \tag{8.5}$$

and using Eqs. (8.4) and (8.2) we have‡

$$\mathrm{d}\boldsymbol{\sigma} = \mathbf{D} \, \mathrm{d}\boldsymbol{\varepsilon} = \mathbf{D}\bar{\mathbf{B}} \, \mathrm{d}\mathbf{a}$$

and if Eq. (8.3) is valid

$$\mathrm{d}\bar{\mathbf{B}} = \mathrm{d}\mathbf{B}_L$$

Therefore

$$\mathrm{d}\boldsymbol{\Psi} = \int_V \mathrm{d}\mathbf{B}_L^{\mathrm{T}} \boldsymbol{\sigma} \, \mathrm{d}V + \bar{\mathbf{K}} \, \mathrm{d}\mathbf{a} \tag{8.6}$$

† It is important to remember here that the stress components defined in Eq. (8.4) are those *corresponding* to the strain component used. In some gross displacement problems such strain components are subject to considerable change of direction from original, fixed axes.

‡ Once again if non-linear stress–strain relations are used $\mathbf{D} = \mathbf{D}(\boldsymbol{\sigma})$ is the incremental elasticity matrix as given in Eq. (7.51).

where

$$\bar{\mathbf{K}} = \int_V \bar{\mathbf{B}}^{\mathrm{T}} \mathbf{D} \bar{\mathbf{B}} \, \mathrm{d}V = \mathbf{K}_0 + \mathbf{K}_L \tag{8.7}$$

in which $\mathbf{K}_0$ represents the usual, small displacements stiffness matrix, i.e.

$$\mathbf{K}_0 = \int_V \mathbf{B}_0^{\mathrm{T}} \mathbf{D} \mathbf{B}_0 \, \mathrm{d}V \tag{8.7a}$$

The matrix $\mathbf{K}_L$ is due to the large displacement and is given by

$$\mathbf{K}_L = \int_V (\mathbf{B}_0^{\mathrm{T}} \mathbf{D} \mathbf{B}_L + \mathbf{B}_L^{\mathrm{T}} \mathbf{D} \mathbf{B}_L + \mathbf{B}_L^{\mathrm{T}} \mathbf{D} \mathbf{B}_0) \, \mathrm{d}V \tag{8.7b}$$

$\mathbf{K}_L$ is variously known as the *initial displacement matrix*,[3] *large displacement matrix*, etc., and contains only terms that are linear and quadratic in $\mathbf{a}$. It will be found that this is a matrix that would alternatively be obtained by using an infinitesimal strain approach but adjusting element coordinates in the computation of the stiffness.

The first term of Eq. (8.6) can generally be written (perhaps less obviously until particular cases are examined) as

$$\int_V \mathrm{d}\mathbf{B}_L^{\mathrm{T}} \boldsymbol{\sigma} \, \mathrm{d}V \equiv \mathbf{K}_\sigma \, \mathrm{d}\mathbf{a} \tag{8.8}$$

where $\mathbf{K}_\sigma$ is a symmetric matrix dependent on the stress level. This matrix is known as the *initial stress matrix*[3,4] or *geometric matrix*.[5,6] Thus

$$\mathrm{d}\boldsymbol{\Psi} = (\mathbf{K}_0 + \mathbf{K}_\sigma + \mathbf{K}_L) \, \mathrm{d}\mathbf{a} = \mathbf{K}_{\mathrm{T}} \, \mathrm{d}\mathbf{a} \tag{8.9}$$

with $\mathbf{K}_T$ being the total, *tangential stiffness*, matrix. Newton-type iteration can once more be applied precisely in the manner of Sec. 7.2.

To summarize, usually:

1. The elastic linear solution is obtained as a first approximation $\mathbf{a}^0$.
2. $\boldsymbol{\Psi}^0$ is found using Eq. (8.1) with an appropriate definition of $\bar{\mathbf{B}}$ and stresses as given by Eq. (8.4) (or any other linear or non-linear law).
3. Matrix $\mathbf{K}_T^0$ is established.
4. Correction is computed as

$$\delta \mathbf{a}^0 = -(\mathbf{K}_T^0)^{-1} \boldsymbol{\Psi}^0$$

   and processes 2, 3 and 4 repeated until $\boldsymbol{\Psi}^i$ becomes sufficiently small.

Again a constant matrix could be used, increasing the number of iterating steps but making use of a semi-inverted, resolve process at small computer cost, providing that at each step $\boldsymbol{\Psi}^i$ is calculated by the correct expressions; however, convergence is sometimes slow by this procedure.

Quite conveniently the tangent stiffness matrix may be made constant after, say, the second iteration of each load increment; this has been used with considerable success.[7]

While all solutions can be accomplished in a one-step operation for a full load, on occasion, as in all non-linear problems, the possibility of a non-unique solution arises and the physically unimportant one may be obtained. In such cases it is wise to proceed by incrementing the load and obtaining the non-linear solution for each increment. This indeed is sometimes computationally cheaper as effects of non-linearity in each step are reduced. Indeed, if load increments of sufficiently small magnitude are taken each incremental solution may be accomplished sufficiently accurately in one step.[5,6,8] It is important, however, to check periodically the total equilibrium by using the full Eq. (8.1).

All the solution techniques described in Chapter 7, Sec. 7.2, have been used in the context of geometrically non-linear analysis problems with success. Such techniques have been extensively evaluated by Haisler *et al.*[9]

8.2.3 *Initial stability problem.* It is of interest to note at this stage that $\mathbf{K}_\sigma$ does not explicitly contain the displacements and is proportional to the stress level $\boldsymbol{\sigma}$. Thus, if at the first step of computation we evaluate $\boldsymbol{\sigma}$ by a linear solution we have from Eq. (8.6)

$$d\boldsymbol{\Psi} = (\mathbf{K}_0 + \mathbf{K}_\sigma)\, d\mathbf{a} \tag{8.10}$$

as $\mathbf{K}_L = 0$ at this stage.

If the loads are increased by a factor λ we may find that neutral stability exists, i.e.

$$d\boldsymbol{\Psi} = (\mathbf{K}_0 + \lambda\mathbf{K}_\sigma)\, d\mathbf{a} \equiv 0 \tag{8.11}$$

From this λ can be obtained by solving the typical *eigenvalue problem* defined above (see Chapter 9 for a discussion of similar eigenvalue problems). This is the classical, 'initial', stability problem such as occurs in the buckling of struts, plates, shells, etc.

Quite frequently in the literature this type of approach is used beyond its limits of applicability. The 'initial stability' expressed can only give physically significant answers if the elastic ($\mathbf{K}_0$) solution gives such deformations that the *large deformation matrix* $\mathbf{K}_L$ *is identically zero.* This only happens in a very limited number of practical situations (e.g. a perfectly straight strut under axial load, complete sphere under uniform pressure, etc.). The preoccupation of such investigators with the subject of 'initial imperfections' is strictly limited to such situations where a true *bifurcation* can occur. In real engineering situations where the qualitative nature of the behaviour is completely unknown such problems should be investigated using the full tangential stiffness matrix.[8] When $\mathbf{K}_T\,\mathbf{da}$ is identically zero, neutral equilibrium is obtained. A step-by-step approach is clearly necessary here.

It is possible to achieve a compromise between the classical stability problem and a full non-linear analysis in a number of ways.[10,11] For example, (*a*) based on a full non-linear analysis the eigenvalue problem may be considered after each increment of load or, alternatively, (*b*) a linear eigenvalue analysis that includes a linearized $\mathbf{K}_L$ as well as the usual $\mathbf{K}_\sigma$ can be used.

8.2.4 *Energy interpretation of the stability criteria.* It was shown in Chapter 2 of Volume 1 that the virtual work done during a displacement variation $\mathbf{da}$ is in fact equal to the variation of total potential energy Π. Thus, for equilibrium,

$$\mathrm{d}\Pi = \mathbf{da}^{\mathrm{T}}\boldsymbol{\Psi} = 0 \tag{8.12}$$

i.e. *the total potential energy is stationary* [which is equivalent to Eq. (8.1)].

The second variation of Π is [by Eq. (8.9)]

$$\mathrm{d}^2\Pi = \mathrm{d}(\mathrm{d}\Pi) = \mathbf{da}^{\mathrm{T}}\,\mathrm{d}\boldsymbol{\Psi} = \mathbf{da}^{\mathrm{T}}\,\mathbf{K}_T\,\mathbf{da} \tag{8.13}$$

The stability criterion is given by a positive value of this second variation and conversely instability by a negative value (as in the first case energy has to be added to the structure while in the second it contains surplus energy). In other words, *if* $\mathbf{K}_T$ *is positive-definite, stability exists.* This criterion is well known[12] and of considerable use when investigating stability during large deformation.†[13–15]

8.2.5 *Forces dependent on deformation.* In the derivation of Eq. (8.5) it was implicitly assumed that the forces $\mathbf{f}$ are not themselves dependent on the deformation. In some instances this is not true. For instance, pressure loads on a grossly deforming structure are in general in this category, as indeed are some aerodynamic forces depending on the deformation (flutter).

If forces vary with displacement then in relation to Eq. (8.5) the variation $\mathbf{df}$ with respect to $\mathbf{da}$ has to be considered. This leads to the introduction of the *load-correction matrix.*[16–17] Stability and large deformation problems under such (non-conservative) loads can be once again studied if proper consideration is given to the above term.

† An alternative test is to investigate the sign of the determinant of $\mathbf{K}_T$.[15]

8.3 Large deflection and 'initial' stability of plates

8.3.1 *Definitions.* As a first example we shall consider the problems associated with the deformation of plates subject to 'in-plane' and 'lateral' forces, when displacements are not infinitesimal but also not excessively large (Fig. 8.1). In this situation the 'change in geometry' effect is less important than the *relative magnitudes* of the linear and non-linear strain–displacement terms, and in fact for 'stiffening' problems the non-linear displacements are always less than the corresponding linear ones (see Fig. 8.2). It is well known that in such situations the lateral displacements will be responsible for development of 'membrane'-type strains and now the two problems of 'in-plane' and 'lateral' deformation can no longer be dealt with separately but are *coupled.*

We shall, as before, describe the plate 'strains' in terms of middle surface displacements, i.e. if the xy plane coincides with the middle surface as in Fig. 8.1(a) we shall have (see Chapters 2 and 4)

$$\boldsymbol{\varepsilon}=\left\{\begin{array}{c}\varepsilon_x\\ \varepsilon_y\\ \gamma_{xy}\\ -\dfrac{\partial^2 w}{\partial x^2}\\ -\dfrac{\partial^2 w}{\partial y^2}\\ 2\dfrac{\partial^2 w}{\partial x\,\partial y}\end{array}\right\}=\left\{\begin{array}{c}\boldsymbol{\varepsilon}^p\\ \boldsymbol{\varepsilon}^b\end{array}\right\}\qquad \boldsymbol{\sigma}=\left\{\begin{array}{c}T_x\\ T_y\\ T_{xy}\\ M_x\\ M_y\\ M_{xy}\end{array}\right\}=\left\{\begin{array}{c}\boldsymbol{\sigma}^p\\ \boldsymbol{\sigma}^b\end{array}\right\} \tag{8.14}$$

The 'stresses' are defined in terms of the usual resultants;† $T_x=\bar{\sigma}_x t$, where $\bar{\sigma}_x$ is the average membrane stress, etc. Now, if the deformed shape is considered as in Fig. 8.1(b), we see that displacement w produces some additional extension in the x and y directions of the middle surface and the length $\mathrm{d}x$ stretches to

$$\mathrm{d}x'=\mathrm{d}x\sqrt{1+\left(\frac{\partial w}{\partial x}\right)^2}=\mathrm{d}x\left\{1+\frac{1}{2}\left(\frac{\partial w}{\partial x}\right)^2+\cdots\right\}$$

i.e. in defining the x elongation we can write (to second approximation)

$$\varepsilon_x=\frac{\partial u}{\partial x}+\frac{1}{2}\left(\frac{\partial w}{\partial x}\right)^2$$

Considering in a similar way the other components[18] we can write as the

† In-plane and bending components have here been separated by appropriate superscripts.

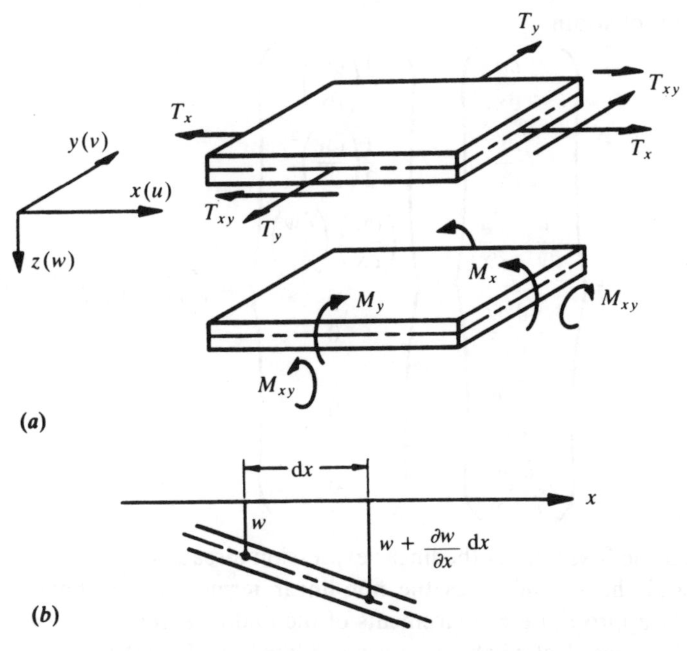

Fig. 8.1 (*a*) 'In-plane' and bending resultants for a flat plate. (*b*) Increase of middle surface length due to lateral displacement

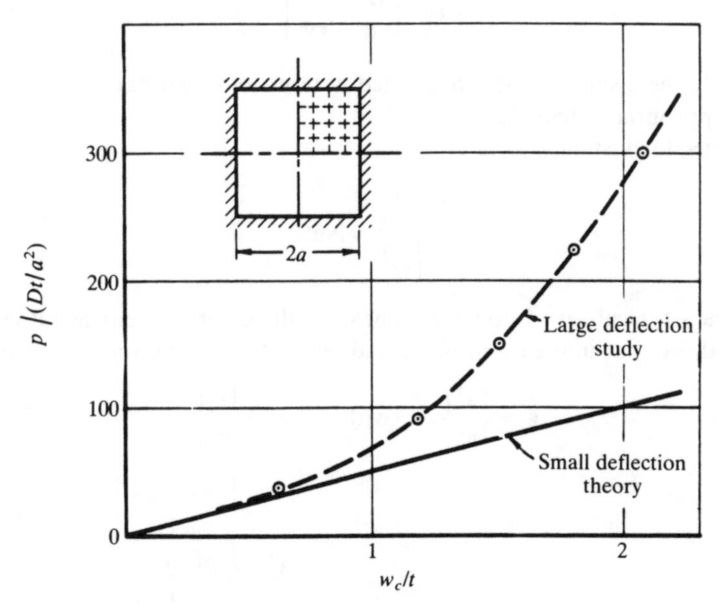

Fig. 8.2 Central deflection w_c of a clamped square plate under uniform load p.[15] $u=v=0$ at edge

definition of strain

$$\boldsymbol{\varepsilon}=\begin{Bmatrix}\dfrac{\partial u}{\partial x}\\[2mm] \dfrac{\partial v}{\partial y}\\[2mm] \dfrac{\partial u}{\partial y}+\dfrac{\partial u}{\partial x}\\[2mm] -\dfrac{\partial^2 w}{\partial x^2}\\[2mm] -\dfrac{\partial^2 w}{\partial y^2}\\[2mm] 2\dfrac{\partial^2 w}{\partial x\,\partial y}\end{Bmatrix}+\begin{Bmatrix}\dfrac{1}{2}\left(\dfrac{\partial w}{\partial x}\right)^2\\[2mm] \dfrac{1}{2}\left(\dfrac{\partial w}{\partial y}\right)^2\\[2mm] \left(\dfrac{\partial w}{\partial x}\right)\left(\dfrac{\partial w}{\partial y}\right)\\[2mm] 0\\[2mm] 0\\[2mm] 0\end{Bmatrix}=\begin{Bmatrix}\boldsymbol{\varepsilon}_0^p\\ \boldsymbol{\varepsilon}_0^b\end{Bmatrix}+\begin{Bmatrix}\boldsymbol{\varepsilon}_L^p\\ \mathbf{0}\end{Bmatrix} \tag{8.15}$$

in which the first term is the linear expression already encountered many times and the second gives the non-linear terms. In the above, u, v, w stand for appropriate displacements of the middle surface.

If only linear elastic behaviour is considered, the $\mathbf{D}$ matrix is composed of plane stress and a bending component (see Chapters 2 and 4):

$$\mathbf{D}=\begin{bmatrix}\mathbf{D}^p & 0\\ 0 & \mathbf{D}^b\end{bmatrix} \tag{8.16}$$

Finally the displacements are defined in terms of nodal parameters using the appropriate shape functions.

Thus, for instance,

$$\begin{Bmatrix}u\\ v\\ w\end{Bmatrix}=\mathbf{N}\mathbf{a}^e \tag{8.17}$$

where a typical set of nodal parameters will be for convenience divided into those that influence in-plane and bending deformation respectively:

$$\mathbf{a}_i=\begin{Bmatrix}\mathbf{a}_i^p\\ \mathbf{a}_i^b\end{Bmatrix}\quad\text{with}\quad \mathbf{a}_i^p=\begin{Bmatrix}u_i\\ v_i\end{Bmatrix},\qquad \mathbf{a}_i^b=\begin{Bmatrix}w_i\\[1mm] \dfrac{\partial w}{\partial x_i}\\[2mm] \dfrac{\partial w}{\partial y_i}\end{Bmatrix} \tag{8.18}$$

Thus the shape function can also be subdivided as

$$\mathbf{N}_i = \begin{bmatrix} \mathbf{N}_i^p & 0 \\ 0 & \mathbf{N}_i^b \end{bmatrix} \tag{8.19}$$

and indeed we shall assume in what follows that the final assembled displacement vector is also subdivided in the manner of Eq. (8.18).

This is convenient as with the exception of the non-linear strain $\boldsymbol{\varepsilon}_L^p$ all the definitions of standard linear analysis apply and therefore do not have to be repeated here.

8.3.2 *Evaluation of* $\bar{\mathbf{B}}$. For further formulation it will be necessary to establish expressions for $\bar{\mathbf{B}}$ and $\mathbf{K}_T$ matrices. Firstly we shall note that

$$\bar{\mathbf{B}} = \mathbf{B}_0 + \mathbf{B}_L \tag{8.20}$$

where

$$\mathbf{B}_0 = \begin{bmatrix} \mathbf{B}_0^p & \mathbf{0} \\ \mathbf{0} & \mathbf{B}_0^b \end{bmatrix} \qquad \mathbf{B}_L = \begin{bmatrix} \mathbf{0} & \mathbf{B}_L^b \\ \mathbf{0} & \mathbf{0} \end{bmatrix}$$

where $\mathbf{B}_0^p$, $\mathbf{B}_0^b$ are the well-defined, standard matrices of appropriate linear in-plane and bending elements and $\mathbf{B}_L^b$ is found by taking a variation of $\boldsymbol{\varepsilon}_L^p$ with respect to the parameters $\mathbf{a}^b$.

This non-linear, strain component of Eq. (8.15) can be written conveniently as

$$\boldsymbol{\varepsilon}_L^p = \frac{1}{2} \begin{bmatrix} \dfrac{\partial w}{\partial x} & 0 \\ 0 & \dfrac{\partial w}{\partial y} \\ \dfrac{\partial w}{\partial y} & \dfrac{\partial w}{\partial x} \end{bmatrix} \begin{Bmatrix} \dfrac{\partial w}{\partial x} \\ \dfrac{\partial w}{\partial y} \end{Bmatrix} = \tfrac{1}{2}\mathbf{A}\boldsymbol{\theta} \tag{8.21}$$

The derivatives (slopes) of w can be related to the nodal parameters $\mathbf{a}^b$ as

$$\boldsymbol{\theta} = \begin{Bmatrix} \dfrac{\partial w}{\partial x} \\ \dfrac{\partial w}{\partial y} \end{Bmatrix} = \mathbf{G}\mathbf{a}^b \tag{8.22}$$

in which we have

$$\mathbf{G}=\begin{bmatrix} \dfrac{\partial \mathbf{N}_1^b}{\partial x}, & \dfrac{\partial \mathbf{N}_2^b}{\partial x}, & \dots \\ \dfrac{\partial \mathbf{N}_1^b}{\partial y}, & \dfrac{\partial \mathbf{N}_2^b}{\partial y} & \dots \end{bmatrix} \tag{8.23}$$

Thus $\mathbf{G}$ is a matrix defined purely in terms of the coordinates.

Taking the variation of Eq. (8.21) we have†

$$\mathrm{d}\boldsymbol{\varepsilon}_L^p=\tfrac{1}{2}\,\mathrm{d}\mathbf{A}\boldsymbol{\theta}+\tfrac{1}{2}\mathbf{A}\,\mathrm{d}\boldsymbol{\theta}=\mathbf{A}\,\mathrm{d}\boldsymbol{\theta}=\mathbf{A}\mathbf{G}\,\mathrm{d}\mathbf{a}^b \tag{8.24}$$

† The manipulation of Eq. (8.24) is due to an interesting property of the matrices $\mathbf{A}$ and $\boldsymbol{\theta}$. It is easy to verify that if

$$\mathbf{x}=\begin{Bmatrix} x_1 \\ x_2 \end{Bmatrix}$$

is an arbitrary vector then

$$\mathrm{d}\mathbf{A}\mathbf{x}=\begin{bmatrix} \mathrm{d}\left(\dfrac{\partial w}{\partial x}\right) & 0 \\ 0 & \mathrm{d}\left(\dfrac{\partial w}{\partial y}\right) \\ \mathrm{d}\left(\dfrac{\partial w}{\partial y}\right) & \mathrm{d}\left(\dfrac{\partial w}{\partial x}\right) \end{bmatrix}\begin{Bmatrix} x_1 \\ x_2 \end{Bmatrix} \equiv \begin{bmatrix} x_1 & 0 \\ 0 & x_2 \\ x_2 & x_1 \end{bmatrix}\mathrm{d}\boldsymbol{\theta}$$

Thus

$$\mathrm{d}\mathbf{A}\boldsymbol{\theta}=\mathbf{A}\,\mathrm{d}\boldsymbol{\theta}$$

Similarly, if

$$\mathbf{y}=\begin{Bmatrix} y_1 \\ y_2 \\ y_3 \end{Bmatrix}$$

then

$$\mathrm{d}\mathbf{A}^{\mathrm{T}}\mathbf{y}=\begin{bmatrix} \mathrm{d}\left(\dfrac{\partial w}{\partial x}\right) & 0 & \mathrm{d}\left(\dfrac{\partial w}{\partial y}\right) \\ 0 & \mathrm{d}\left(\dfrac{\partial w}{\partial y}\right) & \mathrm{d}\left(\dfrac{\partial w}{\partial x}\right) \end{bmatrix}\begin{Bmatrix} y_1 \\ y_2 \\ y_3 \end{Bmatrix}=\begin{bmatrix} y_1 & y_3 \\ y_3 & y_2 \end{bmatrix}\mathrm{d}\boldsymbol{\theta}$$

Use of this second property will be made later.

and hence immediately, by definition,

$$\mathbf{B}_L^b = \mathbf{A}\mathbf{G} \tag{8.25}$$

8.3.3 *Evaluation of* $\mathbf{K}_T$. The linear, small deformation, matrices are written as

$$\mathbf{K}_0 = \begin{bmatrix} \mathbf{K}_0^p & \mathbf{0} \\ \mathbf{0} & \mathbf{K}_0^b \end{bmatrix} \tag{8.26}$$

with appropriate definitions given in Chapters 2 and 4. Using Eq. (8.7b) the large displacement matrices can be defined on substituting Eq. (8.20). Thus after some manipulation

$$\mathbf{K}_L = \int_V \begin{bmatrix} \mathbf{0}, & \mathbf{B}_0^{pT} \quad \mathbf{D}^p \quad \mathbf{B}_L^b \\ \text{sym.}, & \mathbf{B}_L^{bT} \quad \mathbf{D}^{b'} \quad \mathbf{B}_L^b \end{bmatrix} \mathrm{d}V \tag{8.27}$$

Finally $\mathbf{K}_\sigma$ has to be found using the definition of Eq. (8.8). From Eq. (8.20) we have, on taking a variation,

$$\mathrm{d}\mathbf{B}_L^{\mathrm{T}} = \begin{bmatrix} \mathbf{0} & \mathbf{0} \\ \mathrm{d}\mathbf{B}_L^{b\mathrm{T}} & \mathbf{0} \end{bmatrix} \tag{8.28}$$

which on substitution into Eqs (8.8) and (8.25) gives

$$\mathbf{K}_\sigma \,\mathrm{d}\mathbf{a} = \int_V \begin{bmatrix} \mathbf{0} & \mathbf{0} \\ \mathbf{G}^{\mathrm{T}} \,\mathrm{d}\mathbf{A}^{\mathrm{T}} & \mathbf{0} \end{bmatrix} \begin{Bmatrix} T_x \\ T_y \\ T_{xy} \\ M_x \\ M_y \\ M_{xy} \end{Bmatrix} \mathrm{d}V \tag{8.29}$$

However, by the special property described in the footnote of page 294, we can write

$$\mathrm{d}\mathbf{A}^{\mathrm{T}} \begin{Bmatrix} T_x \\ T_y \\ T_{xy} \end{Bmatrix} = \begin{bmatrix} T_x & T_{xy} \\ T_{xy} & T_y \end{bmatrix} \mathrm{d}\boldsymbol{\theta} = \begin{bmatrix} T_x & T_{xy} \\ T_{xy} & T_y \end{bmatrix} \mathbf{G} \,\mathrm{d}\mathbf{a}^b$$

and finally we obtain

$$\mathbf{K}_\sigma = \begin{bmatrix} \mathbf{0} & \mathbf{0} \\ \mathbf{0} & \mathbf{K}_\sigma^b \end{bmatrix} \tag{8.30}$$

with

$$\mathbf{K}_\sigma^b = \int_V \mathbf{G}^\mathrm{T} \begin{bmatrix} T_x & T_{xy} \\ T_{xy} & T_y \end{bmatrix} \mathbf{G}\, \mathrm{d}V \tag{8.31}$$

a well-known *symmetric form of the initial stress* matrix for plate bending.

8.3.4 *Large deflection problem.* All the ingredients necessary for computing the large deflection plate problem are now available.

As a first step displacements $\mathbf{a}^0$ are found according to the small displacement uncoupled solution. This determines the actual strains by considering the non-linear contribution defined by Eq. (8.21) together with the appropriate linear contributions. Corresponding stresses can be found by the elastic expressions and $\mathbf{\Psi}^0$ determined according to Eq. (8.1). For successive iterations $\mathbf{K}_T^i$ is found from Eqs (8.26), (8.27) and (8.30).

A typical solution thus obtained,[15] Fig. 8.2, shows the stiffening of the plate with increasing deformation due to the development of 'membrane' stresses. At the edges of the plate all in-plate and lateral deformations are restrained. The results show excellent agreement with an alternative analytical solution.

The element properties were derived using for the in-plane deformation the simplest bilinear rectangle and for bending deformation the non-conforming shape function for a rectangle (Section 1.3, Chapter 1).

An example of the stress variation with load for a clamped square plate under uniform load is shown in Fig. 8.3.[19] A quarter of the plate is analysed as above with 32 triangular elements, using the 'in-plane' element given in Chapter 3 of Volume 2 together with a modified[20] version of the non-conforming plate bending element of Chapter 1. Many other examples of large plate deformation obtained by finite element methods are available in the literature.[4,19,21-25]

8.3.5 *Bifurcation instability.* In a few practical cases, as in the classical Euler problem, a bifurcation instability is possible. Consider the situation of a plate loaded purely in its own plane. As lateral deflections, w, are not produced, the small deflection theory gives an exact solution. However, even with zero lateral displacements, the initial stress matrix $\mathbf{K}_\sigma^b$ can be found while $\mathbf{K}_L \equiv 0$. If the in-plane stresses are compressive this matrix will generally be such that real eigenvalues of the bending deformation equation can be found:

$$(\mathbf{K}_0^b + \lambda \mathbf{K}_\sigma^b)\mathbf{a}^b = \mathbf{0} \tag{8.32}$$

in which λ denotes the increase factor on in-plane stresses necessary to achieve neutral equilibrium (instability).

At such an increased load incipient buckling occurs and lateral deflections can occur without any lateral load.

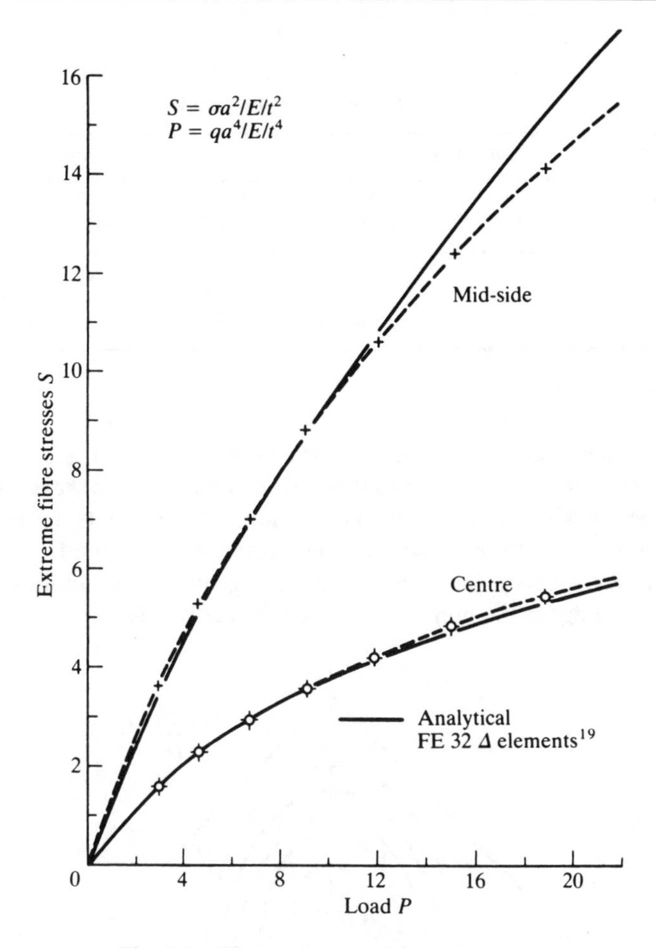

Fig. 8.3 Clamped square plate—stresses

The problem is simply formulated by writing only the bending equation with $\mathbf{K}_0^b$ determined as in Chapter 1 and $\mathbf{K}_\sigma^b$ found from Eq. (8.31).

Points of such incipient stability (buckling) for a variety of plate problems have been determined using various element formulations.[26-31] Some comparative results for a simple problem of a square, simply supported plate under a uniform compression T_x applied in one direction are given in Table 8.1. In this the buckling parameter is defined as

$$C = \frac{T_x a^2}{\pi^2 D}$$

where a is the side of the plate and D the bending rigidity.

TABLE 8.1

VALUES OF C FOR A SQUARE PLATE SIMPLY SUPPORTED AND COMPRESSED UNIAXIALLY

Elements in quarter plate	Non-compatible		Compatible	
	Rectangle[27] 12 DOF	Triangle[29] 9 DOF	Rectangle[30] 16 DOF	Quadrilateral[31] 16 DOF
2×2		3.22		
4×4	3.77	3.72	4.015	4.029
8×8	3.93	3.90	4.001	4.002

Exact $C=4.00$.[18]
DOF = degrees of freedom.

The elements are all of the type described in Chapter 1 and it is of interest to note that all those that are slope compatible always overestimate the buckling factor. The non-conforming elements in this case underestimate it although there is now no theoretical bound.

Figure 8.4 shows a buckling mode for a geometrically more complex case.[29] Here again the non-conforming triangle was used.

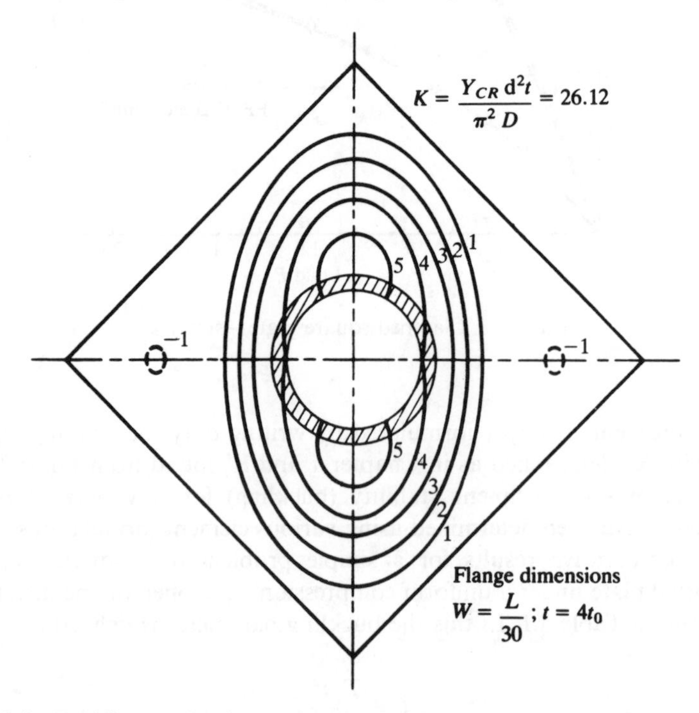

Fig. 8.4 Buckling mode of a square plate under shear—clamped edges, central hole stiffened by flange[29]

Such incipient stability problems in plates are of limited practical importance. As soon as lateral deflection occurs a stiffening of the plate follows and additional loads can be carried. This stiffening was noted in the example of Fig. 8.2. Postbuckling behaviour thus should be studied by the large deformation process generally described in previous sections.[32-34] To avoid the bifurcation difficulty a slight perturbation (or lateral load) should then be imposed.

8.4 Shells

In shells, stability problems are much more relevant than in plates. Here, in general, the problem is one in which the tangential stiffness matrix $\mathbf{K}_T$ should always be determined taking the actual displacements into account, as now the special case of uncoupled membrane and bending effects does not occur under load except in the most trivial cases. If the *initial stability* matrix $\mathbf{K}_\sigma$ is determined for the elastic stresses it is, however, sometimes possible to obtain useful results concerning the stability factor λ, and indeed in the classical work on the subject of shell buckling this initial stability has been almost exclusively considered. The true collapse load may, however, be well *below* the initial stability load and it is important to determine at least approximately the deformation effects.

If shells are assumed to be built up of flat plate elements the same transformations as given in Chapter 3 can be followed with the plate tangential stiffness matrix.[35,36] If curved shell elements are used it is essential to revert to the equation of shell theory and to include in these the non-linear terms.[15,37-39] For the complete formulation required the reader is referred to these references.

In the case of shallow shells the transformations of Chapter 3 may conveniently be avoided by adopting a formulation based on Marguerre shallow shell theory.[25,40,41]

It is extremely important to emphasize again that initial instability calculations are meaningful only in special cases and that they often overestimate the collapse loads considerably. For correct answers a full non-linear process has to be invoked. The progressive 'softening' of a shell under a load well below the one given by linearized buckling is shown in Fig. 8.5, taken from reference 15. Figure 8.6 shows the progressive collapse of an arch at a load much below that given by the linear stability value.[8]

The determination of the actual collapse load of a shell or other slender structure presents obvious difficulties (of a kind already encountered in Chapter 7), as convergence of displacements cannot be obtained when load is increased near the 'peak' carrying capacity.

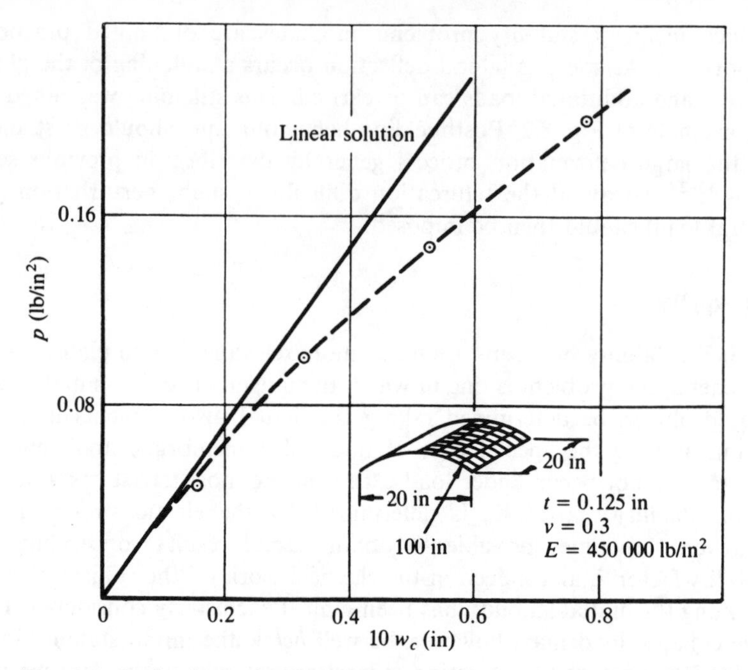

Fig. 8.5 Deflections of cylindrical shell at centre. All edges clamped[15]

It is convenient, then, to proceed immediately prescribing displacement increments and computing the corresponding reactions if one concentrated load is considered. By such processes, Argyris[6] and others[39,42] succeed in following a complete snap-through behaviour of an arch.

Pian and Ping Tong[43] show how the process can be simply generalized when a system of proportional loads is considered and the general procedures are precisely those discussed in Chapter 7, page 221.

Alternative processes of approaching the collapse problem have been described and much work has been accomplished in this important field.[44–48]

8.5 General, large strain and displacement formulation

The non-linear strain displacement relationship for plates, used in Sec. 8.3 [Eq. (8.15)], was derived on an *ad hoc* basis. For shells, alternative relationships may be similarly derived but the possibility of diverse approximations arises at all stages. It is, however, possible to use a general definition of strains which is *valid whether displacements or strains are large or small.* Such a definition was introduced by Green and St Venant and is known as *Green's strain tensor.* Referred to a fixed

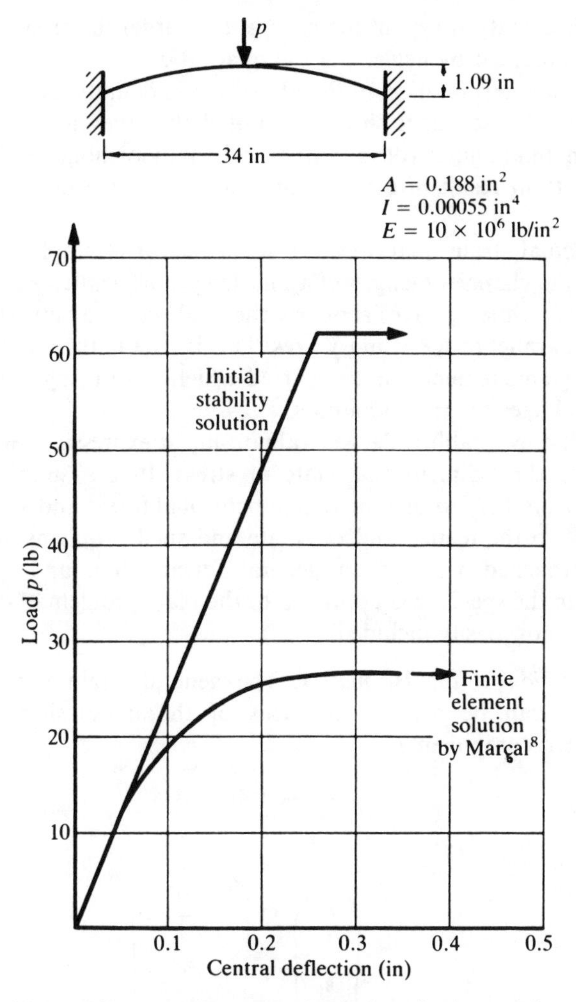

Fig. 8.6 'Initial stability' and incremental solutions for large deformation of an arch under central load p[8]

cartesian system of coordinates, x, y, z, the displacements u, v, w define strains as[49]

$$\varepsilon_x = \frac{\partial u}{\partial x} + \frac{1}{2}\left[\left(\frac{\partial u}{\partial x}\right)^2 + \left(\frac{\partial v}{\partial x}\right)^2 + \left(\frac{\partial w}{\partial x}\right)^2\right]$$

$$\gamma_{xy} = \frac{\partial u}{\partial y} + \frac{\partial v}{\partial x} + \left[\frac{\partial u}{\partial x}\frac{\partial u}{\partial y} + \frac{\partial v}{\partial x}\frac{\partial v}{\partial y} + \frac{\partial w}{\partial x}\frac{\partial w}{\partial y}\right] \tag{8.33}$$

with other components obtained by suitable permutation.

If displacements are small the general first-order linear strain approximation is obtained by neglecting the quadratic terms.

Geometric interpretation of the above strain definitions is not obvious in the general case but it should be noted that they give a measure of elongation and angular rotation of an originally orthogonal element and for small strain result in usual definitions even if the displacements are large.

If the actual strain quantities are small then it is simple to show that ε_x defines the change of length of a unit length *originally oriented parallel to the x axis* while γ_{xy} gives similarly the angle change between two lines *originally* parallel to the x and y axes. This is true in the above definition even if large movements have occurred which have rotated or displaced the original axes by gross amounts.

We shall now establish the general non-linear expressions $\bar{\mathbf{B}}$ and $\mathbf{K}_T$ for a complete three-dimensional state of stress. It is a simple matter to specialize from here to one- or two-dimensional forms and such exercises will be left to the reader. Indeed, plate and shell problems are conveniently approached via such a general formulation, and some terms neglected in the specialized approach to the plate problem of the previous section are now easily included.

8.5.1 *Derivation of the* $\mathbf{B}_L$ *matrix.* The general strain vector in three dimensions can be defined in terms of the infinitesimal and large displacement components

$$\boldsymbol{\varepsilon} = \boldsymbol{\varepsilon}_0 + \boldsymbol{\varepsilon}_L \tag{8.34}$$

where

$$\boldsymbol{\varepsilon}_0 = \begin{Bmatrix} \varepsilon_x \\ \varepsilon_y \\ \varepsilon_z \\ \gamma_{yz} \\ \gamma_{zy} \\ \gamma_{xy} \end{Bmatrix} = \begin{Bmatrix} \dfrac{\partial u}{\partial x} \\ \dfrac{\partial v}{\partial y} \\ \dfrac{\partial w}{\partial z} \\ \dfrac{\partial v}{\partial z} + \dfrac{\partial w}{\partial y} \\ \dfrac{\partial w}{\partial x} + \dfrac{\partial u}{\partial z} \\ \dfrac{\partial u}{\partial y} + \dfrac{\partial v}{\partial x} \end{Bmatrix} \tag{8.35}$$

is the same as defined in Chapter 5 of Volume 1. The non-linear terms of

Eq. (8.33) can be conveniently rewritten as

$$\boldsymbol{\varepsilon}_L = \frac{1}{2}\begin{bmatrix} \boldsymbol{\theta}_x^{\mathrm{T}} & \mathbf{0} & \mathbf{0} \\ \mathbf{0} & \boldsymbol{\theta}_y^{\mathrm{T}} & \mathbf{0} \\ \mathbf{0} & \mathbf{0} & \boldsymbol{\theta}_z^{\mathrm{T}} \\ \mathbf{0} & \boldsymbol{\theta}_z^{\mathrm{T}} & \boldsymbol{\theta}_y^{\mathrm{T}} \\ \boldsymbol{\theta}_z^{\mathrm{T}} & \mathbf{0} & \boldsymbol{\theta}_x^{\mathrm{T}} \\ \boldsymbol{\theta}_y^{\mathrm{T}} & \boldsymbol{\theta}_x^{\mathrm{T}} & \mathbf{0} \end{bmatrix} \begin{Bmatrix} \boldsymbol{\theta}_x \\ \boldsymbol{\theta}_y \\ \boldsymbol{\theta}_z \end{Bmatrix} = \frac{1}{2}\mathbf{A}\boldsymbol{\theta} \tag{8.36}$$

in which

$$\boldsymbol{\theta}_x^{\mathrm{T}} = \left[\frac{\partial u}{\partial x}, \frac{\partial v}{\partial x}, \frac{\partial w}{\partial x}\right], \qquad \text{etc.}$$

and $\mathbf{A}$ is a 6×9 matrix.

The reader can readily verify the validity of the above definition and reestablish the properties of the matrices $\mathbf{A}$ and $\boldsymbol{\theta}$ defined in Sec. 8.3.2 (footnote to p. 294). Once again

$$\mathrm{d}\boldsymbol{\varepsilon}_L = \frac{1}{2}\,\mathrm{d}\mathbf{A}\boldsymbol{\theta} + \frac{1}{2}\mathbf{A}\,\mathrm{d}\boldsymbol{\theta} = \mathbf{A}\,\mathrm{d}\boldsymbol{\theta} \tag{8.37}$$

and as we can determine $\boldsymbol{\theta}$ in terms of the shape function $\mathbf{N}$ and nodal parameters $\mathbf{a}$ we can write

$$\boldsymbol{\theta} = \mathbf{G}\mathbf{a} \tag{8.38}$$

or

$$\mathrm{d}\boldsymbol{\varepsilon}_L = \mathbf{A}\mathbf{G}\,\mathrm{d}\mathbf{a}$$

and

$$\mathbf{B}_L = \mathbf{A}\mathbf{G} \tag{8.39}$$

8.5.2 *Derivation of* $\mathbf{K}_T$ *matrix.* Noting that

$$\bar{\mathbf{B}} = \mathbf{B}_0 + \mathbf{B}_L$$

we can readily form the matrix of Eq. (8.7):

$$\bar{\mathbf{K}} = \mathbf{K}_0 + \mathbf{K}_L = \int_V \bar{\mathbf{B}}^{\mathrm{T}}\mathbf{D}\bar{\mathbf{B}}\,\mathrm{d}V \tag{8.40}$$

To complete the total tangential stiffness matrix it is necessary only to determine the initial stress matrix $\mathbf{K}_\sigma$. Again, by Eq. (8.8) we have

$$\mathbf{K}_\sigma\,\mathrm{d}\mathbf{a} = \int_V \mathrm{d}\mathbf{B}_L^{\mathrm{T}}\boldsymbol{\sigma}\,\mathrm{d}V = \int_V \mathbf{G}^{\mathrm{T}}\,\mathrm{d}\mathbf{A}^{\mathrm{T}}\boldsymbol{\sigma}\,\mathrm{d}V \tag{8.41}$$

Once again we can verify that we can write

$$\mathbf{dA}^{\mathrm{T}}\boldsymbol{\sigma}=\begin{bmatrix}\sigma_x\mathbf{I}_3 & \tau_{xy}\mathbf{I}_3 & \tau_{xz}\mathbf{I}_3\\ & \sigma_y\mathbf{I}_3 & \tau_{yz}\mathbf{I}_3\\ \text{symmetric} & & \sigma_z\mathbf{I}_3\end{bmatrix}\mathbf{d\theta}=\mathbf{MG}\,\mathbf{da} \tag{8.42}$$

in which $\mathbf{I}_3$ is a 3×3 identity matrix.

Substituting Eq. (8.42) into Eq. (8.41) yields[42]

$$\mathbf{K}_\sigma=\int_V \mathbf{G}^{\mathrm{T}}\mathbf{MG}\,\mathrm{d}V \tag{8.43}$$

in which $\mathbf{M}$ is a 9×9 matrix of the six stress components arranged as shown in Eq. (8.42). The *symmetric* form of $\mathbf{K}_\sigma$ is once again demonstrated.

In the above we have omitted element superscripts though in fact all of the above matrices would be obtained element by element and added in the standard manner.

The use of the general expressions is a useful starting point in the analysis of plates and shells if consistent approximations are to be made. In the case of the thick shell formulation of Chapter 4 such expressions are essential. Further, if a suitable stress–strain relation can be found they are valid for large strain analysis. Here, however, it is more usual to define directly a strain energy function in terms of the strain components and to obtain generalized forces by direct minimization. Some examples of such a large strain analysis have been given by Oden[50-53] who discusses large deformation of rubber membranes and continua.

An example of the application of the above formulation for an axisymmetric shell under a central point load and ring loads is shown in Fig. 8.7[42,54,55] in which parabolic–linear isoparametric elements were used. Another typical solution using this two-dimensional formulation is shown in Figs 8.8 and 8.9 for the case of a deep clamped–hinged arch.[54,56] Here extremely large deflections are followed by the lagrangian process described.

It is important to note that the use of isoparametric elements in conjunction with the above general formulation leads to particularly concise representations of $\mathbf{K}_0+\mathbf{K}_L$ and $\mathbf{K}_\sigma$. Furthermore, such formulations result in substantial savings in the computation of $\mathbf{K}_\sigma$.[7,42,55]

8.6 Concluding remarks

This chapter attempted to present a unified approach to all large deformation problems. The various procedures for solution of the non-linear algebraic system have followed those presented in Chapter 7. Again

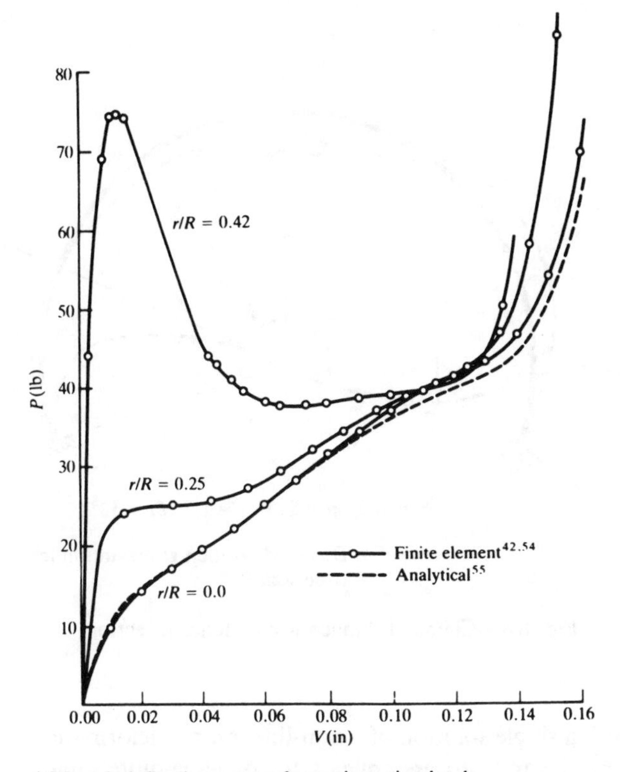

(*a*) Load–deflection curves for various ring loads

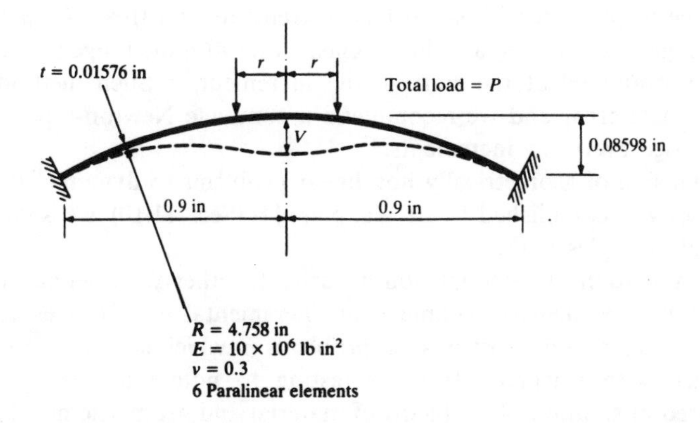

(*b*) Geometry definition and deflected shape

Fig. 8.7 Spherical cap

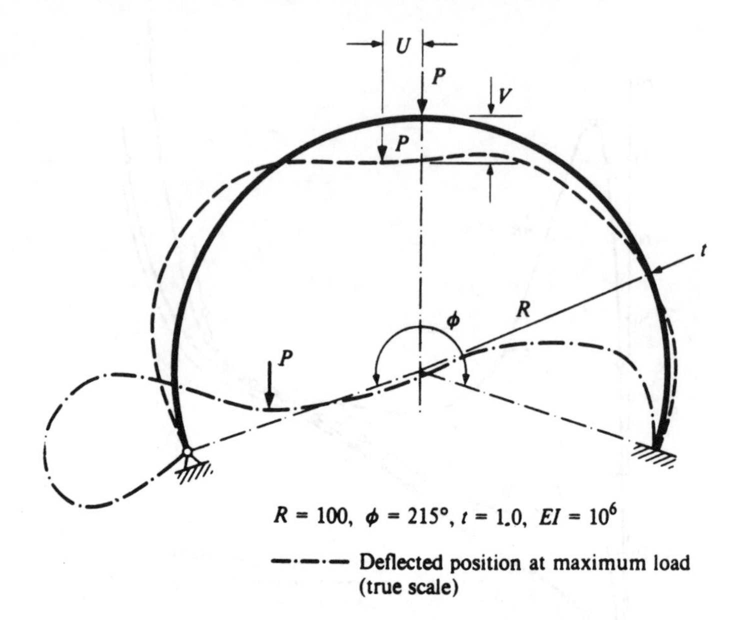

Fig. 8.8 Clamped–hinged arch—load deflection

we find that if a single solution of a non-linear large deformation problem is desired the Newton process appears to converge quite rapidly in most instances. For certain cases, however, the constant matrix methods are more economical.

If a full load deformation study is required it has been the common practice to proceed with small load increments and treat, for each such increment, the problem as a linear elastic one with the tangential stiffness matrix evaluated at the start of the increment.[3,5] Such methods may accumulate error and we recommend a complete Newton-type solution to be used every few increments.

Extension of geometrically non-linear problems to dynamic situations is readily accomplished.[57] In Chapter 11 (Sec. 11.10) we shall show several examples of this.

Combination of material non-linearity together with geometric non-linearity is particularly simple if an incremental matrix is established. Marçal[3] solves a number of such problems in which large deformation is coupled with plasticity. It is interesting to note that the operations required in solution of problems of material and geometric non-linearity are similar, and computer systems capable of dealing with both can be developed.

Finally, two matters should be noted. One is the apparently lengthy

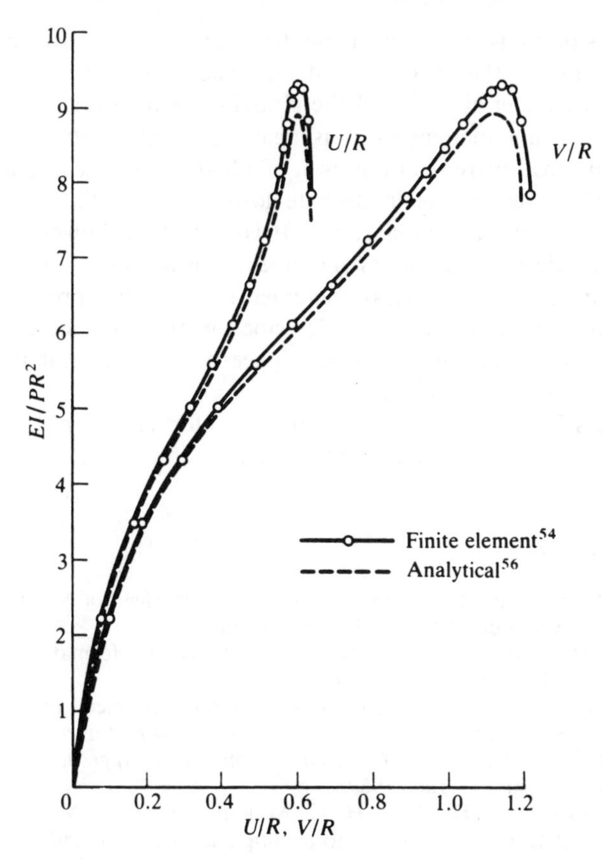

Fig. 8.9 Clamped–hinged arch: load–deflection curves (horizontal and vertical components)

derivation of the initial stress matrix for plates which in previous publications[27,29] is derived in a more direct way. This is due to an attempt at complete generality which hopefully was achieved. The second point is that somewhat involved operations were required in the section dealing with general large strain in order to preserve the convenient matrix formulation used throughout this book. Some simplification would have arisen if tensor notation were used and indeed we introduced this earlier as an alternative. No apology is, however, being made for the choice of the more direct notation.

An alternative to the lagrangian process described in this chapter is the use of the current (updated) element configuration in an eulerian manner. This is advantageous in large strain situations and a clear description of the procedure is given by McMeeking and Rice.[58]

In such situations it is convenient to return to the most obvious of stress components, that of Cauchy stress, which defines stresses in terms of the current material area and the global coordinates. In terms of the infinitesimal strain increments it is again a simple matter to present incremental constitutive relations even of a highly non-linear, plastic type and equilibrium equations in discrete form [viz. Eq. (8.1)] or written precisely as in infinitesimal elasticity. It is important, however, to write in such formulations the stress increment arising due to the rotation alone (Jauman–Zaremba stress increment). With this proviso all the computations follow the patterns described in the previous chapter for each, small, displacement increment. Space does not permit us here to elaborate further on the matter.

References 59 to 62 give the reader further information on this complex subject and introduce some current trends.

References

1. C. TRUESDELL (ed.), *Continuum Mechanics IV: Problems of Non-linear Elasticity*, Vol. 8, Gordon and Beach, London, 1965.
2. K. J. BATHE and H. OZDEMIR, 'Elastic–plastic large deformation static and dynamic analysis', *Comp. Struct.*, **6**, 81–92, 1976.
3. P. V. MARÇAL, 'Finite element analysis of combined problems of material and geometric behaviour', technical report 1, ONR, Brown University, 1969; also *Proc. Am. Soc. Mech. Eng. Conf. on Computational Approaches in Applied Mechanics*, paper 133, June 1969.
4. M. J. TURNER, E. H. DILL, H. C. MARTIN and R. J. MELOSH, 'Large deflection of structures subjected to heating and external loads', *J. Aero. Sci.*, **27**, 97–106, 1960.
5. J. H. ARGYRIS, S. KELSEY and H. KAMEL, *Matrix Methods of Structural Analysis*, AGARD-ograph 72, Pergamon Press, Oxford, 1963.
6. J. H. ARGYRIS, 'Continua and discontinua', in *Proc. Conf. on Matrix Methods in Structural Mechanics*, Air Force Institute of Technology, Wright-Patterson AF Base, Ohio, October 1965.
7. G. C. NAYAK, 'Plasticity and large deformation problems by finite element method', Ph.D. thesis, University of Wales, Swansea, 1971.
8. P. V. MARÇAL, 'Effect of initial displacement on problem of large deflection and stability', technical report ARPA E54, Brown University, 1967.
9. W. E. HAISLER, J. A. STRICKLIN and F. J. STEBBINS, 'Development and evaluation of solution procedures for geometrically non-linear analysis', *JAIAA*, **10**, 264–72, 1972.
10. G. A. DUPUIS, D. D. PFAFFINGER and P. V. MARÇAL, 'Effective use of incremental stiffness matrices in non-linear geometric analysis', in *IUTAM Symp. on High Speed Computing of Elastic Structures*, Liége, August 1970.
11. R. H. GALLAGHER and S. T. MAU, 'A method of limit point calculation in finite element structural analysis', report NASA CR 12115, September 1972.
12. H. L. LANGHAAR, *Energy Methods in Applied Mechanics*, Wiley, New York, 1962.

13. K. MARGUERRE, 'Über die Anwendung der energetischen Methode auf Stabilitätsprobleme', *Hohrb.*, DVL, 252–62, 1938.
14. B. FRAEIJS DE VEUBEKE, 'The second variation test with algebraic and differential contrasts', in *Advanced Problems and Methods for Space Flight Optimisation*, Pergamon Press, Oxford, 1969.
15. C. A. BREBBIA and J. CONNOR, 'Geometrically non-linear finite element analysis', *Proc. Am. Soc. Civ. Eng.*, **95**, EM2, 463–83, 1969.
16. H. D. HIBBITT, P. V. MARÇAL and J. R. RICE, 'A finite element formulation for problems of large strain and large displacement', *Int. J. Solids Struct.*, **6**, 1069–86, 1970.
17. J. T. ODEN, 'Discussion on "Finite element analysis of non-linear structures", by Mallett and Marçal', *Proc. Am. Soc. Civ. Eng.*, **95**, ST6, 1376–81, 1969.
18. S. P. TIMOSHENKO and J. M. GERE, *Theory of Elastic Stability*, 2nd ed., McGraw-Hill, New York, 1961.
19. R. D. WOOD, 'The application of finite element methods to geometrically non-linear analysis', Ph.D. thesis, University of Wales, Swansea, 1973.
20. A. RAZZAQUE, 'Program for triangular bending elements with derivative smoothing', *Int. J. Num. Meth. Eng.*, **6**, 333–5, 1973.
21. L. A. SCHMIT, F. K. BOGNER and R. L. FOX, 'Finite deflection structural analysis using plate and cylindrical shell discrete elements', *JAIAA*, **5**, 1525–7, 1968.
22. T. KAWAI and N. YOSHIMURA, 'Analysis of large deflection of plates by finite element method', *Int. J. Num. Meth. Eng.*, **1**, 123–33, 1969.
23. R. H. MALLETT and P. V. MARÇAL, 'Finite element analysis of non-linear structures', *Proc. Am. Soc. Civ. Eng.*, **94**, ST9, 2081–105, 1968.
24. D. W. MURRAY and E. L. WILSON, 'Finite element large deflection analysis of plates', *Proc. Am. Soc. Civ. Eng.*, **94**, EM1, 143–65, 1968.
25. P. G. BERGAN and R. W. CLOUGH, 'Large deflection analysis of plates and shallow shells using the finite element method', *Int. J. Num. Meth. Eng.*, **5**, 543–56, 1973.
26. H. C. MARTIN, 'On the derivation of stiffness matrices for the analysis of large deflection and stability problems', in *Proc. Conf. on Matrix Methods in Structural Mechanics*, Air Force Institute of Technology, Wright-Patterson AF Base, Ohio, October 1965.
27. K. K. KAPUR and B. J. HARTZ, 'Stability of thin plates using the finite element method', *Proc. Am. Soc. Civ. Eng.*, **92**, EM2, 177–95, 1966.
28. R. H. GALLAGHER and J. PADLOG, 'Discrete element approach to structural instability analysis', *JAIAA*, **1**, 1537–9, 1963.
29. R. G. ANDERSON, B. M. IRONS and O. C. ZIENKIEWICZ, 'Vibration and stability of plates using finite elements', *Int. J. Solids Struct.*, **4**, 1031–55, 1968.
30. W. G. CARSON and R. E. NEWTON, 'Plate buckling analysis using a fully compatible finite element', *JAIAA*, **8**, 527–9, 1969.
31. Y. K. CHAN and A. P. KABAILA, 'A conforming quadrilateral element for analysis of stiffened plates', UNICIV report R-121, University of New South Wales, 1973.
32. D. W. MURRAY and E. L. WILSON, 'Finite element post buckling analysis of thin elastic plates', in *Proc. 2nd Conf. on Matrix Methods in Structural Mechanics*, Wright-Patterson AF Base, Ohio, 1968.
33. K. C. ROCKEY and D. K. BAGCHI, 'Buckling of plate girder webs under partial edge loadings', *Int. J. Mech. Sci.*, **12**, 61–76, 1970.
34. T. M. ROBERTS and D. G. ASHWELL, 'Post-buckling analysis of slightly

curved plates by the finite element method', report 2, Department of Civil and Structural Engineering, University of Wales, Cardiff, 1969.
35. R. G. ANDERSON, 'A finite element eigenvalue solution system', Ph.D. thesis, University of Wales, Swansea, 1968.
36. R. H. GALLAGHER, R. A. GELLATLY, R. H. MALLETT and J. PADLOG, 'A discrete element procedure for thin shell instability analysis', *JAIAA*, **5**, 138–45, 1967.
37. R. H. GALLAGHER and H. T. Y. YANG, 'Elastic instability predictions for doubly curved shells', in *Proc. 2nd Conf. on Matrix Methods in Structural Mechanics*, Air Force Institute of Technology, Wright-Patterson AF Base, Ohio, 1968.
38. J. L. BATOZ, A. CHATTOPADHYAY and G. DHATT, 'Finite element large deflection analysis of shallow shells', *Int. J. Num. Meth. Eng.*, **10**, 35–8, 1976.
39. T. MATSUI and O. MATSUOKA, 'A new finite element scheme for instability analysis of thin shells', *Int. J. Num. Meth. Eng.*, **10**, 145–70, 1976.
40. T. Y. YANG, 'A finite element procedure for the large deflection analysis of plates with initial imperfections', *JAIAA*, **9** (No. 8), 1468–73, 1971.
41. T. M. ROBERTS and D. G. ASHWELL, 'The use of finite element mid-increment stiffness matrices in the post-buckling analysis of imperfect structures', *Int. J. Solids Struct.*, **7**, 805–23, 1971.
42. O. C. ZIENKIEWICZ and G. C. NAYAK, 'A general approach to problems of plasticity and large deformation using isoparametric elements', in *Proc. 3rd Conf. on Matrix Methods in Structural Mechanics*, Wright-Patterson AF Base, Ohio, 1971.
43. T. H. H. PIAN and PING TONG, 'Variational formulation of finite displacement analysis', in *Symp. on High Speed Computing of Elastic Structures*, Liége, 1970.
44. H. C. MARTIN, 'Finite elements and the analysis of geometrically non-linear problems', in *Recent Advances in Matrix Methods and Structural Analysis and Design*, University of Alabama Press, 1971.
45. A. C. WALKER, 'A non-linear finite element analysis of shallow circular arches', *Int. J. Solids Struct.*, **5**, 97–107, 1969.
46. J. M. T. THOMPSON and A. C. WALKER, 'A non-linear perturbation analysis of discrete structural systems', *Int. J. Solids Struct.*, **4**, 757–67, 1968.
47. J. S. PRZEMIENIECKI, 'Stability analysis of complex structures using discrete element techniques', in *Symp. on Structure Stability and Optimisation*, Loughborough University, March 1967.
48. J. CONNOR and N. MORIN, 'Perturbation techniques in the analysis of geometrically non-linear shells', in *Symp. on High Speed Computing of Elastic Structures*, Liége, 1970.
49. Y. C. FUNG, *Foundation of Solid Mechanics*, Prentice-Hall, Englewood Cliffs, N.J., 1965.
50. J. T. ODEN, 'Finite plane strain of incompressible elastic solids by the finite element method', *Aeronaut. Q.*, **19**, 254–64, 1967.
51. J. T. ODEN and T. SATO, 'Finite deformation of elastic membranes by the finite element method', *Int. J. Solids Struct.*, **3**, 471–88, 1967.
52. J. T. ODEN, 'Numerical formulation of non-linear elasticity problems', *Proc. Am. Soc. Civ. Eng.*, **93**, ST3, 235–55, 1967.
53. J. T. ODEN, 'Finite element applications in non-linear structural analysis', in *Proc. ASCE Symp. on Application of Finite Element Methods in Civil Engineering*, Vanderbilt University, 1969.

54. R. D. WOOD and O. C. ZIENKIEWICZ, 'Geometrically non-linear finite element analysis of beams–frames–circles and axisymmetric shells', *Comp. Struct.*, **7**, 725–35, 1977.
55. J. F. MESCALL, 'Large deflections of spherical shells under concentrated loads', *J. Appl. Mech.*, **32**, 936–8, 1965.
56. D. A. DA DEPPO and R. SCHMIDT, 'Instability of clamped–hinged circular arches subjected to a point load', *Trans. Am. Soc. Mech. Eng.*, 894–6, 1975.
57. J. A. STRICKLIN, 'Non-linear dynamic analysis of shells of revolution', in *Symp. on High Speed Computing of Elastic Structures*, Liège, 1970.
58. R. M. MCMEEKING and J. R. RICE, 'Finite element formulations for problems of large elastic–plastic deformation, *Int. J. Solids Struct.*, **11**, 601–16, 1975.
59. M. KLEIBER, *Incremental Finite Element Modelling in Non-linear Solid Mechanics*, Ellis-Horwood, London, 1990.
60. K. J. BATHE, *Finite Element Procedures in Engineering Analysis*, Prentice-Hall, Englewood Cliffs, N.J., 1982.
61. L. H. MALVERN, *Introduction to the Mechanics of a Continuous Medium*, Prentice-Hall, Englewood Cliffs, N.J., 1969.
62. H. KARDESTUNCER and D. H. NORRIE (eds), *Finite Element Handbook*, McGraw-Hill, New York, 1987.

9

The time dimension—semi-discretization of field and dynamic problems and analytical solution procedures

9.1 Introduction

In all the problems considered so far in this text conditions that do not vary with time were generally assumed. There is little difficulty in extending the finite element idealization to situations that are time dependent.

The range of practical problems in which the time dimension has to be considered is great. Transient heat conduction, wave transmission in fluids and dynamic behaviour of structures are typical examples. While it is usual to consider these various problems separately—sometimes classifying them according to the mathematical structure of governing equations as 'parabolic' or 'hyperbolic'[1]—we shall group them in one category to show that the formulation is identical.

In the first part of this chapter we shall formulate, by a simple extension of the methods used so far, matrix differential equations governing such problems for a variety of physical situations. Here a finite element discretization in the space dimension only will be used and a semi-discretization process followed (viz. Chapter 9 of Volume 1). In the remainder of this chapter various analytical procedures of solution for the resulting ordinary linear differential equation system will be dealt with. These form the basic arsenal of steady-state and transient analysis.

Chapter 10 will be devoted to the discretization at the time domain itself.

9.2 Direct formulation of time-dependent problems with spatial finite element subdivision

9.2.1 *The 'quasi-harmonic' equation with time differentials.* In many physical problems the quasi-harmonic equation takes up a form in which time derivatives of the unknown function ϕ occur. In the three-dimensional case typically we might have

$$\frac{\partial}{\partial x}\left(k\frac{\partial \phi}{\partial x}\right)+\frac{\partial}{\partial y}\left(k\frac{\partial \phi}{\partial y}\right)+\frac{\partial}{\partial z}\left(k\frac{\partial \phi}{\partial z}\right)+\left(\bar{Q}-\mu\frac{\partial \phi}{\partial t}-\rho\frac{\partial^2 \phi}{\partial t^2}\right)=0 \tag{9.1}$$

In the above, quite generally, all the parameters may be prescribed functions of time, or in non-linear cases of ϕ, as well as of space $\mathbf{x}$, i.e.

$$k=k(\mathbf{x},\phi,t) \qquad \bar{Q}=\bar{Q}(\mathbf{x},\phi,t), \qquad \text{etc.} \tag{9.2}$$

If a situation at a particular instant of time is considered, the time derivatives of ϕ and all the parameters can be treated as *prescribed functions of space coordinates.* Thus, at that instant the problem is precisely identified with those treated in Chapter 10 of Volume 1 if the whole of the quantity in the last parentheses of Eq. (9.1) is identified as the source term Q.

The finite element discretization of this in terms of *space* elements has already been fully discussed and we found that with the prescription

$$\begin{gathered}\phi=\sum N_i a_i=\mathbf{N}\mathbf{a}\\ \mathbf{N}=\mathbf{N}(x,y,z) \qquad \mathbf{a}=\mathbf{a}(t)\end{gathered} \tag{9.3}$$

for each element, a standard form of assembled equation†

$$\mathbf{K}\mathbf{a}+\bar{\mathbf{f}}=0 \tag{9.4}$$

was obtained. Element contributions to the above matrices are defined in Chapter 10 of Volume 1 and need not be repeated here except for that representing the 'load' term due to Q. This is given by

$$\bar{\mathbf{f}}=-\int_\Omega Q\mathbf{N}^{\mathrm{T}}\,\mathrm{d}\Omega \tag{9.5}$$

Replacing now Q by the last bracketed term of Eq. (9.1) we have

$$\bar{\mathbf{f}}=-\int_\Omega \mathbf{N}^{\mathrm{T}}\left(\bar{Q}-\mu\frac{\partial \phi}{\partial t}-\rho\frac{\partial^2 \phi}{\partial t^2}\right)\mathrm{d}\Omega \tag{9.6}$$

† We have replaced the matrix $\mathbf{H}$ of Chapter 10 of Volume 1 by $\mathbf{K}$ to facilitate comparison with other dynamic equations.

However, from Eq. (9.3) it is noted that ϕ is approximated in terms of the nodal parameters **a**. On substitution of this approximation we have

$$\bar{\mathbf{f}} = -\int_\Omega \mathbf{N}^\mathrm{T}\bar{Q}\,\mathrm{d}\Omega + \left(\int_\Omega \mathbf{N}^\mathrm{T}\mu\mathbf{N}\,\mathrm{d}\Omega\right)\frac{\mathrm{d}}{\mathrm{d}t}\mathbf{a} + \left(\int_\Omega \mathbf{N}^\mathrm{T}\rho\mathbf{N}\,\mathrm{d}\Omega\right)\frac{\mathrm{d}^2}{\mathrm{d}t^2}\mathbf{a} \quad (9.7)$$

and on expanding Eq. (9.4) in its final assembled form we get the following *matrix differential equation*:

$$\mathbf{M}\ddot{\mathbf{a}} + \mathbf{C}\ddot{\mathbf{a}} + \mathbf{K}\mathbf{a} + \mathbf{f} = 0 \quad (9.8)$$

$$\dot{\mathbf{a}} \equiv \frac{\mathrm{d}}{\mathrm{d}t}\mathbf{a} \qquad \ddot{\mathbf{a}} = \frac{\mathrm{d}^2}{\mathrm{d}t^2}\mathbf{a} \quad (9.9)$$

in which all the matrices are assembled from element submatrices in the standard manner with submatrices $\mathbf{K}^e$ and $\mathbf{f}^e$ still given by relations (10.17) in Chapter 10 of Volume 1,† and

$$C^e_{ij} = \int_\Omega N_i \mu N_j \,\mathrm{d}\Omega \quad (9.10)$$

$$M^e_{ij} = \int_\Omega N_i \rho N_j \,\mathrm{d}\Omega \quad (9.11)$$

Once again these matrices are symmetric as seen from the above relations.

Boundary conditions imposed at any time instant are treated in the standard manner.

The variety of physical problems governed by Eq. (9.1) is so large that a comprehensive discussion of them is beyond the scope of this book. A few typical examples will, however, be quoted.

Equation (9.1) *with* $\rho = 0$. This is the standard *transient heat conduction equation*[1,2] which has been discussed in the finite element context by several authors.[3-6] This same equation is applicable in other physical situations—one of these being the *soil consolidation equations*[7] associated with *transient seepage forms.*[8]

Equation (9.1) *with* $\mu = 0$. Now the relationship becomes the famous *Helmholz wave equation* governing a wide range of physical phenomena. Electromagnetic waves,[9] fluid surface waves[10] and compression waves[11] are but a few cases to which the finite element process has been applied.

Equation (9.1) *with* $\mu \neq \rho \neq 0$. This damped wave equation is yet of more

† In Eq. (9.8) it is implied as usual that the loading term $\bar{\mathbf{f}}$ includes also all prescribed values of ϕ (and hence of **a**) on the boundaries. A point which seems to be missed by some is that this implies also a prescription of $\dot{\mathbf{a}}$ and $\ddot{\mathbf{a}}$ on such boundaries and appropriate load terms are contributed from *all the matrices* **K**, **C** and **M**. Only when prescribed **a** values are zero, or when the matrices are diagonal, can such terms be neglected.

general applicability and has particular significance in fluid mechanics (wave) problems.

The reader will recognize that what we have done here is simply an application of the process of partial discretization described in Sec. 9.7 of Chapter 9 of Volume 1. It is convenient, however, to perform the operations in the manner suggested above as all the matrices and discretization expressions obtained from steady-state analysis are immediately available.

9.2.2 *Dynamic behaviour of elastic structures with linear damping.*† While in the previous section we have been concerned with, apparently, a purely mathematical problem, identical reasoning can be applied directly to the wide class of dynamic behaviour of elastic structures following precisely the general lines of Chapter 2 of Volume 1.

When displacements of an elastic body vary with time two sets of additional forces are called into play. The first is the inertia force, which for an acceleration characterized by $\ddot{\mathbf{u}}$ can be replaced by its static equivalent

$$-\rho\ddot{\mathbf{u}}$$

using the well-known d'Alembert principle ($\mathbf{u}$ is here the generalized displacement. This is a force with components in directions identical to those of the displacement $\mathbf{u}$ and (generally) given per unit volume. In this context ρ is simply the mass per unit volume.

The second force is that due to (frictional) resistances opposing the motion. These may be due to microstructure movements, air resistance, etc., and generally are related in a non-linear way to the displacement velocity $\dot{\mathbf{u}}$.

For simplicity of treatment, however, only a linear, viscous-type, resistance will be permitted, resulting again in unit volume forces in an equivalent static problem of magnitude

$$-\boldsymbol{\mu}\dot{\mathbf{u}}$$

In the above $\boldsymbol{\mu}$ is some property which (presumably) can be given numerical values.

The equivalent static problem, at any instant of time, is now discretized precisely in the manner of Chapter 2 of Volume 1, but replacing the distributed body force $\mathbf{b}$ by its equivalent

$$\bar{\mathbf{b}}-\rho\ddot{\mathbf{u}}-\boldsymbol{\mu}\dot{\mathbf{u}}$$

† For simplicity we shall only consider *distributed* inertia and damping effects—concentrated mass and damping forces being a limiting case.

The element (nodal) forces given by Eq. (2.13) in Chapter 2 of Volume 1 now become (excluding initial stress–strain contributions)

$$\mathbf{f}^e = -\int_{V^e} \mathbf{N}^{\mathrm{T}}\mathbf{b}\,\mathrm{d}V = -\int_{V^e} \mathbf{N}^{\mathrm{T}}\bar{\mathbf{b}}\,\mathrm{d}V + \int_{V^e} \mathbf{N}^{\mathrm{T}}\rho\ddot{\mathbf{u}}\,\mathrm{d}V + \int_{V^e} \mathbf{N}^{\mathrm{T}}\boldsymbol{\mu}\dot{\mathbf{u}}\,\mathrm{d}V \tag{9.12}$$

in which the first force is the static one due to an external distributed load and need not be further considered.

Substituting Eq. (9.12) into the general equilibrium equations we obtain finally, on assembly, the following matrix differential equation:

$$\mathbf{M}\ddot{\mathbf{a}} + \mathbf{C}\dot{\mathbf{a}} + \mathbf{K}\mathbf{a} + \mathbf{f} = 0 \tag{9.13}$$

in which $\mathbf{K}$ and $\mathbf{f}$ are assembled stiffness and force matrices obtained by the usual addition of stiffness coefficients of elements and of element forces due to external, specified loads, initial stresses, etc., in the manner fully described before. The new matrices $\mathbf{C}$ and $\mathbf{M}$ are assembled by the usual rule from element submatrices given by

$$\mathbf{C}^e_{ij} = \int_{V^e} \mathbf{N}^{\mathrm{T}}\boldsymbol{\mu}\mathbf{N}\,\mathrm{d}V \tag{9.14}$$

and

$$\mathbf{M}^e_{ij} = \int_{V^e} \mathbf{N}^{\mathrm{T}}\rho\mathbf{N}\,\mathrm{d}V \tag{9.15}$$

The matrix $\mathbf{M}^e$ is known as the *element mass matrix* and the assembled matrix $\mathbf{M}$ as the system mass matrix.

It is of interest to note that in early attempts to deal with dynamic problems of this nature the mass of the elements was usually arbitrarily 'lumped' at nodes, resulting always in a diagonal matrix even if no actual concentrated masses existed. The fact that such a procedure was, in fact, unnecessary and apparently inconsistent was simultaneously recognized by Archer[12] and independently by Leckie and Lindberg[13] in 1963. The general presentation of the results given in Eq. (9.15) is due to Zienkiewicz and Cheung.[14] The name of 'consistent mass matrix' has been coined for the mass matrix here defined, a term which may be considered to be unnecessary since it is the logical and natural consequence of the discretization process.

By analogy the matrices $\mathbf{C}^e$ and $\mathbf{C}$ may be called *consistent damping matrices.*

For many computational processes the lumped mass matrix is, however, more convenient and economical. Many practitioners are today using such matrices exclusively showing sometimes a good accuracy. While with simple elements a physically obvious methodology of lumping is easy to devise, this is not the case with higher-order elements and we shall return to the process of 'lumping' later.

Determination of the damping matrix $\mathbf{C}$ is in practice difficult as knowledge of the viscous matrix $\boldsymbol{\mu}$ is lacking. It is often assumed, therefore, that the damping matrix is a linear combination of stiffness and mass matrices, i.e.

$$\mathbf{C}=\alpha\mathbf{M}+\beta\mathbf{K} \tag{9.16}$$

Here α and β are determined experimentally.[15]

Such damping is known as 'Rayleigh damping' and has certain mathematical advantages which we shall discuss later. On occasion $\mathbf{C}$ may be completely specified and such approximation devices are not necessary.

It is perhaps worth recognizing that on occasion different shape functions need to be used to describe the inertia forces from those specifying the displacements $\mathbf{u}$. For instance, in plates and beams (Chapter 1) the full strain state was prescribed simply by defining w, the lateral displacement, as the additional plate bending assumptions were introduced. When considering, however, the inertia forces it may be desirable not only to include the simple lateral inertia force given by

$$-\rho\frac{\partial^2 w}{\partial t^2}$$

(in which ρ is now the weight per unit area of the plate) but also to consider *rotary inertia couples* of the type

$$\frac{\rho t^2}{12}\frac{\partial^2}{\partial t^2}\left(\frac{\partial w}{\partial x}\right), \qquad \text{etc.}$$

Now it will be necessary to describe a more generalized displacement $\bar{\mathbf{u}}$:

$$\bar{\mathbf{u}}=\begin{Bmatrix} w \\ \dfrac{\partial w}{\partial x} \\ \dfrac{\partial w}{\partial y} \end{Bmatrix}=\bar{\mathbf{N}}\mathbf{a}^e$$

in which $\bar{\mathbf{N}}$ will follow directly from the definition of $\mathbf{N}$ which specifies only the w component. Relations such as Eq. (9.15) still are valid, providing we replace $\mathbf{N}$ by $\bar{\mathbf{N}}$ and put in place of ρ a matrix

$$\begin{bmatrix} \rho & 0 & 0 \\ 0 & \dfrac{\rho t^2}{12} & 0 \\ 0 & 0 & \dfrac{\rho t^2}{12} \end{bmatrix}$$

Such specialized usage is, however, rare.

9.2.3 *'Mass' or 'damping' matrices for some typical elements.* It is impracticable to present in an explicit form all the mass matrices for the various elements discussed in previous chapters. Some selected examples only will be discussed here.

Plane stress and strain. Using triangular elements discussed in Chapter 3 of Volume 1, the matrix $\mathbf{N}$ is defined as

$$\mathbf{N}^e = \mathbf{I}[N_i, N_j, N_k]$$

in which

$$\mathbf{I} = \begin{bmatrix} 1 & 0 \\ 0 & 1 \end{bmatrix}$$

with Eq. (3.8) in Volume 1 giving

$$N_i = \frac{a_i + b_i x + c_i y}{2\Delta}, \qquad \text{etc.}$$

where Δ is the area of a triangle.

If the thickness of the element is t and this is assumed to be constant within the element, we have, for the mass matrix Eq. (9.15),

$$\mathbf{M}^e = \rho t \iint \mathbf{N}^T \mathbf{N} \, dx \, dy$$

$$\text{or} \qquad \mathbf{M}^e_{rs} = \rho t \mathbf{I} \iint N_r N_s \, dx \, dy \tag{9.17}$$

If the relationships of Eq. (3.8) in Volume 1 are substituted, it is easy to show that

$$\int N_r N_s \, dx \, dy = \begin{cases} \frac{1}{12}\Delta & \text{when } r \neq s \\ \frac{1}{6}\Delta & \text{when } r = s \end{cases} \tag{9.18}$$

Thus, taking the mass of the element as

$$\rho t \Delta = W$$

the mass matrix becomes

$$\mathbf{M}^e = \frac{W}{3} \left[\begin{array}{cc|cc|cc} \frac{1}{2} & 0 & \frac{1}{4} & 0 & \frac{1}{4} & 0 \\ 0 & \frac{1}{2} & 0 & \frac{1}{4} & 0 & \frac{1}{4} \\ \hline \frac{1}{4} & 0 & \frac{1}{2} & 0 & \frac{1}{4} & 0 \\ 0 & \frac{1}{4} & 0 & \frac{1}{2} & 0 & \frac{1}{4} \\ \hline \frac{1}{4} & 0 & \frac{1}{4} & 0 & \frac{1}{2} & 0 \\ 0 & \frac{1}{4} & 0 & \frac{1}{4} & 0 & \frac{1}{2} \end{array} \right] \tag{9.19}$$

If the mass had been lumped at the nodes in three equal parts the mass matrix contributed by the element would have been

$$\mathbf{M}^e = \frac{W}{3}\left[\begin{array}{cc:cc:cc} 1 & 0 & 0 & 0 & 0 & 0 \\ 0 & 1 & 0 & 0 & 0 & 0 \\ \hdashline 0 & 0 & 1 & 0 & 0 & 0 \\ 0 & 0 & 0 & 1 & 0 & 0 \\ \hdashline 0 & 0 & 0 & 0 & 1 & 0 \\ 0 & 0 & 0 & 0 & 0 & 1 \end{array}\right] \tag{9.20}$$

Certainly both matrices differ considerably and yet in application the results of analysis are almost identical.

Plate bending. Vibration of plates presents problems of considerable engineering importance. Such practical situations as bridge-deck oscillations, vibration of turbine blades, etc., result in analytically intractable formulations.

The use a consistent mass matrix is illustrated in many references.[15-22]

If the rectangular plate element of Sec. 1.3 is considered, for instance, the displacement function is defined by Eq. (1.29) as

$$\mathbf{N} = \mathbf{P}\mathbf{C}^{-1} \tag{9.21}$$

with notation as defined in Chapter 1.

It will be observed that $\mathbf{C}$ is not dependent on the coordinates and that $\mathbf{P}$ is determined as

$$\mathbf{P} = [1, x, y, x^2, xy, y^2, x^3, x^2y, xy^2, y^3, x^3y, xy^3]$$

Thus, the mass matrix for a plate element of constant thickness t becomes, from Eq. (9.15),

$$\mathbf{M}^e = \rho t \mathbf{C}^{-1^{\mathrm{T}}}\left(\iint \mathbf{P}^{\mathrm{T}}\mathbf{P}\,\mathrm{d}x\,\mathrm{d}y\right)\mathbf{C}^{-1} \tag{9.22}$$

Once again only the central integral needs to be evaluated, thus presenting no difficulty, and the full matrix can be obtained by matrix multiplication. However, an explicit expression has been presented in Dawe[17] and is quoted in Table 9.1.

Similar mass matrices can be obtained for triangular elements discussed in Sec. 1.5 *et seq.* Numerical integration procedures are recommended for use with such elements, though explicit integrals are available in references 20 and 21.

TABLE 9.1

MASS MATRIX OF A RECTANGULAR PLATE ELEMENT

$$\mathbf{M}^e = \mathbf{LML}$$

$$\mathbf{M}^e = \lambda \begin{bmatrix}
3454 & & & & & & & & & & & \\
-461 & 80 & & & & & & & & & & \\
-461 & -63 & 80 & & & & & & & & & \\
1226 & -274 & 199 & 3454 & & & & & & & & \\
274 & -60 & 42 & 461 & 80 & & & & & & & \\
199 & -42 & 40 & 461 & 63 & 80 & & & & & & \\
1226 & -199 & 274 & 394 & 116 & 116 & 3454 & & & & & \\
-199 & 40 & -42 & -116 & -30 & -28 & -461 & 80 & & & & \\
-274 & 42 & -60 & -116 & -28 & -30 & -461 & 63 & 80 & & & \\
394 & -116 & 116 & 1226 & 199 & 274 & 1226 & -274 & -199 & 3454 & & \\
116 & -30 & 28 & 199 & 40 & 42 & 274 & -60 & -42 & 461 & 80 & \\
-116 & 28 & -30 & -274 & -42 & -60 & -199 & 42 & 40 & -461 & -63 & 80
\end{bmatrix}$$

L is defined in Table 1.1 and $\lambda = \dfrac{\rho tab}{6300}$.

Shells. If the mass matrices for the 'in-plane' and 'bending' motions of an element are found, then once again the mass matrices referred to a general coordinate system can be determined. The rules of transformation are, obviously, precisely the same as for forces. The derivation of the mass matrices for each element in general coordinates and the final assembly of the mass matrix associated with a node follow the steps for similar operations with stiffness matrices (see Chapter 1 of Volume 1).

In principle, therefore, shell vibration problems present no special difficulties.

Indeed, the same may be said of a wide variety of matrices that arise in the field of structural dynamics. Performance of numerical integration allows the mass (or damping) matrices to be evaluated in a direct and simple fashion with procedures described in Chapter 7 of Volume 1.

9.2.4 *Mass 'lumping' or diagonalization.* We have referred to the computational convenience of lumping of mass matrices and presenting these in a diagonal form. On some occasions such lumping is physically obvious (viz. the linear triangle for instance), in others this is not the case and a 'rational' procedure is required. For matrices of the type given in Eq. (9.15) several alternative approximations have been developed. These are discussed in detail in Appendix 8 of Volume 1. In all of these the essential requirement of mass preservation is satisfied, i.e.

$$\sum_i \tilde{M}_{ii} = \int_\Omega \rho \, d\Omega \tag{9.23}$$

where $\tilde{M}_{ii}$ is the diagonal of the lumped mass matrix $\tilde{\mathbf{M}}$.

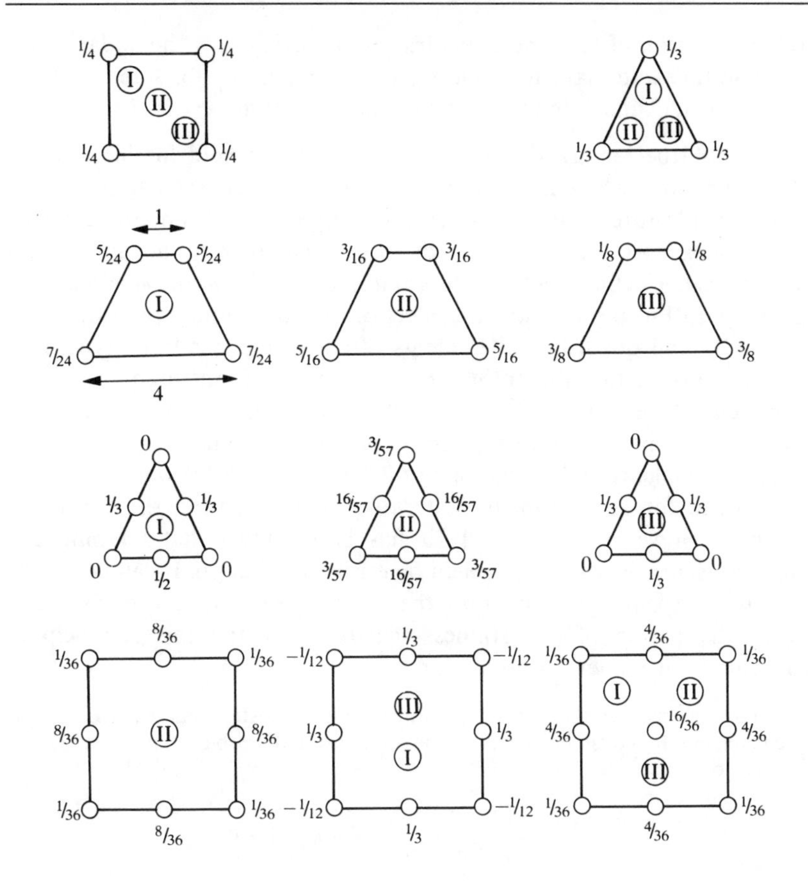

(I) Row sum procedure

(II) Diagonal scaling procedure

(III) Quadrature using nodal points

Fig. 9.1 Mass lumping for some two-dimensional elements

Three main procedures exist (see Fig. 9.1):

I. The row sum method in which

$$\tilde{M}_{ii} = \sum_j M_{ij}$$

II. Diagonal scaling in which

$$\tilde{M}_{ii} = aM_{ii}$$

with a so adjusted that Eq. (9.23) is satisfied,[23,24] and

III. Evaluation of M using a quadrature involving only the nodal points and thus automatically yielding a diagonal matrix for standard finite element shape functions[25,26] in which $N_i=0$ for $x=x_j$, $j\neq i$.

It should be remarked that Eq. (9.23) does not hold for hierarchical shape functions where no lumping procedure appears satisfactory.

The quadrature (numerical integration) process (III) is mathematically most appealing but frequently leads to negative (or zero) masses. Such a loss of positive definiteness is undesirable and cancels out the advantages of lumping. In Fig. 9.1 we show the effect of various lumping procedures on triangular and quadrilateral elements of linear and quadratic type. It is clear from these that the optimal choice is by no means unique.

In general we would recommend the use of lumped matrices only as a convenient numerical device generally paid for by some loss of accuracy. In some problems of fluid mechanics (Chapters 12 to 15) we shall indeed only use it for an intermediate iterative step in getting the consistent solution. However, occasionally it has been shown that lumping can *improve* accuracy of some problem by error cancellation. It can be shown that in transient approximation the lumping process introduces additional dissipation of the 'stiffness matrix' form and this can help in cancelling out numerical oscillation.

To demonstrate the nature of lumped and consistent mass matrices it is convenient to consider a typical one-dimensional problem specified by the equation

$$\frac{\partial\phi}{\partial t}-\frac{\partial}{\partial x}\mu\frac{\partial}{\partial x}\left(\frac{\partial\phi}{\partial t}\right)-\frac{\partial}{\partial x}\left(k\frac{\partial\phi}{\partial x}\right)=0$$

Semi-discretization gives here a typical nodal equation i as

$$(M_{ij}+H_{ij})\dot{a}_j+K_{ij}a_j=0$$

where

$$M_{ij}=\int_\Omega N_iN_j\,\mathrm{d}x$$

$$H_{ij}=\int_\Omega \frac{\mathrm{d}N_i}{\mathrm{d}x}\mu\frac{\mathrm{d}N_j}{\mathrm{d}x}\,\mathrm{d}x$$

$$K_{ij}=\int_\Omega \frac{\mathrm{d}N_i}{\mathrm{d}x}k\frac{\mathrm{d}N_j}{\mathrm{d}x}\,\mathrm{d}x$$

and it is observed that H and K have an identical structure. With linear elements of constant size h the approximating equation at a typical node i (and surrounding nodes $i-1$ or $i+1$) can be written as follows (as the reader can readily verify):

$$M_{ij}\dot{a}_j \equiv h\left(\frac{\dot{a}_{i-1}}{6}+\frac{2\dot{a}_i}{3}+\frac{\dot{a}_{i+1}}{6}\right)$$

$$H_{ij}\dot{a}_j \equiv \frac{\mu}{h}(-\dot{a}_{i-1}+2\dot{a}_i-\dot{a}_{i+1})$$

$$K_{ij}a_j = \frac{k}{h}(-a_{i-1}+2a_i-a_{i+1})$$

If a lumped approximation is used for **M**, that is $\tilde{\mathbf{M}}$, we have simply (by adding coefficients)

$$\tilde{M}_{ij}\dot{a}_j = h\dot{a}_i$$

The difference of the two expressions is

$$\tilde{M}_{ij}\dot{a}_j - M_{ij}\dot{a}_j \equiv \frac{h}{6}(-\dot{a}_{i-1}+2\dot{a}_i-\dot{a}_{i+1})$$

and is clearly identical to that which would be obtained by increasing μ by $h^2/6$. As μ in above example can be considered as a viscous dissipation we note that the effect of using a lumped matrix is that of adding an extra amount of such viscosity and can often result in smoother (though probably less accurate) solutions.

EIGENVALUES AND
ANALYTICAL SOLUTION PROCEDURES

9.3 General classification

We have seen that as a result of semi-discretization many time-dependent problems can be reduced to a system of ordinary differential equations of the characteristic form given by Eq. (9.13):

$$\mathbf{M}\ddot{\mathbf{a}}+\mathbf{C}\dot{\mathbf{a}}+\mathbf{K}\mathbf{a}+\mathbf{f}=\mathbf{0} \tag{9.24}$$

In this, in general, all the matrices are symmetric (some cases involving non-symmetric matrices will be discussed in Chapter 12). This second-order system often becomes one of first order if $\mathbf{M}=0$ as, for instance, in transient heat conduction problems. We shall now discuss some methods of solution of such ordinary differential equation systems. In general, the above equations can be non-linear (if, for instance, stiffness matrices are dependent on non-linear material properties or if large deformation is involved) but initially we shall concentrate on linear cases only.

Systems of ordinary, linear, differential equations can always in principle be solved analytically without the introduction of additional approximations, and the remainder of this chapter will be concerned with such analytical processes. While such solutions are possible they may be so complex that a further recourse has to be taken to the process of approximation; we shall deal with this matter in the next chapter. The analytical approach provides, however, an insight into the behaviour of the system which the investigator always finds helpful.

Some of the matter in this chapter will be an extension of standard well-known procedures used for the solution of differential equations with constant coefficients that have been encountered by most students of dynamics or mathematics. In the following we shall deal successively with

(*a*) determination of free response ($\mathbf{f}=0$),
(*b*) determination of periodic response [$\mathbf{f}(t)$—periodic],
(*c*) determination of transient response [$\mathbf{f}(t)$–arbitrary].

In the first two, initial conditions of the system are of no importance and a general solution is simply sought. The last, most important, phase presents a problem to which considerable attention will be devoted.

9.4 Free response—eigenvalues for second-order problems and dynamic vibrations

9.4.1 *Free dynamic vibration—real eigenvalues.* If no damping or forcing terms exist in the dynamic problem of Eq. (9.24) this is reduced to

$$\mathbf{M}\ddot{\mathbf{a}}+\mathbf{K}\mathbf{a}=0 \tag{9.25}$$

If a general solution of such an equation is written as

$$\mathbf{a}=\bar{\mathbf{a}}\,\mathrm{e}^{i\omega t}$$

(the real part of which represents simply a harmonic response as $\mathrm{e}^{i\omega t}\equiv\cos\omega t+\mathrm{i}\sin\omega t$) then on substitution we find that ω can be determined from

$$(-\omega^2\mathbf{M}+\mathbf{K})\bar{\mathbf{a}}=0 \tag{9.26}$$

This is an *eigenvalue or characteristic value problem* and for non-zero solutions the determinant of the above equation must be zero:

$$|-\omega^2\mathbf{M}+\mathbf{K}|=0 \tag{9.27}$$

Such a determinant will in general give n values of ω^2 (or ω_j, $j=1-n$) when the size of the matrices $\mathbf{K}$ and $\mathbf{M}$ is $n\times n$. Providing the matrices $\mathbf{K}$ and $\mathbf{M}$ are positive definite—which is the usual case with structural problems—it can be shown that all the roots of Eq. (9.27) are real positive numbers (for the proof of this see reference 1). These are known as the natural frequencies of the system.

While the solution of Eq. (9.27) cannot determine the actual values of

a we can find n vectors $\bar{\mathbf{a}}_j$ that give the proportions of the various terms. Such vectors are known as the *natural modes of the system of eigenvectors* and are made unique by normalizing so that

$$\bar{\mathbf{a}}_j^{\mathrm{T}}\mathbf{M}\bar{\mathbf{a}}_j=1 \tag{9.28}$$

At this stage it is useful to note the property of *modal orthogonality*, i.e. that

$$\begin{aligned}\bar{\mathbf{a}}_j^{\mathrm{T}}\mathbf{M}\bar{\mathbf{a}}_i&\equiv\mathbf{0}\\ \bar{\mathbf{a}}_j^{\mathrm{T}}\mathbf{K}\bar{\mathbf{a}}_i&\equiv\mathbf{0}\end{aligned}\qquad (i\neq j) \tag{9.29}$$

The proof of the above statement is simple. As Eq. (9.26) is valid for any mode we can write

$$\omega_i^2\mathbf{M}\bar{\mathbf{a}}_i=\mathbf{K}\bar{\mathbf{a}}_i$$
$$\omega_j^2\mathbf{M}\bar{\mathbf{a}}_j=\mathbf{K}\bar{\mathbf{a}}_j$$

Premultiplying the first by $\bar{\mathbf{a}}_j^{\mathrm{T}}$ and the second by $\bar{\mathbf{a}}_i^{\mathrm{T}}$ we have on subtraction (noting the symmetry of the matrix **M**, that is $\bar{\mathbf{a}}_i^{\mathrm{T}}\mathbf{M}\bar{\mathbf{a}}_j\equiv\bar{\mathbf{a}}_j^{\mathrm{T}}\mathbf{M}\bar{\mathbf{a}}_i$)

$$(\omega_i^2-\omega_j^2)\bar{\mathbf{a}}_i^{\mathrm{T}}\mathbf{M}\bar{\mathbf{a}}_j=0$$

and if $\omega_i\neq\omega_j$ the orthogonality condition for the matrix **M** has been proved. Immediately from this, orthogonality of the matrix **K** follows.

9.4.2 *Determination of eigenvalues.* To find the actual eigenvalues it is seldom practicable to write the polynomial expanding the determinant given in Eq. (9.27) and alternative techniques have to be developed. The discussion of such techniques is best left to specialist texts and indeed many standard computer programs exist today as library routines.

Many new and extremely efficient procedures are being added to the arsenal available. Their description is beyond the scope of this text but the reader can find some interesting matter in references.[27-34]

In most processes the starting point is the *special eigenvalue* problem given by

$$\mathbf{H}\mathbf{x}=\lambda\mathbf{x} \tag{9.30}$$

in which **H** is a symmetric, positive definite, matrix. Equation (9.26) can be written as

$$\mathbf{K}^{-1}\mathbf{M}\bar{\mathbf{a}}=\lambda\bar{\mathbf{a}} \tag{9.31}$$

on inverting **K** and with $\lambda=1/\omega^2$, but symmetry is in general lost.

If, however, we write in a triangular form

$$\mathbf{K}=\mathbf{L}\mathbf{L}^{\mathrm{T}}\qquad\text{and}\qquad\mathbf{K}^{-1}=(\mathbf{L}^{\mathrm{T}})^{-1}\mathbf{L}^{-1}$$

in which $\mathbf{L}$ is a matrix having only zero coefficients above the diagonal we have, on multiplying Eq. (9.31) by $\mathbf{L}^{\mathrm{T}}$,

$$\mathbf{L}^{-1}\mathbf{M}\bar{\mathbf{a}} = \lambda \mathbf{L}^{\mathrm{T}}\bar{\mathbf{a}}$$

Calling

$$\mathbf{L}^{\mathrm{T}}\bar{\mathbf{a}} = \mathbf{x} \tag{9.32}$$

we have finally

$$\mathbf{H}\mathbf{x} = \lambda \mathbf{x} \tag{9.33}$$

in which

$$\mathbf{H} = \mathbf{L}^{-1}\mathbf{M}(\mathbf{L}^{\mathrm{T}})^{-1} \tag{9.34}$$

which is of the standard form of Eq. (9.30), as $\mathbf{H}$ is now symmetric.

Having determined λ (all, or only a few of the selected largest values corresponding to fundamental periods) the modes of $\mathbf{x}$ are found, and hence by use of Eq. (9.32) the modes of $\bar{\mathbf{a}}$.

If the matrix $\mathbf{M}$ is diagonal—as is the case if the masses have been 'lumped'—the procedure of deriving the standard eigenvalue problem is simplified and here appears the first advantage of the diagonalization, which we have discussed in Sec. 9.2.4.

9.4.3 *Free vibration with singular* **K** *matrix.* In static problems we have always introduced a suitable number of *support* conditions to allow the stiffness matrix $\mathbf{K}$ to be inverted, or what is equivalent, to solve the static equations uniquely. If such 'support' conditions are in fact not specified, as may well be the case with a rocket travelling in space, an arbitrary fixing of a minimum number of support conditions allows a static solution to be obtained without affecting the stresses. In dynamic problems such a fixing is not permissible and frequently one is faced with a problem of a free oscillation for which $\mathbf{K}$ is singular and therefore does not possess an inverse.

To preserve the applicability of the general methods described in the previous section a simple artifice is possible. Equation (9.26) is modified to

$$[(\mathbf{K} + \alpha\mathbf{M}) - (\omega^2 + \alpha)\mathbf{M}]\bar{\mathbf{a}} = 0 \tag{9.35}$$

in which α is an arbitrary constant of the same order as the typical ω^2 sought.

The new matrix $(\mathbf{K} + \alpha\mathbf{M})$ is no longer singular and can be inverted and the standard process maintained to find $(\omega^2 + \alpha)$.

This simple but effective sidestepping of otherwise serious difficulties was first suggested by Cox[35] and Jennings.[36] Alternative methods of dealing with the above problem are given in references 37 and 38.

9.4.4 *Reduction of the eigenvalue system.* Whatever technique is used in the process of determining the eigenvalues of eigenmodes of the system, for a given size of the matrices $n \times n$ the effort is at least one order greater than that involved in an equivalent static situation. Further, while the number of eigenvalues of the real system is infinite, in practice, we are generally interested only in a relatively small number of the lower frequencies and it is possible to simplify the computation by reducing the size of the problem.

To achieve a reduced problem we assume that the unknown $\mathbf{a}$ can be expressed in terms of m ($\ll n$) vectors $\mathbf{t}_1, \mathbf{t}_2, \ldots, \mathbf{t}_m$ and corresponding participating factors x_i. We now write

$$\mathbf{a} = \mathbf{t}_1 x_1 + \mathbf{t}_2 x_2 + \cdots + \mathbf{t}_m x_m = \mathbf{T}\mathbf{x} \tag{9.36}$$

Inserting above into Eq. (9.26) and premultiplying by $\mathbf{T}^{\mathrm{T}}$ we have a reduced problem with only m eigenvectors:

$$(-\omega^2 \mathbf{M}^* + \mathbf{K}^*)\mathbf{x} = 0 \tag{9.37}$$

where

$$\mathbf{M}^* = \mathbf{T}^{\mathrm{T}}\mathbf{M}\mathbf{T} \qquad \mathbf{K}^* = \mathbf{T}^{\mathrm{T}}\mathbf{K}\mathbf{T}$$

If by good fortune the trial vectors were to be chosen as eigenvectors of the original matrix the system would become diagonal and all eigenvalues could be determined by a trivial calculation. This of course is very difficult to achieve but it is possible by physical insight to find vectors $\mathbf{t}$ that correspond closely to the principal modes of the movement.

In an early development of an 'economizer' procedure the vector $\mathbf{x}$ was simply a selection of *master* degrees of freedom from the total vector $\mathbf{a}$ with the remainder $\mathbf{a}_s$ (*slave*) being determined by solving the deformations of an otherwise unloaded structure. We can thus write

$$\mathbf{K}\mathbf{a} = \begin{bmatrix} \mathbf{K}_{ss} & \mathbf{K}_{sx} \\ \mathbf{K}_{sx}^{\mathrm{T}} & \mathbf{K}_{xx} \end{bmatrix} \begin{Bmatrix} \mathbf{a}_s \\ \mathbf{x} \end{Bmatrix} = \begin{Bmatrix} 0 \\ f_s \end{Bmatrix} \tag{9.38}$$

Immediately from the equation of the unloaded nodes we have

$$\mathbf{a}_s = -\mathbf{K}_{ss}^{-1}\mathbf{K}_{sx}\mathbf{x} \tag{9.39}$$

and

$$\mathbf{a} = \begin{Bmatrix} -\mathbf{K}_{ss}^{-1}\mathbf{K}_{sx} \\ \mathbf{I} \end{Bmatrix} \mathbf{x} = \mathbf{T}\mathbf{x}$$

thus defining the transformation matrix. This procedure can be quite effective in some problems, providing the master degrees of freedom are appropriately chosen as shown by Irons,[39,40] Guyan[41] and others.[42,43] Indeed, in the last of these references a suggestion is made that modes with the lowest ratio

$$\frac{K_{ii}}{M_{ii}}$$

be retained as masters. However, the procedure is somewhat arbitrary and can on occasion lead to very poor results.

An alternative to the above process is to devise m static loads, $\mathbf{f}_i$, that approximate in the deformed structure to possible vibration modes. These deflected forms or Ritz vectors[44,45] give the vectors $\mathbf{t}_i$ of Eq. (9.36), and this allows a much more effective choice of the transformation matrix to be obtained. The procedure is still far from being automatic and presents difficulties if a large number of eigenvalues is to be retained. However, an eigenvalue iterative solution may be used directly to determine the loads, $\mathbf{f}_i$, the process being known as subspace iteration.[46,47] We shall refer again to this in the chapter dealing with programming (Chapter 16).

9.4.5 *Some examples.* There are a variety of problems for which practical solutions exist, so only a few simple examples will be shown.

Vibration of plates. Figure 9.2 shows the vibration of a rectangular cantilever plate solved using only four triangular elements. The results are compared against an elaborate calculation carried out by Barton.[44] It is seen that the results using the simple non-conforming triangle are here superior to those using the more elaborate formulation and the accuracy is quite remarkable both in frequency and mode shape.

A similar problem is presented in Fig. 9.3 where the effect of using the *eigenvalue economizer* method is examined. It will be seen how very small the changes in the first four frequencies are on restricting the degrees of freedom from 90 through various stages to 6.

So many further examples of plate and shell vibration analysis are included in current literature that a list of references is here impracticable and a survey of such literature should be consulted.[45,46]

Shell vibration. Application of the process to any elastic two- or three-dimensional continuum can obviously be made and shell vibrations are a typical problem of much interest. In Fig. 9.4, by contrast to the previous simple examples, the elaborate thick shell elements described in Chapter 5 are used to solve a problem of turbine blade vibration.[47,48]

Some other dynamic analyses of shells are given in references 49 to 52. Reference 20 also shows some applications utilizing full three-dimensional isoparametric elements.

The 'wave' equation. Electromagnetic and fluid problems. The basic

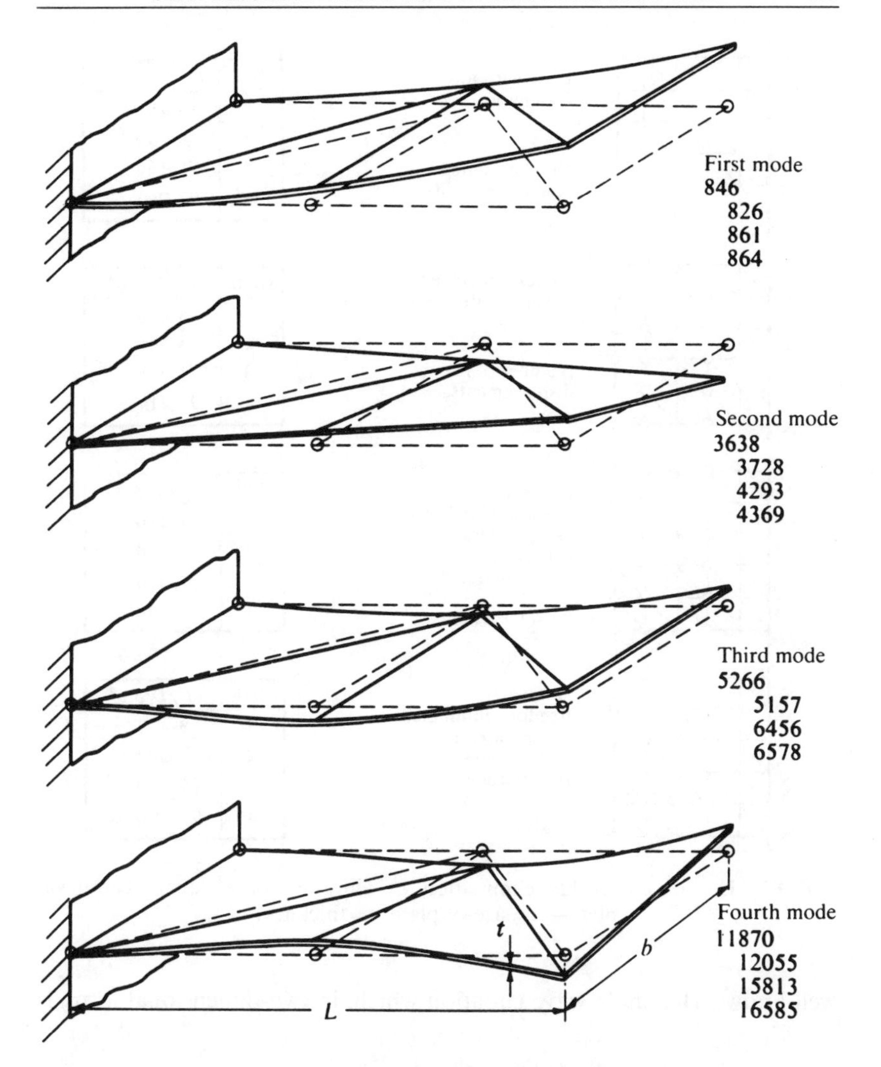

Fig. 9.2 Vibration of a cantilever plate divided into four triangular elements. Modal shapes. Data: $E=30\times10^6$ lb/in^2, $t=0.1$ in, $L=2$ in, $b=1$ in, $\nu=0.3$, density $\rho=0.283$ lb/in^3. The numbers listed show frequencies in cycles per second (Hz) for (1) exact solution;[44] (2) 'non-conforming' triangle (Sec. 1.5); (3) conforming triangle [Sec. 1.10, Eq.(1.52)]; (4) conforming triangle, corrective function [Sec. 1.10, Eq.(1.53)]

dynamic equation (9.8) can be derived from a variety of non-structural problems. The eigenvalue problem once again occurs with 'stiffness' and 'mass' matrices now having alternate physical meanings.

A particular form of the more general equations discussed earlier is the

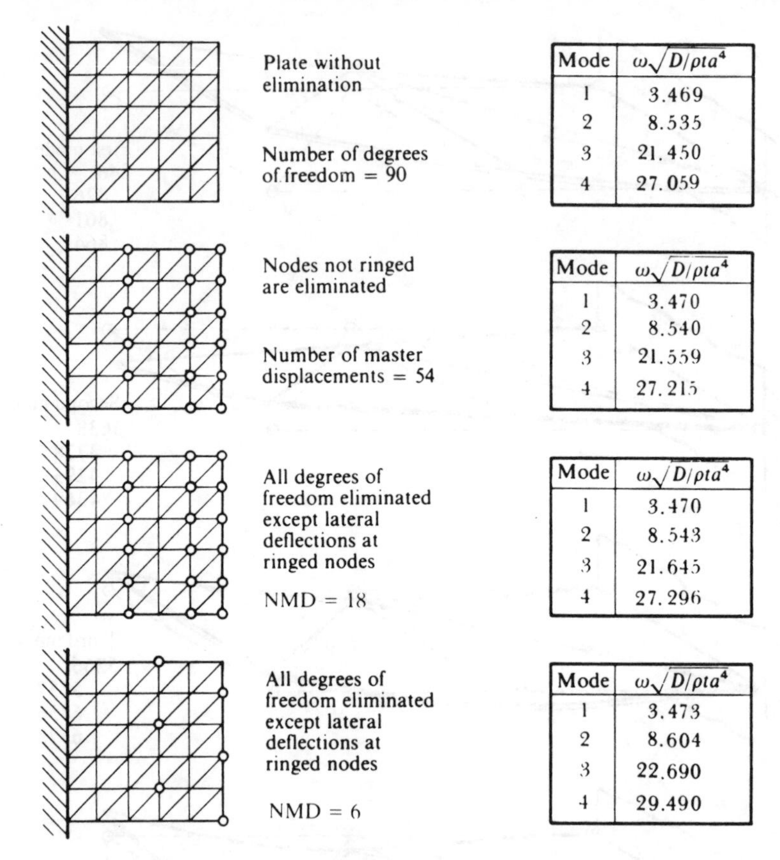

Mode	$\omega\sqrt{D/\rho t a^4}$
1	3.469
2	8.535
3	21.450
4	27.059

Mode	$\omega\sqrt{D/\rho t a^4}$
1	3.470
2	8.540
3	21.559
4	27.215

Mode	$\omega\sqrt{D/\rho t a^4}$
1	3.470
2	8.543
3	21.645
4	27.296

Mode	$\omega\sqrt{D/\rho t a^4}$
1	3.473
2	8.604
3	22.690
4	29.490

Fig. 9.3 Use of eigenvalue elimination in vibration of a square cantilever plate—a = size of plate, t = thickness

well-known Helmholz wave equation which, in two-dimensional form, is

$$\frac{\partial^2\phi}{\partial x^2}+\frac{\partial^2\phi}{\partial y^2}+\frac{1}{\bar{c}^2}\frac{\partial^2\phi}{\partial t^2}=0 \tag{9.40}$$

If the boundary conditions do not force a response, an eigenvalue problem results which has a significance in several fields of physical science.

The first application is to *electromagnetic* fields.[53] Figure 9.5 shows a modal shape of a field for a *waveguide problem.* Simple triangular elements are used here. More complex three-dimensional oscillations are also discussed in reference 53.

A similar equation also describes to a reasonable approximation the

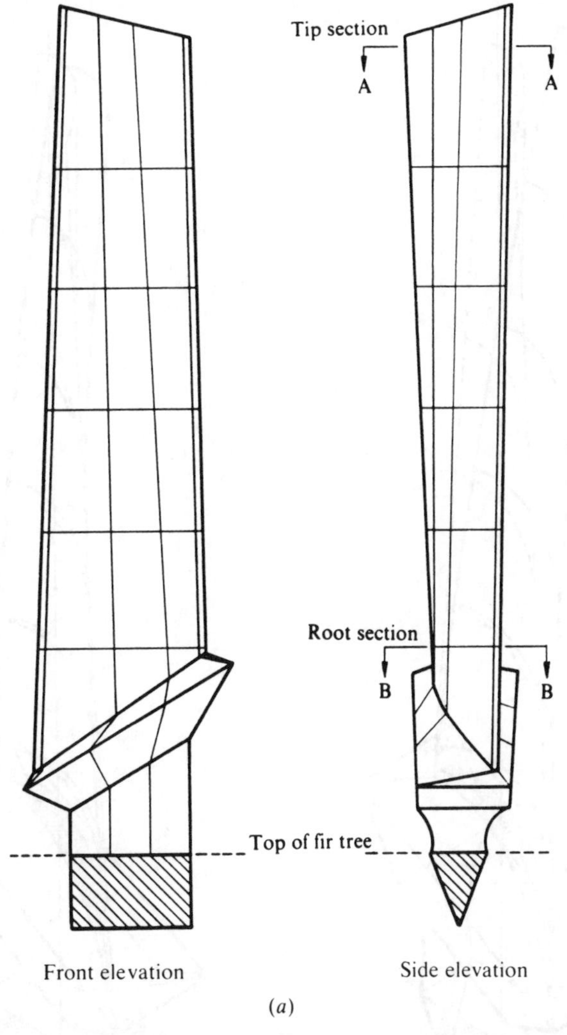

Fig. 9.4 Vibration of turbine blade treated as a thick shell. (*a*) Element subdivision (parabolic type), (*b*) modal shapes and frequencies compared with experiment

behaviour of shallow water waves in a body of water:

$$\frac{\partial}{\partial x}\left(h\frac{\partial \psi}{\partial x}\right)+\frac{\partial}{\partial y}\left(h\frac{\partial \psi}{\partial y}\right)+\frac{1}{g}\frac{\partial^2 \psi}{\partial t^2}=0 \tag{9.41}$$

in which h is the average water depth, ψ the surface elevation above average and g the gravity acceleration (viz. Chapter 15).

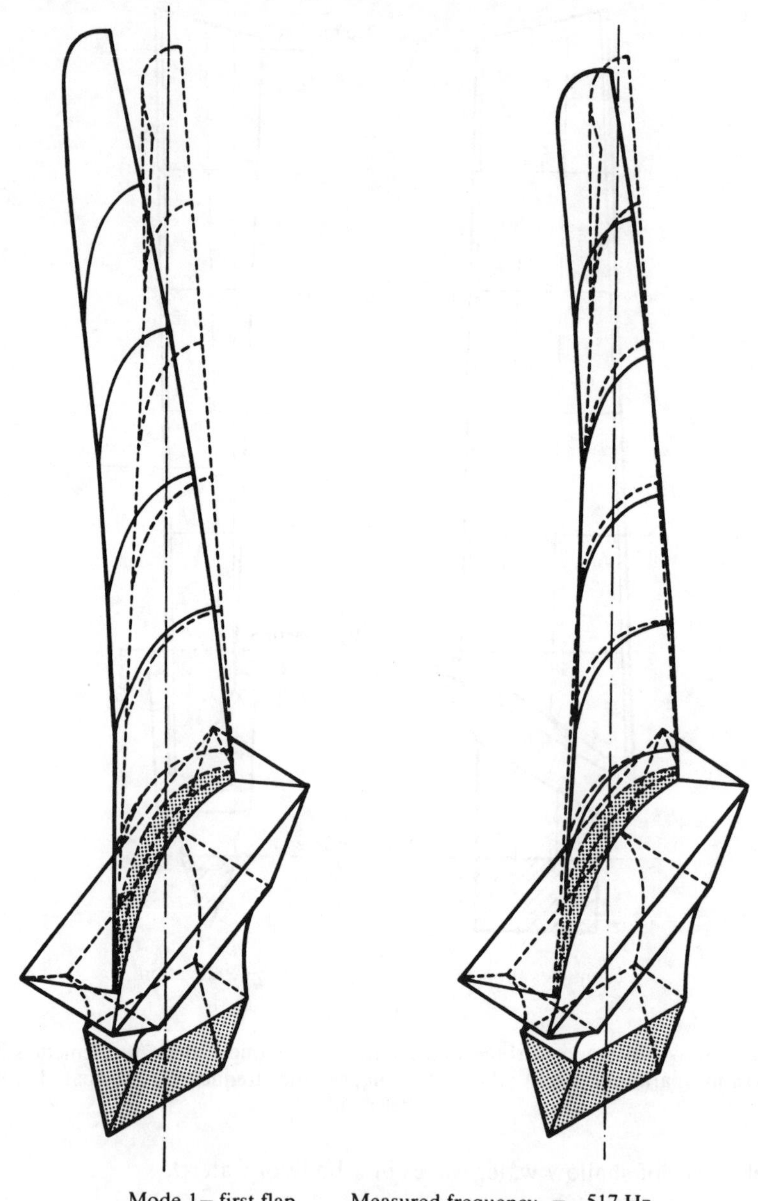

Mode 1 – first flap Measured frequency = 517 Hz
Calculated frequency = 518 Hz

Mode 2 – first edgewise Measured frequency = 1326 Hz
Calculated frequency = 1692 Hz

(*b*) cont.

Fig. 9.4*b* (continued)

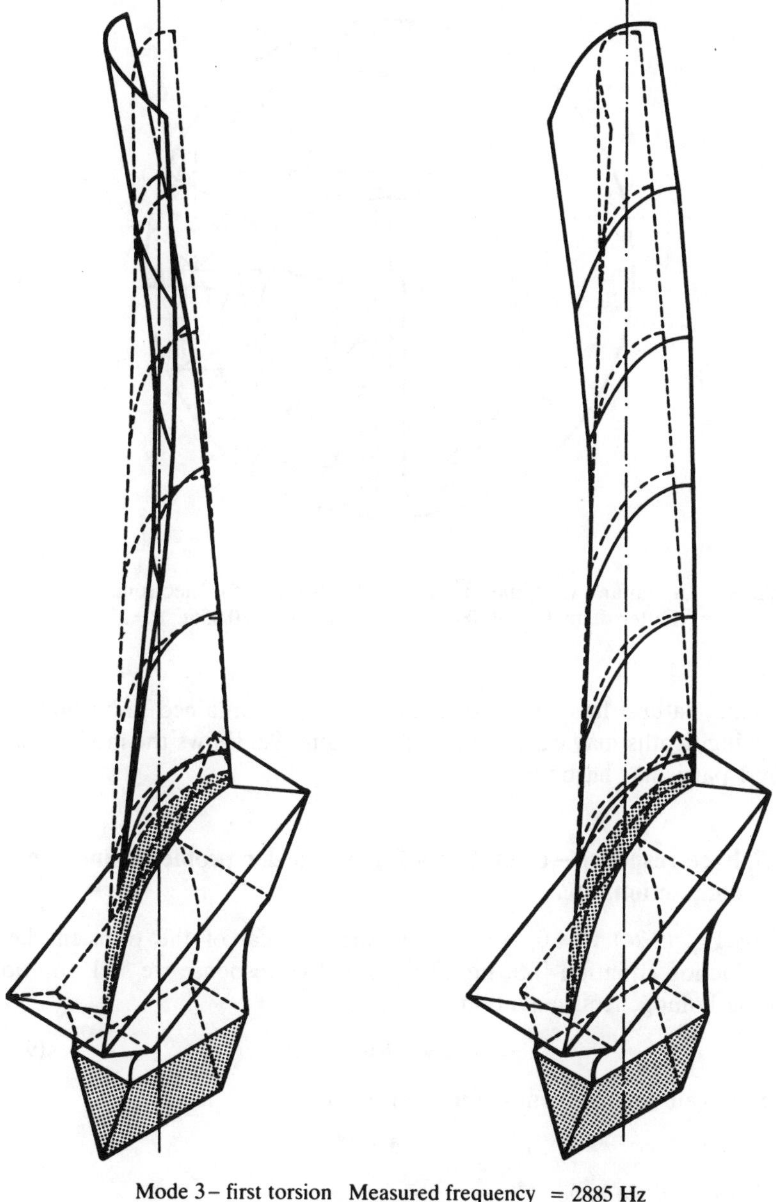

Mode 3 – first torsion	Measured frequency	= 2885 Hz
	Calculated frequency	= 2686 Hz
Mode 4 – second flap	Measured frequency	= 2510 Hz
	Calculated frequency	= 2794 Hz

(*b*)

Fig. 9.4*b* (continued)

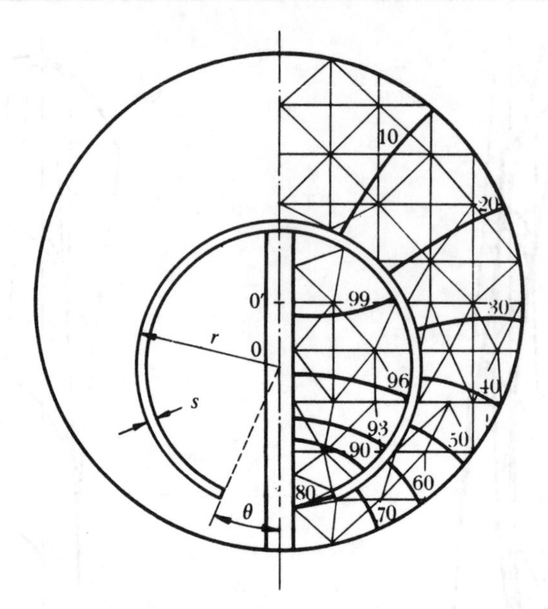

Fig. 9.5 A 'lunar' waveguide;[53] mode of vibration for electromagnetic field. Outer diameter $=d$, $00'=1.3d$, $r=0.29d$, $S=0.055d$, $\theta=22°$

Thus natural frequencies of bodies of water contained in harbours of varying depths may easily be found.[19] Figure 9.6 shows the modal shape for a particular harbour.

9.5 Free response—eigenvalues for first-order problems and heat conduction, etc.

If in Eq. (9.24) $\mathbf{M}=0$, we have a form typical of the transient heat conduction equation [see Eq. (9.1)]. For free response we seek solution of the homogeneous equation

$$\mathbf{C}\dot{\mathbf{a}}+\mathbf{K}\mathbf{a}=0 \tag{9.42}$$

Once again an exponential form can be used:

$$\mathbf{a}=\bar{\mathbf{a}}\,\mathrm{e}^{-\omega t}$$

Substituting we have

$$(-\omega\mathbf{C}+\mathbf{K})\bar{\mathbf{a}}=0 \tag{9.43}$$

which gives again an eigenvalue problem identical to that of Eq. (9.26). As $\mathbf{C}$ and $\mathbf{K}$ are usually positive definite, ω will be positive and real. The solution represents, therefore, simply an exponential decay term and is

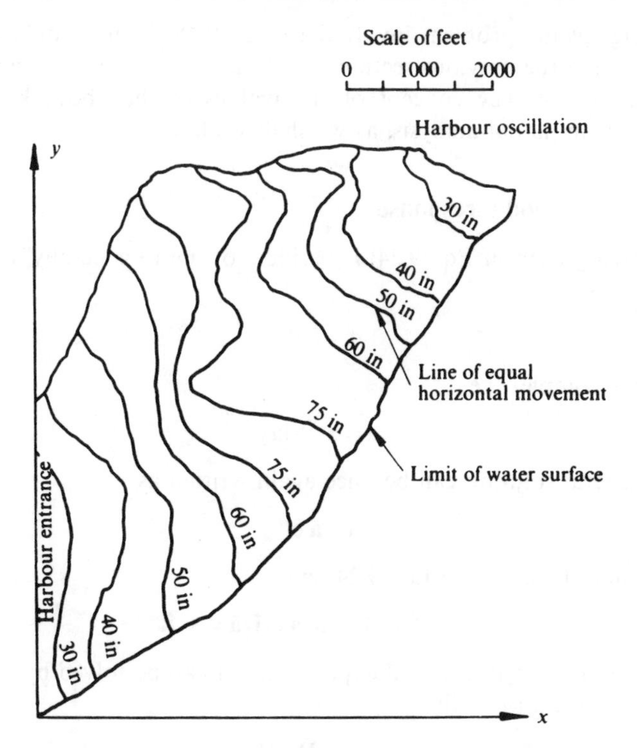

Fig. 9.6 Oscillations of a natural harbour: contours of velocity amplitudes[10]

not really steady state. Combination of such terms, however, can be useful in the solution of initial value transient problems but *per se* is of little value.

9.6 Free response—damped dynamic eigenvalues

We shall now consider the full equation (9.24) for free response conditions. Writing

$$\mathbf{M}\ddot{\mathbf{a}} + \mathbf{C}\dot{\mathbf{a}} + \mathbf{K}\mathbf{a} = 0 \tag{9.44}$$

and substituting

$$\mathbf{a} = \bar{\mathbf{a}}\, e^{\alpha t} \tag{9.45}$$

we have the characteristic equation

$$(\alpha^2\mathbf{M} + \alpha\mathbf{C} + \mathbf{K})\bar{\mathbf{a}} = 0 \tag{9.46}$$

where α and $\bar{\mathbf{a}}$ will in general be found to be complex. The real part of the solution represents a decaying vibration.

The eigenvalue problem involved in Eq. (9.45) is more difficult than that arising in the previous sections. Fortunately this seldom needs to be solved explicitly. The concept of eigenvalues of the above kind is of importance in modal analysis, as we shall see later.

9.7 Forced periodic response

If the forcing term in Eq. (9.24) is periodic or, more generally, if we can express it as

$$\mathbf{f} = \bar{\mathbf{f}}\, e^{\alpha t} \tag{9.47}$$

where α is complex, i.e.

$$\alpha = \alpha_1 + i\alpha_2 \tag{9.48}$$

then a general solution can be once again written as

$$\mathbf{a} = \bar{\mathbf{a}}\, e^{\alpha t} \tag{9.49}$$

Substituting the above in Eq. (9.24) gives

$$(\alpha^2 \mathbf{M} + \alpha \mathbf{C} + \mathbf{K})\bar{\mathbf{a}} \equiv \mathbf{D}\bar{\mathbf{a}} = -\bar{\mathbf{f}} \tag{9.50}$$

which is no longer an eigenvalue problem but can be solved by inverting the matrix **D**, i.e. formally

$$\bar{\mathbf{a}} = -\mathbf{D}^{-1}\bar{\mathbf{f}} \tag{9.51}$$

The solution is thus precisely of the same form as that used for static problems but now, however, has to be determined in terms of complex quantities. Computer programs are available for operation of complex numbers but the computation can be arranged directly, noting that

$$\begin{aligned} e^{\alpha t} &= e^{\alpha_1 t}(\cos \alpha_2 t + i \sin \alpha_2 t) \\ \bar{\mathbf{f}} &= \bar{\mathbf{f}}_1 + i\bar{\mathbf{f}}_2 \\ \bar{\mathbf{a}} &= \bar{\mathbf{a}}_1 + i\bar{\mathbf{a}}_2 \end{aligned} \tag{9.52}$$

in which $\alpha_1, \alpha_2, \mathbf{f}_1, \mathbf{f}_2, \bar{\mathbf{a}}_1$ and $\bar{\mathbf{a}}_2$ are real quantities. Inserting the above into Eq. (9.50) we have

$$\begin{bmatrix} (\alpha_1^2 - \alpha_2^2)\mathbf{M} + \alpha_1 \mathbf{C} + \mathbf{K}, & -2\alpha_1\alpha_2 \mathbf{M} - \alpha_2 \mathbf{C} \\ 2\alpha_1\alpha_2 \mathbf{M} + \alpha_2 \mathbf{C}, & (\alpha_1^2 - \alpha_2^2)\mathbf{M} + \alpha_1 \mathbf{C} + \mathbf{K} \end{bmatrix} \begin{Bmatrix} \bar{\mathbf{a}}_1 \\ \bar{\mathbf{a}}_2 \end{Bmatrix} = -\begin{Bmatrix} \bar{\mathbf{f}}_1 \\ \mathbf{f}_2 \end{Bmatrix} \tag{9.53}$$

Equations (9.53) form a system in which all quantities are real and from which the response to any periodic input can be determined by direct solution. The system is no longer positive definite although it is still symmetric.

With periodic input the solution after an initial transient is not sensitive to the initial conditions and the above solution represents the finally established response. It is valid for problems of dynamic structural response as well as for the problems typical of heat conduction in which we simply put $\mathbf{M}=0$.

9.8 Transient response by analytical procedures

9.8.1 *General.* In the previous sections we have been concerned with steady-state general solutions which took no account of the initial conditions of the system or of the non-periodic form of the forcing terms. The response taking these features into account is essential if we consider, for instance, the earthquake behaviour of structures or the transient behaviour of the heat conduction problem. The solution of such general cases requires either a full-time discretization, which we shall discuss in detail in the next chapter, or the use of special analytical procedures. Here two broad possibilities exist:

(*a*) the frequency response procedure;
(*b*) the modal analysis procedure.

We shall discuss these briefly.

9.8.2 *Frequency response procedures.* In Sec. 9.7 we have shown how the response of the system to any forcing terms of the general periodic type or in particular to a periodic forcing function

$$\mathbf{f}=\bar{\mathbf{f}}\,e^{i\omega t} \tag{9.54}$$

can be obtained by solving a simple equation system. As a completely arbitrary forcing function can be represented approximately by a Fourier series or in the limit, exactly, as a Fourier integral, the response to such an input can be obtained by a synthesis of a curve representing the response of any quantity of interest, e.g. the displacement at a particular point, etc., to all frequencies ranging from zero to infinity. In fact only a limited number of such forcing frequencies has to be considered and the results can be synthesized efficiently by fast Fourier transform techniques.[54] We shall not discuss the mathematical details for such procedures which can be found in standard texts on dynamics.[15]

The technique of frequency response is readily adapted to problems where the damping matrix $\mathbf{C}$ is of an arbitrary specified form. This is not the case with the more widely used modal synthesis procedures which are to be described in the next section.

By way of illustration we show in Fig. 9.7 the frequency response of an artificial harbour [see Eq. (9.41)] to an input of waves with different frequencies and damping due to the radiation of reflected waves which imposes a very particular form on the damping matrix. Details of this

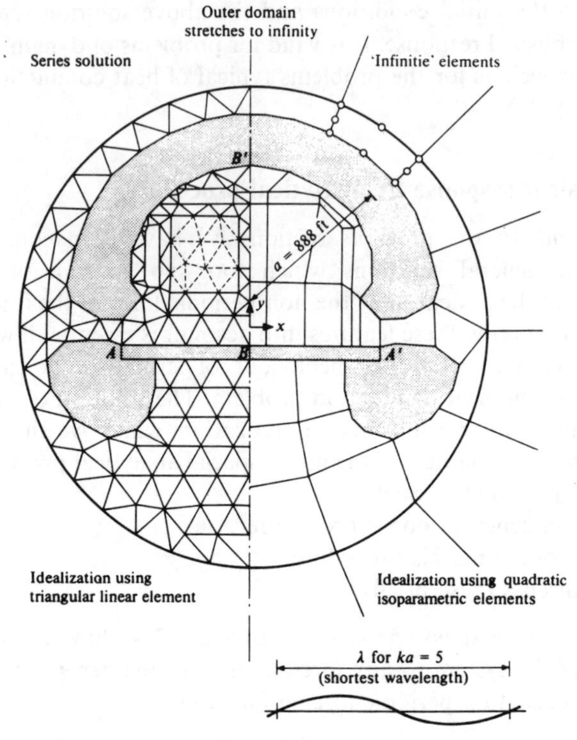

(a) Geometric details and FEM idealization.
Wave forcing frequency $\omega = k\sqrt{gh} = ka$, h = depth of water

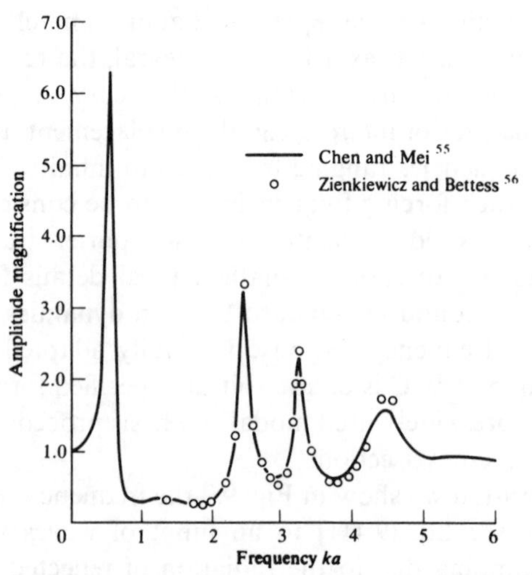

(b) Amplitude magnification response of mean depth in harbour for various frequencies

Fig. 9.7 Frequency response of an artificial harbour to an input of periodic waves

problem are given elsewhere[55,56] (see also Chapter 15). Similar techniques are frequently used in the analysis of the foundation response of structures where radiation of energy occurs.[57]

9.8.3 *Modal decomposition analysis.* This procedure is probably the most important and widely used in practice. Further, it provides an insight into the behaviour of the whole system, which is of value where strictly numerical processes are used, and we shall therefore describe it in detail in the context of the general problem of Eq. (9.24), i.e.

$$\mathbf{M}\ddot{\mathbf{a}}+\mathbf{C}\dot{\mathbf{a}}+\mathbf{K}\mathbf{a}+\mathbf{f}=\mathbf{0} \tag{9.55}$$

where $\mathbf{f}$ is an arbitrary function of time.

We have seen that the general solution for the free response is of the form

$$\mathbf{a}=\bar{\mathbf{a}}\,\mathrm{e}^{\alpha t}=\sum_{i=1}^{n}\bar{\mathbf{a}}_i\,\mathrm{e}^{\alpha_i t} \tag{9.56}$$

where α_i are the eigenvalues and $\bar{\mathbf{a}}_i$ are the eigenvectors (Sec. 9.6). For forced response we shall assume that the solution can be written in a linear combination of modes as

$$\mathbf{a}=\sum\bar{\mathbf{a}}_i\,y_i=[\bar{\mathbf{a}}_1,\bar{\mathbf{a}}_2,\ldots]\,\mathbf{y} \tag{9.57}$$

where the scalar mode participation factor $y_i(t)$ is a function of time. This shows in a clear manner the proportions of each mode occurring. Such a decomposition of an arbitrary vector presents no restriction as all the modes are linearly independent vectors (except for repeated frequencies).

If expression (9.57) is substituted into Eq. (9.55) and the result is premultiplied by $\bar{\mathbf{a}}_i^{\mathrm{T}}$ ($i=1$ to n), then the result is simply a set of scalar, independent, equations

$$m_i\,\ddot{y}_i+c_i\,\dot{y}_i+k_i\,y_i+f_i=0 \tag{9.58}$$

where

$$\begin{aligned} m_i&=\bar{\mathbf{a}}_i^{\mathrm{T}}\mathbf{M}\bar{\mathbf{a}}_i=1 \text{ (if modes normalized)}\\ c_i&=\bar{\mathbf{a}}_i^{\mathrm{T}}\mathbf{C}\bar{\mathbf{a}}_i\\ k_i&=\bar{\mathbf{a}}_i^{\mathrm{T}}\mathbf{K}\bar{\mathbf{a}}_i\\ f_i&=\bar{\mathbf{a}}_i^{\mathrm{T}}\mathbf{f}\end{aligned} \tag{9.59}$$

as for true eigenvectors $\bar{\mathbf{a}}_i$

$$\bar{\mathbf{a}}_i^{\mathrm{T}}\mathbf{M}\bar{\mathbf{a}}_j=\bar{\mathbf{a}}_i^{\mathrm{T}}\mathbf{C}\bar{\mathbf{a}}_j=\bar{\mathbf{a}}_i^{\mathrm{T}}\mathbf{K}\bar{\mathbf{a}}_j=0 \tag{9.60}$$

when

$$i\neq j$$

(This result was proved in Sec. 9.4 for real eigenvalues but is valid generally for complex eigenvalues and vectors, as could be verified by the reader.)

Each scalar equation of (9.58) can be solved by elementary procedures independently and the total vector of response obtained by superposition following Eq. (9.59). In a general case, as we have shown in Sec. 9.6, the eigenvalues and eigenvectors are complex and their determination is not simple.[37] The more usual procedure is to use real eigenvalues corresponding to the solution of Eq. (9.25):

$$(-\omega^2\mathbf{M}+\mathbf{K})\bar{\mathbf{a}}=0 \tag{9.61}$$

Now repetition of procedures using the process described in Eqs (9.57) to (9.60) leads to decoupled equations with real variables $\mathbf{y}$ only if

$$\bar{\mathbf{a}}_i^{\mathrm{T}}\mathbf{C}\mathbf{a}_j=0 \tag{9.62}$$

which generally does not occur as the eigenvectors now guarantee only the orthogonality of $\mathbf{M}$ and $\mathbf{K}$ and not of the damping matrix. However, if the damping matrix $\mathbf{C}$ is of the form of Eq. (9.16), i.e. a linear combination of $\mathbf{M}$ and $\mathbf{K}$, such orthogonality will obviously occur. Unless the damping is of a definite form which requires special treatment, an assumption of orthogonality is made and Eq. (9.58) is assumed valid in terms of such eigenvectors.

From Eq. (9.61) we have

$$\omega_i^2\mathbf{M}\bar{\mathbf{a}}_i=\mathbf{K}\bar{\mathbf{a}}_i \tag{9.63}$$

and on premultiplying by $\bar{\mathbf{a}}_i^{\mathrm{T}}$ we obtain

$$\omega_i^2 m_i=k_i \tag{9.64}$$

Writing

$$c_i=2\omega_i c_i' \tag{9.65}$$

(where c_i' represents the ratio of damping to its critical value) and assuming that the nodes have been normalized so that $m_i=1$ [see Eq. (9.28)], Eq. (9.58) can be rewritten in a standard form of second degree:

$$\ddot{y}_i+2\omega_i c_i'\dot{y}_i+\omega_i^2 y_i+f_i=0 \tag{9.66}$$

The general solution can be then obtained by writing

$$y_i=\int_0^t f_i\,\mathrm{e}^{-v'_i\omega_i(t-\tau)}\sin\omega_i(t-\tau)\,\mathrm{d}\tau \tag{9.67}$$

Such integration can be carried out numerically and the response obtained. In principle, superposition will result in the full transient required. In practice, often a single calculation is carried out for each mode to determine the maximum responses and a suitable addition of these results is used. Such processes are well described in standard texts and are used as standard procedures to calculate the behaviour of structures subject to earthquake shocks.[15,27]

For a first-order equation system, such as arises in heat conduction,

$$\mathbf{C}\dot{\mathbf{a}} + \mathbf{K}\mathbf{a} + \mathbf{f} = 0$$

exactly analogous procedures can be used. Using now the real eigenvalues determined in Sec. 9.5, decomposition into scalar equations can be made yielding a set of decoupled equations

$$c_i \dot{y}_i + k_i y_i + f_i = 0 \tag{9.68}$$

and this can be used again to get the general solution analytically.[58] We leave details of such a solution to the reader as an exercise.

9.8.4 *Damping and participation of modes.* The type of calculation implied in modal decomposition apparently necessitates the determination of all modes and eigenvalues—a task of considerable magnitude. In fact only a limited number of modes usually need to be taken into consideration as often the response to higher frequency is critically damped and insignificant.

To show that this is true consider the form of the damping matrices. In Sec. 9.2 [Eq. (9.16)] we have indicated that the damping matrix is often assumed as

$$\mathbf{C} = \alpha \mathbf{M} + \beta \mathbf{K} \tag{9.69}$$

Indeed a form of this type is necessary for the use of modal decomposition, although other generalizations are possible.[59,60] From the definition of c'_i, the critical damping ratio in Eq. (9.65), we see that this can now be written as

$$c'_i = \frac{1}{2\omega_i} \bar{\mathbf{a}}_i^{\mathrm{T}} (\alpha \mathbf{M} + \beta \mathbf{K}) \bar{\mathbf{a}}_i = \frac{1}{2\omega_i} (\alpha + \beta \omega_i^2) \tag{9.70}$$

Thus if the coefficient β is of larger importance, as is the case with most structural damping, c'_i grows with ω_i and at high frequency an over-damped condition will arise. This is indeed fortunate as, in general, an infinite number of high frequencies exist which are not modelled by any finite element discretizations.

We shall see in the next chapter that in the step-by-step recurrence computation the high frequencies often control the problem, and this effect needs to be filtered out for realistic results.

9.9 Symmetry and repeatability

In concluding this chapter it is worth remarking that in dynamic calculation we have once again encountered all the general principles of assembly, etc., that are applicable to static problems. However, some aspects of symmetry and repeatability which were used previously (see Chapter 8 of Volume 1) need amending. It is obviously possible for a symmetric structure to vibrate in an unsymmetric manner, for instance, and similarly a repeatable structure contains modes which are themselves non-repeatable. However, even here considerable simplification can still be made; details of this are discussed in references 61 to 63.

References

1. S. CRANDALL, *Engineering Analysis*, McGraw-Hill, New York, 1956.
2. H. S. CARSLAW and J. C. JAEGER, *Conduction of Heat in Solids*, 2nd ed., Clarendon Press, Oxford, 1959.
3. W. VISSER, 'A finite element method for the determination of non-stationary temperature distribution and thermal deformation', in *Proc. Conf. on Matrix Methods in Structural Mechanics*, Air Force Institute of Technology, Wright-Patterson AF Base, Ohio, 1965.
4. O. C. ZIENKIEWICZ and Y. K. CHEUNG, *The Finite Element Method in Structural and Continuum Mechanics*, 1st ed., McGraw-Hill, New York, 1967.
5. E. L. WILSON and R. E. NICKELL, 'Application of finite element method to heat conduction analysis', *Nucl. Eng. Des.*, **4**, 1–11, 1966.
6. O. C. ZIENKIEWICZ and C. J. PAREKH, 'Transient field problems—two and three dimensional analysis by isoparametric finite elements', *Int. J. Num. Mech. Eng.*, **2**, 61–71, 1970.
7. K. TERZHAGI and R. B. PECK, *Soil Mechanics in Engineering Practice*, Wiley, New York, 1948.
8. D. K. TODD, *Ground Water Hydrology*, Wiley, New York, 1959.
9. P. L. ARLETT, A. K. BAHRANI and O. C. ZIENKIEWICZ, 'Application of finite elements to the solution of Helmholtz's equation', *Proc. IEE*, **115**, 1962–6, 1968.
10. C. TAYLOR, B. S. PATIL and O. C. ZIENKIEWICZ, 'Harbour oscillation: a numerical treatment for undamped natural modes' *Proc. Inst. Civ. Eng.*, **43**, 141–56, 1969.
11. O. C. ZIENKIEWICZ and R. E. NEWTON, 'Coupled vibrations in a structure submerged in a compressible fluid', in *Int. Symp. on Finite Element Techniques*, Stuttgart, 1968.
12. J. S. ARCHER, 'Consistent mass matrix for distributed systems', *Proc. Am. Soc. Civ. Eng.*, **89**, ST4, 161, 1963.
13. F. A. LECKIE and G. M. LINDBERG, 'The effect of lumped parameters on beam frequencies', *Aero. Q.*, **14**, 234, 1963.
14. O. C. ZIENKIEWICZ and Y. K. CHEUNG, 'The finite element method for analysis of elastic isotropic and orthotropic slabs', *Proc. Inst. Civ. Eng.*, **28**, 471–88, 1964.
15. R. W. CLOUGH and J. PENZIEN, *Dynamics of Structures*, McGraw-Hill, New York, 1975.
16. O. C. ZIENKIEWICZ, B. M. IRONS and B. NATH, 'Natural frequencies of complex free or submerged structures by the finite element method', in *Symp. on Vibration in Civil Engineering, London, April 1965*, Butterworth, 1966.
17. D. J. DAWE, 'A finite element approach to plate vibration problems', *J. Mech. Eng. Sci.*, **7**, 28, 1965.

18. R. J. GUYAN, 'Distributed mass matrix for plate elements in bending', *JAIAA*, **3**, 567, 1965.
19. G. P. BAZELEY, Y. K. CHEUNG, B. M. IRONS and O. C. ZIENKIEWICZ, 'Triangular elements in plate bending—conforming and non-conforming solutions', in *Proc. Conf. on Matrix Methods in Structural Mechanics*, Air Force Institute of Technology, Wright-Patterson AF Base, Ohio, 1965.
20. R. G. ANDERSON, B. M. IRONS and O. C. ZIENKIEWICZ, 'Vibration and stability of plates using finite elements', *Int. J. Solids Struct.*, **4**, 1031–55, 1968.
21. R. G. ANDERSON, 'The application of the non-conforming triangular plate bending element to plate vibration problems', M.Sc. thesis, University of Wales, Swansea, 1966.
22. R. W. CLOUGH, 'Analysis of structure vibrations and response', in *Recent Advances in Matrix Method of Structure Analysis and Design* (eds R. H. Gallagher, Y. Yamada and J. T. Oden), pp. 25–45, First US–Japan Seminar, Alabama Press, 1971.
23. S. W. KEY and Z. E. BEISINGER, 'The transient dynamic analysis of thin shells in the finite element method', in *Proc. 3rd Conf. on Matrix Methods and Structural Mechanics*, Wright-Patterson AF Base, Ohio, 1971.
24. E. HINTON, A. ROCK and O. C. ZIENKIEWICZ, 'A note on mass lumping and related processes in the finite element method', *Int. J. Earthquake Eng. Struct. Dynam.*, **4**, 245–9, 1976.
25. P. TONG, T. H. H. PIAN and L. L. BOCIOVELLI, 'Mode shapes and frequencies by the finite element method using consistent and lumped matrices', *Comp. Struct.*, **1**, 623–38, 1971.
26. I. FRIED and D. S. MALKUS, 'Finite element mass matrix lumping by numerical integration with the convergence rate loss', *Int. J. Solids Struct.*, **11**, 461–5, 1975.
27. K. J. BATHE and E. L. WILSON, *Numerical Methods in Finite Element Analysis*, Prentice-Hall, Englewood Cliffs, N.J., 1976.
28. J. H. WILKINSON, *The Algebraic Eigenvalue Problem*, Clarendon Press, Oxford, 1965.
29. I. FRIED, 'Gradient methods for finite element eigen problems', *JAIAA*, **7**, 739–41, 1969.
30. O. RENFIELD, 'Higher vibration modes by matrix iteration', *JAIAA*, **9**, 505–741, 1971.
31. K. J. BATHE and E. L. WILSON, 'Large eigenvalue problems in dynamic analysis', *Proc. Am. Soc. Civ. Eng.*, **98**, EM6, 1471–85, 1972.
32. K. J. BATHE and E. L. WILSON, 'Solution methods for eigenvalue problems in structural dynamics', *Int. J. Num. Meth. Eng.*, **6**, 213–26, 1973.
33. A. JENNINGS, 'Mass condensation and similarity iterations for vibration problems', *Int. J. Num. Meth. Eng.*, **6**, 543–52, 1973.
34. K. K. GUPTA, 'Solution of eigenvalue problems by Sturm sequence method', *Int. J. Num. Meth. Eng.*, **4**, 379–404, 1972.
35. H. L. COX, 'Vibration of missiles', *Aircraft Eng.*, **33**, 2–7 and 48–55, 1961.
36. A. JENNINGS, 'Natural vibration of a free structure', *Aircraft Eng.*, **34**, 8, 1962.
37. W. C. HURTY and M. F. RUBINSTEIN, *Dynamics of Structures*, Prentice-Hall, Englewood Cliffs, N.J., 1974.
38. A. CRAIG and M. C. C. BAMPTON, 'On the iterative solution of semi definite eigenvalue problems', *Aero. J.*, **75**, 287–90, 1971.
39. B. M. IRONS, 'Eigenvalue economisers in vibration problems', *J. Roy. Aero. Soc.*, **67**, 526, 1963.

40. B. M. IRONS, 'Structural eigenvalue problems: elimination of unwanted variables', *JAIAA*, **3**, 961, 1965.
41. R. J. GUYAN, 'Reduction of stiffness and mass matrices', *JAIAA*, **3**, 380, 1965.
42. J. N. RAMSDEN and J. R. STOKER, 'Mass condensation; a semi-automatic method for reducing the size of vibration problems', *Int. J. Num. Meth. Eng.*, **1**, 333–49, 1969.
43. R. D. HENSHELL and J. H. ONG, 'Automatic masters for eigenvalue economisation', *Int. J. Earthquake Struct. Dynam.*, **3**, 375–83, 1975.
44. M. V. BARTON, 'Vibration of rectangular and skew cantilever plates', *J. Appl. Mech.*, **18**, 129–34, 1951.
45. G. B. WARBURTON, 'Recent advances in structural dynamics', in *Symp. on Dynamic Analysis of Structures*, NEL, East Kilbride, Scotland, October 1975.
46. J. C. MACBAIN, 'Vibratory behaviour of twisted cantilever plates', *J. Aircraft*, **12**, 357–9, 1975.
47. S. AHMAD, R. G. ANDERSON and O. C. ZIENKIEWICZ, 'Vibration of thick, curved, shells with particular reference to turbine blades', *J. Strain Anal.*, **5**, 200–6, 1970.
48. M. A. J. BOSSAK and O. C. ZIENKIEWICZ, 'Free vibration of initially stressed solids with particular reference to centrifugal force effects in rotating machinery', *J. Strain Anal.*, **8**, 245–52, 1973.
49. J. S. ARCHER and C. P. RUBIN, 'Improved linear axi-symmetric shell-fluid model for launch vehicle longitudinal response analysis', in *Proc. Conf. on Matrix Methods in Structural Mechanics*, Air Force Institute of Technology, Wright-Patterson AF Base, Ohio, 1965.
50. J. H. ARGYRIS, 'Continua and discontinua', in *Proc. Conf. on Matrix Methods in Structural Mechanics*, Air Force Institute of Technology, Wright-Patterson AF Base, Ohio, 1965.
51. S. KLEIN and R. J. SYLVESTER, 'The linear elastic dynamic analysis of shells of revolution by the matrix displacement method', in *Proc. Conf. on Matrix Methods in Structural Mechanics*, Air Force Institute of Technology, Wright-Patterson AF Base, Ohio, 1965.
52. R. DUNGAR, R. T. SEVERN and P. R. TAYLOR, 'Vibration of plate and shell structures using triangular finite elements', *J. Strain Anal.*, **2**, 73–83, 1967.
53. P. L. ARLETT, A. K. BAHRANI and O. C. ZIENKIEWICZ, 'Application of finite elements to the solution of Helmholtz's equation', *Proc. IEE*, **115**, 1762–64, 1968.
54. E. O. BRIGHAM, *The Fast Fourier Transform*, Prentice-Hall, Englewood Cliffs, N.J., 1974.
55. H. S. CHEN and C. C. MEI, 'Hybrid-element method for water waves', in *Proc. Modelling Techniques Conf. (Modelling 1975)*, Vol. 1, pp. 63–81, San Francisco, 1975.
56. O. C. ZIENKIEWICZ and P. BETTESS, 'Infinite elements in the study of fluid–structure interaction problems', in *2nd Int. Symp. on Computing Methods in Applied Science and Engineering*, Versailles, France, December 1975.
57. J. PENZIEN, 'Frequency domain analysis including radiation damping and water load coupling', in *Numerical Methods in Offshore Engineering* (eds O. C. Zienkiewicz, R. W. Lewis and K. G. Stagg), Wiley, New York, 1978.
58. R. H. GALLAGHER and R. H. MALLETT, *Efficient Solution Process for Finite Element Analysis of Transient Heat Conduction*, Bell Aero Systems, Buffalo, N.Y., 1969.
59. E. L. WILSON and J. PENZIEN, 'Evaluation of orthogonal damping matrices', *Int. J. Num. Meth. Eng.*, **4**, 5–10, 1972.

60. H. T. THOMSON, T. COLKINS and P. CARAVANI, 'A numerical study of damping', *Int. J. Earthquake Eng. Struct. Dynam.*, **3**, 97–103, 1974.
61. F. W. WILLIAMS, 'Natural frequencies of repetitive structures', *Q. J. Mech. Appl. Math.*, **24**, 285–310, 1971.
62. D. A. EVENSEN, 'Vibration analysis of multi-symmetric structures', *JAIAA*, **14**, 446–53, 1976.
63. D. L. THOMAS, 'Standing waves in rotationally periodic structures', *J. Sound Vibr.*, **37**, 288–90, 1974.

10

The time dimension—discrete approximation in time

10.1 Introduction

In the last chapter we have shown how semi-discretization of dynamic or transient field problems leads in the case of linear problems to sets of ordinary differential equations of the form

$$\mathbf{M\ddot{a}} + \mathbf{C\dot{a}} + \mathbf{Ka} + \mathbf{f} = \mathbf{0} \tag{10.1}$$

for dynamics or

$$\mathbf{C\dot{a}} + \mathbf{Ka} + \mathbf{f} = \mathbf{0} \qquad \text{where } \frac{\mathrm{d}}{\mathrm{d}t}a \equiv \dot{a}, \qquad \text{etc.} \tag{10.2}$$

for heat transfer or similar problems.

In many practical situations non-linearities exist, typically altering the above equations by making

$$\mathbf{M} = \mathbf{M(a)} \qquad \mathbf{C} = \mathbf{C(a)} \qquad \mathbf{Ka} = \mathbf{P(a)} \tag{10.3}$$

The analytical solutions previously discussed, while providing much insight into the behaviour patterns (and indispensable in establishing such properties as natural system frequencies), are in general not economical for the solution of transient problems in linear cases and not applicable when non-linearity exists. In this chapter we shall therefore revert to discretization processes applicable directly to the time domain.

For such discretization the finite element method, including in its definition the finite difference approximation, is of course widely applicable and provides the greatest possibilities, though much of the classical literature on the subject uses only the latter.[1-6] We shall demonstrate here how the finite element method provides a useful generalization unifying many existing algorithms and providing a variety of new ones.

As the time domain is infinite we shall inevitably curtail it to a finite time increment Δt and relate the initial conditions at t_n (and before) to those at time $t_{n+1} = t_n + \Delta t$, obtaining so-called *recurrence relations*. In all of this chapter, the starting point will be that of the semi-discrete equations (10.1) or (10.2), though, of course, the full time–space domain discretization could be considered simultaneously. This, however, in general offers no advantage, for, with the regularity of the time domain, irregular time–space elements are not required. Indeed, if product-type shape functions are chosen, the process will of course be identical to that obtained by using first semi-discretization in space followed by time discretization. An exception here is provided in the convection dominated problems where simultaneous discretizations may be desirable, as we shall show in Chapter 12.

The first concepts of time–space elements were introduced in 1969–70[7–10] and the development of processes involving semi-discretization is presented chronologically in references 11 to 20. Full time–space elements for convection-type equations are described in references 21 and 22.

The presentation of this chapter will be divided into three parts. In the first we shall derive a set of *single-step* recurrence relations for the linear, first- and second-order problems of Eqs (10.2) and (10.1). Such schemes have a very general applicability and are preferable to *multistep schemes* described in the second part as the time step can be adaptively varied. In the final part we shall deal with generalizations necessary for *non-linear problems*.

When discussing stability problems we shall often revert to the concept of modally uncoupled equations introduced in the previous chapter. Here we recall that the equation systems (10.1) and (10.2) can be written as a set of scalar equations:

$$m_i \ddot{y}_i + c_i \dot{y}_i + k_i y_i + f_i = 0 \tag{10.4}$$

or

$$c_i \dot{y}_i + k_i y_i + f_i = 0 \tag{10.5}$$

in the respective eigenvalue participation factors. We shall find that the stability requirements here are dependent on the eigenvalues associated with such equations, ω_i. It turns out, however, fortunately, that it is never necessary to obtain the system eigenvalues or eigenvectors due to a powerful theorem first stated by Irons and Treharne.[23]

The theorem states simply that the system eigenvalues can be bounded by the eigenvalues of individual elements ω^e. Thus

$$\begin{aligned} &(\omega_j)^2_{\min} \geqslant (\omega^e)^2_{\min} \\ \text{and} \quad &(\omega_j)^2_{\max} \leqslant (\omega^e)^2_{\max} \end{aligned} \tag{10.6}$$

The stability limits can thus (as will be shown later) be related to Eqs (10.4) or (10.5) written for a single element.

SINGLE-STEP ALGORITHMS

10.2 Simple time-step algorithms for the first-order equation

10.2.1 *Weighted residual* (*or finite element*) *approach.* We shall now consider Eq. (10.2) which may represent a semi-discrete approximation to a particular physical problem or indeed simply be itself a discrete system. The objective is to obtain an approximation for $\mathbf{a}_{n+1}$ given the value of $\mathbf{a}_n$ and the forcing vector $\mathbf{f}_t$ acting in the interval of time Δt.

In the interval, in the manner used in all finite element approximations, we assume that $\mathbf{a}$ varies as a polynomial and take here the lowest (linear) expansion as shown in Fig. 10.1, writing

$$\mathbf{a} \approx \hat{\mathbf{a}}(\tau) = \mathbf{a}_n + \frac{\tau}{\Delta t}(\mathbf{a}_{n+1} - \mathbf{a}_n) \tag{10.7}$$

with $$\tau = t - t_n$$

This can, of course, be translated to the standard finite element expansion, giving

$$\hat{\mathbf{a}}(\tau) = \sum \mathbf{N}_i \mathbf{a}_1 = \left(1 - \frac{\tau}{\Delta t}\right)\mathbf{a}_n + \left(\frac{\tau}{\Delta t}\right)\mathbf{a}_{n+1} \tag{10.8}$$

in which the unknown parameter is $\mathbf{a}_{n+1}$.

The equation by which this unknown parameter is provided will be the weighted residual approximation to Eq. (10.2), i.e.

$$\int_0^{\Delta t} W(\mathbf{C}\dot{\hat{\mathbf{a}}} + \mathbf{K}\hat{\mathbf{a}} + \mathbf{f})\, \mathrm{d}\tau = \mathbf{0} \tag{10.9}$$

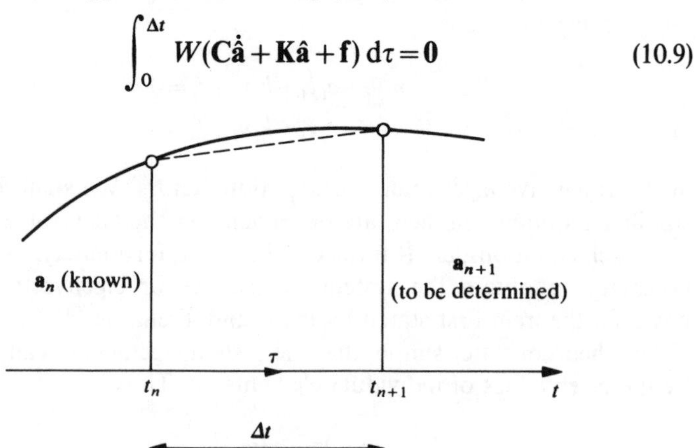

Fig. 10.1 Approximation to $\mathbf{a}$ in the time domain

Introducing Θ as the weighting parameter given by

$$\Theta = \frac{1}{\Delta t} \frac{\int_0^{\Delta t} W\tau \, \mathrm{d}\tau}{\int_0^{\Delta t} W \, \mathrm{d}\tau} \tag{10.10}$$

we can immediately write

$$\frac{\mathbf{C}(\mathbf{a}_{n+1} - \mathbf{a}_n)}{\Delta t} + \mathbf{K}[(\mathbf{a}_n + \Theta(\mathbf{a}_{n+1} - \mathbf{a}_n)] + \bar{\mathbf{f}} = 0 \tag{10.11}$$

where $\bar{\mathbf{f}}$ represents an average value of $\mathbf{f}$ given by

$$\bar{\mathbf{f}} = \frac{\int_0^{\Delta t} \mathbf{f} W \, \mathrm{d}\tau}{\int_0^{\Delta t} W \, \mathrm{d}\tau} \tag{10.12}$$

or

$$\bar{\mathbf{f}} = \bar{\mathbf{f}}_n + \Theta(\bar{\mathbf{f}}_{n+1} - \bar{\mathbf{f}}_n) \tag{10.13}$$

if a linear variation of $\mathbf{f}$ is assumed.

Equation (10.11) is in fact almost identical to a finite difference approximation to the governing equation (10.2) at time $t_n + \Theta\Delta t$, and in this example little advantage is gained by introducing the finite element approximation. However, the averaging of the forcing term is important, as shown in Fig. 10.2, where a constant W (that is $\Theta = \frac{1}{2}$) is used and a finite difference approximation presents difficulties.

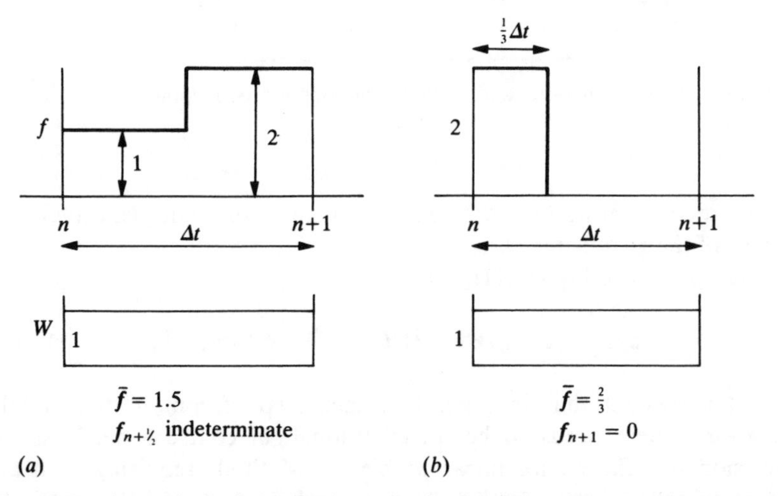

Fig. 10.2 'Averaging' of the forcing term in the finite element–time approach

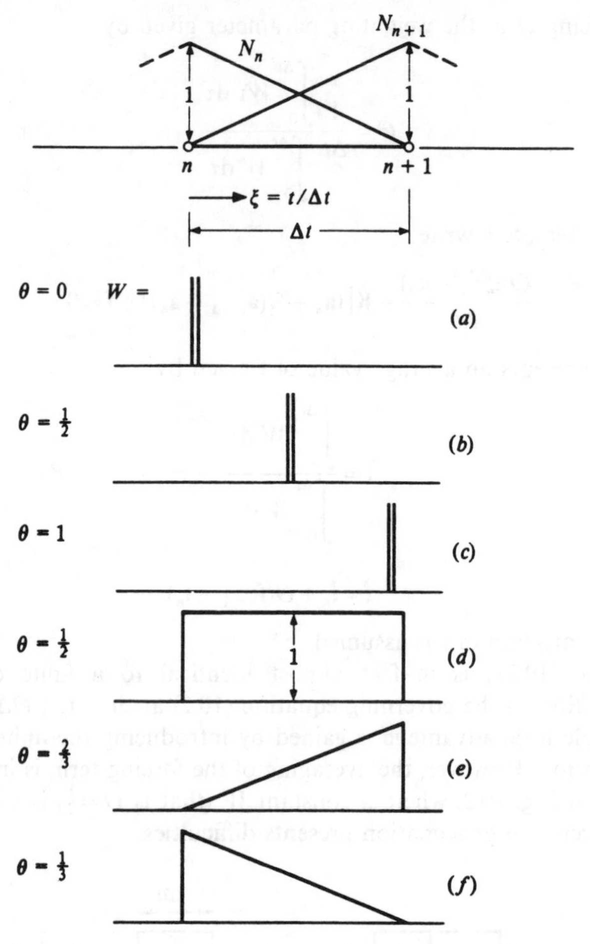

Fig. 10.3 Shape functions and weight functions for two point recurrence formulae

Figure 10.3 shows how different weight functions can yield alternate values of the parameter Θ.

The solution of Eq. (10.11) yields

$$\mathbf{a}_{n+1} = (\mathbf{C} + \Delta t \Theta \mathbf{K})^{-1} [(\mathbf{C} - \Delta t(1-\Theta)\mathbf{K})\mathbf{a}_n - \bar{\mathbf{f}}] \tag{10.14}$$

and it is evident that in general at each step of computation a full equation system needs to be solved (though of course a single semi-inversion is sufficient for linear problems). Methods requiring such an inversion are known as *implicit*. However, when $\Theta = 0$ and the matrix **C** is approximated by its lumped equivalent $\mathbf{C}_L$ the solution, known as *explicit*, is exceedingly cheap.

10.2.2 *Taylor series collocation.* A frequently used alternative to the algorithm presented in the preceding pages is obtained by approximating separately $\mathbf{a}_{n+1}$ and $\dot{\mathbf{a}}_{n+1}$ by truncated Taylor series. We can write, assuming that $\mathbf{a}_n$ and $\dot{\mathbf{a}}_n$ are known:

$$\mathbf{a}_{n+1} \approx \mathbf{a}_n + \Delta t \dot{\mathbf{a}}_n + \Delta t \beta (\dot{\mathbf{a}}_{n+1} - \dot{\mathbf{a}}_n) \tag{10.15}$$

and use collocation satisfying the governing equation at t_{n+1}, or

$$\mathbf{C}\dot{\mathbf{a}}_{n+1} + \mathbf{K}\mathbf{a}_{n+1} + \mathbf{f}_{n+1} = \mathbf{0} \tag{10.16}$$

In the above β is a parameter, $0 \leqslant \beta \leqslant 1$, such that the last term of Eq. (10.15) represents a suitable difference approximation to the truncated expansion.

Substitution of Eq. (10.15) into Eq. (10.16) yields a recurrence relation for $\dot{\mathbf{a}}_{n+1}$:

$$\dot{\mathbf{a}}_{n+1} = (\mathbf{C} + \Delta t \beta \mathbf{K})^{-1} [-\mathbf{K}(\mathbf{a}_n + \Delta t(1-\beta)\dot{\mathbf{a}}_n) - \mathbf{f}_{n+1}] \tag{10.17}$$

where $\mathbf{a}_{n+1}$ is now computed by substitution of Eq. (10.17) into Eq. (10.15).

We remark that:

(*a*) the scheme is not self-starting, requiring the satisfaction of Eq. (10.2) at $t=0$;

(*b*) the computation requires, with identification of the parameters $\beta = \Theta$, an identical equation-solving problem to that in the finite element scheme of Eq. (10.14) and, finally, as we shall see later, stability considerations are identical.

The procedure is introduced here as it has some advantages in non-linear computations which will be shown later.

10.2.3 *Other single-step procedures.* As an alternative to the weighted residual process other possibilities of deriving finite element approximations exist, as discussed in Chapter 9 of Volume 1. For instance, variational principles in time could be established and used for the purpose. This indeed was done in the early approaches to finite element approximation using Hamilton's or Gurtin's variational principles.[8,9,24–27] However, as expected, the final algorithms turn out to be identical. A variant on the above procedures is the use of a least square approximation for minimization of the equation residual.[12,13] This is obtained by insertion of the approximation (10.7) into Eq. (10.2).

The reader can verify that the recurrence algorithm becomes

$$\left(\frac{\mathbf{C}^{\mathrm{T}}\mathbf{C}}{\Delta t}+\frac{\mathbf{K}^{\mathrm{T}}\mathbf{C}+\mathbf{C}^{\mathrm{T}}\mathbf{K}}{2}+\frac{\mathbf{K}^{\mathrm{T}}\mathbf{K}\Delta t}{3}\right)\mathbf{a}_{n+1}$$
$$+\left(\frac{-\mathbf{C}^{\mathrm{T}}\mathbf{C}}{\Delta t}-\frac{\mathbf{K}^{\mathrm{T}}\mathbf{C}-\mathbf{C}^{\mathrm{T}}\mathbf{K}}{2}+\frac{\mathbf{K}^{\mathrm{T}}\mathbf{K}\Delta t}{6}\right)\mathbf{a}_{n}$$
$$+\mathbf{K}^{\mathrm{T}}\frac{1}{\Delta t^2}\int_0^1 \mathbf{f}\tau\,\mathrm{d}\tau+\mathbf{C}^{\mathrm{T}}\frac{1}{\Delta t^2}\int_0^1 \mathbf{f}\,\mathrm{d}\tau=\mathbf{0} \qquad (10.18)$$

requiring a more complex equation solution and always remaining 'implicit'. For this reason the algorithm is largely of purely theoretical interest, though as expected its accuracy is good, as shown in Fig. 10.4, in which a single degree of freedom equation (10.2) is used with

$$\mathbf{K}=\mathbf{C}=1 \qquad \text{and} \qquad \mathbf{f}=0 \qquad \text{with } a_0=1$$

Here, the various algorithms previously discussed are compared. Now we see from this example that the $\Theta=\frac{1}{2}$ algorithm performs almost as well as the least squares one. It is popular for this reason and is known as the Crank–Nicolson scheme after its originators.[28]

10.2.4 *Consistency and approximation error.* For the convergence of any

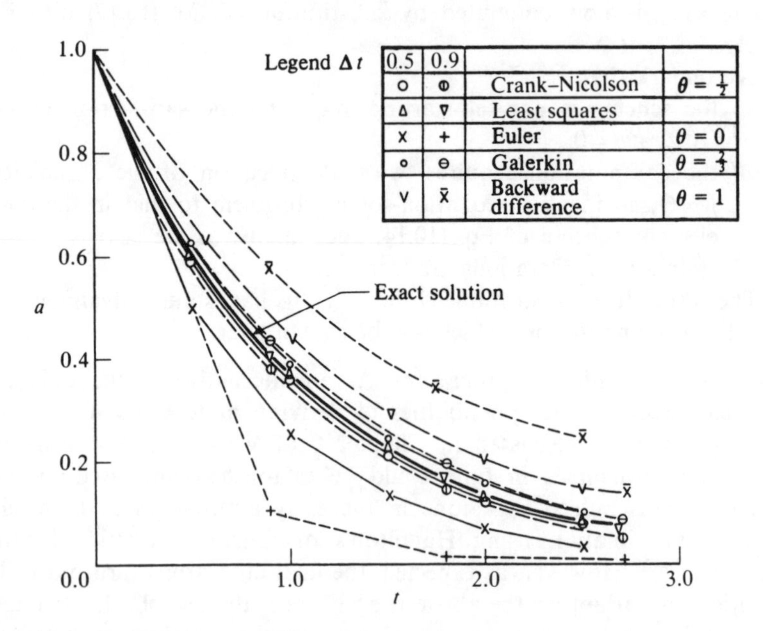

Fig. 10.4 Comparison of various time-stepping schemes on a first-order initial value problem

finite element approximation it is necessary and sufficient that it be *consistent and stable.* We have discussed these two conditions in Volume 1 and introduced appropriate requirements for boundary value problems. In the temporal approximation similar conditions apply though the stability problem is more delicate.

Clearly the function $\mathbf{a}$ itself and its derivatives occurring in the equation have to be approximated with a truncation error of $O(\Delta t)^{\alpha}$, where $\alpha \geqslant 1$ is needed for consistency to be achieved. For the first-order equation (10.2) it is thus necessary to use an approximation polynomial of order $p \geqslant 1$ which is capable of approximating $\dot{\mathbf{a}}$ to $O(\Delta t)$.

The *truncation error in the local approximation* of $\mathbf{a}$ with such an approximation is $O(\Delta t)^2$ and all the algorithms we have presented here using the $p=1$ approximation of Eq. (10.7) will have at least that *local accuracy.*[29] As at a given time, $t = n\Delta t$, the total error can be magnified n times, the final accuracy at a given time for schemes here discussed is of the order $O(\Delta t)$ in general.

We shall see later that the arguments used here lead to $p \geqslant 2$ for the second-order equation (10.1) and that an increase of accuracy can generally be achieved by use of higher-order approximating polynomials.

It would of course be possible to apply such a polynomial increase to the approximating function (10.7) by adding higher-order degrees of freedom. For instance, we could write in place of the original approximation a quadratic expansion:

$$\mathbf{a} \approx \hat{\mathbf{a}}(\tau) = \mathbf{a}_n + \frac{\tau}{\Delta t}(\mathbf{a}_{n+1} - \mathbf{a}_n) + \frac{\tau}{\Delta t}\left(1 - \frac{\tau}{\Delta t}\right)\tilde{\mathbf{a}} \tag{10.19}$$

where $\tilde{\mathbf{a}}$ is a hierarchic internal variable. Obviously now both $\mathbf{a}_{n+1}$ and $\tilde{\mathbf{a}}$ will have to be solved for simultaneously, increasing the size of the problem. The necessary algebraic equations are introduced by using now two sets of weighting functions, W and $\tilde{W}$.

It is of interest to consider the first of these obtained by using the weighting W alone in the manner of Eq. (10.9). The reader will easily verify that we now have to add to Eq. (10.11) a term involving $\tilde{\mathbf{a}}$ which is

$$\left[\frac{\mathbf{C}(1-2\Theta)}{\Delta t} + \mathbf{K}(\Theta - \tilde{\Theta})\right]\tilde{\mathbf{a}} \tag{10.20}$$

where

$$\tilde{\Theta} = \frac{1}{\Delta t^2} \frac{\int_0^{\Delta t} W\tau^2 \, d\tau}{\int_0^{\Delta t} W \, d\tau}$$

It is clear that the choice of $\Theta=\tilde{\Theta}=\frac{1}{2}$ eliminates the quadratic term and regains the previous scheme, thus showing that the values so obtained have a local truncation error of $O(\Delta t)^3$. This explains why the Crank–Nicolson scheme possesses higher accuracy.

In general the addition of higher-order internal variables makes recurrence schemes too expensive and we shall show later how an increase of accuracy can be more economically achieved.

10.2.5 *Stability*. If we consider any of the recurrence algorithms so far derived, we note that for the homogeneous form (i.e. with $\mathbf{f}\equiv 0$) all can be written in the form

$$\mathbf{a}_{n+1}=\mathbf{A}\mathbf{a}_n \tag{10.21}$$

where $\mathbf{A}$ is known as the *amplification matrix*.

The form of this matrix for the first algorithm derived is, for instance, evident from Eq. (10.14) as

$$\mathbf{A}=(\mathbf{C}+\Delta t\Theta\mathbf{K})^{-1}(\mathbf{C}-\Delta t(1-\Theta)\mathbf{K}) \tag{10.22}$$

Any errors present in the solution will of course be subject to amplification by precisely the same factor.

A general solution of any recurrence scheme can be written as

$$\mathbf{a}_{n+1}=\mu\mathbf{a}_n \tag{10.23}$$

and by insertion into Eq. (10.21) we observe that μ is given by eigenvalues of the matrix $\mathbf{A}$ as

$$(\mathbf{A}-\mu\mathbf{I})\mathbf{a}_n=0 \tag{10.24}$$

Clearly if any eigenvalue μ is such that

$$|\mu|>1 \tag{10.25}$$

all initially small errors will increase without limit and the solution will be unstable. In the case of complex eigenvalues the recurrence is modified to the requirement that the modulus of μ satisfy Eq. (10.25).

As the determination of system eigenvalues is a large undertaking it is useful to consider only a scalar equation of the form (10.5) (generally representing, say, one element performance). Thus for the case of the algorithm discussed in Eq. (10.22) we have a scalar A, i.e.

$$A=\frac{c-\Delta t(1-\Theta)k}{c+\Delta t\Theta k}=\frac{1-(\omega\Delta t)(1-\Theta)}{1+\omega\Delta t\Theta}=\mu \qquad \text{where } \omega=\frac{k}{c} \tag{10.26}$$

and μ is evaluated from Eq. (10.24) simply as $\mu=A$ to allow non-trivial $\mathbf{a}_n$. (This is of course equivalent to making the determinant of $[\mathbf{A}-\mu\mathbf{I}]$ zero in a more general case.)

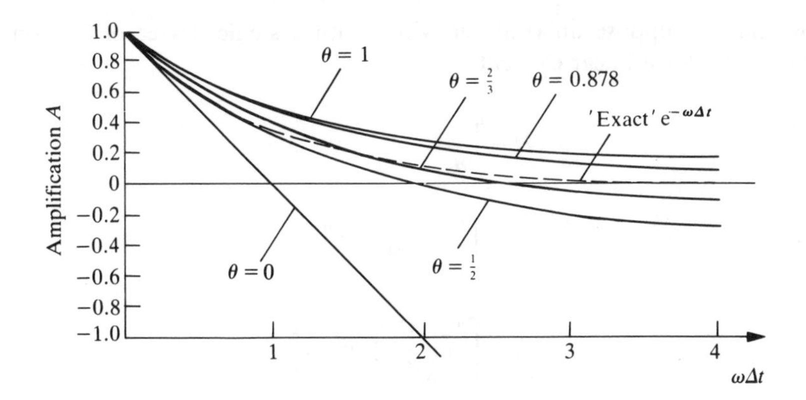

$\theta = 1$ The fully implicit method which decreases too slowly (backward difference)

$\theta = \frac{1}{2}$ Crank–Nicolson or the trapezium rule which gives a higher order of accuracy but can produce oscillations ($\omega \Delta t > 2$)

$\theta = \frac{2}{3}$ A compromise value

$\theta = 0.878$ Liniger algorithm where θ is chosen to minimize the domain error

$\theta = 0$ Explicit, cheap Euler scheme

Fig. 10.5 The amplification A for various versions of the Θ algorithm

In Fig. 10.5 we show how μ (or A) varies with $\omega \Delta t$ for various Θ values. We observe immediately that:

(*a*) for $\Theta \geqslant \frac{1}{2}$:

$$|\mu| \leqslant 1 \tag{10.27a}$$

and such algorithms are *unconditionally stable*;

(*b*) for $\Theta < \frac{1}{2}$ we require

$$\omega \Delta t \leqslant \frac{2}{1 - 2\Theta} \tag{10.27b}$$

for stability. Such algorithms are therefore only *conditionally stable.* Here of course the explicit form with $\Theta = 0$ is typical.

The critical value of Δt below which the scheme is stable with $\Theta < \frac{1}{2}$ needs the determination of the maximum values of μ from a typical element. For instance, in the case of the thermal conduction problem in which we have the coefficients c_{ii} and k_{ii} defined by expressions

$$c_{ii} = \int_{\Omega^i} \tilde{c} N_1^2 \, d\Omega \qquad \text{and} \qquad k_{ii} = \int_{\Omega^i} \nabla N_i \tilde{k} \nabla N_i a \, d\Omega \tag{10.28}$$

we can presuppose uniaxial behaviour with a single degree of freedom and write for a linear element

$$N=\frac{h-x}{h}$$

$$c=\int_0^h \tilde{c}N^2\,\mathrm{d}x=\frac{\tilde{c}h}{3}$$

$$k=\int_0^h \tilde{k}\left(\frac{\mathrm{d}N}{\mathrm{d}x}\right)^2\,\mathrm{d}x=\frac{\tilde{k}}{h}$$

Now

$$\omega=\frac{k}{c}=\frac{3\tilde{k}}{h^2\tilde{c}}$$

This gives

$$\Delta t\leqslant\frac{2}{1-2\Theta}\,\frac{h^2\tilde{c}}{3\tilde{k}}=\Delta t_{\mathrm{crit}} \tag{10.29}$$

which of course means that the smallest element size, $h_{\min}$, dictates overall stability. We note from above that:

(*a*) in first-order problems the critical time step is proportional to h^2 and thus decreases rapidly with element size making explicit computations difficult and

(*b*) if mass lumping is assumed and therefore $c=\tilde{c}h/2$ the critical time step is larger.

In Fig. 10.6 we show the performance of the scheme described in Sec. 10.2.1 for various values of Θ and Δt in the example we have already illustrated in Fig. 10.4, but now using larger values of Δt. We note now that the conditionally stable scheme with $\Theta=0$ and a stability limit of $\Delta t=2$ shows oscillations as this limit is approached ($\Delta t=1.5$) and diverges when exceeded.

Stability computations which were presented for the algorithm of Sec. 10.2.1 can of course be repeated for the other algorithms which we have discussed.

If identical procedures are used, for instance on the algorithm of Sec. 10.2.2, we shall find that the stability conditions, based on the determinant of the amplification matrix $(\mathbf{A}-\mu\mathbf{I})$, are identical with the previous one providing we set $\Theta=\beta$. Algorithms that give such identical determinants will be called, in what follows, *similar*.

In general, it is possible for different amplification matrices $\mathbf{A}$ to have

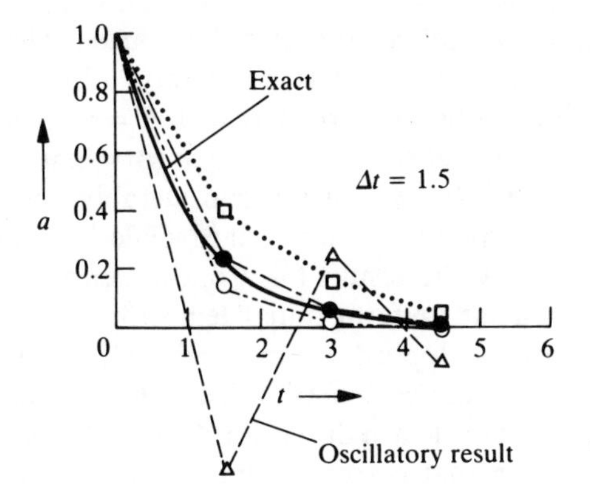

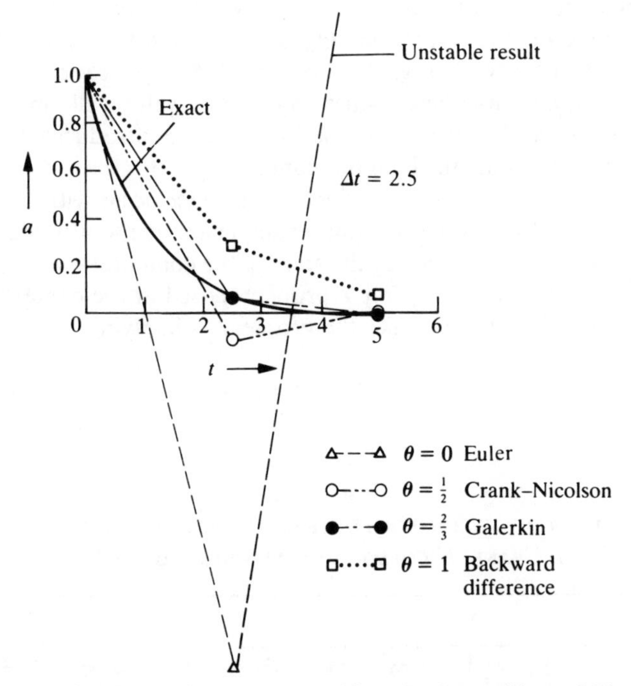

Fig. 10.6 Performance of some Θ algorithms in the problem of Fig. 10.4 and larger time steps. Note oscillation and instability

identical determinants of $(\mathbf{A}-\mu\mathbf{I})$ and hence stability conditions but differ otherwise. If in addition the amplification matrices are the same, schemes are known as *identical*. In the present case, indeed, such an identity can be shown to exist despite different derivations.

10.2.6 *Some further remarks. Initial conditions and examples.* The question of choosing an optimal value of Θ is not always obvious from theoretical accuracy considerations. In particular with $\Theta = \frac{1}{2}$ oscillations are sometimes present,[13] as we observe in Fig. 10.6 ($\Delta t = 2.5$), and for this reason some prefer to use[30] $\Theta = \frac{2}{3}$, which is considerably 'smoothed' (and which incidentally corresponds to a standard Galerkin approximation). In Table 10.1 we show the results for a one-dimensional finite element problem where a bar at uniform initial temperature is subject suddenly to zero temperatures applied at the ends. Here 10 linear elements are used in the space dimension with $L = 1$. The oscillational errors occurring with $\Theta = \frac{1}{2}$ are much reduced for $\Theta = \frac{2}{3}$. The time step used here is much longer than that corresponding to the lowest eigenvalue period, but the main cause of the oscillation is in the abrupt discontinuity of temperature change.

For similar reasons Liniger[31] derives Θ which minimizes the error in the whole time domain and gives $\Theta = 0.878$ for the simple one-dimensional case. We observe in Fig. 10.5 how well the amplification factor fits the exact solution with these values. Again this value will smooth out many oscillations. However, most oscillations are introduced by using simply a physically unrealistic initial condition.

In part at least, the oscillations which for instance occur with $\Theta = \frac{1}{2}$ and $\Delta t = 2.5$ (viz. Fig. 10.6) in the previous example are due to a sudden jump in the forcing term introduced at the start of the computation. This jump is evident if we consider this simple problem posed in the context of the whole time domain. We can take the problem as implying

$$f(t) = -1 \qquad \text{for } t < 0$$

TABLE 10.1

PERCENTAGE ERROR FOR FINITE ELEMENTS IN TIME ($\Theta = \frac{2}{3}$) AND CRANK-NICOLSON ($\Theta = \frac{1}{2}$) SCHEME; $\Delta t = 0.01$

t	$x=0.1$		$x=0.2$		$x=0.3$		$x=0.4$		$x=0.5$	
	$\Theta=\frac{2}{3}$	$\Theta=\frac{1}{2}$	$\Theta=\frac{2}{3}$	$\Theta=\frac{1}{2}$	$\Theta=\frac{2}{3}$	$\Theta=\frac{1}{2}$	$\Theta=\frac{2}{3}$	$\Theta=\frac{1}{2}$	$\Theta=\frac{2}{3}$	$\Theta=\frac{1}{2}$
0.01	10.8	28.2	1.6	3.2	0.5	0.7	0.6	0.1	0.5	0.2
0.02	0.5	3.5	2.1	9.5	0.1	0.0	0.5	0.7	0.7	0.4
0.03	1.3	9.9	0.5	0.7	0.8	3.1	0.5	0.2	0.5	0.6
0.05	0.5	4.5	0.4	0.2	0.5	2.3	0.4	0.8	0.5	1.0
0.10	0.1	1.4	0.1	2.0	0.1	1.5	0.1	1.9	0.1	1.6
0.15	0.3	2.2	0.3	2.1	0.3	2.2	0.3	2.1	0.3	2.2
0.20	0.6	2.6	0.6	2.6	0.6	2.6	0.6	2.6	0.6	2.6
0.30	1.4	3.5	1.4	3.5	1.4	3.5	1.4	3.5	1.4	3.5

giving the solution $u=1$ with a sudden change at $t=0$, resulting in

$$f(t)=0 \qquad \text{for } t \geqslant 0$$

As shown in Fig. 10.7 this represents a discontinuity of the loading function at $t=0$. Although load discontinuities are permitted by the algorithm they lead to sudden discontinuity of u and hence are undesirable. If in place of this discontinuity we assume that f varies linearly in the first time step Δt $(-\Delta t/2 < t < \Delta t/2)$ then smooth results are obtained with a much improved physical representation of the true solution, even for such a long time step as $t=2.5$, as shown in Fig. 10.7.

Similar use of smoothing is illustrated in a multi degree of freedom system (representation of heat conduction in a wall) which is solved using

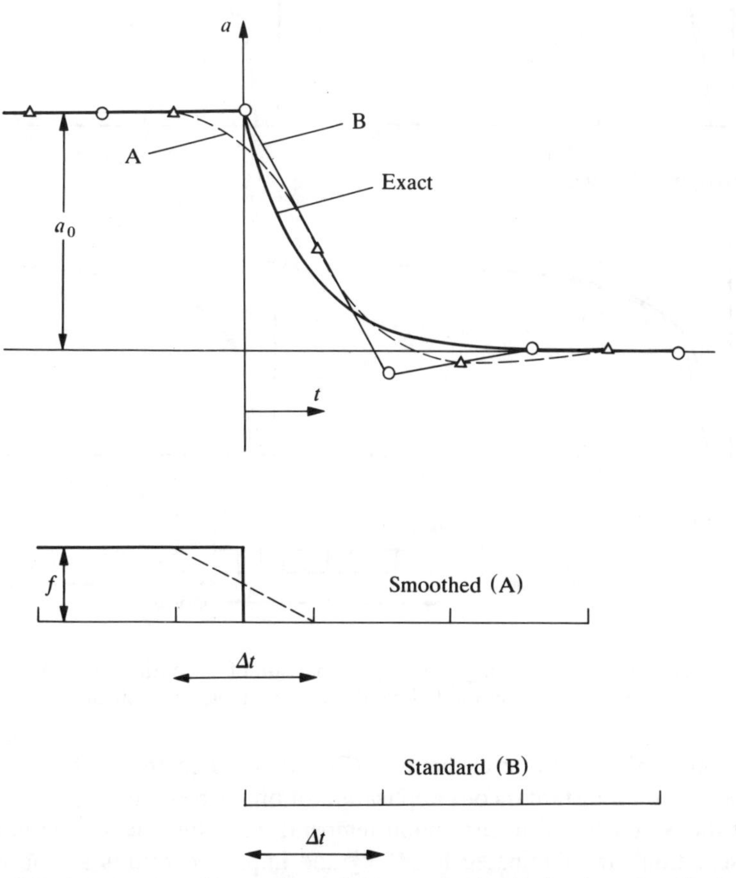

Fig. 10.7 Importance of 'smoothing' the force term in elimination of oscillations in the solution. $\Delta t=2.5$

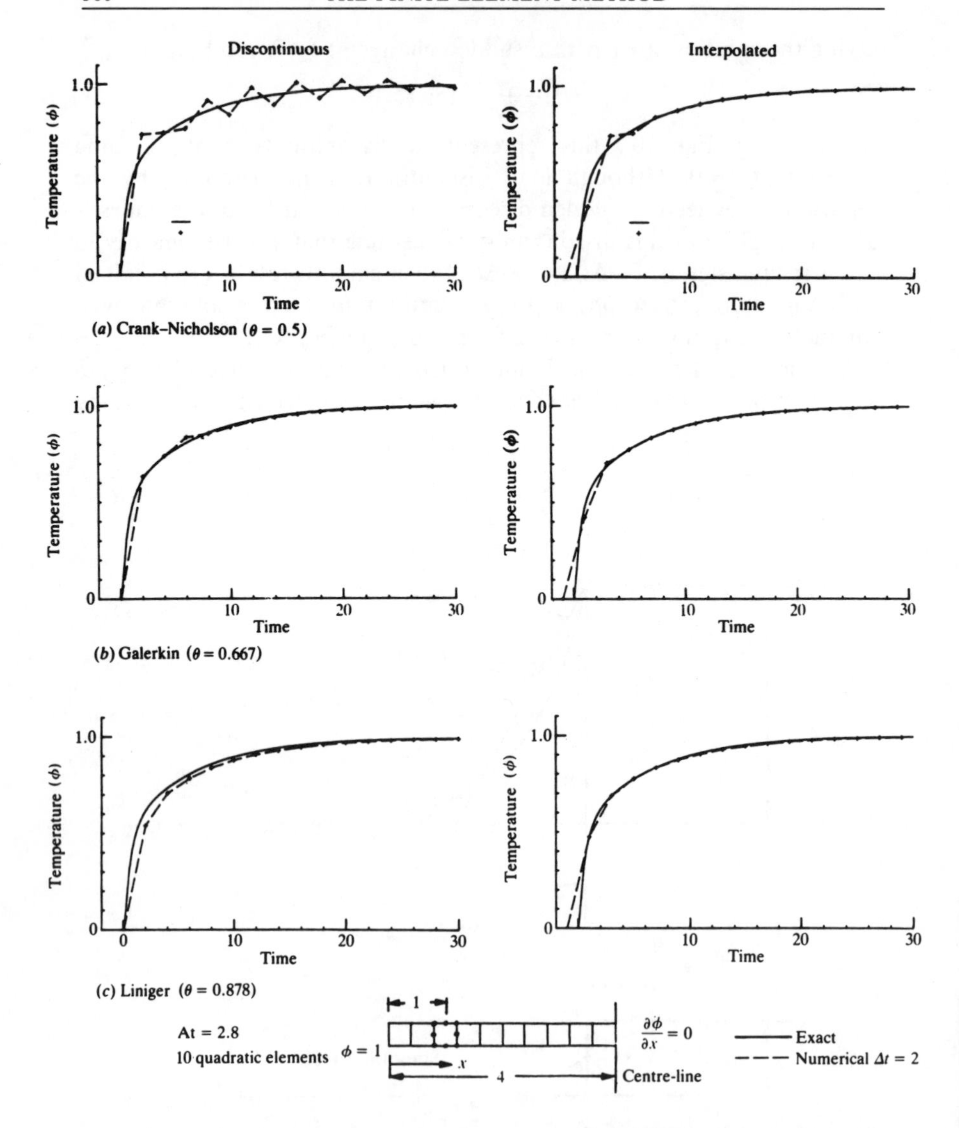

Fig. 10.8 Transient heating of a bar; comparison of discontinuous and interpolated (smoothed) initial conditions for single-step schemes

two-dimensional finite elements[32] (Fig. 10.8). Here the problem corresponds to an instantaneous application of prescribed temperature ($T=1$) at the wall sides with zero initial temperatures. Now again troublesome oscillations are eliminated for $\Theta=\frac{1}{2}$ and improved results are obtained for other values of $\Theta(\frac{2}{3}, 0.878)$ by assuming the step change to be replaced by a continuous one. Such smoothing is always advisable and a continuous representation of the forcing term is important.

We conclude this section by showing a typical example of temperature distribution in an industrial example in which high-order elements are used (Fig. 10.9). Here the equation (10.2) corresponds to the discretized form of the heat conduction equation [viz. Sec. 9.2].

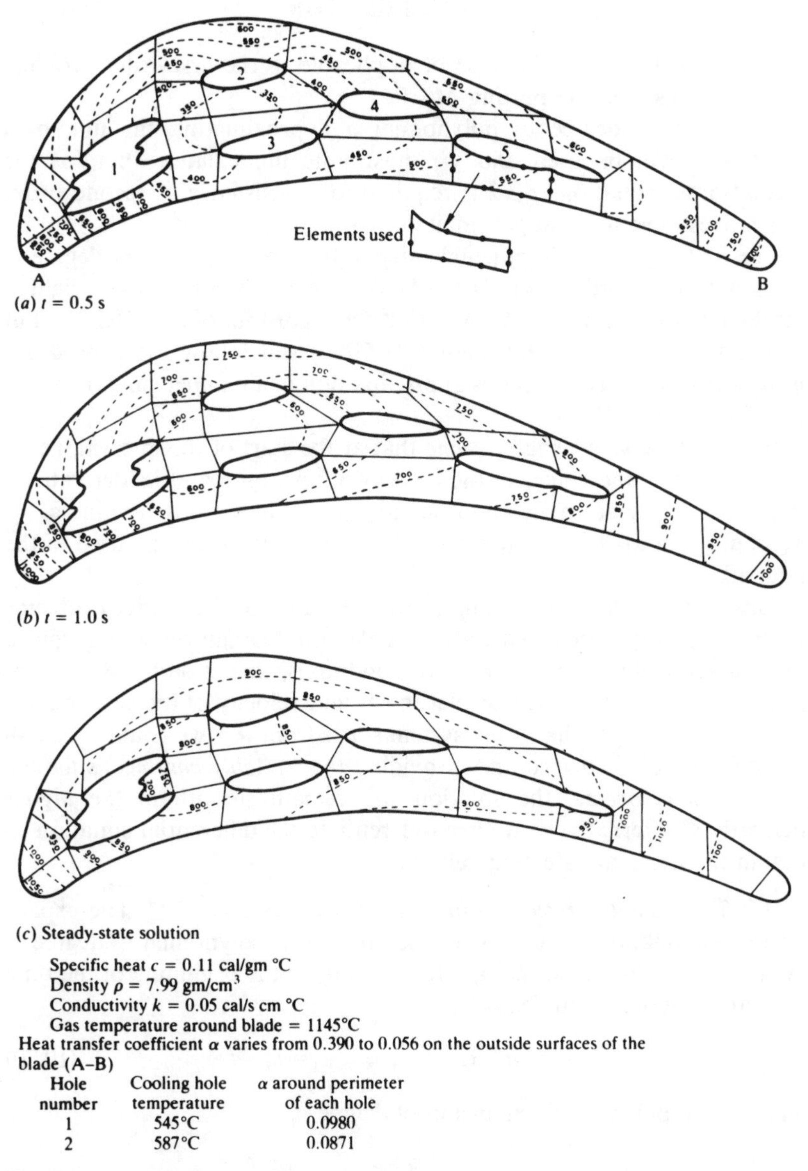

Fig. 10.9 Temperature distribution in a cooled rotor blade, initially at zero temperature

10.3 General single-step algorithms for first- and second-order equations

10.3.1 *Introduction.* We shall introduce in this section two general single-step algorithms applicable to Eq. (10.1):

$$\mathbf{M}\ddot{\mathbf{a}} + \mathbf{C}\dot{\mathbf{a}} + \mathbf{K}\mathbf{a} + \mathbf{f} = \mathbf{0} \tag{10.1}$$

These algorithms will of course be applicable to the first-order problem of Eq. (10.2) simply by putting $M=0$.

An arbitrary degree of polynomial p approximating the unknown function $\mathbf{a}$ will be used and we must note immediately that for the second-order equations $p \geqslant 2$ is required for consistency as second-order derivatives have to be approximated.

The first algorithm SSpj (single step with approximation of degree p, for equations of order $j=1, 2$) will be derived by the use of the weighted residual process and we shall find that the algorithm of Sec. 10.2.1 is but a special case. The second algorithm GNpj will follow the procedures using a truncated Taylor series approximation in a manner similar to that described in Sec. 10.2.2.

In what follows we shall *assume* that at the start of the interval, i.e. at $t=t_n$, we know the values of the unknown function $\mathbf{a}$ and its derivatives, that is $\mathbf{a}_n$, $\dot{\mathbf{a}}_n$, $\ddot{\mathbf{a}}_n$ up to ${}^{p}\bar{\mathbf{a}}_n^{1}$ and our objective will be to determine $\mathbf{a}_{n+1}$, $\dot{\mathbf{a}}_{n+1}$, $\ddot{\mathbf{a}}_{n+1}$ up to ${}^{p}\bar{\mathbf{a}}_{n+1}^{1}$, where p is the order of the expansion used in the interval.

This is indeed a rather strong presumption as for first-order problems we have already stated that only a single initial condition, $\mathbf{a}(0)$, is given, and for second-order problems two conditions, that is $\mathbf{a}(0)$ and $\dot{\mathbf{a}}(0)$, are generally necessary (i.e. the displacement and velocity of the system). We can, however, argue that if the system starts from rest we could take $\mathbf{a}(0)$ to ${}^{p}\bar{\mathbf{a}}^{1}(0)$ as equal to zero and, providing *that suitably continuous forcing of the system occurs*, the solution will remain smooth in the higher derivatives. Alternatively, we can differentiate the differential equation to obtain the necessary starting values.

10.3.2 *The weighted residual (finite element) form SSpj.*[18,19] The expansion of the unknown vector $\mathbf{a}$ will be taken as a polynomial of degree p. With *known* values of $\mathbf{a}_n$, $\dot{\mathbf{a}}_n$, $\ddot{\mathbf{a}}_n$, etc., up to $\mathbf{a}_n$ at the beginning of the time step Δt, we write, as in Sec. 10.2.1,

$$\tau = t - t_n \qquad \Delta t = t_{n+1} - t_n \tag{10.30}$$

and using a polynomial expansion of degree p,

$$\mathbf{a} \approx \mathbf{a}_n + \dot{\mathbf{a}}_n \tau + \frac{\ddot{\mathbf{a}}_n \tau^2}{2!} + \cdots + \frac{\boldsymbol{\alpha}_n^p \tau^p}{p!} \tag{10.31}$$

where the only unknown is the vector $\boldsymbol{\alpha}_n^p$,

$$\boldsymbol{\alpha}_n^p \equiv \overset{p}{\mathbf{a}} \equiv \frac{\mathrm{d}^p}{\mathrm{d}t^p}\mathbf{a} \tag{10.32}$$

which represents some average value of the pth derivative occurring in the interval Δt. The approximation to $\mathbf{a}$ for the case of $p=2$ is shown in Fig. 10.10.

We note that in order to obtain a consistent approximation to all the derivatives that occur in the differential equations (10.1) and (10.2), $p \geqslant 2$ is necessary for the full dynamic equation and $p \geqslant 1$ is necessary for the first-order equation. Indeed, the lowest approximation, that is $p=1$, is the basis of the algorithm derived in the previous section.

The recurrence algorithm will now be obtained by inserting $\mathbf{a}$, $\dot{\mathbf{a}}$ and $\ddot{\mathbf{a}}$ obtained by differentiating Eq. (10.31) into Eq. (10.1) and satisfying the weighted residual equation with a single weighting function $W(t)$. This gives

$$\int_0^{\Delta t} W(t)\left\{\mathbf{M}\left[\ddot{\mathbf{a}}_n + \ddot{\mathbf{a}}\tau + \cdots + \frac{\boldsymbol{\alpha}_n^p \tau^{p-2}}{(p-2)!}\right] + \mathbf{C}\left[\dot{\mathbf{a}}_n + \ddot{\mathbf{a}}\tau + \cdots + \frac{\boldsymbol{\alpha}_n^p \tau^{p-1}}{(p-1)!}\right] + \mathbf{K}\left(\mathbf{a}_n + \dot{\mathbf{a}}_n\tau + \cdots + \frac{\boldsymbol{\alpha}_n^p \tau^p}{p!}\right) + \mathbf{f}\right\} \mathrm{d}\tau = \mathbf{0} \tag{10.33}$$

as the basic equation for determining $\boldsymbol{\alpha}_n^p$.

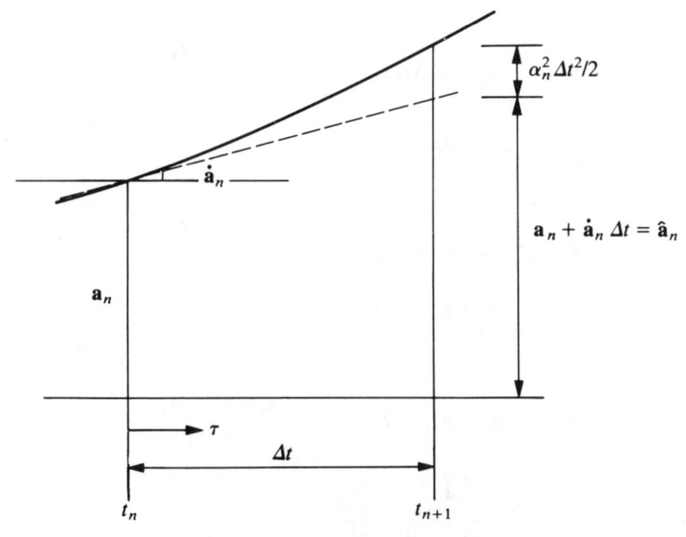

Fig. 10.10 A second-order time approximation

Without specifying the weighting function used we can as in Sec. 10.2.1 generalize its effects by writing [similarly to Eq. (10.10)]

$$\Theta_1 = \frac{\int_0^{\Delta t} W\tau \, \mathrm{d}\tau}{\Delta t \int_0^{\Delta t} W \, \mathrm{d}\tau}$$

to

$$\Theta_p = \frac{\int_0^{\Delta t} W\tau^p \, \mathrm{d}\tau}{\Delta t^p \int_0^{\Delta t} W \, \mathrm{d}\tau}$$

and

$$\bar{\mathbf{f}} = \frac{\int_0^{\Delta t} Wf \, \mathrm{d}\tau}{\int_0^{\Delta t} W \, \mathrm{d}\tau} \qquad \text{with } \Theta_0 \equiv 1 \tag{10.34}$$

Equation (10.33) can now be written more compactly as

$$\mathbf{M}\left[\ddot{\tilde{\mathbf{a}}}_{n+1} + \frac{\boldsymbol{\alpha}_n^p \Theta_{p-2} \Delta t^{p-2}}{(p-2)!}\right] + \mathbf{C}\left[\dot{\tilde{\mathbf{a}}}_{n+1} + \frac{\boldsymbol{\alpha}_n^p \Theta_{p-1} \Delta t^{p-1}}{(p-1)!}\right] + \mathbf{K}\left(\tilde{\mathbf{a}}_{n+1} + \frac{\boldsymbol{\alpha}_n^p \Theta_p \Delta t^p}{p!}\right) + \bar{\mathbf{f}}_n = \mathbf{0} \tag{10.35}$$

where

$$\begin{aligned}
\tilde{\mathbf{a}}_{n+1} &= \mathbf{a}_n + \dot{\mathbf{a}}_n \Delta t \Theta_1 + \cdots + \frac{\overset{p-1}{\mathbf{a}}_n \Delta t^{p-1} \Theta_{p-1}}{(p-1)!} \\
&= \sum_{q=0}^{p-1} \frac{\overset{q}{\mathbf{a}}_n \Delta t^q \Theta_q}{q!} \\
\dot{\tilde{\mathbf{a}}}_{n+1} &= \dot{\mathbf{a}}_n + \ddot{\mathbf{a}}_n \Delta t \Theta_1 + \cdots + \frac{\overset{p-1}{\mathbf{a}}_n \Delta t^{p-2} \Theta_{p-2}}{(p-2)!} \\
&= \sum_{q-1}^{p-1} \frac{\overset{q}{\mathbf{a}}_n \Delta t^{q-1} \Theta_{q-1}}{(q-1)!} \\
\ddot{\tilde{\mathbf{a}}}_{n+1} &= \ddot{\mathbf{a}}_n + \mathbf{a}_n \Delta t \Theta_1 + \cdots + \frac{\overset{p-1}{\mathbf{a}}_n \Delta t^{p-3} \Theta_{p-3}}{(p-3)!} \\
&= \sum_{q=2}^{p-1} \frac{\overset{q}{\mathbf{a}}_n \Delta t^{q-2} \Theta_{q-2}}{(q-2)!}
\end{aligned} \tag{10.36}$$

As $\tilde{\mathbf{a}}_{n+1}$, $\mathring{\mathbf{a}}_{n+1}$ and $\ddot{\mathring{\mathbf{a}}}_{n+1}$ can be computed directly from the initial values we can solve Eq. (10.35) to obtain

$$\boldsymbol{\alpha}_n^p = -\left[\frac{\mathbf{M}\Delta t^{p-2}\Theta_{p-2}}{(p-2)!} + \frac{\mathbf{C}\Delta t^{p-1}\Theta_{p-1}}{(p-1)!} + \frac{\mathbf{K}\Delta t^p\Theta_p}{p!}\right]^{-1}$$
$$\times(\mathbf{M}\ddot{\mathring{\mathbf{a}}}_{n+1} + \mathbf{C}\mathring{\mathbf{a}}_{n+1} + \mathbf{K}\tilde{\mathbf{a}}_{n+1} + \bar{\mathbf{f}}) \qquad (10.37)$$

It is important to observe that $\tilde{\mathbf{a}}_{n+1}$, $\mathring{\mathbf{a}}_{n+1}$ and $\ddot{\mathring{\mathbf{a}}}_{n+1}$ represent here some mean predicted values of $\mathbf{a}$, $\dot{\mathbf{a}}$ and $\ddot{\mathbf{a}}$ in the interval that satisfy the governing Eq. (10.1) in a weighted sense if $\boldsymbol{\alpha}_n^p$ *is chosen as zero.*

The procedure is now complete as the knowledge of the vector $\boldsymbol{\alpha}_n^p$ permits the evaluation of $\mathbf{a}_{n+1}$ to ${}^{p-1}\mathbf{a}_{n+1}$ from the expansion originally used in Eq. (10.31) by putting $\tau = \Delta t$. This gives

$$\mathbf{a}_{n+1} = \mathbf{a}_n + \dot{\mathbf{a}}_n\Delta t + \cdots + \frac{\boldsymbol{\alpha}_n^p\Delta t^p}{p!} = \hat{\mathbf{a}}_{n+1} + \frac{\boldsymbol{\alpha}_n^p\Delta t^p}{p!}$$

$$\dot{\mathbf{a}}_{n+1} = \dot{\mathbf{a}}_n + \ddot{\mathbf{a}}_n\Delta t + \cdots + \frac{\boldsymbol{\alpha}_n^p\Delta t^{p-1}}{(p-1)!} = \hat{\dot{\mathbf{a}}}_{n+1} + \frac{\boldsymbol{\alpha}_n^p\Delta t^{p-1}}{(p-1)!} \qquad (10.38)$$

$$\vdots$$

$${}^{p-1}\mathbf{a}_{n+1} = {}^{p-1}\mathbf{a}_n + \boldsymbol{\alpha}_n^p\Delta t$$

In the above, $\hat{\mathbf{a}}_{n+1}$, $\hat{\dot{\mathbf{a}}}_{n+1}$, etc., are again quantities that can be written down *a priori* (before solving for $\boldsymbol{\alpha}_n^p$). These represent predicted values at the end of the interval with $\boldsymbol{\alpha}_n^p = 0$.

To summarize, the general algorithm necessitates the choice of values of Θ_1 to Θ_p and requires

(*a*) computation of $\tilde{\mathbf{a}}_{n+1}$, $\mathring{\mathbf{a}}_{n+1}$ and $\ddot{\mathring{\mathbf{a}}}_{n+1}$ using the definitions of Eqs (10.36);

(*b*) computation of $\boldsymbol{\alpha}_n^p$ by solution of Eq. (10.37);

(*c*) computation of $\mathbf{a}_{n+1}$ to ${}^{p-1}\mathbf{a}_{n+1}$ by Eqs (10.38).

After completion of stage (*c*) a new time step can be started. In first-order problems computation of $\ddot{\mathring{\mathbf{a}}}_{n+1}$ can obviously be omitted.

If the matrices $\mathbf{C}$ and $\mathbf{M}$ are diagonal the solution of Eq. (10.37) is trivial providing we choose

$$\Theta_p = 0 \qquad (10.39)$$

With this choice the algorithms are *explicit* but, as we shall find later, only sometimes *conditionally stable.*

When $\Theta_p \neq 0$, *implicit* algorithms of various kinds will be available and some of these will be found to be *unconditionally stable.* Indeed, it is such algorithms that are of great practical use.

Important special cases of the general algorithm are the SS11 and SS22 forms given below.

The SS11 algorithm. If we consider the first-order equation (that is $j=1$) it is evident that only the value of $\mathbf{a}_n$ is necessarily specified as the initial value for any computation. For this reason the choice of a linear expansion in the time interval is *natural* ($p=1$) and the SS11 algorithm is for that reason most widely used.

Now the approximation of Eq. (10.31) is simply

$$\mathbf{a}=\mathbf{a}_n+\boldsymbol{\alpha}\tau \qquad (\boldsymbol{\alpha}_n^1=\boldsymbol{\alpha}=\dot{\mathbf{a}}) \tag{10.40}$$

and the approximation to the average satisfaction of Eq. (10.2) is simply

$$\mathbf{C}\boldsymbol{\alpha}+\mathbf{K}(\tilde{\mathbf{a}}_{n+1}+\boldsymbol{\alpha}\Theta\Delta t)+\bar{\mathbf{f}}_n=\mathbf{0} \qquad (\Theta_1=\Theta) \tag{10.41}$$

with

$$\tilde{\mathbf{a}}_{n+1}=\mathbf{a}_n$$

Solution of Eq. (10.41) determines $\boldsymbol{\alpha}$ as

$$\boldsymbol{\alpha}=-(\mathbf{C}+\Delta t\Theta\mathbf{K})^{-1}(\bar{\mathbf{f}}+\mathbf{K}\mathbf{a}_n) \tag{10.42}$$

and finally

$$\mathbf{a}_{n+1}=\mathbf{a}_n+\boldsymbol{\alpha}\Delta t \tag{10.43}$$

The reader will verify that this process is identical to that developed in Eqs (10.7) to (10.11) and hence will not be further discussed except perhaps for noting the more elegant computational form above.

The SS22 algorithm. With Eq. (10.1) we considered a second-order system ($j=2$) in which the necessary initial conditions require the specification of two quantities, $\mathbf{a}_n$ and $\dot{\mathbf{a}}_n$. The simplest and natural choice here is to specify the minimum value of p, that is $p=2$, as this does not require computation of additional derivatives at the start. This algorithm, SS22, is *thus basic for dynamic equations* and we present it here in a full form.

From Eq. (10.31) the approximation is a quadratic

$$\mathbf{a}=\mathbf{a}_n+\dot{\mathbf{a}}_n\tau+\frac{\boldsymbol{\alpha}\tau^2}{2} \qquad (\boldsymbol{\alpha}_n^2=\boldsymbol{\alpha}=\ddot{\mathbf{a}}) \tag{10.44}$$

The approximate form of the ‘average’ dynamic equation is now

$$\mathbf{M}\boldsymbol{\alpha}+\mathbf{C}(\tilde{\dot{\mathbf{a}}}_{n+1}+\boldsymbol{\alpha}\Theta_1\Delta t)+\mathbf{K}\left(\tilde{\mathbf{a}}_{n+1}+\frac{\boldsymbol{\alpha}\Theta_2\Delta t^2}{2}\right)+\bar{\mathbf{f}}=\mathbf{0} \tag{10.45}$$

with predicted, ‘mean’, values

$$\begin{aligned}\tilde{\mathbf{a}}_{n+1}&=\mathbf{a}_n+\dot{\mathbf{a}}_n\Delta t\Theta_1 \\ \tilde{\dot{\mathbf{a}}}_{n+1}&=\dot{\mathbf{a}}_n\end{aligned} \tag{10.46}$$

After evaluation of $\boldsymbol{\alpha}$ from Eq. (10.45), the values of $\mathbf{a}_{n+1}$ are evaluated by Eqs (10.38) which become simply

$$\mathbf{a}_{n+1} = \mathbf{a}_n + \dot{\mathbf{a}}_n \Delta t + \frac{\boldsymbol{\alpha} \Delta t^2}{2}$$
$$\dot{\mathbf{a}}_{n+1} = \dot{\mathbf{a}}_n + \boldsymbol{\alpha} \Delta t \tag{10.47}$$

This completes the algorithm which is of much practical value in solution of dynamic problems.

In many respects it resembles the Newmark algorithm[33] which we shall discuss in the next section and which is widely used in practice. Indeed, its stability properties turn out to be identical with the Newmark algorithm, i.e.

$$\Theta_1 = \gamma \text{ and } \Theta_1 \geqslant \Theta_2 \geqslant \tfrac{1}{2} \qquad \text{for unconditional stability}$$
$$\Theta_2 = 2\beta \tag{10.48}$$

In the above γ and β are conventionally used Newmark parameters.

For $\Theta_2 = 0$ the algorithm is 'explicit' (assuming both $\mathbf{M}$ and $\mathbf{C}$ to be diagonal) and can be made conditionally stable if $\Theta_1 \geqslant \frac{1}{2}$.

The algorithm is clearly applicable to first-order equations described as SS21 and we shall find that the stability conditions are identical.

10.3.3 *Truncated Taylor series collocation algorithm GNpj.* It will be shown that again as in Sec. 10.2.2 a non-self-starting process is obtained, which in most cases, however, gives an algorithm similar to the SSpj one we have derived. The classical Newmark method[33] will be recognized as a particular case together with its derivation process in a form presented generally in existing texts.[34] Because of this similarity we shall term the new algorithm as generalized Newmark (GNpj).

In the derivation we shall now consider the satisfaction of the governing equation (10.1) only at the end-points of the interval Δt (collocation) and write

$$\mathbf{M}\ddot{\mathbf{a}}_{n+1} + \mathbf{C}\dot{\mathbf{a}}_{n+1} + \mathbf{K}\mathbf{a}_{n+1} + \mathbf{f}_{n+1} = \mathbf{0} \tag{10.49}$$

with appropriate approximations for the values of $\mathbf{a}_{n+1}$, $\dot{\mathbf{a}}_{n+1}$ and $\ddot{\mathbf{a}}_{n+1}$.†

† An alternative is to use a weighted residual approach on the governing equation which gives a generalization to Eq. (10.11) of

$$\frac{\mathbf{M}(\dot{\mathbf{a}}_{n+1} - \dot{\mathbf{a}}_n)}{\Delta t} + \mathbf{C}[\dot{\mathbf{a}}_n + \Theta_1(\dot{\mathbf{a}}_{n+1} - \dot{\mathbf{a}}_n)] + \mathbf{K}[\mathbf{a}_n + \Theta_1(\mathbf{a}_{n+1} - \mathbf{a}_n)] + \mathbf{f}_n + \Theta_2(\mathbf{f}_{n+1} - \mathbf{f}_n) = \mathbf{0}$$

This form may be combined with a weighted residual approach as described in reference 16 or with a collocation algorithm as generalized from reference 63. This is similar to the algorithm described in this section although the optimal parameters are usually different.

If we consider a truncated Taylor series expansion similar to Eq. (10.15) for the function $\mathbf{a}$ and its derivatives, we can write

$$
\begin{aligned}
\mathbf{a}_{n+1} &= \mathbf{a}_n + \Delta t \dot{\mathbf{a}}_n + \cdots + \frac{\Delta t^p}{p!} \overset{p}{\mathbf{a}}_n + \beta_p \frac{\Delta t^p}{p!} (\overset{p}{\mathbf{a}}_{n+1} - \overset{p}{\mathbf{a}}_n) \\
\dot{\mathbf{a}}_{n+1} &= \dot{\mathbf{a}}_n + \Delta t \ddot{\mathbf{a}}_n + \cdots + \frac{\Delta t^{p-1}}{(p-1)!} \overset{p}{\mathbf{a}}_n + \beta_{p-1} \frac{\Delta t^{p-1}}{(p-1)!} (\overset{p}{\mathbf{a}}_{n+1} - \overset{p}{\mathbf{a}}_n) \\
&\vdots \\
\overset{p-1}{\mathbf{a}}_{n+1} &= \overset{p-1}{\mathbf{a}}_n + \Delta t \overset{p}{\mathbf{a}}_n + \beta_1 \Delta t (\overset{p}{\mathbf{a}}_{n+1} - \overset{p}{\mathbf{a}}_n) \qquad (\beta_0 = 1)
\end{aligned}
\tag{10.50}
$$

In Eqs (10.50) we have effectively allowed for a polynomial of degree p plus a Taylor series remainder term in each of the expansions for the function and its derivatives with parameters β_j, $j = 1, 2, \ldots, p$, which can be chosen to give good approximation properties to the algorithm.

Insertion of the first three expressions of (10.50) into Eq. (10.49) gives a single equation from which $\overset{p}{\mathbf{a}}_{n+1}$ can be found. When this is determined, $\mathbf{a}_{n+1}$ to $\overset{p-1}{\mathbf{a}}_{n+1}$ can be evaluated using Eqs (10.50). Satisfying Eq. (10.49) is almost a 'collocation' which could be obtained by inserting the expressions (10.50) into a weighted residual form (10.33) with $W = \delta(t_{n+1})$ (the Dirac delta). However, the expansion does not correspond to a unique function for $\mathbf{a}$.

In detail we can write the first three expansions of Eqs (10.50) as

$$
\begin{aligned}
\mathbf{a}_{n+1} &= \bar{\mathbf{a}}_{n+1} + \frac{\Delta t^p}{p!} \beta_p \overset{p}{\mathbf{a}}_{n+1} \\
\dot{\mathbf{a}}_{n+1} &= \dot{\bar{\mathbf{a}}}_{n+1} + \frac{\Delta t^{p-1}}{(p-1)!} \beta_{p-1} \overset{p}{\mathbf{a}}_{n+1} \\
\ddot{\mathbf{a}}_{n+1} &= \ddot{\bar{\mathbf{a}}}_{n+1} + \frac{\Delta t^{p-2}}{(p-2)!} \beta_{p-2} \overset{p}{\mathbf{a}}_{n+1}
\end{aligned}
\tag{10.51}
$$

where the contractions $\bar{\mathbf{a}}$, $\dot{\bar{\mathbf{a}}}$ and $\ddot{\bar{\mathbf{a}}}$ are self-evident. Inserting above into Eq. (10.49) gives

$$
\mathbf{M}\left[\ddot{\bar{\mathbf{a}}}_{n+1} + \beta_{p-2} \frac{\Delta t^{p-2}}{(p-2)!} \overset{p}{\mathbf{a}}_{n+1}\right] + \mathbf{C}\left[\dot{\bar{\mathbf{a}}}_{n+1} + \beta_{p-1} \frac{\Delta t^{p-1}}{(p-1)!} \overset{p}{\mathbf{a}}_{n+1}\right] + \mathbf{K}\left(\bar{\mathbf{a}}_{n+1} + \beta_p \frac{\Delta t^p}{p!} \overset{p}{\mathbf{a}}_{n+1}\right) + \mathbf{f}_{n+1} = 0 \tag{10.52}
$$

Solving the above equation for $\overset{p}{\mathbf{a}}_{n+1}$ we have

$$\overset{p}{\mathbf{a}}_{n+1} = -\left[\frac{\mathbf{M}\Delta t^{p-2}\beta_{p-2}}{(p-2)!} + \frac{\mathbf{C}\Delta t^{p-1}\beta_{p-1}}{(p-1)!} + \frac{\mathbf{K}\Delta t^{p}\beta_{p}}{p!}\right]^{-1} \times (\mathbf{M}\ddot{\bar{\mathbf{a}}}_{n+1} + \mathbf{C}\dot{\bar{\mathbf{a}}}_{n+1} + \mathbf{K}\bar{\mathbf{a}}_{n+1} + \mathbf{f}_{n+1}) \tag{10.53}$$

We note immediately that the above expression is formally identical to that of the SSpj algorithm, Eq. (10.37), if we make

$$\beta_{p-2} = \Theta_{p-2} \qquad \beta_{p-1} = \Theta_{p-1} \qquad \beta_p = \Theta_p \tag{10.54}$$

However, $\bar{\mathbf{a}}_{n+1}$, $\dot{\bar{\mathbf{a}}}_{n+1}$, etc., in the generalized Newmark, GNpj, are not identical to $\tilde{\mathbf{a}}_{n+1}$, $\dot{\tilde{\mathbf{a}}}_{n+1}$, etc., in SSpj. In the SSpj algorithm these represent predicted mean values in the interval Δt while in the GN algorithms they represent predicted values at t_{n+1}.

The computation procedure for the GN algorithms is very similar to that for the SS algorithms, starting now with known values of $\mathbf{a}_n$ to $\overset{p}{\mathbf{a}}_n$. As before we have the given initial conditions and we can usually arrange to use the differential equation and its derivatives to generate higher derivatives of $\mathbf{a}$ at $t=0$. However, the GN algorithm requires more storage and computer operations because of the necessity of retaining and using $\overset{p}{\mathbf{a}}_0$ in the computation of the next time step.

The most important member of this family is of course the GN22 algorithm.

The Newmark algorithm (*GN22*).[33] We have already mentioned this algorithm as it is one of the most popular for dynamic analysis and is known as the Newmark scheme. It is indeed a special case of the general algorithm of the preceding section in which a quadratic ($p=2$) expansion is used, this being the minimum required for second-order problems. We spell out here the details in view of its widespread use.

The expansion of Eq. (10.50) gives

$$\begin{aligned} \mathbf{a}_{n+1} &= \mathbf{a}_n + \Delta t\dot{\mathbf{a}}_n + \frac{\Delta t^2}{2}(1-\beta_2)\ddot{\mathbf{a}}_n + \frac{\Delta t^2}{2}\beta_2\ddot{\mathbf{a}}_{n+1} \\ \dot{\mathbf{a}}_{n+1} &= \dot{\mathbf{a}}_n + \Delta t(1-\beta_1)\ddot{\mathbf{a}}_n + \Delta t\beta_1\ddot{\mathbf{a}}_{n+1} \end{aligned} \tag{10.55}$$

This, together with the dynamic equation (10.49),

$$\mathbf{M}\ddot{\mathbf{a}}_{n+1} + \mathbf{C}\dot{\mathbf{a}}_{n+1} + \mathbf{K}\mathbf{a}_{n+1} + \mathbf{f}_{n+1} = \mathbf{0} \tag{10.56}$$

allows the three unknowns $\ddot{\mathbf{a}}_{n+1}$, $\dot{\mathbf{a}}_{n+1}$ and $\mathbf{a}_{n+1}$ to be determined.

It is convenient to proceed as we have already indicated and solve first for $\ddot{\mathbf{a}}_{n+1}$ by substituting (10.55) into (10.56). This yields as the first step the equivalent of Eq. (10.53):

$$\ddot{\mathbf{a}}_{n+1} = -\left(\mathbf{M} + \mathbf{C}\beta_1 \Delta t + \frac{\mathbf{K}\beta_2 \Delta t^2}{2}\right)^{-1} \times \left\{\mathbf{f}_{n+1} + \mathbf{C}[\dot{\mathbf{a}}_n + \Delta t(1-\beta_1)\ddot{\mathbf{a}}_n] + \mathbf{K}\left[\mathbf{a}_n + \Delta t \dot{\mathbf{a}}_n + \frac{\Delta t^2}{2}(1-\beta_2)\ddot{\mathbf{a}}_n\right]\right\} \tag{10.57}$$

After this step the values of $\dot{\mathbf{a}}_{n+1}$ and $\mathbf{a}_{n+1}$ can be found by simple vector operations by using Eqs (10.55).

Some users prefer a slightly more complicated procedure in which the first unknown determined is $\mathbf{a}_{n+1}$. This results in

$$\mathbf{a}_{n+1} = \left(\mathbf{M} + \mathbf{C}\beta_1 \Delta t + \frac{\mathbf{K}\beta_2 \Delta t^2}{2}\right)^{-1} \times \left\{\frac{-\beta_2 \Delta t^2}{2}\mathbf{f}_{n+1} - \mathbf{C}\left[\frac{\beta_2 \Delta t^2}{2}(\dot{\mathbf{a}}_n + \Delta t(1-\beta_1)\ddot{\mathbf{a}}_n) - \beta_1 \Delta t\left(\mathbf{a}_n + \Delta t \dot{\mathbf{a}}_n + \frac{\Delta t^2}{2}(1-\beta_2)\ddot{\mathbf{a}}_n\right)\right] + \mathbf{M}\left[\mathbf{a}_n + \Delta t \dot{\mathbf{a}}_n + \frac{\Delta t^2}{2}(1-\beta_2)\ddot{\mathbf{a}}_n\right]\right\} \tag{10.58}$$

which again on using Eqs (10.55) gives $\dot{\mathbf{a}}_{n+1}$ and $\ddot{\mathbf{a}}_{n+1}$. The inversion is here identical but vector operations are slightly simplified.†

We shall return to the problem of stability later but will anticipate here the result, which is that stability is unconditional for precisely the same values as those of Eq. (10.48) if we identify $\beta_1 = \Theta_1$ and $\beta_2 = \Theta_2$. Indeed, the stability requirements for SS22 in GN22 prove to be identical though the algorithms are only similar.

As in the general case, $\beta_2 = 0$ produces an explicit algorithm whose solution is very simple if **M** and **C** are assumed diagonal.

† Some programs write Eq. (10.58) in yet another alternative form as

$$\mathbf{a}_{n+1} = \left(\frac{2\mathbf{M}}{\beta_2 \Delta t^2} + \frac{2\mathbf{C}\beta_1}{\beta_2 \Delta t} + \mathbf{K}\right)^{-1} \left\{\mathbf{f}_{n+1} - \cdots\right.$$

This has the merit of yielding immediately the static solutions as the inversion concerns a 'stiffness' matrix. However, this form is not desirable as it does not permit the explicit, $\beta_2 = 0$, form to be used.

It is of interest to remark that accuracy can be slightly improved and yet the advantages of the explicit form preserved for algorithms SS/GN by a simple iterative process. In this (for the GN algorithm for instance) we predict $\mathbf{a}_{n+1}^i$, $\dot{\mathbf{a}}_{n+1}^i$ and $\ddot{\mathbf{a}}_{n+1}^i$ using expressions (10.51) with

$$(\overset{p}{\mathbf{a}}_{n+1})^{i-1}$$

setting for $i=1$,

$$(\overset{p}{\mathbf{a}}_{n+1})^0 = 0$$

This is followed by rewriting the governing equation (10.52) as

$$\mathbf{M}\left[(\ddot{\mathbf{a}}_{n+1})^{i-1} + \beta_{p-2}\frac{\Delta t^{p-2}}{(p-2)!}(\overset{p}{\mathbf{a}}_{n+1})^i\right] + \mathbf{C}\mathbf{a}_{n+1}^{i-1} + \mathbf{K}\ddot{\mathbf{a}}_{n+1}^{i-1} + \mathbf{f}_{n+1} = \mathbf{0} \quad (10.59)$$

and solving for $(\overset{p}{\mathbf{a}}_{n+1})^i$.

This predictor/corrector iteration has been successfully used for various algorithms, though of course the stability conditions remain unaltered from those of a fully explicit scheme.[35]

10.3.4 *Stability of general algorithms. Consistency* of the general algorithms of SS and GN type is self-evident and assured by their formulation.

In a similar manner to that used in Sec. 10.2.5 we can conclude from this that the *local truncation error* is $O(\Delta t)^{p+1}$ as the expansion contains all terms up to τ^p. However, the total truncation error after n steps is $O(\Delta t)^p$ only for the first-order equation system, becoming $O(\Delta t)^{p-1}$ for the second. Details of accuracy discussions and reasons for this can be found in reference 6.

The question of stability is paramount, and in this section we shall discuss it in detail for the SS type of algorithms. The establishment of similar conditions for the GN algorithm follows precisely the same pattern and can be left as an exercise to the reader. It is, however, important to remark here that it can be shown that

(*a*) the SS and GN are similar generally in performance and
(*b*) *their stability conditions are identical* (assuming $\Theta_p \equiv \beta_p$).

The proof of the last statement requires some elaborate algebra and is given in reference 6.

The determination of stability requirements follows precisely the pattern outlined in Sec. 10.2.5. However, for practical reasons we shall

(*a*) avoid writing explicitly the amplification matrix $\mathbf{A}$;
(*b*) immediately consider the scalar equation system implying modal decomposition and no forcing, i.e.

$$m\ddot{a} + c\dot{a} + ka = 0 \quad (10.60)$$

Equations (10.35), (10.36) and (10.38) written in scalar terms define the recurrence algorithms. For the homogeneous case the general solution can be written down as

$$
\begin{aligned}
a_{n+1} &= \mu a_n \\
\dot{a}_{n+1} &= \mu \dot{a}_n \\
&\vdots \\
\overset{p-1}{a}_{n+1} &= \mu^{p} \overset{p-1}{a}_n
\end{aligned}
\tag{10.61}
$$

and substitution of the above into the equations governing the recurrence can be written quite generally as

$$
\mathbf{S}\mathbf{X}_n = \mathbf{0} \tag{10.62}
$$

where

$$
\mathbf{X}_n = \begin{Bmatrix} a_n \\ \Delta t \dot{a}_n \\ \vdots \\ \Delta t^{p-1} \overset{p-1}{a}_n \\ \Delta t^{p} \alpha_n \end{Bmatrix} \tag{10.63}
$$

The matrix $\mathbf{S}$ is given below in a compact form which can be verified by the reader:

$$
\mathbf{S} = \begin{bmatrix}
b_0, & b_1, & b_2, \ldots, & b_{p-1}, & b_p \\
1-\mu, & 1, & 1/2!, \ldots, & \dfrac{1}{(p-1)!}, & \dfrac{1}{p!} \\
0, & 1-\mu, & 1, \ldots, & \dfrac{1}{(p-2)!}, & \dfrac{1}{(p-1)!} \\
\cdots & \cdots & \cdots & \cdots & \cdots \\
\cdots & \cdots & \cdots & \cdots & \cdots \\
\cdots & \cdots & \cdots & \cdots & \cdots \\
0, & 0, & 0, \ldots, 1-\mu, & 1, & 1/2! \\
0, & 0, & 0, \ldots, 0, & 1-\mu, & 1
\end{bmatrix} \tag{10.64}
$$

where $b_0 = \Theta_0 k \Delta t^2$

$$b_1 = \Theta_0 c \Delta t + \Theta_1 k \Delta t^2, \quad \Theta_0 = 1$$

$$b_q = \frac{\Theta_{q-2} m}{(q-2)!} + \frac{\Theta_{q-1} c \Delta t}{(q-1)!} + \frac{\Theta_q k \Delta t^2}{q!}, \qquad q = 2, 3, \ldots, p$$

For non-trivial solutions for the vector $\mathbf{X}_n$ to exist it is necessary for the determinant of $\mathbf{S}$ to be zero:

$$\det \mathbf{S}=0 \tag{10.65}$$

This provides the *characteristic polynomial* for μ of order p, yielding the eigenvalues of the amplification matrix, and for stability it is sufficient and necessary that the moduli of all eigenvalues [viz. Eq. (10.25)]

$$|\mu|\leqslant 1 \tag{10.66}$$

We remark that in the case of repeated roots, the equality sign does not apply. The reader will have noticed that the direct derivation of the determinant of $\mathbf{S}$ is much simpler than writing down matrix $\mathbf{A}$ and finding its eigenvalues, but the results are, of course, identical.

The calculation of stability limits, even with the scalar (modal) equation system, is non-trivial and for this reason we shall only do it for $p=2$ and $p=3$ in what follows. However, two general procedures will be introduced here.

The first of these is the so-called *z transformation.* In this we use a change of the variable in the polynomial, putting

$$\mu=\frac{1+z}{1-z} \tag{10.67}$$

where z as well as μ are in general complex numbers. It is easy to show that the requirement of Eq. (10.66) is identical to that demanding the *real part of z to be negative* (viz. Fig. 10.11).

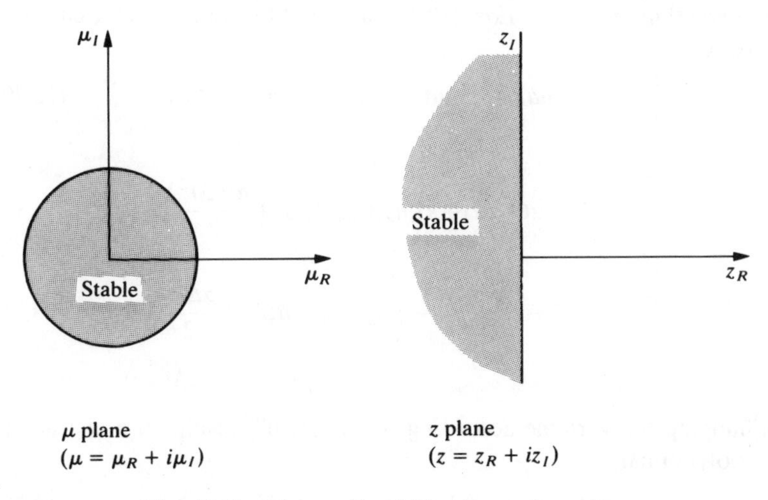

Fig. 10.11 The $\mu=(1+z)/(1-z)$ transformation

The second is the well-known Routh–Hurwitz condition[36–38] which states that for a polynomial with $c_0 > 0$

$$c_0 z^n + c_1 z^{n+1} + \cdots + c_n = 0 \tag{10.68}$$

the real part of all roots will be negative if, for $c_1 > 0$,

$$\det\begin{bmatrix} c_1 & c_3 \\ c_0 & c_2 \end{bmatrix} > 0$$

$$\det\begin{bmatrix} c_1 & c_3 & c_5 \\ c_0 & c_2 & c_4 \\ 0 & c_1 & c_3 \end{bmatrix} > 0 \tag{10.69a}$$

and generally

$$\det\begin{bmatrix} c_1 & c_3 & c_5 & \cdots & \\ c_0 & c_2 & c_4 & \cdots & \\ 0 & c_1 & c_3 & \cdots & \\ 0 & 0 & c_2 & \cdots & \\ \cdots & \cdots & \cdots & \cdots & \\ 0 & & & & c_n \end{bmatrix} > 0 \tag{10.69b}$$

With these tools in hand we can discuss in detail the stability of specific algorithms.

10.3.5 *Stability of the SS22/SS21 algorithms.* The recurrence relations for the algorithm given in Eqs (10.45) and (10.47) can be written after inserting

$$a_{n+1} = \mu a_n \qquad \text{and} \qquad \dot{a}_{n+1} = \mu \dot{a}_n \qquad f = 0 \tag{10.70}$$

as

$$\begin{aligned} m\alpha + c(\dot{a}_n + \alpha\Theta_1 \Delta t) + k\left(a_n + \dot{a}_n \Theta_1 \Delta t + \frac{\alpha\Theta\Delta t^2}{2}\right) &= 0 \\ -\mu a_n + a_n + \dot{a}_n \Delta t + \frac{\alpha \Delta t^2}{2} &= 0 \\ -\mu \dot{a}_n + \dot{a}_n + \alpha\Delta t &= 0 \end{aligned} \tag{10.71}$$

Changing the variable according to Eq. (10.67) results in a characteristic polynomial

$$c_0 z^2 + c_1 z + c_2 = 0 \tag{10.72}$$

with

$$
\begin{aligned}
c_0 &= 4m + 2(2\Theta_1 - 1)c\Delta t + 2(\Theta_2 - \Theta_1)k\Delta t^2 \\
c_1 &= 2c\Delta t + k\Delta t^2(2\Theta_1 - 1) \\
c_2 &= k\Delta t^2
\end{aligned}
$$

The Routh–Hurwitz requirement for stability is simply that

$$
c_0 > 0 \qquad c_1 \geqslant 0 \qquad \det\begin{bmatrix} c_1 & 0 \\ c_0 & c_2 \end{bmatrix} > 0
$$

or simply

$$
c_0 \geqslant 0 \qquad c_1 \geqslant 0 \qquad c_2 \geqslant 0 \tag{10.73}
$$

These inequalities give for *unconditional stability* the condition that

$$
\Theta_2 \geqslant \Theta_1 \geqslant 0.5 \tag{10.74}
$$

This condition is also generally valid when $m=0$, i.e. for the SS21 algorithm (the first-order equation) though now $\Theta_2 = \Theta_1$ must be excluded.

It is possible of course to satisfy the inequalities (10.73) only at some values of Δt yielding conditional stability. For the explicit process $\Theta_2 = 0$ with SS22/SS21 algorithms the inequalities (10.73) demand that

$$
\begin{aligned}
2m + (2\Theta_1 - 1)c\Delta t - \Theta_1 k\Delta t^2 &\geqslant 0 \\
2c + (2\Theta_1 - 1)k\Delta t &\geqslant 0
\end{aligned} \tag{10.75}
$$

The second one is satisfied whenever

$$
\Theta_1 \geqslant 0.5 \tag{10.76}
$$

and for $\Theta_1 = 0.5$ the first supplies the requirement that

$$
\Delta t^2 \leqslant \frac{4m}{k} \tag{10.77}
$$

The last condition does not permit an explicit scheme for SS21, i.e. when $m=0$. Here, however, if we take $\Theta_1 > 0.5$ we have from the first equation of Eq. (10.75)

$$
\Delta t < \frac{c}{k}\,\frac{2\Theta_1 - 1}{\Theta_1} \tag{10.78}
$$

It is of interest for problems of structural dynamics to consider the nature of the bounds in an elastic situation. Here we can use the same process as that described on page 355 for thermal, first-order, problems of conduction. Looking at a single element with a single degree of freedom yields in place of condition (10.77) (for consistent mass matrices)

$$t \leqslant \left(\frac{2}{\sqrt{3}}\right)\frac{h}{C} = \Delta t_{\text{crit}}$$

where h is the element size and

$$C = \sqrt{\frac{E}{\xi}}$$

is the speed of elastic wave propagation. (For lumped matrices the factor becomes $2/\sqrt{2}$.)

Once again the smallest element size governs the stability but it is interesting to note that in problems of dynamics the critical time step is proportional to h while, as shown in Eq. (10.29), for first-order problems it is proportional to h^2. Clearly for decreasing mesh size explicit schemes in dynamics are more realistic than in thermal analysis and, as we shall see later, are exceedingly popular.

10.3.6 *Stability of various higher-order schemes and equivalence with some known alternatives.* Identical stability considerations as those described in previous sections can be applied to SS32/31 and higher-degree approximations. We omit here the algebra and simply quote some results.[6]

SS32/31. Here for zero damping ($C=0$) in SS32 we require for unconditional stability that

$$\Theta_1 > \tfrac{1}{2} \qquad \Theta_2 \geqslant \Theta_1 + \tfrac{1}{6} \qquad \Theta_3 \geqslant \tfrac{3}{2} \qquad \Theta_2 \geqslant \tfrac{1}{4} \tag{10.79}$$

and

$$3\Theta_1\Theta_2 - 3\Theta_1^2 + \Theta_1 \geqslant \Theta_3$$

For first-order problems ($m=0$), i.e. SS31, the first requirements as in dynamics are identical but the last one becomes

$$3\Theta_1\Theta_2 - 3\Theta_1^2 + \Theta_1 \geqslant \Theta_3 - \frac{[6\Theta_1(\Theta_1 - 1) + 1]^2}{9(2\Theta_1 - 1)} \tag{10.80}$$

With $\Theta_3 = 0$, i.e. an explicit scheme when $c=0$,

$$\Delta t^2 \leqslant \frac{m}{k}\,\frac{12(2\Theta_1 - 1)}{6\Theta_2 - 1} \tag{10.81a}$$

and when $m=0$,

$$\Delta t \leqslant \frac{c}{k}\,\frac{\Theta_2 - \Theta_1}{6\Theta_2 - 1} \tag{10.81b}$$

SS42/41. For this (and indeed higher orders) unconditional stability in dynamic problems $m \neq 0$ does not exist. This is a consequence of a theorem of Dahlquist.[39] The SS41 scheme can have unconditional stability but the general expressions for this are cumbersome. We quote one example that is unconditionally stable:

$$\Theta_1 = \tfrac{5}{2} \qquad \Theta_2 = \tfrac{35}{6} \qquad \Theta_3 = \tfrac{25}{2} \qquad \Theta_4 = 24$$

This set of values corresponds to a backward difference four-step algorithm of Gear.[40]

It is of general interest to remark that certain members of the SS (or GN) families of algorithms are similar in performance and identical in the stability (and hence recurrence) properties to others published in the large literature on the subject. Each algorithm claims particular advantages and properties. In Tables 10.2 to 10.4 we show some members of this

TABLE 10.2
SS21 EQUIVALENTS

Algorithms	Theta values
Zlamal[30]	$\Theta_1 = \frac{5}{6}$, $\Theta_2 = 2$
Gear[40]	$\Theta_1 = \frac{3}{2}$, $\Theta_2 = 2$
Liniger[41]	$\Theta_1 = 1.0848$, $\Theta_2 = 1$
Liniger[41]	$\Theta_1 = 1.2184$, $\Theta_2 = 1.292$

TABLE 10.3
SS31 EQUIVALENTS

Algorithm	Theta values
Gear[40]	$\Theta_1 = 2$, $\Theta_2 = \frac{11}{3}$, $\Theta_3 = 6$
Liniger[41]	$\Theta_1 = 1.84$, $\Theta_2 = 3.07$, $\Theta_3 = 4.5$
Zlamal[30]	$\Theta_1 = 0.8$, $\Theta_2 = 1.03$, $\Theta_3 = 1.29$

TABLE 10.4
SS32 EQUIVALENTS[47]

Algorithm	Theta values
Houbolt[42]	$\Theta_1 = 2$, $\Theta_2 = \frac{11}{3}$, $\Theta_3 = 6$
Wilson[43,44]	$\Theta_1 = \Theta$, $\Theta_2 = \Theta^2$, $\Theta_3 = \Theta^3$ ($\Theta = 1.4$ common)
Bossak–Newmark[45] ($m\ddot{x} + kx = 0$, $\gamma_B = \frac{1}{2} - \alpha_B$)	$\Theta_1 = 1 - \alpha_B$ $\Theta_2 = \frac{2}{3} - \alpha_B + 2\beta_B$ $\Theta_3 = 6\beta_B$
Bossak–Newmark[45] ($m\ddot{x} + c\dot{x} + kx = 0$, $\gamma_B = \frac{1}{2} - \alpha_B$, $\beta_B = \frac{1}{6} - \alpha_B/2$)	$\Theta_1 = 1 - \alpha_B$ $\Theta_2 = 1 - 2\alpha_B$ $\Theta_3 = 1 - 3\alpha_B$
Hilber–Hughes–Taylor[46] ($m\ddot{x} + kx = 0$, $\gamma_H = \frac{1}{2} - \alpha_H$)	$\Theta_1 = 1$ $\Theta_2 = \frac{2}{3} + 2\beta_H - 2\alpha_H^2$ $\Theta_3 = 6\beta_H(1 + \alpha_H)$

family.[30,39–46] Clearly many more algorithms that are applicable are present in the general formulae and a study of their optimality is yet to be undertaken.

We remark here that identity of stability and recurrence always occurs with multistep algorithms, which we shall discuss in the next section.

MULTISTEP METHODS

10.4 Multistep recurrence algorithms

10.4.1 *Introduction.* In the previous sections we have been concerned with recurrence algorithms valid within a single time step and relating the values of $\mathbf{a}_{n+1}$, $\dot{\mathbf{a}}_{n+1}$, $\ddot{\mathbf{a}}_{n+1}$ to $\mathbf{a}_n$, $\dot{\mathbf{a}}_n$, $\ddot{\mathbf{a}}_n$, etc. It is possible to derive, using very similar procedures to those previously introduced, multistep algorithms in which we relate $\mathbf{a}_{n+1}$ to values of $\mathbf{a}_n$, $\mathbf{a}_{n-1}$, $\mathbf{a}_{n-2}$, etc., without introducing explicitly the derivatives and assuming that each set is separated by an equal interval Δt. Such algorithms are in general less convenient to use than the single-step processes as they do not permit an easy change of the time step magnitude. However, much classical work on stability and accuracy has been introduced on such multistep algorithms and hence they deserve a mention.

We shall show in this section that a series of such algorithms may be simply derived using the weighted residual process and that this set possesses identical stability and accuracy properties as the SSpj procedures.

10.4.2 *The approximation procedure for a general multistep algorithm.* As in Sec. 10.3.2 we shall approximate the function $\mathbf{a}$ of the second-order equation

$$\mathbf{M}\ddot{\mathbf{a}} + \mathbf{C}\ddot{\mathbf{a}} + \mathbf{K}\mathbf{a} = \mathbf{f} \tag{10.82}$$

by a polynomial expansion of the order p, containing now a single unknown $\mathbf{a}_{n+1}$. This polynomial assumes the value of $\mathbf{a}_n$, $\mathbf{a}_{n-1}, \ldots, \mathbf{a}_{n-p+1}$ at appropriate times (Fig. 10.12). We can write this polynomial as

$$\mathbf{a} = \sum_{j=-p+1}^{1} N_j \mathbf{a}_{n+j} \tag{10.83}$$

where

$$N_j = N_j(\xi) = \prod_{\substack{k=-p+1 \\ k \neq j}}^{1} \frac{\xi - k}{i - k} \tag{10.84}$$

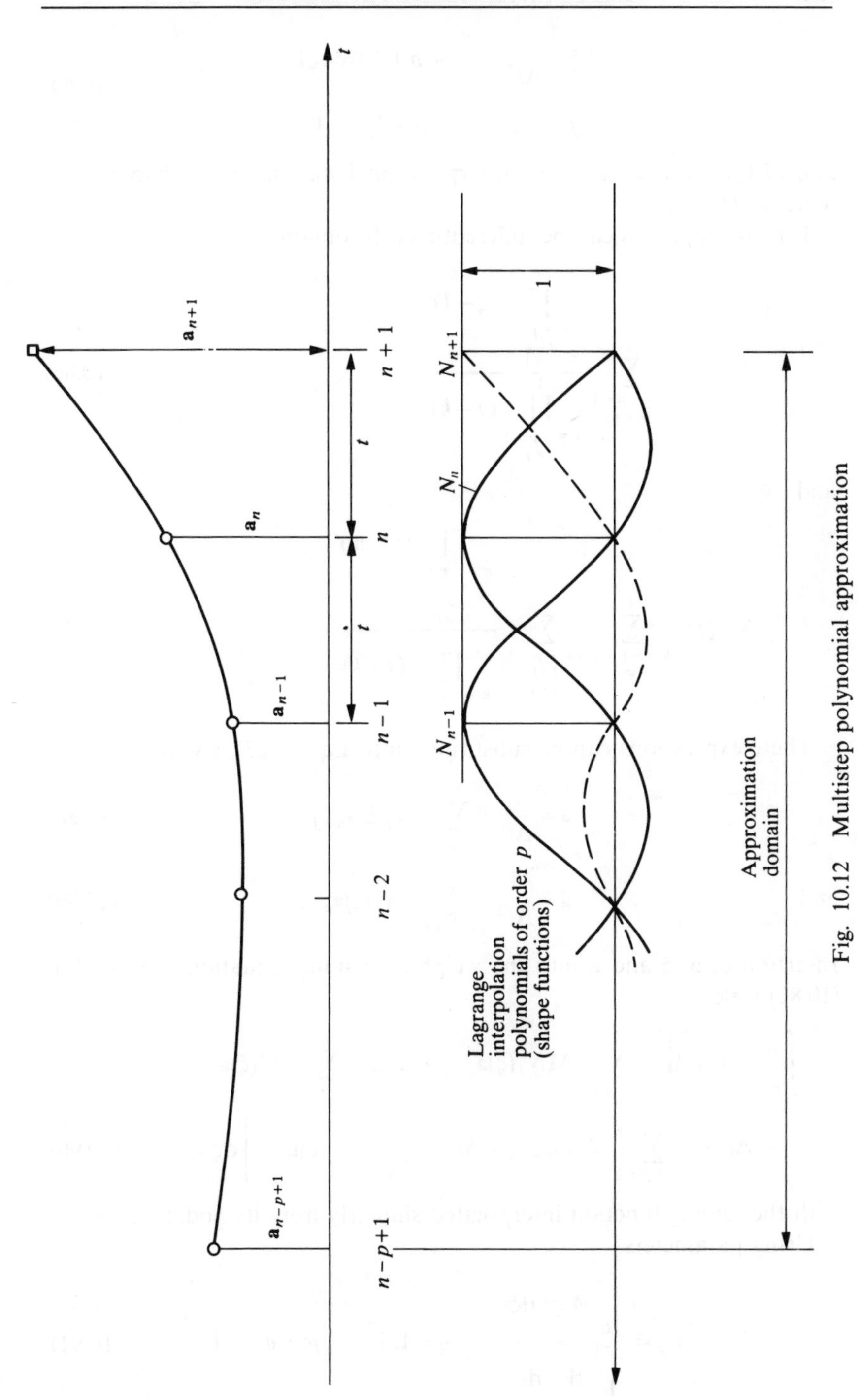

Fig. 10.12 Multistep polynomial approximation

$$\xi = \frac{\tau}{\Delta t}, \qquad -p+1 \leqslant \xi \leqslant 1$$
$$j = -p+1,\ -p+2, \ldots, 0,\ 1 \tag{10.85}$$

and $N_j(\xi)$ are the Lagrange interpolation functions (viz. Chapter 7 of Volume 1).

Equations (10.84) can be differentiated to obtain

$$N'_j(\xi) = \sum_{\substack{k=-p+1 \\ k \neq j}}^{1} \frac{\prod\limits_{\substack{l=-p+1 \\ l \neq k \\ l \neq j}}^{1} (\xi - l)}{\prod\limits_{\substack{k=-p+1 \\ k \neq j}}^{1} (j-k)} \tag{10.86}$$

and

$$N''_j(\xi) = \sum_{\substack{k=-p+1 \\ k \neq j}}^{1} \sum_{\substack{l=-p+1 \\ l \neq k \\ l \neq j}}^{1} \frac{\prod\limits_{\substack{m=-p+1 \\ m \neq k \\ m \neq j \\ m \neq l}}^{1} (\xi - m)}{\prod\limits_{\substack{k=-p+1 \\ k \neq j}}^{1} (j-k)} \tag{10.87}$$

These expressions can be substituted into Eq. (10.83), giving

$$\dot{\mathbf{a}} = \frac{1}{\Delta t} \sum_{j=-p+1}^{1} N'_j(\xi) \mathbf{a}_{n+j} \tag{10.88}$$

and

$$\ddot{\mathbf{a}} = \frac{1}{\Delta t^2} \sum_{j=-p+1}^{1} N''_j(\xi) \mathbf{a}_{n+j} \tag{10.89}$$

Insertion of $\mathbf{a}$, $\dot{\mathbf{a}}$ and $\ddot{\mathbf{a}}$ into the weighted residual equation form of Eq. (10.82) yields

$$\int_{-p+1}^{1} W(\xi) \Bigg[\sum_{j=-p+1}^{1} \mathbf{M} N''_j(\xi) \mathbf{a}_{n+j} + \Delta t \mathbf{C} \sum_{j=-p+1}^{1} N'_j(\xi) \mathbf{a}_{n+j} + \Delta t^2 \mathbf{K} \sum_{j=-p+1}^{1} N_j(\xi) \mathbf{a}_{n+j} + \Delta t^2 \sum_{j=-p+1}^{1} N_j(\xi) \mathbf{f}_{n+j} \Bigg] \mathrm{d}\xi = 0 \tag{10.90}$$

with the forcing function interpolated similarly from its nodal values.

Using parameters

$$\phi_q = \frac{\int_0^1 W \xi^q \, \mathrm{d}\xi}{\int_0^1 W \xi \, \mathrm{d}\xi}, \qquad q = 1, 2, \ldots, p; \quad \phi_0 = 1 \tag{10.91}$$

we now have an algorithm that enables us to compute $\mathbf{a}_{n+1}$ from known values $\mathbf{a}_{p+1}, \mathbf{a}_{p+2}, \ldots, \mathbf{a}_n$. [*Note*: so long as the limits of $-p+1$, $-p+2$ integration are the *same* in Eqs (10.90) and (10.91) it makes no difference what we choose them to be.]

For example, for $p=3$, Eq. (10.84) gives

$$
\begin{aligned}
N_{-2}(\xi) &= \frac{(\xi+1)\xi(\xi-1)}{(2)(-3)} = -\tfrac{1}{6}(\xi^3-\xi) \\
N_{-1}(\xi) &= \frac{(\xi+2)\xi(\xi-1)}{(-2)} = \tfrac{1}{2}(\xi^3+\xi^2-2\xi) \\
N_0\ \ (\xi) &= \frac{(\xi+2)(\xi+1)(\xi-1)}{(-2)} = \tfrac{1}{2}(\xi^3+2\xi^2-\xi-2) \\
N_1\ \ (\xi) &= \frac{(\xi+2)(\xi+1)\xi}{(3)(2)(1)} = \tfrac{1}{6}(\xi^3+3\xi^2+2\xi)
\end{aligned}
\tag{10.92a}
$$

Similarly, from Eqs (10.86) and (10.87),

$$
\begin{aligned}
N'_{-2}(\xi) &= \tfrac{1}{6}[\xi(\xi-1)+(\xi+1)(\xi-1)+(\xi+1)] = \tfrac{1}{6}(3\xi^2-1) \\
N'_{-1}(\xi) &= \tfrac{1}{2}[\xi(\xi-1)+(\xi+2)(\xi-1)+(\xi+2)] = \tfrac{1}{2}(3\xi^2+2\xi-2) \\
N'_{-0}(\xi) &= -\tfrac{1}{2}[(\xi+1)(\xi-1)+(\xi+2)(\xi-1)+(\xi+2)(\xi+1)] \\
&= -\tfrac{1}{2}(3\xi^2+4\xi-1) \\
N'_{-1}(\xi) &= \tfrac{1}{6}[(\xi+1)\xi+(\xi+2)\xi+(\xi+2)(\xi+1)] \\
&= \tfrac{1}{6}(3\xi^2+6\xi+2)
\end{aligned}
\tag{10.92b}
$$

and

$$
\begin{aligned}
N''_{-2}(\xi) &= -\xi \\
N''_{-1}(\xi) &= 3\xi+1 \\
N''_0\ \ (\xi) &= -3\xi-2 \\
N''_1\ \ (\xi) &= \xi+1
\end{aligned}
\tag{10.92c}
$$

We now have a three-step algorithm for the solution of Eq. (10.82) of the form (taking $f=0$)

$$
\sum_{j=-2}^{1} [\alpha_{j+2}\mathbf{M} + \gamma_{j+2}\Delta t\mathbf{C} + \beta_{j+2}\Delta t^2\mathbf{K}]\mathbf{a}_{n+j} = 0 \tag{10.93}
$$

where

$$
\alpha_{j+2} = \int_0^1 W(\xi)N''_j(\xi)\,\mathrm{d}\xi
$$

$$
\gamma_{j+2} = \int_0^1 W(\xi)N'_j(\xi)\,\mathrm{d}\xi \tag{10.94}
$$

and

$$
\beta_{j+2} = \int_0^1 W(\xi)N_j(\xi)\,\mathrm{d}\xi
$$

above gives after integration

$$
\begin{aligned}
\alpha_0 &= -\phi_1 \\
\alpha_1 &= 3\phi_1 + 1 \\
\alpha_2 &= -3\phi_1 - 2 \\
\alpha_3 &= \phi_1 + 1 \\
\gamma_0 &= -\tfrac{1}{2}\phi_2 + \tfrac{1}{6} \\
\gamma_1 &= \tfrac{3}{2}\phi_2 + \phi_1 - 1 \\
\gamma_2 &= -\tfrac{3}{2}\phi_2 - 2\phi_1 + \tfrac{1}{2} \\
\gamma_3 &= \tfrac{1}{2}\phi_2 + \phi_1 + \tfrac{1}{3} \\
\beta_0 &= -\tfrac{1}{6}\phi_3 + \tfrac{1}{6}\phi_1 \\
\beta_1 &= \tfrac{1}{2}\phi_3 + \tfrac{1}{2}\phi_2 - \phi_1 \\
\beta_2 &= -\tfrac{1}{2}\phi_3 - \phi_2 + \tfrac{1}{2}\phi_1 + 1 \\
\beta_3 &= \tfrac{1}{6}\phi_3 + \tfrac{1}{2}\phi_2 + \tfrac{1}{3}\phi_1
\end{aligned}
\tag{10.95}
$$

An algorithm of the form given in Eq. (10.93) is called a linear three-step method. The general p-step form is

$$\sum_{j=-p+1}^{1} (\alpha_{j+p-1}\mathbf{M} + \gamma_{j+p-1}\Delta t\mathbf{C} + \beta_{j+p-1}\Delta t^2\mathbf{K})\mathbf{a}_{n+j} = \mathbf{0} \tag{10.96}$$

This is the form generally given in mathematics texts; it is an extension of the form given by Lambert[2] for $\mathbf{C}=\mathbf{0}$. The weighted residual approach described here derives the α's, β's and γ's in terms of the parameters ϕ_0, $\phi_1, \ldots, \phi_p$ and thus ensures consistency.

From Eq. (10.96) the unknown $\mathbf{a}_{n+1}$ is obtained in the form

$$\mathbf{a}_{n+1} = (\alpha_3\mathbf{M} + \gamma_3\Delta t\mathbf{C} + \beta_3\Delta t^2\mathbf{K})^{-1} \quad \text{(known values)} \tag{10.97}$$

For example, for $p=3$ the matrix to be inverted is

$$[(\phi_1 + 1)\mathbf{M} + (\tfrac{1}{2}\phi_2 + \phi_1 + \tfrac{1}{3})\Delta t\mathbf{C} + (\tfrac{1}{6}\phi_3 + \tfrac{1}{2}\phi_2 + \tfrac{1}{3}\phi_1)\Delta t^2\mathbf{K}]$$

Comparing this with the matrix to be inverted in the SSpj algorithm given in Eq. (10.37) suggests a correspondence between SSpj and the p-step algorithm above which we explore further in the next section.

10.4.3 *The relationship between SSpj and the weighted residual p-step algorithm.* For simplicity we now consider the p-step algorithm described in the previous section applied to the scalar equation

$$m\ddot{a} + c\dot{a} + ka = 0 \tag{10.98}$$

As in previous stability considerations we can obtain the general solution of the recurrence relation

$$\sum_{j=-p+1}^{1} (\alpha_{j+p-1}m + \Delta t \gamma_{j+p-1}c + \Delta t^2 \beta_{j+p-1}k)a_{n+j} = 0 \tag{10.99}$$

by putting $a_{n+1} = \mu$, $a_n = \mu^2$, $a_{n-1} = \cdots$ where the values of μ are the roots μ_k of the stability polynomial of the p-step algorithm:

$$\sum_{j=-p+1}^{1} (\alpha_{j+p-1}m + \Delta t c \gamma_{j+p-1} + \Delta t^2 \beta_{j+p-1}k)\mu^{p-1+j} = 0 \tag{10.100}$$

This stability polynomial can be quite generally identified with the one resulting from the determinant of Eq. (10.64) as shown in reference 6, by using a suitable set of relations linking Θ and ϕ. Thus, for instance, in the case $p=3$ discussed we shall have the identity of stability and indeed of the algorithm when

$$\begin{aligned} \Theta_1 &= 1 + \phi_1 \\ \Theta_2 &= \tfrac{2}{3} + 2\phi_1 + \phi_2 \\ \Theta_3 &= \phi_3 + 3\phi_2 + 2\phi_1 \end{aligned} \tag{10.101}$$

Table 10.5 summarizes the identities of $p=2$, 3 and 4.

Many results obtained previously with p-step methods[15,47] can be used to give the accuracy and stability properties of the solutions produced by the SSpj algorithms. Tables 10.6 and 10.7 give the accuracy of stable algorithms from the SSp1 and SSp2 families respectively for $p=2$, 3, 4. Algorithms that are only conditionally stable (i.e. only stable for values of the time step less than some critical value) are marked CS. Details are given in reference 2.

TABLE 10.5

Algorithm	Identities
SS22/21	
	$\Theta_1 = \phi_1 + \frac{1}{2}$
	$\Theta_2 = \phi_1 + \phi_2$
SS32/31	
	$\Theta_1 = \phi_1 + 1$
	$\Theta_2 = \frac{2}{3} + 2\phi_1 + \phi_2$
	$\Theta_3 = 2\phi_1 + 3\phi_2 + \phi_3$
SS42/41	
	$\Theta_1 = \frac{3}{2} + \phi_1$
	$\Theta_2 = \frac{11}{6} + 3\phi_1 + \phi_2$
	$\Theta_3 = \frac{3}{2} + 11\phi_1/2 + 9\phi_2/2 + \phi_3$
	$\Theta_4 = 6\phi_1 + 11\phi_2 + 6\phi_3 + \phi_4$

TABLE 10.6

Method	Parameters	Error
SS11	Θ_1	$O(\Delta t)$
	$\Theta_1=0.5$	$O(\Delta t^2)$
SS21	Θ_1, Θ_2	$O(\Delta t^2)$
	$\Theta_1-\Theta_2=\frac{1}{6}$	$O(\Delta t^2)$CS
SS31	$\Theta_1, \Theta_2, \Theta_3$	$O(\Delta t^3)$
	$\Theta_1-3\Theta_2+2\Theta_3=0$	$O(\Delta t^4)$CS
SS41	$\Theta_1, \Theta_2, \Theta_3, \Theta_4$	$O(\Delta t^4)$

TABLE 10.7

Method	Parameters	Error	
		$C=0$	$C\neq 0$
SS22	Θ_1, Θ_2	$O(\Delta t)$	$O(\Delta t)$
	$\Theta_1, \Theta_2=0.5$	$O(\Delta t^2)$	$O(\Delta t^2)$
	$\Theta_1=0.5, \Theta_2=\frac{1}{6}$	$O(\Delta t^4)$CS	$O(\Delta t^2)$CS
SS32	$\Theta_1, \Theta_2, \Theta_3$	$O(\Delta t^2)$	$O(\Delta t^2)$
	$\Theta_2=\Theta_1-\frac{1}{6}$	$O(\Delta t^3)$CS	$O(\Delta t^3)$CS
	$\Theta_3=\Theta_1/2$	$O(\Delta t^4)$CS	—
SS42	$\Theta_1, \Theta_2, \Theta_3, \Theta_4$	$O(\Delta t^3)$CS	$O(\Delta t^3)$CS

We conclude this section by writing in full the second degree (two-step) algorithm that corresponds precisely to SS22 and GN22 methods. Indeed, it is written below in the form originally derived by Newmark:

$$\begin{aligned}(\mathbf{M}+\Delta t\mathbf{C}+\beta\Delta t^2\mathbf{K})\mathbf{a}_{n+1}\\ +[-2\mathbf{M}+(1-2\gamma)\Delta t\mathbf{C}+(\tfrac{1}{2}-2\beta+\gamma)\Delta t^2\mathbf{K}]\mathbf{a}_n\\ +[\mathbf{M}-(1-\gamma)\Delta t\mathbf{C}+(\tfrac{1}{2}+\beta-\gamma)\Delta t^2\mathbf{K}]\mathbf{a}_n+\bar{\mathbf{f}}\Delta t^2=\mathbf{0}\end{aligned} \tag{10.102}$$

Here, of course, we have the original Newmark parameters, which can be changed to correspond with SS22/GN22 form as follows:

$$\gamma=\Theta_1=\beta_1 \qquad \beta=\tfrac{1}{2}\Theta_2=\tfrac{1}{2}\beta_2$$

The explicit form of this algorithm ($\Theta_2=0$) is frequently used as an alternative to the single-step explicit form. It is then known as the *central difference approximation* obtained by direct differencing. The reader can easily verify that the simplest finite difference approximation of Eq. (10.1) in fact corresponds to above with $\Theta_1=\frac{1}{2}$, $\Theta_2=0$.

10.5 Some remarks on general performance of numerical algorithms

In Secs 10.2.5 and 10.3.3 we have considered the *exact* solution of the approximate recurrence algorithm given in the form

$$\mathbf{a}_{n+1} = \mu \mathbf{a}_n, \qquad \text{etc.} \tag{10.103}$$

for the modally decomposed, single degree of freedom systems typical of Eqs (10.4) and (10.5). The evaluation of μ was important to ensure that its modulus does not exceed unity and stability is preserved.

However, analytical solution of the linear homogeneous equations is easy to obtain and comparison of μ with such a solution which is always of the form

$$a = \tilde{a}\, e^{\lambda t} \tag{10.104a}$$

or

$$a_{n+1} = a_n\, e^{\lambda \Delta t} \tag{10.104b}$$

is instructive and provides information on the performance of the algorithm in the particular range of eigenvalues.

In Fig. 10.5 we have indeed plotted the exact solution $e^{\omega \Delta t}$ and compared it with values of μ for various Θ algorithms approximating the first-order equation, noting that here

$$\lambda \equiv -\omega = -\frac{k}{c} \tag{10.105}$$

and is real.

Immediately we see that here the performance error is very different for various values of Δt and obviously deteriorates at large values of it. Such values in a real multivariable problem correspond of course to the 'high-frequency' responses which are less important, and for smooth solutions we favour algorithms where μ tends to zero for such values. However, response through the whole time range is important and attempts to choose, for instance, the optimal value of Θ by Liniger consider various time ranges.[41] Table 10.1 of Sec. 10.2.6 illustrates how an algorithm $\Theta = \frac{2}{3}$ with a higher truncation error than that of $\Theta = \frac{1}{2}$ can perform better in a multidimensional system because of such properties.

Similar analysis can of course be applied to the second-order equation. Here to simplify matters we frequently consider only the undamped equation in the form

$$m\ddot{a} + ka = 0 \tag{10.106}$$

in which the value of λ is purely imaginary and corresponds to a simple oscillation. By examining μ we can find not only the amplitude ratio (which for high accuracy should be unity) but also the phase retardation.

In Fig. 10.13(*a*) and (*b*) we show both the variation of the modulus μ

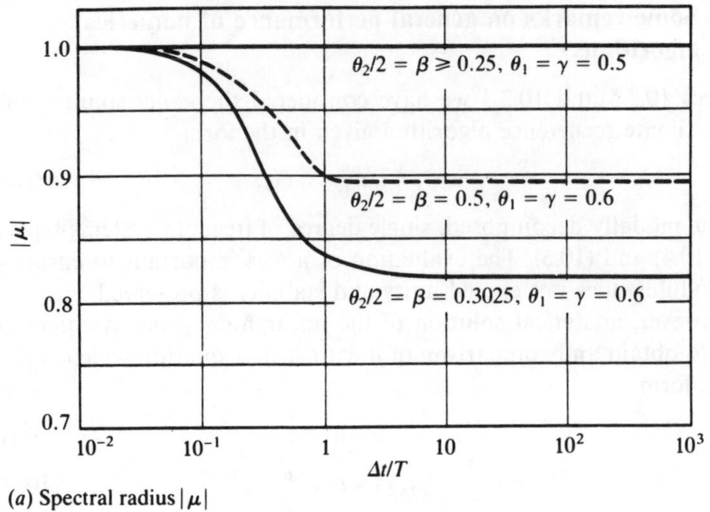

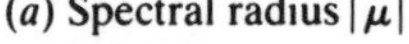

(*a*) Spectral radius $|\mu|$

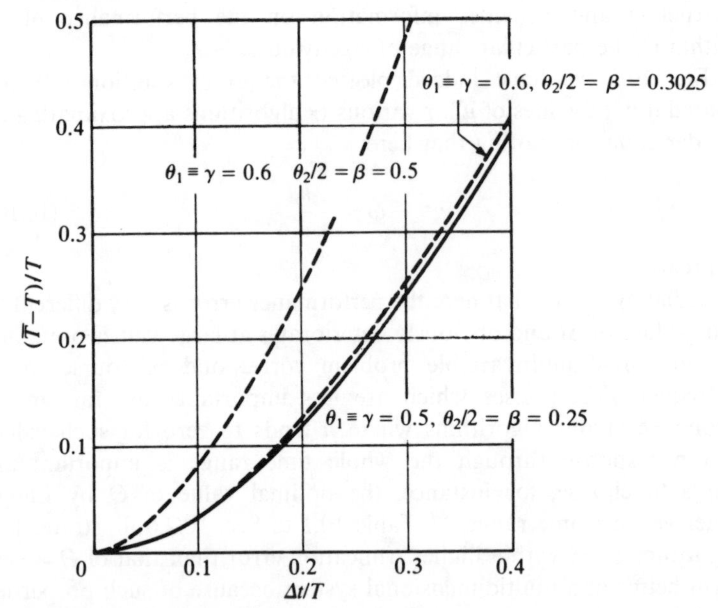

(*b*) Relative period elongation

Fig. 10.13 SS22, GN22 (Newmark) or their two-step equivalent

(i.e. the spectral radius) and of the relative period for the SS22/GN22 schemes, which of course are also applicable to their two-step equivalents. The results are plotted against

$$\Delta t/T \qquad \text{where } T = 2\pi/\omega;\ \omega^2 = k/m \tag{10.107}$$

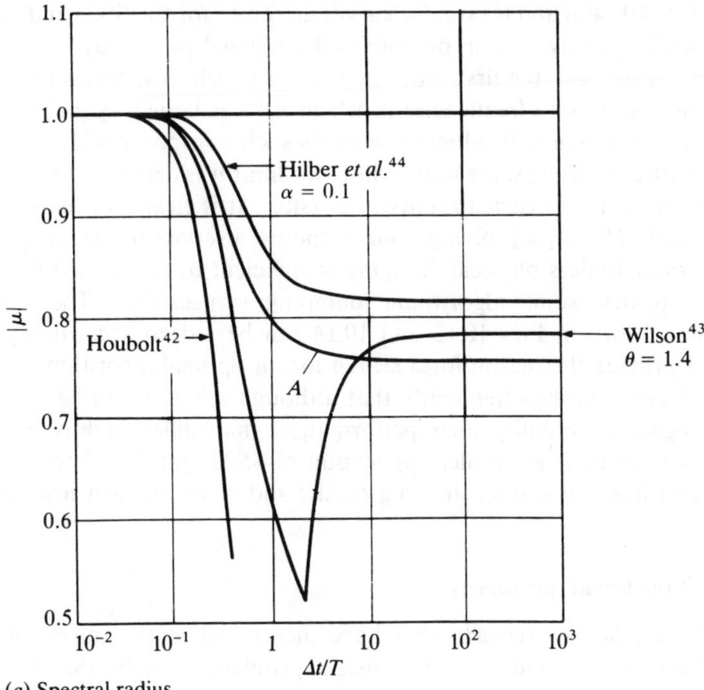

(*a*) Spectral radius

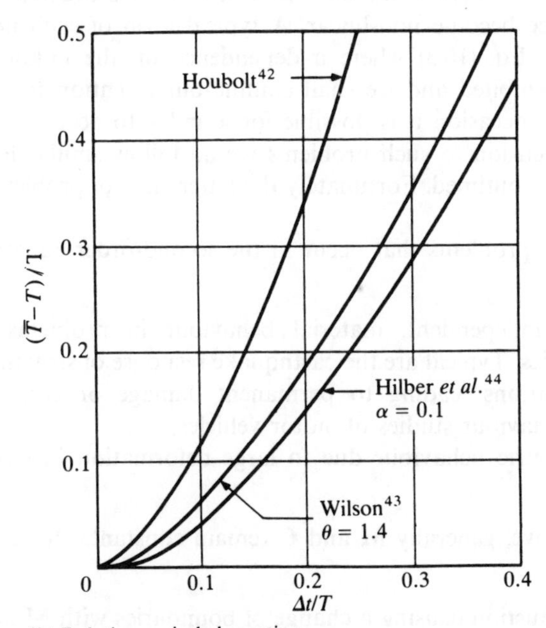

(*b*) Relative period elongation

Fig. 10.14 SS23,GN23 and their two-step equivalent

In Fig. 10.14(*a*) and (*b*) similar curves are given for the SS23 and GN23 schemes frequently used in practice and discussed previously.

Here again as in the first-order problem we wish to suppress (or damp out) the response to frequencies in which $\Delta t/T$ is large (say greater than $\frac{1}{10}$) in multi degree of freedom systems, as such a response will invariably be inaccurate. At the same time below this limit it is desirable to have the amplitude ratios as close to unity as possible. It is clear that the stability limit with $\Theta=\Theta_2=\frac{1}{2}$ giving unit response everywhere is in general undesirable (unless physical damping is sufficient to damp high frequencies) and that some algorithmic damping is necessary. The various schemes shown in Figs 10.13 and 10.14 can be judged accordingly and indeed provide the reason for a search for an optimal algorithm.

We have remarked frequently that although schemes can be identical with regard to stability their performances may differ slightly. In Fig. 10.15 we illustrate a parallel application of SS22 and GN22 to a single degree of freedom system showing results and errors in each scheme.

10.6 Non-linear problems

10.6.1 *Introductory remarks.* We have mentioned at the outset of this chapter that the typical linear transient problem given by the ordinary differential equation (10.1) (valid for both first- and second-order problems) may in practice become non-linear. A typical form of such non-linearity is given by Eq. (10.3) where a dependence on the unknown function $\mathbf{a}$ itself is exhibited and we shall confine our attention to that class of problem. On occasion it is possible for $\dot{\mathbf{a}}$ and $\ddot{\mathbf{a}}$ to occur in the dependence, and extension to such problems would follow similar lines to those which will be outlined. Fortunately the latter class of problem is rare.

Typical non-linear problems that occur in the second-order equation ($\mathbf{M}\neq 0$) are:

1. Non-linear, rate-independent, material behaviour in problems of structural dynamics. Typical are the earthquake response of structures and their foundations leading to permanent damage or collapse; impact (crash) behaviour studies of motor vehicles; etc.
2. Non-linear geometric behaviour due to large deformation in structural dynamics.

In both of the above, generally $\mathbf{M}$ and $\mathbf{C}$ remain constant. However, these can vary as in:

3. Ablation or combustion causing a change of boundaries with $\mathbf{M}$ now variable.

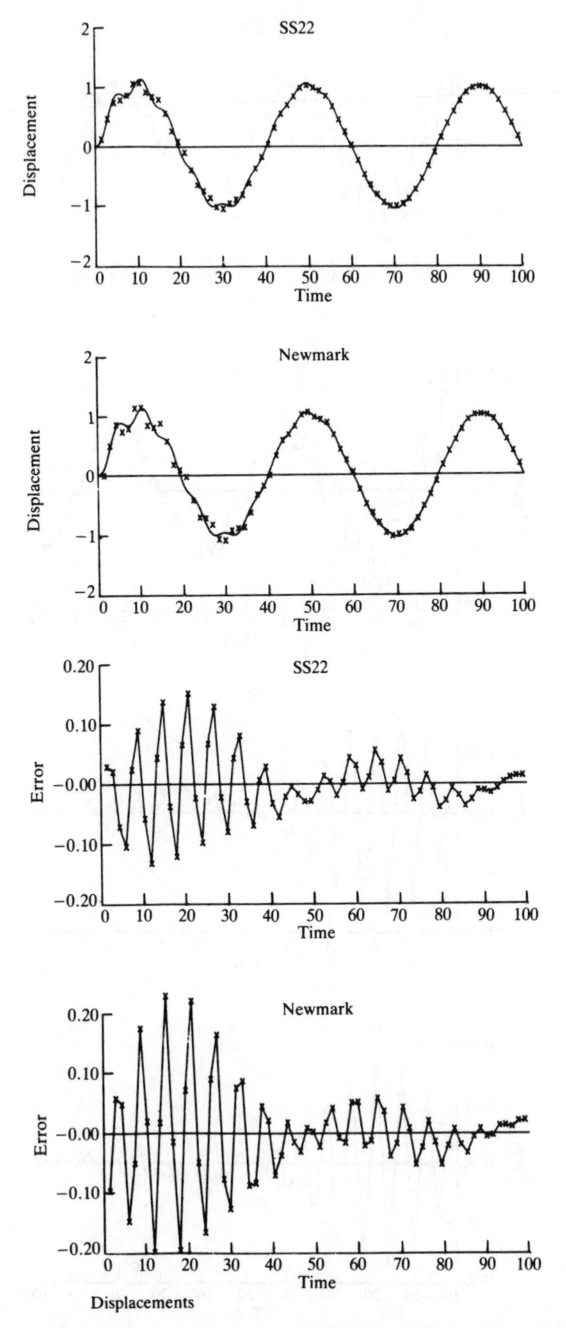

Fig. 10.15 Comparison of SS22 and Newmark (GN22) algorithms. A single DOF dynamic equation with periodic forcing term, $\Theta_1=\beta_1=\frac{1}{2}, \Theta_2=\beta_2=0$

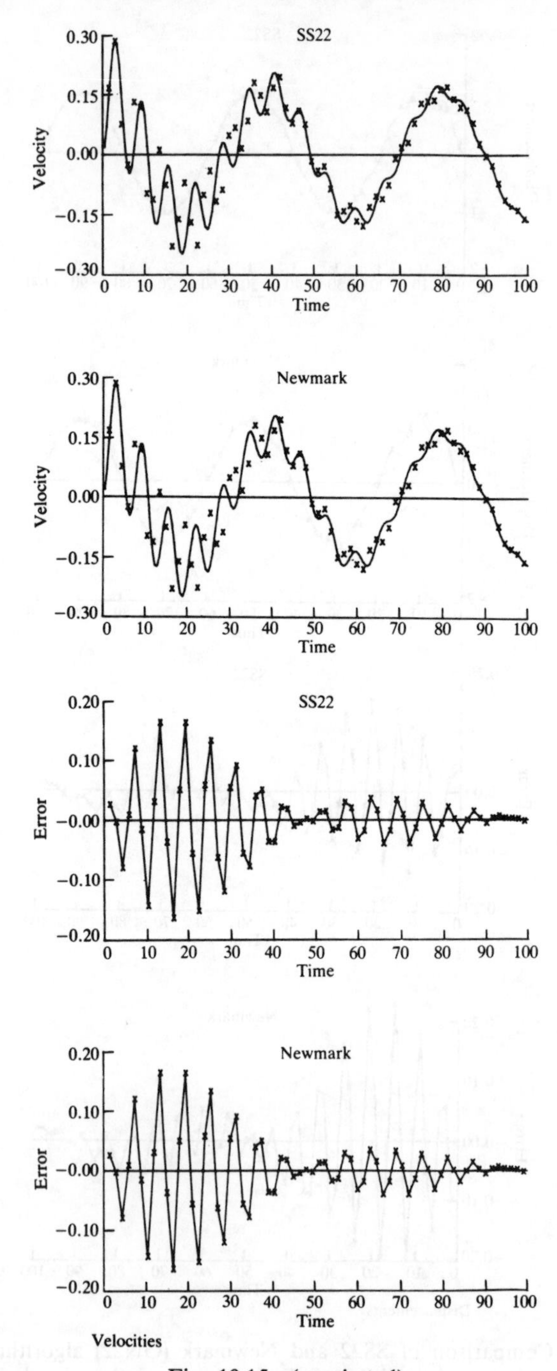

Fig. 10.15 (continued)

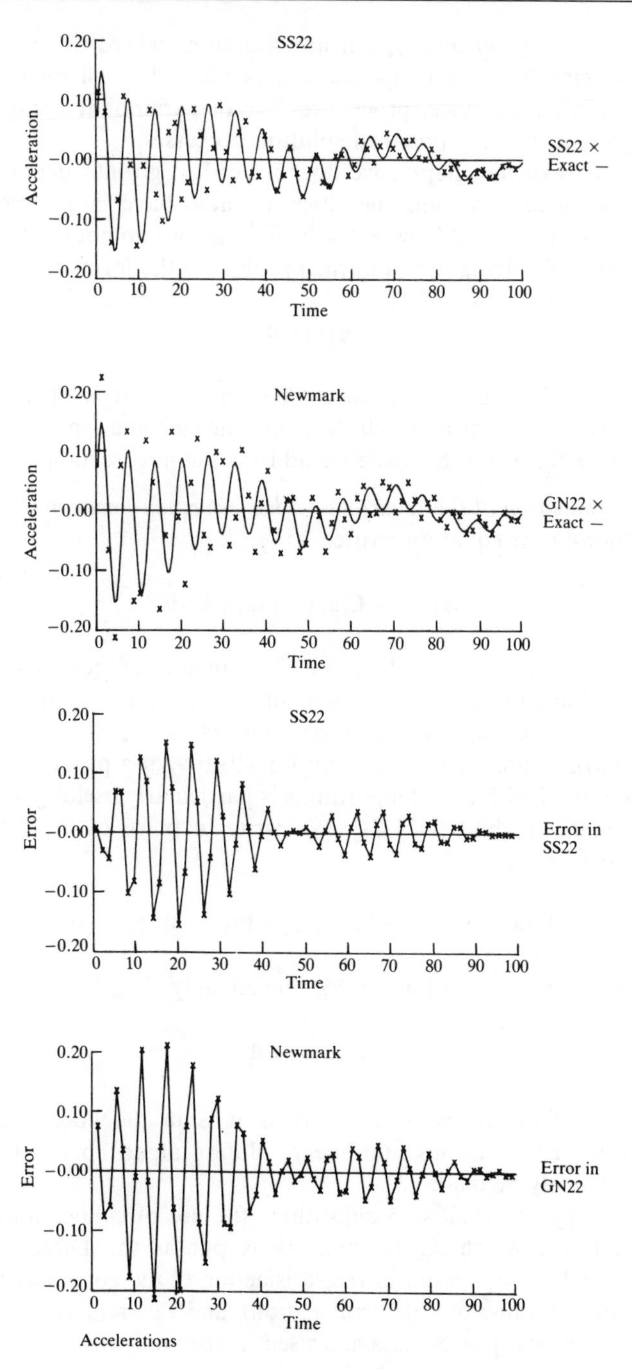

Fig. 10.15 (continued)

In first-order problems typical are situations where the conductivity and/or specific heat are temperature dependent. For all such problems time stepping, recurrence, procedures are of paramount importance as they represent the only practical solution processes.

In this part of the chapter we do not need to go into detail of either material or geometric non-linearities as these have been described in detail in Chapters 7 and 8 respectively. Further, we do not need to discuss the methods of solving a non-linear problem of the form

$$\boldsymbol{\psi}(\mathbf{x}) = \mathbf{0} \tag{10.108}$$

This again is fully documented in Chapter 7. What is of importance, however, is the manner in which the recurrence scheme can be formulated in the above form. To this issue we address the next section.

10.6.2 *Formulation of the recurrence scheme for non-linear problems.* The general non-linear equation written as

$$\mathbf{M}(\mathbf{a})\ddot{\mathbf{a}} + \mathbf{C}(\mathbf{a})\dot{\mathbf{a}} + \mathbf{P}(\mathbf{a}) + \mathbf{f} = \mathbf{0} \tag{10.109}$$

could of course be discretized by any of the preceding processes using the weighted residual method or its particular, collocation, form.

The *collocation form* is most obvious as the non-linear algebraic equation arises directly from writing Eq. (10.109) at a particular point of time. Here the GN series of algorithms is particularly useful [for instance the Newmark method (GN22)]. We write in place of Eq. (10.56) the non-linear form as

$$\mathbf{M}(\mathbf{a}_{n+1})\ddot{\mathbf{a}}_{n+1} + \mathbf{C}(\mathbf{a}_{n+1})\dot{\mathbf{a}}_{n} + \mathbf{P}(\mathbf{a}_{n+1}) + \mathbf{f}_{n+1} = \mathbf{0} \tag{10.110}$$

which on substitution of Eqs (10.55) immediately gives the required form

$$\boldsymbol{\psi}(\ddot{\mathbf{a}}_{n+1}) = \mathbf{0} \tag{10.111}$$

where the unknown vector is $\ddot{\mathbf{a}}_{n+1}$, as it is in the linear case. Here application of the various non-linear solution algorithms of Chapter 7 can immediately be made.

In the SSpj and multistep algorithms the use of collocation is again possible. If the weighting function W is put as the Dirac delta, the weighting will always result in the satisfaction of the governing equation at a particular value of τ (i.e. collocation), and this replaces Eq. (10.33). Now the identical process is again used in solution with appropriate Θ values substituted. We note that sampling at

$\tau = \Delta t$ gives $\Theta_1 = 1$, $\Theta_2 = 1$, $\Theta_3 = 1$, etc.

$\tau = \frac{3}{4}\Delta t$ gives $\Theta_1 = \frac{3}{4}$, $\Theta_2 = \frac{9}{16}$, $\Theta_3 = \frac{27}{64}$, etc.

$\tau = \frac{1}{2}\Delta t$ gives $\Theta_1 = \frac{1}{2}$, $\theta_2 = \frac{1}{4}$, $\theta_3 = \frac{1}{16}$, etc.

For SS22 only the first two processes are unconditionally stable. All will give the non-linear equation in the form

$$\boldsymbol{\psi}(\boldsymbol{\alpha}_n) = \mathbf{0} \tag{10.112}$$

which again is in a standard non-linear shape.

Identical remarks apply to the multistep algorithm if Dirac delta weighting is used in Eq. (10.90). We shall not pursue the point but will remark that:

1. When the collocation occurs at point $t = t_n$ in the two-step algorithm we have precisely the central difference (explicit) scheme frequently used in dynamic analysis of non-linear problems.[48-54]
2. When the collocation occurs as $t = t_{n+1}$ and a three-step algorithm is used we have the Houboldt scheme.[42]

At this stage an additional remark on *explicit schemes* is in order. As we have remarked earlier, for the GN22 scheme an explicit form arises with $2\beta = \beta_2 = \Theta_2 = 0$ and it would appear from Eq. (10.111) that a set of non-linear equations still needs to be solved. However, inspection of Eq. (10.55) shows that the first of these now gives

$$\mathbf{a}_{n+1} = \mathbf{a}_n + \Delta t \dot{\mathbf{a}}_n + \frac{\Delta t^2}{2} \ddot{\mathbf{a}}_n \tag{10.113}$$

which can be directly evaluated, but the second one remains as

$$\dot{\mathbf{a}}_{n+1} = \dot{\mathbf{a}}_n + \Delta t(1 - \beta_1)\ddot{\mathbf{a}}_n + \Delta t \beta_1 \ddot{\mathbf{a}}_{n+1} \tag{10.114}$$

However, if no dependence of $\mathbf{M}$, $\mathbf{C}$ or $\mathbf{P}$ on $\dot{\mathbf{a}}$ exists then the scheme does not need any iteration and is truly explicit for a diagonal $\mathbf{M}$. This direct calculation is an obvious advantage for explicit schemes as *practically no additional cost is involved for non-linear problems over that for linear ones.* For this reason such schemes are extremely popular.

Unfortunately the SS22 scheme with a Dirac weighting at t_n cannot be so simply interpreted. Indeed, for arbitrary weighting resulting in all the possible Θ combinations it is not easy to use the SSpj family. The most logical procedure appears to write the non-linear equation as the equivalent of Eq. (10.35) for the general case as

$$\boldsymbol{\psi}(\boldsymbol{\alpha}_n) \equiv \mathbf{M}(\tilde{\mathbf{a}}_{n+1})\ddot{\tilde{\mathbf{a}}}_{n+1} + \mathbf{C}(\tilde{\mathbf{a}}_{n+1})\dot{\tilde{\mathbf{a}}}_{n+1} + \mathbf{P}(\tilde{\mathbf{a}}_{n+1}) + \bar{\mathbf{f}} = \mathbf{0} \tag{10.115a}$$

with

$$\tilde{\ddot{\mathbf{a}}}_{n+1} = \tilde{\ddot{\mathbf{a}}}_{n+1} + \frac{\alpha_n \Theta_{p-2} \Delta t^{p-2}}{(p-2)!}$$

$$\dot{\tilde{\mathbf{a}}}_{n+1} = \dot{\tilde{\mathbf{a}}}_{n+1} + \frac{\alpha_n \Theta_{p-1} \Delta t^{p-1}}{(p-1)!} \tag{10.115b}$$

$$\tilde{\mathbf{a}}_{n+1} = \tilde{\mathbf{a}}_{n+1} + \frac{\alpha_n \Theta_p \Delta t^p}{p!}$$

and here $\tilde{\mathbf{a}}$, $\dot{\tilde{\mathbf{a}}}$ and $\ddot{\tilde{\mathbf{a}}}$ are defined by Eq. (10.36). Now of course any choice of Θ is possible again.

The application to first-order problems will follow similar lines and here all the single-step algorithms can again be used. Quite popular is the two-step algorithm due to Lees[55] which corresponds to that of Eq. (10.102) with $\gamma = \frac{1}{2}$, $\beta = \frac{1}{6}$ and all the non-linear parameters evaluated at t_n.

10.6.3 *The non-linear solution process.* As we have remarked, the solution process at each recurrent step between t_n and t_{n+1} requires the (iterative) solution of a non-linear equation set unless explicit form is used with its conditional stability. Here the standard methods of Chapter 7 are used with appropriate convergence limits prescribed. Each of the methods is applicable and most can be used in practical problems. Clearly, however, it is not desirable to use many iterations in each time step and if convergence does not occur within three or four iterations it is advantageous to reduce the time step size. An economic balance between iteration number and the time step size needs to be struck and the precise limits are somewhat subjective. Indeed, the convergence criteron should not be too tightly set and must be compatible with the error in each time step due to temporal discretization.

10.6.4 *Some examples.* The reader will find numerous examples of time stepping methods applied to the solution of non-linear problems of first and second orders in the current literature. We shall quote some of these here.

Transient heat conduction. The governing equation for this set of physical problems is reported here although we have already shown some linear applications. Thus we have

$$\frac{\partial}{\partial x}\left(k\frac{\partial T}{\partial x}\right) + \frac{\partial}{\partial y}\left(k\frac{\partial T}{\partial y}\right) + \frac{\partial}{\partial z}\left(k\frac{\partial T}{\partial z}\right) + \rho c\frac{\partial T}{\partial t} - Q = 0 \tag{10.116}$$

with boundary conditions

$$T = \bar{T} \quad \text{or} \quad k\frac{\partial T}{\partial n} = -\alpha(T - T_a)^n \tag{10.117}$$

where α is known as the heat transfer coefficient.

Non-linearity clearly can arise due to c, k and Q being temperature dependent or from the radiation boundary condition with $n \neq 1$. We shall show two examples to illustrate the above.

The first concerns the *freezing of ground* in which the latent heat of freezing is represented by varying the specific heat with temperature in a narrow zone, as shown in Fig. 10.16. Further, in the transition from the fluid to the frozen state a variation in conductivity occurs. We now thus have a problem in which both matrices **C** and **K** are variable and the solution of Fig. 10.17 illustrates the progression of a freezing front which was derived using the three-point (Lees) algorithm[55,56] with $\mathbf{C}=\mathbf{C}_n$ and $\mathbf{K}=\mathbf{K}_n$.

A computational feature of some significance arises in this problem as values of the specific heat become very high in the transition zone and in time stepping can be missed if the temperature step *straddles* the freezing point. To avoid this difficulty and keep the heat balance correct the concept of enthalpy is introduced, defining

$$H = \int_0^T \rho c \, \mathrm{d}T \tag{10.118}$$

Now, whenever a change of temperature is considered, an appropriate value of ρc is calculated that gives the correct change of H.

The heat conduction problem involving phase change is of considerable importance in welding and casting technology. Some very useful

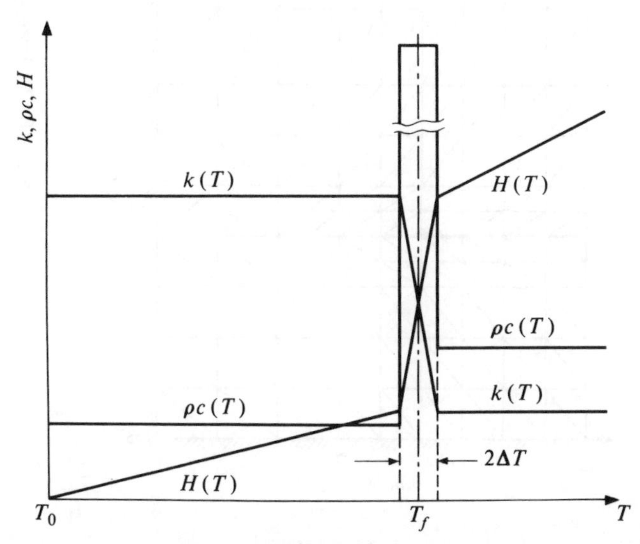

Fig. 10.16 Estimation of thermophysical properties in phase change problems. Latent heat effect is approximated by a large capacity over a small temperature interval $2\Delta T$

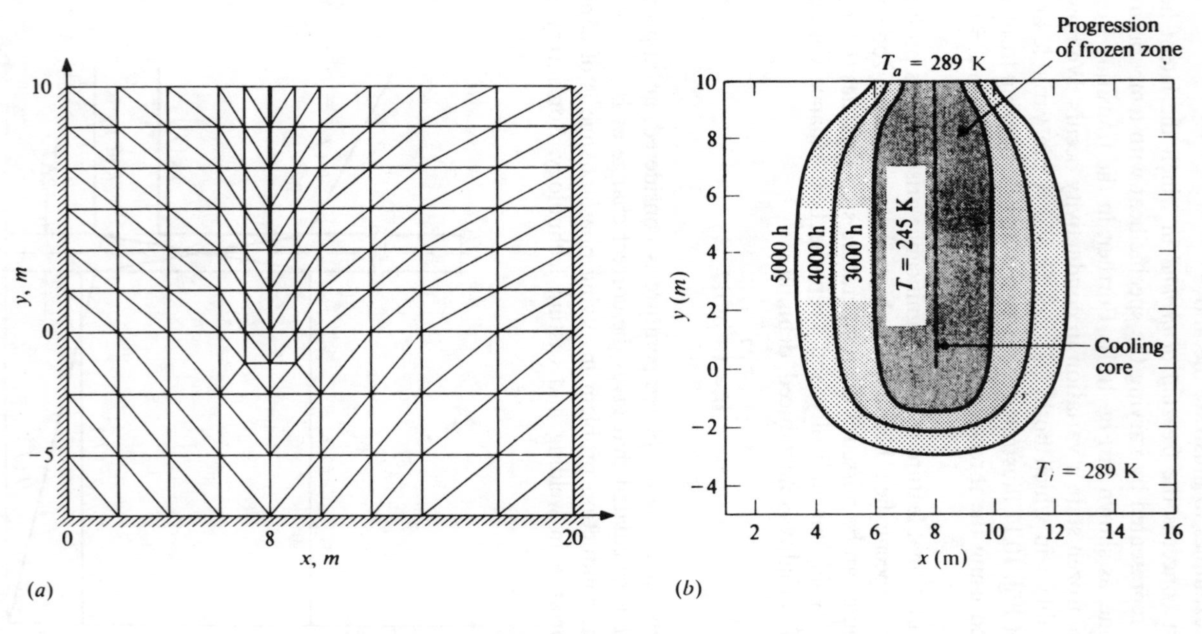

Fig. 10.17 Freezing of a moist soil (sand), (a) and (b)

finite element solutions of these problems have been obtained.[57] Further elaboration of the procedure described above is given in reference 58.

The second non-linear example concerns the problem of *spontaneous ignition*.[59] We have discussed the steady-state case of this problem on page 274 and now will be concerned with the transient. Here the heat generated depends on the temperature

$$Q = \delta \, e^{T} \tag{10.119}$$

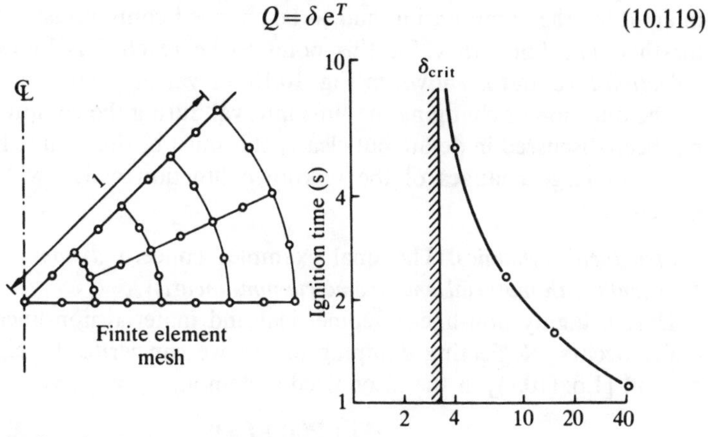

(*a*) Induction time versus Frank–Kamenetskii parameter

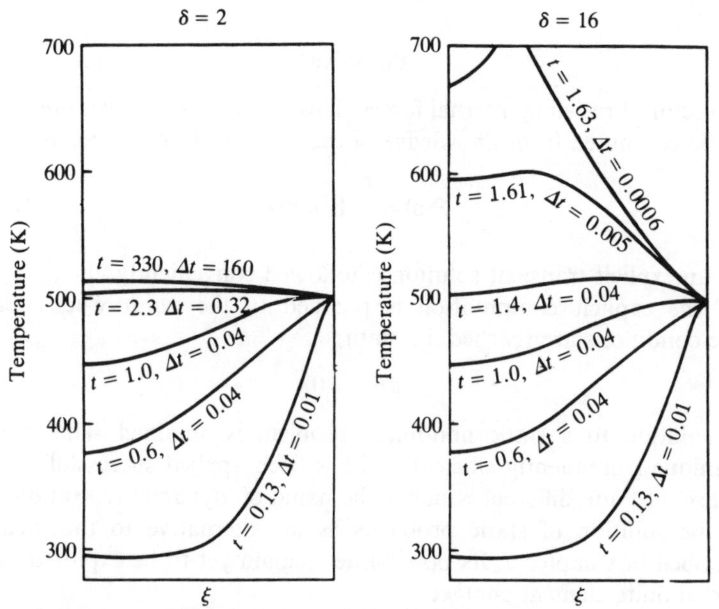

(*b*) Temperature profiles for ignition ($\delta = 16$) and non-ignition ($\delta = 2$) transient behaviour of a reactive sphere

Fig. 10.18 Reactive sphere

and the situation can become *physically unstable* with temperatures rising continuously to extreme values. In Fig. 10.18 we show a transient solution for a sphere at an initial temperature of $T=290$ K immersed in a bath of 500 K. The solution is given for two values of the parameter δ with $k=\rho c=1$, and the non-linearities are now so severe that an iterative solution in each time increment was necessary. For the larger value of δ the temperature increases to an infinite value in *a finite time* and the time interval for the computation had to be changed continuously to account for this. The finite time for this point to be reached is known as the *induction time* and is shown in Fig. 10.18 for various values of δ.

The question of changing the time interval during the computation has not been discussed in detail, but clearly this must be done quite frequently to avoid large changes of the unknown function which will result in inaccuracies.

Structural dynamics. The final examples concern *dynamic structural transients with material and geometric non-linearity.*

Here a highly non-linear geometrical and material non-linearity generally occurs. Neglecting damping forces, we can write the equation of motion [Eq. (10.1)] in the discretized system as

$$\mathbf{M}\ddot{\mathbf{a}}+\mathbf{P}(\mathbf{a})+\mathbf{f}=\mathbf{0} \tag{10.120}$$

where

$$\mathbf{P}(\mathbf{a})\equiv\mathbf{K}\mathbf{a} \tag{10.121}$$

is a vector of resisting internal forces. This, as has been shown previously, can be computed from knowledge of the stresses at any stage, as

$$\mathbf{P}(\mathbf{a})=\int_{\Omega}\bar{\mathbf{B}}^{\mathrm{T}}\sigma\,\mathrm{d}\Omega \tag{10.122}$$

and an explicit transient solution is followed conveniently.

If the explicit computation is pursued to the point when steady-state conditions are reached, i.e. until

$$\ddot{\mathbf{a}}=\dot{\mathbf{a}}=0$$

the solution to a static non-linear problem is obtained. This type of technique is frequently efficient and has been applied successfully in the context of finite differences under the name of 'dynamic relaxation'[60–62] for the solution of static problems as an alternative to the methods described in Chapter 7. Its possibilities remain yet to be explored in the general finite element context.

Two examples of explicit dynamic analysis will be given here. The first problem, illustrated in Plate 1 (see frontispiece), is a large three-dimensional one and its solution was obtained with the use of an explicit

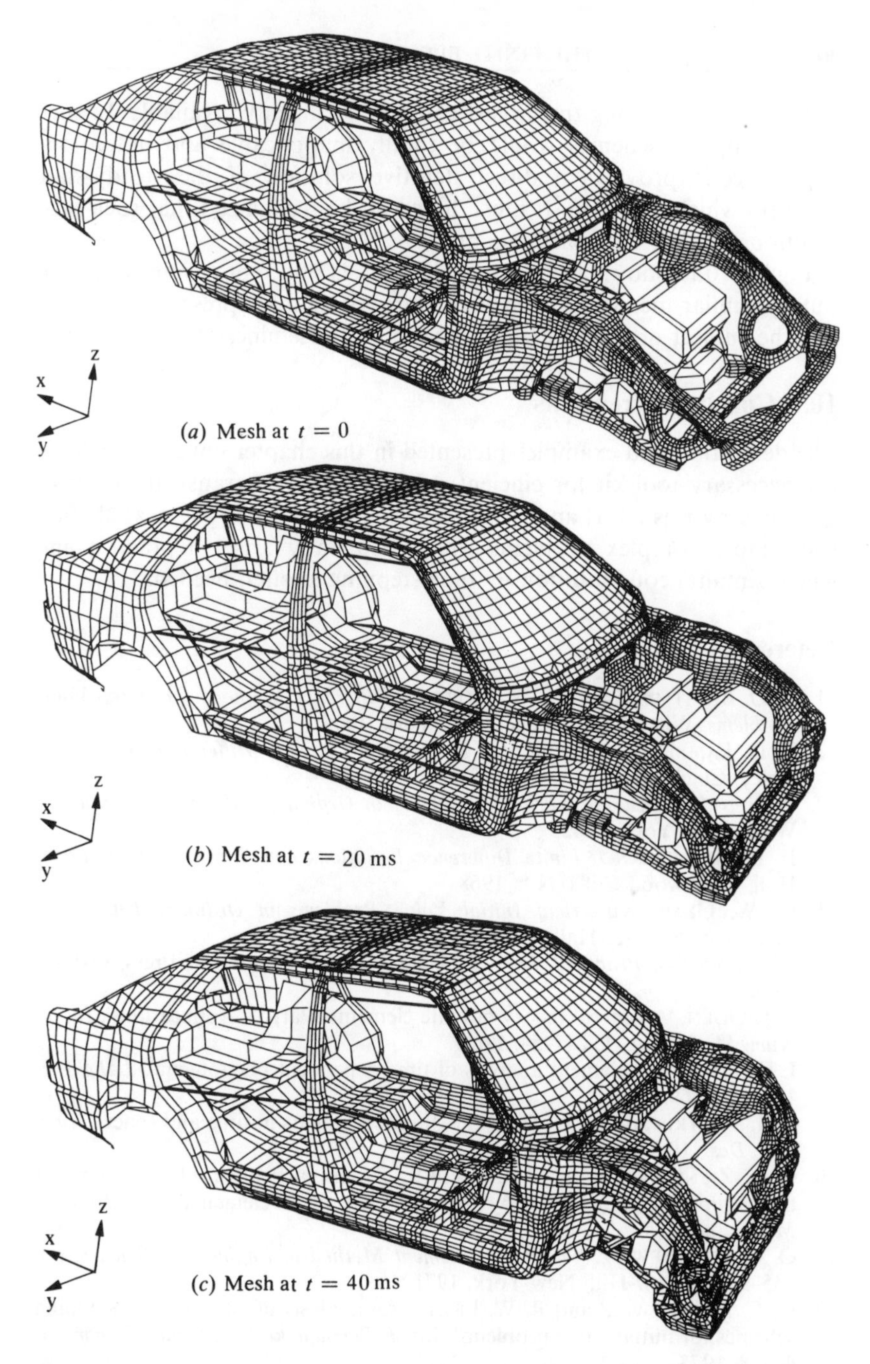

Fig. 10.19 Crash analysis of a SAAB 9000. The analysis was carried out using the explicit code LS—DYNA 3D (J. Hallquist) 14805 shell and 6 beam elements with average $\Delta t = 0.8 \times 10^{-8}$ s, CPU time 17 h and CRAY x −MP/48. (Reproduced by kind permission of SAAB Automobile AB)

dynamic scheme using the largest computers currently available. In such a case implicit schemes would be totally inapplicable and indeed the explicit code provides a direct iterative solution of the steady-state problem which is shown. It must, however, be recognized that such final solutions are not necessarily unique.

Figure 10.19 shows a typical crash analysis of a motor vehicle carried out by similar means. Explicit procedures of analysis presented here are still the subject of much research due to their usefulness.[63]

10.7 Concluding remarks

The derivation and examples presented in this chapter cover, we believe, the necessary tool-kit for efficient solution of many transient problems governed by Eqs (10.1) and (10.2). We shall elaborate in the next chapter some more complex problems which frequently arise in practice and where simultaneous solution by time stepping is often needed.

References

1. R. D. RICHTMYER and K. W. MORTON, *Difference Methods for Initial Value Problems*, Wiley (Interscience), New York, 1967.
2. T. D. LAMBERT, *Computational Methods in Ordinary Differential Equations*, Wiley, Chichester, 1973.
3. P. HENRICI, *Discrete Variable Methods in Ordinary Differential Equations*, Wiley, New York, 1962.
4. F. B. HILDEBRAND, *Finite Difference Equations and Simulations*, Prentice-Hall, Englewood Cliffs, N.J., 1968.
5. G. W. GEAR, *Numerical Initial Value Problems in Ordinary Differential Equations*, Prentice-Hall, Englewood Cliffs, N.J., 1971.
6. W. L. WOOD, *Practical Time Stepping Schemes*, Clarendon Press, Oxford, 1990.
7. J. T. ODEN, 'A general theory of finite elements, Part II. Applications', *Int. J. Num. Meth. Eng.*, **1**, 247–54, 1969.
8. I. FRIED, 'Finite element analysis of time-dependent phenomena', *JAIAA*, **7**, 1170–3, 1969.
9. J. H. ARGYRIS and D. W. SCHARPF, 'Finite elements in time and space', *Nucl. Eng. Des.*, **10**, 456–69, 1969.
10. O. C. ZIENKIEWICZ and C. J. PAREKH, 'Transient field problems—two and three dimensional analysis by isoparametric finite elements', *Int. J. Num. Meth. Eng.*, **2**, 61–71, 1970.
11. O. C. ZIENKIEWICZ, *The Finite Element Method in Engineering Science*, pp. 335–9, McGraw-Hill, New York, 1971.
12. O. C. ZIENKIEWICZ and R. W. LEWIS, 'An analysis of various time stepping schemes for initial value problems', *Int. J. Earthquake Eng. Struct. Dynam.*, **1**, 407–8, 1973.
13. W. L. WOOD and R. W. LEWIS, 'A comparison of time marching schemes for the transient heat conduction equation', *Int. J. Num. Meth. Eng.*, **9**, 679–89, 1975.
14. O. C. ZIENKIEWICZ, 'A new look at the Newmark, Houbolt and other time

stepping formulas. A weighted residual approach', *Int. J. Earthquake Eng. Struct. Dynam.*, **5**, 413–18, 1977.

15. W. L. WOOD, 'On the Zienkiewicz four-time-level scheme for the numerical integration of vibration problems', *Int. J. Num. Meth. Eng.*, **11**, 1519–28, 1977.
16. O. C. ZIENKIEWICZ, W. L. WOOD and R. L. TAYLOR, 'An alternative single-step algorithm for dynamic problems', *Int. J. Earthquake Eng. Struct. Dynam.*, **8**, 31–40, 1980.
17. W. L. WOOD, 'A further look at Newmark, Houbolt, etc., time-stepping formulae', *Int. J. Num. Meth. Eng.*, **20**, 1009–17, 1984.
18. O. C. ZIENKIEWICZ, W. L. WOOD, N. W. HINE and R. L. TAYLOR, 'A unified set of single-step algorithms, Part 1: general formulation and applications', *Int. J. Num. Meth. Eng.*, **20**, 1529–52, 1984.
19. W. L. WOOD, 'A unified set of single-step algorithms, Part 2: theory', *Int. J. Num. Meth. Eng.*, **20**, 2303–9, 1984.
20. M. G. KATONA and O. C. ZIENKIEWICZ, 'A unified set of single-step algorithms, Part 3: the beta-*m* method, a generalisation of the Newmark scheme', *Int. J. Num. Meth. Eng.*, **21**, 1345–59, 1985.
21. E. VAROGLU and N. D. L. FINN, 'A finite element method for the diffusion convection equations with concurrent coefficients', *Adv. Water Resources*, **1**, 337–41, 1973.
22. C. J. JOHNSON, V. NAVERT and J. PITKÄRANTA, 'Finite element methods for linear, hyperbolic problems', *Comp. Meth. Appl. Mech. Eng.*, **45**, 285–312, 1984.
23. B. M. IRONS and C. TREHARNE, 'A bound theorem for eigen values and its practical applications', in *2nd Conf. on Matrix Methods in Structural Mechanics*, pp. 245–54, Wright-Patterson Air Force Base, Ohio, 1971.
24. K. WASHIZU, *Variational Methods in Elasticity and Plasticity*, 2nd ed., Pergamon Press, Oxford, 1975.
25. M. GURTIN, 'Variational principles for linear elastodynamics', *Arch. Nat. Mech. Anal.*, **16**, 34–50, 1969.
26. M. GURTIN, 'Variational principles for linear initial-value problems', *Q. Appl. Math.*, **22**, 252–6, 1964.
27. E. L. WILSON and R. E. NICKELL, 'Application of finite element method to heat conduction analysis', *Nucl. Eng. Des.*, **4**, 1–11, 1966.
28. J. CRANK and P. NICOLSON, 'A practical method for numerical integration of solutions of partial differential equations of heat conduction type', *Proc. Camb. Phil. Soc.*, **43**, 50, 1947.
29. R. L. TAYLOR and O. C. ZIENKIEWICZ, 'A note on the "order of approximation"', *Int. J. Solids Struct.*, **21**, 793–8, 1985.
30. M. ZLAMAL, 'Finite element methods in heat conduction problems', in *The Mathematics of Finite Elements and Applications* (ed. J. Whiteman), pp. 85–104, Academic Press, London, 1977.
31. W. LINIGER, 'Optimisation of a numerical integration method for stiff systems of ordinary differential equations', IBM research report RC2198, 1968.
32. J. M. BETTENCOURT, O. C. ZIENKIEWICZ and G. CANTIN, 'Consistent use of finite elements in time and the performance of various recurrence schemes for the heat diffusion equation', *Int. J. Num. Meth. Eng.*, **17**, 931–8, 1981.
33. N. M. NEWMARK, 'A method for computation of structural dynamics', *Proc. Am. Soc. Civ. Eng.*, **85**, EM3, 67–94, 1959.
34. T. BELYTSCHKO and T. J. R. HUGHES (eds), *Computational Methods for Transient Analysis*, North Holland, Dordrecht, 1983.

35. I. Miranda, R. M. Ferencz and T. J. R. Hughes, 'An improved implicit–explicit time integration method for structural dynamics', *Int. J. Earthquake Eng. Struct. Dynam.*, **18**, 643–55, 1989.
36. E. J. Routh, *A Treatise on the Stability of a Given State or Motion*, Macmillan, London, 1877.
37. A. Hurwitz, 'Uber die Bedingungen, unter welchen eine Gleichung nur Würzeln mit negativen reellen teilen besitzt', *Math. Ann.*, **46**, 273–84, 1895.
38. F. R. Gantmacher, *The Theory of Matrices*, Chelsea, New York, 1959.
39. G. G. Dahlquist, 'A special stability problem for linear multistep methods', *BIT*, **3**, 27–43, 1963.
40. C. W. Gear, 'The automatic integration of stiff ordinary differential equations', in *Information Processing 68* (ed. A. J. H. Morrell), pp. 187–93, North Holland, Dordrecht, 1969.
41. W. Linger, 'Global accuracy and *A*-stability of one and two step integration formulae for stiff ordinary differential equations', in *Proc. Conf. on Numerical Solution of Differential Equations*, Dundee University, 1969.
42. J. C. Houbolt, 'A recurrence matrix solution for the dynamic response of elastic aircraft', *J. Aero. Sci.*, **17**, 540–50, 1950.
43. E. L. Wilson, 'A computer program for the dynamic stress analysis of underground structures', SEL report, 68–1, University of California, Berkeley, 1968.
44. K. J. Bathe and E. L. Wilson, 'Stability and accuracy analysis of direct integration methods', *Int. J. Earthquake Eng. Struct. Dynam.*, **1**, 283–91, 1973.
45. W. L. Wood, M. Bossak and O. C. Zienkiewicz, 'An alpha modification of Newmark's method', *Int. J. Num. Meth. Eng.*, **15**, 1562–6, 1980.
46. H. M. Hilber, T. J. R. Hughes and R. L. Taylor, 'Improved numerical dissipation for time integration algorithms in structural dynamics, *Int. J. Earthquake Eng. Struct. Dynam.*, **5**, 283–92, 1977.
47. W. L. Wood, 'On the Zienkiewicz three- and four-time level schemes applied to the numerical integration of parabolic equations', *Int. J. Num. Meth. Eng.*, **12**, 1717–26, 1978.
48. T. Belytschko, R. L. Chiapetta and H. D. Bartel, 'Efficient large scale non-linear transient analysis by finite elements', *Int. J. Num. Meth. Eng.*, **10**, 579–96, 1976.
49. D. Shantaram, D. R. J. Owen and O. C. Zienkiewicz, 'Dynamic transient behaviour of two and three dimensional structures including plasticity, large deformation and fluid interaction', *Int. J. Earthquake Eng. Struct. Dynam.*, **4**, 561–78, 1976.
50. R. D. Krieg and S. W. Key, 'Transient shock response by numerical time integration', *Int. J. Num. Meth. Eng.*, **7**, 273–86, 1973.
51. C. C. Fu, 'On the stability of explicit methods for numerical integration of the equations of matrices in finite elements', *Int. J. Num. Meth. Eng.*, **4**, 95–107, 1972.
52. J. O. Hallquist, 'A numerical treatment of sliding interfaces and impact', *ASME*, **AMD-30**, 117–33, 1978.
53. G. L. Goudreau and J. O. Hallquist, 'Recent developments in large-scale finite element lagrangian hydrocode technology', *Comp. Meth. Appl. Mech. Eng.*, **33**, 725–57, 1982.
54. J. O. Hallquist, G. L. Goudreau and D. J. Benson, 'Sliding interfaces with contact-impact in large-scale lagrangian computations', *Comp. Meth. Appl. Mech. Eng.*, **51**, 107–37, 1985.

55. M. LEES, 'A linear three level difference scheme for quasilinear parabolic equations', *Maths. Comp.*, **20**, 516–622, 1966.
56. G. COMINI, S. DEL GUIDICE, R. W. LEWIS and O. C. ZIENKIEWICZ, 'Finite element solution of non-linear conduction problems with special reference to phase change', *Int. J. Num. Meth. Eng.*, **8**, 613–24, 1974.
57. H. D. HIBBITT and P. V. MARCAL, 'Numerical thermo-mechanical model for the welding and subsequent loading of a fabricated structure', *Comp. Struct.*, **3**, 1145–74, 1973.
58. K. MORGAN, R. W. LEWIS and O. C. ZIENKIEWICZ, 'An improved algorithm for heat convection problems with phase change', *Int. J. Num. Meth. Eng.*, **12**, 1191–5, 1978.
59. C. A. ANDERSON and O. C. ZIENKIEWICZ, 'Spontaneous ignition: finite element solutions for steady and transient conditions', *Trans. ASME, J. Heat Transfer*, 398–404, 1974.
60. J. R. M. OTTER, 'Dynamic relaxation', *Proc. Inst. Civ. Eng.*, **35**, 633–56, 1966.
61. O. C. ZIENKIEWICZ and R. LÖHNER, 'Accelerated "relaxation" or direct solution? Future prospects for finite element methods', *Int. J. Num. Mech. Eng.*, **21**, 1–11, 1985.
62. P. UNDERWOOD, 'Dynamic relaxation', in *Computer Methods for Transient Analysis* (eds T. Belytchko and T. J. R. Hughes), chap. 5, pp. 245–65, North Holland, Dordrecht, 1983.
63. J. C. SIMO and K. WONG, 'Unconditionally stable algorithms for rigid body dynamics that exactly conserve energy and momentum', *Int. J. Num. Meth. Eng.* **31**, 19–52, 1991.

11

Coupled systems

11.1 Coupled problems—definition and classification

Frequently two or more physical systems interact with each other, with the independent solution of any one system being impossible without simultaneous solution of the others. Such systems are known as coupled and of course such coupling may be weak or strong depending on the degree of interaction.

An obvious 'coupled' problem is that of dynamic fluid–structure interaction. Here neither the fluid nor the structural system can be solved independently of the other due to the unknown interface forces.

A definition of coupled systems may be generalized to include a wide range of problems and their numerical discretization as:[1]

Coupled systems and formulations are those applicable to multiple domains and dependent variables which usually (but not always) describe different physical phenomena and in which

(a) neither domain can be solved while separated from the other;

(b) neither set of dependent variables can be explicitly eliminated at the differential equation level.

The reader may well contrast this with definitions of *mixed* and *irreducible* formulations given in Chapter 12 of Volume 1 and find some similarities. Clearly 'mixed' and 'coupled' formulations are analogous, with the main difference being that in the former elimination of some dependent variables is possible at the governing differential equation level. In the coupled system a full analytical solution or inversion of a (discretized) single system is necessary before such elimination is possible.

Indeed, a further distinction can be made. In coupled systems the solution of any single system is a well-posed problem and is possible when the variables corresponding to the other system are prescribed. This is not always the case in mixed formulations.

It is convenient to classify coupled systems in two categories:

Class I. This class contains problems in which coupling occurs on domain interfaces via the boundary conditions imposed there. Generally the domains describe different physical situations but it is possible to consider coupling between domains that are physically similar in which different discretization processes have been used.

Class II. This class contains problems in which the various domains overlap (totally or partially). Here the coupling occurs through the differential governing equations describing different physical phenomena.

Typical of the first category are the problems of fluid–structure interaction illustrated in Fig. 11.1(*a*) where physically different problems interact and also structure–structure interactions of Fig. 11.1(*b*) where the interface simply divides arbitrarily chosen regions in which different numerical discretizations are used.

The need for the use of different discretization may arise from different causes. Here for instance:

1. Different finite element meshes may be advantageous to describe the subdomains.
2. Different procedures such as the combination of boundary method and finite elements in respective regions may be computationally desirable.
3. Domains may simply be divided by the choice of different time-stepping procedures, e.g. of an implicit and explicit kind.

In the second category typical problems are illustrated in Fig. 11.2. One of these is that of metal extrusion where the plastic flow is strongly coupled with the temperature field while at the same time the latter is influenced by the heat generated in the plastic flow. The other is that of soil dynamics (earthquake response of a dam) in which the seepage flow and pressures interact with the dynamic behaviour of the soil 'skeleton'.

We observe that, in the examples illustrated, motion invariably occurs. Indeed, the vast majority of coupled problems involve such dynamic behaviour and for this reason the present chapter will only consider this area. It will thus follow and expand the analysis techniques of Chapters 9 and 10.

As the problems encountered in coupled analysis of various kinds are similar, we shall focus the presentation on three examples:

1. Fluid–structure interaction (confined to small amplitudes);
2. Soil–fluid interaction;

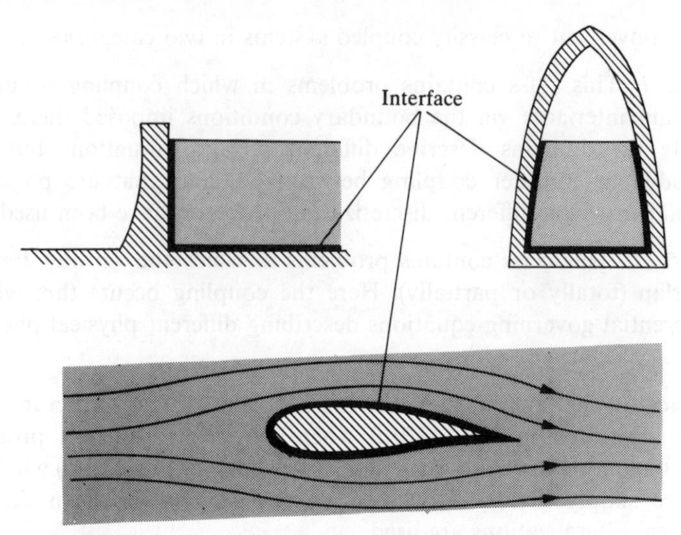

(*a*) Fluid–structure interaction (physically different domains)

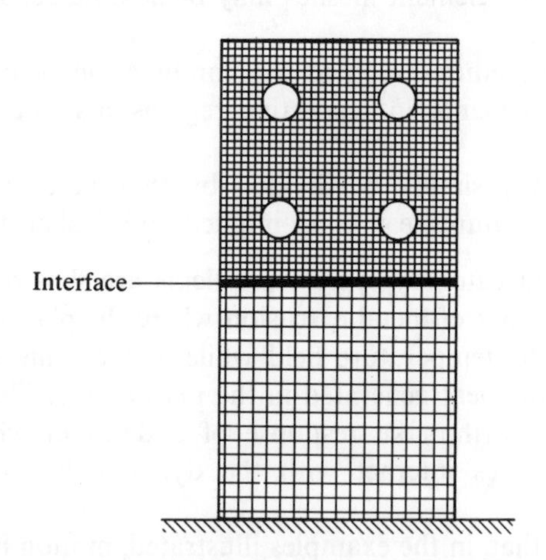

(*b*) Structure–structure interaction (physically identical domains)

Fig. 11.1 Class I problems with coupling via interfaces (shown in thick line)

3. Implicit–explicit dynamic analysis of a structure where the separation involves the process of temporal discretization.

In these problems all the typical features of coupled analysis will be found and extension to others will follow similar lines. In Chapter 13 we

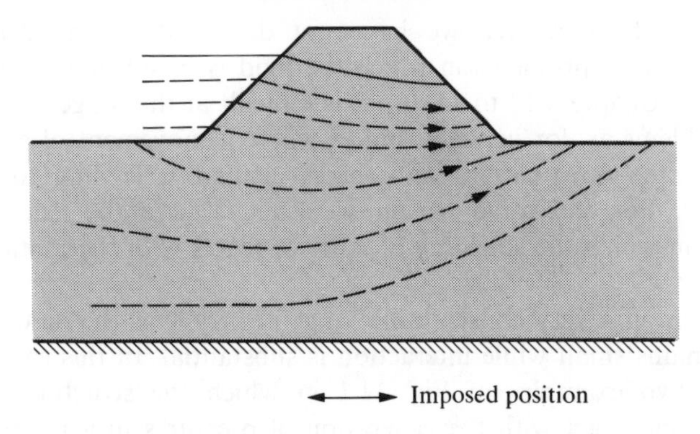

(*a*) Seepage through a porous medium interacts with its dynamic, structural behaviour

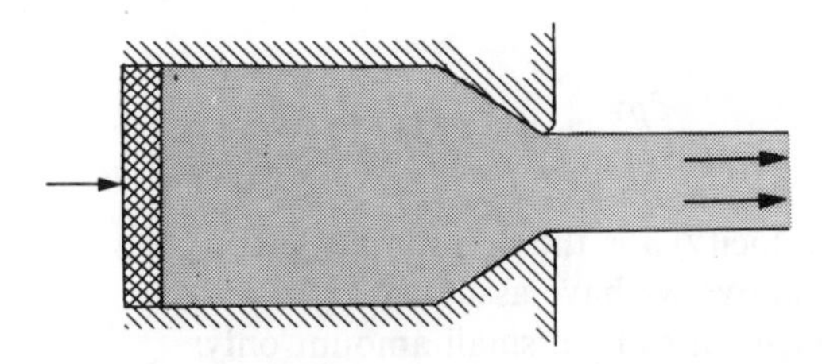

(*b*) Problem of metal extrusion in which the plastic flow is coupled with the thermal field

Fig. 11.2 Class II coupled problems with coupling in overlapping domains

shall, for instance, deal in more detail with the problem of coupled metal forming[2] and the reader will discover the similarities.

As a final remark, it is worth while mentioning that such problems as thermal stress analysis to which we have referred frequently in this book are not coupled in the terms defined here. In this the stress analysis problem requires a knowledge of temperature fields but the temperatures, in general, can be determined by a solution independent of the stress field. Thus the problem decouples in one direction. Many examples of truly coupled problems will be found in available books.[3,4]

11.2 Fluid–structure interaction (class I problem)

11.2.1 *General remarks and fluid behaviour equations.* The problem of fluid–structure interaction is a wide one and covers many forms of fluid behaviour which, as yet, we have not discussed in any detail. The consideration of problems in which the fluid is in substantial motion is deferred to Chapters 13 to 15 and it is difficult at this stage to introduce such problems as, for instance, flutter where a movement of an aerofoil influences the flow pattern and forces around it leading to possible instability. For the same reason we exclude here the 'singing wire' problem in which the shedding of vortices reacts with the motion of the wire.

However, in a very considerable range of problems the fluid displacement remains small while interaction is substantial. In this category fall the first two examples of Fig. 11.1 in which the structural motions influence and react with the generation of pressures in a reservoir or a container. A number of symposia have been entirely devoted to this class of problem which is of considerable engineering interest, and here fortunately considerable simplifications are possible in the description of the fluid phase. References 5 to 17 give some typical studies.

In such problems, it is possible to write the dynamic equations of fluid behaviour simply as

$$\frac{\partial \rho \mathbf{v}}{\partial t} \approx \rho \dot{\mathbf{v}} = -\nabla p \tag{11.1}$$

where $\mathbf{v}$ is the fluid velocity, ρ is the density and p the pressure.

In postulating the above we have assumed

(*a*) that the density ρ varies by a small amount only;

(*b*) that velocities are small enough for convective effects to be omitted; and

(*c*) that viscous effects by which deviatoric stresses are introduced can be neglected in the fluid.

The reader can in fact note that with the preceding assumption Eq. (11.1) is a special form of a more general relation described in Chapter 13 [viz. Eq. (13.21)].

The continuity equation based on the same assumptions is

$$\rho \operatorname{div} \mathbf{v} \equiv \rho \nabla^{\mathrm{T}} \mathbf{v} = \frac{\partial \rho}{\partial t} \tag{11.2}$$

and noting that

$$\mathrm{d}\rho = \frac{\rho \, \mathrm{d}p}{K} \tag{11.3}$$

where K is the bulk modulus, we can write

$$\nabla^{\mathrm{T}} \mathbf{v} = \frac{1}{K} \frac{\partial p}{\partial t} \tag{11.4}$$

Elimination of $\mathbf{v}$ between (11.1) and (11.4) gives the well-known Helmholtz equation governing the pressure p:

$$\nabla^2 p + \frac{1}{c^2}\frac{\partial^2 p}{\partial t^2} = 0 \tag{11.5a}$$

where

$$c = \sqrt{\frac{K}{\rho}} \tag{11.5b}$$

denotes the speed of sound in the fluid.

The equations described above are in fact the basis of all *acoustic* problems.

11.2.2 *Boundary conditions for the fluid. Coupling and radiation.* In Fig. 11.3 we focus on the class I problem illustrated in Fig. 11.1(*a*) and on the boundary conditions possible for the fluid part described by the governing equation (11.5). As we know well, either normal gradients or values of p need now to be specified.

Interface with solid (boundary ①, ② in Fig. 11.3). On these boundaries the normal velocities (or their time derivatives) are prescribed. Considering the pressure gradient in the normal direction to the face n we can thus write, by Eq. (11.1),

$$\frac{\partial p}{\partial n} = -\rho \dot{\bar{v}}_n \tag{11.6}$$

where $\dot{\bar{v}}_n$ is prescribed.

Thus, for instance, on boundary ① coupling with the motion of the

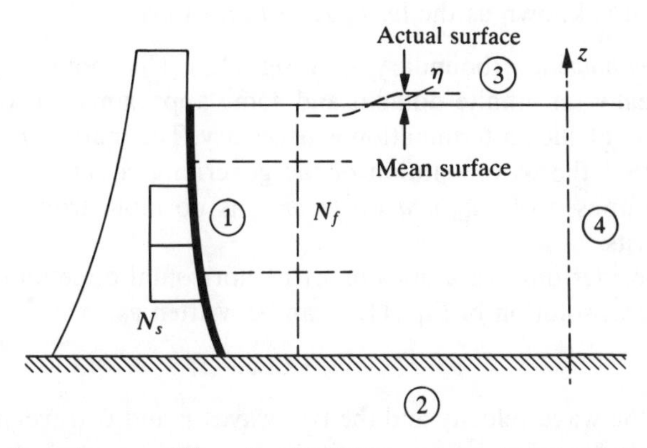

Fig. 11.3 Boundary conditions for the fluid component of structure–fluid interaction

structure described by displacements $\mathbf{u}$ occurs. Here we put

$$\dot{\bar{\mathbf{v}}}_n = \ddot{\mathbf{u}}_n \tag{11.7}$$

while on boundary ② where only horizontal motion exists we have

$$\dot{\mathbf{v}}_z = 0 \tag{11.8}$$

Coupling with the structure motion clearly occurs only via boundary ①.

Free surface (boundary ③ in Fig. 11.3). On the free surface the simplest assumption is that

$$p = 0 \tag{11.9}$$

However, this does not allow for any possibility of surface gravity waves. These can be approximated by assuming the actual surface to be at an elevation η relative to the mean surface. Now

$$p = \rho g \eta \tag{11.10a}$$

where ρ is the density and g the gravity acceleration. However, by Eq. (11.1), we have, on noting that $v_z = \partial\eta/\partial t$ and assuming ρ to be constant,

$$\rho \frac{\partial^2 \eta}{\partial t^2} = -\frac{\partial p}{\partial z} \tag{11.10b}$$

and on elimination of η, by using Eq. (11.10a) we have a specified normal gradient condition

$$\frac{\partial p}{\partial z} = -\frac{1}{g}\frac{\partial^2 p}{\partial t^2} = -\frac{1}{g}\ddot{p} \tag{11.11}$$

This allows for gravity waves to be approximately incorporated in the analysis and is known as the *linearized surface wave condition.*

Radiation boundary (boundary ④ in Fig. 11.3). This 'boundary' terminates physically an infinite domain and some approximation to account for the effect of such a termination is necessary. The main dynamic effect is simply that the wave solution of the governing equation (11.5) must here be composed of *outgoing waves only* as no input from the infinite domain exists.

If we consider only variations in x (the horizontal direction) we know that a general solution of Eq. (11.5) can be written as

$$p = F(x - ct) + G(x + ct) \tag{11.12}$$

where c is the wave velocity and the two waves F and G travel in positive and negative directions of x respectively.

The absence of the incoming wave G means that on boundary ④ we have only

$$p = F(x - ct) \tag{11.13}$$

Thus

$$\frac{\partial p}{\partial n} \equiv \frac{\partial p}{\partial x} = F' \tag{11.14a}$$

and

$$\frac{\partial p}{\partial t} = -cF' \tag{11.14b}$$

where F' is the derivative of F with respect to $(x - ct)$. We can therefore eliminate the unknown function F' and write

$$\frac{\partial p}{\partial n} = -\frac{1}{c}\dot{p} \tag{11.15}$$

which is a condition very similar to that of Eq. (11.11). This boundary condition was first presented in reference 5 for radiating boundaries and has an analogy with a damping element placed there.

11.2.3 *The discrete coupled problem.* We shall now consider the structure discretized in the standard (displacement) manner with the displacement vector approximated as

$$\mathbf{u} \approx \hat{\mathbf{u}} = \mathbf{N}_u \bar{\mathbf{u}} \tag{11.16a}$$

and the fluid similarly approximated by

$$\mathbf{p} \approx \hat{\mathbf{p}} = \mathbf{N}_p \bar{\mathbf{p}} \tag{11.16b}$$

where $\bar{\mathbf{u}}$ and $\bar{\mathbf{p}}$ are the (nodal) parameters of each field and $\mathbf{N}_u$ and $\mathbf{N}_p$ appropriate shape functions.

The discrete structural equations thus become

$$\mathbf{M}\ddot{\bar{\mathbf{u}}} + \mathbf{C}\dot{\bar{\mathbf{u}}} + \mathbf{K}\bar{\mathbf{u}} - \mathbf{Q}\bar{\mathbf{p}} + \mathbf{f} = \mathbf{0} \tag{11.17}$$

where the coupling term arises due to the pressures specified on the boundary and is

$$\int_{\Gamma_1} \mathbf{N}_u^{\mathrm{T}} \, \mathbf{n} p \, \mathrm{d}\Gamma \equiv \left(\int_{\Gamma_1} \mathbf{N}_u^{\mathrm{T}} \mathbf{n} \mathbf{N}_p \, \mathrm{d}\Gamma \right) \bar{\mathbf{p}} = \mathbf{Q}\bar{\mathbf{p}} \tag{11.18}$$

In the above $\mathbf{n}$ is the direction vector of the normal to the interface. The terms of the other matrices are already well known to the reader as mass, damping, stiffness and force.

Standard (Galerkin) discretization applied to the fluid equation (11.5) and its boundary conditions leads to

$$\mathbf{S}\ddot{\bar{\mathbf{p}}} + \tilde{\mathbf{C}}\dot{\bar{\mathbf{p}}} + \mathbf{H}\bar{\mathbf{p}} + \mathbf{Q}^{\mathrm{T}}\ddot{\bar{\mathbf{u}}} + \mathbf{q} = \mathbf{0} \tag{11.19}$$

Here we shall for completeness quote the form of some of the matrices of the above. Thus

$$\mathbf{S} = -\int_{\Omega} \mathbf{N}_p^{\mathrm{T}} \frac{1}{c^2} \mathbf{N}_p \, \mathrm{d}\Omega + \int_{\Gamma_3} \mathbf{N}_p^{\mathrm{T}} \frac{1}{g} \mathbf{N}_p \, \mathrm{d}\Omega \tag{11.20}$$

$$\tilde{\mathbf{C}} = \int_{\Gamma_4} \mathbf{N}_p^{\mathrm{T}} \frac{1}{c^2} \mathbf{N}_p \, \mathrm{d}\Omega$$

$$\mathbf{H} = \int_{\Omega} \nabla \mathbf{N}^{\mathrm{T}} \nabla \mathbf{N} \, \mathrm{d}\Omega$$

and $\mathbf{Q}$ is identical to that of Eq. (11.18).

11.2.4 *Free vibrations.* If we consider free vibrations and omit all damping terms (noting that in the fluid component the damping is strictly that due to radiation energy loss) we can write the two equations (11.17) and (11.19) as a set:

$$\begin{bmatrix} \mathbf{M} & \mathbf{O} \\ \mathbf{Q}^{\mathrm{T}} & \mathbf{S} \end{bmatrix} \begin{Bmatrix} \ddot{\bar{\mathbf{u}}} \\ \ddot{\bar{\mathbf{p}}} \end{Bmatrix} + \begin{bmatrix} \mathbf{K} & -\mathbf{Q} \\ \mathbf{O} & \mathbf{H} \end{bmatrix} \begin{Bmatrix} \bar{\mathbf{u}} \\ \bar{\mathbf{p}} \end{Bmatrix} = \mathbf{0} \tag{11.21}$$

and attempt to proceed and establish the eigenvalues corresponding to natural frequencies. However, we note immediately that the system is not symmetric (nor positive definite) and that standard eigenvalue computation methods are not directly applicable. Physically it is, however, clear that the eigenvalues are real and that free vibration modes exist.

The above problem is similar to that arising in vibration of rotating solids and special solution methods are available, though costly.[18] It is possible by various manipulations to arrive at a symmetric form and reduce the problem to a standard eigenvalue one.[14,15,19–21]

A simple method recently suggested[22] proceeds to achieve the symmetrization objective as follows.

Putting $\bar{\mathbf{u}} = \tilde{\mathbf{u}}\, \mathrm{e}^{i\omega t}$ $\bar{\mathbf{p}} = \tilde{\mathbf{p}}\, \mathrm{e}^{i\omega t}$ we rewrite Eq. (11.21) as

$$\mathbf{K}\tilde{\mathbf{u}} - \mathbf{Q}\tilde{\mathbf{p}} - \omega^2 \mathbf{M}\tilde{\mathbf{u}} = \mathbf{0} \tag{11.22a}$$

$$\mathbf{H}\tilde{\mathbf{p}} - \omega^2 \mathbf{S}\tilde{\mathbf{p}} - \omega^2 \mathbf{Q}^{\mathrm{T}}\tilde{\mathbf{u}} = \mathbf{0} \tag{11.22b}$$

and introduce an additional variable $\mathbf{q}$ such that

$$\tilde{\mathbf{p}} = \omega^2 \mathbf{q} \tag{11.22c}$$

After some manipulation and substitution we can write the new system as

$$\left\{ \begin{bmatrix} \mathbf{K} & 0 & 0 \\ 0 & \mathbf{S} & 0 \\ 0 & 0 & 0 \end{bmatrix} - \omega^2 \begin{bmatrix} \mathbf{M} & 0 & \mathbf{Q} \\ 0 & 0 & \mathbf{S} \\ \mathbf{Q}^{\mathrm{T}} & \mathbf{S}^{\mathrm{T}} & \mathbf{H} \end{bmatrix} \right\} \begin{Bmatrix} \tilde{\mathbf{u}} \\ \tilde{\mathbf{p}} \\ \tilde{\mathbf{q}} \end{Bmatrix} = 0 \tag{11.23}$$

which is a symmetric standard form. Further, the variable $\mathbf{q}$ can now be eliminated by static condensation and the final system becomes symmetric and still contains only the basic variables.

An alternative that has frequently been used is to introduce a new symmetrizing variable at the governing equation level, but this is clearly not necessary.[14,15]

As an example of a simple problem in the present category we show an analysis of a three-dimensional flexible wall vibrating with a fluid encased in a 'rigid' container[23] (Fig. 11.4).

11.2.5 *Forced vibrations and transient step-by-step algorithms.* The reader can easily verify that steady-state, linear, response to periodic input can be readily computed in the complex frequency domain by the procedures described in Chapter 9. Here no difficulties arise due to the non-symmetric nature of equations and standard procedures can be applied. Chopra and coworkers have, for instance, done many studies of dam/reservoir interaction using such methods.[24,25] However, such methods are not generally economical and fail in non-linear response studies. Here time-stepping procedures are required in the manner discussed in the previous chapter. However, simple application of methods developed there is not possible for the combined system (with $\tilde{\mathbf{u}}$ and $\tilde{\mathbf{p}}$ as variables) due to the non-symmetric matrices and a modified approach is necessary.[26] In this each of the equations (11.17) and (11.19) is *discretized in time separately* using the general approaches of Chapter 10.

Thus in the time interval Δt we can approximate $\bar{\mathbf{u}}$ using, say, the general SS22 procedure as follows. First we write

$$\bar{\mathbf{u}} = \bar{\mathbf{u}}_n + \dot{\bar{\mathbf{u}}}_n \tau + \frac{\boldsymbol{\alpha}\tau^2}{2} \tag{11.24a}$$

with a similar expression for $\bar{\mathbf{p}}$,

$$\bar{\mathbf{p}} = \bar{\mathbf{p}}_n + \dot{\bar{\mathbf{p}}}_n \tau + \frac{\boldsymbol{\beta}\tau^2}{2} \tag{11.24b}$$

where $\tau = t - t_n$ and the time interval is Δt.

Insertion of the above into Eqs (11.17) and (11.19) and weighting with two *separate weighting functions* results in two relations in which $\boldsymbol{\alpha}$ and $\boldsymbol{\beta}$ are the unknowns. These are

$$\begin{aligned} \mathbf{M}\boldsymbol{\alpha} + \mathbf{C}(\dot{\bar{\mathbf{u}}}_n + \boldsymbol{\alpha}\Theta_1 \Delta t) + \mathbf{K}\left(\bar{\mathbf{u}}_n + \mathbf{u}_n \Theta_1 \Delta t + \frac{\boldsymbol{\alpha}\Theta_2 \Delta t^2}{2}\right) \\ - \mathbf{Q}\left(\bar{\mathbf{p}}_n + \dot{\bar{\mathbf{p}}}_n \Theta_1 \Delta t + \frac{\boldsymbol{\beta}\Theta_2 \Delta t^2}{2}\right) + \bar{\mathbf{f}} = \mathbf{0} \end{aligned} \tag{11.25a}$$

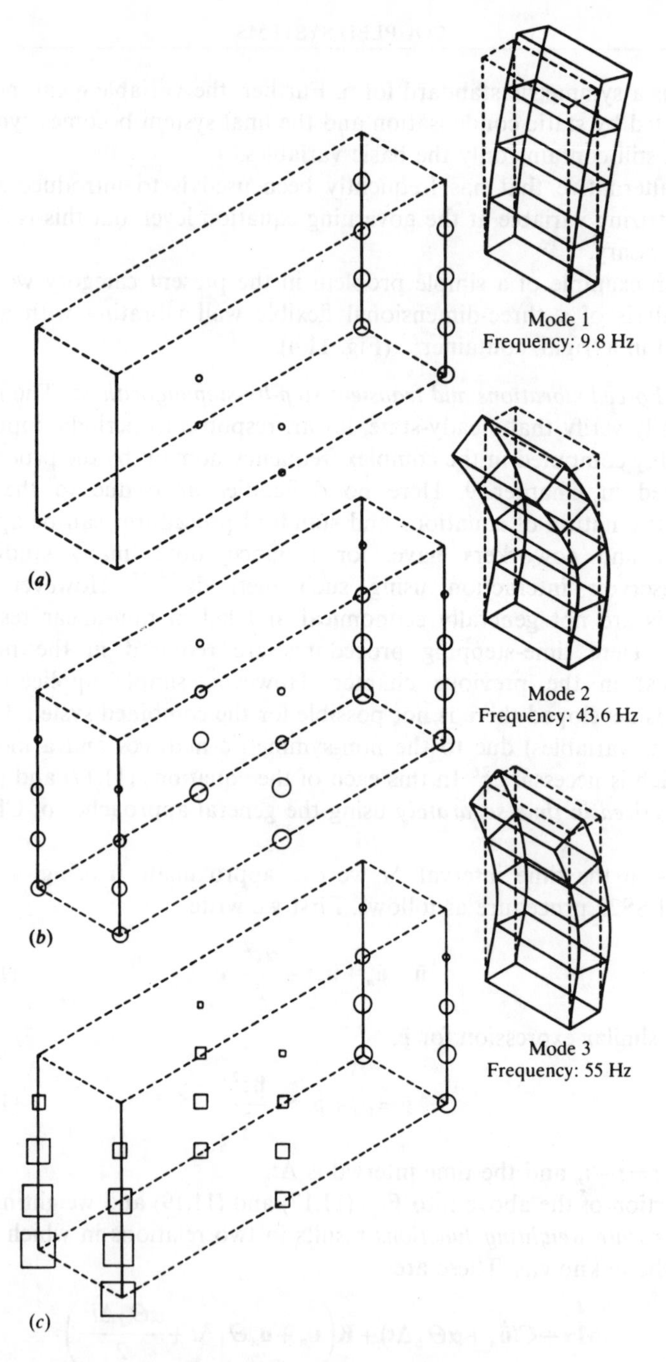

Fig. 11.4 Body of fluid with a free surface oscillating with a wall. Circles show pressure amplitudes and squares indicate opposite signs. A three-dimensional approach using parabolic elements[23]

and $$\mathbf{S}\boldsymbol{\beta}+\mathbf{Q}^{\mathrm{T}}\boldsymbol{\alpha}+\mathbf{H}\left(\bar{\mathbf{p}}_n+\mathbf{p}_n\bar{\Theta}_1\Delta t+\frac{\boldsymbol{\beta}\Theta_2\Delta t^2}{2}\right)+\mathbf{q}=\mathbf{b} \tag{11.25b}$$

In the above the parameters $\Theta_1, \Theta_2, \bar{\Theta}_1$ and $\bar{\Theta}_2$ are similar to those of Eq. (10.35) and can be chosen by the user. It is interesting to observe that the equation system written as below becomes symmetric of the form

$$\begin{bmatrix} \mathbf{M}+\mathbf{C}\Theta_1\Delta t, & +\mathbf{K}\Theta_2\Delta t^2/2, & -\mathbf{Q}\Theta_2\Delta t^2/2 \\ -\mathbf{Q}\Theta_2\Delta t^2/2, & -\mathbf{S}\Theta_2\Delta t^2/2, & -\mathbf{H}\bar{\Theta}_2\Theta_2\Delta t^4/2 \end{bmatrix}\begin{Bmatrix}\boldsymbol{\alpha}\\ \boldsymbol{\beta}\end{Bmatrix}=\begin{Bmatrix}\mathbf{F}_1\\ \mathbf{F}_2\end{Bmatrix} \tag{11.26}$$

where $\mathbf{F}_1$ and $\mathbf{F}_2$ are composed of known quantities. In deriving above, the second equation of (11.25) was multiplied by $-\Theta_2\Delta t^2/2$.

It is not necessary to go into detail of computation as these follow the usual patterns of determining $\boldsymbol{\alpha}$ and $\boldsymbol{\beta}$ and then evaluation of the problem variables, that is $\bar{\mathbf{u}}_{n+1}, \bar{\mathbf{p}}_{n+1}, \dot{\bar{\mathbf{u}}}_{n+1}$ and $\dot{\bar{\mathbf{p}}}_{n+1}$ at $\mathbf{t}_{n+1}$ before proceeding with the next step.

Non-linearity of structural behaviour can readily be accommodated again using procedures described in the previous chapter.

It is, however, important to consider the stability of the system which of course will depend on the choice of Θ_i and $\bar{\Theta}_i$. Here we find, by using procedures described in Chapter 10, that unconditional stability is obtained when

$$\begin{aligned} \Theta_2 &\geqslant \Theta_1 \qquad & \Theta_1 &\geqslant \tfrac{1}{2} \\ \bar{\Theta}_2 &\geqslant \Theta_2 \qquad & \bar{\Theta}_1 &\geqslant \tfrac{1}{2} \end{aligned} \tag{11.27}$$

It is instructive to note that precisely the same result would be obtained if GN22 approximations were used in Eq. (11.24).

The derivation of such stability conditions is straightforward and follows precisely the lines of Sec. 10.3.4 of the previous chapter. However, the algebra is sometimes tedious and in general we shall only present the results in the present chapter. Nevertheless, to allow the reader to repeat such calculations for any case encountered we shall outline the calculations for the present example.

Stability of the structure–fluid time-stepping scheme.[26] For stability evaluations is it always advisable to consider the modally decomposed system with scalar values. We thus rewrite Eq. (11.25) omitting the forcing terms and putting $\Theta_1=\bar{\Theta}_1$, etc., as

$$m\alpha+c(\dot{u}_n+\alpha\Theta_1\Delta t)+k\left(u_n+\dot{u}_n\Theta_1\Delta t+\frac{\alpha\Theta_1\Delta t^2}{2}\right)$$
$$-q\left(p_n+\dot{p}_n\Theta_1\Delta t+\frac{\alpha\Theta_2\Delta t^2}{2}\right)=0 \tag{A}$$

and $$s\beta+q\alpha+h\left(p_n+\dot{p}_n\Theta_1\Delta t+\frac{\beta\Theta_2\Delta t^2}{2}\right)=0 \tag{B}$$

To complete the recurrence relations we have

$$
\begin{aligned}
u_{n+1} &= u_n + \dot{u}_n \Delta t + \frac{\alpha \Delta t^2}{2} \\
\dot{u}_{n+1} &= \dot{u}_n + \alpha \Delta t \\
p_{n+1} &= p_n + \dot{p}_n \Delta t + \frac{\beta \Delta t^2}{2} \\
\dot{p}_{n+1} &= \dot{p}_n + \beta \Delta t
\end{aligned} \tag{C}
$$

The exact solution of the above system will always be of the form

$$
\begin{aligned}
u_{n+1} &= \mu u_n \\
\dot{u}_{n+1} &= \mu \dot{u}_n \\
p_{n+1} &= \mu p_n \\
\dot{p}_{n+1} &= \mu \dot{p}_n
\end{aligned} \tag{D}
$$

and immediately we put

$$\mu = \frac{1-z}{1+z}$$

knowing that for stability we require a negative real part of z.

Eliminating all $n+1$ values from Eqs (C) and (D) leads to

$$
\begin{aligned}
\dot{u}_n &= \frac{2z}{\Delta t} u_n & \dot{p}_n &= \frac{2z}{\Delta t} p_n \\
\alpha &= \frac{4z^2}{(1-z)\Delta t^2} u_n & \beta &= \frac{4z^2}{(1-z)\Delta t^2} p_n
\end{aligned} \tag{E}
$$

Inserting (E) into the system (A) gives

$$
\begin{aligned}
\{[4m' - 2(1-2\Theta_1)c' - 2k(\Theta_1 - \Theta_2)]z^2 + [2c' - k(1-2\Theta_1)]z + k\}u_n \\
+ \{2q(\Theta_1 - \Theta_2)z^2 + q(1-2\Theta_1)z - q\}p_n = 0
\end{aligned} \tag{F}
$$

$$\text{and } 4qz^2 u_n + \{[4s - 2h'(\Theta_1 - \Theta_2)]z^2 - h'(1-2\Theta_1)z + h'\}p_n = 0$$

where

$$m' = \frac{m}{\Delta t^2} \qquad c' = \frac{c}{\Delta t} \qquad h' = h\,\Delta t^2$$

For non-trivial solutions to exist the determinant of Eq. (F) has to be zero. This determinant provides the characteristic equation for z which is a polynomial of the fourth order in the form

$$a_0 z^4 + a_1 z^3 + a_2 z^2 + a_4 = 0$$

The use of the Routh–Hurwitz conditions (viz. page 374) ensures that stability requirements are satisfied, i.e. that the roots of z have negative real parts.

The requirements are the following:

$$a_0 > 0, \qquad a_1 \geqslant 0, \qquad a_2 \geqslant 0, \qquad a_3 \geqslant 0 \quad \text{and} \quad a_4 \geqslant 0$$

$$
\begin{aligned}
a_0 = [4m' - 2(1-2\Theta_1)c' - 2k(\Theta_1 - \Theta_2)]\,[4s - 2h'(\Theta_1 - \Theta_2)] - 2q(\Theta_1 - \Theta_2) + 4q \\
> 0 \quad \text{for } m', c', k, s, h' \geqslant 0
\end{aligned} \tag{G}
$$

The inequality is satisfied if

$$\Theta_1 \geqslant \tfrac{1}{2} \qquad \Theta_2 \geqslant \Theta_1$$

If all the equalities hold then $m's > 0$ has to be satisfied. If $m's = 0$ and $c' = 0$ then $\Theta_2 > \Theta_1$ has to be satisfied.

$$\begin{aligned} a_1 &= [4m' - 2(1-2\Theta_1)c' - 2k(\Theta_1 - \Theta_2)]\,[-h'(1-2\Theta_1)] \\ &\quad + [2c' - k(1-2\Theta_1)]\,[4s - 2h'(\Theta_1 - \Theta_2)] - 4q^2(1-2\Theta_1) \\ &\geqslant 0 \end{aligned} \tag{H}$$

The inequality is satisfied if

$$\Theta_1 \geqslant \tfrac{1}{2} \qquad \Theta_2 \geqslant \Theta_1$$

$$\begin{aligned} a_2 &= [4m' - 2(1-2\Theta_1)c' - 2k(\Theta_1 - \Theta_2)]h' + [2c' - k(1-2\Theta_1)]\,[-h'(1-2\Theta_1)] \\ &\quad + k[4s - 2h'(\Theta - \Theta_2)] + 4q^2 \\ &\geqslant 0 \end{aligned} \tag{I}$$

The inequality is satisfied if (G) and (H) are satisfied.

$$a_3 = [2c' - k(1-2\Theta_1)]h' - kh'(1-2\Theta_1) \geqslant 0 \tag{J}$$

The inequality is satisfied if (G) and (H) are satisfied.

$$a_4 = kh' \geqslant 0 \tag{K}$$

The inequality is satisfied automatically.

$$\left.\begin{aligned} a_1 a_2 - a_0 a_3 &\geqslant 0 \\ a_1 a_2 a_3 - a_0 a_3^2 - a_4 a_1^2 &\geqslant 0 \end{aligned}\right\} \tag{L}$$

The two inequalities are satisfied if (G) and (H) are satisfied.

11.2.6 *Special case of incompressible fluids.* If the fluid is incompressible as well as being inviscid, its behaviour is described by a simple laplacian equation

$$\nabla^2 p = 0 \tag{11.28}$$

obtained by putting $c = \infty$ in Eq. (11.5).

In the absence of surface wave effects and of non-zero prescribed pressures the discrete equation (11.19) becomes simply

$$\mathbf{H}\bar{\mathbf{p}} = -\mathbf{Q}^{\mathrm{T}}\ddot{\bar{\mathbf{u}}} \tag{11.29a}$$

as wave radiation disappears. It is simple to obtain now

$$\bar{\mathbf{p}} = -\mathbf{H}^{-1}\mathbf{Q}^{\mathrm{T}}\ddot{\bar{\mathbf{u}}} \tag{11.29b}$$

and substitution of the above into the structure equation (11.17) results in

$$(\mathbf{M} + \mathbf{Q}\mathbf{H}^{-1}\mathbf{Q}^{\mathrm{T}})\ddot{\bar{\mathbf{u}}} + \mathbf{C}\dot{\bar{\mathbf{u}}} + \mathbf{K}\bar{\mathbf{u}} + \mathbf{f} = \mathbf{0} \tag{11.30}$$

This is now a standard structural system in which the mass matrix has been augmented by an *added mass matrix*

$$\mathbf{M}_a = \mathbf{Q}\,\mathbf{H}^{-1}\,\mathbf{Q}^{\mathrm{T}} \tag{11.31}$$

and its solution follows standard patterns of previous chapters.

We have to remark that

1. In general the complete inverse of **H** is not required as pressures at interface nodes only are needed.
2. In general the question of when compressibility effects can be ignored is a difficult one and will depend much on the frequencies that have to be considered in the analysis. For instance, in the analysis of reservoir–dam interaction much debate on the subject has been recorded.[27] Here the fundamental compressible period may be of the order H/c where H is a typical dimension (such as the height of the dam). If this period is of the same order as that of, say, earthquake forcing motion then of course compressibility must be taken into account. If it is much shorter then its neglect can be justified.

11.2.7 *Cavitation effects in fluids.* In such fluids as water the linear behaviour under volumetric strain ceases when pressures fall below a certain threshold. This is the vapour pressure limit. When this is reached cavities or distributed bubbles form and the pressure remains almost constant. To follow such behaviour a non-linear constitutive law has to be introduced. A convenient variable useful in this formulation was defined by Newton:[28]

$$s = \text{div}(\rho \mathbf{u}) \equiv \nabla^T(\rho \mathbf{u}) \tag{11.32}$$

where $\mathbf{u}$ is the fluid displacement. The non-linearity now is such that

$$\begin{aligned} & p = -Ku_{i,i} = c^2 s \qquad \text{in linear behaviour} \\ \text{if} \quad & s < (p_a - p_v)/c^2 \\ \text{and} \quad & \\ & p = p_a - p_v \\ \text{if} \quad & s > (p_a - p_v)/c^2 \qquad \text{below the threshold} \end{aligned} \tag{11.33}$$

Here p_a is the atmospheric pressure (at which $u = 0$ is assumed), p_v is the vapour pressure and c is the sound velocity.

Clearly monitoring strains is a difficult problem in the formulation using the velocity and pressure variables [Eqs (11.1) to (11.5)]. Here therefore a change of variable is desirable and it is convenient to introduce the displacement potential ψ such that

$$\rho \mathbf{u} = -\nabla \psi \tag{11.34}$$

From the momentum equation (11.1) we see that

$$\rho \ddot{\mathbf{u}} = -\nabla p = -\nabla \ddot{\psi} \tag{11.35}$$

Thus

$$\ddot{\psi} = p$$

The continuity equation (11.2) now gives

$$s=\rho \operatorname{div} \mathbf{u}=-\nabla^2\psi=\left(\frac{1}{c^2}\right)p=\left(\frac{1}{c^2}\right)\ddot{\psi} \tag{11.36}$$

in the linear case [with appropriate change according to conditions (11.33) during cavitation].

It is interesting to note that Eq. (11.36) has the same form as that of Eq. (11.5) though the variable has been changed.

Details of boundary conditions, discretization and coupling are fully described in reference 29 and follow the standard methodology previously given. Figure 11.5, taken from that reference, illustrates results of a

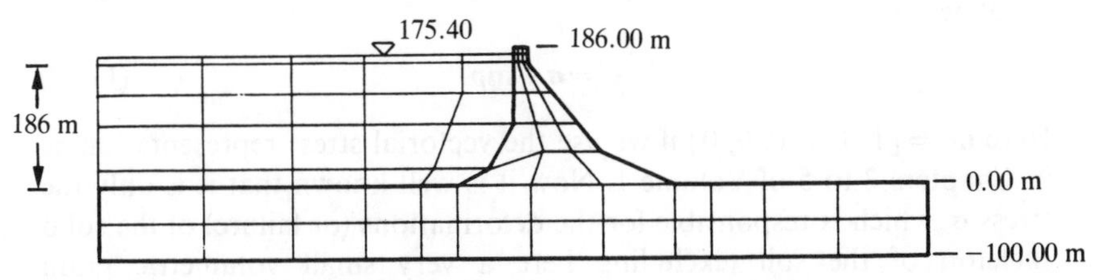

(*a*) Structure–fluid mesh (quadratic elements)

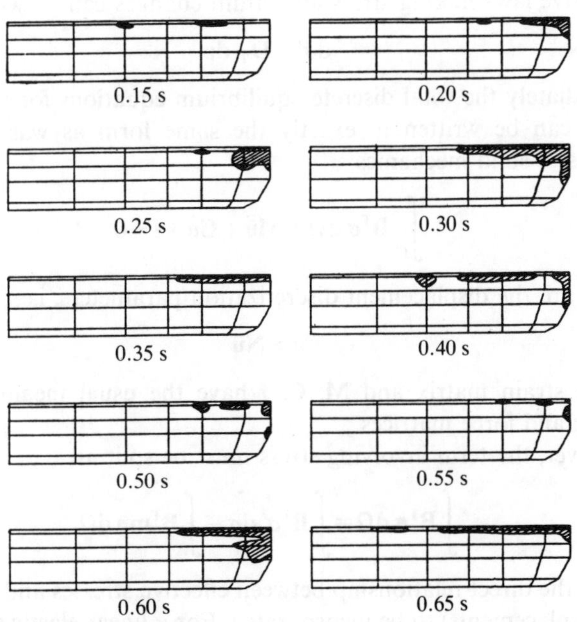

(*b*) Zones in which cavitation develops

Fig. 11.5 The Bhakra Dam–reservoir system.[29] Interaction during the first second of earthquake motion showing the development of cavitation

non-linear analysis showing the development of cavity zones in a reservoir.

11.3 Soil–pore fluid interaction (class II problem)

11.3.1 *The problem and the governing equation. Discretization.* It is well known that the behaviour of soils (and indeed other geomaterials) is strongly influenced by the pressures of the fluid present in the pores of the material. Indeed, the concept of *effective stress* is here of paramount importance. Thus if $\boldsymbol{\sigma}$ is the tensor describing the total stress (positive in tension) acting on the total area of the solid and the pores, and p is the pressure of the fluid in the pores (generally of water), the effective stress is defined as

$$\boldsymbol{\sigma}' = \boldsymbol{\sigma} + \mathbf{m}p \tag{11.37}$$

Here $\mathbf{m}^{\mathrm{T}} = [1, 1, 1, 0, 0, 0]$ if we use the vectorial stress representation as in Chapters 2 to 5 of Volume 1. Now it is well known that it is only the stress $\boldsymbol{\sigma}'$ which is responsible for the deformations (or failure) of the solid skeleton of the soil (excluding here a very small volumetric grain compression which has to be included in some cases). Thus incremental constitutive laws linking stress and strain changes can be written as

$$\mathrm{d}\boldsymbol{\sigma}' = \mathbf{D}_T \, \mathrm{d}\boldsymbol{\varepsilon} \tag{11.38}$$

Immediately the total discrete equilibrium equations for the soil–fluid mixture can be written in exactly the same form as was done in all problems of solid mechanics:

$$\int_{\Omega} \mathbf{B}^{\mathrm{T}} \boldsymbol{\sigma} \, \mathrm{d}\Omega + \mathbf{M}\ddot{\bar{\mathbf{u}}} + \mathbf{C}\dot{\bar{\mathbf{u}}} + \mathbf{f} = 0 \tag{11.39}$$

where $\bar{\mathbf{u}}$ are the displacement discretization parameters, i.e.

$$\mathbf{u} \approx \mathbf{N}\bar{\mathbf{u}} \tag{11.40}$$

$\mathbf{B}$ is the strain matrix and $\mathbf{M}$, $\mathbf{C}$, $\mathbf{f}$ have the usual meaning of mass, damping and force matrices.

However, the term involving stress must be split as

$$\int \mathbf{B}^{\mathrm{T}} \boldsymbol{\sigma} \, \mathrm{d}\Omega \equiv \int \mathbf{B}^{\mathrm{T}} \boldsymbol{\sigma}' \, \mathrm{d}\sigma - \int \mathbf{B}^{\mathrm{T}} \mathbf{m} p \, \mathrm{d}\Omega \tag{11.41}$$

to allow the direct relationship between effective stresses and strains (and hence displacements) to be incorporated. For a linear elastic soil skeleton where $\mathbf{D}_T$ is constant we have immediately

$$\mathbf{M}\ddot{\bar{\mathbf{u}}} + \mathbf{C}\dot{\bar{\mathbf{u}}} + \mathbf{K}\bar{\mathbf{u}} - \mathbf{Q}\bar{\mathbf{p}} + \mathbf{f} = \mathbf{0} \tag{11.42}$$

where $\mathbf{K}$ is the standard stiffness matrix written as

$$\int_\Omega \mathbf{B}^T\boldsymbol{\sigma}'\,d\Omega=\left(\int_\Omega \mathbf{B}^T\mathbf{D}\mathbf{B}\,d\Omega\right)\bar{\mathbf{u}}=\mathbf{K}\bar{\mathbf{u}} \tag{11.43}$$

and $\mathbf{Q}$ couples the field of pressures and the equilibrium equations assuming that these are discretized as

$$p\approx\mathbf{N}_p\bar{\mathbf{p}} \tag{11.44}$$

Thus

$$\mathbf{Q}=\int_\Omega \mathbf{B}^T\mathbf{m}\mathbf{N}_p\,d\Omega \tag{11.45}$$

In the above discretization conveniently the same element shapes are used for $\bar{\mathbf{u}}$ and $\bar{\mathbf{p}}$ variables, though not necessarily identical interpolations. With the dynamic equations coupled to the pressure field an additional equation is clearly needed from which the pressure field can be derived. This is provided by the transient seepage equation of the form

$$-\nabla(k\,\nabla p)+\frac{\dot{p}}{Q}+\dot{\varepsilon}_{ii}=0 \tag{11.46}$$

where Q is related to the compressibility of the fluid, k is the permeability and the last term represents the volumetric strain rate of the soil skeleton, which of course is given on discretization of displacements as

$$\varepsilon_{ii}\approx\mathbf{m}^T\mathbf{B}\bar{\mathbf{u}} \tag{11.47}$$

The equation of seepage can now of course be discretized in the standard Galerkin manner as

$$\mathbf{Q}^T\dot{\bar{\mathbf{u}}}+\mathbf{S}\dot{\bar{\mathbf{p}}}+\mathbf{H}\bar{\mathbf{p}}+\mathbf{q}=\mathbf{0} \tag{11.48}$$

where $\mathbf{Q}$ is precisely that of Eq. (11.45), and

$$\mathbf{S}=\int_\Omega \mathbf{N}_p^T\frac{1}{Q}\mathbf{N}_p\,d\Omega \qquad \mathbf{H}=\int_\Omega \nabla\mathbf{N}_p^T k\,\nabla\mathbf{N}_p\,d\Omega \tag{11.49}$$

with $\mathbf{q}$ containing the forcing and boundary terms. The derivation of coupled flow–soil equations was first introduced by Biot[30] but the present formulation is elaborated in references 31 to 42 where the various approximations as well as the effect of various constitutive laws with non-linear responses are discussed.

We shall not comment in detail on any of the boundary conditions as these are of standard type and are well documented in previous chapters.

11.3.2 *The format of the coupled equations.* Although the main interest in the solution of the coupled equations is in their non-linear form, it is instructive to consider the linear version of Eqs (11.42) and (11.48). This can be written as

$$\begin{bmatrix} \mathbf{M} & \mathbf{0} \\ \mathbf{0} & \mathbf{0} \end{bmatrix} \begin{Bmatrix} \ddot{\bar{\mathbf{u}}} \\ \ddot{\bar{\mathbf{p}}} \end{Bmatrix} + \begin{bmatrix} \mathbf{C} & \mathbf{0} \\ \mathbf{Q}^{\mathrm{T}} & \mathbf{S} \end{bmatrix} \begin{Bmatrix} \dot{\bar{\mathbf{u}}} \\ \dot{\bar{\mathbf{p}}} \end{Bmatrix} + \begin{bmatrix} \mathbf{K} & -\mathbf{Q} \\ \mathbf{0} & \mathbf{H} \end{bmatrix} \begin{Bmatrix} \bar{\mathbf{u}} \\ \bar{\mathbf{p}} \end{Bmatrix} = -\begin{Bmatrix} \mathbf{f} \\ \mathbf{q} \end{Bmatrix} \tag{11.50}$$

Once again, like in the structural–fluid interaction problem, overall dissymmetry occurs despite the inherent symmetry of **M**, **C**, **S**, **K** and **H** matrices. However, as the free vibration problem is of no great interest here we shall not discuss its symmetrization. In the transient solution algorithm we shall proceed in a similar manner to that described in Sec. 11.2.5 and again symmetry will be observed.

11.3.3 *Transient, step-by-step, algorithm.* Time-stepping procedures can be derived in a manner analogous to that presented in Sec. 11.2.5. We choose here to use the GNpj algorithm of lowest order (viz. Sec. 10.3.3) to approximate each variable.

Thus for $\bar{\mathbf{u}}$ we shall use GN22, writing

$$\begin{aligned} \dot{\bar{\mathbf{u}}}_{n+1} &= \dot{\bar{\mathbf{u}}}_n + \ddot{\bar{\mathbf{u}}}_n \Delta t + \beta_1 \Delta \ddot{\bar{\mathbf{u}}}_n \Delta t \\ &\equiv \dot{\bar{\mathbf{u}}}_{n+1}^p + \beta_1 \Delta \ddot{\bar{\mathbf{u}}}_n \Delta t \\ \bar{\mathbf{u}}_{n+1} &= \bar{\mathbf{u}}_n + \dot{\bar{\mathbf{u}}}_n \Delta t + \frac{\ddot{\bar{\mathbf{u}}}_n \Delta t^2}{2} + \frac{\beta_2 \ddot{\bar{\mathbf{u}}}_n \Delta t^2}{2} \\ &\equiv \bar{\mathbf{u}}_{n+1}^p + \frac{\beta_2 \Delta \ddot{\bar{\mathbf{u}}}_n \Delta t^2}{2} \end{aligned} \tag{11.51a}$$

For the variables p that occur in first order we shall use GN11, as

$$\begin{aligned} \bar{\mathbf{p}}_{n+1} &= \bar{\mathbf{p}}_n + \dot{\bar{\mathbf{p}}}_n \Delta t + \Theta \Delta \dot{\bar{\mathbf{p}}}_n \Delta t \\ &= \bar{\mathbf{p}}_{n+1}^p + \Theta \Delta \dot{\bar{\mathbf{p}}}_n \Delta t \end{aligned} \tag{11.51b}$$

In the above $\bar{\mathbf{u}}_{n+1}^p$, etc., stand for values that can be immediately 'predicted' from known parameters at time t_n and

$$\Delta \ddot{\bar{\mathbf{u}}}_n = \ddot{\bar{\mathbf{u}}}_{n+1} - \ddot{\bar{\mathbf{u}}}_n \qquad \Delta \dot{\bar{\mathbf{p}}}_n = \dot{\bar{\mathbf{p}}}_{n+1} - \dot{\bar{\mathbf{p}}}_n \tag{11.52}$$

are the unknowns.

To complete the recurrence algorithm it is necessary to insert the above into the coupled governing equations [(11.39) and (11.48)] written at time t_{n+1}. Thus we require the following equalities:

$$\mathbf{M}\ddot{\bar{\mathbf{u}}}_{n+1} + \mathbf{C}\dot{\bar{\mathbf{u}}}_{n+1} + \int \mathbf{B}^{\mathrm{T}} \boldsymbol{\sigma}'_{n+1} \, d\Omega + -\mathbf{Q}\bar{\mathbf{p}}_{n+1} + \mathbf{f} = \mathbf{0}$$

and

$$\mathbf{Q}^{\mathrm{T}}\dot{\bar{\mathbf{u}}}_{n+1} + \mathbf{S}\dot{\bar{\mathbf{p}}}_{n+1} + \mathbf{H}\bar{\mathbf{p}}_{n+1} + \mathbf{q}_{n+1} = \mathbf{0} \tag{11.53}$$

in which quite generally $\boldsymbol{\sigma}'_{n+1}$ is evaluated using the constitutive relation (11.38) and knowledge of $\boldsymbol{\sigma}'_n$.

The system is non-linear and indeed on many occasions the $\mathbf{H}$ matrix itself may be dependent on the values of $\bar{\mathbf{u}}_{n+1}$ due to permeability variations with strain. Solution methods of such non-linear systems have been discussed in Chapter 7 and we shall not repeat them here. However, it is of interest to look at the linear form as the non-linear system will solve similar equations iteratively.

Here insertion of Eq. (11.51) into (11.53) will result in the equation system

$$\begin{bmatrix} \mathbf{M}+\mathbf{C}\beta_1\Delta t+\mathbf{K}\beta_2\Delta t^2/2, & -\mathbf{Q}\Theta\Delta t \\ -\mathbf{Q}^{\mathrm{T}}\beta_1\Delta t, & -\mathbf{S}\beta_1\Delta t-\mathbf{H}\beta_1\Theta\Delta t^2 \end{bmatrix} \begin{Bmatrix} \Delta\ddot{\bar{\mathbf{u}}}_n \\ \Delta\dot{\bar{\mathbf{p}}}_n \end{Bmatrix} = \begin{Bmatrix} \mathbf{F}_1 \\ \mathbf{F}_2 \end{Bmatrix} \tag{11.54}$$

where symmetry was obtained by multiplying the second by $-\beta_1\Delta t$ and where $\mathbf{F}_1$ and $\mathbf{F}_2$ stand for vectors that can be evaluated from initial, known, values.

The solution of Eq. (11.54) and the use of Eqs (11.51) complete the recurrence relation.

The stability of the scheme can be found by following identical procedures to those used in Sec. 11.2.5 and the result is[26] that stability is unconditional when

$$\beta_2 \geqslant \beta_1 \qquad \beta_1 \geqslant \tfrac{1}{2} \qquad \Theta \geqslant \tfrac{1}{2} \tag{11.55}$$

11.3.4 *Special cases and robustness requirements.* Frequently the compressibility of the fluid phase which results in the matrix $\mathbf{S}$ is such that

$$\mathbf{S} \approx 0$$

compared with other terms. Further, the permeability k may on occasion also be very small (as, say, in clays) and

$$\mathbf{H} \approx 0$$

leading to so-called 'undrained' behaviour.

Now the equation matrix in (11.54) becomes of the lagrangian constrained form (viz. Chapter 12 of Volume 1), i.e.

$$\begin{bmatrix} \mathbf{A} & \mathbf{B} \\ \mathbf{B}^{\mathrm{T}} & 0 \end{bmatrix} \begin{Bmatrix} \Delta\ddot{\bar{\mathbf{u}}} \\ \Delta\dot{\bar{\mathbf{p}}} \end{Bmatrix} = \begin{Bmatrix} \mathbf{F}_1 \\ \mathbf{F}_2 \end{Bmatrix} \tag{11.56}$$

and it is solvable only if

$$n_u \geqslant n_p$$

where n_u and n_p stand for the numbers of $\bar{\mathbf{u}}$ and $\bar{\mathbf{p}}$ parameters.

The problem is indeed identical to that encountered in incompressible

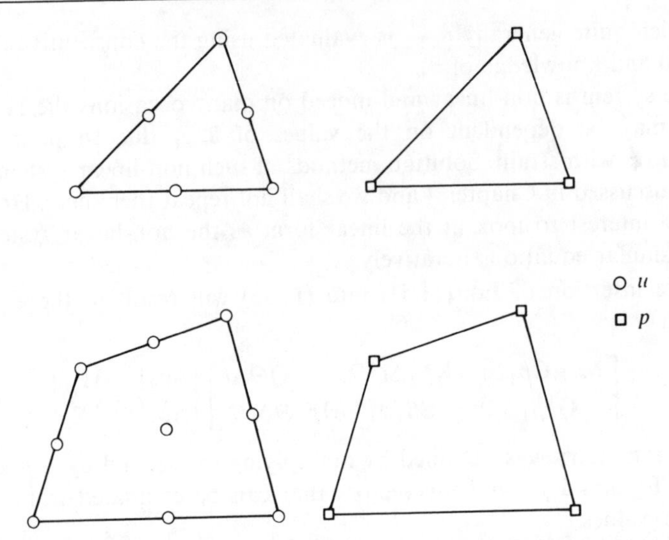

Fig. 11.6 'Robust' interpolations for the coupled soil–fluid problem

behaviour and the interpolations used for the **u** and p variables have to satisfy the identical criteria. As C_0 interpolation for both variables is now necessary, suitable element forms are shown in Fig. 11.6 and can be used with confidence.

The formulation can of course be used for steady-state solution but it must be remarked that in such cases an uncoupling occurs as the seepage equation can be solved independently.

Finally, it is worth remarking that the formulation also solves the well-known soil consolidation problem where the phenomena are so slow that the dynamic term

$$\mathbf{M}\ddot{\mathbf{u}} \to 0$$

However, no special modifications are necessary and the algorithm is again applicable.

11.3.5 *Examples–soil liquefaction.* As we have already mentioned, the most interesting application of the coupled soil–fluid behaviour is when the non-linear soil properties are taken into account. In particular, it is a well-known fact that repeated straining of a granular, soil-like material in the absence of the pore fluid results in a decrease of volume (densification) due to particle rearrangement. This, when a pore fluid is present, will (via the coupling terms) tend to increase the fluid pressures and hence reduce the soil strength. This, as is well known, decreases with the mean, compressive, effective, stress.

It is not surprising therefore that under dynamic action the soil frequently loses all of its strength and behaves almost like a fluid, leading occasionally to catastrophic failures of structural foundations in earthquakes. The reproduction of such phenomena with computational models is not easy as constitutive behaviour description is not perfect for soils. However, much effort devoted to the subject has produced fruit,[34-42] and a reasonable confidence in predictions achieved by comparison with experimental studies exists. One such study is illustrated in Fig. 11.7 where a comparison with tests carried out in a centrifuge is made.[42] In particular the close correlation between computed pressure and displacement with experiments should be noted.

11.3.6 *Biomechanics, oil recovery and other applications.* The interaction between a porous medium and interstitial fluid is not confined to soils. The same equations describe, for instance, the biomechanics problem of bone–fluid interaction in living bodies. Applications in this field have been documented.[43,44]

On occasions two (or more) fluids are present in the pores and here similar equations again can be written[45,46] to describe the interaction. Problems of ground settlement in oil fields due to oil extraction, or flow of water/oil mixtures in oil recovery, etc., are good examples of the application of techniques here described.

11.4 Partitioned single-phase systems—implicit–explicit partitions (class I problems)

In Fig. 11.1(*b*), describing problems coupled by an interface, we have already indicated the possibility of a structure being partitioned into substructures and linked along an interface only. Here the substructures will in general be of a similar kind but may differ in the manner (or simply size) of discretization used in each or even in the transient recurrence algorithms employed. In Chapter 13 of Volume 1 we have described special kinds of mixed formulations allowing the linking of domains in which, say, boundary-type approximations are used in one and standard finite elements in the other. We shall not return to this phase and will simply assume that the total system can be described using such procedures by a single set of equations in time. Here we shall only consider a first-order problem (but a similar approach can be extended to the second-order dynamic system):

$$\mathbf{C\dot{a}} + \mathbf{Ka} + \mathbf{f} = \mathbf{0} \tag{11.57}$$

which can be partitioned into two (or more) components, writing

$$\begin{bmatrix} \mathbf{C}_{11} & \mathbf{C}_{12} \\ \mathbf{C}_{21} & \mathbf{C}_{22} \end{bmatrix} \begin{Bmatrix} \dot{\mathbf{a}}_1 \\ \dot{\mathbf{a}}_2 \end{Bmatrix} + \begin{bmatrix} \mathbf{K}_{11} & \mathbf{K}_{12} \\ \mathbf{K}_{21} & \mathbf{K}_{22} \end{bmatrix} \begin{Bmatrix} \mathbf{a}_1 \\ \mathbf{a}_2 \end{Bmatrix} + \begin{Bmatrix} \mathbf{f}_1 \\ \mathbf{f}_2 \end{Bmatrix} = \mathbf{0} \tag{11.58}$$

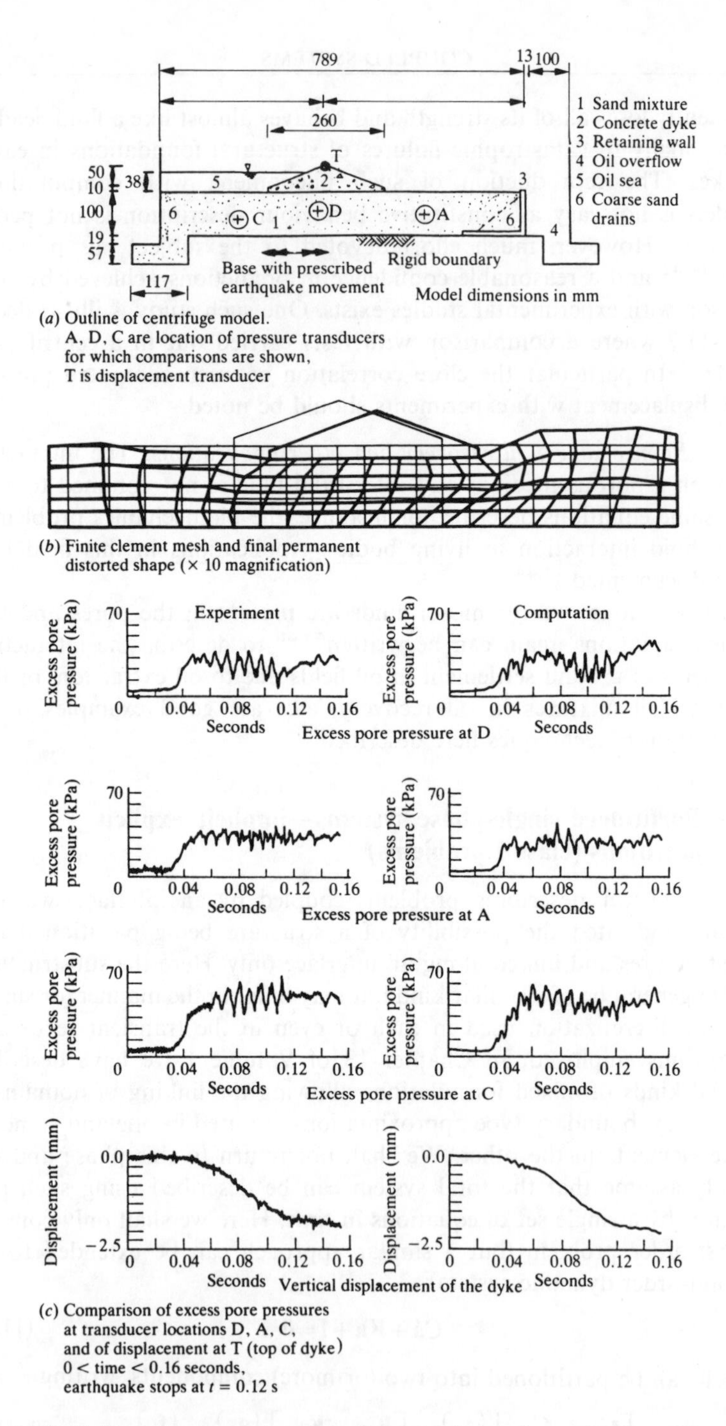

Fig. 11.7 Soil–pore water interaction. Computation and centrifuge model results compared on the problem of a dyke foundation subject to a simulated earthquake

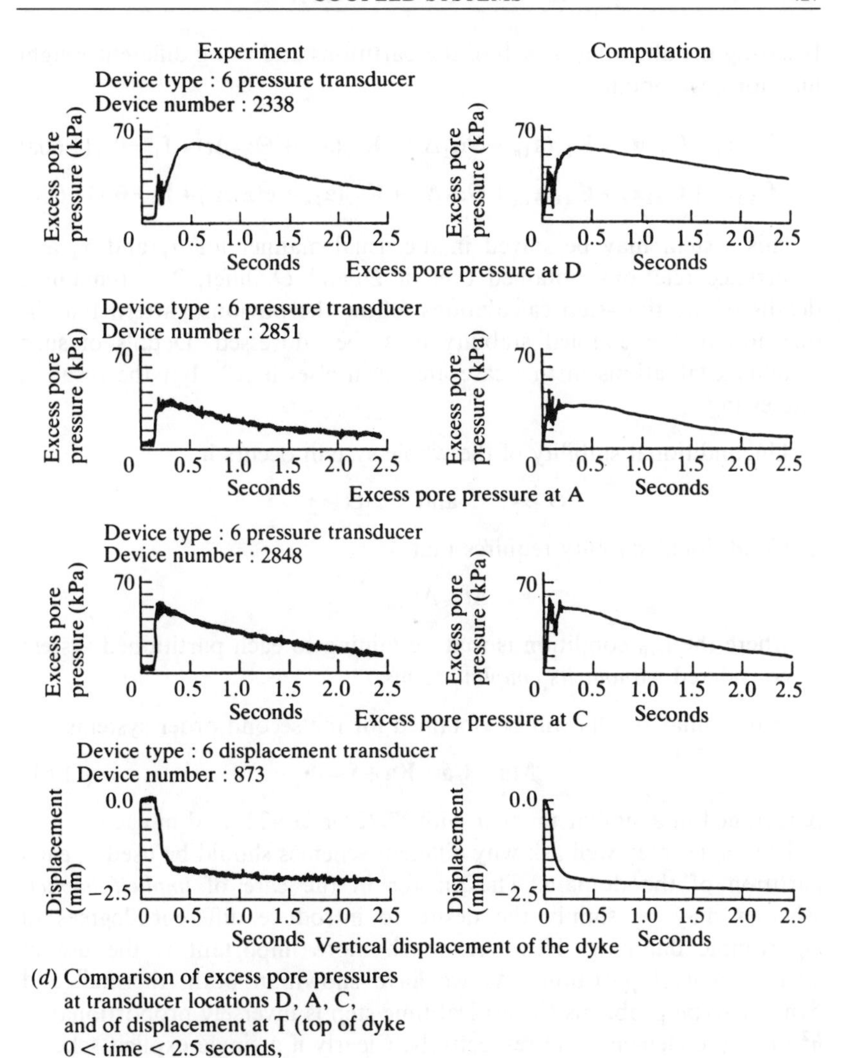

(*d*) Comparison of excess pore pressures at transducer locations D, A, C, and of displacement at T (top of dyke $0 < \text{time} < 2.5$ seconds, Note consolidation process

Fig. 11.7 (continued)

Now for various reasons it may be desirable to use in each partition a different time-step algorithm. Here we shall assume the same structure of the algorithm (SS11) and the same time step (Δt) but simply a different parameter Θ in each. Proceeding thus as in the other coupled analyses we write

$$\begin{aligned} \mathbf{a}_1 &= \mathbf{a}_{1n} + \boldsymbol{\alpha}_1 \tau \\ \mathbf{a}_2 &= \mathbf{a}_{2n} + \boldsymbol{\alpha}_2 \tau \end{aligned} \tag{11.59}$$

Inserting the above into each of the partitions and using different weight functions, we obtain

$$\mathbf{C}_{11}\boldsymbol{\alpha}_1+\mathbf{C}_{12}\boldsymbol{\alpha}_2+\mathbf{K}_{11}(\mathbf{a}_{1n}+\Theta\boldsymbol{\alpha}_1\Delta t)+\mathbf{K}_{12}(\mathbf{a}_{2n}+\Theta\boldsymbol{\alpha}_2\Delta t)+\mathbf{f}_1=\mathbf{0} \quad (11.60\text{a})$$

$$\mathbf{C}_{12}\boldsymbol{\alpha}_1+\mathbf{C}_{22}\boldsymbol{\alpha}_2+\mathbf{K}_{21}(\mathbf{a}_{1n}+\bar{\Theta}\boldsymbol{\alpha}_1\Delta t)+\mathbf{K}_{22}(\mathbf{a}_{2n}+\bar{\Theta}\boldsymbol{\alpha}_2\Delta t)+\mathbf{f}_1=\mathbf{0} \quad (11.60\text{b})$$

This system may be solved in the usual manner for α_1 and α_2 and recurrence relations obtained even if Θ and $\bar{\Theta}$ differ. The remaining details of the time-step calculations follow the obvious pattern but the question of the coupled stability must be addressed. Details of such stability evaluations in this case are given elsewhere[47] but the result is interesting.

1. Unconditional stability of the whole system occurs if

$$\Theta\geqslant\tfrac{1}{2} \quad \text{and} \quad \bar{\Theta}\geqslant\tfrac{1}{2}$$

2. Conditional stability requires that

$$\Delta t\leqslant\Delta t_{\text{crit}}$$

   where the t_{crit} condition is that pertaining to each partitioned system considered *without its coupling terms.*

Indeed, similar results will be obtained for the second-order systems

$$\mathbf{M}\ddot{\mathbf{a}}+\mathbf{C}\dot{\mathbf{a}}+\mathbf{K}\mathbf{a}+\mathbf{f}=\mathbf{0} \quad (11.61)$$

partitioned in a similar manner with SS22 or GN22 used in each.

The reader may well ask why different schemes should be used in each partition of the domain. The answer in the case of *implicit–implicit* schemes may be simply the desire to introduce different degrees of algorithmic damping. However, much more important is the use of *explicit–implicit* partitions. As we have shown in both 'thermal' and dynamic-type problems the critical time step is inversely proportional to h^2 or h (the element size) respectively. Clearly if a single explicit scheme were to be used with very small elements occurring in one partition, this time step may become too short for economy to be preserved in its use. In such cases it may be advantageous to use an *explicit* scheme (with $\Theta=0$ in the first-order problem, $\Theta_2=0$ in dynamics) for a part of the domain with larger elements while maintaining unconditional stability with $\bar{\Theta}>\frac{1}{2}$, using the same time step in the partition in which the elements are small. For this reason such explicit–implicit partitions are frequently used in practice.

Indeed, with a lumped representation of matrices **C** or **M** such schemes are in effect *staggered* as the explicit part can be advanced independently of the implicit and immediately provides the boundary values for the implicit partition. We shall return to such staggered solutions in the next section.

The first use of explicit–implicit partition was first recorded in 1978.[48,49] In the first reference the process is given on an identical partition as presented here. In the second, the different algorithms are associated with elements necessitating a slightly augmented partitioning.

11.5 Staggered solution processes

11.5.1 *General remarks.* We have just observed that in the explicit–implicit partitioning of time stepping it was possible to proceed in the *staggered* fashion, achieving a complete solution of the explicit scheme independently of the implicit one and then using the results to progress with the second. It is tempting to examine the possibility of such staggered procedures generally even if each uses an independent algorithm.

In such procedures the first equation would be solved with some assumed (predicted) values for the variable of the other. Once the solution for the first system was obtained its values could be substituted in the second, allowing again its independent treatment. If such procedures can be made stable and reasonably accurate many possibilities are immediately open, for instance:

1. Completely different methodologies could be used in each part of the coupled system.
2. Independently developed codes dealing efficiently with single systems could be combined.
3. Parallel computation with its inherent advantages could be used.
4. Finally, in systems of the same physics, efficient iterative solvers could easily be developed.

The problems of such staggered solutions have been frequently discussed[29,50–53] and on occasion unconditional stability could not be achieved without substantial modification. In the following we shall indicate some options available.

11.5.2 *Staggered process of solution in single-phase systems.* We shall look at this possibility first, having already mentioned it as a special form arising naturally in explicit–implicit processes of Sec. 11.4. We return here to consider the problem of Eq. (11.57) and the partitioning given in Eq. (11.58). Further, for simplicity, we shall assume a diagonal form of the **C** matrix, i.e. that the problem is posed as

$$\begin{bmatrix} \mathbf{C}_{11} & \mathbf{0} \\ \mathbf{0} & \mathbf{C}_{22} \end{bmatrix} \begin{Bmatrix} \dot{\mathbf{a}}_1 \\ \dot{\mathbf{a}}_2 \end{Bmatrix} + \begin{bmatrix} \mathbf{K}_{11} & \mathbf{K}_{12} \\ \mathbf{K}_{21} & \mathbf{K}_{22} \end{bmatrix} \begin{Bmatrix} \mathbf{a}_1 \\ \mathbf{a}_2 \end{Bmatrix} + \begin{Bmatrix} \mathbf{f}_1 \\ \mathbf{f}_2 \end{Bmatrix} = \mathbf{0} \qquad (11.62)$$

As we have already remarked, the use of $\Theta=0$ in the first equation and $\bar{\Theta}\geqslant\frac{1}{2}$ in the second [viz. Eqs (11.60)] allowed the explicit part to be solved independently of the implicit. Now, however, we shall use the same Θ in both equations but in the first of the approximations, analogous to Eq. (11.60a), we shall insert a predicted value for the second variable:

$$\mathbf{a}_2=\mathbf{a}_2^p=\mathbf{a}_{2n} \tag{11.63}$$

This gives in place of the first equation of (11.60),

$$\mathbf{C}_{11}\boldsymbol{\alpha}_1+\mathbf{K}_{11}(\mathbf{a}_{1n}+\Theta\boldsymbol{\alpha}_1\Delta t)=-\mathbf{K}_{12}\mathbf{a}_{2n}-\mathbf{f}_1 \tag{11.64a}$$

allowing the direct solution for $\boldsymbol{\alpha}_1$.

Following this step, the second equation can be solved, of course, for $\boldsymbol{\alpha}_2$ with the previous value of $\boldsymbol{\alpha}_1$ inserted, i.e.

$$\mathbf{C}_{22}\boldsymbol{\alpha}_2+\mathbf{K}_{22}(\mathbf{a}_{2n}+\Theta\boldsymbol{\alpha}_2\Delta t)=-\mathbf{K}_{21}(\mathbf{a}_{1n}+\Theta\boldsymbol{\alpha}_1\Delta t)-\mathbf{f}_2 \tag{11.64b}$$

It turns out that this scheme is unconditionally stable if $\Theta\geqslant\frac{1}{2}$, i.e. as before that its stability is unconditional providing each component is stable. It turns out indeed that similar conditions pertain in second-order dynamic problems.

Obviously, however, some accuracy will be lost as the approximation of Eq. (11.64a) is that of the explicit form in $\mathbf{a}_2$, but the approximation is consistent and hence convergence will occur.

The advantage of using the staggered process in the above is clear as the equation solving, even though not explicit, is now confined to the magnitude of each partition and computational economy occurs.

Further, it is obvious that precisely the same procedures can be used for any number of partitions and that again the same stability conditions will apply. Consider a partition of the form

$$\begin{bmatrix} \mathbf{C}_{11} & & & & \\ & \mathbf{C}_{22} & & \mathbf{0} & \\ & & \ddots & & \\ & \mathbf{0} & & \mathbf{C}_{ii} & \\ & & & & \ddots \ \mathbf{C}_{kk} \end{bmatrix} \begin{Bmatrix} \mathbf{a}_1 \\ \mathbf{a}_2 \\ \vdots \\ \mathbf{a}_i \\ \mathbf{a}_k \end{Bmatrix} + \begin{bmatrix} \mathbf{K}_{11} & \mathbf{K}_{12} & \cdots & & \\ \mathbf{K}_{22} & & & & \\ & \ddots & & & \\ & & \mathbf{K}_{ii} & & \\ & & & \ddots & \mathbf{K}_{kk} \end{bmatrix} \begin{Bmatrix} \mathbf{a}_1 \\ \mathbf{a}_2 \\ \vdots \\ \mathbf{a}_i \\ \mathbf{a}_k \end{Bmatrix} + \begin{Bmatrix} \mathbf{f}_1 \\ \mathbf{f}_2 \\ \vdots \\ \mathbf{f}_i \\ \mathbf{f}_k \end{Bmatrix} = \mathbf{0} \tag{11.65}$$

Now in approximating the first equation it is necessary to use predicted values for $\mathbf{a}_2, \mathbf{a}_3, \ldots, \mathbf{a}_k$, writing in place of Eq. (11.64a),

$$\mathbf{C}_{11}\boldsymbol{\alpha}_1+\mathbf{K}_{11}(\mathbf{a}_{1n}+\Theta\boldsymbol{\alpha}_1\Delta t)=-\mathbf{K}_{12}\mathbf{a}_{2n}-\mathbf{K}_{13}\mathbf{a}_{3n}-\cdots-\mathbf{f}_1 \tag{11.66}$$

and continue similarly to (11.64b), with the predicted values now continually being replaced by better approximations as the solution progresses.

The partitioning of Eq. (11.65) can be continued until each includes only single degrees of freedom. Then at each step the equation that requires solving for α_i is of the form

$$(C_{ii}+\Theta\Delta t K_{ii})\alpha_i = F_1 \qquad (11.67)$$

This is a scalar and computation is thus *fully explicit and yet preserves unconditional stability* for $\theta \geqslant \frac{1}{2}$. This type of partitioning and the derivation of an unconditionally stable explicit scheme was first proposed by Zienkiewicz *et al.*[53] An alternative and somewhat more limited scheme of a similar kind was given by Trujillo.[54]

Clearly the error in the approximation in the time step decreases as the solution sweeps through the partitions and hence it is advisable to alter the sweep directions during the computation. For instance, in Fig. 11.8 we show quite reasonable accuracy for a one-dimensional heat-conduction problem in which the *explicit–split* process was used with alternating direction of sweeps. Of course the accuracy is much inferior to that exhibited by a standard implicit scheme with the same time step, though the process could be used quite effectively as an iteration to obtain steady-state solutions. Here many options are possible.

It is, for instance, of interest to consider the equation system given in Eq. (11.65) as originating from a simple finite difference approximation

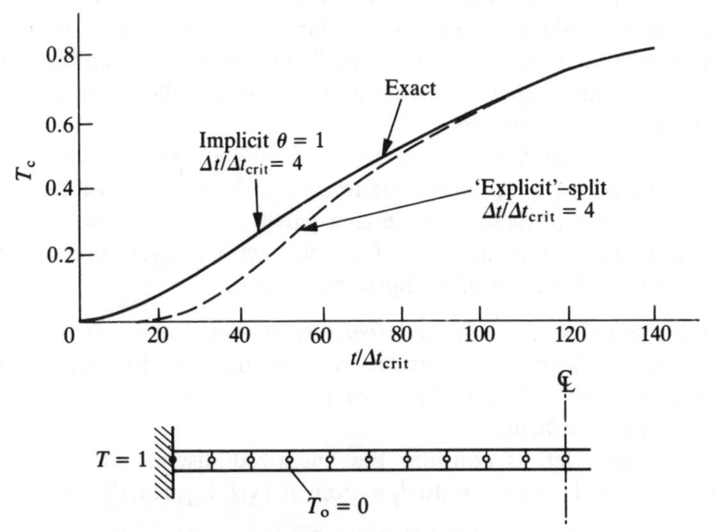

Fig. 11.8 Accuracy of an explicit–split procedure compared with standard implicit process for heat conduction in a bar

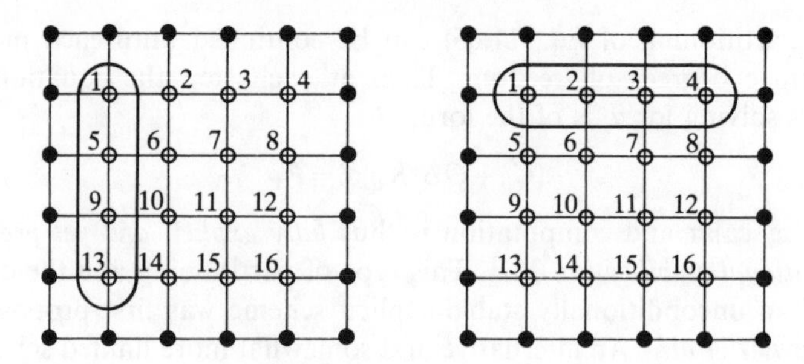

Fig. 11.9 Partitions corresponding with the well-known ADI (alternating direction implicit) finite difference scheme

to, say, a heat-conduction equation on a rectangular mesh of Fig. 11.9.

Here it is well known that the so-called alternating direction implicit (ADI) scheme[55] presents an efficient solution for both transient and steady-state problems. It is fairly obvious that that scheme simply represents the procedure just outlined with partitions representing lines of nodes such as (1, 5, 9, 13), (2, 6, 10, 14), etc., of Fig. 11.9 alternating with partitions (1, 2, 3, 4), (5, 6, 7, 8), etc.

Obviously the bigger the partition, the more accurate the scheme becomes, though of course at the expense of computational costs. The concept of the staggered partitioning clearly allows easy adoption of such procedures in the finite element context. Here irregular partitions arbitrarily chosen could be made but so far applications have only been recorded in regular mesh subdivisions.[56] The field of possibilities obviously is large and can yet be explored. Use of parallel computation is obvious in such procedures.

A further possibility which has many advantages is to use hierarchical variables based on, say, linear, quadratic and higher expansions and to consider each set of these variable as a partition.[57] Such procedures are particularly efficient in iteration if coupled with suitable preconditioning[58] and form the basis of *multigrid procedures.*

11.5.3 *Staggered schemes in fluid–structure systems and stabilization processes.* The application of staggered solution methods in coupled problems representing different phenomena is more obvious—though, as it turns out, more difficult.

For instance, let us consider the linearized discrete structure–fluid equations (with damping omitted) written as [viz. Eqs (11.17) and (11.19)]

$$\begin{bmatrix} \mathbf{M} & \mathbf{0} \\ \mathbf{Q}^{\mathrm{T}} & \mathbf{S} \end{bmatrix} \begin{Bmatrix} \ddot{\mathbf{u}} \\ \ddot{\mathbf{p}} \end{Bmatrix} + \begin{bmatrix} \mathbf{K} & -\mathbf{Q} \\ \mathbf{0} & \mathbf{H} \end{bmatrix} \begin{Bmatrix} \mathbf{u} \\ \mathbf{p} \end{Bmatrix} + \begin{Bmatrix} \mathbf{f} \\ \mathbf{q} \end{Bmatrix} = \mathbf{0} \tag{11.68}$$

where we have omitted the bar superscript for simplicity.

For simplicity we shall use the GN22 type of approximation of both variables and write

$$\dot{\mathbf{u}}_{n+1} = \dot{\mathbf{u}}_n + \ddot{\mathbf{u}}_n \Delta t + \beta_1 \Delta \ddot{\mathbf{u}}_n \Delta t = \dot{\mathbf{u}}^p_{n+1} + \beta_1 \Delta \ddot{\mathbf{u}}_n \Delta t$$

$$\mathbf{u}_{n+1} = \mathbf{u}_n + \dot{\mathbf{u}}_n \Delta t + \frac{\ddot{\mathbf{u}}_n \Delta t^2}{2} + \beta_2 \frac{\Delta \ddot{\mathbf{u}}_n \Delta t^2}{2} = \mathbf{u}^p_{n+1} + \frac{\beta_2 \Delta \ddot{\mathbf{u}}_n \Delta t^2}{2} \tag{11.69}$$

$$\dot{\mathbf{p}}_{n+1} = \dot{\mathbf{p}}_n + \ddot{\mathbf{p}}_n \Delta t + \bar{\beta}_1 \Delta \ddot{\mathbf{p}}_n \Delta t = \dot{\mathbf{p}}^p_{n+1} + \bar{\beta}_1 \Delta \ddot{\mathbf{p}}_n \Delta t$$

$$\mathbf{p}_{n+1} = \mathbf{p}_n + \dot{\mathbf{p}}_n \Delta t + \frac{\ddot{\mathbf{p}} \Delta t^2}{2} + \bar{\beta}_2 \frac{\Delta \ddot{\mathbf{p}}_n \Delta t^2}{2} = \mathbf{p}^p_{n+1} + \frac{\bar{\beta}_2 \Delta \ddot{\mathbf{p}}_n \Delta t^2}{2}$$

which together with Eq. (11.68) written at $t = t_{n+1}$ completes the system of equations requiring simultaneous solution for $\Delta \ddot{\mathbf{u}}_n$ and $\Delta \ddot{\mathbf{p}}_n$.

Now a staggered solution of a fairly obvious kind would be to write the first set of equations (11.68) corresponding to the structural behaviour with a predicted (approximate) value of $\mathbf{p}_{n+1} = \mathbf{p}^p_{n+1}$, as this would allow an independent solution for $\Delta \ddot{\mathbf{u}}_n$ writing

$$\mathbf{M}\ddot{\mathbf{u}}_{n+1} + \mathbf{K}\mathbf{u}_{n+1} = -\mathbf{f}_{1,n+1} + \mathbf{Q}\mathbf{p}^p_{n+1} \tag{11.70}$$

This would then be followed by the solution of the fluid problem for $\Delta \ddot{\mathbf{p}}_n$, writing

$$\mathbf{S}\ddot{\mathbf{p}}_{n+1} + \mathbf{H}\mathbf{p}_{n+1} = -\mathbf{q} - \mathbf{Q}^{\mathrm{T}}\ddot{\mathbf{u}}_{n+1} \tag{11.71}$$

This scheme turns out, however, to be only conditionally stable,[47] even if β_1 and β_2 are chosen so that unconditional stability of a simultaneous solution is achieved. (The stability limit is indeed the same as if a fully explicit scheme were chosen for *the fluid phase.*)

Various stabilization schemes can be used here.[20,47] One of these is given below. In this Eq. (11.70) is augmented to

$$\mathbf{M}\mathbf{u}_{n+1} + (\mathbf{K} + \mathbf{Q}\mathbf{S}^{-1}\mathbf{Q}^{\mathrm{T}})\mathbf{u}_{n+1} = -\mathbf{f}_{n+1} + \mathbf{Q}\mathbf{p}^p_{n+1} + \mathbf{Q}\mathbf{S}^{-1}\mathbf{Q}^{\mathrm{T}}\mathbf{u}^p_{n+1} \tag{11.72}$$

before solving for $\mathbf{u}_{n+1}$. It turns out that the scheme is now unconditionally stable provided the usual conditions

$$\beta_2 \geqslant \beta_1 \qquad \beta_1 \geqslant \tfrac{1}{2}$$

are satisfied.

Such stabilization involves the inverse of $\mathbf{S}$ but again it should be noted that this needs to be obtained only for the coupling nodes on the interface. Another stable scheme involves a similar inversion of $\mathbf{H}$ and is useful as incompressible behaviour is automatically given.

Similar stabilization processes can and have been applied with success to the soil–fluid system.[59,60]

References

1. O. C. ZIENKIEWICZ, 'Coupled problems and their numerical solution', in *Numerical Methods in Coupled Systems* (eds R. W. Lewis, P. Bettess and E. Hinton), chap. 1, pp. 35–68, Wiley, Chichester, 1984.
2. O. C. ZIENKIEWICZ, E. ONATE and J. C. HEINRICH, 'A general formulation for coupled thermal flow of metals using finite elements', *Int. J. Num. Meth. Eng.*, **17**, 1497–514, 1981.
3. R. W. LEWIS, P. BETTESS and E. HINTON (eds), *Numerical Methods in Coupled Systems*, Wiley, Chichester, 1984.
4. R. W. LEWIS, E. HINTON, P. BETTESS and B. A. SCHREFLER (eds), *Numerical Methods in Transient and Coupled Problems*, Wiley, Chichester, 1987.
5. O. C. ZIENKIEWICZ and R. E. NEWTON, 'Coupled vibrations of a structure submerged in a compressible fluid', in *Proc. Int. Symp. on Finite Element Techniques*, Stuttgart, 1–15 May 1969.
6. P. BETTESS and O. C. ZIENKIEWICZ, 'Diffraction and refraction of surface waves using finite and infinite elements', *Int. J. Num. Meth. Eng.*, **11**, 1271–90, 1977.
7. O. C. ZIENKIEWICZ, D. W. KELLY and P. BETTESS, 'The Sommerfield (radiation) condition on infinite domains and its modelling in numerical procedures', *Proc. IRIA 3rd Int. Symp. on Computing Methods in Applied Science and Engineering*, Versailles, December 1977.
8. O. C. ZIENKIEWICZ, P. BETTESS and D. W. KELLY, 'The finite element method for determining fluid loadings on rigid structures. Two- and three-dimensional formulations', in *Numerical Methods in Offshore Engineering* (eds O. C. Zienkiewicz, R. W. Lewis and K. G. Stagg), chap. 4, pp. 141–83, Wiley, Chichester, 1978.
9. O. C. ZIENKIEWICZ and P. BETTESS, 'Dynamic fluid–structure interaction. Numerical modelling of the coupled problem', in *Numerical Methods in Offshore Engineering* (eds O. C. Zienkiewicz, R. W. Lewis and K. G. Stagg), chap. 5, pp. 185–93, Wiley, Chichester, 1978.
10. O. C. ZIENKIEWICZ and P. BETTESS, 'Fluid–structure dynamic interaction and wave forces. An introduction to numerical treatment', *Int. J. Num. Meth. Eng.*, **13**, 1–16, 1978.
11. O. C. ZIENKIEWICZ, H. HARA and P. BETTESS, 'Application of finite elements to determination of wave effects on offshore structures', in *BOSS* (Proc. 2nd Int. Conf. Behaviour of Offshore Structures), pp. 383–90, 1979.
12. O. C. ZIENKIEWICZ and P. BETTESS, 'Fluid structure dynamic interaction and some "unified" approximation processes', in *5th Int. Symp. on Unification of Finite Elements, Finite Differences and Calculus of Variations*, University of Connecticut, May 1980.
13. O. C. ZIENKIEWICZ and P. BETTESS, 'Fluid structure interaction', in *Proc. Ocean Structural Dynamics Symp.*, Oregon State University, pp. 65–102, September 1982.
14. R. OHAYON and R. VALID, 'True symmetric formulation of free vibrations for fluid–structure interaction in bounded media', in *Numerical Methods in Coupled Systems* (eds R. W. Lewis, P. Bettess and E. Hinton), Wiley, Chichester, 1984.
15. R. OHAYON, 'Symmetric variational formulations for harmonic vibration problems by coupling primal and dual variables—applications to fluid–structure coupled systems', *La Recherche Aerospatiale*, **3**, 69–77, 1979.

16. M. GERADIN, G. ROBERTS and J. HUCK, 'Eigenvalue analysis and transient response of fluid structure interaction problems', *Eng. Comp.*, **1**, 152–60, 1984.
17. G. SANDBERG and P. GORENSSON, 'A symmetric finite element formulation of acoustic fluid–structure interaction analysis', *J. Sound Vibr.*, **123**, 507–15, 1988.
18. K. K. GUPTA, 'On a numerical solution of the supersonic panel flutter eigen problem', *Int. J. Num. Meth. Eng.*, **10**, 637–45, 1976.
19. B. M. IRONS, 'The role of part inversion in fluid–structure problems with mixed variables', *JAIAA*, **7**, 568, 1970.
20. W. J. T. DANIEL, 'Modal methods in finite element fluid–structure eigenvalue problems', *Int. J. Num. Meth. Eng.*, **15**, 1161–75, 1980.
21. C. FELIPPA, 'Symmetrization of coupled eigenproblems by eigenvector augmentation', *Comm. Appl. Num. Meth.*, **4**, 561–3, 1988.
22. R. OHAYON, personal communication, 1990.
23. J. HOLBECHE, Ph.D. thesis, University of Wales, Swansea, 1971.
24. A. K. CHOPRA and S. GUPTA, 'Hydrodynamic and foundation interaction effects in earthquake response of a concrete gravity dam', *J. Struct. Div. Am. Soc. Civ. Eng.*, **578**, 1399–412, 1981.
25. J. F. HALL and A. K. CHOPRA, 'Hydrodynamic effects in the dynamic response of concrete gravity dams', *J. Earthquake Eng. Struct. Dynam.*, **10**, 333–95, 1982.
26. O. C. ZIENKIEWICZ and R. L. TAYLOR, 'Coupled problems—a simple time-stepping procedure', *Comm. Appl. Num. Meth.*, **1**, 233–9, 1985.
27. O. C. ZIENKIEWICZ, R. N. CLOUGH and H. B. SEED, 'Earthquake analysis procedures for concrete and earth dams—state of the art', Int. Commission on Large Dams, Bulletin 32, Paris, 1986.
28. R. E. NEWTON, 'Finite element study of shock induced cavitation', *ASCE Spring Convention*, Portland, Oregon, 1980.
29. O. C. ZIENKIEWICZ, D. K. PAUL and E. HINTON, 'Cavitation in fluid–structure response (with particular reference to dams under earthquake loading', *J. Earthquake Eng. Struct. Dynam.*, **11**, 463–81, 1983.
30. M. A. BIOT, 'Theory of propagation of elastic waves in a fluid saturated porous medium, Part I: low frequency range; Part II: high frequency range', *J. Acoust. Soc. Am.*, **28**, 168–91, 1956.
31. O. C. ZIENKIEWICZ, C. T. CHANG and P. BETTESS, 'Drained, undrained, consolidating and dynamic behaviour assumptions in soils. Limits of validity', *Geotechnique*, **30**, 385–95, 1980.
32. O. C. ZIENKIEWICZ, C. T. CHANG, E. HINTON and K. H. LEUNG, 'Earth dam analysis for earthquakes. Numerical solution and constitutive relations for non-linear (damage) analysis', in *Dams and Engineering*, pp. 170–94, Institute of Civil Engineering, 1980.
33. O. C. ZIENKIEWICZ and T. SHIOMI, 'Dynamic behaviour of saturated porous media, the generalised Biot formulation and its numerical solution', *Int. J. Num. Anal. Meth. Geomech.*, **8**, 71–96, 1984.
34. O. C. ZIENKIEWICZ, C. T. CHANG and E. HINTON, 'Non linear seismic response and liquefaction', *Int. J. Num. Anal., Meth. Geomech.*, **2**, 381–404, 1978.
35. O. C. ZIENKIEWICZ, 'Non linear problems of soil statics and dynamics', in *Non-linear Finite Element Analysis in Structural Mechanics* (eds W. Wunderlich, E. Stein and K. J. Bathe), pp. 259–73, Springer-Verlag, Berlin, 1980.
36. O. C. ZIENKIEWICZ, K. H. LEUNG and M. PASTOR, 'Simple model for

transient soil loading in earthquake analysis: Part I—basic model and its application', *Int. J. Num. Anal. Meth. Geomech.*, **9**, 453–76, 1985.
37. O. C. Zienkiewicz, K. H. Leung and M. Pastor, 'Simple model for transient soil loading in earthquake analysis: Part II—non-associative models for sands', *Int. J. Num. Anal. Meth. Geomech.*, **9**, 477–98, 1985.
38. M. Pastor and O. C. Zienkiewicz, 'A generalised plasticity, hierarchical model for sand under monotonic and cyclic loading', in *Proc. Int. Symp. on Numerical Models in Geomechanics* (eds G. N. Pande and W. F. Van Impe), Ghent, pp. 131–50, Jackson, England, 1986.
39. O. C. Zienkiewicz, A. H. C. Chan, M. Pastor and T. Shiomi, 'Computational approach to soil dynamics', in *Soil Dynamics and Liquefaction* (ed. A. S. Czamak), Developments in Geotechnical Engineering 42, Elsevier, Amsterdam, 1987.
40. O. C. Zienkiewicz, M. Pastor and A. H. C. Chan, 'Simple models for soil behaviour and applications to problems of soil liquefaction', in *Inst. Num. Meth. Eng.*, report CR/598/87, University College of Swansea, December 1987.
41. O. C. Zienkiewicz, A. H. C. Chan, M. Pastor, D. K. Paul and T. Shiomi, 'Static and dynamic behaviour of soils: a rational approach to quantitative solutions: Part I—fully saturated problems', *Proc. Roy. Soc. A.*, **429**, 285–309, 1990.
42. O. C. Zienkiewicz, Y. M. Xie, B. A. Schrefler, A. Ladesme and N. Bicanic, 'Static and dynamic behaviour of soils: a rational approach to quantitative solutions: Part II—semi saturated problems', *Proc. Roy. Soc. A.*, **429**, 311–21, 1990.
43. B. R. Simon, J. S-S. Wu, M. W. Carlton, L. E. Kazarian, E. P. France, J. H. Evans and O. C. Zienkiewicz, 'Poroelastic dynamic structural models of rhesus spinal motion segments', *Spine*, **10** (6), 494–507, 1985.
44. B. R. Simon, J. S-S. Wu and O. C. Zienkiewicz, 'Higher order, mixed and Hermitean finite element procedures for dynamic analysis of saturated porous media', *Int. J. Num. Anal. Meth. Geomech.*, **10**, 483–99, 1986.
45. R. L. Lewis and B. A. Schrefler, *The Finite Element Method in the Deformation and Consolidation of Porous Media*, Wiley, Chichester, 1987.
46. X. K. Li, O. C. Zienkiewicz and Y. M. Xie, 'A numerical model for immiscible two-phase fluid flow in porous medium and its time domain solution', *Inst. Num. Meth. Eng.*, **30**, 1195–212, 1990.
47. O. C. Zienkiewicz and A. H. C. Chan, 'Coupled problems and their numerical solution', in *Advances in Computational Non Linear Mechanics* (ed. I. S. Doltsinis), chap. 3, pp. 109–76, Springer-Verlag, Berlin, 1988.
48. T. Belytchko and R. Mullen, 'Stability of explicit–implicit time integration', *Int. J. Num. Meth. Eng.*, **12**, 1575–1586, 1978.
49. T. J. R. Hughes and W. K. Liu, 'Implicit–explicit finite elements in transient analysis', *J. Appl. Mech.*, pt I, **45**, 371–4; pt II, **45**, 375–8, 1978.
50. C. A. Felippa and K. C. Park, 'Staggered transient analysis procedures for coupled mechanics systems; formulation', *Comp. Meth. Appl. Mech. Eng.*, **24**, 61–111, 1980.
51. K. C. Park, 'Partitioned transient analysis procedures for coupled field problems: stability analysis', *J. Appl. Mech.*, **47**, 370–6, 1980.
52. K. D. Park and C. A. Felippa, 'Partitioned transient analysis procedures for coupled field problems: accuracy analysis', *J. Appl. Mech.*, **47**, 919–26, 1980.

53. O. C. ZIENKIEWICZ, E. HINTON, K. H. LEUNG and R. L. TAYLOR, 'Staggered, time marching schemes in dynamic soil analysis and selective explicit extrapolations algorithms', in *Proc. Conf. on Innovative Numerical Analysis for the Engineering Sciences* (eds R. Shaw *et al.*), University of Virginia Press, 1980.
54. D. M. TRUJILLO, 'An unconditionally stable explicit scheme of structural dynamics', *Int. J. Num. Meth. Eng.*, **11**, 1579–92, 1977.
55. A. R. MITCHELL, *Computational Methods in Partial Differential Equations*, Wiley, New York, 1969.
56. L. J. HAYES, 'Implementation of finite element alternating-direction methods on non-rectangular regions', *Int. J. Num. Meth. Eng.*, **16**, 35–49, 1980.
57. A. W. CRAIG and O. C. ZIENKIEWICZ, 'A multigrid algorithm using a hierarchical finite element basis', in *Multigrid Methods in Integral and Differential Equations* (eds D. J. Pedolon and H. Holstein), pp. 310–12, Clarendon Press, Oxford, 1985.
58. I. BABUSKA, A. W. CRAIG, J. MANDEL and J. PITKÄRANTA, 'Efficient preconditioning for the *p*-inversion finite element method in two dimensions', to be published in *SIAM J. Num. Anal.*
59. K. C. PARK, 'Stabilization of partitioned solution procedures for pore fluid–soil interaction analysis', *Int. J. Num. Meth. Eng.*, **19**, 1669–73, 1983.
60. O. C. ZIENKIEWICZ, D. K. PAUL and A. H. C. CHAN, 'Unconditionally stable staggered solution procedure for soil–pore fluid interaction problems', *Int. J. Num. Meth. Eng.*, **26**, 1039–55, 1988.

12

Convection dominated problems

12.1 Introduction

In this chapter we are concerned with the steady-state and transient solutions of equations of the type

$$\frac{\partial \mathbf{U}}{\partial t}+\frac{\partial \mathbf{F}_i}{\partial x_i}+\frac{\partial \mathbf{G}_i}{\partial x_i}+\mathbf{Q}=0 \tag{12.1}$$

where in general **U** is the basic dependent, vector-valued variable, **Q** is a source term vector and the *flux* matrices **F** and **G** are such that

$$\mathbf{F}_i=\mathbf{F}_i(\mathbf{U}) \tag{12.2a}$$

and in general

$$\mathbf{G}_i=\mathbf{G}_i\left(\frac{\partial \mathbf{U}}{\partial x_j}\right)$$

$$\mathbf{Q}=\mathbf{Q}(x_i, \mathbf{U}) \tag{12.2b}$$

In the above, x_i and i refer in the indicial manner to cartesian coordinates and quantities associated with these (see Chapter 6 of Volume 1 for details).

Equations (12.1) and (12.2) are *conservation laws* arising from a balance of quantity **U** with its fluxes **F** and **G** entering a control volume. Such equations are typical of fluid mechanics and will form the basis of Chapters 13 to 15 dealing with specific applications. As such equations may also arise in other physical situations this chapter is devoted to the general discussion of their approximate solution.

The simplest form of Eqs (12.1) and (12.2) is one in which **U** is a scalar and the fluxes are linear functions. Thus

$$\mathbf{U}=U \qquad \mathbf{Q}=Q$$
$$\mathbf{F}_i=F_i=A_iU \qquad \mathbf{G}_i=-k\frac{\partial U}{\partial x_i} \tag{12.3}$$

We now have in cartesian coordinates a scalar equation of the form

$$\frac{\partial U}{\partial t}+\frac{\partial(A_iU)}{\partial x_i}-\frac{\partial}{\partial x_i}\left(k\frac{\partial U}{\partial x_i}\right)+Q$$
$$\equiv\frac{\partial U}{\partial t}+\frac{\partial(A_xU)}{\partial x}+\frac{\partial(A_yU)}{\partial y}-\frac{\partial}{\partial x}\left(k\frac{\partial U}{\partial x}\right)-\frac{\partial}{\partial y}\left(k\frac{\partial U}{\partial y}\right)+Q$$
$$=0 \tag{12.4}$$

which will serve as the basic model for most of the present chapter.

The above can also be written as

$$\frac{\partial U}{\partial t}+A_x\frac{\partial U}{\partial x}+A_y\frac{\partial U}{\partial y}-\frac{\partial}{\partial x}\left(k\frac{\partial U}{\partial x}\right)-\frac{\partial}{\partial y}\left(k\frac{\partial U}{\partial y}\right)+Q=0 \tag{12.5}$$

if A_x and A_y are such that

$$\nabla^{\mathrm{T}}\begin{Bmatrix}A_x\\A_y\end{Bmatrix}\equiv 0$$

which is more restrictive but is often associated with the transport of the quantity U by convection in a field of velocity with components A_x and A_y in two dimensions (though of course the equations are similar for one or three space dimensions).

We have encountered this equation in Volume 1 [Eq. (9.11), Sec. 9.1] in connection with heat transport, and indeed the general equation (12.1) can be termed the *transport equation* with **F** standing for the *convective* and **G** for *diffusive* flux quantities. Indeed it is applicable to diverse problems such as transport of pollutants, etc.

With the variable **U** being approximated in the usual way

$$\mathbf{U}\approx\hat{\mathbf{U}}=\mathbf{N}\bar{\mathbf{U}}=\sum\mathbf{N}_k\bar{\mathbf{U}}_k \tag{12.6}$$

the problem could be presented following the usual (weighted residual) semi-discretization process as

$$\mathbf{M}\dot{\bar{\mathbf{U}}}+\mathbf{H}\bar{\mathbf{U}}+\mathbf{f}=0 \tag{12.7}$$

but now even with standard Galerkin (Bubnov) weighting the matrix **H** will not be symmetric. However, this is a relatively minor computational problem compared with inaccuracies and instabilities in the solution which follow the arbitrary use of this weighting function.

This chapter will discuss the manner in which these difficulties can be overcome and the approximation improved.

We shall in the main address the problem of solving Eq. (12.4), i.e. the scalar form, and to simplify matters further we shall often start with the idealized one-dimensional equation:

$$\frac{\partial U}{\partial t}+A\frac{\partial U}{\partial x}-k\frac{\partial^2 U}{\partial x^2}+Q=0 \tag{12.8}$$

reduced in steady state to an ordinary differential equation:

$$A\frac{dU}{dx}-k\frac{d^2U}{dx^2}+Q=0 \tag{12.9}$$

in which we shall assume A, k and Q to be constant. The basic concepts will be evident from above which will later be extended to multidimensional problems, still treating U as a scalar variable.

Indeed the methodology of dealing with the first space derivatives occurring in differential equations governing a problem, which as shown in Chapter 8 of Volume 1 lead to non-self-adjointness, opens the way for many new physical situations.

The present chapter will be divided into two parts. Part I deals with *steady-state situations* starting from Eq. (12.9) and Part II with *transient solutions* starting from Eq. (12.8). Although the scalar problem will mainly be dealt with here in detail, the discussion of the procedures can indicate the choice of optimal ones which will have much bearing on the solution of the general case of Eq. (12.1). Indeed we shall discuss the full form of this equation when introducing certain procedures so that the basis for solution of problems of Chapters 13 to 15 is available.

PART I: STEADY STATE

12.2 The steady-state problem in one dimension—some preliminaries and Petrov–Galerkin methods

We shall consider the discretization of Eq. (12.9) with

$$U\approx\sum N_i a_i=\mathbf{N}\mathbf{a} \tag{12.10}$$

where N_i are shape functions and $\mathbf{a}$ represents a set of still unknown parameters. Here we shall take these to be the nodal values of U and write

$$\mathbf{a}=\bar{\mathbf{U}} \qquad U=\mathbf{N}\bar{\mathbf{U}} \tag{12.11}$$

using a general weighting procedure (Sec. 9.2 of Chapter 9 of Volume 1). This gives for a typical internal node i the approximating equation

$$K_{ij}\bar{U}_j + f_i = 0 \tag{12.12}$$

where

$$K_{ij} = \int_0^L W_i A \frac{\partial N_j}{\partial x}\,dx + \int_0^L \frac{\partial W_i}{\partial x} k \frac{\partial N_j}{\partial x}\,dx$$

$$f_i = \int_0^L W_i Q\,dx \tag{12.13}$$

and the domain of the problem is $0 \leqslant x \leqslant L$.

For linear shape functions, Galerkin weighting ($W_i = N_i$) and equal elements of size h, we have for *constant* values of A, k and Q (Fig. 12.1) a typical assembled equation

$$(-Pe-1)\bar{U}_{i-1} + 2\bar{U}_i + (Pe-1)\bar{U}_{i+1} + \frac{Qh^2}{k} = 0 \tag{12.14}$$

where

$$Pe = \frac{Ah}{2k} \tag{12.15}$$

is the mesh *Peclet* number. The above is, incidently, identical to the usual central finite difference approximation obtained by putting

$$\frac{dU}{dx} \approx \frac{\bar{U}_{i+1} - \bar{U}_{i-1}}{2h} \tag{12.16a}$$

and

$$\frac{d^2U}{dx} \approx \frac{U_{i+1} - 2U_i + U_{i-1}}{h^2} \tag{12.16b}$$

The algebraic equations are obviously non-symmetric and in addition their accuracy deteriorates as the parameter Pe increases. Indeed as $Pe \to \infty$, i.e. when only convective terms are of importance, the solution

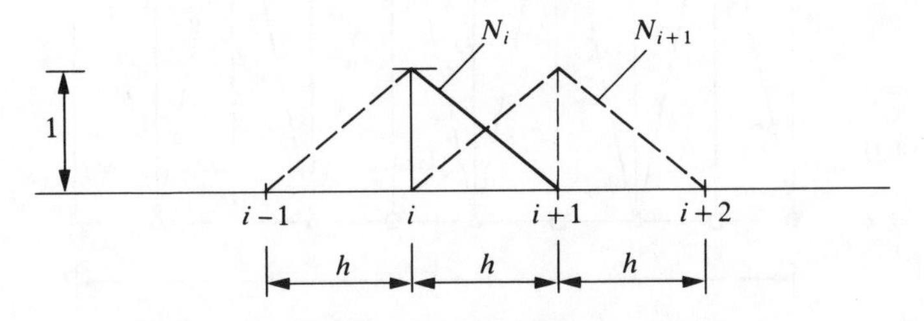

Fig. 12.1 A one-dimensional problem shape function

is purely oscillatory and bears no relation to the underlying problem, as shown in the simple example of Fig. 12.2 and curves labelled $\alpha = 0$. (Indeed the solution for this problem is now only possible for an odd number of elements and not for even.)

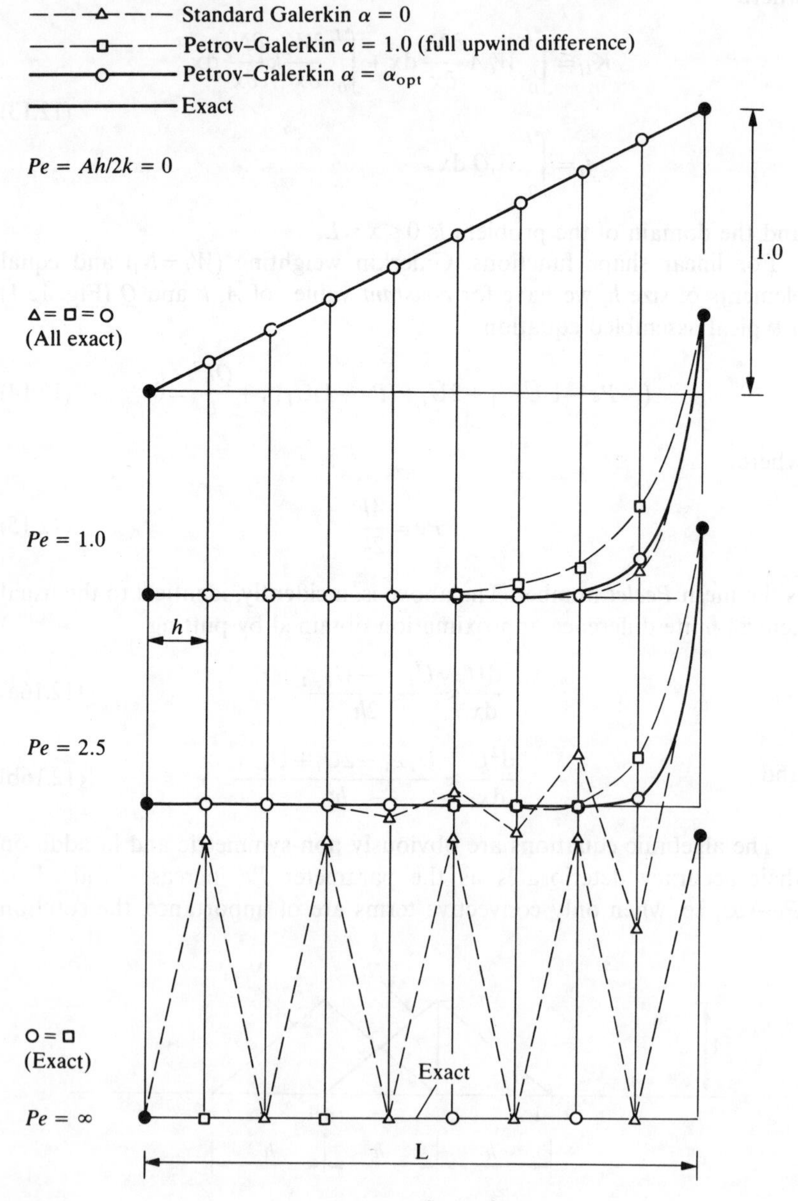

Fig. 12.2 Approximations to $dU/dx + k d^2U/dx^2 = 0$ for $U = 0$, $x = 0$ and $U = 1$, $x = L$ for various Peclet numbers

Of course the above is partly a problem of boundary conditions. When diffusion is omitted only a single boundary condition can be imposed and when the diffusion is small we note that the downstream boundary condition ($U=1$) is felt in only a very small region of a *boundary layer* evident from the exact solution[1]

$$U=\frac{1-e^{Ax/k}}{1-e^{AL/k}} \tag{12.17}$$

Motivated by the fact that the propagation of information is in the direction of velocity A, the finite difference practitioners were the first to overcome the bad approximation problem by using *one-sided* finite differences for the first derivative approximation.[2-5] Thus in place of Eq. (12.16a) and with positive A, the approximation was put as

$$\frac{dU}{dx}\approx\frac{U_i-U_{i-1}}{h} \tag{12.18}$$

changing the central finite difference form of the approximation to the governing equation as given by Eq. (12.14) to

$$(-2Pe-1)\bar{U}_{i-1}+(2+2Pe)\bar{U}_i-\bar{U}_{i+1}+\frac{Qh^2}{k}=0 \tag{12.19}$$

With this *upwind* difference approximation, realistic (though not always accurate) solutions can be obtained through the whole range of Pe numbers of the example of Fig. 12.2 as shown there by curves labelled $\alpha=1$. However, now exact nodal solutions are obtained for pure convection ($Pe=\infty$), as shown in Fig. 12.2, in a similar way as the Galerkin finite element form gives exact nodal answers for pure diffusion.

How can such upwind differencing be introduced into the finite element scheme and generalized to more complex situations? This is the problem that we shall now address, and indeed will show that again, as in self-adjoint equations, the finite element solution can result in exact nodal values for the one-dimensional approximation for all Peclet numbers.

The first possibility is that of the use of a Petrov–Galerkin type of weighting in which $W_i \neq N_i$.[6-9] In particular, again for elements with linear shape functions N_i, shown in Fig. 12.1, we shall take, as shown in Fig. 12.3, weighting functions so constructed that

$$W_i=N_i+\alpha\tilde{W}_i \tag{12.20}$$

where W_i is such that

$$\int_{\Omega_e}\tilde{W}_i\,dx=\pm\frac{h}{2} \tag{12.21}$$

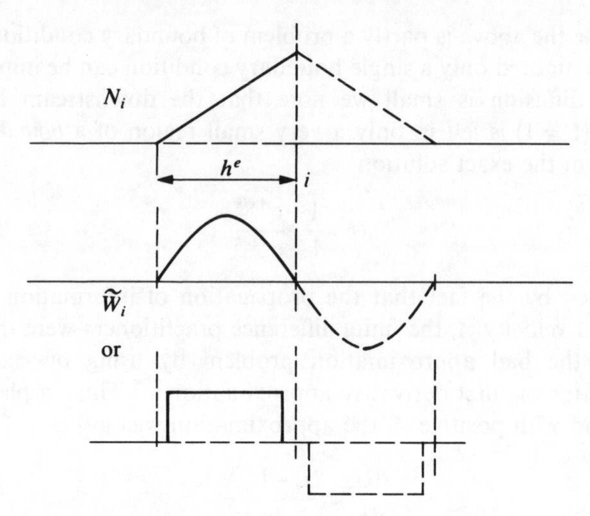

Fig. 12.3 Petrov–Galerkin weight function $W_i = N_i + \alpha\widetilde{W}_i$. Continuous and discontinuous definitions

the sign depending on whether A is a velocity directed towards or away from the node.

Various forms of W_i are possible, but the most convenient is the following simple definition which is, of course, a discontinuous function (see the note at the end of this section):

$$\alpha\widetilde{W}_i = \alpha\frac{h}{2}\frac{\mathrm{d}N_i}{\mathrm{d}x} \quad (\text{sign } A) \tag{12.22}$$

With the above weighting functions the approximation equivalent to that of Eq. (12.14) becomes

$$[-Pe(\alpha+1)-1]\bar{U}_{i-1} + [2+2\alpha(Pe)]\bar{U}_i + [-Pe(\alpha-1)-1]\bar{U}_{i+1} + \frac{Qh^2}{k} = 0 \tag{12.23}$$

Immediately we see that with $\alpha = 0$ the standard Galerkin approximation is recovered [Eq. (12.14)] and that with $\alpha = 1$ the full upwinded discrete equation (12.19) is available, each giving exact nodal values for purely diffusive or purely convective cases.

Now if the value of α is chosen as

$$|\alpha| = \alpha_{\text{opt}} = \coth|Pe| - \frac{1}{|Pe|} \tag{12.24}$$

then exact nodal values will be given *for all values of Pe*. The proof of this is given in reference 7 for the present, one-dimensional, case where it is also shown that if

$$|\alpha| > \alpha_{crit} = 1 - \frac{1}{|Pe|} \qquad (12.25)$$

oscillatory solutions will never arise. The results of Fig. 12.2 show indeed that with $\alpha = 0$, i.e. the Galerkin procedure, oscillations will occur when

$$Pe > 1 \qquad (12.26)$$

Figure 12.4 shows the variation of α_{opt} and α_{crit} with Pe.

Although the proof of optimality for the upwinding parameter was given for the case of constant coefficients and constant size elements, nodally exact values will also be given if $\alpha = \alpha_{opt}$ is chosen for each element individually. We show some typical solutions in Fig. 12.5[10] for a variable source term $Q = Q(x)$, convection coefficients $A = A(x)$ and element sizes. Each of these is compared with a standard Galerkin solution, showing that even when the latter does not result in oscillations the accuracy is improved. Of course in the above examples the Petrov–Galerkin weighting must be applied to all terms of the equation. When this is not done (as in simple finite difference upwinding) totally wrong results will be obtained, as shown in the example of Fig. 12.6, which was used in reference 11 to discredit upwinding methods. The effect of α on the source term is not apparent in Eq. (12.23) where Q is constant in the whole domain, but its influence is strong when $Q = Q(x)$.

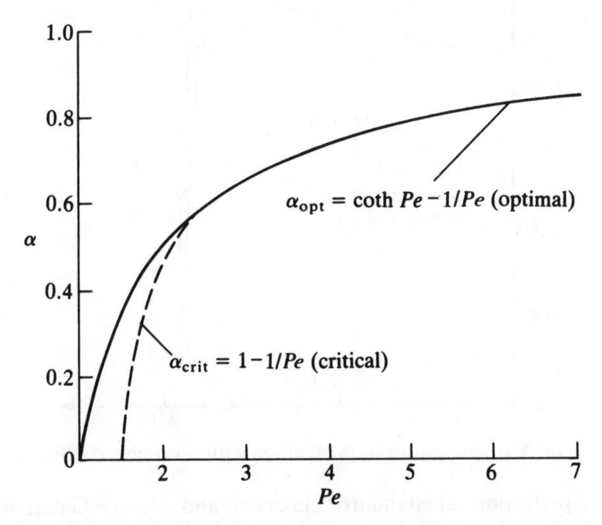

Fig. 12.4 Critical (stable) and optimal values of the 'upwind' parameter α for different values of $Pe = Ah/2k$

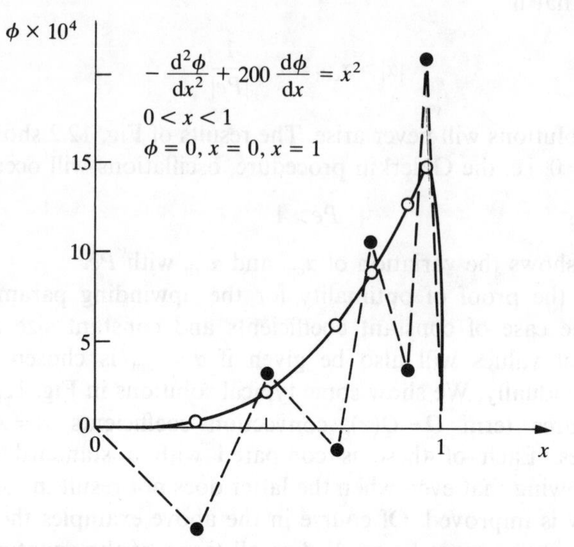

(*a*) Variable source term equation with constant coefficient

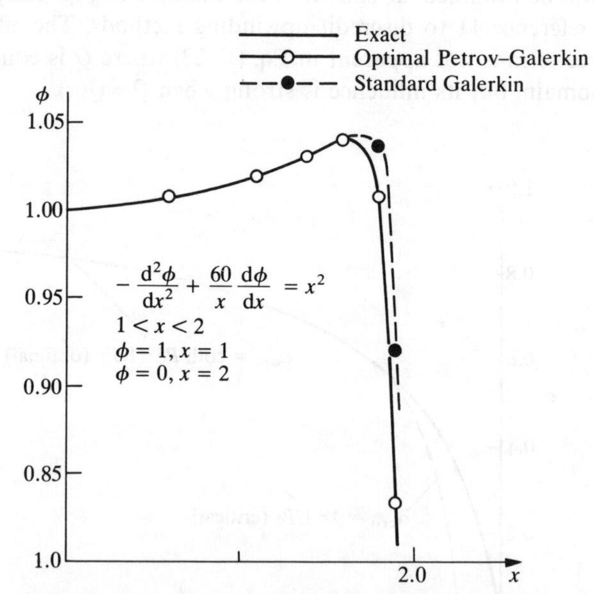

(*b*) Variable source term equation with a variable coefficient

Fig. 12.5 Application of standard Galerkin and Petrov–Galerkin (optimal) approximation

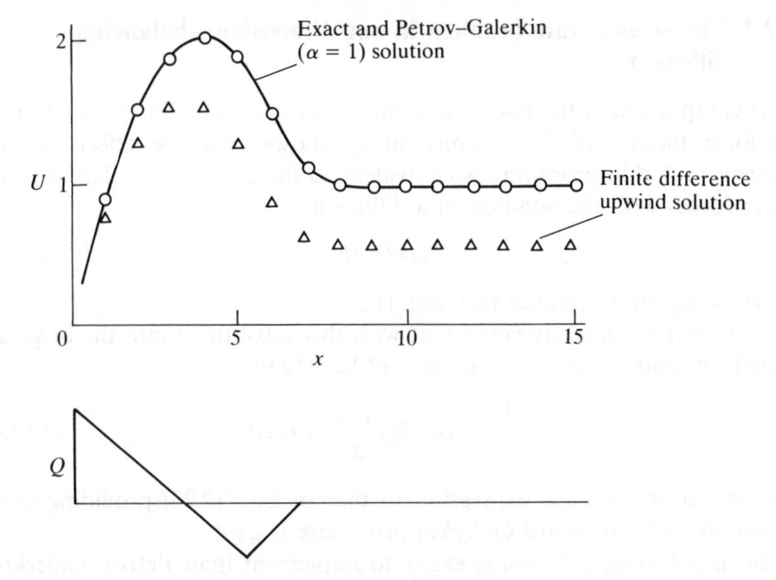

Fig. 12.6 A one-dimensional convective problem with a variable source term Q. Petrov–Galerkin procedure results in an exact solution but simple upwindng gives substantial error

Continuity requirements for weighting functions. The weighting function W_i (or $\tilde{W}_i$) introduced in Fig. 12.3 can of course be discontinuous as far as the contributions to the convective terms are concerned [viz. Eq. (12.13)], i.e.

$$\int_0^L W_i \frac{\partial F}{\partial x} \mathrm{d}x \qquad \text{or} \qquad \int_0^L W_i A \frac{\partial N_i}{\partial x} \mathrm{d}x$$

Clearly no difficulty arises at the discontinuity in evaluation of the above integrals. However, when evaluating the diffusion term, we generally introduce integration by parts and evaluate such terms as

$$\int \frac{\partial W_i}{\partial x} k \frac{\partial N_j}{\partial x} \mathrm{d}x$$

in place of the form

$$\int W_i \frac{\partial}{\partial x}\left(k \frac{\partial N_j}{\partial x}\right) \mathrm{d}x$$

Here a local infinity will occur with discontinuous W_i. To avoid this difficulty we consider the discontinuity of the $\tilde{W}_i$ part of the weighting function to occur *within* the element[1] rather than at the node in the manner shown in Fig. 12.3. Now direct integration can be used, showing in the present case zero contributions to the diffusion term, as indeed happens with C_0 continuous functions for $\tilde{W}$ used in earlier references.

12.3 The steady-state problem in one dimension—balancing diffusion

The comparison of the nodal equations (12.14) and (12.23) obtained on a uniform mesh and for a constant Q shows that the effect of the Petrov–Galerkin procedure is equivalent to the use of a standard Galerkin process with the addition of a diffusion

$$k_b = \tfrac{1}{2}\alpha A h \tag{12.27}$$

to the original differential equation (12.9).

The reader can easily verify that with this substituted into the original equation, thus writing now in place of Eq. (12.9)

$$A\frac{\mathrm{d}U}{\mathrm{d}x} - (k+k_b)\frac{\mathrm{d}^2 U}{\mathrm{d}x^2} + Q = 0 \tag{12.28}$$

we obtain an identical expression to that of Eq. (12.23) providing Q is constant and a standard Galerkin procedure is used.

Such *balancing diffusion* is easier to implement than Petrov–Galerkin weighting, particularly in two or three dimensions, and has some physical merit in interpretation of the Petrov–Galerkin methods. However, it does not provide the modification of source terms required, and for instance in the example of Fig. 12.6 will give erroneous results identical with a simple finite difference, upwind, approximation.

The concept of *artificial diffusion* introduced frequently in finite difference models suffers of course from the same drawbacks and in addition cannot be logically justified.

It is of interest to observe that a central difference approximation, when applied to the original equations (or the use of the standard Galerkin process), fails by introducing a *negative diffusion* into the equations. This 'negative' diffusion is countered by the present, balancing, one.

12.4 The steady-state problem in one dimension—a variational principle

Equation (12.9) which we are here considering is not self-adjoint and hence is not derivable from any variational principle. However, it was shown by Guymon *el al.*[12] that it is a simple matter to derive a variational principle (or ensure self-adjointness which is equivalent) if the operator is premultiplied by a suitable function p. Thus we write a weak form of Eq. (12.9) as

$$\int_0^L Wp\left[A\frac{\mathrm{d}U}{\mathrm{d}x} - \frac{\mathrm{d}}{\mathrm{d}x}\left(k\frac{\mathrm{d}U}{\mathrm{d}x}\right) + Q\right]\mathrm{d}x = 0 \tag{12.29}$$

where $p=p(x)$ is as yet undetermined. This gives, on integration by parts,

$$\int_0^L \left[W\frac{\mathrm{d}U}{\mathrm{d}x}\left(pA+k\frac{\mathrm{d}p}{\mathrm{d}x}\right)+\frac{\mathrm{d}W}{\mathrm{d}x}(kp)\frac{\mathrm{d}U}{\mathrm{d}x}+WpQ\right]\mathrm{d}x+\left|Wpk\frac{\mathrm{d}U}{\mathrm{d}x}\right|_0^L=0 \quad (12.30)$$

Immediately we see that the operator can be made self-adjoint and a symmetric approximation achieved if the first term in square brackets is made zero (viz. also Chapter 9 of Volume 1, Sec. 9.11.2, for this derivation). This requires that p be so chosen that

$$pA+k\frac{\mathrm{d}p}{\mathrm{d}x}=0 \quad (12.31a)$$

or that

$$p=\text{constant}\times \mathrm{e}^{-Ax/k}=\text{constant}\times \mathrm{e}^{-2(Pe)x/h} \quad (12.31b)$$

For such a form corresponding to the existence of a variational principle the 'best' approximation is that of the Galerkin method with

$$W=N_i \qquad U=\sum N_j U_j \quad (12.32)$$

Indeed, as shown in Volume 1, such a formulation will, in one dimension, yield answers exact at nodes (viz. Appendix 7 of Volume 1). It must therefore be equivalent to that obtained earlier by weighting in the Petrov–Galerkin manner. Inserting the approximation of Eq. (12.32) into Eq. (12.30), with Eqs (12.31) defining p using an origin at $x=x_i$, we have for the ith equation of the uniform mesh

$$\int_{-h}^{h}\left[\frac{\mathrm{d}N_i}{\mathrm{d}x}(k\,\mathrm{e}^{-2(Pe)x/h})\frac{\mathrm{d}N_j}{\mathrm{d}x}\bar{U}_j+N_i\mathrm{e}^{-2(Pe)x/h}Q\right]\mathrm{d}x=0 \quad (12.33)$$

with $j=i-1,\ i,\ i+1$. This gives, after some algebra, a typical nodal equation:

$$(1-\mathrm{e}^{-2(Pe)})\bar{U}_{i+1}+(\mathrm{e}^{-2(Pe)}-\mathrm{e}^{-2(Pe)})\bar{U}_i-(1-\mathrm{e}^{-2(Pe)})\bar{U}_{i+1} - \frac{Qh^2}{2(Pe)k}(\mathrm{e}^{Pe}-\mathrm{e}^{-Pe})^2=0 \quad (12.34)$$

which can be shown to be identical with the expression (12.23) into which $\alpha=\alpha_{\text{opt}}$ *given by Eq. (12.24) has been inserted.*

Here we have a somewhat more convincing proof of the optimality of the Petrov–Galerkin weighting proposed.[13,14] However, serious drawbacks exist. The numerical evaluation of the integrals is difficult and the equation system, though symmetric overall, is not well conditioned if p is taken as a continuous function of x through the whole domain. The second point is easily overcome by taking p to be *discontinuously* defined, for instance taking the origin of x at point i for *all assemblies* as we did in deriving Eq. (12.34). This is permissible by arguments given in Sec. 12.2, page 447, and is equivalent to scaling the full equation system row by row.[13] Now of course the total equation system ceases to be symmetric.

The numerical integration difficulties disappear, of course, if the simple weighting functions previously derived are used. However, the proof of equivalence is important, as the problem of determining the optimal weighting disappears.

An interesting alternative to the local use of exponential functions has recently been presented by Idelsohn.[15] In this we note that the Galerkin weighting of the form given by Eq. (12.29) is self-adjoint in *the discrete sense* for any, locally defined, function $\hat{p}$ that gives [viz. Eq. (12.30)]

$$\int_{-h}^{h} N_i \frac{\mathrm{d}U}{\mathrm{d}x}\left(A\hat{p}+k\frac{\mathrm{d}\hat{p}}{\mathrm{d}x}\right)\mathrm{d}x=0$$

for the element assembly.

If locally $\mathrm{d}U/\mathrm{d}x$ is assumed as a constant (average) value then we simply require that

$$\int_{-h}^{h} N_i\left(A\hat{p}+k\frac{\mathrm{d}\hat{p}}{\mathrm{d}x}\right)\mathrm{d}x=0$$

where $\hat{p}$ is *any continuous function in the assembly domain.* For instance, if we take a linear distribution as shown in Fig. 12.7,

$$\hat{p}=1+\gamma\frac{x}{h}$$

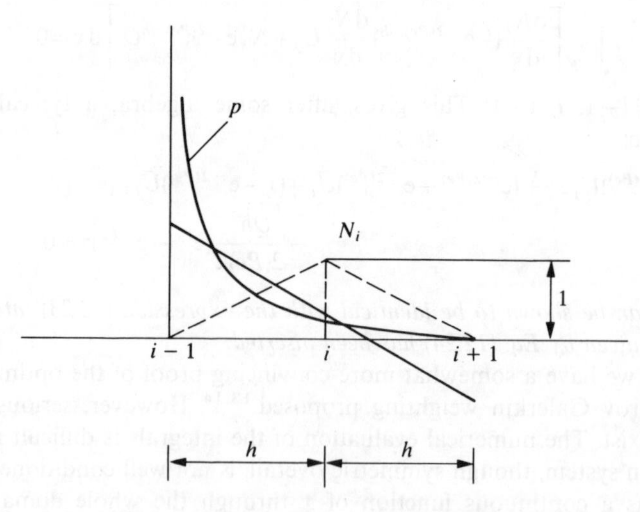

Fig. 12.7 A local linear approximation to modifying function p in the form $\hat{p}=1+\gamma x/h$

it follows immediately that

$$\gamma = -\frac{Ah}{k}$$

satisfies the above, but of course the definition of the approximation $\hat{p}$ is not unique. Almost any one parameter function can be made to satisfy the above equations.

Further, the results of this approximation are disappointing. If

$$\frac{Ah}{k} \equiv 2Pe > 1$$

the function $\hat{p}$ becomes negative over part of the domain and positive definitiveness is lost. An alternative approach is suggested by Sampaio.[16] He introduces piecewise constant weighting functions to approximate $\hat{p}$ so that

$$\int_{-h}^{h} N_i(p-\hat{p})\,\mathrm{d}x = \int_{-h}^{h} N_i(\mathrm{e}^{-2(Pe)x/h} - \hat{p})\,\mathrm{d}x = 0$$

This yields, as shown in Fig. 12.8,

$$\hat{p} = 1 \pm \alpha$$

where α is determined by the expression (12.24).

It is not surprising that this approximation yields in the simple case precisely the optimal expression. However, the use of the simple functions is advantageous if element sizes are not identical and is a procedure easily extendible to two or three dimensions.

12.5 Least square–Galerkin approximation

In the preceding sections we have shown that several, apparently different, approaches have resulted in identical (or almost identical) approximations. Here yet another procedure is presented which again will produce similar results. In this a combination of the standard Galerkin and least square approximations is made.[17,18]

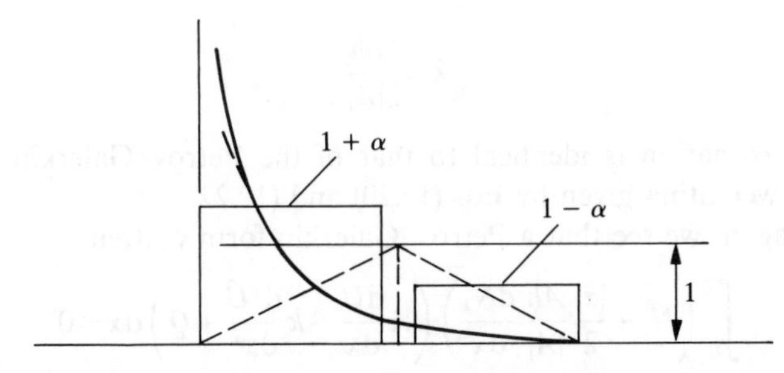

Fig. 12.8 Piecewise constant, local, approximation to modifying function

If Eq. (12.9) is rewritten as

$$LU+Q=0 \qquad U\approx\hat{U}=\mathbf{N}\bar{\mathbf{U}} \tag{12.35a}$$

with

$$L=A\frac{\mathrm{d}}{\mathrm{d}x}-k\frac{\mathrm{d}^2}{\mathrm{d}x^2} \tag{12.35b}$$

the standard Galerkin approximation gives for the kth equation

$$\int_0^L N_k L(\mathbf{N})\bar{\mathbf{U}}\,\mathrm{d}x+\int_0^L N_k Q\,\mathrm{d}x=0 \tag{12.36}$$

with boundary conditions omitted for clarity.

Similarly, a least square residual minimization (viz. Chapter 9 of Volume 1, Sec. 9.14.2) results in

$$R=L\hat{U}+Q$$

and

$$\frac{1}{2}\frac{\mathrm{d}}{\mathrm{d}\bar{U}_k}\int_0^L R^2\,\mathrm{d}x=\int_0^L\frac{\mathrm{d}(L\hat{U})}{\mathrm{d}\bar{U}_i}(L\hat{U}+Q)\,\mathrm{d}x=0 \tag{12.37}$$

or

$$\int_0^L\left(A\frac{\mathrm{d}N_k}{\mathrm{d}x}-k\frac{\mathrm{d}^2}{\mathrm{d}x^2}N_k\right)(L\hat{U}+Q)=0 \tag{12.38}$$

If the final approximation is written as a linear combination of Eqs (12.36) and (12.38), we have

$$\int_0^L\left(N_k+\lambda A\frac{\mathrm{d}N_k}{\mathrm{d}x}-\lambda k\frac{\mathrm{d}^2}{\mathrm{d}x^2}N_k\right)(L\hat{U}+Q)\,\mathrm{d}x=0 \tag{12.39}$$

This is of course, a Petrov–Galerkin approximation with an undetermined parameter λ. If the second-order term is omitted (as could be done assuming linear N_k and a curtailment as in Fig. 12.3) and further if we take

$$\lambda=\frac{|\alpha|h}{2|A|} \tag{12.40}$$

the approximation is identical to that of the Petrov–Galerkin method with the weighting given by Eqs (12.20) and (12.22).

Once again we see that a Petrov–Galerkin form written as

$$\int_0^L\left(N_i+\frac{|\alpha|}{2}\frac{Ah}{|A|}\frac{\mathrm{d}N_k}{\mathrm{d}x}\right)\left(A\frac{\mathrm{d}\hat{U}}{\mathrm{d}x}-k\frac{\mathrm{d}^2\hat{U}}{\mathrm{d}x^2}+Q\right)\mathrm{d}x=0 \tag{12.41}$$

is a result that follows from diverse approaches, though only the variational form of Sec. 12.4 explicitly determines the value of α that should optimally be used. In all the other derivations this value is determined by an *a posteriori* analysis.

12.6 Higher-order approximations

The derivation of accurate Petrov–Galerkin procedures for the convective diffusion equation is of course possible for any order of finite element expansion. In reference 9 Heinrich and Zienkiewicz show how the procedure of studying exact discrete solutions can yield optimal upwind parameters for quadratic shape functions. However, here the simplest approach involves the procedures of Sec. 12.4, which are available of course for any element expansion and, as shown before, will always give an optimal approximation.

We thus recommend the reader to pursue the example discussed in that section and, by extending Eq. (12.33), to arrive at an appropriate equation linking the two quadratic elements of Fig. 12.9.

For practical purposes for such elements it is possible to extend the Petrov–Galerkin weighting of the type given in Eqs (12.20) to (12.22) now using

$$\alpha_{\text{opt}} = \coth Pe - \frac{1}{Pe}$$

and

$$\alpha \tilde{W}_i = \alpha \frac{h}{4} \frac{\mathrm{d}N_i}{\mathrm{d}x} \qquad (\text{sign } A) \tag{12.42}$$

This procedure, though not as exact as that for linear elements, is very effective and has been used with success for solution of Navier–Stokes equations.[19]

12.7 Extension to two (or three) dimensions

12.7.1 *General remarks.* It is clear that the application of standard Galerkin discretization to the steady-state scalar convection–diffusion

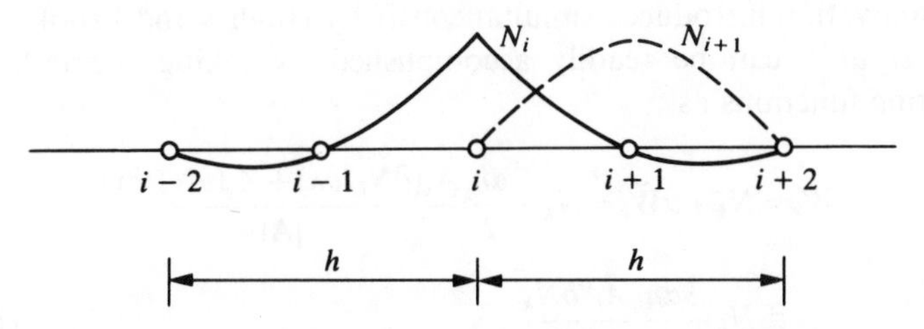

Fig. 12.9 Assembly of one-dimensional quadratic elements

equation in several space dimensions is similar to the problem discussed previously in one dimension and will again yield unsatisfactory answers with high oscillation for local Peclet numbers greater than unity.

The equation now considered is the steady-state version of Eq. (12.5), i.e.

$$A_x \frac{\partial U}{\partial x} + A_y \frac{\partial U}{\partial x} - \frac{\partial}{\partial x}\left(k \frac{\partial U}{\partial x}\right) - \frac{\partial}{\partial y}\left(k \frac{\partial U}{\partial y}\right) + Q = 0 \tag{12.43}$$

in two dimensions (with an addition of appropriate terms with z derivatives for three).

Obviously the problem is now of greater practical interest than the one-dimensional case so far discussed, and a satisfactory solution is important. Again, all of the possible approaches we have discussed are applicable.

12.7.2 *Streamline Petrov–Galerkin weighting.* The most obvious procedure is to use again some form of Petrov–Galerkin method of the type introduced in Sec. 12.2 and Eqs (12.20) to (12.24), seeking optimality of α in some heuristic manner. We note immediately that the Peclet parameter

$$\mathbf{Pe} = \frac{\mathbf{A}h}{2k} \qquad \mathbf{A} = \begin{Bmatrix} A_v \\ A_y \end{Bmatrix} \tag{12.44}$$

is now a 'vector' quantity and hence that upwinding needs to be 'directional'.

The first reasonably satisfactory attempt to do this consisted of determining the optimal Petrov–Galerkin formulation using $\alpha\tilde{W}$ based on components of $\mathbf{A}$ associated to the *sides of elements* and of obtaining the final weight functions by a blending procedure.[8,9]

A better method was soon realized when the analogy between balancing diffusion and upwinding was established, as shown in Sec. 12.3. In two (or three) dimensions the convection is only active in the direction of the resultant element velocity $\mathbf{A}$, and hence the corrective, or *balancing, diffusion* introduced by upwinding should be anisotropic with a coefficient different from zero only in the direction of the velocity resultant. This innovation introduced simultaneously by Hughes and Brooks[20] and Kelly *et al.*[10] can be readily accomplished by taking the individual weighting functions as

$$W_k = N_k + \alpha \tilde{W}_k = N_k + \frac{\alpha h}{2} \frac{A_x(\partial N_k/\partial x) + A_y(\partial N_k/\partial y)}{|\mathbf{A}|}$$

$$\equiv N_k + \frac{\alpha h}{2} \frac{A_i}{|\mathbf{A}|} \frac{\partial N_k}{\partial x_i} \tag{12.45}$$

where α is determined for each element by the previously found expression (12.21) written as follows:

$$\alpha = \alpha_{\text{opt}} = \coth Pe - \frac{1}{Pe} \tag{12.46}$$

with

$$Pe = \frac{|\mathbf{A}|h}{2k} \tag{12.47a}$$

and

$$|\mathbf{A}| = (A_x^2 + A_y^2)^{1/2} \tag{12.47b}$$

The above expressions presuppose that the velocity components A_x and A_y in a particular element are substantially constant and that the element size h can be reasonably defined.

Figure 12.10 shows an assembly of linear triangles and bilinear quadrilaterals for each of which the mean resultant velocity $\mathbf{A}$ is indicated. Determination of the element size h to use in expression (12.47) is of course somewhat arbitrary. In Fig. 12.10 we show it simply as the maximum size in the direction of the velocity vector.

The form of Eq. (12.45) is such that the 'non-standard' weighting $\tilde{W}$ has a zero effect in the direction in which the velocity component is zero. Thus the balancing diffusion is only introduced in the direction of the resultant velocity (convective) vector $\mathbf{A}$. This can be verified if Eq. (12.43) is written in a tensorial (index) notation as

$$A_i \frac{\partial U}{\partial x_i} - \frac{\partial}{\partial x_i}\left(k \frac{\partial U}{\partial x_i}\right) + Q = 0 \tag{12.48a}$$

In the discretized form the 'balancing diffusion' term [obtained from

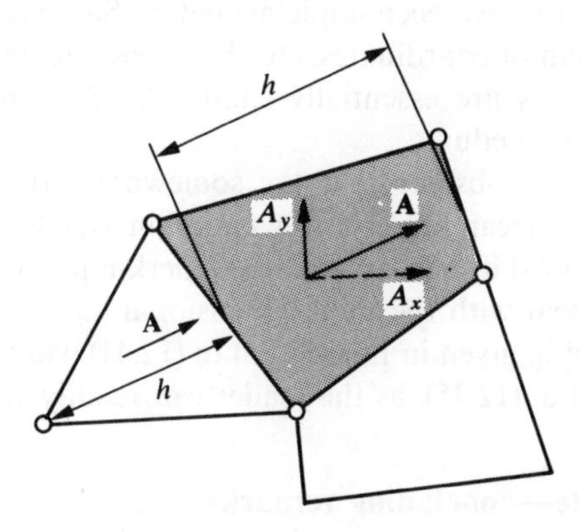

Fig. 12.10 A two-dimensional, streamline assembly. Element size h and streamline directions

weighting the first term of the above with W of Eq. (12.45)] becomes

$$\int_{\Omega} \frac{\partial N}{\partial x_i} \tilde{k}_{ij} \frac{\partial N}{\partial x_j} \, d\Omega \tag{12.48b}$$

with

$$\tilde{k}_{ij} = \frac{\alpha A_i A_j}{|\mathbf{A}|} \tag{12.48c}$$

This indicates a highly anisotropic diffusion with zero coefficients normal to the convective velocity vector directions. It is therefore named the *streamline balancing diffusion*[10,20,21] or streamline Petrov–Galerkin process.

The mathematical validity of the procedures introduced in this section has been established by Johnson *et al.*[22] for $\alpha = 1$, showing convergence improvement over the standard Galerkin process. However, the proof does not include any optimality in the selection of α values as shown by Eq. (12.46).

Figure 12.11 shows a typical solution of Eq. (12.43), indicating the very small amount of 'cross wind diffusion', i.e. allowing discontinuities to propagate in the direction of flow without substantial smearing.[23]

A more convincing 'optimality' can be achieved by applying the exponential modifying function, making the problem self-adjoint. This of course follows precisely the procedures of Sec. 12.4 and is easily accomplished if the velocities are constant in the element assembly domain. If velocities vary from element to element, again the exponential functions

$$p = e^{-Ax'/k} \tag{12.49}$$

with x' orientated in the velocity direction in each element can be taken. This appears first to have been implemented by Sampaio[23] but problems regarding the origin of coordinates, etc., have once again to be addressed. However, the results are essentially similar here to those achieved by Petrov–Galerkin procedures.

It is of interest to observe that the somewhat intuitive approach to generation of the 'streamline' Petrov–Galerkin weight functions of Eq. (12.45) can be avoided if the least square–Galerkin procedures of Sec. 12.4 are extended to deal with the multidimensional equation. Simple extension of the reasoning given in Eqs (12.35) to (12.41) will immediately yield the weighting of Eq. (12.45), as the reader can readily verify.

12.8 Steady state—concluding remarks

In Secs 12.2 to 12.7 we presented several currently used procedures for dealing with the steady-state convection–diffusion equation with a scalar

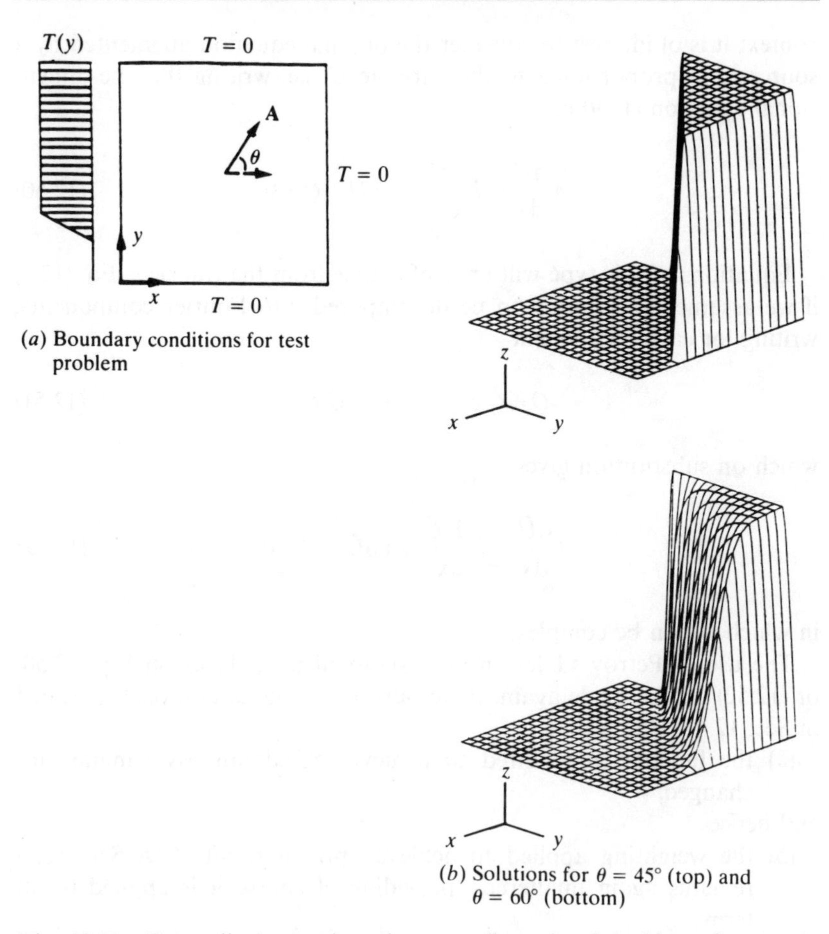

(*a*) Boundary conditions for test problem

(*b*) Solutions for $\theta = 45°$ (top) and $\theta = 60°$ (bottom)

Fig. 12.11 'Streamline' procedures in a two-dimensional problem of pure convection. Bilinear elements[23]

variable. All of these translate essentially to the use of streamline Petrov–Galerkin discretization, though of course the modification of the basic equations to a self-adjoint form given in Sec. 12.4 produces the *full justification* of the special weighting. Which of the procedures is best used in practice is largely a matter of taste, as all will give excellent results. However, we shall see from the second part of this chapter, in which transient problems are dealt with, that other methods can be adopted if time-stepping procedures are used as an iteration to derive steady-state algorithms. Further, the identity of the procedures in the case of a multicomponent, **U**, does not appear to exist and indeed some are not applicable. We shall defer consideration of these matters till later. In this context it is of interest to consider the original equation augmented by a source term proportional to the variable U, i.e. writing the one-dimensional equation (12.9) as

$$A\frac{dU}{dx} - k\frac{d^2U}{dx^2} + mU + Q = 0 \tag{12.50}$$

Equations of this type will arise of course from the transient Eq. (12.8) if we assume the solution to be decomposed into Fourier components, writing for each component

$$Q = \tilde{Q}\, e^{i\omega t} \qquad U = \tilde{U}\, e^{i\omega t} \tag{12.51}$$

which on substitution gives

$$A\frac{d\tilde{U}}{dx} - k\frac{d^2\tilde{U}}{dx^2} + i\omega\tilde{U} + \tilde{Q} = 0 \tag{12.52}$$

in which $\tilde{U}$ can be complex.

The use of Petrov–Galerkin or variational procedures on Eq. (12.50) or (12.52) can be made again. If we pursue the line of approach outlined in Sec. 12.4 we note that

(*a*) the function p required to achieve self-adjointness remains unchanged;

and hence

(*b*) the weighting applied to achieve optimal results (viz. Sec. 12.3) remains again unaltered—providing of course it is applied to all terms.

Although the above result is encouraging and permits the solution in frequency domain for transient problems, it does not readily 'transplant' to the problems in which time-stepping procedures are required.

Some further points require mentioning at this stage. These are simply that:

1. When pure convection is considered (that is $k = 0$) only one boundary condition—generally that giving the value of U at the inlet—can be specified, and in such a case the violent oscillations observed in Fig. 12.2 with standard Galerkin procedures will not occur generally.
2. Specification of no boundary condition at the outlet edge in the case when $k > 0$, which is equivalent to imposing a zero conduction flux there, generally results in quite acceptable solutions with standard Galerkin weighting even for quite high Peclet numbers.

PART II: TRANSIENTS

12.9 Transients—introductory remarks

The objective of this section is to develop procedures of general applicability for the solution by direct time-stepping methods of Eq. (12.1) written for scalar values of U, F_i and G_i:

$$\frac{\partial U}{\partial t}+\frac{\partial F_i}{\partial x_i}+\frac{\partial G_i}{\partial x_i}+Q=0 \tag{12.53}$$

though consideration of the procedure for dealing with a vector-valued function will be included in Sec. 12.13. However, to allow a simple interpretation of the various methods and of behaviour patterns the scalar equation in one dimension [viz. Eq. (12.8)], i.e.

$$\frac{\partial U}{\partial t}+A\frac{\partial U}{\partial x}-\frac{\partial}{\partial x}\left(k\frac{\partial U}{\partial x}\right)+Q=0 \tag{12.54a}$$

will be considered. This of course is a particular case of Eq. (12.53) in which $F=F(U)$, $A=\partial F/\partial U$ and $Q=Q(U, x)$ and therefore

$$\frac{\partial F}{\partial x}=\frac{\partial F}{\partial U}\frac{\partial U}{\partial x}=A\frac{\partial U}{\partial x} \tag{12.54b}$$

The problem so defined is non-linear unless A is constant. However, the non-conservative equations (12.54) admit a spatial variation of A and are quite general.

The main behaviour patterns of the above equations can be determined by a change of the independent variable x to x' such that

$$\mathrm{d}x_i'=\mathrm{d}x_i-A_i\,\mathrm{d}t \tag{12.55}$$

Noting that for $U=U(x_i', t)$ we have

$$\left.\frac{\partial U}{\partial t}\right|_{x\text{ const}}\frac{\partial U}{\partial x_i'}\frac{\partial x_i'}{\partial t}+\left.\frac{\partial U}{\partial t}\right|_{x'\text{ const}}=-A_i\frac{\partial U}{\partial x_i'}+\left.\frac{\partial U}{\partial t}\right|_{x'\text{ const}} \tag{12.56}$$

The one-dimensional equation (12.54a) becomes now simply

$$\frac{\partial U}{\partial t}-\frac{\partial}{\partial x'}\left(k\frac{\partial U}{\partial x'}\right)+Q=0 \tag{12.57}$$

and equations of this type with self-adjoint spatial operators can be readily discretized and solved by procedures developed previously (viz. Chapter 10).

The coordinate system of Eq. (12.55) describes *characteristic directions* and the moving nature of the coordinates must be noted. A further corollary of the coordinate change is that with no conduction or heat generation terms, i.e. when $k=0$ and $Q=0$, we have simply

$$\frac{\partial U}{\partial t}=0$$

or

$$U(x')=U(x-At)=\text{constant} \tag{12.58}$$

along a characteristic [assuming A to be constant, which will be the case if $A=A(U)$]. This is a typical equation of a wave propagating with a velocity A in the x direction, as shown in Fig. 12.12. The wave nature is evident in the problem even if the conduction (diffusion) is not zero, and in this case we shall have solutions showing a wave that attenuates with the distance travelled.

The discretization and solution procedures which are available for the solution of the problem must include the wave characteristic nature of the problem, but in addition should allow the same type of approximation that we have arrived at in Part I as, obviously, steady state is a particular case of the transient. Indeed, frequently transient solutions are used simply as an iteration process of arriving at the solution of steady state, as we shall illustrate in the case of fluid mechanics in Chapters 13 to 15. Here obviously an explicit type of time marching procedure will be of paramount interest.

The simultaneous achievement of all the objectives of the discretization is difficult and a measure of compromise is unfortunately necessary. In what follows we shall outline main lines of attack that are feasible and discuss their merits and shortcomings. These are:

1. *Petrov–Galerkin methods* (Sec. 12.10) in which spatial discretization proceeds according to the methods developed in Part I of this chapter and is followed by 'standard' time stepping as in Chapter 10.
2. *Least square and space–time Petrov–Galerkin methods* (Sec. 12.11) in which an attempt is made to combine (with severe limitations) the optimal approximations procedures.
3. *Characteristic-based methods* (Sec. 12.12) in which *a priori* the wave nature of the equations is taken into account.

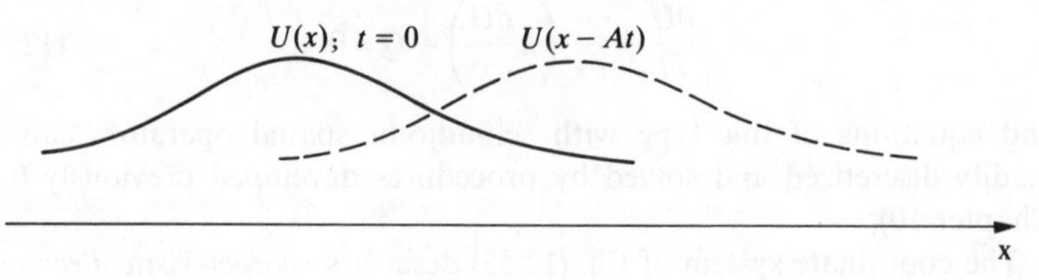

Fig. 12.12 The wave nature of solution with no conduction. Constant wave velocity A

4. *Time-stepping schemes of higher order followed by standard Galerkin approximations.* Although such procedures do not take any account of the wave nature of the problem they will often be found to reproduce other methods in a simpler and more general manner (Sec. 12.13). Here steady-state solutions will depend on the size of the time step Δt used which acts as an upwinding parameter.

12.10 Transient formulation of the variational and Petrov–Galerkin procedures

12.10.1 *General forms.* In Secs 12.2 and 12.4 we have shown that, in the absence of time derivatives, the approximation derived by a heuristic use of 'upwinded' weighting functions and that based on the variational principle obtained by premultiplication by an exponential function in Eq. (12.33) were identical. It is therefore logical to consider first the extension of the latter procedure to the time-dependent problem.

Thus if Eq. (12.54) is premultiplied by the function p given by Eq. (12.31) and a standard Galerkin space approximation is used, we obtain

$$\int_\Omega \mathbf{N}^{\mathrm{T}} p \left[\frac{\partial \hat{U}}{\partial t} + A \frac{\partial \hat{U}}{\partial x} - \frac{\partial}{\partial x} k \left(\frac{\partial \hat{U}}{\partial x} \right) \right] + Q \, \mathrm{d}\Omega$$

$$\equiv \left(\int_\Omega \mathbf{N}^{\mathrm{T}} p \mathbf{N} \, \mathrm{d}\Omega \right) \dot{\bar{\mathbf{U}}} + \left(\int_\Omega (\nabla \mathbf{N})^{\mathrm{T}} k p \nabla \mathbf{N} \, \mathrm{d}\Omega \right) \bar{\mathbf{U}} + \int_\Omega p \mathbf{N}^{\mathrm{T}} Q \, \mathrm{d}\Omega$$

$$\equiv \tilde{\mathbf{M}} \dot{\bar{\mathbf{U}}} + \tilde{\mathbf{H}} \bar{\mathbf{U}} - \tilde{\mathbf{f}} = 0 \tag{12.59}$$

In this we approximate the spatial distribution of U in the usual way as

$$U \approx \hat{U} = \mathbf{N} \bar{\mathbf{U}} \tag{12.60}$$

The above equation system is of course symmetric and capable of solution in the time domain by any of the methods described in Chapter 10. We can, of course, scale each equation, changing the origin of the function

$$p = \mathrm{e}^{-2(Pe)x/h} \tag{12.61}$$

in the same manner as we used in Sec. 12.4 to avoid conditioning difficulties but now losing symmetry. With such scaling a typical nodal equation derived for steady-state conditions [viz. Eq. (12.34)] is augmented by time derivatives. These add to the left-hand side of Eq. (12.34) the following terms:

$$\begin{aligned} &h/A[(1-1/Pe)\,\mathrm{e}^{2Pe} + 1 + 1/Pe]\dot{\bar{U}}_{i-1} \\ &+ h/A[1/(Pe)\,\mathrm{e}^{2Pe} - 4 - 1/(Pe)\,\mathrm{e}^{-2Pe}]\dot{\bar{U}}_i \\ &+ h/A[1 - 1/Pe + (1 + 1/Pe)\,\mathrm{e}^{-2Pe}]\dot{\bar{U}}_{i+1} \end{aligned} \tag{12.62}$$

Although the application of the Petrov–Galerkin procedure with optimal upwinding to Eq. (12.54a) gives indentical terms in the steady state, the transient terms are different and become

$$h/A[(\tfrac{5}{12}-\tfrac{1}{4}Pe)\,e^{2Pe}-\tfrac{1}{3}+\tfrac{1}{2}Pe-(\tfrac{1}{12}+\tfrac{1}{4}Pe)\,e^{-2Pe}]\dot{\bar{U}}_{i-1}+\tfrac{2}{3}h/A[e^{Pe}-e^{-Pe}]^2\dot{\bar{U}}_i$$
$$+h/A[(-\tfrac{1}{12}+\tfrac{1}{4}Pe)\,e^{2Pe}-\tfrac{1}{3}-\tfrac{1}{2}Pe+(\tfrac{5}{12}+\tfrac{1}{4}Pe)\,e^{-2Pe}]\dot{\bar{U}}_{i-1} \qquad (12.63)$$

As we shall observe later, the performance of both schemes is different. For comparison we give below the expression that is obtained using a standard Galerkin procedure applied to Eq. (12.54):

$$\left(\frac{h}{6k}\right)(\dot{\bar{U}}_{i-1}+4\dot{\bar{U}}_i+\dot{\bar{U}}_{i+1})-(1+Pe)\bar{U}_{i-1}$$
$$+2\bar{U}_i-(1-Pe)\bar{U}_{i+1}+\frac{Qh^2}{k}=0 \qquad (12.64)$$

Each of the equations (12.62), (12.63) and (12.64) could be solved in the time domain by standard time-stepping procedures. Writing the above equations in the form

$$\tilde{\mathbf{M}}\dot{\bar{\mathbf{U}}}+\tilde{\mathbf{H}}\bar{\mathbf{U}}+\tilde{\mathbf{f}}=0 \qquad (12.65)$$

we can, for instance, approximate in time as [Eq. (10.11) of Chapter 10]

$$\frac{\tilde{\mathbf{M}}[\bar{\mathbf{U}}^{n+1}-\bar{\mathbf{U}}^n]}{\Delta t}+\tilde{\mathbf{H}}[(1-\Theta)\bar{\mathbf{U}}^n+\Theta\bar{\mathbf{U}}^{n+1}]+\mathbf{f}^n=0 \qquad (12.66)$$

and solve for $\bar{\mathbf{U}}^{n+1}$. We would expect that as usual $\Theta \geqslant \frac{1}{2}$ would give unconditional stability and that with $\Theta=0$ and substitution of a lumped matrix

$$\tilde{\mathbf{M}}\approx\tilde{\mathbf{M}}_L \qquad (12.67)$$

an explicit procedure would be available. However, this is not the case generally as the stability proofs (viz. Chapter 10) have been restricted to situations arising from self-adjoint problems in which matrices of the type occurring in Eq. (12.65) were symmetric and positive definite. With convective terms dominating the problem, more elaborate procedures for assessment of stability and performance are needed. A very convenient process is that of comparing exact and numerical solutions on a uniform spatial mesh and obtaining so-called *amplitude* and *relative celerity* ratios. The process is described below.

Consider a typical equation

$$\frac{\partial U}{\partial t}+A\frac{\partial U}{\partial x}=0 \qquad \text{(A)}$$

arising from Eq. (12.54) with $k=0$ and $Q=0$. This has a general analytical solution

$$U = U\,e^{i\sigma(x-At)} \tag{B}$$

where U and σ are arbitrary constants.

With a real value of σ, the above gives periodic solutions of wavelength

$$L = \frac{2\pi}{\sigma} \tag{C}$$

and this allows a convenient Fourier expansion to be used spatially.

At a given x coordinate it is of interest to assess the 'growth' or *amplitude* ratio

$$\lambda = \frac{U(t+\Delta t)}{U(t)} = e^{-iA\sigma\Delta t} \tag{D}$$

This has a modulus $|\lambda|$ such that

$$|\lambda| = 1$$

and an argument β,

$$\arg \equiv \beta = A\sigma\Delta t \equiv pC$$

where we define

$$p \equiv \sigma\,\Delta x \qquad C \equiv A\frac{\Delta t}{\Delta x} \equiv \text{Courant number} \tag{E}$$

We can now obtain a similar ratio, λ^*, for any numerical scheme that links discrete values of U at two time levels t and $t+\Delta t$, assuming that in x the exact solution is satisfied, i.e. that

$$\frac{U_{i+1}}{U_i} = \frac{U_i}{U_{i-1}} = e^{i\sigma\Delta x}$$

where $\Delta x = h$ is the node spacing.

For assessment of the numerical solution the ratio

$$\bar{\lambda} = \frac{\lambda^*}{\lambda} \tag{F}$$

is of particular importance. The modulus of $\bar{\lambda}$ is known as the *amplitude ratio* and the *relative celerity* is the ratio of arguments of λ^* and λ. Both determine the type of performance we can expect from the numerical scheme. Thus the modulus of $\bar{\lambda}$ must be close to unity for accuracy, and in the present case should never exceed it if stability is to be achieved. Similarly, the relative celerity should, for good performance, be close to unity.

To illustrate the procedure consider the discrete approximation by standard Galerkin given by Eq. (12.64) but reduced to the purely convective case and with the temporal approximation of Eq. (12.66) used.

The reader can now verify that we have

$$\frac{1}{6C}(\bar{U}_{i-1}^{n+1}+4\bar{U}_i^{n+1}+\bar{U}_{i+1}^{n+1})-\frac{1}{6C}(\bar{U}_{i-1}^n+4\bar{U}_i^n+\bar{U}_{i+1}^n)$$
$$+\frac{1-\Theta}{2}(-\bar{U}_{i+1}^n+\bar{U}_{i-1}^n)+\frac{\Theta}{2}(-\bar{U}_{i+1}^n+\bar{U}_{i-1}^{n+1})=0$$

with $C=A\,\Delta t/h$ being the Courant number. Putting

$$\bar{U}_{i+1}^n=\bar{U}_i^n\,\mathrm{e}^{i\sigma h}=\bar{U}_i^n\,\mathrm{e}^{ip}, \qquad \text{etc.}$$

and inserting into the previous equation we obtain the ratio λ^* as

$$\lambda^*\equiv\frac{\bar{U}_i^{n+1}}{\bar{U}_i^n}=\frac{(\cos p+2)-3Ci(1+\Theta)\sin p}{(\cos p+2)+3Ci\,\Theta\ \sin p}$$

noting that

$$\frac{\mathrm{e}^{ip}+\mathrm{e}^{-ip}}{2}\equiv\cos p \qquad \frac{\mathrm{e}^{ip}-\mathrm{e}^{-ip}}{2i}=\sin p$$

From the above and Eq. (D), both the modulus and argument of $\bar{\lambda}$ can be obtained, giving the desired amplitude and celerity ratios for this scheme.

12.10.2 *The performance characteristics of variational, Petrov–Galerkin and Galerkin approximations.* In Fig. 12.13(*a*), (*b*) and (*c*) we show the performance characteristics, i.e. the amplitude and celerity ratios, for the three schemes given by Eqs (12.62) to (12.64) combined with the time-step algorithm of Eq. (12.66). Only the pure convection case is considered ($k=0$) for simplicity.

These plots should ideally give unity for both the amplitude and celerity ratios; as mentioned before, values of the first exceeding unity will lead to instability or progressive growth of errors and thus are not acceptable.

The parameters used in the above plots are the Courant number

$$C=A\frac{\Delta t}{\Delta x}\equiv A\frac{\Delta t}{h} \tag{12.68}$$

and the Θ used in time stepping.

It is immediately evident that:

1. The variational form is generally unstable.
2. The Petrov–Galerkin is unconditionally stable for $\Theta\geqslant\frac{1}{2}$ but that its conditional stability with $\Theta=0$ is rather limited.
3. The standard (Bubnov) Galerkin process with $\Theta=0.5$ is stable and accurate but is unconditionally unstable for $\Theta=0$.

The last result is not surprising as we have already demonstrated for the steady state that the standard Galerkin process is not applicable.

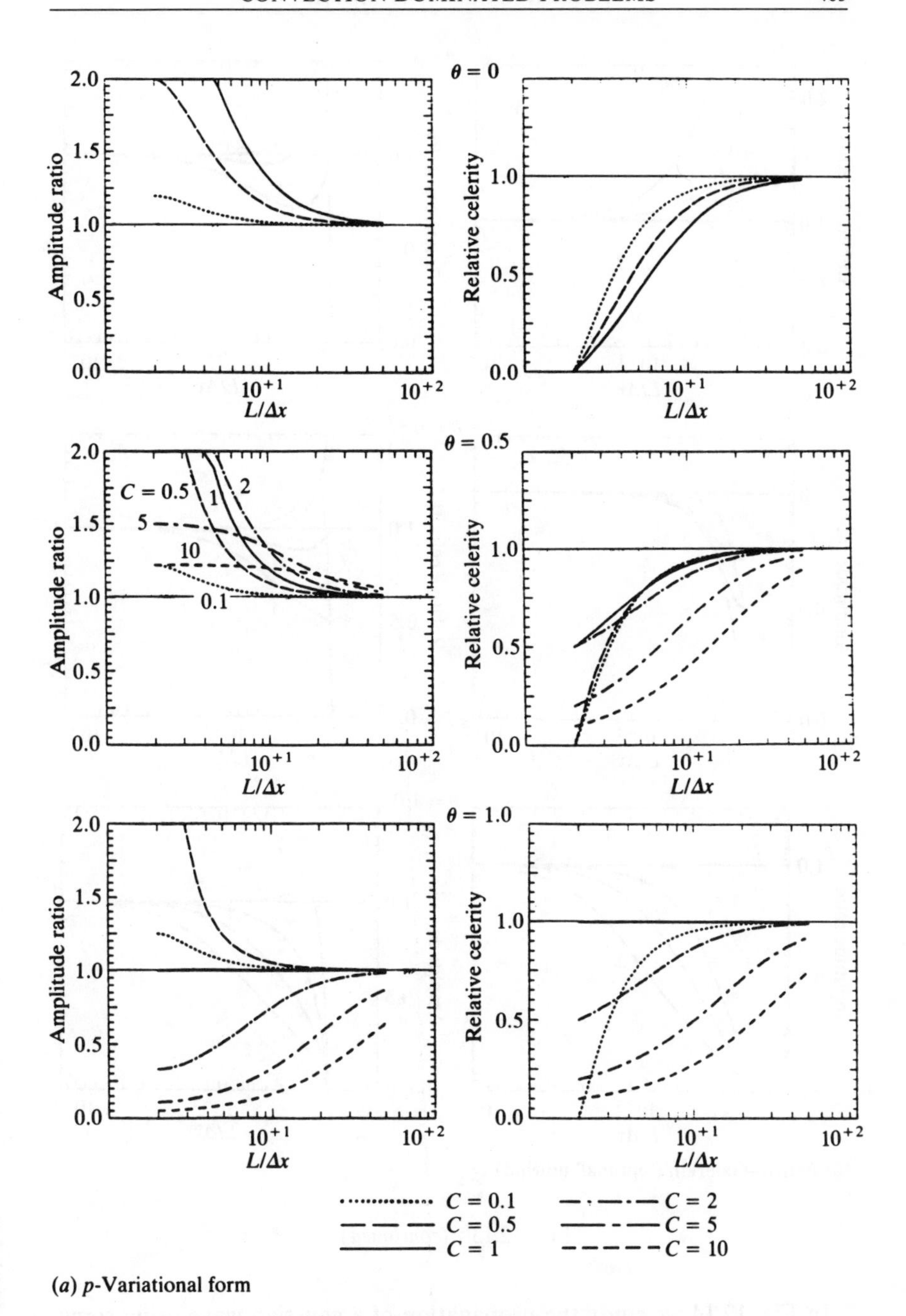

(*a*) *p*-Variational form

Fig. 12.13 Amplitude ratio and relative celerity for various algorithms. Pure convection at various Courant numbers

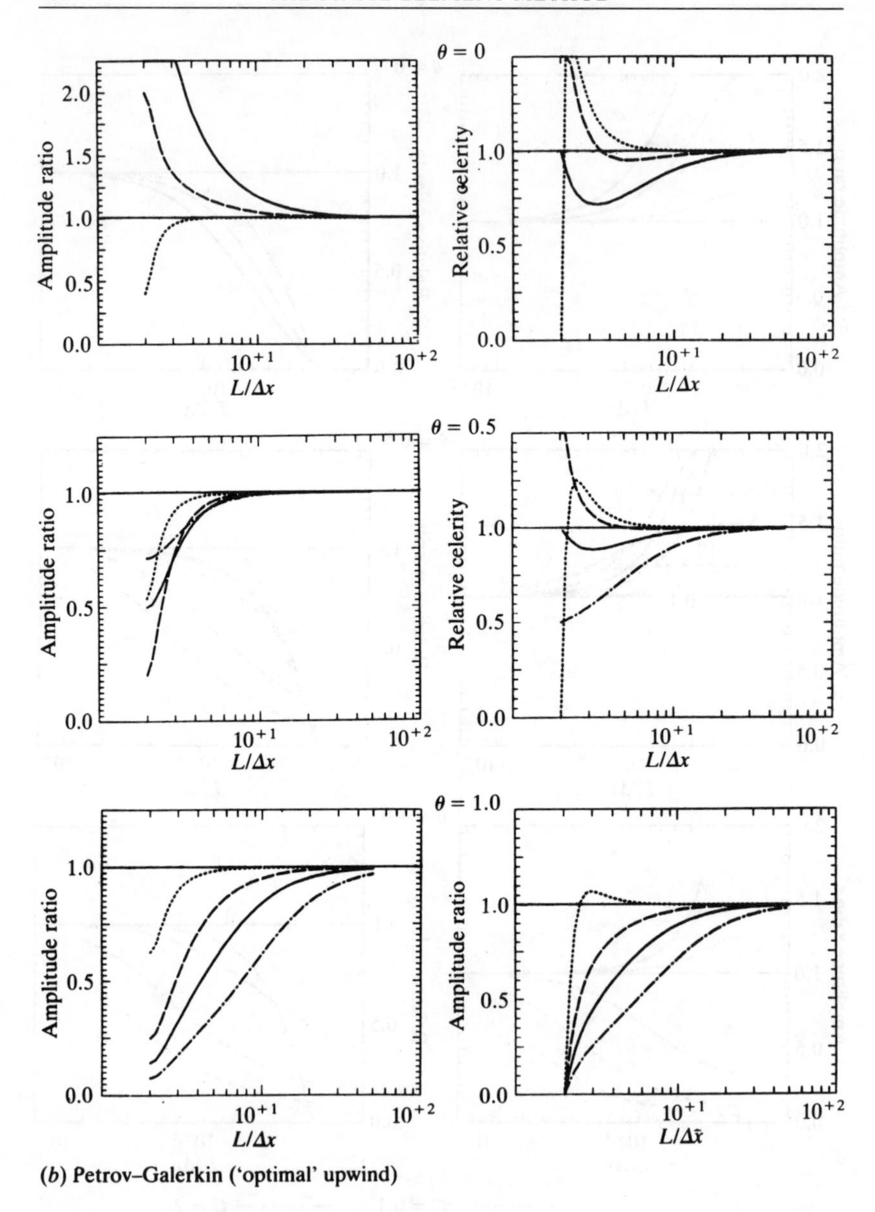

Fig. 12.13 *(continued)*

In Fig. 12.14 we study the propagation of a gaussian wave using some of the implicit methods derived in this chapter. The performance of the methods described so far is poor except for very small Courant numbers ($C = 0.1$) showing excessive damping.

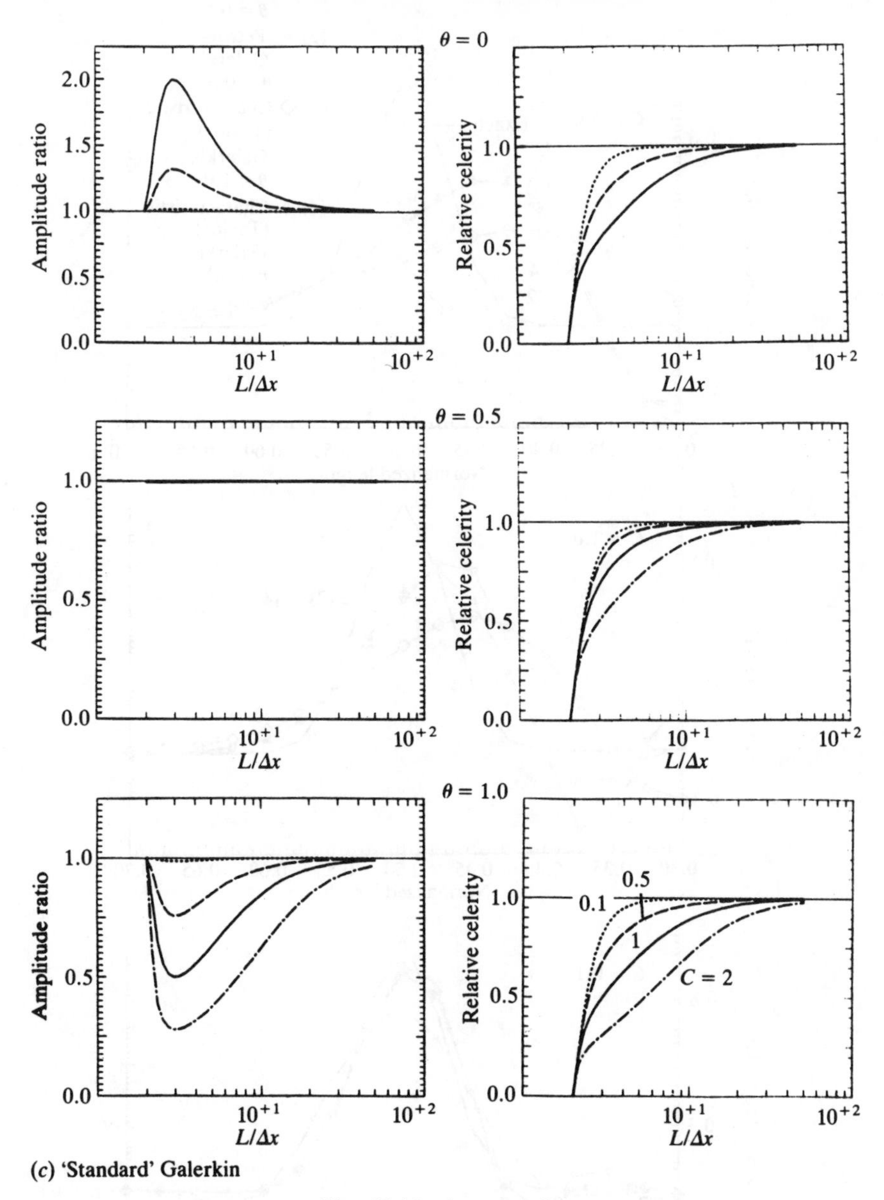

Fig. 12.13 (*continued*)

12.11 Least squares and space–time Petrov–Galerkin methods

The successful and simple application of a Galerkin–least square residual minimization for the steady-state problem of Sec. 12.5 suggests that perhaps this procedure applied now in the space–time domain of Fig. 12.15

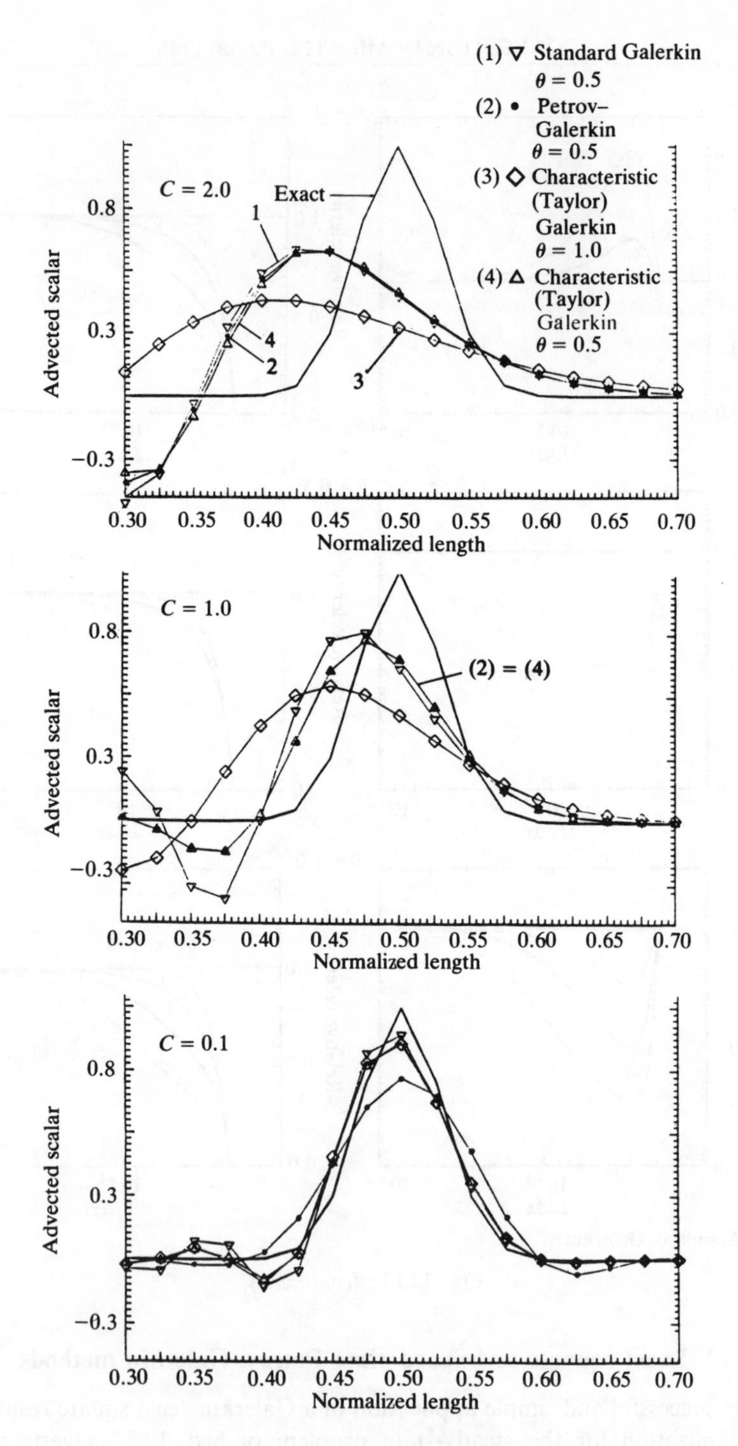

Fig. 12.14 Propagation of a gaussian wave. Profile after $t=0.5$ $(A=1.0)$, $h=0.025$ (20 steps of $C=1.0$), implicit methods

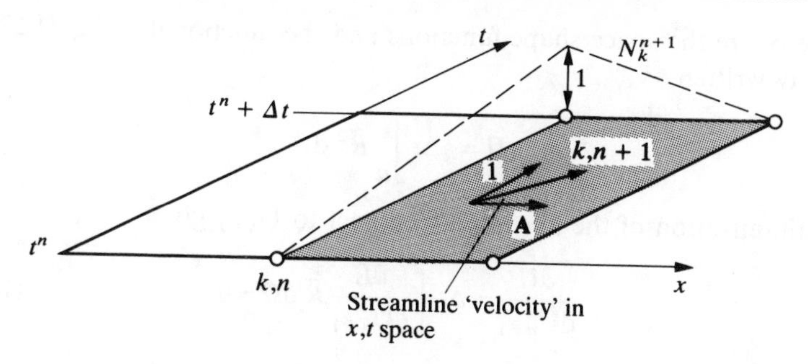

Fig. 12.15 Space–time elements in a 'streamline Petrov–Galerkin' approximation

may again be useful. This indeed turns out to be the case as we shall demonstrate below.

Consider the problem of Eq. (12.53) in which we attempt to approximate U in the space–time domain, writing

$$U \approx \hat{U} = \tilde{\mathbf{N}}^n(x_i, t)\,\bar{\mathbf{U}}_n + \tilde{\mathbf{N}}^{n+1}(x_i, t)\,\bar{\mathbf{U}}_{n+1} \tag{12.69}$$

where subscripts n and $n+1$ refer to times t and $t+\Delta t$. The residual R is defined by

$$R \equiv \frac{\partial \hat{U}}{\partial t} + A_i \frac{\partial \hat{U}}{\partial x_i} + \frac{\partial \hat{G}_i}{\partial x_i} + Q \tag{12.70}$$

and we shall attempt to minimize the functional with respect to $\bar{\mathbf{U}}_{n+1}$:

$$\Pi = \frac{1}{2}\int_t^{t+\Delta t}\int_\Omega R^2\, \mathrm{d}\Omega\, \mathrm{d}t \tag{12.71}$$

thus using now a pure least squares approach.[24]

At this stage it is convenient to approximate further to eliminate the time integration and to write

$$R \approx \tilde{R} \equiv \frac{\hat{U}_{n+1} - \hat{U}_n}{\Delta t} + \left(A_i \frac{\partial \hat{U}_{n+1}}{\partial x_i} + \frac{\partial \hat{G}_i^{n+1}}{\partial x_i}\right)\Theta + \left(A_i \frac{\partial \hat{U}_n}{\partial x_i} + \frac{\partial \hat{G}_i^n}{\partial x_i}\right)(1-\Theta) + Q^{n+\theta} \tag{12.72}$$

where

$$\hat{U}_{n+1} = \mathbf{N}\bar{\mathbf{U}}_{n+1}$$
$$\hat{U}_n = \mathbf{N}\bar{\mathbf{U}}_n$$
$$G_i^{n+1} = -k\frac{\partial \mathbf{N}}{\partial x_i}\bar{\mathbf{U}}_{n+1}$$
$$G_i^n = -k\frac{\partial \mathbf{N}}{\partial x_i}\bar{\mathbf{U}}_n$$

Here $\mathbf{N}$ are the space shape functions and the functional of Eq. (12.71) is simply written as

$$\Pi = \tfrac{1}{2}\Delta t \int_{\Omega} \tilde{R}^2 \, d\Omega \tag{12.73}$$

Minimization of the above with respect to $\bar{\mathbf{U}}_{n+1}$ gives

$$\frac{\partial \Pi}{\partial \bar{\mathbf{U}}_{n+1}} = \Delta t \int \frac{\partial \tilde{R}}{\partial \bar{\mathbf{U}}_{n+1}} \tilde{R} \, d\Omega = 0 \tag{12.74}$$

or

$$\int_{\Omega} \left[\mathbf{N}^{\mathrm{T}} + \Theta \Delta t A_i \left(\frac{\partial \mathbf{N}}{\partial x_i} \right)^{\mathrm{T}} - \Theta \Delta t \frac{\partial}{\partial x_i} \left(k \frac{\partial \mathbf{N}}{\partial x_i} \right)^{\mathrm{T}} \right] \tilde{R} \, d\Omega = 0 \tag{12.75}$$

This form causes some difficulties due to the second-order term in the weight function. However, if this term is omitted the expression is similar to that which is obtainable using the Petrov–Galerkin procedure as in Sec. 12.10 with the simple time-stepping scheme of Eq. (12.66).

If the time increment Δt is chosen so that the Courant number is unity, i.e.

$$C = |A| \frac{\Delta t}{h} = 1 \tag{12.76}$$

and $\Theta = \frac{1}{2}$, then the coefficient in the second term of Eq. (12.75) becomes

$$\Theta A_i \Delta t = \tfrac{1}{2} h \frac{A_i}{|A|} \tag{12.77}$$

and identity between the Petrov–Galerkin scheme with $\alpha = 1$ and above results.

Indeed, one could argue that this scheme is in some manner optimal, corresponding to time increments necessary for the disturbance to propagate through elements 'naturally', though of course in a real multidimensional problem the achievement of the condition of Eq. (12.76) would be generally impossible.

It is of interest to observe that in the steady state, i.e. when $\bar{\mathbf{U}}_n = \bar{\mathbf{U}}_{n+1}$, the equation (12.75) reduces to

$$\int_{\Omega} \left(\mathbf{N}^{\mathrm{T}} + \Theta \Delta t A_i \frac{\partial \mathbf{N}^{\mathrm{T}}}{\partial x_i} \right) \left(A_i \frac{\partial \hat{U}}{\partial x_i} + \frac{\partial \hat{G}}{\partial x_i} + Q \right) d\Omega \tag{12.78}$$

or the Petrov–Galerkin weighting, which is identical to that of Eq. (12.45) if Δt is independently chosen for each element to satisfy Eq. (12.77).

The time-stepping scheme and approximations here described follow the work of Carey and Jiang.[25,26] The scheme of Eqs (12.72) and (12.75) is of course of no consequence in its explicit mode ($\theta=0$) as then it is simply identical to the standard Galerkin scheme of Eq. (12.64) which we have then found to be unconditionally unstable. However, for $\Theta \geqslant \frac{1}{2}$, the process is unconditionally stable and demonstrated to be reasonably accurate in many transient problems with C values in the range 0.5–20.[25,26]

A somewhat more complex approach involving time–space elements has been evolved by Johnson and coworkers[22,27,28] and later by Yu and Heinrich.[29,30] Here we observe that the transient problem of Eq. (12.53) bears much similarity to the multidimensional steady-state case if time t is simply considered as a coordinate. Procedures of Secs 12.2 to 12.8 therefore appear to be applicable. For instance using the Galerkin–least squares approach of Sec. 12.5 and shape functions defined by Eq. (12.69), we could write in the space–time domain a weighted form [analogous to Eq. (12.39)]

$$\int_0^t \int_\Omega \left[\tilde{\mathbf{N}}^{\mathrm{T}} + \lambda \left(-\frac{\partial \tilde{\mathbf{N}}^{\mathrm{T}}}{\partial t} + A_i \frac{\partial \tilde{\mathbf{N}}}{\partial x_i} \right) \right] R \, \mathrm{d}\Omega \, \mathrm{d}t = 0 \tag{12.79}$$

where R is the residual defined in Eq. (12.70).

Some ambiguity in determination of the optimal λ exists as the 'size' of the element in the space–time domain cannot be easily defined. It is therefore suggested that expression (12.40) is used with $\alpha=1$ and a space dimension h, i.e.

$$\lambda = \frac{h}{2|A|} \tag{12.80}$$

The 'streamline' nature of the effective diffusion in the time–space domain is, however, still preserved.

The algorithm resulting from Eq. (12.79) is, like that of Eq. (12.75), always implicit and reduces for the steady state to the Petrov–Galerkin approximation for all Δt values. This is not the case with the least squares approximation of (Eq. 12.75), which gives the correct steady-state 'diffusion' only for $C=1$. Indeed, of all the algorithms presented here this one is alone in having this very desirable property.

12.12 Characteristic-based methods

12.12.1 *Mesh updating and interpolation methods.* We have already observed that, if the spatial coordinate is 'convected' in the manner implied by Eq. (12.55), i.e. along the problem *characteristics*, then convective, first-order, terms disappear and the remaining problem is that of simple diffusion for which standard discretization procedures with the Galerkin spatial approximation apply.

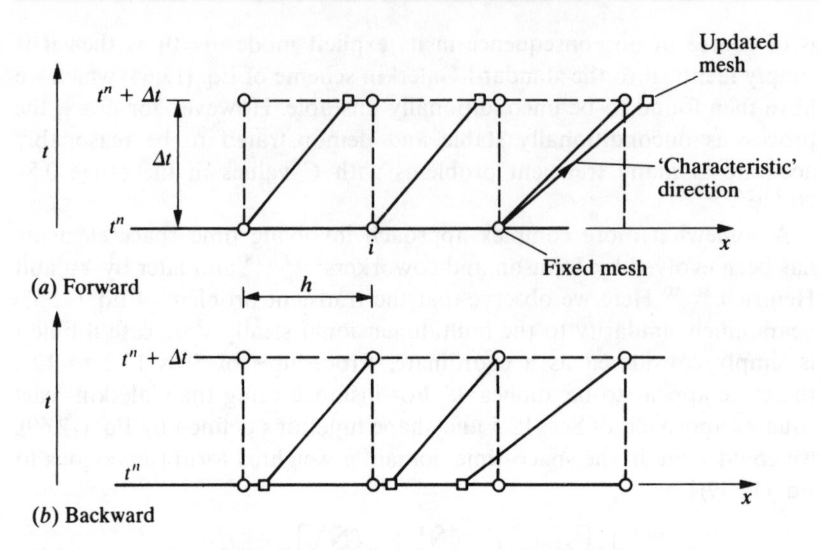

Fig. 12.16 Mesh updating and interpolation

The most obvious use of this in the finite element context is to update the position of the mesh points in a lagrangian manner. In Fig. 12.16(*a*) we show such an update for the one-dimensional problem of Eq. (12.54) occurring in an interval Δt.

For a constant x' coordinate

$$dx = A\,dt \tag{12.81}$$

and for a typical nodal point i, we have

$$x_i^{n+1} = x_i^n + \int_0^t A\,dt \tag{12.82}$$

where in general the 'velocity' A may be dependent on x. However, if $F = F(U)$ and $A = \partial F/\partial U = A(U)$ then the wave velocity is constant along a characteristic in the absence of diffusion or source terms by virtue of Eq. (12.58) and the characteristics are straight lines.

For such a constant A we have simply

$$x_i^{n+1} = x_i^n + A\Delta t \tag{12.83}$$

for the updated mesh position.

On the updated mesh only the time-dependent diffusion problem needs to be solved, using the methods of Chapter 10. This we need not discuss in detail here.

The process of continuously updating the mesh and solving the diffusion problem on the new mesh is, of course, impracticable. When applied to two- or three-dimensional configurations very distorted elements would result and difficulties will always arise on boundaries of the domain. For that reason it seems obvious that after completion of a single step a return to the original mesh should be made by interpolating from the updated values, as shown in Fig. 12.16(*a*).

This procedure can of course be reversed and characteristic origins traced backwards, as shown in Fig. 12.16(*b*) using appropriate interpolated starting values.

The method described is somewhat intuitive but has been used with success by Adye and Brebbia[31] and others for solution of transport equations. The procedure can be formalized and presented more generally and gives the basis of so-called characteristic–Galerkin methods.[32]

The diffusion part of the computation is carried out either on the original or on the final mesh, each representing a certain approximation. Intuitively we imagine in the updating scheme that the *operator is split* with the diffusion changes occurring separately from those of convection. This idea is explained in the procedures of the next section.

12.12.2 *Characteristic–Galerkin procedures.* We shall consider that the equation of convective diffusion in its one-dimensional form (12.54) is split into two parts such that

$$U = U^* + U^{**} \tag{12.84}$$

and

$$\frac{\partial U^*}{\partial t} + A\frac{\partial U}{\partial x} = 0 \tag{12.85a}$$

is a purely convective system while

$$\frac{\partial U^{**}}{\partial t} - \frac{\partial}{\partial x}\left(k\frac{\partial U}{\partial x}\right) + Q = 0 \tag{12.85b}$$

represents the diffusion.

Both U^* and U^{**} are to be approximated by standard expansions

$$\hat{U}^* = \mathbf{N}\bar{\mathbf{U}}^* \qquad \hat{U}^{**} = \mathbf{N}\bar{\mathbf{U}}^{**} \tag{12.86}$$

and in a single time step t^n to $t^n + \Delta t = t^{n+1}$ we shall assume that the initial conditions are

$$t = t^n \qquad U^* = 0 \qquad U^{**} = U^n \tag{12.87}$$

Standard Galerkin discretization of the diffusion equation allows $\bar{\mathbf{U}}^{**}_{n+1}$ to be determined on the given fixed mesh by solving an equation of the form

$$\mathbf{M}\Delta\bar{\mathbf{U}}^{**}_n = \Delta t\mathbf{H}(\bar{\mathbf{U}}_n + \Delta\bar{\mathbf{U}}^{**}_n\Theta) + \mathbf{f} \tag{12.88}$$

with

$$\bar{\mathbf{U}}^{**}_{n+1} = \bar{\mathbf{U}}^{**}_n + \Delta\bar{\mathbf{U}}^{**}_n$$

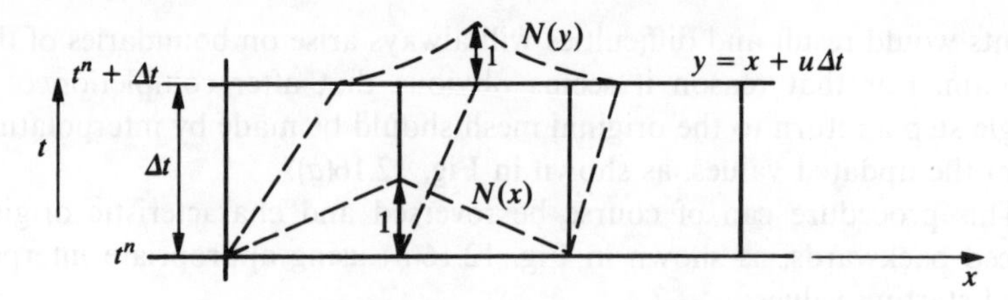

Fig. 12.17 Distortion of convected shape function

In solving the convective problem we assume that U^* remains unchanged along the characteristic. However, Fig. 12.17 shows how the initial value of U_n^* interpolated by standard linear shape functions at time n [viz. Eq. (12.86)] becomes shifted and distorted. The new value is given by

$$U_{n+1}^* = \mathbf{N}(y)\bar{\mathbf{U}}_n^* \qquad y = x + A\Delta t \tag{12.89}$$

As we require U_{n+1}^* to be approximated by standard shape functions, we shall write a projection for smoothing of these values as

$$\int_\Omega \mathbf{N}^\mathrm{T}(\mathbf{N}\bar{\mathbf{U}}_{n+1}^* - \mathbf{N}(y)\bar{\mathbf{U}}_n^*)\,\mathrm{d}x = 0 \tag{12.90}$$

giving

$$\mathbf{M}\bar{\mathbf{U}}_{n+1}^* = \int_\Omega [\mathbf{N}^\mathrm{T}(y)\,\mathrm{d}x]\bar{\mathbf{U}}_n \tag{12.91a}$$

where $\mathbf{N} = \mathbf{N}(x)$ and $\mathbf{M}$ is

$$\mathbf{M} = \int_\Omega \mathbf{N}^\mathrm{T}\mathbf{N}\,\mathrm{d}x \tag{12.91b}$$

The evaluation of the above integrals is of course still complex, especially if the procedure is extended to two or three dimensions. This is generally performed numerically and the stability of the formulation is dependent on the accuracy of such integration.[32] The scheme is stable and indeed exact as far as the convective terms are concerned if the integration is performed exactly (which of course is an unreachable goal). However, stability and indeed accuracy will even then be controlled by the diffusion terms where several approximations have been involved.

12.12.3 *A simple explicit characteristic–Galerkin procedure.* Many variants of the schemes described in the previous section are possible and have been introduced quite early. References 31 to 40 present some successful versions. However, all methods are somewhat complex in programming and time consuming. For this reason a simpler alternative was developed in which the difficulties are avoided at the expense of

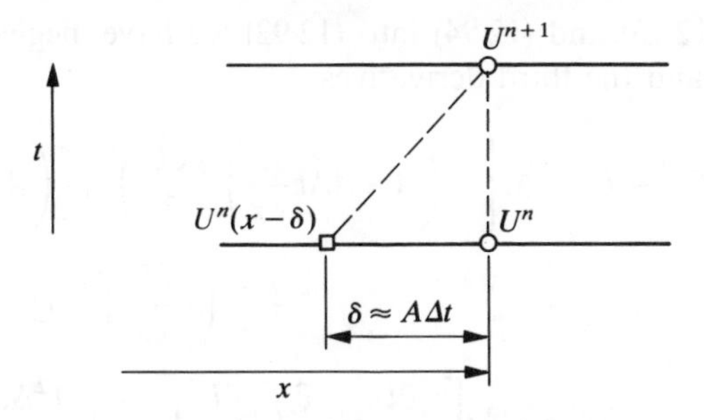

Fig. 12.18 A simple characteristic–Galerkin procedure

conditional stability. This method was first published in 1984[41] and is fully described in numerous publications.[42-45] Its derivation involves a local Taylor expansion and we illustrate this in Fig. 12.18.

We shall here discretize Eq. (12.54) *in time* before attempting the spatial discretization. Using an explicit scheme *along the characteristic*, we can write

$$U^{n+1}(x)-U^n(x-\delta)=\Delta t\left[\frac{\partial}{\partial x}\left(k\frac{\partial \tilde{U}}{\partial x}\right)-\tilde{Q}\right]_n \tag{12.92}$$

where $(x-\delta)$ is the origin of the characteristic in Fig. 12.18 and $\tilde{U},\tilde{Q}$ represent some average values in the interval, δ. As before, the convective terms have been dropped, as they disappear along the characteristic.

Now we note that by Taylor expansion we can approximate to second order using values defined at x at time n:

$$U^n(x-\delta)\approx U^n-\delta\frac{\partial U^n}{\partial x}+\frac{\delta^2}{2}\frac{\partial^2 U^n}{\partial x^2} \tag{12.93a}$$

Further, we shall approximate

$$\frac{\partial}{\partial x}\left(k\frac{\partial \tilde{U}}{\partial x}\right)\approx\frac{\partial}{\partial x}\left(k\frac{\partial U}{\partial x}\right)^n \tag{12.93b}$$

$$\tilde{Q}^n(x-\delta)\approx Q^n-\tfrac{1}{2}\delta\left(\frac{\partial Q}{\partial x}\right)^n \tag{12.93c}$$

where $$\delta=\tilde{A}\Delta t$$

Similarly, we can evaluate

$$\begin{aligned}\tilde{A}&=\left[A-\delta\left(\frac{\partial A}{\partial x}\right)+\cdots\right]^n\\ &\approx\left[A-A\Delta t\left(\frac{\partial A}{\partial x}\right)\right]^n\end{aligned} \tag{12.94}$$

Inserting (12.93) and (12.94) into (12.92) we have, neglecting terms of higher order and the third derivatives,

$$\Delta U^n = U^{n+1} - U^n = \Delta t\left[-\left(A - A\Delta t \frac{\partial A}{\partial x}\right)^n \left(\frac{\partial U}{\partial x}\right)^n + \frac{1}{2}\left(A^2 \Delta t \frac{\partial^2 U}{\partial x^2}\right)^n + \frac{\partial}{\partial x}\left(k\frac{\partial U}{\partial x}\right)^n - Q^n + \left(\frac{A\Delta t}{2}\frac{\partial Q}{\partial x}\right)^n \right]$$

$$= -\Delta t\left[A\frac{\partial U}{\partial x} - \frac{\partial}{\partial x}\left(k\frac{\partial U}{\partial x}\right) - \frac{\partial}{\partial x}\left(\frac{A^2\Delta t}{2}\frac{\partial U}{\partial x}\right) + Q - \frac{A\Delta t}{2}\frac{\partial Q}{\partial x}\right]^n \tag{12.95}$$

This appears as a direct temporal approximation to the original equation (12.54) to which an artificial diffusivity $A^2\Delta t/2$ has been added—recalling the simple 'upwinding' procedures.

At this stage space discretization can proceed by a *standard Galerkin process as the equation is self-adjoint along the characteristic.* Approximating thus in the usual finite element manner we write

$$U \approx \mathbf{N}\bar{\mathbf{U}} \tag{12.96}$$

and after premultiplying by $\mathbf{N}^{\mathrm{T}}$ and integrating by parts we obtain

$$\mathbf{M}\Delta\bar{\mathbf{U}}^n \equiv \mathbf{M}(\bar{\mathbf{U}}^{n+1} - \bar{\mathbf{U}}^n) = -\Delta t[(\mathbf{V} + \mathbf{H} + \mathbf{H}_D)\bar{\mathbf{U}} + \mathbf{f}]^n \tag{12.97}$$

where

$$\mathbf{M} = \int_\Omega \mathbf{N}^{\mathrm{T}}\mathbf{N}\,\mathrm{d}\Omega \tag{12.98a}$$

is the standard (consistent) mass matrix;

$$\mathbf{V} = \int_\Omega \mathbf{N}^{\mathrm{T}} A \frac{\partial \mathbf{N}}{\partial x}\,\mathrm{d}\Omega \tag{12.98b}$$

is the non-symmetric matrix obtained from convective terms and a Galerkin approximation;

$$\mathbf{H}_D = \int_\Omega \frac{\partial \mathbf{N}^{\mathrm{T}}}{\partial x}\left(\frac{A^2\Delta t}{2}\right)\frac{\partial \mathbf{N}}{\partial x}\,\mathrm{d}\Omega \tag{12.98c}$$

is a balancing diffusion term similar to that arising in steady-state problems due to a Petrov–Galerkin approximation (viz. Sec. 12.2);

$$\mathbf{H} = \int_\Omega \frac{\partial \mathbf{N}^{\mathrm{T}}}{\partial x} k \frac{\partial \mathbf{N}}{\partial x}\,\mathrm{d}\Omega \tag{12.98d}$$

is the standard diffusion matrix; and, finally,

$$\mathbf{f}=\int_{\Omega}\mathbf{N}^{\mathrm{T}}\left(Q-\frac{A\Delta t}{2}\frac{\partial Q}{\partial x}\right)\mathrm{d}\Omega+\text{boundary terms}$$

$$=\int_{\Omega}\left(\mathbf{N}^{\mathrm{T}}+\frac{A\Delta t}{2}\frac{\partial \mathbf{N}^{\mathrm{T}}}{\partial x}\right)Q\,\mathrm{d}\Omega+\text{boundary terms} \qquad (12.98\mathrm{e})$$

gives the modified force vector.

The algorithm developed is conditionally stable and its critical (or maximum allowable) time step requires that

$$C\leqslant\sqrt{\frac{1}{Pe^2}+\frac{1}{3}}-\frac{1}{Pe} \qquad (12.99\mathrm{a})$$

or, if $\mathbf{M}$ is replaced by $\mathbf{M}_L$,

$$C\leqslant\sqrt{\frac{1}{Pe^2}+1}-\frac{1}{Pe} \qquad (12.99\mathrm{b})$$

where $\mathbf{M}_L$ is the diagonalized or 'lumped' form of $\mathbf{M}$.

In the above $C=A\Delta t/h$ is the element Courant number and

$$Pe=\frac{Ah}{2k}$$

is the element Peclet number [viz. Eq. (12.15)]. Figure 12.19 shows the stability limit variation prescribed by Eq. (12.99) with a lumped mass matrix.

It is of considerable interest to examine the behaviour of the solution when steady state is reached—for instance, if we use the time-stepping algorithm of Eq. (12.97) as an iterative process. Now the final solution is given by taking

$$\bar{\mathbf{U}}^{n+1}=\bar{\mathbf{U}}^{n}=\bar{\mathbf{U}}$$

which gives

$$(\mathbf{V}+\mathbf{H}_D+\mathbf{H})\bar{\mathbf{U}}+\mathbf{f}=0 \qquad (12.100)$$

Inspection of Secs 12.2 and 12.3 shows that the above is identical in form with the use of the Petrov–Galerkin approximation. In the latter the matrix $\mathbf{V}$ is identical and the matrix $\mathbf{H}_D$ includes balancing diffusion of

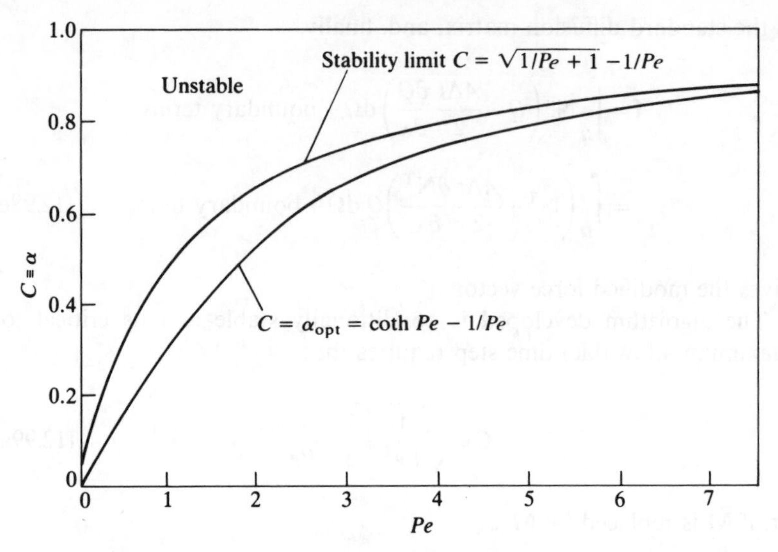

Fig. 12.19 Stability limit for lumped mass approximation and optimal upwind parameter

the amount given by $\frac{1}{2}\alpha Ah$. However, if we take

$$\tfrac{1}{2}\alpha Ah = \frac{A^2 \Delta t}{2} \tag{12.101}$$

the identity of the two schemes results. This can be written as a requirement that

$$\alpha = \frac{A\Delta t}{h} = C \tag{12.102}$$

In Fig. 12.19 we therefore plot the optimal value of α as given in Eq. (12.24) against *Pe*. *We note immediately that if the time-stepping scheme is operated at or near the critical stability limit of the lumped scheme the steady-state solution reached will be close to that resulting from the optimal Petrov–Galerkin process for the steady state.* However, if smaller time steps than the critical ones are used, the final solution, though stable, will tend towards the standard Galerkin steady-state discretization and may show oscillations if boundary conditions are such that boundary layers are created. Nevertheless, such small time steps result in very accurate transients so we can conclude that it is unlikely that optimality for transients and steady state can be reached simultaneously.

It is important to extend the scheme for the scalar variable U defined in more than one dimension. Here, if we start with Eq. (12.53) and apply precisely identical arguments to those used in one dimension we shall arrive at the form of Eqs (12.97) and (12.98) in which

$$\mathbf{V}=\int_{\Omega}\mathbf{N}^{\mathrm{T}}A_i\frac{\partial\mathbf{N}}{\partial x_i}\,\mathrm{d}\Omega \tag{12.103a}$$

$$\mathbf{H}_D=\int_{\Omega}\frac{\partial\mathbf{N}^{\mathrm{T}}}{\partial x_i}\left(\frac{A_iA_j\Delta t}{2}\right)\frac{\partial\mathbf{N}}{\partial x_j}\,\mathrm{d}\Omega \tag{12.103b}$$

$$\mathbf{H}=\int_{\Omega}\frac{\partial\mathbf{N}^{\mathrm{T}}}{\partial x_i}k\frac{\partial\mathbf{N}}{\partial x_i}\,\mathrm{d}\Omega \tag{12.103c}$$

$$\mathbf{f}=\int_{\Omega}\left(\mathbf{N}^{\mathrm{T}}+\frac{A_i\Delta t}{2}\frac{\partial\mathbf{N}^{\mathrm{T}}}{\partial x_i}\right)\mathbf{Q}+\text{boundary terms} \tag{12.103d}$$

and therefore the balancing diffusion introduced by the scheme is of the form

$$\frac{A_iA_j\Delta t}{2} \tag{12.104}$$

and hence is completely analogous to the *streamline diffusion* of Eq. (12.48).

Examination of Eqs (12.97), (12.98) and (12.103) shows that the characteristic Galerkin algorithm could have been obtained by applying a Petrov–Galerkin weighting

$$\mathbf{N}^{\mathrm{T}}+\frac{\Delta t}{2}A_i\frac{\partial\mathbf{N}^{\mathrm{T}}}{\partial x_i}$$

to the various terms of the governing equation (12.53) excluding the time derivative $\partial U/\partial t$ to which the standard Galerkin weighting of $\mathbf{N}^{\mathrm{T}}$ is attached. Comparing the above with the steady-state problem and the weighting given in Eq. (12.45) the connection is obvious.

The *characteristic–Galerkin process* introduced in this section is extremely versatile and presents near-optimal behaviour patterns. In Figs 12.20 and 12.21 we show its performance in the one-dimensional problem previously presented in Figs 12.13 and 12.14.

A two-dimensional application is illustrated in Fig. 12.22 in which we show pure convection of a disturbance in a circulating flow. It is remarkable to note that almost no dispersion occurs after a complete revolution. The present scheme is here contrasted with the solution obtained by the finite difference scheme of Lax and Wendroff[46] which for a regular one-dimensional mesh gives a scheme identical to the characteristic–Galerkin except for the mass matrix, which is lumped in the finite difference scheme.

It seems here that the difference is entirely due to the proper form of the mass matrix **M** now used and we note that for transient response

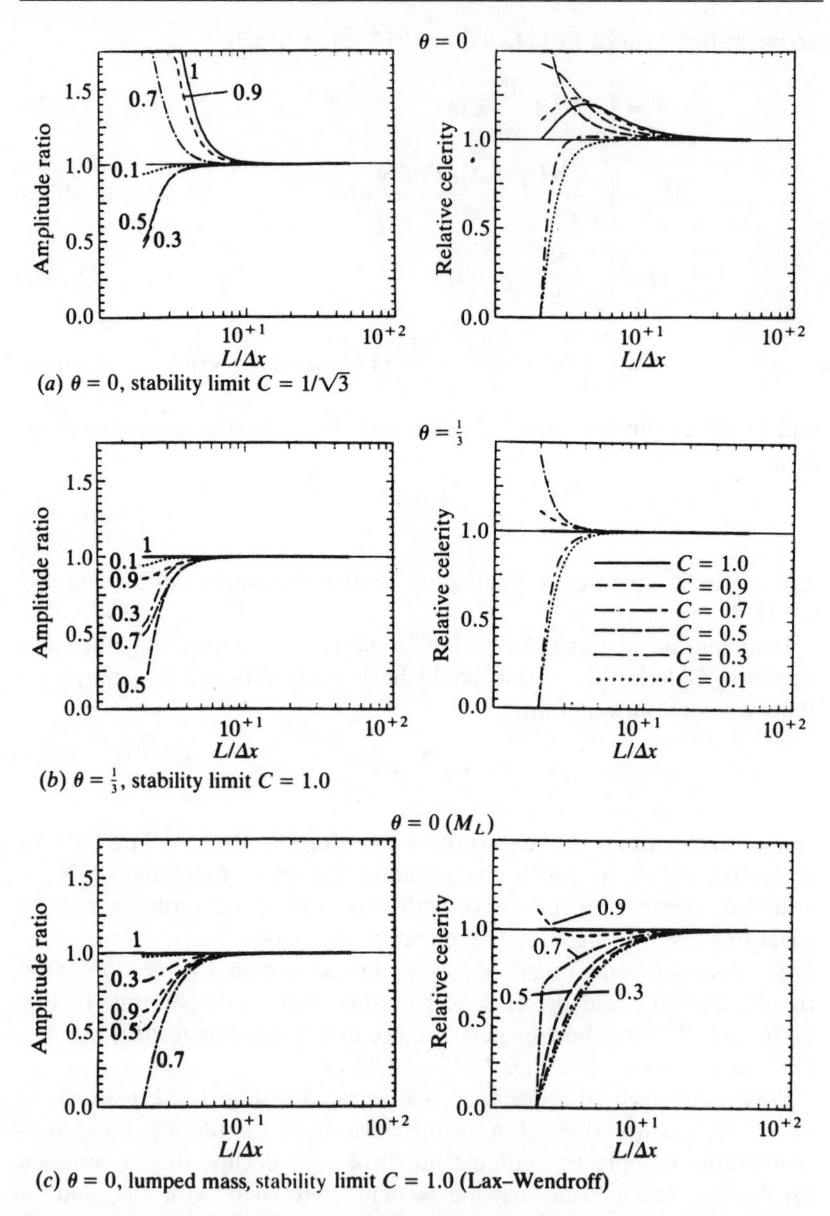

Fig. 12.20 Explicit characteristic–Galerkin methods. Amplitude ratio and relative celerity

importance of the consistent mass matrix is crucial. However, the numerical convenience of using the lumped form is over-whelming in an explicit scheme. It is easy to recover the performance of the consistent mass matrix by using a simple iteration. In this we write

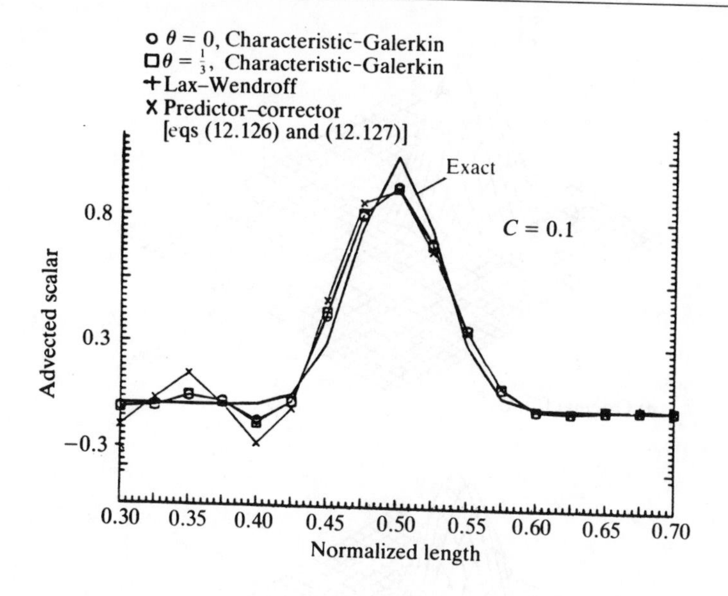

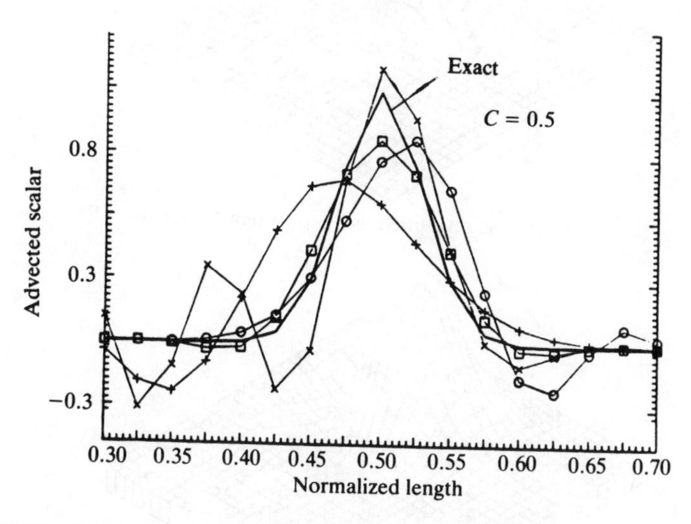

Fig. 12.21 Propagation of a gaussian wave. Profile after $t=0.5 (A=1.0)$, $h=0.025$, explicit methods

Eq. (12.97) as

$$\mathbf{M}\Delta\bar{\mathbf{U}}^n = \Delta t\mathbf{S}^n \tag{12.105}$$

with

$$\bar{\mathbf{U}}^{n+1} = \bar{\mathbf{U}}^n + \Delta\bar{\mathbf{U}}^n$$

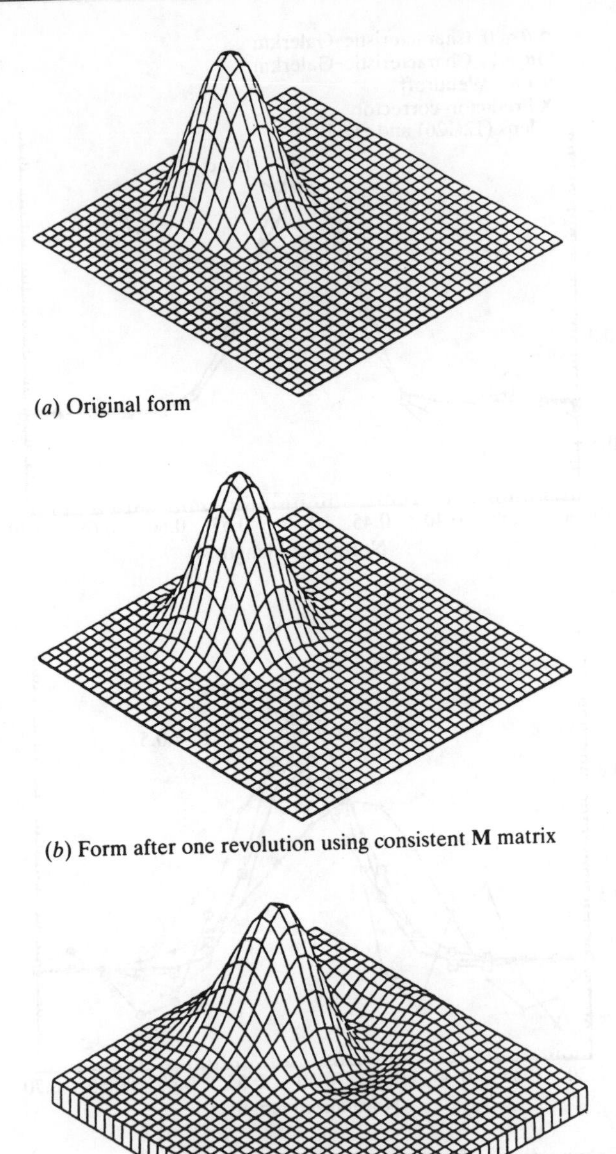

(*a*) Original form

(*b*) Form after one revolution using consistent **M** matrix

(*c*) Form after one revolution using lumped mass (Lax–Wendroff)

Fig. 12.22 Advection of a gaussian cone in a rotating fluid by characteristic–Galerkin method

Substituting a lumped mass matrix $\mathbf{M}_L$ to ease the solution problem we can iterate as follows:

$$(\Delta\bar{\mathbf{U}})_l^n = \mathbf{M}_L^{-1}[\Delta t\mathbf{S}^n - \mathbf{M}(\Delta\bar{\mathbf{U}})_{l-1}^n] + (\Delta\bar{\mathbf{U}})_{l-1}^n \tag{12.106}$$

where l is the iteration number. The process converges very rapidly and in Fig. 12.23 we show the dramatic improvements of results in the solution of a wave propagation with three such iterations done at each time step.

12.12.4 *Boundary conditions—radiation.* As we have already indicated the convection–diffusion problem allows a single boundary condition of the type

$$U = \bar{\bar{U}} \text{ on } \Gamma_u \tag{12.107a}$$

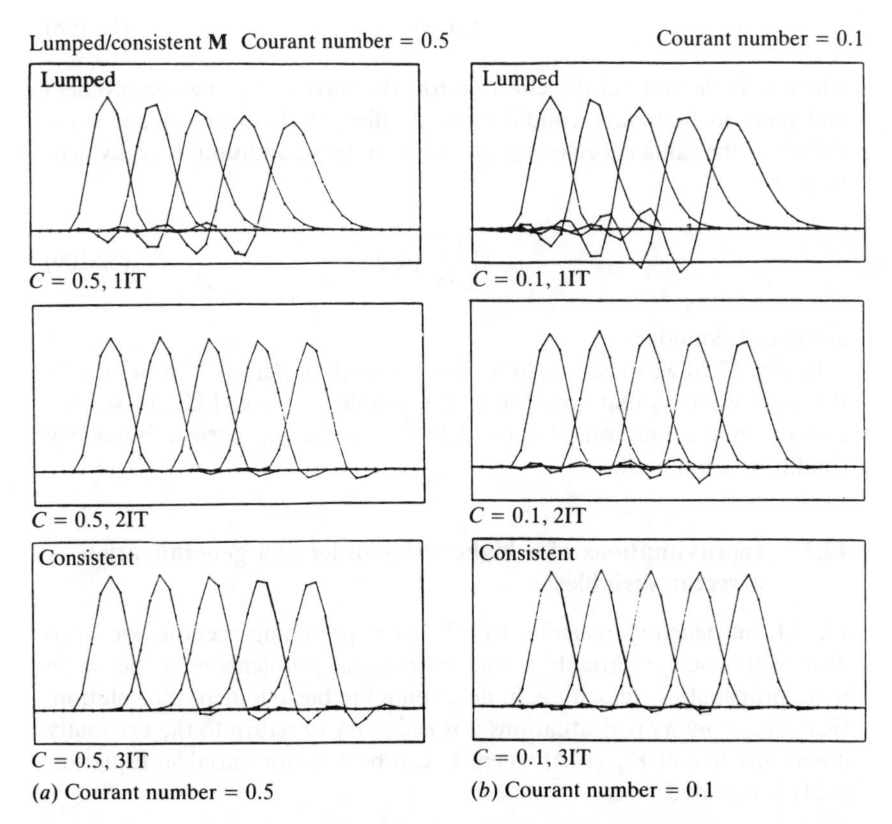

(*a*) Courant number = 0.5 (*b*) Courant number = 0.1

Fig. 12.23 Characteristic–Galerkin method in the solution of a one-dimensional wave progression. Effect of using a lumped mass matrix and of consistent iteration

and $$-k\left(\frac{\partial U}{\partial h}\right)=\bar{q} \text{ on } \Gamma_q \tag{12.107b}$$

(where $\Gamma=\Gamma_u \cup \Gamma_q$) to be imposed, providing the equation is of second order and diffusion is present.

In the case of pure convection this is no longer the case as the differential equation is of the first order. Indeed this was responsible for the difficulty of obtaining a solution in the example of Fig. 12.1 when $Pe \to \infty$ and an exit boundary condition of the type given by Eq. (12.107a) was imposed. In this one-dimensional case for pure convection only the inlet boundary condition can be given; at the exit no boundary condition needs to be prescribed if A, the wave velocity, is positive.

For multidimensional problems of pure convection the same wave specification depends on the value of the normal component of A. Thus if

$$A_i n_i \geqslant 0 \tag{12.108}$$

where n_i is the normal direction vector, the wave is leaving the problem and then no boundary condition is specified. If the problem has some diffusion, the same specification of 'no boundary condition' is equivalent to putting

$$-k\left(\frac{\partial U}{\partial n}\right)=0 \tag{12.109}$$

at the exit boundary.

In Fig. 12.24 we illustrate, following the work of Peraire,[47] how cleanly the same wave as that specified in the problem of Fig. 12.22 leaves the domain in the uniform velocity field[45,47] when the correct boundary condition is imposed.

12.13 Approximations of a higher time order and generalization to vector variables

12.13.1 *Introductory remarks.* In all of the preceding sections we have dealt with a scalar variable U and in transient problems used the single wave propagation velocity $\mathbf{A}$ in describing the behaviour of the solution. However, in many real situations it is necessary to return to the originally posed problem of Eq. (12.1) where $\mathbf{U}$ can be a vector variable. Equation (12.1) is repeated below:

$$\frac{\partial \mathbf{U}}{\partial t}+\frac{\partial \mathbf{F}_i}{\partial x_i}+\frac{\partial \mathbf{G}_i}{\partial x_i}+\mathbf{Q}=0 \tag{12.1}$$

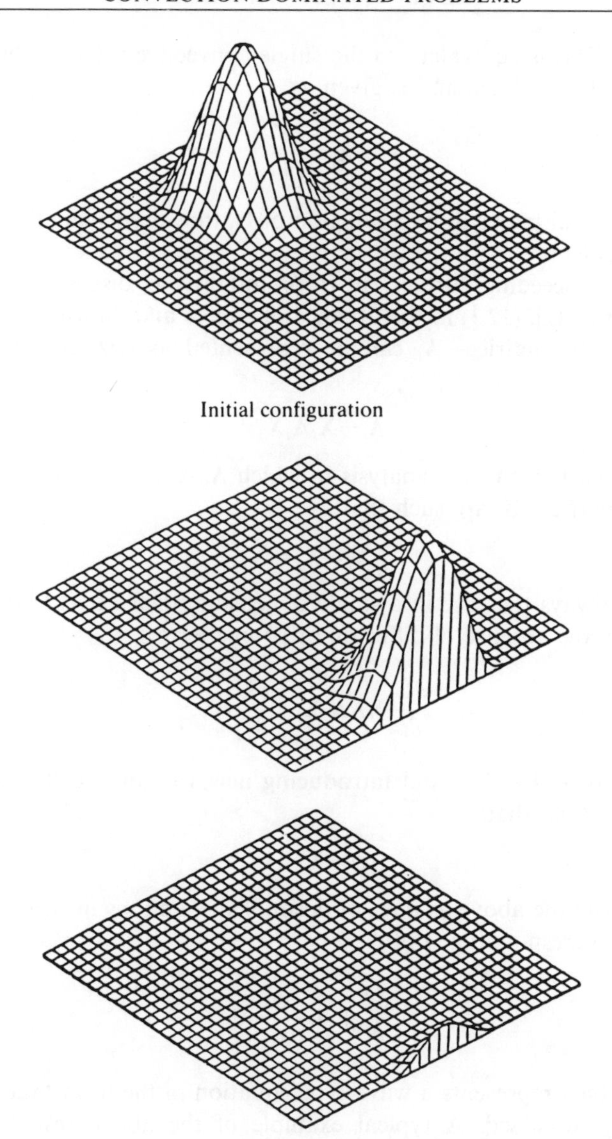

Fig. 12.24 A gaussian distribution advected in a constant velocity field. Boundary condition causes no reflection

and this can be alternatively written as

$$\frac{\partial \mathbf{U}}{\partial t}+\mathbf{A}_i\frac{\partial \mathbf{U}}{\partial x_i}+\frac{\partial \mathbf{G}_i}{\partial x_i}+\mathbf{Q}=0 \tag{12.110}$$

where $\mathbf{A}_i$ is a matrix of the size corresponding to the variables in the vector $\mathbf{U}$. This is equivalent to the single convective velocity component A_i in a scalar problem and is given as

$$\mathbf{A}_i \equiv \frac{\partial \mathbf{F}_i}{\partial \mathbf{U}} \tag{12.111}$$

This in general may still be a function of $\mathbf{U}$, thus destroying the linearity of the problem.

Before proceeding further, it is of interest to discuss the general behaviour of Eq. (12.1) in the absence of source and diffusion terms. We note that the matrices $\mathbf{A}_i$ can be represented as (viz. Appendix 1 of Volume 1)

$$\mathbf{A}_i = \mathbf{X}_i \mathbf{\Lambda}_i \mathbf{X}_i^{-1} \tag{12.112}$$

by a standard eigenvalue analysis in which $\mathbf{\Lambda}_i$ is a diagonal matrix.

If the matrices $\mathbf{X}_i$ are such that

$$\mathbf{X}_i = \mathbf{X}_j = \mathbf{X} \tag{12.113}$$

which is always the case in a single dimension, then Eq. (12.110) can be written (in the absence of diffusion or source terms) as

$$\frac{\partial \mathbf{U}}{\partial t} + \mathbf{X}\mathbf{\Lambda}_i \mathbf{X}^{-1} \frac{\partial \mathbf{U}}{\partial x_i} = 0 \tag{12.114}$$

Premultiplying by $\mathbf{X}^{-1}$ and introducing new variables (called Riemann invariants) such that

$$\boldsymbol{\phi} = \mathbf{X}^{-1}\mathbf{U} \tag{12.115}$$

we can write the above as a set of decoupled equations in components ϕ of $\boldsymbol{\phi}$ and corresponding Λ of $\mathbf{\Lambda}$:

$$\frac{\partial \phi}{\partial t} + \Lambda_i \frac{\partial \phi}{\partial x_i} = 0 \tag{12.116}$$

each of which represents a wave-type equation of the form that we have previously discussed. A typical example of the above results from a one-dimensional elastic, dynamics problem describing stress waves in a bar in terms of stresses (σ) and velocities (v) as

$$\frac{\partial \sigma}{\partial t} - E\frac{\partial v}{\partial x} = 0$$

$$\frac{\partial v}{\partial t} - \frac{1}{\rho}\frac{\partial \sigma}{\partial x} = 0$$

This can be written in the standard form of Eq. (12.1) with

$$U = \begin{Bmatrix} \sigma \\ v \end{Bmatrix} \qquad F = \begin{Bmatrix} Ev \\ \sigma/g \end{Bmatrix}$$

The two variables of Eq. (12.115) become

$$\phi_1 = \sigma - cv \qquad \phi_2 = \sigma + cv$$

where $c = \sqrt{E/\rho}$ and the equations corresponding to (12.116) are

$$\frac{\partial \phi_1}{\partial t} + c\frac{\partial \phi_1}{\partial t} = 0$$

$$\frac{\partial \phi_2}{\partial t} - c\frac{\partial \phi_2}{\partial t} = 0$$

representing respectively two waves moving with velocities $\pm c$.

Unfortunately the condition of Eq. (12.113) seldom pertains and hence the determination of general characteristics and hence of decoupling is not usually possible for more than one space dimension. For this reason the procedures developed in the preceding section are not obviously applicable and we have to proceed differently, as we shall discuss later in Sec. 12.13.4.[47]

12.13.2 *The Taylor–Galerkin procedure.* A very general procedure following the methodology originally developed by Lax and Wendroff[46] was introduced by Donea and coworkers.[48,49] Here they combine the high-order expansion of the variable **U** in time using a Taylor series with the application of the standard Galerkin finite element discretization introduced without any consideration of optimality. The resulting *Taylor–Galerkin* scheme is simple and, as we shall demonstrate, recovers the previously derived characteristic–Galerkin procedure [viz. Eqs (12.97) and (12.98)] as a special case for a scalar variable U.

The derivation proceeds as follows. Firstly, we expand the variable **U** of Eq. (12.1) by a Taylor series in time, retaining terms of the second order, giving in the interval Δt

$$\mathbf{U}^{n+1} = \mathbf{U}^n + \Delta t \left.\frac{\partial U}{\partial t}\right|_n + \left.\frac{\Delta t^2}{2}\frac{\partial^2 U}{\partial t^2}\right|_{n+\Theta} \tag{12.117}$$

where Θ is a number such that $0 \leqslant \Theta \leqslant 1$.

From Eq. (12.1),

$$\left[\frac{\partial \mathbf{U}}{\partial t}\right]_n = -\left[\frac{\partial \mathbf{F}_i}{\partial x_i} + \frac{\partial \mathbf{G}_i}{\partial x_i} + \mathbf{Q}\right]_n \tag{12.118a}$$

and differentiating

$$\left[\frac{\partial^2 \mathbf{U}}{\partial t^2}\right]_{n+\Theta} = -\frac{\partial}{\partial t}\left[\frac{\partial \mathbf{F}_i}{\partial x_i} + \frac{\partial \mathbf{G}_i}{\partial x_i} + \mathbf{Q}\right]_{n+\Theta} \tag{12.118b}$$

In the above we can write

$$\frac{\partial}{\partial t}\left(\frac{\partial \mathbf{F}_i}{\partial x_i}\right) \equiv \frac{\partial}{\partial x_i}\left(\frac{\partial \mathbf{F}_i}{\partial \mathbf{U}}\frac{\partial \mathbf{U}}{\partial t}\right) = -\frac{\partial}{\partial x_i}\left[\mathbf{A}_i\left(\frac{\partial \mathbf{F}_j}{\partial x_j} + \frac{\partial \mathbf{G}_j}{\partial x_j} + \mathbf{Q}\right)\right] \tag{12.118c}$$

and if $\mathbf{Q} = \mathbf{Q}(U, x)$ and $\partial \mathbf{Q}/\partial \mathbf{U} = \mathbf{S}$,

$$\frac{\partial \mathbf{Q}}{\partial t} = \frac{\partial \mathbf{Q}}{\partial \mathbf{U}}\frac{\partial \mathbf{U}}{\partial t} = -\mathbf{S}\left(\frac{\partial \mathbf{F}_i}{\partial x_i} + \frac{\partial \mathbf{G}_i}{\partial x_i} + \mathbf{Q}\right) \tag{12.118d}$$

We can therefore approximate Eq. (12.117) as

$$\begin{aligned}\Delta \mathbf{U}^n &\equiv \mathbf{U}^{n+1} - \mathbf{U}^n \\ &= -\Delta t\left[\frac{\partial \mathbf{F}_i}{\partial x_i} + \frac{\partial \mathbf{G}_i}{\partial x_i} + \mathbf{Q}\right]_n + \frac{\Delta t^2}{2}\left\{\frac{\partial}{\partial x_i}\left[\mathbf{A}_i\left(\frac{\partial \mathbf{F}_j}{\partial x_j} + \frac{\partial \mathbf{G}_j}{\partial x_j} + \mathbf{Q}\right)\right]\right. \\ &\qquad \left. + \frac{\partial}{\partial t}\frac{\partial \mathbf{G}_i}{\partial x_i} + \mathbf{S}\left(\frac{\partial \mathbf{F}_j}{\partial x_j} + \frac{\partial \mathbf{G}_j}{\partial x_j} + \mathbf{Q}\right)_{n+\Theta}\right\}\end{aligned} \tag{12.119}$$

Omitting the second derivatives of $\mathbf{G}_i$ and approximating the $n+\Theta$ values we have

$$\begin{aligned}\Delta \mathbf{U} &\equiv \mathbf{U}^{n+1} - \mathbf{U}^n \\ &= -\Delta t\left[\frac{\partial \mathbf{F}_i}{\partial x_i} + \mathbf{Q}\right]_n - \Delta t\left(\left[\frac{\partial \mathbf{G}_i}{\partial x_i}\right]_{n+1}\Theta + \left[\frac{\partial \mathbf{G}_i}{\partial x_i}\right]_n(1-\Theta)\right) \\ &\quad + \frac{\Delta t^2}{2}\left[\frac{\partial}{\partial x_i}\left\{\mathbf{A}_i\left(\frac{\partial \mathbf{F}_j}{\partial x_j} + \mathbf{Q}\right)\right\} + \mathbf{S}\left(\frac{\partial \mathbf{F}_j}{\partial x_j} + \mathbf{Q}\right)\right]_{n+1}\Theta \\ &\quad + \frac{\Delta t^2}{2}\left[\frac{\partial}{\partial x_j}\left\{\mathbf{A}_i\left(\frac{\partial \mathbf{F}_j}{\partial x_j} + \mathbf{Q}\right)\right\} + \mathbf{S}\left(\frac{\partial \mathbf{F}_j}{\partial x_j} + \mathbf{Q}\right)\right]_n(1-\Theta)\end{aligned} \tag{12.120}$$

At this stage a standard Galerkin approximation is applied which will result in a discrete, implicit, time-stepping scheme that is unconditionally stable if $\Theta \geqslant \frac{1}{2}$. As the explicit form is of particular interest we shall only give the details of the discretization process for $\Theta = 0$. Writing as usual

$$\mathbf{U} \approx \mathbf{N}\bar{\mathbf{U}}$$

we have

$$\left(\int_\Omega \mathbf{N}^\mathrm{T}\mathbf{N}\,\mathrm{d}\Omega\right)\Delta\bar{\mathbf{U}} = -\Delta t\left[\int_\Omega \mathbf{N}^\mathrm{T}\left(\frac{\partial \mathbf{F}_i}{\partial x_i}+\frac{\partial \mathbf{G}_i}{\partial x_i}+\mathbf{Q}\right)\mathrm{d}\Omega\right.$$

$$-\frac{\Delta t}{2}\int_\Omega \mathbf{N}^\mathrm{T}\frac{\partial}{\partial x_i}\left\{\mathbf{A}_i\left(\frac{\partial \mathbf{F}_j}{\partial x_j}+\frac{\partial \mathbf{G}_j}{\partial x_j}+\mathbf{Q}\right)\right\}\mathrm{d}\Omega$$

$$\left.-\frac{\Delta t}{2}\int_\Omega \mathbf{N}^\mathrm{T}\mathbf{S}\left(\frac{\partial \mathbf{F}_j}{\partial x_j}+\frac{\partial \mathbf{G}_j}{\partial x_j}+\mathbf{Q}\right)\mathrm{d}\Omega\right]_n \tag{12.121}$$

This can be written in a compact matrix form similar to Eq. (12.97) as

$$\mathbf{M}\,\Delta\bar{\mathbf{U}} = -\Delta t\,[(\mathbf{V}+\mathbf{H}_D+\mathbf{H})\bar{\mathbf{U}}+\mathbf{f}]^n \tag{12.122a}$$

in which, with

$$\mathbf{G}_i = -k_{ij}\frac{\partial \mathbf{U}}{\partial x_j}$$

we have (on omitting the third derivative terms and the effect of **S**) matrices of the form of Eq. (12.103), i.e.

$$\mathbf{V} = \int_\Omega \mathbf{N}^\mathrm{T}\mathbf{A}_i\frac{\partial \mathbf{N}}{\partial x_i}\,\mathrm{d}\Omega$$

$$\mathbf{H}_D = \int_\Omega \frac{\partial \mathbf{N}^\mathrm{T}}{\partial x_i}\left(\mathbf{A}_i\mathbf{A}_j\frac{\Delta t}{2}\right)\frac{\partial \mathbf{N}}{\partial x_j}\,\mathrm{d}\Omega$$

$$\mathbf{H} = \int_\Omega \frac{\partial \mathbf{N}^\mathrm{T}}{\partial x_i}k_{ij}\frac{\partial \mathbf{N}}{\partial x_j}\,\mathrm{d}\Omega \tag{12.122b}$$

$$\mathbf{f} = \int_\Omega\left(\mathbf{N}^\mathrm{T}+\frac{\Delta t}{2}\mathbf{A}_i\frac{\partial \mathbf{N}^\mathrm{T}}{\partial x_i}\right)\mathbf{Q}\,\mathrm{d}\Omega + \text{boundary terms}$$

$$\mathbf{M} = \int_\Omega \mathbf{N}^\mathrm{T}\mathbf{N}\,\mathrm{d}\Omega$$

Clearly the algorithm is identical in form to the characteristic–Galerkin method of Eq. (12.97) for the particular case of a scalar **U**, and again its form as a Petrov–Galerkin weighting form can be observed.

It is surprising that a mere improvement of the temporal approximation has resulted in the correct, streamline, diffusion obtained previously by completely different means which have, however, justified the use of standard Galerkin discretization. Such a justification is lacking in the present derivation but has the merit of extending (with success) the procedure to problems in which the characteristics are not available.

The Taylor–Galerkin derivation has, however, further advantages. With $\Theta=\frac{1}{3}$ it can be shown that the order of approximation increases and for this scheme a simple iterative solution is possible.[50] In Fig. 12.20 we show the amplitude and relative celerity ratios for the Taylor–Galerkin in a purely convective equation for $\Theta=0$ and $\Theta=\frac{1}{3}$. We note that with the consistent mass matrix $\mathbf{M}$ the stability limit for $\Theta=\frac{1}{3}$ is increased to $C=1$.

Use of $\Theta=\frac{1}{3}$ apparently requires an implicit solution. However, similar iteration to that used in Eq. (12.106) is rapidly convergent and the scheme can be used very economically. In Fig. 12.21, showing again the propagation of a gaussian wave, the excellent results of $\Theta=\frac{1}{3}$ should be noted.

12.13.3 *Two-step predictor–corrector methods. Two-step Taylor–Galerkin operation.* There are of course various alternative procedures for improving the temporal approximation other than the Taylor expansion used in the previous section. Such procedures will be particularly useful if the evaluation of the derivative matrix $\mathbf{A}$ can be avoided. In this section we shall consider two predictor–corrector schemes (of Runge–Kutta type) that avoid the evaluation of this matrix and are explicit.

The first starts with a standard Galerkin space approximation being applied to the basic equation (12.1). This results in the form

$$\mathbf{M}\frac{\mathrm{d}\bar{\mathbf{U}}}{\mathrm{d}t}\equiv\mathbf{M}\dot{\bar{\mathbf{U}}}=\mathbf{P}_C+\mathbf{P}_D+\mathbf{f}=\boldsymbol{\psi} \tag{12.123}$$

where again $\mathbf{M}$ is the standard mass matrix, $\mathbf{f}$ are the prescribed 'forces' and

$$\mathbf{P}_C(\bar{\mathbf{U}})=\int_\Omega \mathbf{N}^{\mathrm{T}}\frac{\partial \mathbf{F}_i}{\partial x_i}\,\mathrm{d}\Omega \tag{12.124a}$$

represents the convective 'forces', while

$$\mathbf{P}_D(\bar{\mathbf{U}})=\int_\Omega \mathbf{N}^{\mathrm{T}}\frac{\partial \mathbf{G}_i}{\partial x_i}\,\mathrm{d}\Omega \tag{12.124b}$$

are the diffusive ones.

If an explicit time integration scheme is used, i.e.

$$\mathbf{M}\,\Delta\bar{\mathbf{U}}\equiv\mathbf{M}(\bar{\mathbf{U}}^{n+1}-\bar{\mathbf{U}}^n)=\Delta t\,\boldsymbol{\psi}^n(\bar{\mathbf{U}}^n) \tag{12.125}$$

the evaluation of the right-hand side does not require the matrix product representation and $\mathbf{A}_i$ does not have to be computed.

Of course the scheme presented is not accurate for the various reasons previously discussed, and indeed becomes *unconditionally unstable* in the absence of diffusion and external force vectors.

The reader can easily verify that in the case of the linear one-dimensional problem the right-hand side is equivalent to a central difference scheme with $\bar{U}^n_{i-1}$ and $\bar{U}^n_{i+1}$ only being used to find the value of U^{n+1}_i, as shown in Fig. 12.25(*a*).

The scheme can, however, be recast as a two-step, predictor–corrector operation and conditional stability regained. Now we proceed as follows:

Step 1. Compute $\bar{\mathbf{U}}^{n+1/2}$ using an explicit approximation of Eq. (12.125), i.e.

$$\bar{\mathbf{U}}^{n+1/2} = \bar{\mathbf{U}}^n + \frac{\Delta t}{2}\mathbf{M}^{-1}\boldsymbol{\psi}^n \tag{12.126}$$

and

Step 2. Compute $\bar{\mathbf{U}}^{n+1}$ inserting the improved value of $\bar{\mathbf{U}}^{n+1/2}$ in the right-hand side of Eq. (12.125), giving

$$\bar{\mathbf{U}}^{n+1} = \mathbf{U}^n + \Delta t\,\mathbf{M}^{-1}\boldsymbol{\psi}^{n+1/2} \tag{12.127}$$

This is precisely equivalent to the second-order Runge–Kutta scheme being applied to the ordinary system of differential equations (12.123).

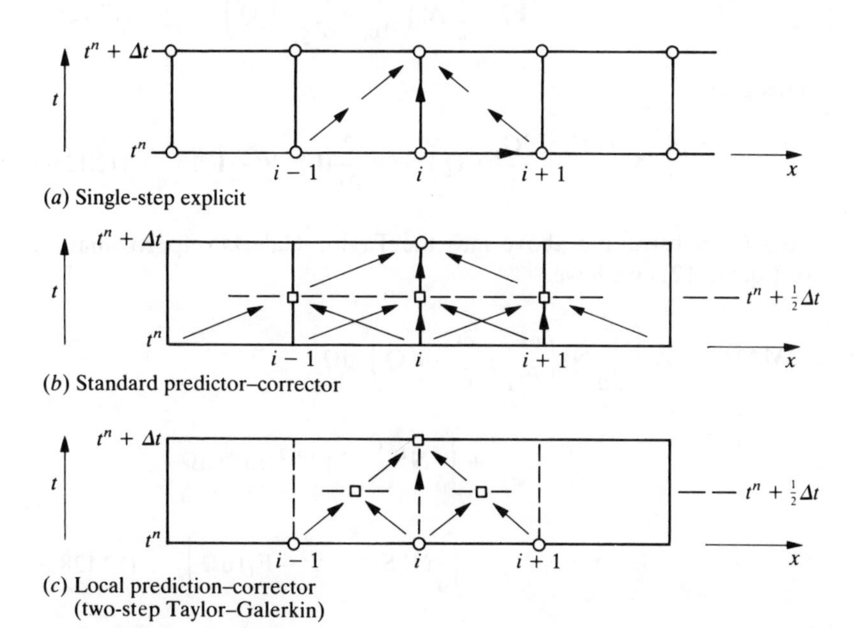

Fig. 12.25 Progression of information in explicit one- and two-step schemes

Figure 12.25(*b*) shows in the one-dimensional example how the information 'spreads', i.e. that now $\bar{U}_i^{n+1}$ will be dependent on values at nodes $i-2, \ldots, i+2$.

It is found that the scheme, though stable, is overdiffusive and numerical results are poor (viz. Fig. 12.21).

An alternative is possible, however, using a two-step Taylor–Galerkin operation. Here we return to the original equation (12.1) and proceed as follows:

Step I. Find an improved value of $\mathbf{U}^{n+1/2}$ *using only the convective and and source parts.* Thus

$$\mathbf{U}^{n+1/2} = \mathbf{U}^n - \frac{\Delta t}{2}\left(\frac{\partial \mathbf{F}_i^n}{\partial x_i} + \mathbf{Q}^n\right) \tag{12.128a}$$

which of course allows the evaluation of $\mathbf{F}_i^{n+1/2}$

We note, however, that we can also write an approximate expansion as

$$\begin{aligned}\mathbf{F}_i^{n+1/2} &= \mathbf{F}_i^n + \frac{\Delta t}{2}\frac{\partial \mathbf{F}_i^n}{\partial t} = \mathbf{F}_i^n - \frac{\Delta t}{2}\mathbf{A}_i^n \frac{\partial \mathbf{U}^n}{\partial t} \\ &= \mathbf{F}_i^n - \frac{\Delta t}{2}\mathbf{A}_i^n\left(\frac{\partial \mathbf{F}_j}{\partial x_j} + \frac{\partial \mathbf{G}_j}{\partial x_j} + \mathbf{Q}\right)^n\end{aligned} \tag{12.128b}$$

This gives

$$\mathbf{A}_i^n\left(\frac{\partial \mathbf{F}_j}{\partial x_j} + \frac{\partial \mathbf{G}_j}{\partial x_j} + \mathbf{Q}\right)^n = -\frac{2}{\Delta t}(\mathbf{F}_i^{n+1/2} - \mathbf{F}_i^n) \tag{12.128c}$$

Step II. Substituting above into the Taylor–Galerkin approximation of Eq. (12.121) we have

$$\begin{aligned}\mathbf{M}\Delta\bar{\mathbf{U}} = -\Delta t\Bigg[&\int_\Omega \mathbf{N}^\mathrm{T}\left(\frac{\partial \mathbf{F}_i}{\partial x_i} + \frac{\partial \mathbf{G}_i}{\partial x_i} + \mathbf{Q}\right)^n \mathrm{d}\Omega \\ &+ \int_\Omega \mathbf{N}^\mathrm{T}\frac{\partial}{\partial x_i}(\mathbf{F}_i^{n+1/2} - \mathbf{F}_i^n)\,\mathrm{d}\Omega \\ &+ \int_\Omega \mathbf{N}^\mathrm{T}\,\mathbf{S}(\mathbf{F}_i^{n+1/2} - \mathbf{F}_i^n)\,\mathrm{d}\Omega\Bigg]\end{aligned} \tag{12.128d}$$

and after integration by parts of the terms with respect to the x_i

derivatives we obtain simply

$$\mathbf{M}\,\Delta\bar{\mathbf{U}} = -\Delta t \left\{ \int_\Omega \frac{\partial \mathbf{N}^{\mathrm{T}}}{\partial x_i} (\mathbf{F}_i^{n+1/2} + \mathbf{G}_i^n)\, \mathrm{d}\Omega + \int \mathbf{N}^{\mathrm{T}} [\mathbf{Q} + \mathbf{S}(\mathbf{F}^{n+1/2} - \mathbf{F}^n)]\, \mathrm{d}\Omega + \int_\Gamma \mathbf{N}^{\mathrm{T}} (\mathbf{F}_i^{n+1/2} + \mathbf{G}_i^n) \mathbf{n}_i \, \mathrm{d}\Gamma \right\} \tag{12.129}$$

We note immediately that:

1. The above expression is identical to *using a standard Galerkin approximation on Eq. (12.1)* and an explicit step with $\mathbf{F}_i$ values updated by the simple equation (12.128a).
2. The final form of Eq. (12.129) does not require the evaluation of the matrices $\mathbf{A}_i$ resulting in substantial computation savings as well as yielding essentially the same results. Indeed, some omissions made in deriving Eqs (12.122) did not occur now and presumably the accuracy is improved.

A further practical point must be noted:

3. In non-linear problems it is convenient to interpolate $\mathbf{F}_i$ directly in the finite element manner as

$$\mathbf{F}_i = \mathbf{N}\bar{\mathbf{F}}_i$$

rather than to compute it as $\mathbf{F}_i(\bar{\mathbf{U}})$.

Thus the evaluation of $\mathbf{F}_i^{n+1/2}$ need only be made at the quadrature (integration) points within the element, and the evaluation of $\bar{\mathbf{U}}^{n+1/2}$ by Eq. (12.128a) is only done on such points. For a linear triangle element this reduces to a single evaluation of $\mathbf{U}^{n+1/2}$ and $\mathbf{F}^{n+1/2}$ for each element at its centre, taking of course $\mathbf{U}^n$ and $\mathbf{F}^n$ as the appropriate interpolation average there.

In the simple one-dimensional linear example the information progresses in the manner shown in Fig. 12.25(*c*). The scheme, which originated at Swansea, can be appropriately called the *Swansea two step*,[41,47,51-55] and has shown itself to be applicable successfully to a wide range of problems, some of which will be discussed in the following chapters.

We conclude this section noting the rather puzzling situation in which different computational schemes proposed on the basis of diverse arguments converge frequently to a single formula. With simplicity being, at least in the eyes of the author, a virtue, the most direct derivation has an obvious merit.

12.13.4 *Boundary condition–radiation in the generalized equation.* In Sec. 12.12.4 we discussed the question of boundary conditions which can be imposed on simple linear problems with a scalar variable U. For a vector variable $\mathbf{U}$ the wave nature of the problem is less clear and even if it exists more than one wave velocity may be present, as shown in Sec. 12.13.1.

The problem of determining such wave speeds can be approximated in the multidimensional case for which diagonalization is not generally possible (viz. page 486) by considering plane disturbances propagating in the direction normal to a boundary. Now the determination of a single vector $\mathbf{X}$ and a single eigenvalue in Eq. (12.112) is always possible. Thus, starting from the general equation in which the variables are assumed not to be dependent on the tangential direction to the boundary, we write (for the convective part in the direction of the outward normal)

$$\frac{\partial \mathbf{U}}{\partial t}+\frac{\partial \mathbf{F}_n}{\partial x_n}=\frac{\partial \mathbf{U}}{\partial t}+\mathbf{A}_n\frac{\partial \mathbf{U}}{\partial x_n}=0 \tag{12.130}$$

and using procedures of Eqs (12.112) to (12.116) we can establish a new variable set

$$\boldsymbol{\phi}=\mathbf{X}^{-1}\mathbf{U} \qquad \boldsymbol{\phi}=(\phi_1,\phi_2,\ldots)^{\mathrm{T}} \tag{12.131}$$

Now a set of decoupled equations can be written as

$$\begin{aligned}\frac{\partial \phi_1}{\partial t}+\lambda_1\frac{\partial \phi_1}{\partial x}&=0\\ \frac{\partial \phi_2}{\partial t}+\lambda_2\frac{\partial \phi_2}{\partial x}&=0\\ \text{etc.}\end{aligned} \tag{12.132}$$

On an open radiation boundary the only components of ϕ_k that can be specified are those corresponding to *incoming waves*. This procedure is of great importance in allowing the free radiation of waves and was used by Peraire[47] and others in the context of compressible flow equations. It necessitates determination of local eigenvalues and a change of boundary variables.

12.14 Non-linear waves and shocks

The procedures developed in the previous sections are in principle of course available for both linear and non-linear problems (with explicit procedures of time stepping being particularly efficient for the latter).

Quite generally the convective part of the equation, i.e.

$$\frac{\partial \mathbf{U}}{\partial t}+\frac{\partial \mathbf{F}_i}{\partial x_i}\equiv\frac{\partial \mathbf{U}}{\partial t}+\mathbf{A}_i\frac{\partial \mathbf{U}}{\partial x_i}=0 \tag{12.133}$$

will have the matrix $\mathbf{A}_i$ dependent on $\mathbf{U}$. Thus

$$\mathbf{A}_i\equiv\frac{\partial \mathbf{F}_i}{\partial x_i}=\mathbf{A}_i(\mathbf{U}) \tag{12.134}$$

In a one-dimensional case with a scalar variable (resulting, say, from decomposition) we shall have equations of the type

$$\frac{\partial U}{\partial t}+\frac{\partial F}{\partial x}\equiv\frac{\partial U}{\partial t}+A(U)\frac{\partial U}{\partial x}=0 \tag{12.135}$$

corresponding to waves moving with a non-uniform velocity A. A typical problem in this category is that due to Burger, which is defined by

$$\frac{\partial U}{\partial t}+\frac{\partial}{\partial x}\left(\frac{1}{2}U^2\right)=\frac{\partial U}{\partial t}+U\frac{\partial U}{\partial x}=0 \tag{12.136}$$

In Fig. 12.26 we illustrate qualitatively how different parts of the wave moving with velocities proportional to their amplitude cause it to steepen and finally develop into a shock form. This behaviour is typical of many non-linear systems and in Chapter 14 we shall see how shocks develop in compressible flow at transsonic speeds.

To illustrate the necessity for the development of the shock, consider the propagation of a wave with an originally smooth profile illustrated in Fig. 12.27(*a*). Here as we know the characteristics along which U is constant are straight lines shown in Fig. 12.27(*b*). These show different propagation speeds intersecting at time $t=2$ when a discontinuous shock appears. This shock propagates at a finite speed (which here is the average of the two extreme values).

In such a shock the differential equation is no longer valid but the conservation integral is. We can thus write for a small length Δs around

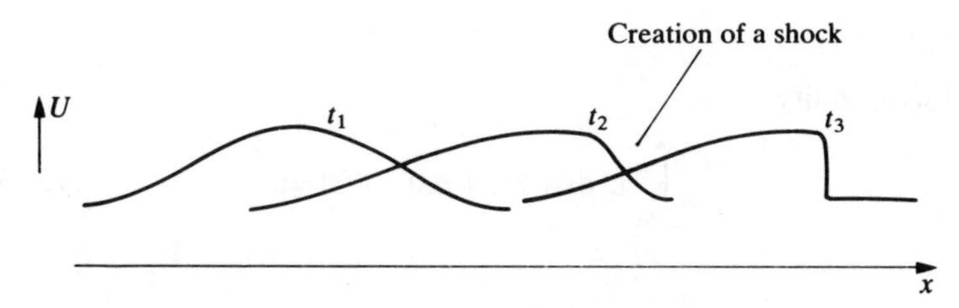

Fig. 12.26 Progression of a wave with velocity $A=U$

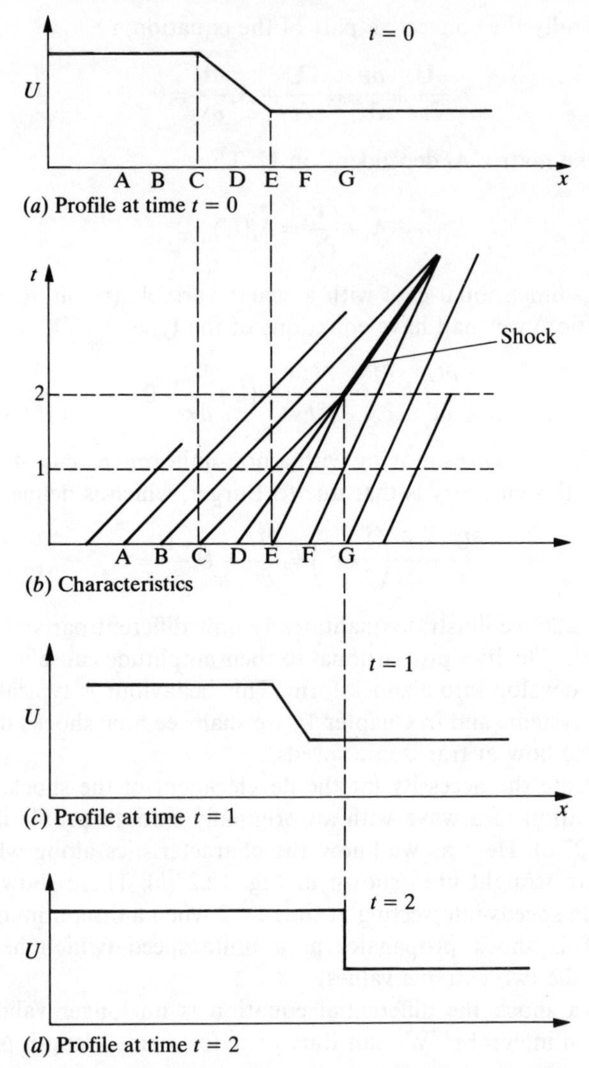

(*a*) Profile at time $t = 0$

(*b*) Characteristics

(*c*) Profile at time $t = 1$

(*d*) Profile at time $t = 2$

Fig. 12.27 Development of a shock (Burger equation)

the discontinuity

$$\frac{\partial}{\partial t} \int_{\Delta s} U \, \mathrm{d}s + F(s + \Delta s) - F(s) = 0 \tag{12.137}$$

or

$$C \Delta U - \Delta F = 0 \tag{12.138a}$$

where $C = \lim \Delta s/\Delta t$ is the speed of shock propagation and ΔU and ΔF are the discontinuities in U and F.

Equation (12.138a) is known as the Rankine–Hugoniot condition which for the general case of a vector-valued function and several space dimensions can be written as

$$C_i \Delta \mathbf{U} - n_i \Delta \mathbf{F}_j = 0 \tag{12.138b}$$

for a discontinuity surface with a normal $\mathbf{n}$, moving with a velocity C_i.

We shall find that such shocks develop frequently in the context of compressible flow and shallow water flow (Chapters 14 and 15) and often can exist even in the presence of diffusive terms in the equation. Indeed, such shocks are not specific to transients but can persist in the steady state. Clearly, approximation of the finite element kind in which we have postulated in general a C_0 continuity for the vector 2 $\mathbf{U}$ can at best *smear* such a discontinuity over an element length, and generally oscillations near such a discontinuity arise even when the best algorithms of the preceding sections are used.

Figure 12.28 illustrates the difficulties of modelling such steep waves occurring even in linear problems in which the physical dissipation contained in the equations is incapable of smoothing the solution out reasonably, and to overcome this problem artificial diffusivity is frequently used. This artificial diffusivity must have the following characteristics:

1. It must vanish as the element size tends to zero.
2. It must not affect substantially the smooth domain of the solution.

A typical diffusivity often used is a finite element version of that introduced by Lapidus[56] for finite differences, but many other forms of local smoothing have been proposed.[57–59] The additional diffusivity is of the form

$$\tilde{k} = C_{\text{Lap}} h^2 \left| \frac{\partial U}{\partial x} \right| \tag{12.139}$$

where the last term gives the maximum gradient.

In Fig. 12.29 we show a problem of discontinuous propagation in the Burger equation and how a progressive increase of the C_{Lap} coefficient kills spurious oscillation, but at the expense of rounding of a steep wave.

For a multidimensional problem with a vector value U a degree of anisotropy can be introduced and a possible expression generalizing (12.139) is

$$\tilde{k}_{ij} = C_{\text{Lap}} h^2 \frac{|V_i V_j|}{|V|} \tag{12.140}$$

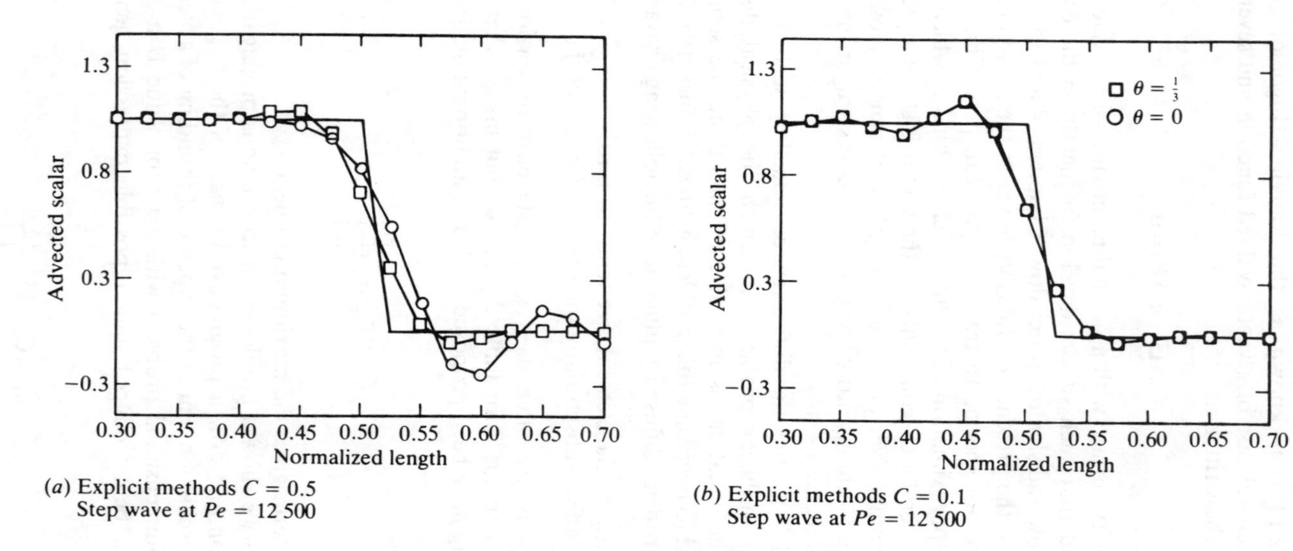

(a) Explicit methods $C = 0.5$
Step wave at $Pe = 12\,500$

(b) Explicit methods $C = 0.1$
Step wave at $Pe = 12\,500$

Fig. 12.28 Propagation of a steep wave by Taylor–Galerkin process

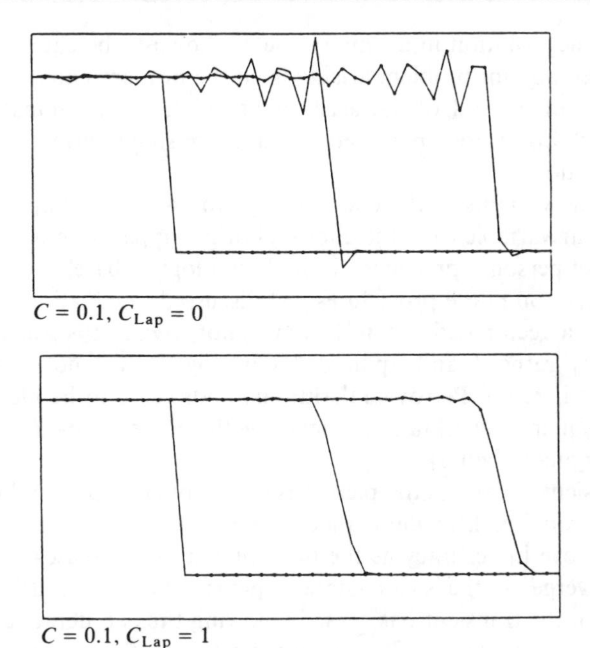

Fig. 12.29 Propagation of a steep front in Burger's equation with solution obtained using different values of Lapidus $C_v = C_{\text{Lap}}$

where

$$V_i = \frac{\partial}{\partial x_i}(\mathbf{U}^{\mathrm{T}}\,\mathbf{U})^{1/2}$$

Other possibilities are open here and much current work is focused on the subject of 'shock capture'. We shall return to the problems in Chapter 14 where its importance in the high-speed flow of gases is paramount.

12.15 Summary and concluding remarks

The reader may well be confused by the variety of apparently unrelated approaches given in this chapter. This may be excused by the fact that

(*a*) optimality guaranteed by the finite element approaches in elliptic problems does not automatically transfer to hyperbolic ones;

(*b*) the problems are still under active research stimulated by the practical needs of aeronautics, meteorology, etc.

In steady-state applications with a scalar variable the situation is reasonably clear. The original Petrov–Galerkin approach arrived at somewhat heuristically, with a 'free parameter' determined *a posteriori* to give an exact solution in a simple case, is shown to be equivalent to the optimal self-adjoint problem arising from the primary equations. Indeed, the same can be said of the least square–Galerkin combination which again contains a free parameter and surprisingly gives an identical approximation.

As all the variants in the end converge to an identical approximation in the scalar variable case, the choice of origin appears immaterial and is a matter of personal preference in finding a logical base.

The extension of the procedures just discussed to equation systems, i.e. when $\mathbf{U}$ is a vector variable, is, however, not trivial. It is generally made 'by analogy' rather than by pure deductive logic. It will now, however, be found that here the Petrov–Galerkin methods cease to be identical with the least square–Galerkin approaches (with the performance of the first being marginally better).

In transient situations the picture is even more complex and confusing. Here we should seek methods that:

(*a*) improve in accuracy as the time interval Δt decreases to zero and
(*b*) converge to the steady-state approximation previously obtained when the transient has ceased (allowing thus an iterative solution).

Only two of the approaches presented satisfy both requirements for all Δt values. The first is the simple space Petrov–Galerkin discretization of Sec. 12.10. The second is the space–time application of the Petrov–Galerkin methods suggested by Johnson *et al.*[22] and outlined in Sec. 12.12. However, neither of the schemes can be used in a simple explicit mode necessary for large equation systems, and indeed where the transient computation is merely a device for obtaining the steady-state solution iteratively.

What of the other alternatives? Here the characteristic basis of the Taylor–Galerkin scheme and its excellent time-domain performance in transient computations (viz. Sec. 12.13) makes it a favourite contender for transient computation, especially as explicit (or nearly explicit[50]) procedures are efficient. However, it only converges to stable and accurate, steady-state, solution if the Courant number C is approximately unity in every element, thus requiring [viz. Eq. (12.102)]

$$\Delta t \approx \frac{h}{|A|} = \Delta t_{\text{crit}} \tag{12.141}$$

for purely convective problems. Indeed, as we have shown, this reproduces here the Petrov–Galerkin procedure.

The same is true of the space–time least squares approximation of Sec. 12.12, though here explicit formulation is not applicable (though other merits exist).

What is the practitioners' guide here? Taking a pragmatic view we have generally pursued the following line:

1. If the problem is *truly transient* the Taylor–Galerkin process is used, putting the Courant number as close to unity as possible in the smallest element of the domain—and occasionally splitting the domain into parts in which different Δt's are used (to avoid the oscillations inherent in very small Courant numbers).[60,61]
2. If the transient solution is used as a device for obtaining the steady state then for each element a local value of $\Delta t = \Delta t_{crit}$ is chosen to ensure that when this is reached the solution coincides as nearly as possible with the optimal Petrov–Galerkin form. This Δt_{crit} is then used to compute such matrices as $\mathbf{H}_D$ of Eq. (12.98c) in the characteristic–Galerkin scheme [or (12.103b) in the Taylor–Galerkin scheme]. However, the value of Δt outside the square bracket of Eq. (12.97) is maintained at that required for general stability of computation.

The approach outlined above will allow, of course, any of the successful implicit procedures to be used in an iterative mode, providing that a correct steady-state 'upwinded' form is available.[62]

The logical basis of the characteristic–Galerkin approach provides the optimal procedure for the scalar variable problem. It is applicable to vector-valued variables providing only one specie of wave propagation is present and does not extend easily to the general case where several wave propagation speeds exist. In such cases it is not possible to arrive at the unique upwind damping terms necessary for an accurate steady-state solution. For this reason it is convenient to use operator split procedures in which the effects with dominant wave speed are dealt with separately.[63] Here, much current research is still in progress.

References

1. A. N. BROOKS and T. J. R. HUGHES, 'Streamline upwind/Petrov–Galerkin formulation for convection dominated flows with particular emphasis on the incompressible Navier Stokes equation', *Comp. Meth. Appl. Mech. Eng.*, **32**, 199–259, 1982.
2. R. COURANT, E. ISAACSON and M. REES, 'On the solution of non-linear hyperbolic differential equations by finite differences', *Comm. Pure Appl. Math.*, **V**, 243–55, 1952.
3. A. K. RUNCHALL and M. WOLFSTEIN, 'Numerical integration procedure for the steady state Navier–Stokes equations', *J. Mech. Eng. Sci.*, **11**, 445–53, 1969.
4. D. B. SPALDING, 'A novel finite difference formulation for differential equations involving both first and second derivatives', *Int. J. Num. Meth. Eng.*, **4**, 551–9, 1972.

5. K. E. BARRETT, 'The numerical solution of singular perturbation boundary value problems', *Q. J. Mech. Appl. Math.*, **27**, 57–68, 1974.
6. O. C. ZIENKIEWICZ, R. H. GALLAGHER and P. HOOD, 'Newtonian and non-Newtonian viscous incompressible flow. Temperature induced flows and finite element solutions', in *The Mathematics of Finite Elements and Applications* (ed. J. Whiteman), Vol. II, Academic Press, London, 1976 (Brunel University, 1975).
7. I. CHRISTIE, D. F. GRIFFITHS, A. R. MITCHELL and O. C. ZIENKIEWICZ, 'Finite element methods for second order differential equations with significant first derivatives', *Int. J. Num. Meth. Eng.*, **10**, 1389–96, 1976.
8. O. C. ZIENKIEWICZ, J. C. HEINRICH, P. S. HUYAKORN and A. R. MITCHELL, 'An upwind finite element scheme for two dimensional convective transport equations', *Int. J. Num. Meth. Eng.*, **11**, 131–44, 1977.
9. J. C. HEINRICH and O. C. ZIENKIEWICZ, 'Quadratic finite element schemes for two dimensional convective-transport problems', *Int. J. Num. Meth. Eng.*, **11**, 1831–44, 1977.
10. D. W. KELLY, S. NAKAZAWA and O. C. ZIENKIEWICZ, 'A note on anisotropic balancing dissipation in the finite element method approximation to convective diffusion problems', *Int. J. Num. Meth. Eng.*, **15**, 1705–11, 1980.
11. B. P. LEONARD, 'A survey of finite differences of opinion on numerical muddling of the incomprehensible defective confusion equation', in *Finite Elements for Convection Dominated Flows* (ed. T. J. R. Hughes), AMD Vol. 34, ASME, 1979.
12. G. L. GUYMON, V. H. SCOTT and L. R. HERRMANN, 'A general numerical solution of the two dimensional diffusion–convection equation by the finite element method', *Water Resouces Res.*, **6**, 1611–17, 1970.
13. T. J. R. HUGHES and J. D. ATKINSON, 'A variational basis of "upwind" finite elements', in *Variational Methods in the Mechanics of Solids* (ed. S. Nemat-Nasser), pp. 387–91, Pergamon Press, Oxford, 1980.
14. G. F. CAREY, 'Exponential upwinding and integrating factors for symmetrization', *Comm. Appl. Num. Meth.*, **1**, 57–60, 1985.
15. S. R. IDELSOHN, 'Upwind techniques via variational principles', *Int. J. Num. Meth. Eng.*, **28**, 769–84, 1989.
16. P..A. B. DE SAMPAIO, 'A Petrov–Galerkin modified operator formulation for convection diffusion problems', *Int. J. Num. Mech. Eng.* (to be published).
17. T. J. R. HUGHES, L. P. FRANCA, G. M. HULBERT, Z. JOHAN and F. SAKHIB, 'The Galerkin least square method for advective diffusion equations', in *Recent Developments in Computational Fluid Mechanics* (eds T. E. Tezduyar and T. J. R. Hughes), AMD 95, ASME, 1988.
18. J. DONEA, T. BELYTSCHKO and P. SMOLINSKI, 'A generalized Galerkin method for steady state convection–diffusion problems with application to quadratic shape function', *Comp. Meth. Appl. Mech. Eng.*, **48**, 25–43, 1985.
19. S. NAKAZAWA, J. F. PITTMAN and O. C. ZIENKIEWICZ, 'Numerical solution of flow and heat transfer in polymer melts', in *Finite Elements in Fluids* (eds R. H. Gallagher *et al.*), Vol. 4, chap. 13, pp. 251–83, Wiley, Chichester, 1982.
20. T. J. R. HUGHES and A. BROOKS, 'A multi-dimensional upwind scheme with no cross wind diffusion', in *Finite Elements for Convection Dominated Flows* (ed. T. J. R. Hughes), AMD 34, ASME, 1979.
21. T. J. R. HUGHES and A. N. BROOKS, 'A theoretical framework for Petrov–Galerkin methods with discontinuous weighting function', in *Finite Elements in Fluids* (eds R. H. Gallagher *et al.*), Vol. 4, pp. 47–65, Wiley, Chichester, 1982.

22. C. Johnson, V. Nävert and J. Pitkäranta, 'Finite element methods for linear, hyperbolic problems', *Comp. Meth. Appl. Mech. Eng.*, **45**, 285–312, 1984.
23. P. A. B. de Sampaio, 'A modified operator analysis of convection diffusion problems', in *Proc. II National Meeting on Thermal Sciences*, Aguas de Lindoia (Brazil), pp. 180–3, 1988.
24. N. Nguen and J. Reynen, 'A space–time least square finite element scheme for advection–diffusion equations', *Comp. Meth. Appl. Mech. Eng.*, **42**, 331–42, 1984.
25. G. F. Carey and B. N. Jiang, 'Least square finite elements for first order hyperbolic systems', *Int. J. Num. Meth. Eng.*, **26**, 81–93, 1988.
26. B. N. Jiang and G. F. Carey, 'A stable least-square finite element method for non-linear hyperbolic problems', *Int. J. Num. Meth. Fluids*, **8**, 933–42, 1988.
27. C. Johnson, 'Streamline diffusion methods for problems in fluid mechanics', in *Finite Elements in Fluids* (eds R. H. Gallagher *et al.*), Vol. 6, pp. 251–61, Wiley, Chichester, 1986.
28. C. Johnson, *Numerical Solution of Partial Differential Equations by the Finite Element Method*, Cambridge University Press, Cambridge, 1987.
29. C. C. Yu and J. C. Heinrich, 'Petrov–Galerkin methods for the time dependent convective transport equation', *Int. J. Num. Meth. Eng.*, **23**, 883–901, 1986.
30. C. C. Yu and J. C. Heinrich, 'Petrov–Galerkin method for multidimensional, time dependent convective diffusion equation', *Int. J. Num. Meth. Eng.*, **24**, 2201–15, 1987.
31. R. A. Adey and C. A. Brebbia, 'Finite element solution of effluent dispersion', in *Numerical Methods in Fluid Mechanics* (eds C. A. Brebbia and J. J. Connor), pp. 325–54, Pentech Press, Southampton, 1974.
32. K. W. Morton, 'Generalised Galerkin methods for hyperbolic problems', *Comp. Meth. Appl. Mech. Eng.*, **52**, 847–71, 1985.
33. R. E. Ewing and T. F. Russell, 'Multistep Galerkin methods along characteristics for convection–diffusion problems', in *Advances in Computational Methods for PDEs* (eds R. Vichnevetsky and R. S. Stepleman), Vol. IV, IMACS, pp. 28–36, Rutgers University, Brunswick, N.J., 1981.
34. J. Douglas, Jr and T. F. Russell, 'Numerical methods for convection dominated diffusion problems based on combining the method of characteristics with finite element or finite difference procedures', *SIAM J. Num. Anal.*, **19**, 871–85, 1982.
35. O. Pironneau, 'On the transport diffusion algorithm and its application to the Navier Stokes equation', *Num. Math.*, **38**, 309–32, 1982.
36. M. Bercovier, O. Pironneau, Y. Harbani and E. Livne, 'Characteristics and finite element methods applied to equations of fluids', in *The Mathematics of Finite Elements and Applications* (ed. J. R. Whiteman), Vol. V, pp. 471–8, Academic Press, London, 1982.
37. J. Goussebaile, F. Hecht, C. Labadie and L. Reinhart, 'Finite element solution of the shallow water equations by a quasi-direct decomposition procedure', *Int. J. Num. Meth. Fluids*, **4**, 1117–36, 1984.
38. M. Bercovier, O. Pironneau and V. Sastri, 'Finite elements and characteristics for some parabolic–hyperbolic problems', *Appl. Math. Modelling*, **7**, 89–96, 1983.
39. J. P. Benque, J. P. Gregoire, A. Hauguel and M. Maxant, 'Application des Methodes du decomposition aux calculs numeriques en hydraulique

industrielle', in *INRIA, 6th Coll. Int. Methodes de Calcul Sci. et Techn.*, Versailles, 12–16 Dec. 1983.
40. A. BERMUDEZ, J. DURANY, M. POSSE and C. VAZQUEZ, 'An upwind method for solving transport–diffusion–reaction systems', *Int. J. Num. Meth. Eng.*, **28**, 2021–40, 1984.
41. O. C. ZIENKIEWICZ, R. LÖHNER, K. MORGAN and S. NAKAZAWA, 'Finite elements in fluid mechanics—a decade of progress', in *Finite Elements in Fluids* (eds R. H. Gallagher *et al.*), Vol. 5, chap. 1, pp. 1–26, Wiley, Chichester, 1984.
42. R. LÖHNER, K. MORGAN and O. C. ZIENKIEWICZ, 'The solution of non-linear hyperbolic equation systems by the finite element method', *Int. J. Num. Meth. Fluids*, **4**, 1043–63, 1984.
43. O. C. ZIENKIEWICZ, R. LÖHNER, K. MORGAN and J. PERAIRE, 'High speed compressible flow and other advection dominated problems of fluid mechanics', in *Finite Elements in Fluids* (eds R. H. Gallagher *et al.*), Vol. 6, chap. 2, pp. 41–88, Wiley, Chichester, 1986.
44. R. LÖHNER, K. MORGAN and O. C. ZIENKIEWICZ, 'An adaptive finite element procedure for compressible high speed flows', *Comp. Meth. Appl. Mech. Eng.*, **51**, 441–65, 1985.
45. O. C. ZIENKIEWICZ, R. LÖHNER and K. MORGAN, 'High speed inviscid compressive flow by the finite element method', in *The Mathematics of Finite Elements and Applications* (ed. J. R. Whiteman), Vol. VI, pp. 1–25, Academic Press, London, 1985.
46. P. D. LAX and B. WENDROFF, 'Systems of conservative laws', *Comm. Pure Appl. Math.*, **13**, 217–37, 1960.
47. J. PERAIRE, 'A finite element method for convection dominated flows', Ph.D. thesis, University of Wales, Swansea, 1986.
48. J. DONEA, 'A Taylor–Galerkin method for convective transport problems', *Int. J. Num. Meth. Eng.*, **20**, 101–19, 1984.
49. V. SELMIN, J. DONEA and L. QUATRAPELLE, 'Finite element method for non-linear advection', *Comp. Meth. Appl. Mech. Eng.*, **52**, 817–45, 1985.
50. L. BOTTURA and O. C. ZIENKIEWICZ, 'Experiments on iterative solution of the semi-implicit characteristic Galerkin algorithm', *Comm. Appl. Num. Meth.*, **6**, 387–93, 1990.
51. R. LÖHNER, K. MORGAN, J. PERAIRE, O. C. ZIENKIEWICZ and L. KONG, 'Finite element methods for compressible flow', in *Numerical Methods in Fluid Dynamics* (eds K. W. Morton and M. J. Baines), Vol. II, pp. 27–52, Clarendon Press, Oxford, 1986.
52. J. PERAIRE, K. MORGAN and O. C. ZIENKIEWICZ, 'Convection dominated problems', in *Numerical Methods for Compressible Flows—Finite Difference, Element and Volume Techniques*, AMD 78, pp. 129–47, ASME, 1987.
53. O. C. ZIENKIEWICZ, J. Z. ZHU, Y. C. LIU, K. MORGAN and J. PERAIRE, 'Error estimates and adaptivity. From elasticity to high speed compressible flow', in *The Mathematics of Finite Elements and Applications* (ed. J. R. Whiteman), Vol. VII, Academic Press, London, 1988.
54. J. PERAIRE, J. PEIRO, L. FORMAGGIA, K. MORGAN and O. C. ZIENKIEWICZ, 'Finite element Euler computations in three dimensions', *AIAA 26th Aerospace Sciences Meeting*, paper AIAA-87-0032, Reno, USA, January 1988.
55. J. PERAIRE, J. PEIRO, L. FORMAGGIA, K. MORGAN and O. C. ZIENKIEWICZ, 'Finite element Euler computations in 3-D', *Int. J. Num. Meth. Eng.*, **26**, 2135–59, 1988.

56. A. LAPIDUS, 'A detached shock calculation by second order finite differences', *J. Comp. Phys.*, **2**, 154–77, 1967.
57. J. P. BORIS and D. L. BROOK, 'Flux corrected transport I Shasta = a fluid transport algorithm that works', *J. Comp. Phys.*, **11**, 38–69, 1973.
58. R. LÖHNER, K. MORGAN, J. PERAIRE and M. VAHDATI, 'Finite element, flux corrected transport (FEM–FCT) for the Euler and Navier–Stokes equations', *Int. J. Num. Meth. Fluids*, **7**, 1093–109, 1987.
59. S. T. ZALESIAK, 'Fully multidimensional flux corrected transport algorithm for fluids', *J. Comp. Phys.*, **31**, 335–62, 1979.
60. R. LÖHNER, K. MORGAN and O. C. ZIENKIEWICZ, 'The use of domain splitting with an explicit hyperbolic solver', *Comp. Meth. Appl. Mech. Eng.*, **45**, 313–29, 1984.
61. R. LÖHNER and K. MORGAN, 'An unstructured multigrid method for elliptic problems', *Int. J. Num. Meth. Eng.*, **24**, 101–15, 1987.
62. J. L. SOHN and J. C. HEINRICH, 'Pressure calculations in penalty finite element approximations to the Navier–Stokes equations', *Int. J. Num. Meth. Eng.* (to be published, 1990).
63. O. C. ZIENKIEWICZ, 'Explicit (or semiexplicit) general algorithm for compressible and incompressible flows with equal finite element interpolation', Report 90.5, Chalmers Technical University, Gothenborg, 1990.

13

Fluid mechanics—governing equations and incompressible flow; newtonian and non-newtonian viscous flows

13.1 Introduction

The application of finite element method problems of fluid mechanics dates back to the early seventies.[1–12] However, the early success with this methodology was confined to flows in which either viscosity was dominant or inviscid problems (in which a potential formulation was used). The extension to problems in which convection was of major importance, i.e. to high Reynolds number flows or compressible flows, had to await the extension of finite element processes which we discussed in the previous chapter (Chapter 12). Today it appears that finite element procedures are available for all fluid mechanics problems and that these indeed, once again, demonstrate the merits vis-à-vis alternatives of finite differences and finite volume methods.[13,14] We shall therefore commence this chapter, to avoid repetition, with a full version of the governing differential equations applicable to viscous, compressible, flow and then deal with the various specialized forms. In particular, the remainder of this chapter will concentrate on *incompressible* or *nearly incompressible* flows where frequently the convective terms are negligible and standard Galerkin procedures remain admissible. The introduction of Petrov–Galerkin methods will only be indicated in Sec. 13.7 dealing with viscous, high Reynolds number, flows.

The consideration of high-speed compressible flows and of shallow water problems with high convective terms will be deferred to Chapters 14 and 15. However, the basic equations derived here will continue to form the foundation.

13.2 The governing equations of fluid mechanics

13.2.1 *Stresses in fluids.* The essential characteristic of a fluid is its inability to sustain shear stresses when at rest. Here only hydrostatic 'stress' or pressure is possible. Any analysis must therefore concentrate on the motion, and the essential independent variable is thus the velocity **u** or, if we adopt the indicial notation (with the *xyz* axis referred to as x_i, $i=1, 2, 3$),

$$u_i \tag{13.1}$$

This replaces the displacement variable which was of primary importance in solid mechanics.

The rates of strain are thus the primary cause of the general stresses, σ_{ij}, and these are defined in a manner analogous to that of infinitesimal strain as

$$\dot{\varepsilon}_{ij}=\frac{\partial u_i/\partial x_j+\partial u_j/\partial x_i}{2}\equiv\mathbf{Su} \tag{13.2}$$

where **S** is the same operator as that introduced in solid mechanics for defining strain.

The stress–strain relations for a linear (newtonian) isotropic fluid require the definition of two constants.

The first of these links the *deviatoric stresses* τ_{ij} to *deviatoric strain rates*:

$$\tau_{ij}\equiv\frac{\sigma_{ij}-\delta_{ij}\sigma_{ii}}{3}=2\mu\left(\frac{\dot{\varepsilon}_{ij}-\delta_{ij}\dot{\varepsilon}_{ii}}{3}\right) \tag{13.3}$$

where δ_{ij} is the Kroneker delta and repeated index means summation; thus

$$\sigma_{ii}\equiv\sigma_{11}+\sigma_{22}+\sigma_{33} \quad \text{and} \quad \dot{\varepsilon}_{ii}=\dot{\varepsilon}_{11}+\dot{\varepsilon}_{22}+\dot{\varepsilon}_{33} \tag{13.4}$$

The coefficient μ is known as the shear viscosity or simply viscosity and is analogous to the shear modulus G in linear elasticity.

The second relation is that between the mean stress changes and the volumetric strain rates. This defines the pressure as

$$p=\frac{\sigma_{ii}}{3}=-\kappa\dot{\varepsilon}_{ii}+p_0 \tag{13.5}$$

where κ is a *volumetric viscosity* coefficient analogous to the bulk modulus K in linear elasticity and p_0 is the initial hydrostatic pressure independent of the strain rate (note that p and p_0 are invariably defined as positive when compressive).

Immediately we can write the 'constitutive' relation for fluids from Eqs

(13.3) and (13.5) as

$$\sigma_{ij}=2\mu\left(\dot{\varepsilon}_{ij}-\frac{\delta_{ij}\dot{\varepsilon}_{ii}}{3}\right)+\delta_{ij}\kappa\dot{\varepsilon}_{ii}-\delta_{ij}p_0$$
$$=\tau_{ij}-\delta_{ij}p \tag{13.6a}$$

or
$$\sigma_{ij}=2\mu\dot{\varepsilon}_{ij}+\delta_{ij}(\kappa-\tfrac{2}{3}\mu)\dot{\varepsilon}_{ii}+\delta_{ij}p_0 \tag{13.6b}$$

Traditionally the Lamé notation is used putting

$$\kappa-\tfrac{2}{3}\mu\equiv\lambda \tag{13.7}$$

but this has little to recommend it and the relation (13.6a) is basic. There is little evidence about the existence of volumetric viscosity and we shall take

$$\kappa\dot{\varepsilon}_{ii}\equiv 0 \tag{13.8}$$

in what follows, giving the essential constitutive relation as (now dropping the suffix p_0)

$$\sigma_{ij}=2\mu\left(\dot{\varepsilon}_{ij}-\frac{\delta_{ij}\dot{\varepsilon}_{ii}}{3}\right)-\delta_{ij}p\equiv\tau_{ij}-\delta_{ij}p \tag{13.9a}$$

without necessarily implying incompressibility $\varepsilon_{ii}=0$.

In the above,

$$\tau_{ij}=2\mu\left(\dot{\varepsilon}_{ij}-\frac{\delta_{ij}\varepsilon_{ii}}{3}\right)=\mu\left[\left(\frac{\partial u_i}{\partial x_j}+\frac{\partial u_j}{\partial x_i}\right)-\delta_{ij}\frac{2}{3}\frac{\partial u_i}{\partial x_i}\right] \tag{13.9b}$$

All of the above relationships are analogous to those of elasticity, as we shall note again later for incompressible flow.

Non-linearity of some fluid flows is observed with a coefficient μ depending on strain rates. Such flows we shall term 'non-newtonian'.

13.2.2 *Mass conservation.* If ρ is the fluid density then the balance of mass flow ρu_i entering and leaving a unit volume requires that

$$\frac{\partial\rho}{\partial t}+\frac{\partial}{\partial x_i}(\rho u_i)\equiv\frac{\partial\rho}{\partial t}+\nabla^{\mathrm{T}}(\rho\mathbf{u})=0 \tag{13.10a}$$

or in cartesian coordinates

$$\frac{\partial\rho}{\partial t}+\frac{\partial}{\partial x}(\rho u)+\frac{\partial}{\partial y}(\rho v)+\frac{\partial}{\partial z}(\rho w)=0 \tag{13.10b}$$

13.2.3 *Momentum conservation—or dynamic equilibrium.* Now the balance of momentum in the jth direction, this is $(\rho u_j)\ u_i$ leaving and entering a control volume, has to be an equilibrium with the stresses σ_{ij}

and body forces ρf_j giving a typical component equation

$$\frac{\partial(\rho u_j)}{\partial t}+\frac{\partial}{\partial x_i}[(\rho u_j)u_i]-\frac{\partial}{\partial x_i}(\sigma_{ij})-\rho f_j=0 \tag{13.11}$$

or using (13.9a),

$$\frac{\partial(\rho u_j)}{\partial t}+\frac{\partial}{\partial x_i}[(\rho u_j)u_i]-\frac{\partial(\tau_{ij})}{\partial x_i}+\frac{\partial p}{\partial x_j}-\rho f_j=0 \tag{13.12a}$$

with (13.9b) implied.

Once again the above can, of course, be written as three sets of equations in a cartesian form:

$$\frac{\partial}{\partial t}(\rho u)+\frac{\partial}{\partial x}(\rho u^2)+\frac{\partial}{\partial y}(\rho uv)+\frac{\partial}{\partial z}(\rho uw)$$

$$-\frac{\partial \tau_{xx}}{\partial x}-\frac{\partial \tau_{xy}}{\partial y}-\frac{\partial \tau_{xz}}{\partial z}+\frac{\partial p}{\partial x}-\rho f_x=0 \tag{13.12b}$$

etc.

13.2.4 *Energy conservation and equation of state.* We note that in the equations of Secs 13.2.2 and 13.2.3 the independent variables are u_i (the velocity), p (the pressure) and ρ (the density). The stresses, of course, were defined by Eq. (13.9) in terms of velocities and hence are not independent.

Obviously, there is one variable too many for this equation system to be capable of solution. However, if density is assumed constant (as in incompressible fluids) or if a single relationship linking pressure and density can be established (as in isothermal flow with small compressibility) the system becomes complete and is solvable.

More generally, the pressure (p), density (ρ) and temperature (T) are related by an equation of state of the form

$$\rho=\rho(p, T) \tag{13.13}$$

For an ideal gas this takes, for instance, the form

$$\rho=\frac{p}{RT} \tag{13.14}$$

where R is the universal gas constant.

In such a general case, it is necessary to supplement the governing equation system by the equation of *energy conservation*. This equation is indeed of interest even if it is not coupled, as it provides additional information about the behaviour of the system.

Before proceeding with the derivation of the energy conservation equation we must define some further quantities. Thus we introduce e, the *intrinsic energy* per unit mass. This is dependent on the state of the fluid, i.e. its pressure and temperature or

$$e = e(T, p) \tag{13.15}$$

The total energy per unit mass, E, includes of course the kinetic energy per unit mass and thus

$$E = e + \frac{u_i u_i}{2} \tag{13.16}$$

Finally, we can define *enthalpy* as

$$h = e + \frac{p}{\rho} \qquad \text{or} \qquad H = h + \frac{u_i u_i}{2} = E + \frac{p}{\rho} \tag{13.17}$$

and these variables are found to be convenient.[13]

The energy transfer can take place by convection and by conduction. The conductive heat flux q_i is defined as

$$q_i = -k \frac{\partial}{\partial x_i} T \tag{13.18}$$

where k is an isotropic heat conduction coefficient.

To complete the relationship it is necessary to determine heat source terms. These can be specified per unit volume as q_H due to chemical reaction (if any) and must include the energy dissipation due to internal stresses, i.e. using Eq. (13.9),

$$\frac{\partial}{\partial x_i}(\sigma_{ij} u_j) = \frac{\partial}{\partial x_i}(\tau_{ij} u_j) - \frac{\partial}{\partial x_j}(p u_j) \tag{13.19}$$

The balance of energy in a unit volume can now thus be written as

$$\frac{\partial(\rho E)}{\partial t} + \frac{\partial}{\partial x_i}(\rho u_i E) - \frac{\partial}{\partial x_i}\left(k \frac{\partial T}{\partial x_i}\right) + \frac{\partial}{\partial x_i}(p u_i)$$

$$- \frac{\partial}{\partial x_i}(\tau_{ji} u_i) - \rho f_i u_i - q_H = 0 \tag{13.20a}$$

or more simply

$$\frac{\partial(\rho E)}{\partial t} + \frac{\partial}{\partial x_i}(\rho u_i H) - \frac{\partial}{\partial x_i}\left(k \frac{\partial T}{\partial x_i}\right) + \frac{\partial}{\partial x_i}(\tau_{ji} u_i) - \rho f_i u_i - q_H = 0 \tag{13.20b}$$

Here, the penultimate term represents the work done by body forces.

13.2.5 *Navier–Stokes and Euler equations.* The governing equations derived in the preceding sections can be written in the general conservative form which we have introduced in Chapter 12 [viz. Eq. (12.1)], i.e.

$$\frac{\partial \mathbf{U}}{\partial t}+\nabla\mathbf{F}+\nabla\mathbf{G}+\mathbf{Q}=0 \qquad (13.21a)$$

or

$$\frac{\partial \mathbf{U}}{\partial t}+\frac{\partial \mathbf{F}_i}{\partial x_i}+\frac{\partial \mathbf{G}_i}{\partial x_i}+\mathbf{Q}=0 \qquad (13.21b)$$

in which Eqs (13.10), (13.12) or (13.20) provide the particular vectors.

Thus, the vector of independent unknowns is, using both indicial and cartesian notation,

$$\mathbf{U}=\begin{Bmatrix}\rho\\ \rho u_1\\ \rho u_2\\ \rho u_3\\ \rho E\end{Bmatrix} \text{ or, in cartesian notation, } \mathbf{U}=\begin{Bmatrix}\rho\\ \rho u\\ \rho v\\ \rho w\\ \rho E\end{Bmatrix} \qquad (13.22a)$$

$$\mathbf{F}_i=\begin{Bmatrix}\rho u_i\\ \rho u_1 u_i+p\delta_{1i}\\ \rho u_2 u_1+p\delta_{2i}\\ \rho u_3 u_i+p\delta_{3i}\\ \rho H u_i\end{Bmatrix} \quad \text{or } \mathbf{F}_x=\begin{Bmatrix}\rho u\\ \rho u^2+p\\ \rho uv\\ \rho uw\\ \rho H u\end{Bmatrix}, \text{ etc.} \qquad (13.22b)$$

$$\mathbf{G}_i=\begin{Bmatrix}0\\ -\tau_{1i}\\ -\tau_{2i}\\ -\tau_{3i}\\ -\dfrac{\partial}{\partial x_i}(\tau_{ij}u_i)-k\dfrac{\partial T}{\partial x_i}\end{Bmatrix} \text{ or } \mathbf{G}_x=\begin{Bmatrix}0\\ -\tau_{xx}\\ -\tau_{yx}\\ -\tau_{zx}\\ -\dfrac{\partial}{\partial x}(\tau_{xx}u+\tau_{xy}v+\tau_{xz}w)-k\dfrac{\partial T}{\partial x}\end{Bmatrix},$$

etc. (13.22c)

$$\mathbf{Q}=\begin{Bmatrix}0\\ -\rho f_1\\ -\rho f_2\\ -\rho f_3\\ -\rho f_i u_i-q_H\end{Bmatrix} \quad \text{or } \mathbf{Q}=\begin{Bmatrix}0\\ -\rho f_x\\ -\rho f_y\\ -\rho f_z\\ -\rho(f_x u+f_y v+f_z w)-q_H\end{Bmatrix}, \text{ etc.}$$

(13.22d)

with

$$\tau_{ij}=\mu\left[\left(\frac{\partial u_i}{\partial x_j}+\frac{\partial u_j}{\partial x_i}\right)-\delta_{ij}\frac{2}{3}\frac{\partial u_i}{\partial x_i}\right]$$

The complete set of equations is known as the *Navier–Stokes equation* and a particular case of an inviscid fluid with no heat conduction ($\tau_{ij}=0$, $k=0$) as the *Euler equation.*

The above equations are the basis from which all fluid mechanics studies start and it is not surprising that many alternative forms are given in literature obtained by combinations of the various equations.[13] The above set is, however, convenient and physically meaningful, defining the conservation of important quantities. Further, we note immediately why the discussions of the preceding Chapter 12 are important and why some modification of the standard Galerkin process will be necessary in discretization.

In many actual situations one or other feature of flow is predominant. For instance, frequently the viscosity is only of importance close to the boundaries at which velocities are specified, i.e.

$$\Gamma_u \qquad \text{where } u_i=\bar{u}_i$$

or on which tractions are prescribed:

$$\Gamma_\tau \qquad \text{where } n_i\sigma_{ij}=t_i$$

In the above n_i are the direction cosines of the normal.

In such cases the problem can be considered separately in two parts: one as the *boundary layer* near such boundaries and another as *inviscid flow* outside the boundary layer.

Further, in many cases a steady-state solution is not available with the fluid exhibiting *turbulence*, i.e. a random fluctuation of velocity. Here it is still possible to use the general Navier–Stokes equations now written in terms of mean flow but with a *Reynolds viscosity* replacing the molecular one. The subject is dealt with elsewhere in detail and in this text we shall limit ourselves to merely showing its onset when stable solutions cease to exist. The turbulent instability is inherent in the simple Navier–Stokes equations and it is in principle always possible to obtain the transient, turbulent, solution modelling of the flow, providing the mesh size is capable of reproducing the random eddies. Such computations, though possible, are extremely costly and hence the Reynolds averaging is of practical importance.

Two important points have to be made concerning *inviscid flow* (ideal fluid flow as it is sometimes known).

Firstly, the Euler equations are of a purely convective form:

$$\frac{\partial \mathbf{U}}{\partial t}+\frac{\partial \mathbf{F}_i}{\partial x_i}=0 \qquad \mathbf{F}_i=\mathbf{F}_i(\mathbf{U}) \tag{13.23}$$

and hence very special methods for their solutions will be necessary, as we have already mentioned in Chapter 12. These methods are applicable and useful mainly in *compressible flow*, as we shall discuss in Chapter 14. Secondly, for incompressible (or nearly incompressible) flows it is of interest to introduce a *potential* that converts the Euler equations to a simple self-adjoint form. We shall confine this procedure to the present chapter. Although potential forms are applicable also to compressible flows we shall not discuss them later as they fail in high-speed, supersonic, cases.

INCOMPRESSIBLE (OR NEARLY INCOMPRESSIBLE) FLOWS

13.3 Governing equations

We observed earlier that the Navier–Stokes equations are completed by the existence of a state relationship giving [Eq. (13.13)]

$$\rho=\rho(p, T)$$

In (nearly) incompressible relations we shall frequently assume that:

1. The problem is isothermal.
2. Variation of ρ with p is very small, i.e. such that in product terms of velocity and density the latter can be assumed constant.

The first assumption will be relaxed, as we shall see later, allowing some thermal coupling via the dependence of the fluid properties on temperature. In such cases we shall introduce the coupling iteratively. Here the problem of density-induced currents (viz. page 541) or temperature-dependent viscosity (viz. page 528) will be typical.

If assumptions introduced above are used we can still allow for small compressibility, noting that density changes are, as a consequence of elastic deformability, related to pressure changes. Thus we can write

$$\mathrm{d}\rho=\frac{\rho}{K}\,\mathrm{d}p \tag{13.24a}$$

where K is the bulk modulus. This can be written as

$$\mathrm{d}\rho=\frac{1}{c^2}\,\mathrm{d}p \tag{13.24b}$$

or
$$\frac{\partial \rho}{\partial t}=\frac{1}{c^2}\frac{\partial p}{\partial t} \tag{13.24c}$$

with $c=\sqrt{K/\rho}$ being the acoustic wave velocity.

Equations (13.21) and (13.22) can now be rewritten omitting the energy transport (and condensing the general form) as

$$\frac{1}{c^2}\frac{\partial p}{\partial t}+\rho\,\frac{\partial u_i}{\partial x_i}=0 \tag{13.25a}$$

$$\frac{\partial u_j}{\partial t}+\frac{\partial}{\partial x_i}(u_j u_i)+\frac{1}{\rho}\frac{\partial p}{\partial x_j}-\frac{1}{\rho}\frac{\partial}{\partial x_i}\,\tau_{ji}-f_j=0 \tag{13.25b}$$

With $j=1$, 2, 3 this represents a system of four equations in which the variables are u_j and p.

Written in terms of cartesian coordinates we have, in place of Eq. (13.25a),

$$\frac{1}{c^2}\frac{\partial p}{\partial t}+\rho\,\frac{\partial u}{\partial x}+\rho\,\frac{\partial v}{\partial y}+\rho\,\frac{\partial w}{\partial z}=0 \tag{13.26a}$$

where the first term is dropped for complete incompressibility ($c=\infty$) and

$$\frac{\partial u}{\partial t}+\frac{\partial}{\partial x}(u^2)+\frac{\partial}{\partial y}(uv)+\frac{\partial}{\partial z}(uw)+\frac{1}{\rho}\frac{\partial p}{\partial x}$$
$$-\frac{1}{\rho}\left(\frac{\partial}{\partial x}\,\tau_{xx}+\frac{\partial}{\partial y}\,\tau_{xy}+\frac{\partial}{\partial z}\,\tau_{xz}\right)-f_x=0 \tag{13.26b}$$

with similar forms for y and z. In both forms

$$\frac{1}{\rho}\tau_{ij}=\nu\left(\frac{\partial u_i}{\partial x_j}+\frac{\partial u_j}{\partial x_i}-\delta_{ij}\frac{2}{3}\frac{\partial u_i}{\partial x_i}\right)$$

where $\nu=\mu/\rho$ is the kinematic viscosity.

The reader will note that the above equations, with the exception of the convective acceleration terms, are *identical to those governing the problem of incompressible* (*or slightly compressible*) *elasticity*, which we have discussed in Chapter 12 of Volume 1.

13.4 Inviscid, incompressible flow—potential formulation

In the absence of viscosity and compressibility flow equations, Eqs (13.26) can be written as

$$\frac{\partial u}{\partial x}+\frac{\partial v}{\partial y}+\frac{\partial w}{\partial z}=0 \tag{13.27a}$$

and
$$\frac{\partial u}{\partial t}+\frac{\partial}{\partial x}(u^2)+\frac{\partial}{\partial y}(uv)+\frac{\partial}{\partial z}(uw)+\frac{1}{\rho}\frac{\partial p}{\partial x}-f_x=0 \tag{13.27b}$$

with similar forms for y and z components.

These Euler equations are not convenient for numerical solution, and it is of interest to introduce a potential, ϕ, defining velocities as

$$u=-\frac{\partial \phi}{\partial x} \qquad v=-\frac{\partial \phi}{\partial y} \qquad w=-\frac{\partial \phi}{\partial z} \tag{13.28}$$

or
$$\mathbf{u}=-\nabla\phi$$

If such a potential exists then insertion of (13.28) into (13.27a) gives a single governing equation

$$\frac{\partial^2 \phi}{\partial x^2}+\frac{\partial^2 \phi}{\partial y^2}+\frac{\partial^2 \phi}{\partial z^2}\equiv\nabla^2\phi=0 \tag{13.29}$$

which, with appropriate boundary conditions, can be readily solved in the manner described in Chapter 10 of Volume 1. For contained flow we can of course impose the normal velocity u_n on the boundaries:

$$u_n=-\frac{\partial \phi}{\partial n} \tag{13.30}$$

and, as we know from discussions in Volume 1, this provides a *natural* boundary condition.

Indeed, at this stage it is not necessary to discuss the application of finite elements to this particular equation, which was discussed at length in Volume 1 and for which many codes are available.[15,16] In Fig. 13.1 an example of a typical potential solution is given.

Of course we must be assured that the potential function ϕ exists, and indeed determine what conditions are necessary for its existence. Here we observe that so far we have not used in the definition of the problem the important momentum conservation equations (13.27b), and we shall now return to these. However, we first note that a single-valued potential function implies that

$$\frac{\partial^2 \phi}{\partial x \partial y}=\frac{\partial^2 \phi}{\partial y \partial x}, \qquad \text{etc.} \tag{13.31}$$

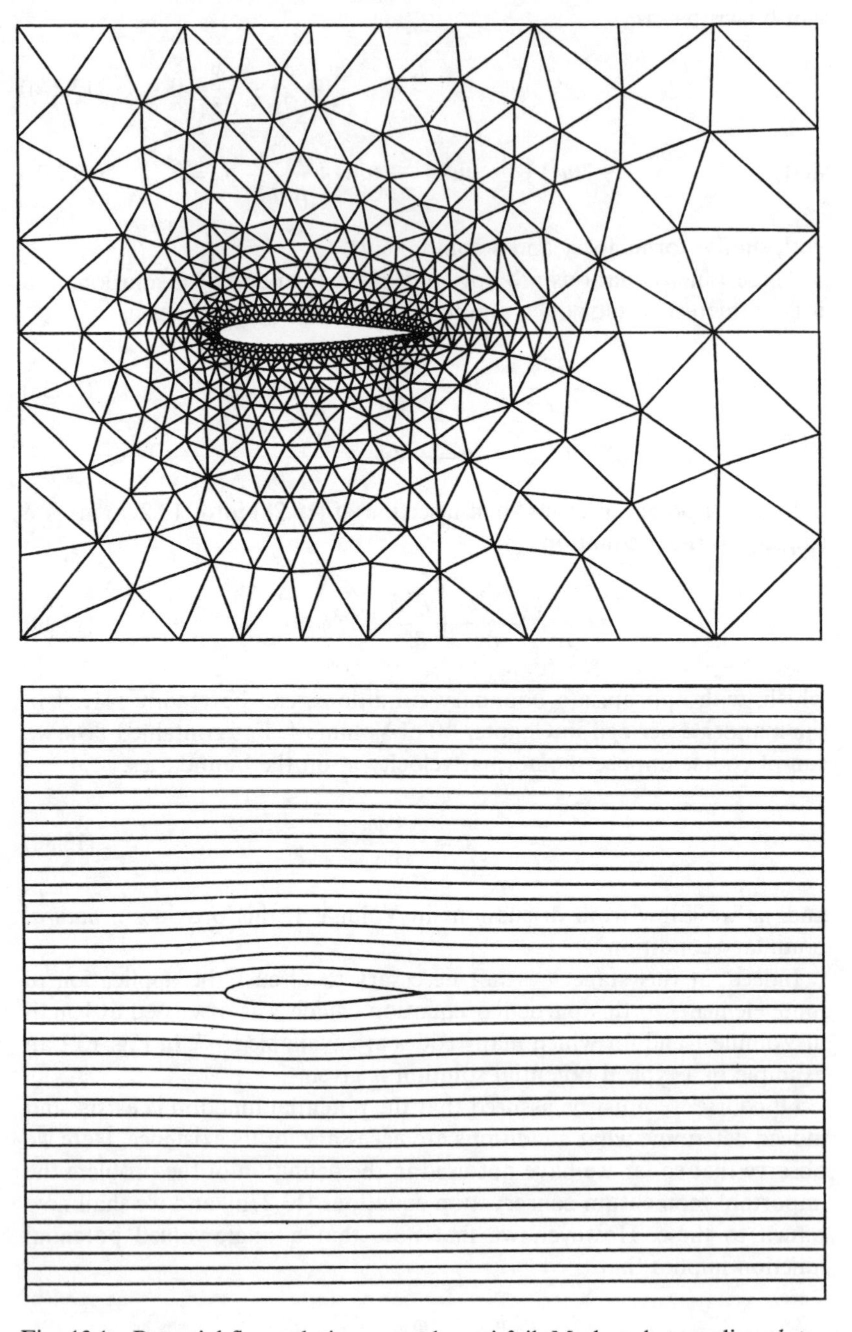

Fig. 13.1 Potential flow solution around an airfoil. Mesh end streamline plots

and hence that, using the definition (13.28),

$$\omega_x \equiv \frac{\partial u}{\partial y} - \frac{\partial v}{\partial x} = 0 \quad \omega_y \equiv \frac{\partial v}{\partial z} - \frac{\partial w}{\partial y} = 0 \quad \text{and} \quad \omega_z \equiv \frac{\partial w}{\partial x} - \frac{\partial u}{\partial z} = 0 \tag{13.32}$$

This is a statement of *irrotationality* of flow which is implied by the existence of the potential.

Inserting the definition of potential into the first term of Eq. (13.27b) and using Eq. (13.27a) and (13.32) we can rewrite this equation as

$$\frac{\partial}{\partial x}\left(\frac{\partial \phi}{\partial t}\right) + \frac{\partial}{\partial x}\left[\frac{1}{2}(u^2 + v^2 + w^2) + \frac{p}{\rho} + P\right] = 0 \tag{13.33}$$

(with similar ones in the y and z directions) in which P is the potential of the body forces giving these as

$$f_i = -\frac{\partial P}{\partial x_i} \tag{13.34}$$

This is alternatively written as

$$\nabla\left(\frac{\partial \phi}{\partial t} + H + P\right) = 0 \tag{13.35}$$

where H is the *enthalpy* introduced in Eq. (13.17).

If isothermal conditions pertain, the specific energy is constant and the above implies that

$$\frac{\partial \phi}{\partial t} + \frac{1}{2}(u^2 + v^2 + w^2) + \frac{p}{\rho} + P = \text{constant} \tag{13.36}$$

for the whole domain. This can be taken as a corollary of the existence of the potential and indeed is a condition for its existence. In steady-state flows it provides the well-known Bernoulli equation that allows the pressures to be determined throughout the whole potential field when the value of the constant is established.

Some problems of specific interest are those of flow with a free surface.[16–18] Here the governing Laplace equation for potential remains identical, but the free surface position has to be found iteratively. In Fig. 13.2 an example of such a free surface flow solution is given.[17]

In problems involving gravity the body force potential is simply

$$P = gz$$

representing gravity forces, and the free surface condition requires that (in two dimensions)

$$\tfrac{1}{2}(u^2 + v^2) - gz = 0$$

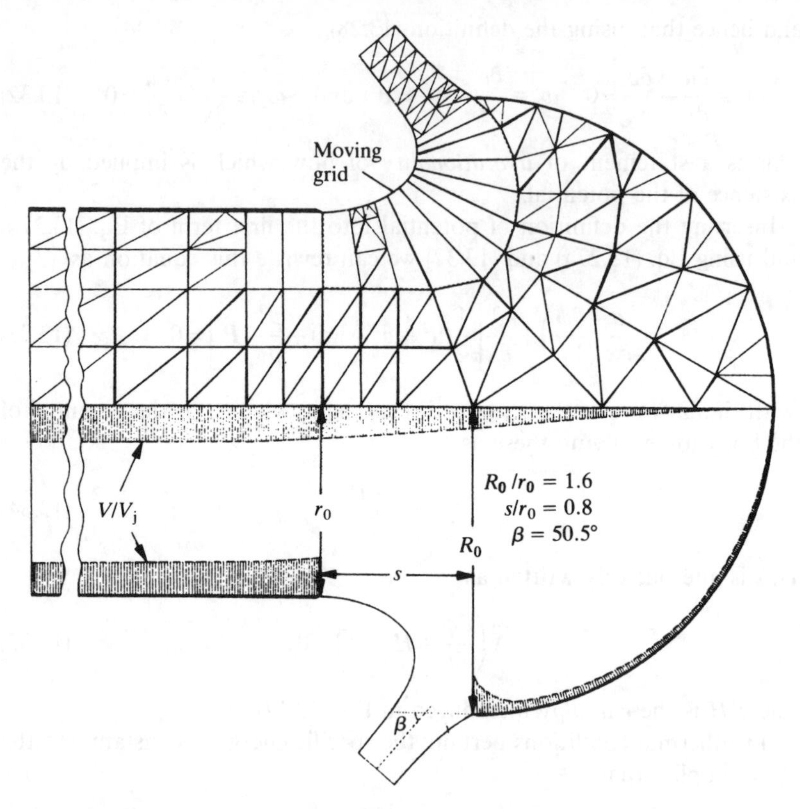

Fig. 13.2 Free surface potential flow, illustrating an axisymmetric jet impinging on a hemispherical thrust reverser. (From Sarpkaya and Hiriart[17])

Such conditions involve an iterative, non-linear solution, as illustrated by examples of overflows in reference 16.

It is interesting to observe that the governing potential equation is self-adjoint and that the introduction of the potential has side-stepped the difficulties of dealing with convective terms.

13.5 Slow (creeping) fully incompressible viscous flow—the Stokes problem

13.5.1 *Analogy with incompressible elasticity.* Slow, viscous incompressible flow represents the extreme situation at the other end of the scale from the inviscid problem of the previous section. Here all dynamic (acceleration) forces are, *a priori*, neglected and Eqs (13.26) reduce, in cartesian coordinates, to

$$\frac{\partial u}{\partial x}+\frac{\partial v}{\partial y}+\frac{\partial w}{\partial z}\equiv\dot{\varepsilon}_v=0 \tag{13.37a}$$

and

$$-\left(\frac{\partial}{\partial x}\tau_{xx}+\frac{\partial}{\partial y}\tau_{xy}+\frac{\partial}{\partial z}\tau_{xz}\right)+\frac{\partial p}{\partial x}-\rho f_x=0 \tag{13.37b}$$

with similar forms in the y and z directions.

The above are completed of course by the constitutive relation (13.26b):

$$\tau_{ij}=\mu\left(\frac{\partial u_i}{\partial x_j}+\frac{\partial u_j}{\partial x_i}-\delta_{ij}\frac{2}{3}\frac{\partial u_i}{\partial x_i}\right) \tag{13.38}$$

which is identical to the problem of incompressible elasticity in which we replace:

(*a*) the displacements by velocities and

(*b*) the shear modulus G by the viscosity μ.

The identity is perhaps clearer if, for consistency with Volume 1, we reintroduce the vectorial notation:

$$\dot{\boldsymbol{\varepsilon}}^{\mathrm{T}}=[\dot{\varepsilon}_x,\dot{\varepsilon}_y,\dot{\varepsilon}_z,2\dot{\varepsilon}_{xy},2\dot{\varepsilon}_{yz},2\dot{\varepsilon}_{zx}] \tag{13.39}$$

for strain rates, and for deviatoric stresses

$$\boldsymbol{\tau}^{\mathrm{T}}=[\tau_{xx}\tau_{yy}\tau_{zz}\tau_{xy}\tau_{yz}\tau_{zx}] \tag{13.40}$$

Now we can write, in place of Eqs (13.37b) and (13.38),

$$\mathbf{S}^{\mathrm{T}}(\mu\mathbf{D}_0)\mathbf{S}\mathbf{u}-\nabla p-\rho\mathbf{f}=0 \tag{13.41a}$$

and in place of Eq. (13.37a),

$$\mathbf{m}^{\mathrm{T}}\mathbf{S}\mathbf{u}=0 \tag{13.41b}$$

where

$$\mathbf{D}_0=\begin{bmatrix} \frac{4}{3}, & -\frac{2}{3}, & -\frac{2}{3}, & 0, & 0, & 0\\ & \frac{4}{3}, & -\frac{2}{3}, & 0, & 0, & 0\\ & & \frac{4}{3}, & 0, & 0, & 0\\ & \text{symmetric} & & 1, & 0, & 0\\ & & & & 1, & 0\\ & & & & & 1 \end{bmatrix} \tag{13.42}$$

and $\qquad \mathbf{m}^{\mathrm{T}}=[1,1,1,0,0,0]$

recovering Eqs (12.44) and (12.45) in Chapter 12 of Volume 1.

13.5.2 *Mixed and penalty discretization.* The discretization can be started from the *mixed form* with independent approximations of $\mathbf{u}$ and p, i.e.

$$\mathbf{u} = \mathbf{N}_u \bar{\mathbf{u}} \qquad p = \mathbf{N}_p \bar{\mathbf{p}} \tag{13.43}$$

or by a penalty form in which Eq. (13.41b) is augmented by p/γ where γ is a large penalty parameter

$$\mathbf{m}^{\mathrm{T}} \mathbf{S} \mathbf{u} - \frac{p}{\gamma} = 0 \tag{13.44}$$

allowing p to be eliminated from the computation. Such penalty forms are only applicable with *reduced integration* and their general equivalence with the mixed form in which p is discretized by discontinuous choice of $\mathbf{N}_p$ between elements has been demonstrated.[19]

As computationally it is advantageous to use the mixed form and introduce the penalty parameter only to eliminate the p values at the element levels, we shall presume such penalization to be done after the mixed discretization.

The use of penalty forms in fluid mechanics was introduced early in the seventies[20-22] and is fully discussed elsewhere.[23-25]

The discretized equations will always be of the form

$$\begin{bmatrix} \mathbf{K} & \mathbf{Q} \\ \mathbf{Q}^{\mathrm{T}} & h\mathbf{I}/\gamma \end{bmatrix} \begin{Bmatrix} \bar{\mathbf{u}} \\ \bar{\mathbf{p}} \end{Bmatrix} = \begin{Bmatrix} \bar{\mathbf{f}} \\ 0 \end{Bmatrix} \tag{13.45}$$

where h is a typical element size, $\mathbf{I}$ an identity matrix,

$$\begin{aligned} \mathbf{K} &= \int_\Omega \mathbf{B}^{\mathrm{T}} \mu \, \mathbf{D}_0 \mathbf{B} \, \mathrm{d}\Omega \qquad \text{where } \mathbf{B} \equiv \mathbf{S} \mathbf{N}_u \\ \mathbf{Q} &= -\int_\Omega \mathbf{N}_u^{\mathrm{T}} \nabla \mathbf{N}_p \, \mathrm{d}\Omega \\ \bar{\mathbf{f}} &= \int_\Omega \mathbf{N}_u^{\mathrm{T}} \rho \, \mathbf{f} \, \mathrm{d}\Omega \qquad \text{(assuming zero boundary tractions)} \end{aligned} \tag{13.46}$$

and the penalty number, γ, is introduced purely as a numerical convenience. This is taken generally as[23,25]

$$\gamma = 10^7 - 10^{10} \mu$$

There is little more to be said about the solution procedures or indeed results of creeping, incompressible, flow solutions for linear problems. The range of applicability is of course limited to low velocities of flow or high viscosity fluids such as oil, blood in biomechanics applications, etc. It is, however, important to recall here that the mixed form allows only certain combinations of $\mathbf{N}_u$ and $\mathbf{N}_p$ interpolations to be used without violating the convergence conditions. This was discussed in detail in Chapter 12 of Volume 1, but for completeness Fig. 13.3 lists some of the

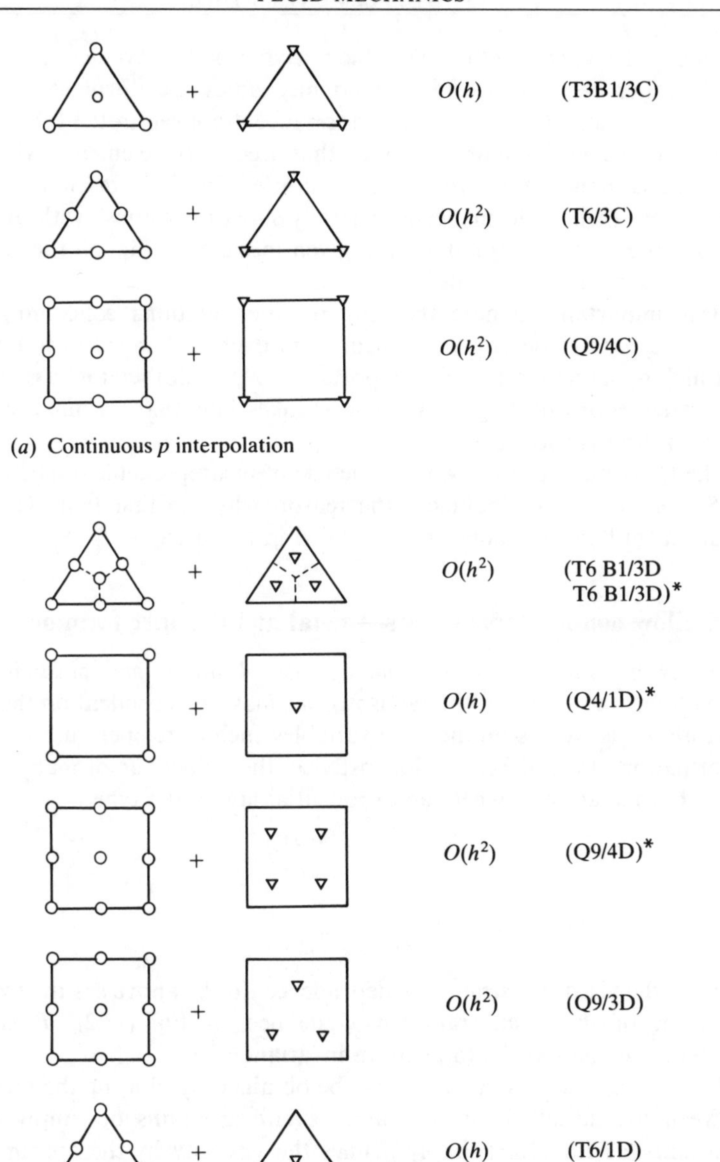

Fig. 13.3 Some useful velocity–pressure interpolations and their asymptotic, energy norm, convergence rates

available elements together with their asymptotic convergence rates.[26] Many other elements useful in fluid mechanics are documented elsewhere,[27-29] but those of proven performance are given in the table.

It is of general interest to note that frequently elements with C_0 continuous pressure interpolations are used in fluid mechanics and indeed that their performance is generally superior to those with discontinuous pressure interpolation on a given mesh, even though the cost of solution is marginally greater.

It is important to note that the recommendations concerning the element types for the Stokes problem carry over unchanged to situations in which dynamic terms are of importance. We shall therefore assume in the later sections dealing with Navier–Stokes flow that a similar spatial discretization is used.

The fairly obvious extension of the use of incompressible elastic codes to Stokes flow is undoubtedly the reason why the first finite element solutions of fluid mechanics were applied in this area.

13.6 Slow non-newtonian flows—metal and polymer forming

13.6.1 *Non-newtonian flows including viscoplasticity and plasticity.* In many fluids the viscosity, though isotropic, may be dependent on the rate of strain $\dot{\varepsilon}_{ij}$ as well as on the state variables such as temperature or total deformation. Typical here is, for instance, the behaviour of many polymers, hot metals, etc., where an exponential law of the type

$$\mu = \mu_0 \dot{\bar{\varepsilon}}^{(m-1)} \tag{13.47}$$

with

$$\mu_0 = \mu_0(T, \bar{\varepsilon})$$

governs the viscosity–strain rate dependence. In the above $\dot{\bar{\varepsilon}}$ is the second invariant of the strain rate tensor defined in Eq. (13.2), T is the temperature and $\bar{\varepsilon}$ is the total strain invariant.

This *secant* viscosity can of course be obtained by plotting the relation between the deviatoric stresses and deviatoric strains or simply their invariants, as Eq. (13.3) simply defines the viscosity by the appropriate ratio of the stress to strain rate. Such plots are shown in Fig. 13.4. The above exponential relation of Eq. (13.47) is known as the Ostwald de Waele law and is illustrated in Fig. 13.4(*b*).

In a similar manner viscosity laws can be found for viscoplastic and indeed purely plastic behaviour of an incompressible kind. For instance, in Fig. 13.4(*c*) we show a viscoplastic Bingham fluid in which a threshold or yield value of the second stress invariant has to be exceeded before any strain rate is observed. Thus for the viscoplastic fluid illustrated it is

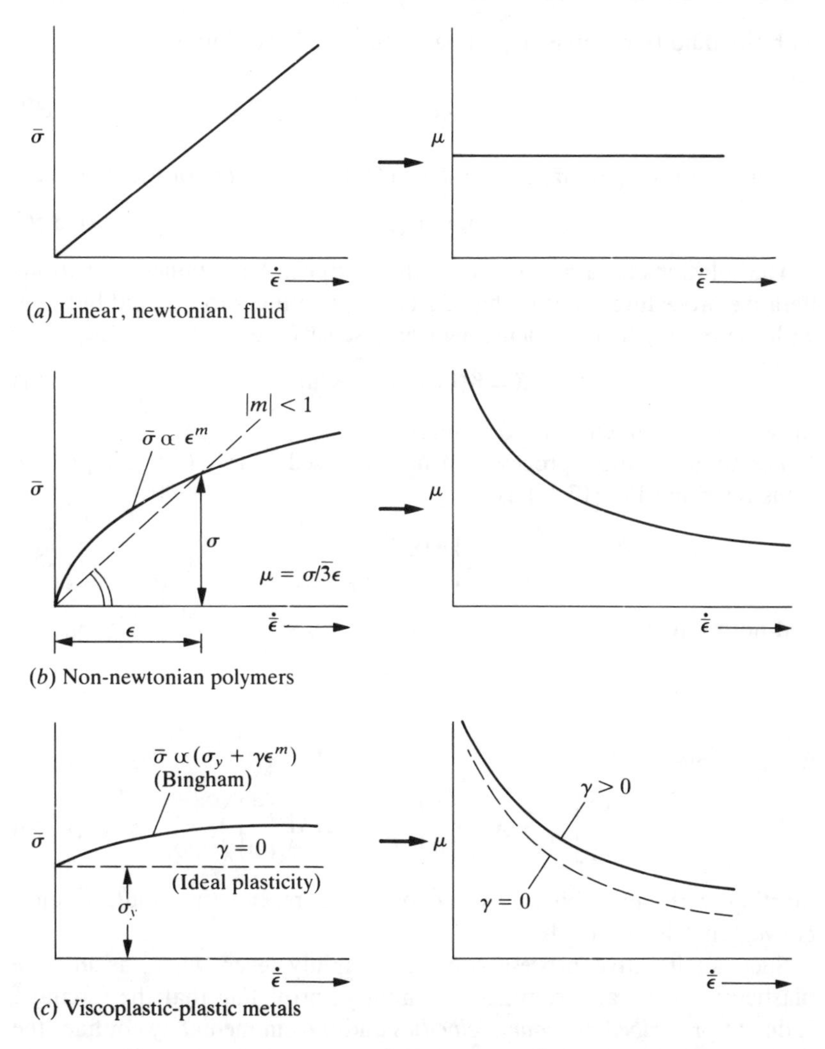

Fig. 13.4 Stress $\bar{\sigma}$, viscosity μ and strain rate $\dot{\bar{\varepsilon}}$ relationships for various materials

evident that a highly non-linear viscosity relation is obtained. This can be written as

$$\mu = \frac{\bar{\sigma}_y + \gamma\,\dot{\bar{\varepsilon}}^m}{\dot{\bar{\varepsilon}}} \qquad (13.48)$$

where $\bar{\sigma}_y$ is the value of the second stress invariant at yield.

The special case of pure plasticity follows of course as a limiting case with the fluidity parameter $\gamma=0$, and now we have simply

$$\mu=\frac{\bar{\sigma}_y}{\dot{\bar{\varepsilon}}} \tag{13.49}$$

Of course, once again $\bar{\sigma}_y$ can be dependent on the *state* of the fluid, i.e.

$$\bar{\sigma}_y=\bar{\sigma}_y(T,\bar{\varepsilon}) \tag{13.50}$$

The solutions (at a given state of the fluid) can be obtained by various iterative procedures, noting that Eq. (13.45) continues to be valid but now with the matrix $\mathbf{K}$ being dependent on viscosity, i.e.

$$\mathbf{K}=\mathbf{K}(\mu)=\mathbf{K}(\dot{\bar{\varepsilon}})=\mathbf{K}(\mathbf{u}) \tag{13.51}$$

thus being dependent on the solution.

The total iteration process can here be used simply (viz. Chapter 7). Thus rewriting Eq. (13.45) as

$$\mathbf{A}\begin{Bmatrix}\bar{\mathbf{u}}\\ \bar{\mathbf{p}}\end{Bmatrix}=\begin{Bmatrix}\bar{\mathbf{f}}\\ 0\end{Bmatrix} \tag{13.52}$$

and noting that

$$\mathbf{A}=\mathbf{A}\left(\begin{Bmatrix}\bar{\mathbf{u}}\\ \bar{\mathbf{p}}\end{Bmatrix}\right)$$

we can write

$$\begin{Bmatrix}\bar{\mathbf{u}}\\ \bar{\mathbf{p}}\end{Bmatrix}^{i+1}=\mathbf{A}_i^{-1}\begin{Bmatrix}\bar{\mathbf{f}}\\ 0\end{Bmatrix}\qquad \mathbf{A}_i=\mathbf{A}\left(\begin{Bmatrix}\bar{\mathbf{u}}\\ \mathbf{p}\end{Bmatrix}\right)^i \tag{13.53}$$

Starting with an arbitrary value of μ we repeat the solution until convergence is obtained.

Such an iterative process converges rapidly (even when, as in pure plasticity, μ can vary from zero to infinity), providing that the forcing $\bar{\mathbf{f}}$ is due to *prescribed boundary velocities* and thus immediately confines the variation of all velocities in a narrow range. In such cases, five to seven iterations are generally required to bring the difference of the ith and $(i+1)$th solutions to within the 1 per cent (euclidian) norm.

The first non-newtonian flow solutions were applied to polymers and to hot metals in the early seventies.[30-32] Application of the same procedures to the forming of metals was introduced at the same time and has subsequently been widely developed.[33-50]

It is perhaps difficult to visualize steel or aluminium behaving as a fluid, being conditioned to use these materials as structural members. If, however, we note that during the forming process the elastic strains are of the order of 10^{-6} while the plastic strain can reach or exceed a value of unity, neglect of the former (which is implied in the viscosity definition) seems justifiable. This is indeed borne out by comparison of computations based on what we now call *flow formulation* with elastoplastic computation or experiment. The process has alternatively been introduced as a 'rigid-plastic' form,[38,39] though such modelling is more complex and less descriptive.

Today the methodology is widely accepted for the solution of metal and polymer forming processes, and only a limited selection of references of application can be cited. The reader would do well to consult reference 48 for a complete survey of the field.

13.6.2 *Steady-state problems of forming.* Two categories of problems arise in forming situations. *Steady-state flow* is the first of these. In this, a real, continuing, flow is modelled, as shown in Fig. 13.5(*a*), and here

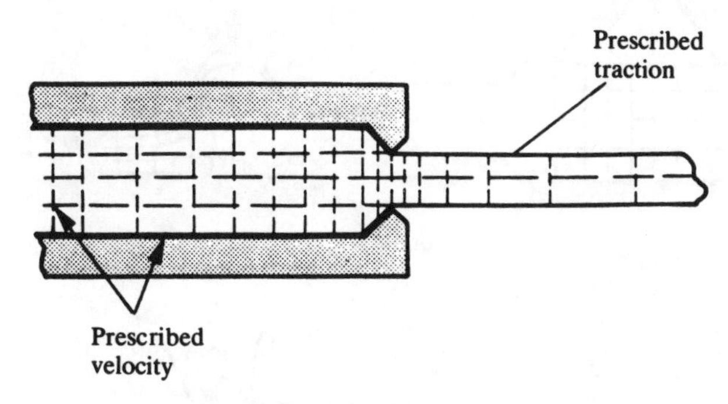

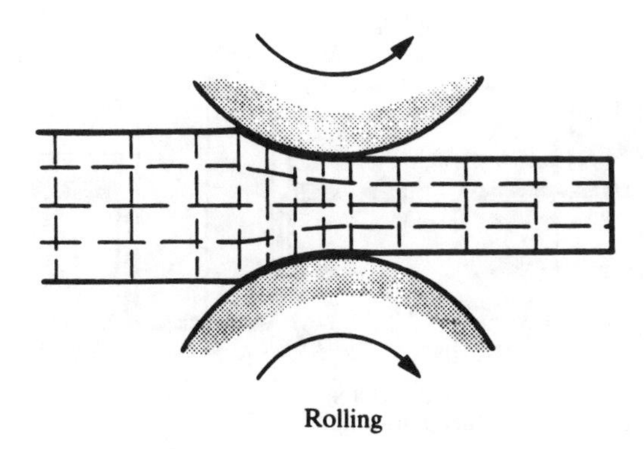

(*a*) Steady state

Fig. 13.5 Forming processes typically used in manufacture

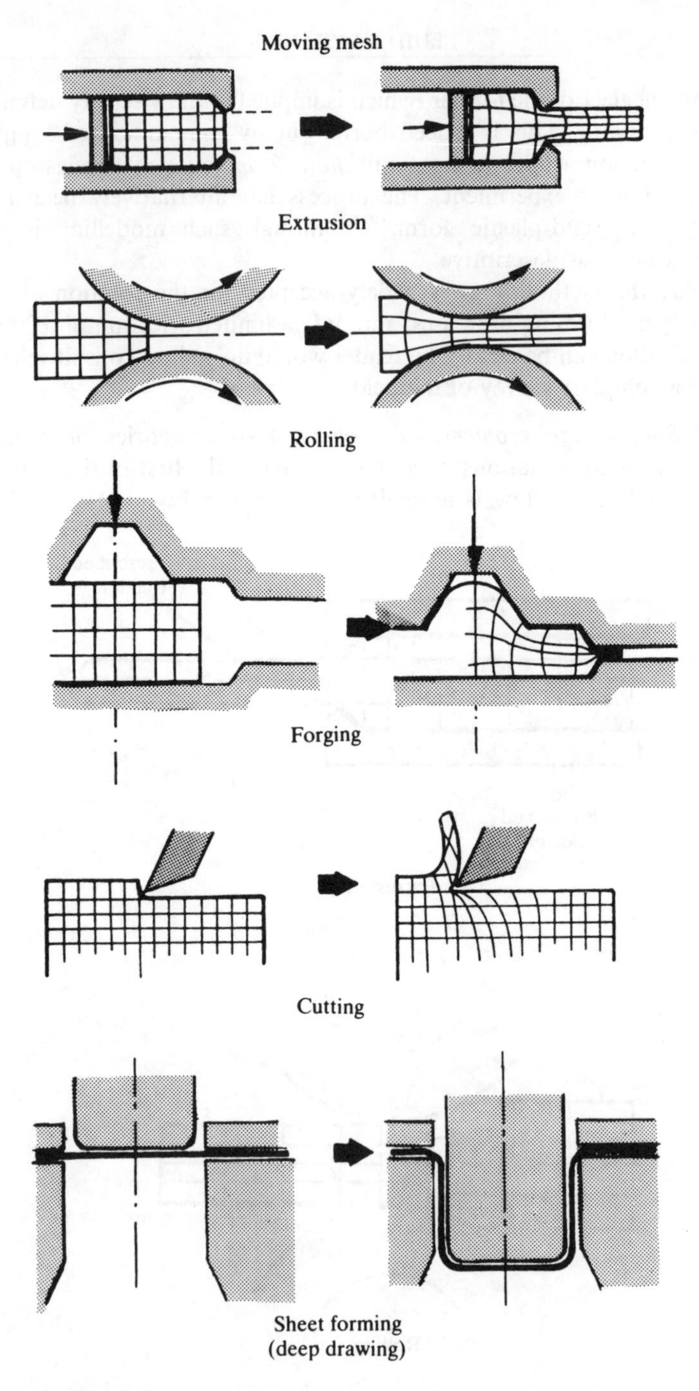

(*b*) Transient

Fig. 13.5 (continued)

velocity and other properties can be assumed to be fixed in a particular point of space. In Fig. 13.5(*b*) the more usual *transient* processes of forming are illustrated and we shall deal with these later. In a typical steady-state problem if the state parameters T and $\bar{\varepsilon}$ defining the temperature and viscosity are known in the whole field, solution can be carried out in the manner previously described. We could, for instance, assume that the 'viscous' flow of the problem of Fig. 13.6 is that of an ideally plastic material under isothermal conditions modelling an extrusion process and obtain the solution shown in Table 13.1. For such a material exact extrusion forces can be calculated[51] and the table shows the errors obtained with the flow formulation using different triangular elements of Fig. 13.3 and two meshes.[26] The fine mesh here was arrived at using error estimates and a single adaptive remeshing.

In general the problem of steady-state flow is accompanied by evolution of temperature (and other state parameters such as the total strain

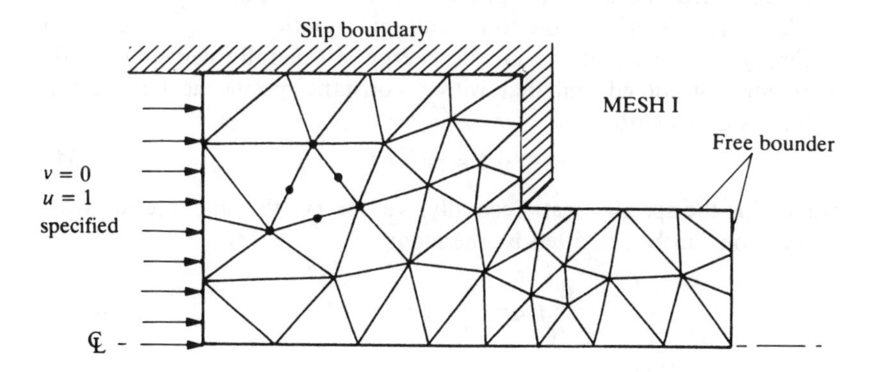

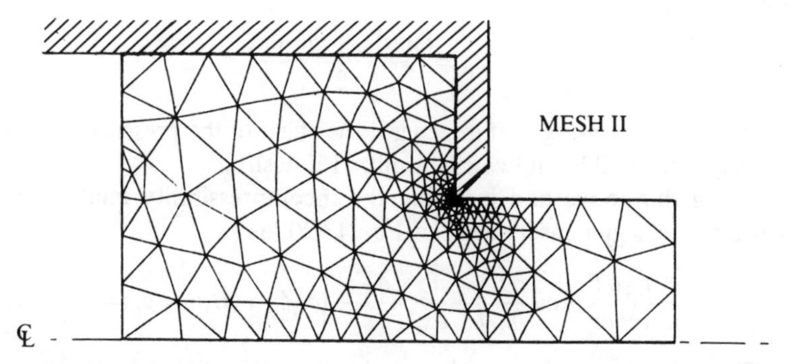

Fig. 13.6 Plane strain extrusion (extrusion ratio 2:1) with ideal plasticity assumed

TABLE 13.1

COMPARISONS OF PERFORMANCE OF SEVERAL TRIANGULAR MIXED ELEMENTS OF FIG. 13.3 IN A PLANE EXTRUSION PROBLEM (IDEAL PLASTICITY ASSUMED[26])

Element type	Mesh 1 (coarse)			Mesh 2 (fine)		
	Ext. force	Force error %	CPU(s)	Ext. force	Force error %	CPU(s)
T6/1D	28 901.0	12.02	67.81	25 990.0	0.73	579.71
T6B1/3D	31 043.0	20.32	74.76	26 258.0	1.78	780.13
T6B1/3D*	29 031.0	12.52	73.08	26 229.0	1.66	613.92
T6/3C	27 902.5	8.15	87.62	25 975.0	0.67	855.38
Exact	25 800.0	0.00	—	25 800.0	0.00	—

invariant $\bar{\varepsilon}$) and here it is necessary to couple the solution with the heat balance and other evolution equations. The evolution of heat has already been discussed and the appropriate conservation equations given as Eq. (13.20). It is convenient now to rewrite this equation in a modified form.

Firstly, we note that the kinetic energy is generally negligible in the problems considered and that with a constant specific heat c per unit volume we can write

$$\rho E \approx \rho e = \hat{c} T \tag{13.54a}$$

where $\hat{c}$ is the specific heat. Secondly, we observe that the internal work dissipation can be rewritten by the identity

$$\frac{\partial}{\partial x_i} p u_i - \frac{\partial}{\partial x_i}(\tau_{ji} u_i) \equiv -\sigma_{ji} \dot{\varepsilon}_{ji} \tag{13.54b}$$

where, by Eq. (13.9),

$$\sigma_{ji} = \tau_{ji} - \delta_{ji} p \tag{13.54c}$$

and, by Eq. (13.2),

$$\dot{\varepsilon}_{ji} = \frac{\partial u_i/\partial x_i + \partial u_i/\partial x_j}{2} \tag{13.54d}$$

We note in passing that in general the effect of the pressure term in Eqs (13.54) is negligible and can be omitted if desired.

Using the above and inserting the incompressibility relation we can write the energy conservation of Eq. (13.20) as

$$\left(\hat{c}\frac{\partial T}{\partial t} + \hat{c} u_i \frac{\partial T}{\partial x_i}\right) - \frac{\partial}{\partial x_i} k \frac{\partial T}{\partial x_i} - (\sigma_{ij}\dot{\varepsilon}_{ij} + \rho f_i u_i + g_h) = 0 \tag{13.55}$$

The solution of the coupled problem can be carried out iteratively. Here the term in the last bracket can be evaluated repeatedly from the known velocities and stresses from the flow solution. We note that the first bracketed term represents a total derivative of the convective kind which, even in the steady state, requires the use of the special weighting procedures discussed in Chapter 13.

Such coupled solutions have been carried out for the first time relatively recently,[35,36] but are today practised routinely.[42-44] Figure 13.7 shows a typical thermally coupled solution for a steady-state rolling problem from reference 36.

It is of interest to note that in this problem boundary friction plays an important role and that this is modelled by using thin elements near the boundary, making the viscosity coefficient in that layer pressure dependent.[49] This procedure is very simple and although not exact gives results of sufficient practical accuracy.

13.6.3 *Transient problems with changing boundaries.* These represent the second, probably larger, category of forming problems. Typical examples here are those of forging, indentation, etc., and again thermal coupling can be included if necessary. Figures 13.8 and 13.9 illustrate typical applications.

The solution for velocities and internal stresses can be readily accomplished at a given configuration providing the temperatures and other state variables are known at that instant. This allows the new configuration to be obtained both for the boundaries and for the mesh by writing explicitly

$$\Delta x_i = u_i \Delta t \tag{13.56}$$

as the incremental relation.

If thermal coupling is important increments of temperature need also to be evaluated. However, we note that for convected coordinates Eq. (13.55) is simplified as the convected terms disappear. We can now write

$$\hat{c}\frac{\partial T}{\partial t} - \frac{\partial}{\partial x_i} k \frac{\partial T}{\partial x_i} - \left(\sigma_{ij}\dot{\varepsilon}_{ij} + \rho f_i u_i + g_h \right) = 0 \tag{13.57}$$

where the last term is the heat input known at the start of the interval and computation of temperature increments is made using either explicit or implicit procedures discussed in Chapter 10.

Indeed, both the coordinate and thermal updating can make use iteratively of the solution on the updated mesh to increase accuracy. However, it must be noted that any continuous mesh updating will soon lead to unacceptable meshes and some form of remeshing is necessary.

In the example of Fig. 13.8,[22] in which ideal plasticity was assumed together with isothermal behaviour, it is necessary only to keep track of boundary movements. As temperature and other state variables do not

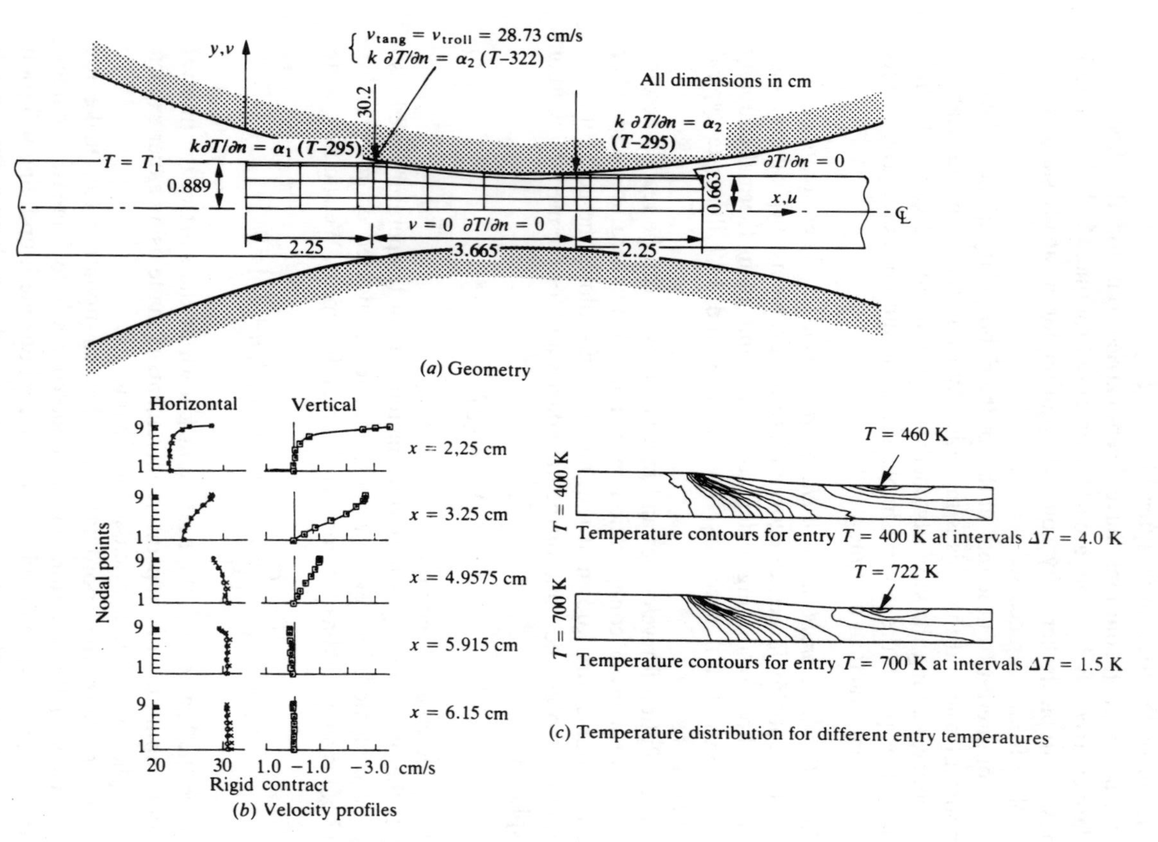

Fig. 13.7 Steady-state rolling process with thermal coupling[36]

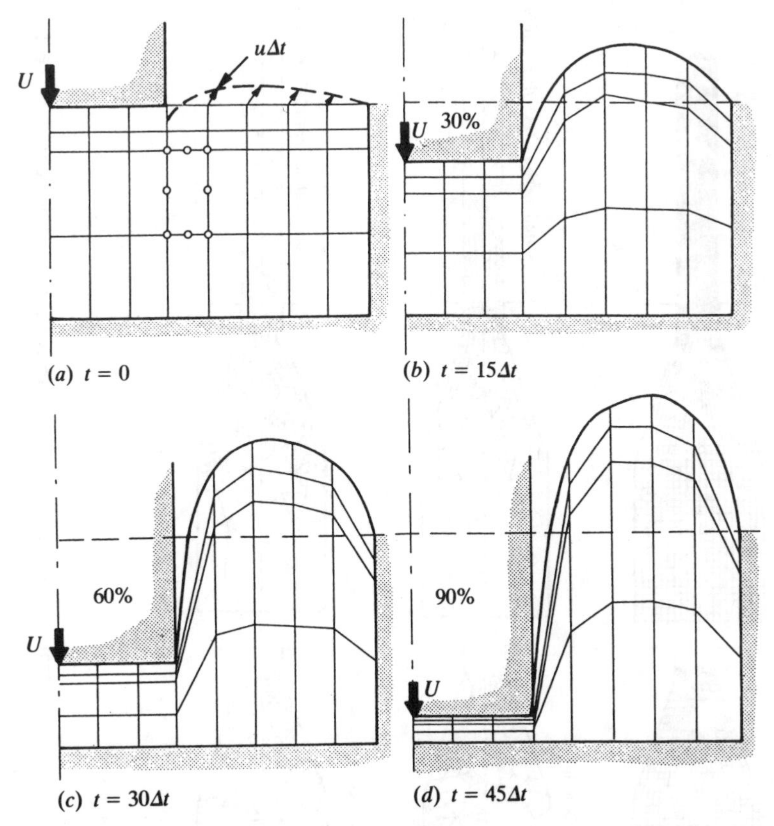

Fig. 13.8 Punch indentation problem (penalty function approach).[22] Updated mesh and surface profile with 24 isoparametric elements. Ideally plastic material: (*a*), (*b*), (*c*) and (*d*) show various depths of indentation. (Reduced integration is used here)

enter the problem the remeshing can be done simply—in the case shown by keeping the same vertical lines for the mesh position.

However, in the example of Fig. 13.9 showing a more realistic problem,[52,53] when a new mesh is created an interpolation of all the state parameters from the old to the new mesh positions is necessary. In such problems it is worth while to strive to obtain discretization errors within specified bounds and to remesh adaptively when these errors are too large.

We have discussed the problem of adaptive remeshing for linear problems in Chapter 14 of Volume 1. In the present examples similar methods have been adopted with success[54,55] and in Fig. 13.9 we show how remeshing proceeds during the forming process. It is of interest simply to observe that once again the *energy norm* of the error is the measure used.

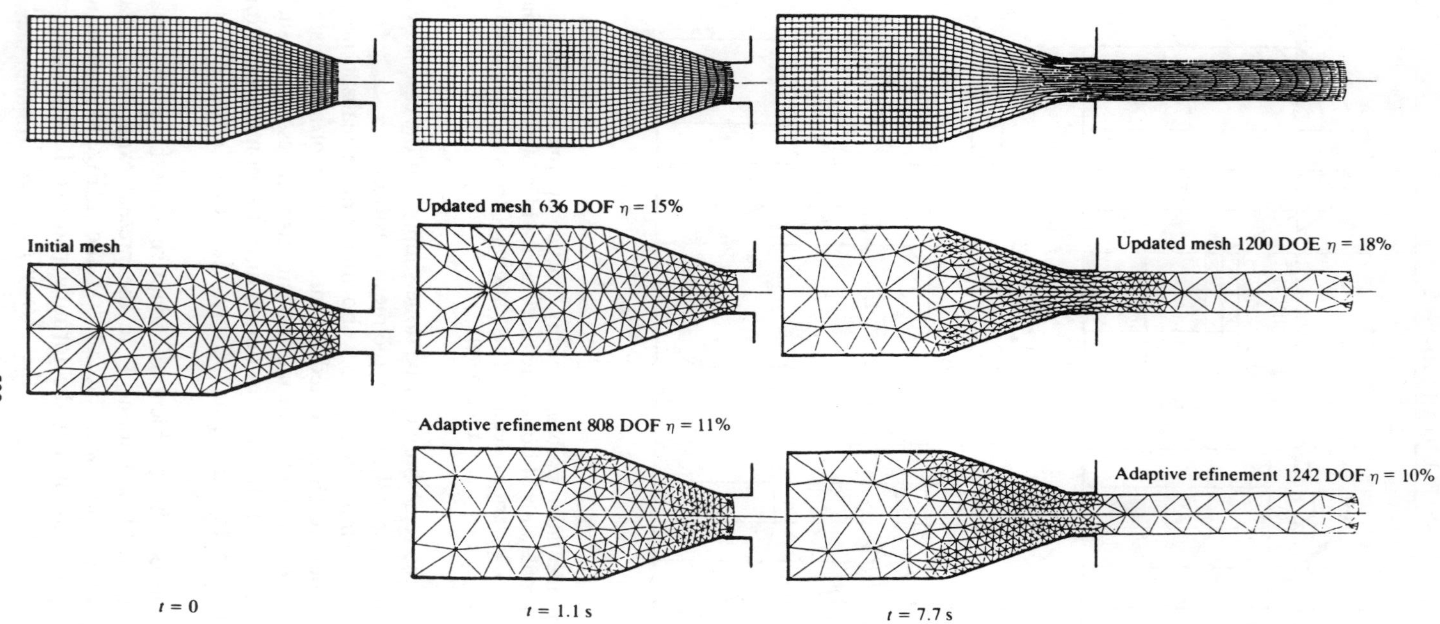

(*a*) **A material grid and updated and adapted meshes with material deformation**
(η percentage error in energy norm)

Fig. 13.9 A transient extrusion problem with temperature and strain-dependent yield.[53] Adaptive mesh refinement uses T6/1D elements of Fig. 13.3

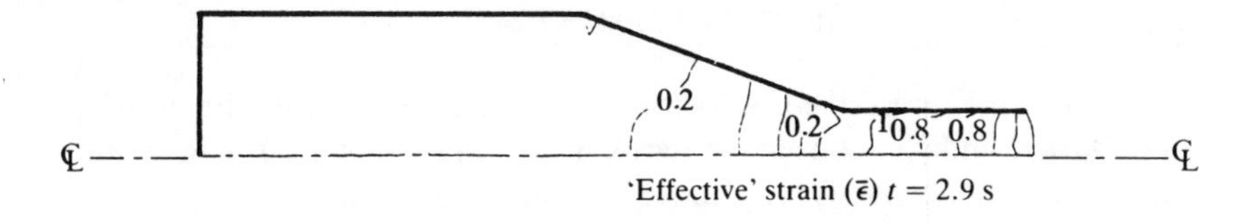

'Effective' strain ($\bar{\epsilon}$) $t = 2.9$ s

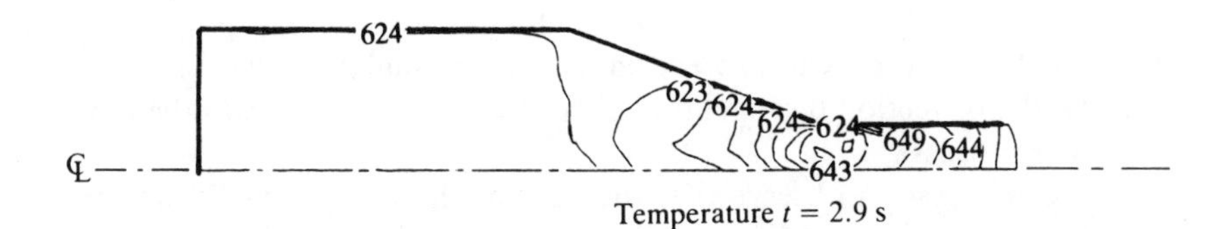

Temperature $t = 2.9$ s

(b) Contours of state parameters at $t = 2.9$ s

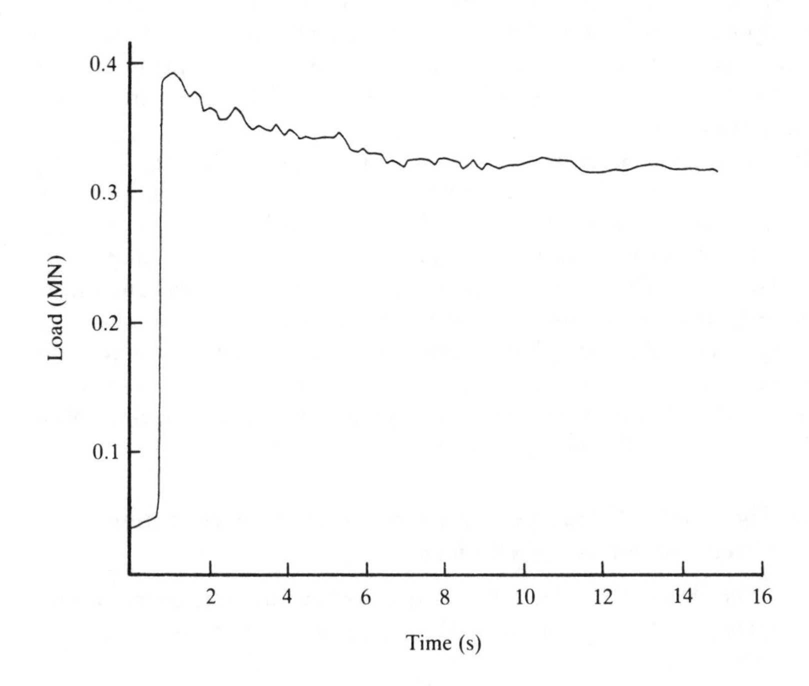

(c) Load versus time

Fig. 13.9 (continued)

The details of various applications can be found in the extensive literature on the subject. This also deals with various sophisticated mesh updating procedures. One particularly successful method is the so-called ALE (arbitrary lagrangian–eulerian) method.[56–60] Here the original mesh is given some prescribed velocity $\bar{\mathbf{v}}$ in a manner fitting the moving boundaries, and the convective terms in the equations are retained with reference to this velocity. In Eq. (13.56), for instance, in place of

$$\hat{c}u_i \frac{\partial T}{\partial x_i} \quad \text{we write} \quad \hat{c}(u_i - \bar{v}_i)\frac{\partial T}{\partial x_i}$$

etc., and the solution can proceed in a manner similar to that of steady state (with convection disappearing of course when $\bar{v}_i = u_i$; i.e. in the pure updating process).

It is of interest to observe that the flow methods can equally well be applied to the forming of thin sections resembling shells. Here of course all the assumptions of shell theory and corresponding finite element technology are applicable. Because of this, incompressibility constraints are no longer a problem but other complications arise. The literature of such applications is large, but much relevant information can be found in references 61 to 67. Practical applications ranging from the forming of beer cans to car bodies abound. Figures 13.10 and 13.11 illustrate some typical problems.

In this section we have confined our attention to flow in which no elastic effects are included, i.e. in which only viscosity varies. However, there is considerable interest in the flow of viscoelastic or viscoplastic–elastic fluids. Such flows are of some interest in metal forming where residual elastic effects always remain, but the major impetus comes from chemical processes involving flow or rubber-like fluids. We shall not discuss this problem here, but we refer the reader to the extension of the present formulation to such cases in references 68 to 70. We must note, however, that if movements are relatively small, procedures of elastoplasticity discussed in Chapter 7 are applicable.

13.7 The Navier–Stokes problem and convective acceleration effects—steady-state solutions

13.7.1 *The basis.* For the full solution of incompressible or nearly incompressible flow we return to Eqs (13.25) and (13.26) and retain the acceleration effects.

To emphasize the convective terms we abandon here the conservative notation used in Eq. (13.26b) and write, for comparison with the slow flow of Eq. (13.37),

$$\frac{1}{c^2}\frac{\partial p}{\partial t} + \rho\frac{\partial u}{\partial x} + \rho\frac{\partial v}{\partial y} + \rho\frac{\partial w}{\partial z} = 0 \tag{13.58a}$$

and
$$\left[\frac{\partial u}{\partial t} + u\frac{\partial u}{\partial x} + v\frac{\partial u}{\partial y} + w\frac{\partial u}{\partial z}\right]\rho - \left[\frac{\partial}{\partial x}\tau_{xx} + \frac{\partial}{\partial y}\tau_{xy} + \frac{\partial}{\partial z}\tau_{xz}\right] + \frac{\partial p}{\partial x} - \rho f_x = 0 \tag{13.58b}$$

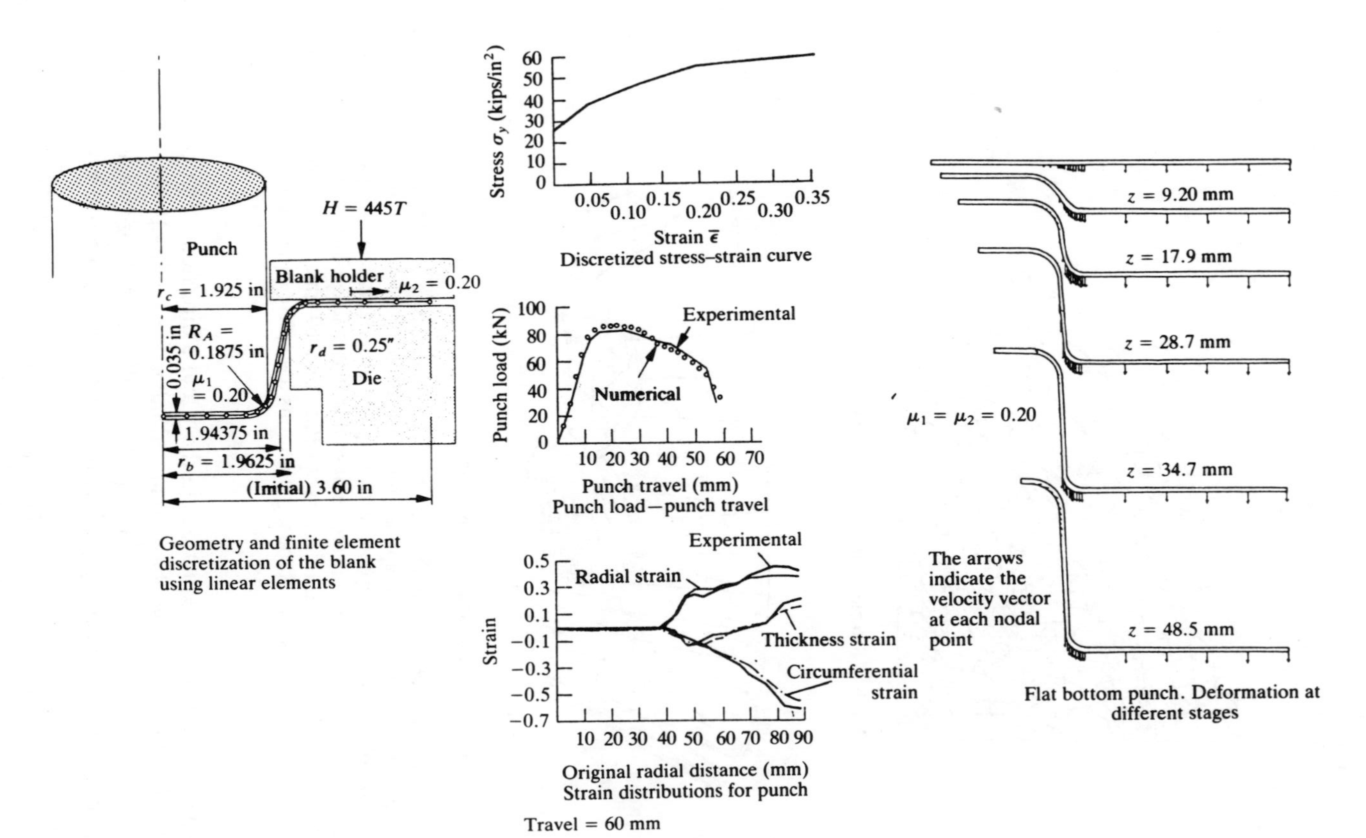

Fig. 13.10 Deep drawing by a flat nosed punch[61]

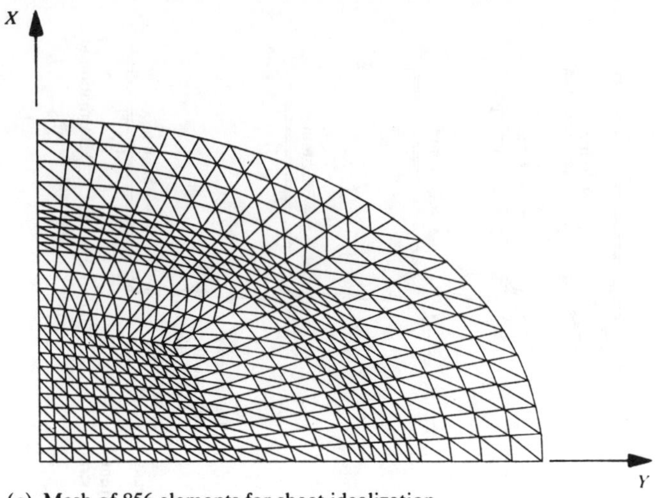

(a) Mesh of 856 elements for sheet idealization

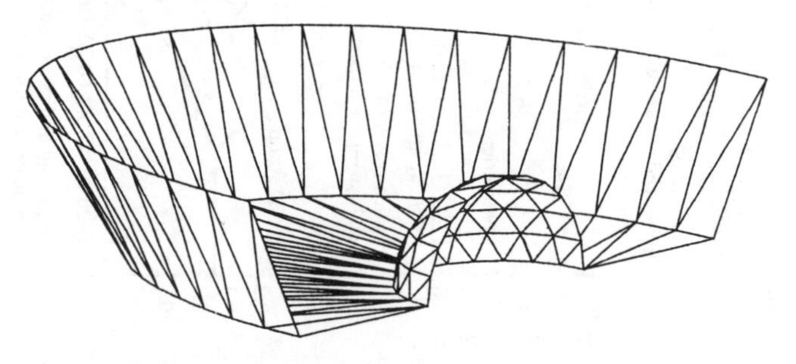

(b) Mesh for establishing die geometry

Fig. 13.11 Finite element simulation of the superplastic forming of a thin sheet component by air pressure application. This example considers the superplastic forming of a truncated ellipsoid with a spherical indent. The original flat blank was 150 by 100 mm. The truncated ellipsoid is 20 mm deep. The original thickness was 1 mm. Minimum final thickness was 0.53 mm. 69 time steps were used with a total of 285 Newton–Raphson iterations (complete equation solutions)[64]

together with similar equations in the y and z directions. (In deriving the second of the above we have assumed incompressibility.)

In the steady state the above equations differ from those we used in the solution of slow creeping problems only by the convective acceleration terms

$$u\frac{\partial u}{\partial x}+v\frac{\partial u}{\partial y}+w\frac{\partial u}{\partial z}$$

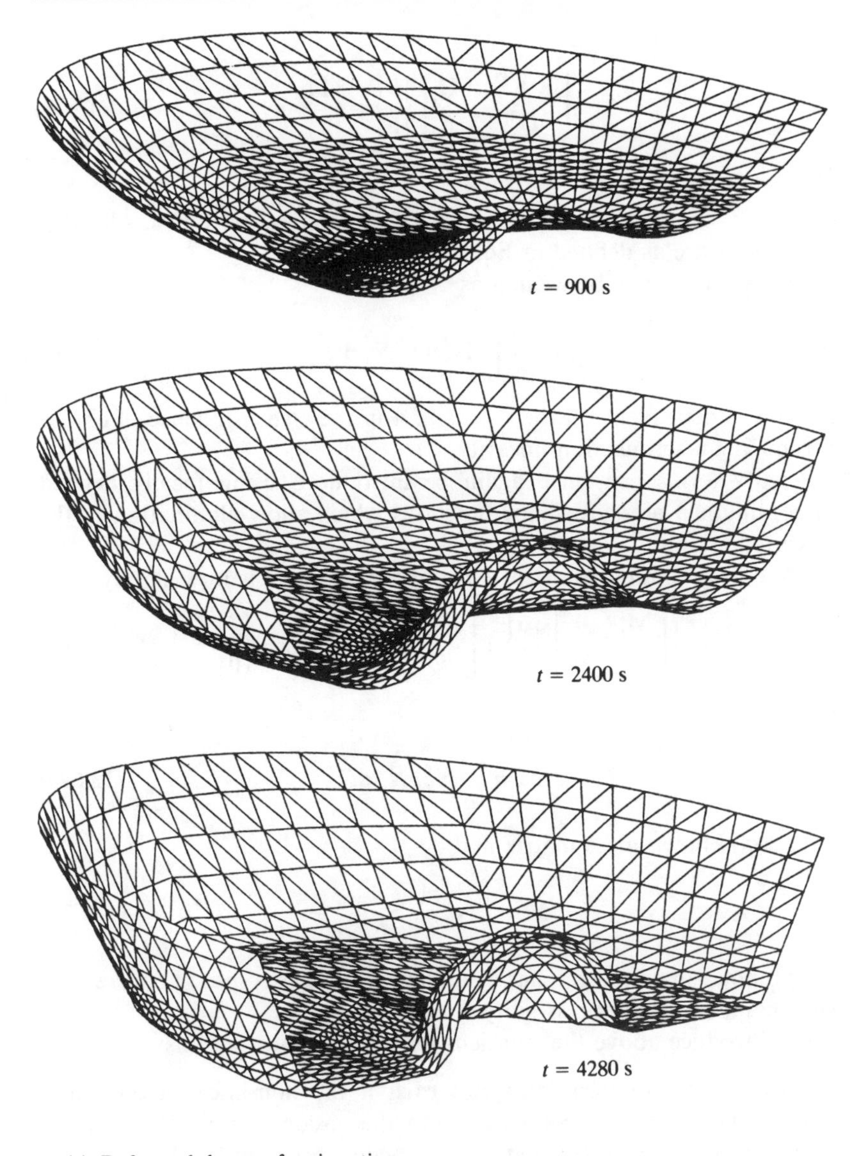

(c) Deformed shapes of various times

Fig. 13.11 (continued)

and hence if these terms are relatively small, the standard Galerkin procedure used in arriving at the discretized forms of Eq. (13.45) can once again be applied.

Now without repeating the previous algebra we can write the steady-state mixed form as

$$\begin{bmatrix} \mathbf{K}+\tilde{\mathbf{K}}, & \mathbf{Q} \\ \mathbf{Q}^{\mathrm{T}}, & \mathbf{0} \end{bmatrix} \begin{Bmatrix} \bar{\mathbf{u}} \\ \bar{\mathbf{p}} \end{Bmatrix} = \begin{Bmatrix} \mathbf{f} \\ 0 \end{Bmatrix} \tag{13.59}$$

with $\qquad \mathbf{u}=\mathbf{N}_u\bar{\mathbf{u}} \qquad \mathbf{p}=\mathbf{N}_p\bar{\mathbf{p}}$

where $\tilde{\mathbf{K}}$ is the additional term arising from convection terms. All other expressions are as defined in Eqs (13.46).

The new term can be written as

$$\tilde{\mathbf{K}} = \int_{\Omega} \mathbf{N}_u^{\mathrm{T}} \mathbf{u} \rho \nabla \mathbf{N}_u \, \mathrm{d}\Omega \tag{13.60}$$

and this is dependent on the values of $\mathbf{u}$ in the element, hence of course introducing a serious non-linearity.

Penalty forms can once again be introduced using locally defined (discontinuous) pressure shape functions which allow elimination of the pressure variable.

The corresponding transient form of the equation is now simply

$$\begin{bmatrix} \mathbf{M}, & \mathbf{0} \\ \mathbf{0}, & \tilde{\mathbf{M}} \end{bmatrix} \begin{Bmatrix} \dot{\bar{\mathbf{u}}} \\ \dot{\bar{\mathbf{p}}} \end{Bmatrix} + \begin{bmatrix} \mathbf{K}+\tilde{\mathbf{K}}, & \mathbf{Q} \\ \mathbf{Q}^{\mathrm{T}}, & \mathbf{0} \end{bmatrix} \begin{Bmatrix} \bar{\mathbf{u}} \\ \bar{\mathbf{p}} \end{Bmatrix} = \begin{Bmatrix} \mathbf{f} \\ \mathbf{0} \end{Bmatrix} \tag{13.61}$$

where

$$\mathbf{M} = \int_{\Omega} \mathbf{N}_u^{\mathrm{T}} \rho \mathbf{N}_u \, \mathrm{d}\Omega \tag{13.62a}$$

is the well-known mass matrix and

$$\tilde{\mathbf{M}} = \int \mathbf{N}_p^{\mathrm{T}} \frac{1}{c^2} \mathbf{N}_p \, \mathrm{d}\Omega \tag{13.62b}$$

though not a mass matrix, has a similar form. $\tilde{\mathbf{M}}$ becomes of course zero when complete incompressibility is assumed.

We introduce above the transient form for various reasons:

1. Steady-state solutions may not exist if the non-linear acceleration terms become large compared with the viscous terms. The ratio of such terms is characterized in a particular problem by the Reynolds number

   $$Re = \frac{UL\rho}{\mu} \equiv \frac{UL}{\nu} \tag{13.63}$$

   where U and L are some typical values of velocity and size and ν is the kinematic viscosity. It is well known that as Re increases, laminar, steady-state flows become unstable, and shedding of eddies, and indeed turbulence, will develop.

2. Even if the steady state exists the transient solution may (and does) provide a convenient, computationally efficient, procedure for obtaining iteratively the steady-state solutions.
3. The problem may indeed be transient and solutions of that type required.

13.7.2 *Direct, steady-state, solution and the Petrov–Galerkin weighting.* Steady-state solutions of Eq. (13.59) were of course the first concern in finite element applications. Proceeding as in the slow Stokes problem the solution of Eq. (13.59) can be obtained, though of course even for newtonian flow matrix non-linearity due to acceleration effects will require an iterative approach. Further, if the non-symmetric nature of $\tilde{\mathbf{K}}$ is noted the system of Eq. (13.59) written as

$$\mathbf{A}\begin{Bmatrix}\bar{\mathbf{u}}\\ \bar{\mathbf{p}}\end{Bmatrix}=\begin{Bmatrix}\bar{\mathbf{f}}\\ 0\end{Bmatrix} \tag{13.64}$$

will require non-symmetric solvers whatever iterative method is adopted.

Direct iteration of the type used in non-newtonian problems [viz. Eq. (13.53)] converges only for very small values of *Re* and for this reason Newton methods are generally preferable, as discussed in Chapter 7.

For relatively low Reynolds numbers satisfactory convergence and accuracy are achieved using the standard Galerkin approximation and many practical solutions have so been obtained.[5-10,71,72] However, as we have already remarked, the convective terms can lead to oscillations in velocity if the Reynolds number is relatively high, even if stable solutions exist. Figure 13.12(*a*)[73] illustrates this by a rather chaotic velocity distribution in a problem solved for a relatively higher Reynolds number. To overcome such oscillations one of the procedures introduced in Chapter 12 needs to be used and, at least for steady-state approximation, the application of Petrov–Galerkin methods is recommended.

We have observed that the Petrov–Galerkin methods involve the use of a weighting function [viz. Eq. (12.45)] of the form

$$\mathbf{W}_u=\mathbf{N}_u+\frac{\alpha}{\|\mathbf{u}\|}u_i\frac{\partial \mathbf{N}_u}{\partial x_i} \tag{13.65}$$

where α is the 'upwind' parameter determined by the value of the element Peclet number.

In the present case the role of the Peclet number is played by the

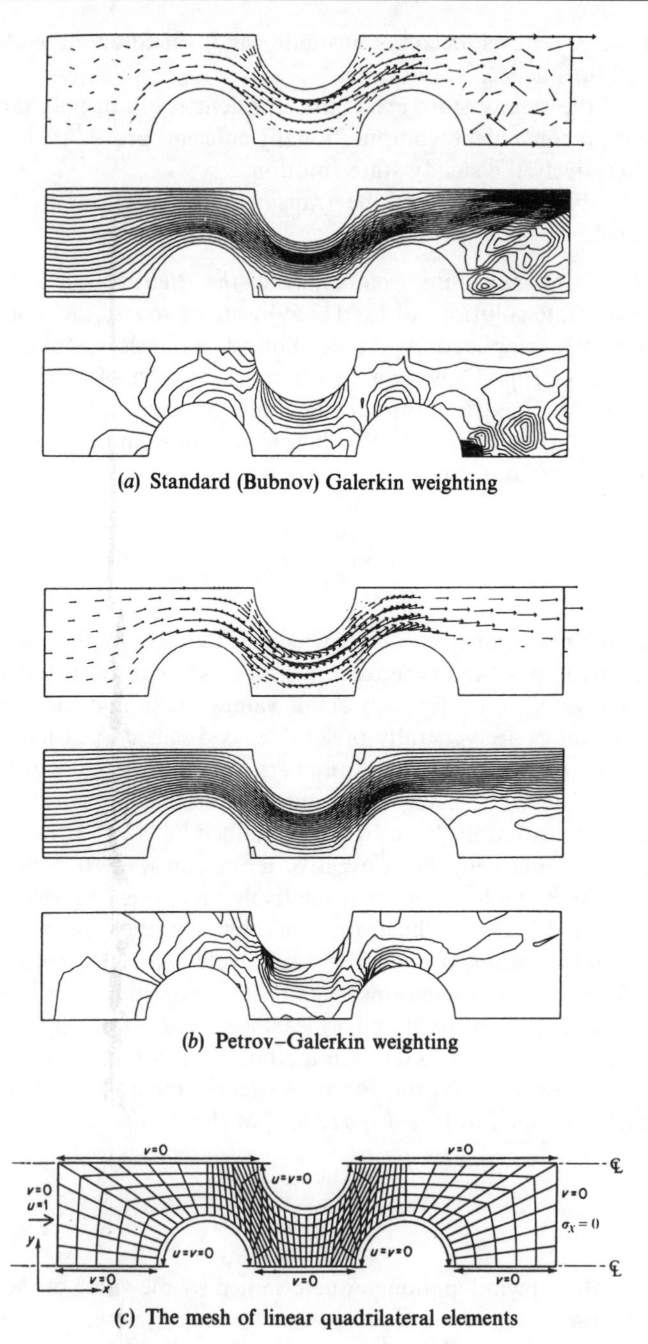

(c) The mesh of linear quadrilateral elements

Fig. 13.12 Viscous flow around a 'tube bundle' at $Re = 500$.[73] Velocity vectors, streamlines and pressure distributions are shown

element Reynolds number:

$$Re_e = \frac{\|u\|h}{2\nu} \qquad \nu = \frac{\mu}{\rho} \tag{13.66}$$

where h is the element size.

Once again the optimal value of the parameter α can be determined by heuristic use of Eq. (12.46):

$$\alpha_{opt} = \coth Re_e - \frac{1}{Re_e} \tag{13.67}$$

Application of the Petrov–Galerkin procedures in the Navier–Stokes problems is advantageous and Fig. 13.12(*b*) shows the improvement of results obtained and the elimination of oscillations.

We note that:

1. Once again only the matrix $\tilde{\mathbf{K}}$ is affected, becoming in place of Eq. (13.60)

$$\tilde{\mathbf{K}} = \int_\Omega \mathbf{W}_u^T \mathbf{u} \rho \nabla \mathbf{N}_u \, d\Omega \tag{13.68}$$

   and the contribution of the weighting of viscous terms is either identically zero (for linear elements) or insignificant for higher-order elements.[73]
2. The upwinding parameter depends now on the unknown values of $\mathbf{u}$ and needs to be updated in every iteration.

It is now generally accepted that Petrov–Galerkin procedures are essential for satisfactory solutions and most codes adopt this practice.[74]

Of course, steady-state solutions obtained by the Navier–Stokes formulation can be used satisfactorily for Stokes solutions, and in most examples previously discussed such general codes were applied. However, the solution cost is increased due to the non-symmetry of matrices which is now included.

13.7.3 *Buoyant convective flows (thermal coupling).* So far, we have considered only the uncoupled (isothermal) incompressible flows in terms of the Navier–Stokes solution. However, it is well known that small density changes due to thermal expansion can be responsible for convective currents such as occur in lakes, atmosphere, etc. To solve such problems one can use the energy equation (13.55) expressed in terms of the temperature variable and solve this either in a staggered manner or simultaneously with the flow equations.

The flow equations (13.58), in which incompressibility is enforced by putting $c^2 = \infty$, need only be modified by introducing a temperature-dependent density in the last term of Eq. (13.58b) (that is ρf_i) and putting

$$\rho = \rho(T) \tag{13.69a}$$

which, with a constant thermal expansion coefficient β, can be written specifically as

$$\rho = \rho_0[1 + \beta(T - T_0)] \tag{13.69b}$$

This particular form of the coupling is known as the Boussinesq approximation.

Many useful studies of thermal circulation have been performed using the above approach.[75-78] Figure 13.13 taken from reference 77 illustrates such thermally induced circulation in a hollow cavity.

13.8 The Navier–Stokes problem—transient solution procedures

13.8.1 *General remarks.* In all the preceding sections we have dealt with steady-state solutions, and generally the direct equation solution process (gaussian elimination in its various forms) has been used even though non-linear iteration was necessary.

The solution of the transient systems given by Eq. (13.61) can of course be obtained by using directly any of the schemes of Chapter 10, but it will immediately be noted that:

1. If the fluid is totally incompressible and the matrix $\tilde{\mathbf{M}}$ is zero only implicit procedures are practical.
2. If the fluid compressibility is finite but small (that is c^2 is large) then explicit schemes are possible but the critical time steps will be very small, necessitating many repeated computations.

For large problems neither procedure is practicable due to prohibitive computation costs. Repeated non-linear iterations at each time step in the case of implicit procedures are not acceptable.

In the case of small compressibility it can easily be shown that the governing critical flow step is of the order of

$$\Delta t_{\text{crit}} = \frac{h}{c} \tag{13.70}$$

i.e. the time necessary for the sound speed wave to traverse the element. With real compressibility of such a fluid as water, for instance, over a million time steps were needed by Hirano *et al.* to obtain a steady-state solution of flow past a sphere[79]—at a large computational expense.

For the above reasons we shall concentrate here on solution schemes which are only in part explicit but where the cost can be greatly reduced as compared with fully implicit procedures. Such schemes are very

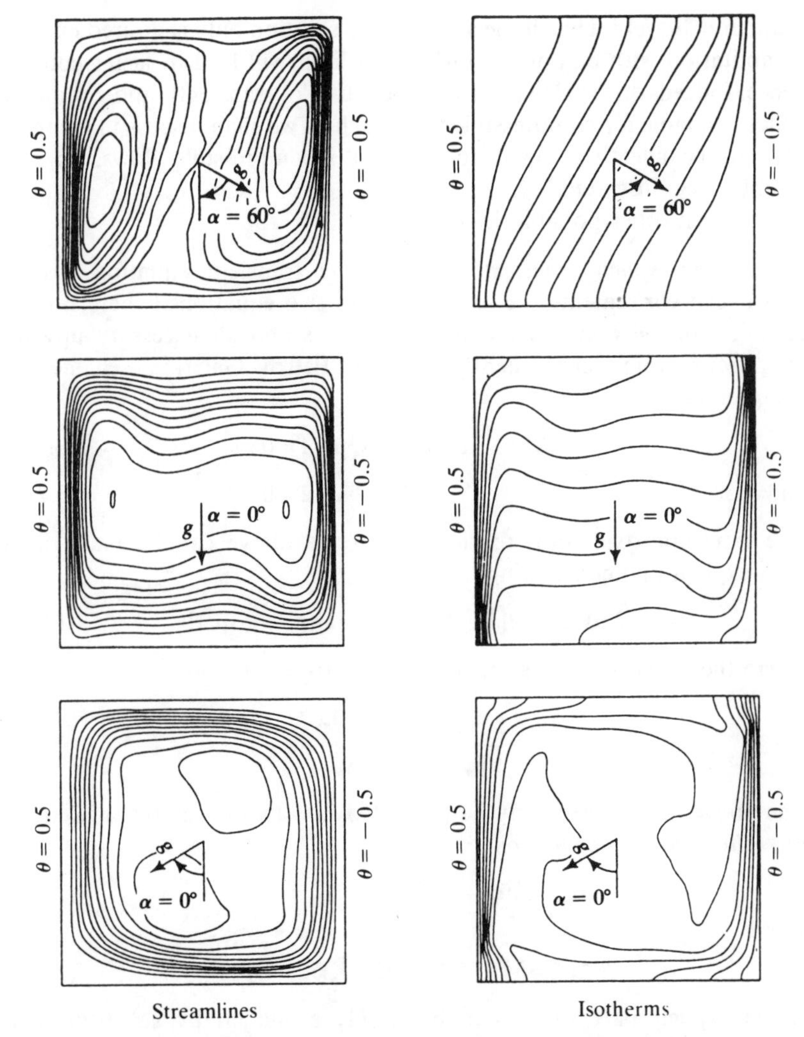

Fig. 13.13 Thermally induced (buoyant) circulation in a rectangular cavity with different gravity directions.[77] Top and bottom of enclosure not conducting. Side temperatures prescribed as shown (Rayleigh number 10^6)

successful and indeed for large problems provide a very economical way of solving steady-state equations in an iterative manner.

In the above context a further remark should be made, which concerns the possible use of *artificial compressibility*, i.e. an artifically changed value of sound velocity c that makes explicit computation feasible. It is found in general that the flow in the steady state is affected only slightly by compressibility providing the Mach number is less than ~ 0.1 (Mach number $= \|u\|/c$).† Thus if the steady-state solution only is desired explicit computation with c approaching such a limit will lead to faster convergence. Indeed, this is also approximately true of transients, and computer codes available for compressible flow, such as we shall discuss in Chapter 15, can be effectively used for solving incompressible flows, as was recently demonstrated.[80,81]

However, other possibilities exist.

13.8.2 *Semi-implicit transient solution.* The transient equation system (13.61) can conveniently be separated into two equations (where now $\tilde{\mathbf{K}}$ contains all the non-linear convection terms and all necessary upwind operations which conveniently use the Petrov–Galerkin schemes of Chapter 12):

$$\mathbf{M}\dot{\bar{\mathbf{u}}} + (\mathbf{K} + \tilde{\mathbf{K}})\bar{\mathbf{u}} + \mathbf{Q}\bar{\mathbf{p}} - \bar{\mathbf{f}} = 0 \tag{13.71a}$$

and

$$\tilde{\mathbf{M}}\dot{\bar{\mathbf{p}}} + \mathbf{Q}^{\mathrm{T}}\bar{\mathbf{u}} = 0 \tag{13.71b}$$

The time approximation to the first of the above can be made in a semi-explicit manner, writing

$$\mathbf{M}\Delta\bar{\mathbf{u}}_n = \Delta t[\bar{\mathbf{f}} - (\mathbf{K} + \tilde{\mathbf{K}})\bar{\mathbf{u}}_n - \mathbf{Q}(\bar{\mathbf{p}}_n + \theta\Delta\bar{\mathbf{p}}_n)] \tag{13.72}$$

where the suffix n denotes the values at t_n, $0 < \theta \leqslant 1$, and

$$\bar{\mathbf{u}}_{n+1} = \bar{\mathbf{u}}_n + \Delta\bar{\mathbf{u}}_n \tag{13.73a}$$

$$\bar{\mathbf{p}}_{n+1} = \bar{\mathbf{p}}_n + \Delta\bar{\mathbf{p}}_n \tag{13.73b}$$

The above can conveniently be separated into two parts (reminiscent of the *split operator application*)

$$\Delta\bar{\mathbf{u}}_n = \Delta\bar{\mathbf{u}}_n^* + \Delta\bar{\mathbf{u}}_n^{**} \tag{13.74}$$

where

$$\Delta\bar{\mathbf{u}}^* = \mathbf{M}^{-1}\Delta t[\bar{\mathbf{f}} - (\mathbf{K} + \tilde{\mathbf{K}})\bar{\mathbf{u}}_n - \mathbf{Q}\bar{\mathbf{p}}_n] \tag{13.75}$$

is the explicit part which can be simply evaluated by substituting a lumped form of $\mathbf{M}$ [and iterating if required for better accuracy as discussed earlier, viz. Eq. (12.106)].

This leaves the determination of

$$\Delta\bar{\mathbf{u}}^{**} = -\mathbf{M}^{-1}\Delta t\theta\mathbf{Q}\Delta\bar{\mathbf{p}}_n \tag{13.76}$$

which will be solved together with an implicit approximation of Eq. (13.71b). This is written as

$$\tilde{\mathbf{M}}\Delta\bar{\mathbf{p}}_n + \mathbf{Q}^{\mathrm{T}}\Delta t[(1-\theta)\bar{\mathbf{u}}_n + \theta\bar{\mathbf{u}}_{n+1}] = 0 \tag{13.77}$$

† If compressibility is so small that equations of the form (13.25) can be adapted it is readily seen that steady-state compressible and incompressible solutions are identical.

Substituting Eqs (13.73a), (13.74) and (13.76) into the above, we have

$$(\tilde{\mathbf{M}} - \Delta t^2 \theta^2 \mathbf{Q}^{\mathrm{T}} \mathbf{M}^{-1} \mathbf{Q}) \Delta \bar{\mathbf{p}}_n = -\mathbf{Q}^{\mathrm{T}} \Delta t (\bar{\mathbf{u}}_n + \theta \Delta \bar{\mathbf{u}}_n^*) \tag{13.78}$$

which can be solved for $\Delta \bar{\mathbf{p}}_n$ even if the fluid is totally incompressible, i.e. when $\tilde{\mathbf{M}} = 0$.

For a totally incompressible fluid Eq. (13.78) simplifies to evaluating

$$\Delta \bar{\mathbf{p}}_n = \frac{1}{\theta^2 \Delta t} [\mathbf{Q}^{\mathrm{T}} \mathbf{M}^{-1} \mathbf{Q}]^{-1} \mathbf{Q}^{\mathrm{T}} (\bar{\mathbf{u}}_n + \theta \Delta \bar{\mathbf{u}}_n^*) \tag{13.79}$$

The corresponding $\Delta \bar{\mathbf{u}}^{**}$ is now computed by expression (13.76) and this completes the calculation cycle in a time step.

The semi-implicit scheme presented above is still conditionally stable but the most severe critical time-step limit imposed by the compressibility (wave speed) [viz. Eq. (13.70)] is no longer applicable providing that $\theta \geqslant \frac{1}{2}$. The limitation of the time step arises from the convective terms contained in $\tilde{\mathbf{K}}$ and the viscous terms contained in $\tilde{\mathbf{K}}$, giving

$$\frac{\|u\| \Delta t}{h} \leqslant \sqrt{\frac{1}{Re} + 1} - \frac{1}{Re} \tag{13.80}$$

where Re is the element Reynolds number of Eq. (13.66). Equation (13.80) is valid for a lumped mass matrix (see page 477, where modification of a consistent mass matrix is used).

This limit is much less severe then that of Eq. (13.70) and indeed may well need to be imposed in order to obtain the desired accuracy.[82]

The computation in a single time step is now very much simpler than with a fully implicit scheme applied to the equation set. It now involves only explicit computation and a repeated solution of a linear, constant, matrix $\mathbf{Q}^{\mathrm{T}} \mathbf{M}^{-1} \mathbf{Q}$. It must, however, be noted that this matrix is of a wider connection than that of the $\mathbf{K}$ matrices and its solution is not trivial in large problems. In the following section we shall present means that can be used to further economize on the solution computation.

The methodology of the above scheme was first introduced in the context of the finite difference approximation of Chorin[82] and was later used extensively in finite element formulation by Donea *et al.*,[83] Hughes *et al.*[84,85] and others.[86-90]

The procedure provides an iterative base for solving steady-state problems described in Sec. 13.7 in an alternative manner.

13.8.3 *An alternative semi-implicit procedure—the pressure laplacian.* The previous procedure constitutes effectively an *operator split*. Starting from a known value of $\bar{\mathbf{u}}_n$ we computed $\Delta \bar{\mathbf{u}}_n^*$ assuming a constant value of $\bar{\mathbf{p}}_n$ in the increment with $\Delta \bar{\mathbf{u}}_n^{**}$ requiring to be determined. In the alternative process we proceed identically up to the point of determining $\Delta \bar{\mathbf{u}}_n^*$ but return to the original differential equations for determination of $\Delta\bar{\mathbf{u}}_n^{**}$.

Thus, omitting the subscript, we return to the not spatially discretized variables and note that Eq. (13.76) stems from (13.58b), and can be written as

$$\frac{\partial \Delta \mathbf{u}^{**}}{\partial t} + \nabla(\Delta p) = 0 \tag{13.81a}$$

Similarly noting that Eq. (13.71b) originated from Eq. (13.58a) we can write

$$\frac{1}{c^2}\frac{\partial \Delta p}{\partial t} + \nabla^{\mathrm{T}}(\mathbf{u}) = 0 \tag{13.81b}$$

noting that $p = p_n + \Delta p$ where p_n is time independent.

Discretization of the above in *time* only gives

$$\Delta \mathbf{u}^{**} + \Delta t \nabla(\theta \Delta p_n) = 0 \tag{13.82a}$$

and

$$\frac{1}{c^2}\Delta p_n + \Delta t \nabla^{\mathrm{T}}(\mathbf{u}_n + \theta \Delta \mathbf{u}_n^{*} + \theta \Delta \mathbf{u}_n^{**}) = 0 \tag{13.82b}$$

From the above we can eliminate $\Delta \mathbf{u}_n^{**}$ and write (noting that $\nabla^{\mathrm{T}} \cdot \nabla \equiv \nabla^2$ is the laplacian operator)

$$\frac{1}{c^2}\Delta p_n + \Delta t \nabla^{\mathrm{T}}(\mathbf{u}_n + \theta \Delta \mathbf{u}_n^{*}) - \Delta t^2 \theta^2 \nabla^2(\Delta p_n) = 0 \tag{13.83}$$

At this stage we perform the spatial discretization and apply standard Galerkin procedure to the above. We put here

$$\Delta p = \mathbf{N}_p \Delta \bar{\mathbf{p}} \quad \mathbf{u}_n = \mathbf{N}_u \bar{\mathbf{u}}_n \quad \Delta \mathbf{u}_n^{*} = \mathbf{N}_u \Delta \bar{\mathbf{u}}_n^{*} \quad \text{and} \quad \Delta \mathbf{u}_n^{**} = \mathbf{N}_u \Delta \bar{\mathbf{u}}_n^{**} \tag{13.84}$$

and obtain, from Eq. (13.83),

$$\left(\tilde{\mathbf{M}} - \Delta t^2 \theta^2 \int_\Omega \mathbf{N}_p^{\mathrm{T}} \nabla^2 \mathbf{N}_p \, \mathrm{d}\Omega\right) \Delta \bar{\mathbf{p}}_n = -\mathbf{Q}^{\mathrm{T}} \Delta t (\bar{\mathbf{u}}_n + \theta \Delta \bar{\mathbf{u}}_n^{*}) \tag{13.85}$$

where $\tilde{\mathbf{M}}$ and $\mathbf{Q}$ are precisely the matrices previously obtained but now a new matrix resulting from the laplacian operator is introduced. Now of course the interpolation of Δp must have C_0 continuity, and integrating above by parts we can write

$$(\tilde{\mathbf{M}} + \Delta t^2 \theta^2 \mathbf{H}) \Delta \bar{\mathbf{p}}_n = -\mathbf{Q}^{\mathrm{T}} \Delta t (\bar{\mathbf{u}}_n + \theta \Delta \bar{\mathbf{u}}_n^{*}) \tag{13.86}$$

in which we have omitted the boundary integral resulting from integration by parts:

$$\left(\int_\Gamma \mathbf{N}_p^{\mathrm{T}} \frac{\partial \mathbf{N}_p}{\partial n} \, \mathrm{d}\Gamma\right) \Delta \bar{\mathbf{p}}_n \tag{13.87}$$

Equation (13.86) is very similar to that of Eq. (13.78) but has the advantage that the matrix **H**, i.e.

$$\mathbf{H}=\int_{\Omega} \nabla \mathbf{N}_p^{\mathrm{T}} \nabla \mathbf{N}_p \, \mathrm{d}\Omega \tag{13.88}$$

has a narrower bandwidth than the corresponding matrix $\mathbf{Q}^{\mathrm{T}}\mathbf{M}^{-1}\mathbf{Q}$ and is much easier to compute.

To complete the computation for a time step it is necessary to determine $\Delta\bar{\mathbf{u}}^{**}$ by discretizing Eq. (13.81a). We now obtain

$$\Delta\bar{\mathbf{u}}^{**}=-\mathbf{M}^{-1}\Delta t\theta \mathbf{Q}\Delta\bar{\mathbf{p}}_n \tag{13.89}$$

which is, not surprisingly, identical to Eq. (13.76).

The computation using this alternative process is simpler than that originally proposed and it appears no less accurate. It is puzzling, however, why it is permissible to ignore the boundary integrals of Eq. (13.87) which is equivalent to assuming that on the boundary Γ either

$$\Delta p=0 \qquad \text{or} \qquad \frac{\partial \Delta p}{\partial n}\equiv 0 \tag{13.90}$$

The reason for this is apparent on examination of Eq. (13.81a) which represents equilibrium conditions in the absence of both viscous and convective forces. We observe that on boundaries where the velocities are prescribed $\Delta\mathbf{u}^{**}\equiv 0$ and hence

$$\nabla(\Delta p)=0 \tag{13.91a}$$

and on traction boundaries neglect of the integral corresponds by simply assuming that no change of tractions occurs. This results in the requirement that

$$\Delta p=0 \tag{13.91b}$$

The use of the laplacian pressure operator in the above context was first introduced in finite differences[91] and later adapted to finite element computation.[92] A generalization to compressible flow is presented in references 93 and 94.

Although the application of the 'laplacian Δp' procedure of this section requires a C_0 pressure continuity to be used in the computation while that of the previous section does not, it is possible to use it in conjunction with discontinuous assumptions in two ways. In the first Δp only is assumed to be C_0 continuous and is interpolated back to the internal Gauss points for the remainder of the computation. In the second the new matrix **H** is used iteratively to provide the solution of Eq. (13.78).

We note that in the computations outlined in the two preceding sections:

1. The time step used is governed purely by the matrix **M** of Eq. (13.62a) and is subject to the restrictions of Eq. (13.80). In steady-state iteration arbitrary values of **M** can, of course, be used to accelerate convergence.
2. The computation is not affected by the presence or absence of compressibility.

13.8.4 *Some examples—exit boundary conditions.* Figures 13.14 and 13.15 show a steady and transient flow of incompressible flow around a cylinder in which the above time-stepping procedures have been used together with adaptive mesh refinement.[95] As we have mentioned before, such procedures are now considered optimal to the solution of both steady and transient flows. The problem incidentally shows the non-existence of steady states at higher Reynolds numbers to which we referred earlier.

In the examples both the inlet and exit boundary conditions need to

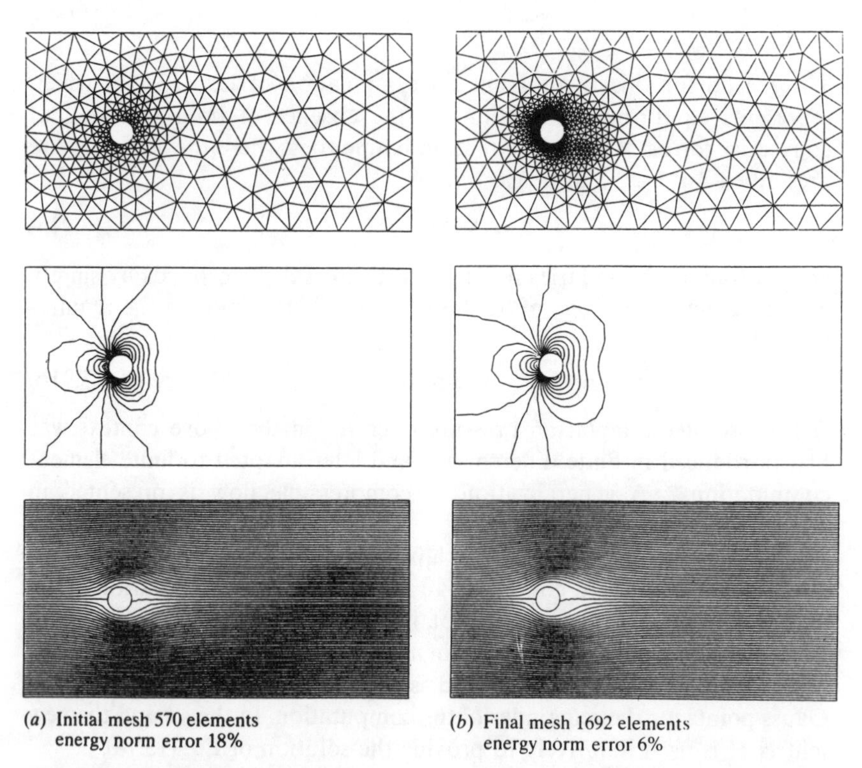

(*a*) Initial mesh 570 elements
energy norm error 18%

(*b*) Final mesh 1692 elements
energy norm error 6%

Fig. 13.14 Steady-state incompressible flow around a cylinder at $Re=60$. Mesh, pressure contours and streamlines with adaptive refinement are shown[95]

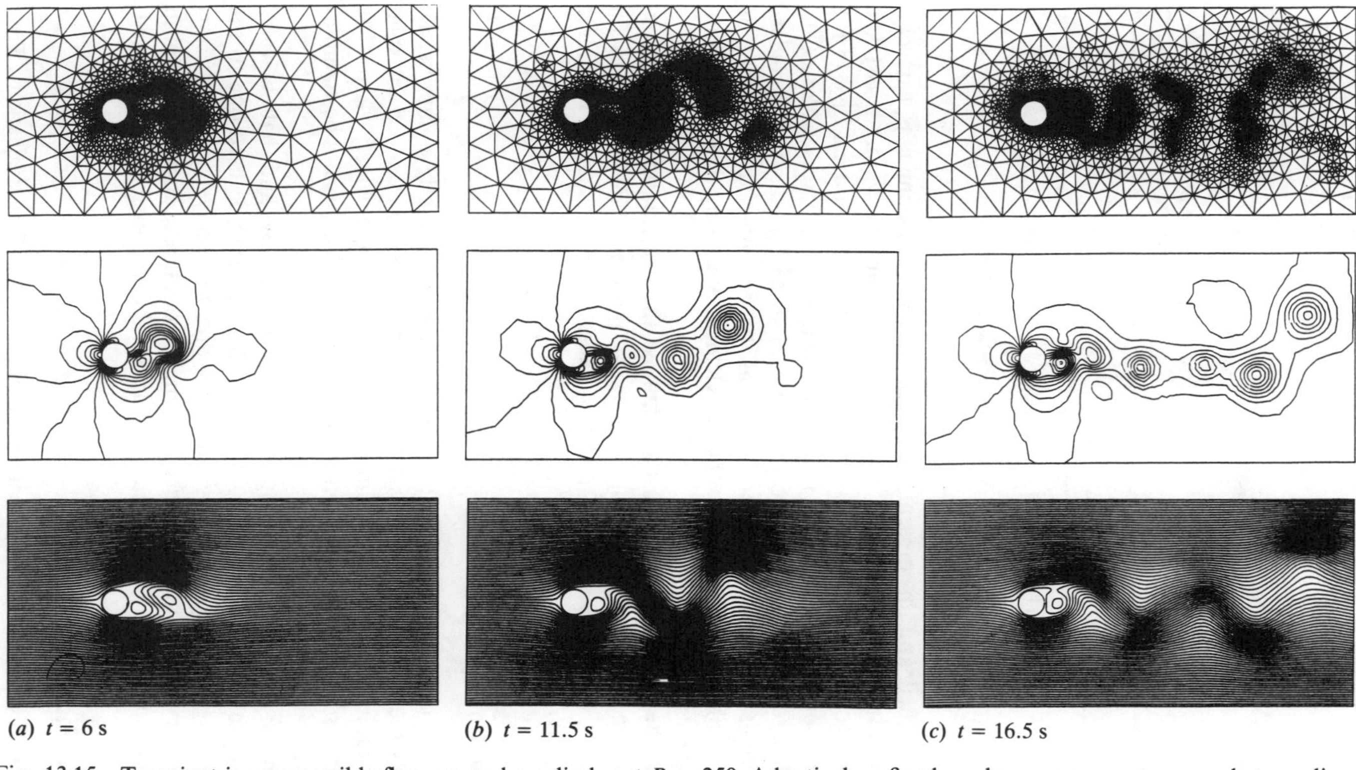

Fig. 13.15 Transient incompressible flow around a cylinder at $Re = 250$. Adaptively refined mesh; pressure contours and streamlines are shown at various times after initiation of motion. Note development of 'vortex street'

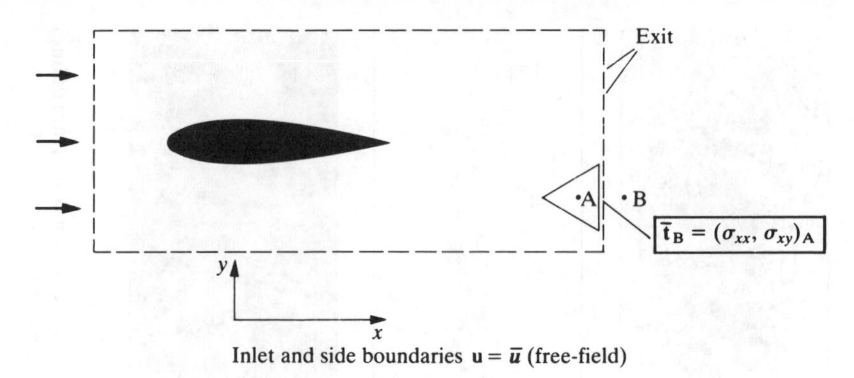

Fig. 13.16 Arbitrary free-field boundaries

be specified, and this presents some difficulties as the domain is 'external'. At the inlet the situation is in general simple and providing it is taken away from the obstacle free-field velocities can be used (viz. Fig. 13.16). At the outlet, however, this is not possible if an oscillatory wake exists, and here it is convenient to assume that a *zero gradient of boundary tractions exists.* Such a condition has to be imposed iteratively by computing the stresses inside the last element and using these as traction on the exit boundary.[93]

In the present chapter we have limited the presentation to discretization methods that are currently most popular. Many other useful alternative approaches exist and the reader will find some in references 96 to 99.

13.9 Concluding remarks

In this chapter dealing with incompressible (or slightly compressible) flows the formulation followed the processes largely developed for solid mechanics and gaussian elimination methods were widely used. The semi-explicit iterative solution is a departure which proves to be economical in larger problems.

In the next chapters dealing with compressible flows and shallow water equations fully explicit computations will invariably be used due to the high degree of non-linearity and also due to relative economy. It is necessary to remind the reader that such procedures can also be used in the context of the present chapter if artificial compressibility is assumed.

References

1. J. T. ODEN, 'A finite element analog of the Navier–Stokes equations', *Proc. Am. Soc. Civ. Eng.*, **96**, EM4, 529–34, 1970.

2. J. T. ODEN, 'The finite element in fluid mechanics, in *Lectures on Finite Element Method in Continuum Mechanics, 1970, Lisbon* (eds J. T. Oden and E. R. A. Oliveira), pp. 151–86, University of Alabama Press, Huntsville, 1973.
3. O. C. ZIENKIEWICZ and C. TAYLOR, 'Weighted residual processes in finite elements with particular reference to some transient and coupled problems', in *Lectures on Finite Element Method in Continuum Mechanics, 1970, Lisbon* (eds J. T. Oden and E. R. A. Oliveira), pp. 415–58, University of Alabama Press, Huntsville, 1973.
4. J. T. ODEN and D. SOMOGYI, 'Finite element applications in fluid dynamics', *Proc. Am. Soc. Civ. Eng.*, **95**, EM4, 821–6, 1969.
5. C. TAYLOR and P. HOOD, 'A numerical solution of the Navier–Stokes equations using the finite element techniques', *Comp. Fluids*, **1**, 73–100, 1973.
6. M. KAWAHARA, N. YOSHIMURA and K. NAKAZAWA, 'Analysis of steady incompressible viscous flow', in *Finite Element Methods for Flow Problems* (eds J. T. Oden, O. C. Zienkiewicz, R. H. Gallagher and C. Taylor), pp. 107–20, University of Alabama Press, Huntsville, 1974.
7. J. T. ODEN, O. C. ZIENKIEWICZ, R. H. GALLAGHER and C. TAYLOR (eds), *Finite Elements in Fluids*, Vols 1 and 2, Wiley, Chichester, 1975.
8. R. H. GALLAGHER, O. C. ZIENKIEWICZ, J. T. ODEN and M. MORANDI CECCHI (eds), *Finite Elements in Fluids*, Vol. 3, Wiley, Chichester, 1978.
9. J. T. ODEN and L. C. WELLFORD, Jr, 'Analysis of viscous flow by the finite element method', *JAIAA*, **10**, 1590–9, 1972.
10. A. J. BAKER, 'Finite element solution algorithm for viscous incompressible fluid dynamics', *Int. J. Num. Meth. Eng.*, **6**, 89–101, 1973.
11. M. D. OLSON, 'Variational finite element methods for two dimensional and Navier–Stokes equations', in *Finite Elements in Fluids* (eds J. T. Oden *et al.*), Vol. 1, pp. 57–72, Wiley, Chichester, 1975.
12. P. HOOD and C. TAYLOR, 'Navier–Stokes equations using mixed interpolation', in *Finite Element Methods for Flow Problems* (eds J. T. Oden, O. C. Zienkiewicz, R. H. Gallagher and C. Taylor), pp. 121–32, University of Alabama Press, Huntsville, 1974.
13. C. HIRSCH, *Numerical Computation of Internal and External Flows*, Vol. 1, Wiley, Chichester, 1988.
14. P. J. ROACH, *Computational Fluid Mechanics*, Hermosa Press, Albuquerque, New Mexico, 1972.
15. J. H. ARGYRIS and G. MARECZEK, 'Potential flow analysis by finite elements', *Ing. Archiv*, **41**, 1–25, 1972.
16. P. BETTESS and J. A. BETTESS, 'Analysis of free surface flows using isoparametric finite elements', *Int. J. Num. Meth. Eng.*, **19**, 1675–89, 1983.
17. T. SARPKAYA and G. HIRIART, 'Finite element analysis of jet impingement on axisymmetric curved deflectors', in *Finite Elements in Fluids* (eds J. T. Oden, O. C. Zienkiewicz, R. H. Gallagher and C. Taylor), Vol. 2, chap. 14, pp. 265–79, Wiley, Chichester, 1975.
18. M. J. O'CARROLL, 'A variational principle for ideal flow over a spillway', *Int. J. Num. Meth. Eng.*, **15**, 767–89, 1980.
19. D. S. MALKUS and T. J. R. HUGHES, 'Mixed finite element method reduced and selective integration techniques: a unification of concepts', *Comp. Meth. Appl. Mech. Eng.*, **15**, 63–81, 1978.
20. O. C. ZIENKIEWICZ and P. N. GODBOLE, 'Viscous, incompressible flow with special reference to non-Newtonian (plastic) fluids, in *Finite Elements in Fluids* (eds J. T. Oden *et al.*), Vol. 2, pp. 25–55, Wiley, Chichester, 1975.
21. T. J. R. HUGHES, R. L. TAYLOR and J. F. LEVY, 'A finite element method for

incompressible flows', in *Proc. 2nd Int. Symp. on Finite Elements in Fluid Problems ICCAD*, Sta Margharita Ligure, Italy, pp. 1–6, 1976.
22. O. C. ZIENKIEWICZ and P. N. GODBOLE, 'A penalty function approach to problems of plastic flow of metals with large surface deformation', *J. Strain Anal.*, **10**, 180–3, 1975.
23. O. C. ZIENKIEWICZ and S. NAKAZAWA, 'The penalty function method and its applications to the numerical solution of boundary value problems', *ASME, AMD*, **51**, 157–79, 1982.
24. J. T. ODEN, 'R.I.P. methods for Stokesian flow', in *Finite Elements in Fluids* (eds R. H. Gallagher, D. N. Norrie, J. T. Oden and O. C. Zienkiewicz), Vol. 4, chap. 15, pp. 305–18, Wiley, Chichester, 1982.
25. O. C. ZIENKIEWICZ, J. P. VILOTTE, S. TOYOSHIMA and S. NAKAZAWA, 'Iterative method for constrained and mixed approximation. An inexpensive improvement of FEM performance', *Comp. Meth. Appl. Mech. Eng.*, **51**, 3–29, 1985.
26. O. C. ZIENKIEWICZ, Y. C. LIU and G. C. HUANG, 'Error estimates and convergence rates for various incompressible elements', *Int. J. Num. Meth. Eng.*, **28**, 2191–202, 1989.
27. M. FORTIN, 'Old and new finite elements for incompressible flow', *Int. J. Num. Meth. Fluids*, **1**, 347–64, 1981.
28. M. FORTIN, and N. FORTIN, 'Newer and newer elements for incompressible flow', in *Finite Elements in Fluids* (eds. R. H. Gallagher, G. F. Carey, J. T. Oden and O. C. Zienkiewicz), Vol. 6, chap. 7, pp. 171–88, Wiley, Chichester, 1985.
29. M. S. ENGLEMAN, R. L. SANI, P. M. GRESHO and H. BERCOVIER, 'Consistent v. reduced integration penalty methods for incompressible media using several old and new elements', *Int. J. Num. Meth. Fluids*, **4**, 25–42, 1982.
30. K. PALIT and R. T. FENNER, 'Finite element analysis of two dimensional slow non-Newtonian flows', *AIChE J.*, **18**, 1163–9, 1972.
31. K. PALIT and R. T. FENNER, 'Finite element analysis of slow non-Newtonian channel flow', *AIChE J.*, **18**, 628–33, 1972.
32. B. ATKINSON, C. C. M. CARD and B. M. IRONS, 'Application of the finite element method to creeping flow problems', *Trans. Inst. Chem. Eng.*, **48**, 276–84, 1970.
33. G. C. CORNFIELD and R. H. JOHNSON, 'Theoretical prediction of plastic flow in hot rolling including the effect of various temperature distributions', *J. Iron and Steel Inst.*, **211**, 567–73, 1973.
34. C. H. LEE and S. KOBAYASHI, 'New solutions to rigid plastic deformation problems using a matrix method', *Trans. ASME, J. Eng. for Ind.*, **95**, 865–73, 1973.
35. J. T. ODEN, D. R. BHANDARI, G. YAGEWA and T. J. CHUNG, 'A new approach to the finite element formulation and solution of a class of problems in coupled thermoelastoviscoplasticity of solids', *Nucl. Eng. Des.*, **24**, 420, 1973.
36. O. C. ZIENKIEWICZ, E. OÑATE and J. C. HEINRICH, *Plastic Flow in Metal Forming*, pp. 107–20, ASME, San Francisco, December 1978.
37. O. C. ZIENKIEWICZ, E. OÑATE and J. C. HEINRICH, 'A general formulation for coupled thermal flow of metals using finite elements', *Int. J. Num. Meth. Eng.*, **17**, 1497–514, 1981.
38. N. REBELO and S. KOBAYASHI, 'A coupled analysis of viscoplastic deformation and heat transfer: I. Theoretical consideration; II. Application', *Int. J. Mech. Sci.*, **22**, 699–705, 1980 and **22**, 707–718, 1980.

39. P. R. DAWSON and E. G. THOMPSON, 'Finite element analysis of steady state elastoviscoplastic flow by the initial stress rate method', *Int. J. Num. Meth. Eng.*, **12**, 47–57, 382–3, 1978.
40. Y. SHIMIZAKI and E. G. THOMPSON, 'Elasto-visco-plastic flow with special attention to boundary conditons', *Int. J. Num. Meth. Eng.*, **17**, 97–112, 1981.
41. O. C. ZIENKIEWICZ, P. C. JAIN and E. OÑATE, 'Flow of solids during forming and extrusion: some aspects of numerical solutions', *Int. J. Solids Struct.*, **14**, 15–38, 1978.
42. S. NAKAZAWA, J. F. T. PITTMAN and O. C. ZIENKIEWICZ, 'Numerical solution of flow and heat transfer in polymer melts', in *Finite Elements in Fluids* (ed. R. H. Gallagher *et al.*), Vol. 4, chap. 13, pp. 251–83, Wiley, Chichester, 1982.
43. S. NAKAZAWA, J. F. T. PITTMAN and O. C. ZIENKIEWICZ, 'A penalty finite element method for thermally coupled non-Newtonian flow with particular reference to balancing dissipation and the treatment of history dependent flows', in *Int. Symp. of Refined Modelling of Flows*, 7–10 Sept. 1982, Paris.
44. R. E. NICKELL, R. I. TANNER and B. CASWELL, 'The solution of viscous incompressible jet and free surface flows using finite elements', *J. Fluid Mech.*, **65**, 189–206, 1974.
45. R. I. TANNER, R. E. NICKELL and R. W. BILGER, 'Finite element method for the solution of some incompressible non-Newtonian fluids mechanics problems with free surface', *Comp. Meth. Appl. Mech. Eng.*, **6**, 155–74, 1975.
46. J. M. ALEXANDER and J. W. H. PRICE, 'Finite element analysis of hot metal forming', in *18th MTDR Conf.*, pp. 267–74, 1977.
47. S. I. OH, G. D. LAHOTI and A. T. ALTAN, 'Application of finite element method to industrial metal forming processes', in *Proc. Conf. on Industrial Forming Processes*, pp. 146–53, Pineridge Press, Swansea, 1982.
48. J. F. T. PITTMAN, O. C. ZIENKIEWICZ, R. D. WOOD and J. M. ALEXANDER (eds), *Numerical Analysis of Forming Processes*, Wiley, Chichester, 1984.
49. O. C. ZIENKIEWICZ, 'Flow formulation for numerical solutions of forming problems', in *Numerical Analysis of Forming Processes* (eds J. F. T. Pittman *et al.*), chap. 1, pp. 1–44, Wiley, Chichester, 1984.
50. S. KOBAYASHI, 'Thermoviscoplastic analysis of metal forming problems by the finite element method', in *Numerical Analysis of Forming Processes* (eds J. F. T. Pittman *et al.*), chap. 2, pp. 45–70, Wiley, Chichester, 1984.
51. W. JOHNSON and P. B. MELLOR, *Engineering Plasticity*, Van Nostrand-Reinhold, London, 1973.
52. G. C. HUANG, 'Error estimates and adaptive remeshing in finite element analysis of forming processes', Ph.D. thesis, University of Wales, Swansea, 1989.
53. G. C. HUANG, Y. C. LIU and O. C. ZIENKIEWICZ, 'Error control, mesh updating schemes and automatic adaptive remeshing for finite element analysis of unsteady extrusion processes', in *Modelling of Metal Forming Processes* (eds J. L. Chenot and E. Oñate), pp. 75–83, Kluwer Academic, Dordrecht, 1988.
54. O. C. ZIENKIEWICZ, Y. C. LIU and G. C. HUANG, 'An error estimate and adaptive refinement method for extrusion and other forming problems,' *Int. J. Num. Meth. Eng.*, **25**, 23–42, 1988.
55. O. C. ZIENKIEWICZ, Y. C. LIU, J. Z. ZHU and S. TOYOSHIMA, 'Flow formulation for numerical solution of forming processes. II—Some new directions', in *Proc. 2nd Int. Conf. on Numerical Methods in Industrial Forming Processes, NUMIFORM 86* (eds K. Mattiasson, A. Samuelsson, R.

D. Wood and O. C. Zienkiewicz), A. A. Balkema, Rotterdam, 1986.
56. T. BELYTCHKO, D. P. FLANAGAN and J. M KENNEDY, 'Finite element methods with user controlled mesh for fluid structure interaction', *Comp. Meth. Appl. Mech. Eng.*, **33**, 669–88, 1982.
57. J. DONEA, S. GIULIANI and J. I. HALLEUX, 'An arbitrary Lagrangian–Eulerian finite element method for transient dynamic fluid-structure interaction', *Comp. Meth. Appl. Mech. Eng.*, **75**, 195–214, 1989.
58. P. J. G. SCHREURS, F. E. VELDPAUS and W. A. M. BRAKALMANS, 'An Eulerian and Lagrangian finite element model for the simulation of geometrical non-linear hyper elastic and elasto-plastic deformation processes', in *Proc. Conf. on Industrial Forming Processes*, pp. 491–500, Pineridge Press, Swansea, 1983.
59. J. DONEA, 'Arbitrary Lagrangian–Eulerian finite element methods', in *Computation Methods for Transient Analysis* (eds T. Belytchko and T. J. R. Hughes), chap. 10, pp. 474–516, Elsevier, Amsterdam, 1983.
60. J. VAN DER LUGT and J. HUETNIK, 'Thermo-mechanically coupled finite element analysis in metal forming processes', *Comp. Meth. Appl. Mech. Eng.*, **54**, 145–60, 1986.
61. E. OÑATE and O. C. ZIENKIEWICZ, 'A viscous sheet formulation for the analysis of thin sheet metal forming', *Int. J. Mech. Sci.*, **25**, 305–35, 1983.
62. N. M. WANG and B. BUDIANSKY, 'Analysis of sheet metal stamping by a finite element method', *Trans. ASME, J. Appl. Mech.*, **45**, 73, 1976.
63. A. S. WIFI, 'An incremented complete solution of the stretch forming and deep drawing of a circular blank using a hemispherical punch', *Int. J. Mech. Sci.*, **18**, 23–31, 1976.
64. J. BONET, R. D. WOOD and O. C. ZIENKIEWICZ, 'Time stepping schemes for the numerical analysis of superplastic forming of thin sheets', in *Modelling of Metal Forming Processes* (eds J. L. Chenot and E. Oñate), pp. 179–86, Kluwer Academic, Dordrecht, 1988.
65. E. MASSONI, M. BELLET and J. L. CHENOT, 'Thin sheet forming numerical analysis with membrane approach', in *Modelling of Metal Forming Processes* (eds J. L. Chenot and E. Oñate), Kluwer Academic, Dordrecht, 1988.
66. R. D. WOOD, J. BONET and A. H. S. WARGEDIPURA, 'Simulation of the superplastic forming of thin sheet components using the finite element method', in *NUMIFORM 89 Proc.*, pp. 85–94, Balkhema Press, 1989.
67. J. BONET, R. D. WOOD and O. C. ZIENKIEWICZ, 'Finite element modelling of the superplastic forming of a thin sheet', *Proc. Conf. on Superquality and Superplastic Forming* (eds C. H. Hamilton and N. E. Paton), The Minerals, Metals and Materials Society, USA, 1988.
68. R. KEUNINGS and A. J. CROCHET, 'Numerical simulation of the flow of a viscoelastic fluid through an abrupt contraction', *J. Non-Newtonian Fluid Mechanics*, **14**, 279–99, 1984.
69. Y. SHIMAZAKI and E. G. THOMPSON, 'Elasto-viscoplastic flow with special attention to boundary conditions', *Int. J. Num. Meth. Eng.*, **17**, 97–112, 1981.
70. R. I. TANNER, *Engineering Rheology*, Clarendon Press, Oxford, 1988.
71. P. M. GRESHO and R. LEE, 'Don't suppress the wiggles, they're telling you something', *Computers and Fluids*, **9**, 223–55, 1981.
72. R. LEE, P. M. GRESHO, S. CHAN, R. SANI and M. CULLEN, 'Conservation law for primitive variable formulations of the incompressible flow equation using the Galerkin finite element methods', in *Finite Elements in Fluids* (eds R. H. Gallagher, D. H. Norrie, J. T. Oden and O. C. Zienkiewicz), Vol. 4, chap. 2, pp. 21–46, Wiley, Chichester, 1982.

73. O. C. Zienkiewicz, R. Löhner, K. Morgan and S. Nakazawa, 'Finite elements in fluid mechanics—a decade of progress', in *Finite Elements in Fluids* (eds R. H. Gallagher *et al.*), Vol. 5, chap. 1, pp. 1–26, Wiley, Chichester, 1984.
74. A. N. Brooks and T. J. R. Hughes, 'Streamline upwind Petrov–Galerkin formulation for convection dominated flows with particular references to the incompressible Navier Stokes equation', *Comp. Meth. Appl. Mech. Eng.*, **32**, 119–259, 1982.
75. O. C. Zienkiewicz, R. H. Gallagher and P. Hood, 'Newtonian and non-Newtonian viscous incompressible flow. Temperature induced flows. Finite element solutions', in *Mathematics of Finite Elements and Applications* (ed. J. Whiteman), Vol. II, chap. 20, pp. 235–67, Academic Press, London, 1976.
76. J. C. Heinrich, R. S. Marshall and O. C. Zienkiewicz, 'Penalty function solution of coupled convective and conductive heat transfer', in *Numerical Methods in Laminar and Turbulent Flows* (eds. C. Taylor, K. Morgan and C. A. Brebbia), pp. 435–47, Pentech Press, 1978.
77. M. Strada and J. C. Heinrich, 'Heat transfer rates in natural convection at high Rayleigh numbers in rectangular enclosures', *Num. Heat Transfer*, **5**, 81–92, 1982.
78. J. C. Heinrich and C. C. Yu, 'Finite element simulation of bouyancy driven flow with emphasis on natural convection in a horizonal circular cylinder', *Comp. Meth. Appl. Mech. Eng.*, **69**, 1–27, 1988.
79. H. Hirano, H. Hara and M. Kawahara, 'Two step finite element methods for high Reynolds number viscous flows, in *Proc. 4th Int. Symp. on Finite Elements in Flow Analysis*, pp. 121–8, University of Tokyo Press, 1982.
80. E. Gode and I. L. Rhyming, '3-D computation of the flow in a Francis runner', *Sulzer Tech. Rev.*, **4**, 31–5, 1987.
81. A. Saxer, H. Felici, C. Newry and I. L. Rhyming, 'Euler flows in hydraulic machinery and ducts related to boundary condition formulation', Swiss Federation Institute of Technology, Lausanne, Switzerland, 1987.
82. A. J. Chorin, 'A numerical method for solving incompressible viscous problems', *J. Comp. Phys.*, **2**, 12–26, 1967.
83. J. Donea, S. Giuliani and H. Laval, 'Finite element solution of the unsteady Navier–Stokes equation by the fractional step method', *Comp. Meth. Appl. Mech. Eng.*, **33**, 53–73, 1982.
84. T. J. R. Hughes, W. K. Liu and T. K. Zimmerman, 'Lagrangian–Eulerian finite element formulations for incompressible viscous flows', *Comp. Meth. Appl. Mech. Eng.*, **29**, 329–49, 1981.
85. T. J. R. Hughes, W. K. Liu and A. Brooks, 'Review of finite element analysis of incompressible viscous flows by the penalty function formulation', *J. Comp. Phys.*, **30**, 1–60, 1979.
86. J. Kim and P. Moin, 'Applications of fractional step methods to incompressible Navier–Stokes equations', *J. Comp. Phys.*, **5**, 308, 1985.
87. P. M. Gresho, R. Chan, C. Upson and R. Lee, 'A modified finite element method for solving the time dependent incompressible Navier–Stokes equations', *Int. J. Num. Meth. Fluids*, pt. I, **4**, 557–618, pt. II, **4**, 619–640, 1984.
88. P. M. Gresho and R. Sani, 'On pressure boundary conditions for the incompressible Navier–Stokes equation', *Int. J. Num. Meth. Fluids*, **7**, 1111–48, 1987.
89. G. Schneider, G. Rathby and M. M. Yovanovich, 'Finite element analysis of incompressible fluid incorporating equal order pressure and velocity

interpolation', in *Numerical Methods in Laminar and Turbulent Flow* (eds C. Taylor, K. Morgan and C. A. Brebbia), Pentech Press, Plymouth, 1978.
90. M. KAWAHARA and K. OHMIYA, 'Finite element analysis of density flow using the velocity correction method', *Int. J. Num. Meth. Fluids*, **5**, 981–93, 1985.
91. S. V. PATANKAR and D. B. SPALDING, 'A calculation procedure for heat mass and momentum transfer in 3-D parabolic flows', *Int. J. Heat and Mass Transfer*, **15**, 1787–806, 1972.
92. G. COMINI and S. DEL GUIDICE, 'Finite element solution of the incompressible Navier–Stokes equation', *Num. Heat Transfer*, **5**, 463–78, 1972.
93. O. C. ZIENKIEWICZ, J. SZMELTER and J. PERAIRE, 'Compressible and incompressible flow. An algorithm for all seasons', *Comp. Mech. Appl. Mech. Eng.*, **78**, 105–21, 1990.
94. J. SZMELTER, 'Computational methods for compressible and incompressible fluid mechanics', Ph.D. thesis, University of Wales, Swansea, 1989.
95. J. WU, Z. J. ZHU, J. SZMELTER and O. C. ZIENKIEWICZ, 'Error estimation and adaptivity in Navier–Stokes incompressible flows', *Comp. Mech.*, **6**, 259–70, 1990.
96. R. TEMAM, *Navier–Stokes Equation*, North Holland, 1977.
97. M. O. BRISTEAU, R. GLOWINSKI and J. PERIAUX, 'Numerical method for the Navier–Stokes equation. Application to the simulation of compressible and incompressible viscous flow', *Comp. Phys. Reports*, **6**, 73–187, 1987.
98. R. GLOWINSKI, B. MANTEL and J. PERIAUX, 'Numerical solution of the time dependent Navier–Stokes equation for incompressible viscous fluids by finite elements and alternating direction methods', in *Numerical Methods in Aeronautics; Fluid Dynamics* (ed. P. L. Roe), pp. 309–36, Academic Press, London, 1982.
99. T. E. TEZDUYAR and J. LIOU, 'Adaptive implicit–explicit finite element algorithms for fluid mechanics problems', in *Recent Developments in Computational Fluid Mechanics* (eds T. E. Tezduyar and T. J. R. Hughes), AMD 95, pp. 163–84, American Society of Mechanical Engineers, 1988.

14

Compressible high-speed gas flow

14.1 Introduction

Problems posed by high-speed gas flow are of obvious importance. Applications range from the *exterior flows* associated with flight to *interior flows* typical of turbomachinery. As the cost of physical experiments is high, the possibilities of computations have been explored early and the development concentrated on the use of finite difference and associated finite volume methods. It is only in recent years that the potential offered by the finite element forms have been realized and the field is expanding rapidly.

One of the main advantages in the use of finite element approximation here is its capability of fitting complex forms and permitting local refinement where required. However, the improved approximation is also of substantial importance as practical problems will often involve three-dimensional discretization with the number of degrees of freedom much larger than those encountered in typical structural problems (10^5–10^6 DOF are here quite typical).

For such large problems direct solution methods are obviously not practicable and iterative methods based generally on transient computation forms are invariably used. Here of course we follow and accept much that has been established by the finite difference applications but generally will lose some computationable efficiency associated with *structured meshes* typically used there. However, the reduction of problem size which, as we shall see, can be obtained by local refinement and adaptivity will more than compensate for this loss (though of course structured meshes are not excluded in the finite element forms).

In the previous chapters we have introduced the basic equations governing the flow of compressible gases and indeed have shown how small amounts of compressibility can be incorporated into the procedures developed there specifically for incompressible flow. Here we shall deal with high-speed flows with Mach numbers generally in excess of 0.6. Such flows will usually involve the formation of shocks with characteristic discontinuities. For this reason we shall concentrate on the use of low-order elements and of explicit methods, following the approximation techniques developed in Chapter 12.

Here the pioneering work of the first author's colleagues Morgan, Löhner and Peraire must be acknowledged. It was this work that opened the doors to practical finite element analysis in the field of aeronautics. We shall refer to their work frequently and indeed the numerous examples presented in this chapter illustrate the effectiveness of this work.

14.2 The governing equations

The Navier–Stokes governing equations for compressible flow were derived in the previous chapter [viz. Eqs (13.21) and (13.22)] and we shall repeat these here using the indicial notation form only. We thus write, for $i=1,2,3,$

$$\frac{\partial \mathbf{U}}{\partial t}+\frac{\partial \mathbf{F}_i}{\partial x_i}+\frac{\partial \mathbf{G}_i}{\partial x_i}+\mathbf{Q}=0 \tag{14.1}$$

with

$$\mathbf{U}^{\mathrm{T}}=[\rho, \rho u_1, \rho u_2, \rho u_3, \rho E] \tag{14.2a}$$

$$\mathbf{F}_i^{\mathrm{T}}=[\rho u_i, \rho u_1 u_i+p\delta_{1i}, \rho u_2 u_1+p\delta_{2,i}, \rho u_3 u_i+p\delta_{3i}, \rho H u_i] \tag{14.2b}$$

$$\mathbf{G}_i^{\mathrm{T}}=\left[0, -\tau_{1i}, -\tau_{2i}, -\tau_{3i}, -\frac{\partial}{\partial x_i}(\tau_{ij}u_i)-k\left(\frac{\partial T}{\partial x_i}\right)\right] \tag{14.2c}$$

and $$\mathbf{Q}^{\mathrm{T}}=[0, -\rho f_1, -\rho f_2, -\rho f_3, -\rho f_i u_i-q_H] \tag{14.2d}$$

In the above $$\tau_{ij}=\mu\left[\left(\frac{\partial u_i}{\partial x_j}+\frac{\partial u_j}{\partial x_i}\right)-\delta_{ij}\frac{2}{3}\frac{\partial u_i}{\partial x_i}\right] \tag{14.2e}$$

The above equations need to be 'closed' by addition of a constitutive law relating the pressure, density and energy [viz. Eqs (13.13) and (13.14)]. For many flows the ideal gas law[1] suffices and this is

$$\rho=\frac{p}{RT} \tag{14.3}$$

where R is the universal gas constant.

In terms of specific heats

$$R(c_p-c_v)=(\gamma-1)c_v \tag{14.4}$$

where

$$\gamma = \frac{c_p}{c_v}$$

is the ratio of constant pressure and constant volume specific heat.

The internal energy e is given as

$$e = c_v T = \left(\frac{1}{\gamma - 1}\right)\frac{p}{\rho} \tag{14.5}$$

and hence

$$\rho E = \left(\frac{1}{\gamma - 1}\right) p + \frac{u_i u_i}{2} \tag{14.6a}$$

$$\rho H = \rho E + p = \left(\frac{\gamma}{\gamma - 1}\right) p + \frac{u_i u_i}{2} \tag{14.6b}$$

The variables for which we shall solve are usually taken as the set of Eq. (14.2a), i.e.

$$\rho, \ \rho u_i \text{ and } \rho E$$

but of course other sets could be used, though then the conservative form of Eq. (14.1) would be lost.

In many of the problems discussed in this section inviscid behaviour will be assumed, with

$$\mathbf{G}_i = 0$$

and we shall deal then with Euler equations.

In many problems the Euler solution will provide information about the main features of the flow and will suffice for some purposes, especially if augmented by separate boundary layer calculations. However, in principle it is possible to include the viscous effects without much apparent complication. Here in general steady-state conditions will never arise as the high speed of flow will be associated with turbulence and this will usually be of a small scale capable of resolution with very small sized elements only. If a 'finite' size of element mesh is used then such turbulence will often be suppressed and steady-state answers will be obtained in areas of no flow separation. We shall in some examples include such full Navier–Stokes solutions using a viscosity dependent on the temperature according to Sutherland's law.[1] In the SI system of units for air this gives

$$\mu = \frac{1.45 \, T^{3/2}}{T + 110} \times 10^{-6} \tag{14.7}$$

where T is in kelvin.

14.3 Boundary conditions—subsonic and supersonic flow

The question of boundary conditions which can be prescribed for Euler and Navier–Stokes equations in compressible flow is by no means trivial and has been addressed in a general sense by Demkowicz *et al.*,[2] determining their influence on existence and uniqueness of solution. In the following we shall discuss the case of inviscid Euler form and of the full Navier–Stokes problem separately.

14.3.1 *Euler equation.* Here only first-order derivatives occur and the number of boundary conditions is less than that for the full Navier–Stokes problem.

For a *solid wall boundary*, Γ_u, only the normal component of velocity u needs to be specified (zero if the wall is stationary). Further, with lack of conductivity the energy flux across the boundary is zero and hence ρE (and ρ) remain unspecified. In general the analysis domain will be limited by some arbitrarily chosen *external boundaries*, Γ_s, for exterior or internal flows, as shown in Fig. 14.1.

Here, as discussed in Sec. 12.13.4, it will in general be necessary to perform a linearized Rieman analysis in the direction of the outward normal to the boundary x_n to determine the speeds of wave propagation of Eq. (12.132). For such linearization of Euler equations three values of propagation speeds will be found

$$\begin{aligned} \lambda_1 &= u_n \\ \lambda_2 &= u_n + c \\ \text{and} \qquad \lambda_3 &= u_n - c \end{aligned} \tag{14.8}$$

where u_n is the normal velocity component and c is the compressible wave

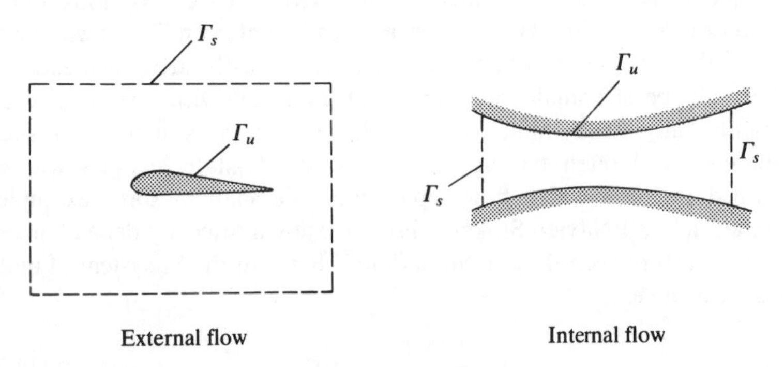

Fig. 14.1 Boundaries of a computation domain. Γ_u, wall boundary; Γ_s, free boundary

celerity given by

$$c=\sqrt{\frac{\gamma p}{\rho}} \tag{14.9}$$

As of course no disturbances can propagate at velocities greater than those of Eqs (14.8) and in the case of supersonic flow, i.e. when the local Mach number

$$M=\frac{|u_n|}{c}\geqslant 1 \tag{14.10}$$

we shall have to distinguish two possibilities:

(*a*) *supersonic inflow boundary* where

$$u_n < -c$$

and the analysis domain cannot influence the exterior; for such boundaries all components of the vector **U** must be specified; and

(*b*) supersonic outflow boundaries where

$$u_n > c$$

and here no components of **U** are prescribed.

For subsonic boundaries the situation is more complex and here the values of **U** that can be specified are the components of the incoming Rieman variables. However, this may frequently present difficulties as the incoming wave may not be known and usual compromises may be necessary as in the treatment of elliptic problems possessing infinite boundaries.

14.3.2 *Navier–Stokes equations.* Here, due to the presence of second derivatives, additional boundary conditions are required.

For the *solid wall boundary*, Γ_u, all the velocity components are prescribed assuming, as in the previous chapter for incompressible flow, that the fluid is attached to the wall. Thus for a stationary boundary we put

$$u_i=0$$

Further, if conductivity is not negligible, boundary temperatures or heat fluxes will generally be given in the usual manner.

For *exterior boundaries* Γ_s of the supersonic inflow kind, the treatment is identical to that used for Euler equations. However, for outflow boundaries a further approximation must be made, either specifying tractions as zero or making their gradient zero in the manner described in Sec. 13.8.4.

14.4 Numerical approximation and solution algorithms

14.4.1 *The two-step algorithm.* The essential features of discretization of transport equations of the form of Eq. (14.1) have been presented in Chapter 12. Here we shall exclusively use the explicit, two-step form of the Taylor–Galerkin scheme described in Sec. 12.13.3 due to its economy and simplicity. We summarize the procedure of Eqs (12.128) and (12.129) below, using always an identical interpolation of $\mathbf{U}$ and $\mathbf{F}_i$ as

$$\mathbf{U}=\mathbf{N}\bar{\mathbf{U}} \qquad \mathbf{F}_i=\mathbf{N}\bar{\mathbf{F}}_i \tag{14.11}$$

Step I. Compute at each integration point of an element

$$\mathbf{U}^{n+1/2}=\mathbf{U}^n-\frac{\Delta t}{2}\left(\frac{\partial \mathbf{F}_i^n}{\partial x_1}+\mathbf{Q}^n\right) \tag{14.12}$$

with

$$\frac{\partial \mathbf{F}_i^n}{\partial x_i}=\frac{\partial \mathbf{N}}{\partial x_i}\bar{\mathbf{F}}_i \qquad \text{and} \qquad \mathbf{U}^n=\mathbf{N}\bar{\mathbf{U}}^n$$

and evaluate

$$\mathbf{F}_i^{n+1/2}=\mathbf{F}_i(\mathbf{U}^{n+1/2})$$

For linear triangles this has to be done at the centroidal coordinates (i.e. a single point only).

Step II. Compute $\Delta\bar{\mathbf{U}}^n$ such that

$$\bar{\mathbf{U}}^{n+1}=\bar{\mathbf{U}}^n+\Delta\bar{\mathbf{U}}^n$$

using Eq. (12.129), i.e.

$$\Delta\bar{\mathbf{U}}^n=-\mathbf{M}^{-1}\Delta t\left[\int_\Omega \frac{\partial \mathbf{N}^{\mathrm{T}}}{\partial x_i}(\mathbf{F}_i^{n+1/2}+\mathbf{G}_i^n)\,\mathrm{d}\Omega + \int_\Omega \mathbf{N}^{\mathrm{T}}\mathbf{Q}^n\,\mathrm{d}\Omega+\int_\Gamma \mathbf{N}^{\mathrm{T}}(\mathbf{F}_i^{n+1/2}+\mathbf{G}_i^n)\,\mathrm{d}\Gamma\right] \tag{14.13}$$

Here, of course, the iteration discussed in Chapter 12 [Eq. (12.106)] with the lumped form of the matrix $\mathbf{M}_L$ will be used to obtain the optimal performance for transients.

In the case of the algorithm being used to obtain a steady-state solution in which the iteration stops when $\Delta\bar{\mathbf{U}}^n=0$, there is no need to use the consistent $\mathbf{M}$ form; only the lumped $\mathbf{M}_L$ is used in Eq. (14.13) which allows a slight extension of the critical time step for stability. For pure convection problems this gives, following Eq. (12.99), the Courant number

$$C=1$$

or in detail

$$\Delta t \leqslant \Delta t_{\text{crit}} = \frac{h}{|u|+c} \tag{14.14}$$

The algorithm described above is developed in references 3 to 6 where the use of domain splitting to ensure the avoidance of low Courant numbers locally is described.

The approximation algorithm can of course use any C_0 continuous set of discretization functions. In the examples that will follow we have used only the simplest linear approximations on triangles in two dimensions and tetrahedra in three dimensions. This is deliberate as such elements are most adapted to deal with possible discontinuities inherent in the wave nature of the problem and also permit the easiest achievement of lumping of the mass matrix **M**.

14.4.2 *Local time-step size for steady-state problems.* We have remarked in Chapter 12 that the Taylor–Galerkin scheme of Eq. (12.102) gives the correct amount of balancing diffusion in steady-state problems if

$$\Delta t = \Delta t_{\text{crit}}$$

Indeed, insertion of the above value can be made to the balancing diffusion term alone without affecting stability to a substantial degree. It can be shown that this is equivalent to using the above Δt in the expression of the first step, that is Eq. (14.12), *independently for each element* while using any appropriate, stable, Δt outside the brackets of Eq. (14.13) of the second step in the general algorithm. To ensure the optimal steady-state solution we recommend that this is always done.

With regard to this second value of Δt which is controlled by stability requirements, we shall still find that if element sizes differ widely in the domain the condition of Eq. (14.14) will give slow convergence due to uneconomically low Courant numbers at some locations. Here again we note that the mass matrix $\mathbf{M}_L$ is fictitious for steady-state problems and its coefficients can be *locally scaled* to ensure that at each node the Courant number is as close to unity as possible.

This is done by determining the highest Courant number of any element joining the particular node and adjusting Δt to the critical value corresponding there. The procedure is known as 'local time stepping' and of course will not result in a true transient but will ensure fastest convergence to the steady state.

14.5 Shock formation and artificial diffusion

The non-linear nature of Euler equations will lead frequently to shock fronts in which an abrupt discontinuity arises. If such shocks are stationary, the fluxes $\mathbf{F}_i$ remain there unchanged but the individual components of the vector $\mathbf{U}$ will be discontinuous [viz. Eqs (12.138)].

Different forms of such discontinuities can occur (as discussed in reference 1, p. 90). The most common are *shock surfaces* through which mass flow occurs. Here discontinuities or jumps will be found in the density ρ, pressure p, energy E and the velocity component u_n in the direction normal to the shock surface. Only the tangential velocity components remain continuous in such shocks.

Another form is that of a *contact discontinuity*, which exists with a zero velocity component normal to the discontinuity surface. Here the pressure p will remain continuous but discontinuity of ρ and E can occur. The tangential velocity components again remain continuous.

A variant on the above is a *slip surface* (or vortex sheet) discontinuity, in which again the velocity component normal to the surface is zero in the steady state. Now, however, only the pressure remains continuous and jumps in tangential velocities exist.

The most serious and difficult to treat is the general shock surface and it is important to add in the formulation some additional diffusivity in the manner discussed in Sec. 12.14 of Chapter 12 to avoid oscillatory solutions. Various forms of such artificial diffusivity have been used in the context of the Euler/Navier–Stokes equation. The most obvious is the application of the procedure due to Lapidus[7–9] in which the diffusion coefficient is based [viz. Eq. (12.139)] on the maximum gradient of some scalar component of $\mathbf{U}$. MacCormack and Baldwin[10] have suggested the use of a similar form but one based on the second derivatives of some scalar quantity—choosing this as the pressure variable.

The diffusion coefficient is thus of the form

$$\tilde{k} = C_a h^\alpha \left| \frac{\partial^2 p}{\partial x_i \, \partial x_i} \right| \equiv C_a h^\alpha \, \nabla^2 p \tag{14.15}$$

where C_a is a coefficient with appropriate dimensionality and α is chosen as 3 to ensure rapid decrease with element size.

The choice of diffusion proportional to second derivatives has considerable logic. We shall see later that these give us some information about errors and indeed tend to be highly localized near shocks.

The question of estimating the laplacian of p in the formulation by the finite element method is not easy due to the nature of interpolation. A relatively inexpensive procedure is suggested by a remark made in Sec. 9.2.4 of Chapter 9, where we have noted, in a one-dimensional context, that the difference between the consistent mass matrix $\mathbf{M}$ and its 'lumped' equivalent $\mathbf{M}_L$ was precisely equal to a laplacian operator, i.e. a diffusion. Extending this reasoning to a multidimensional context and noting that in the computation process both these quantities have been evaluated, we can write

$$h^2|\nabla^2 p| \propto |(\mathbf{M}-\mathbf{M}_L)|\bar{\mathbf{p}} \tag{14.16}$$

The correction due to this artificial diffusion is best carried out using the predicted values of $\mathbf{U}^{n+1}$ [viz. Eq. (14.13)] and results in addition of a 'smoothing' increment

$$\Delta\bar{\mathbf{U}}_s^n = -\Delta t\, \mathbf{M}_L \tilde{k}\left(\int_\Omega \frac{\partial \mathbf{N}^{\mathrm{T}}}{\partial x_i}\frac{\partial \mathbf{N}}{\partial x_i}\,\mathrm{d}\Omega\right)\mathbf{U}^{n+1} \tag{14.17}$$

However, noting that this is also a laplacian operator, we can simplify and write[11]

$$\Delta\bar{\mathbf{U}}_s^n = \tilde{C}_p \Delta t\, \mathbf{M}_L^{-1}|(\mathbf{M}-\mathbf{M}_L)\bar{\mathbf{p}}|(\mathbf{M}-\mathbf{M}_L)\bar{\mathbf{U}}^{n+1} \tag{14.18}$$

In the above, $\tilde{C}_p$ is a non-dimensional coefficient.

Other methods of preventing oscillations have been suggested. One of these is the so-called flux corrected transport (FCT) procedure,[12-15] and this is widely used. Yet another possibility is to return to artificial viscosity but to make this proportional to the residuals of the governing equation.[16,17]

14.6 Some preliminary examples for the Euler equation

The computation procedures outlined can be applied with success to many transient and steady-state problems. In this section we illustrate its performance on a few relatively simple examples.

14.6.1 *Rieman shock tube—a transient problem in one dimension.*[3] This is treated as a one-dimensional problem. Here an initial pressure difference between two sections of the tube is maintained by a diaphram which is destroyed at $t=0$. Figure 14.2 shows the pressure, velocity and energy contours at the seventieth time increment, and the effect of including consistent and lumped mass matrices is illustrated. The problem has an analytical, exact, solution presented by Sod[18] and the numerical solution is from reference 3.

14.6.2 *Isothermal flow through a nozzle—a steady-state problem in one dimension.*[3] Here a variant of the Euler equation is used in which isothermal conditions are assumed and in which the density is replaced

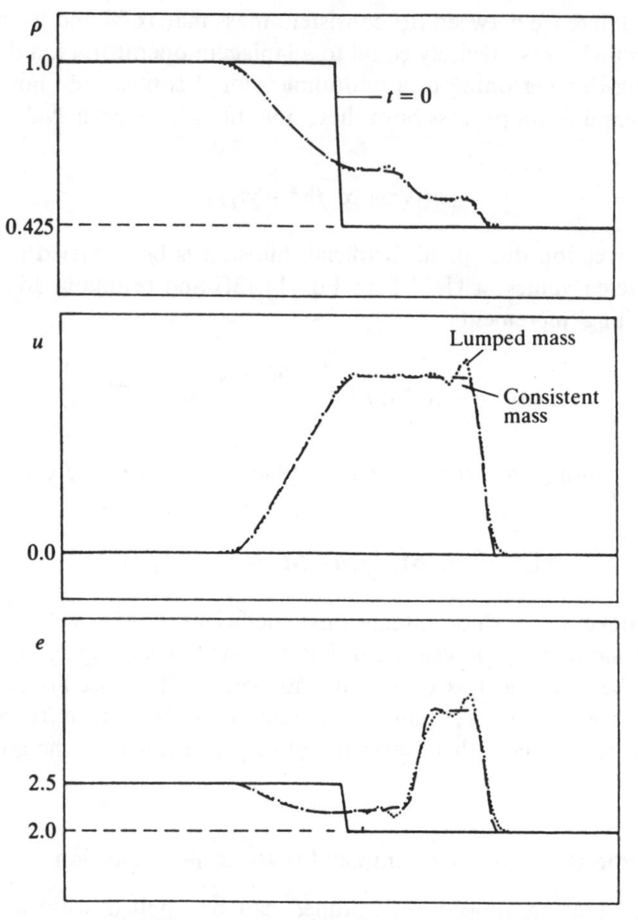

Fig. 14.2 The Riemann shock tube problem.[3,18] The total length is divided into 100 elements. Profile illustrated corresponds to 70 time steps ($\Delta t = 0.250$). Lapidus constant $C_{\text{Lap}} = 1.0$

by ρa where a is the cross-sectional area[3] assumed to vary as[19]

$$a = 1.0 + \frac{(x-2.5)^2}{12.5} \qquad \text{for } 0 \leqslant x \leqslant 5 \tag{14.19}$$

The speed of sound is constant as the flow is isothermal and various conditions at inflow and outflow limits were imposed as shown in Fig. 14.3. In all problems steady state was reached after some 500 time steps. For the case with supersonic inflow and subsonic outflow, a shock forms and Lapidus-type artificial diffusion was used to deal with it, showing in

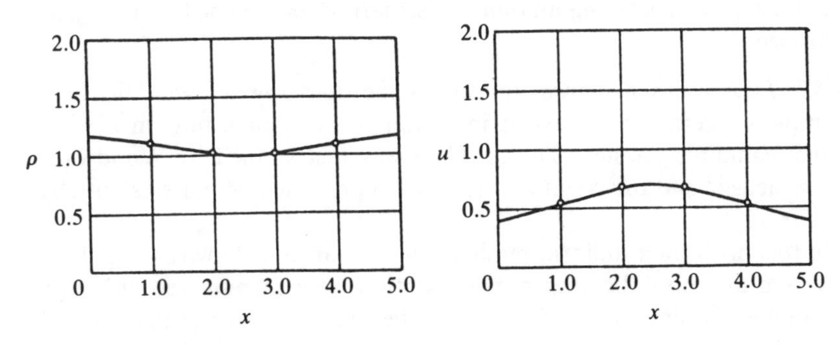

(*a*) Subsonic inflow and outflow

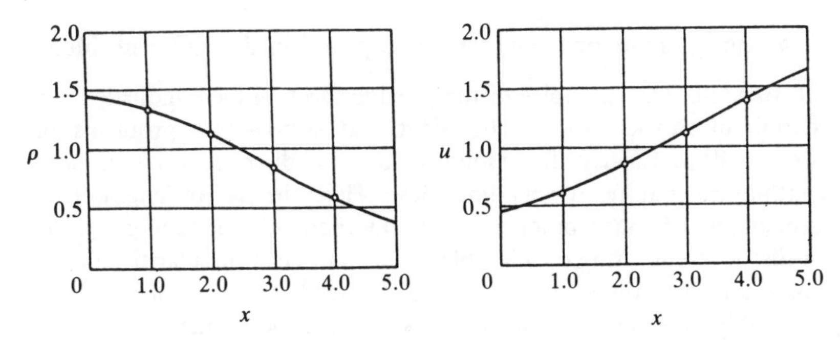

(*b*) Supersonic inflow and outflow

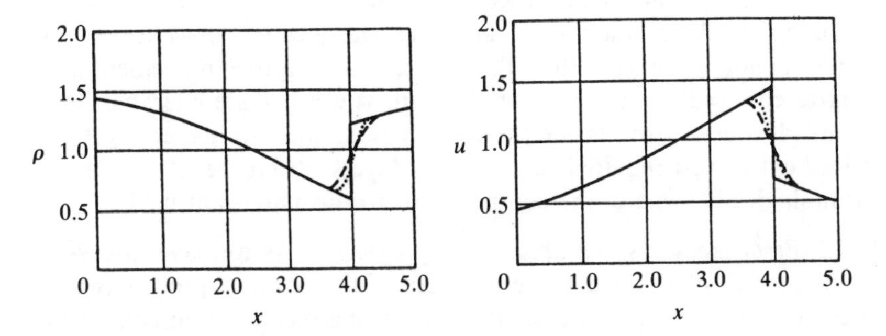

(*c*) Supersonic inflow–subsonic outflow with shock

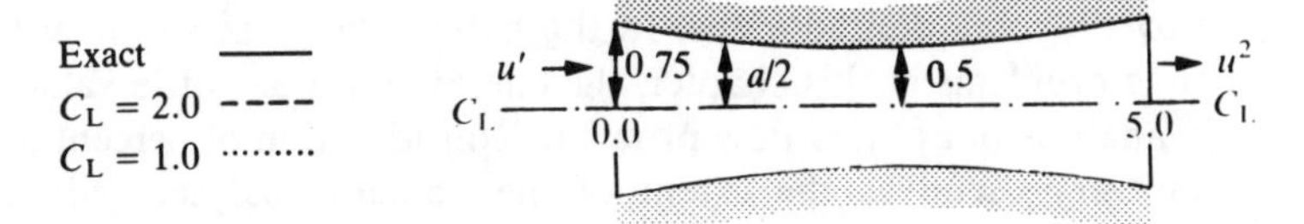

Fig. 14.3 Isothermal flow through a nozzle. 40 elements of equal size used

Fig. 14.3(*c*) the increasing amount of 'smearing' as the coefficient C_{Lap} is increased.

14.6.3 *Two-dimensional transient supersonic flow over a step.* This final example concerns the transient initiation of supersonic flow in a wind tunnel containing a step. The problem was first studied by Woodward and Colella[20] and the results of reference 6 presented here are essentially similar.

In this problem a uniform mesh of linear triangles, shown in Fig. 14.4, was used and no difficulties of computation were encountered although a Lapidus constant $C_{\text{Lap}}=2.0$ had to be used due to the presence of shocks.

14.7 Adaptive refinement and 'shock capture' in the Euler problem

14.7.1 *General.* The examples of the previous section have indicated the formation of shocks both in transient and steady-state problems of high-speed flow. Clearly the resolution of such discontinuities or near discontinuities requires a very fine mesh. Here the use of 'engineering judgement', which many practice in solid mechanics by designing *a priori* mesh distributions with considerable refinement near singularities posed by corners in the boundary, etc., can no longer be used. In problems of compressible flow the position of shocks, where the refinement is most needed, is not known in advance. For this and other reasons, the use of adaptive mesh refinement based on error estimates is essential for obtaining good accuracy and 'capturing' the location of shocks. It is therefore not surprising that the science of adaptive refinement has progressed rapidly in this area and indeed, as we shall see later, has been extended to deal with Navier–Stokes equations where a higher degree of refinement is also required in boundary layers. References 21 to 39 list chronologically some of the contributions to this important field.

14.7.2 *Error measures and accuracy requirements.* As we have discussed in Chapter 12 of Volume 1, the first step of the adaptive process is that of estimating the errors after a stage of the analysis has been computed. In the context of elliptic problems this error is estimated in the energy norm, and the aim is generally to reduce the percentage energy norm error to some desired value. In the hyperbolic or predominantly hyperbolic problems of this chapter, the energy norm has little value. Further, in analysis of exterior flow problems consideration of percentage error is not very useful as the extent of the domain analysed will cause this percentage to vary. Indeed, here the interest focuses on local quantities such as, say, the pressure distribution around an airfoil, skin friction, etc. For this reason, we proceed somewhat differently at the error analysis

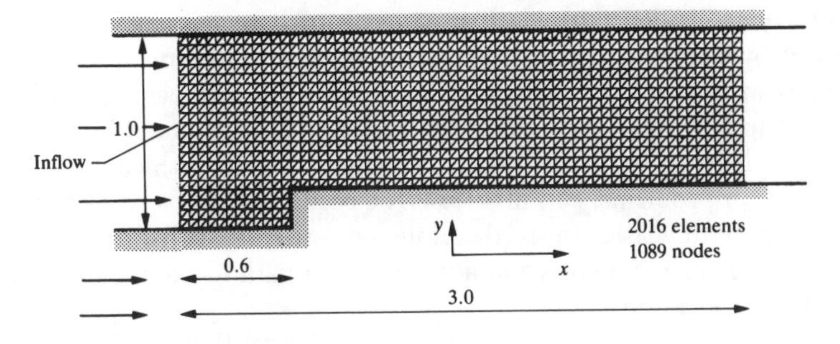

(*a*) Structured uniform mesh

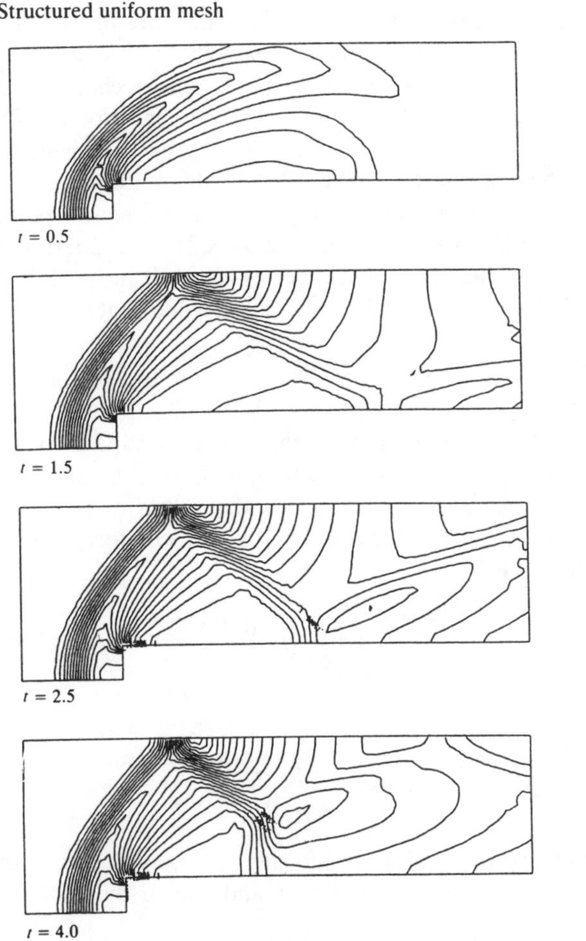

(*b*) Solution — contours of pressure at various times

Fig. 14.4 Transient supersonic flow over a step in a wind tunnel[6] (problem of Woodward and Colella[20]). Inflow Mach 3 uniform flow

stage, trying to limit the *local error value* to a permissible limit in an element-by-element manner. Further, in the refinement process we shall proceed so as to make this local error identical in each element as far as possible to achieve an optimal performance. This has much in common with the errors introduced in time-stepping procedures which, being also propagation problems, can involve error accumulation as shown in Sec. 10.2.4. In both classes it is therefore necessary to consider the total number of elements involved in a propagation path to bound an error in a particular location.

In what follows we shall only consider first-order (linear) elements and the so-called 'h' refinement process in which the increased accuracy is achieved by variation of the element size. The 'p' refinement in which the order of the element polynomial expression is changed is of course possible (as indeed is the h–p combined refinement process). This has been given some attention by Oden *et al.*[28,33] but we believe that such procedures have a limited applicability in shock formation, etc.

The estimation of local error in linear shape function elements is conveniently achieved by consideration of so-called interpolation error. Thus if we take a one-dimensional element of length h and a scalar function u, it is clear that the error in u is of order $O(h^2)$ and that it can be written as

$$e = U - U^b = Ch^2 \frac{\mathrm{d}^2 U}{\mathrm{d}x^2} \approx Ch^2 \frac{\mathrm{d}^2 U^h}{\mathrm{d}x^2} \tag{14.20}$$

where U^h is the computed finite element value and C a constant.

If, for instance, we further assume that $U = U^h$ at nodes, i.e. that the nodal error is zero, then e represents the values on a parabola with a curvature of $\mathrm{d}^2U^h/\mathrm{d}x^2$. This allows C, the unknown constant, to be determined, giving

$$e_{\max} = \tfrac{1}{8} h^2 \frac{\mathrm{d}^2 U^h}{\mathrm{d}x^2}$$

or a RMS departure error as

$$e_{\mathrm{RMS}} = \frac{1}{\sqrt{120}} h^2 \frac{\mathrm{d}^2 U^h}{\mathrm{d}x^2}$$

For simplicity we shall use uniformly $C = 1$, as of course the assumption of exact nodal values is not true, and aim to keep

$$h^2 \frac{\mathrm{d}^2 U^h}{\mathrm{d}x^2} \leqslant e_p \tag{14.21}$$

where e_p is some specified value set by the user. In this, consideration of the possible cost has to be taken into account in setting the limit to be 'reasonable'.

The extension of the process to two or more dimensions is of course needed. Here three, second-order, derivatives

$$\frac{\partial^2 U^h}{\partial x_1^2} \qquad \frac{\partial^2 U^h}{\partial x_1 \, \partial x_2} \qquad \text{and} \qquad \frac{\partial^2 U^h}{\partial x_2^2}$$

exist in two dimensions (and six in three space dimensions) and it is first necessary to determine their *principal values* and directions $\mathbf{x}'$ as shown in Fig. 14.5.

Now if x'_2 corresponds to the direction of the maximum curvature, the element size, measured in that direction as shown, would be given using eq. (14.21) as

$$h_{\min} \leqslant \frac{\sqrt{e_p}}{|(\mathrm{d}^2 U^h/\mathrm{d}x_2^2)|_{\max}} \tag{14.22}$$

Of course if the minimum curvature is much smaller, elements could be elongated and an economy in the required number of degrees of freedom achieved. We could thus make in direction x'_1 (of minimum curvature)

$$h_{\max} = s\, h_{\min}$$

where

$$s = \frac{\sqrt{|(\mathrm{d}^2 U^h/\mathrm{d}x_2^2)|_{\max}}}{|(\mathrm{d}^2 U^h/\mathrm{d}x_1^2)|_{\min}} \tag{14.23}$$

which is known as the stretch ratio.

The determination of the curvatures (or second derivatives) of U^h needs of course some elaboration. With linear elements (e.g. simple triangles or tetrahedra) the curvatures of U^h which is interpolated as

$$U^h = \bar{\mathbf{N}}\bar{\mathbf{U}} \tag{14.24}$$

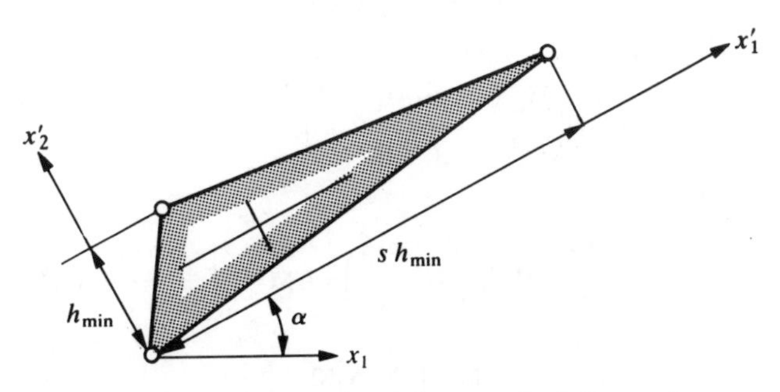

Fig. 14.5 Element elongation s and minimum size

are zero within the elements and take an infinite value at their boundaries. We proceed here first by evaluating the gradients, i.e.

$$\frac{\partial U^h}{\partial x_i} = \frac{\partial \mathbf{N}}{\partial x_i} \bar{\mathbf{U}}, \qquad \text{etc.} \tag{14.25}$$

in each element (where these are constant) and obtaining a projection of these onto the same interpolation as used for U, writing

$$\left(\frac{\partial U^h}{\partial x_i}\right)^* = \mathbf{N}\overline{\left(\frac{\partial U}{\partial x_i}\right)^*}, \qquad \text{etc.} \tag{14.26}$$

where the bar denotes nodal values.

This projection can be achieved by the same procedures as we advocated for stress smoothing in Chapter 12 of Volume 1, where we used a Galerkin projection of the difference between $\partial U^h/\partial x_i$ and $(\partial U^h/\partial x_i)^*$. However, simple averaging of the element values or nodes is quite satisfactory and less expensive.

Once the smoothed values are obtained it is a simple method to estimate the second derivatives such for each element as

$$\left(\frac{\partial^2 U}{\partial x_i^2}\right)^* = \frac{\partial \mathbf{N}}{\partial x_i}\overline{\left(\frac{\partial U}{\partial \mathbf{x}_i}\right)^*} \tag{14.27}$$

Here of course some ambiguity arises from the above process since the equality

$$\frac{\partial^2 U^h}{\partial x_i \partial x_j} = \frac{\partial^2 U^h}{\partial x_j \partial x_i} \tag{14.28}$$

is not automatically preserved. Simple averaging of such cross-derivatives is recommended for best accuracy.†

† It is often more convenient to obtain the second derivatives at nodes in a single operation. Thus, assuming again that we interpolate

$$\left(\frac{\partial^2 U}{\partial x^2}\right)^* = \mathbf{N}\overline{\left(\frac{\partial^2 U}{\partial \mathbf{x}^2}\right)^*}$$

we obtain by projection

$$\int_\Omega \mathbf{N}^\mathrm{T}\left[\mathbf{N}\overline{\left(\frac{\partial^2 U}{\partial \mathbf{x}^2}\right)^*} - \frac{\partial^2 U^h}{\partial x^2}\right] \mathrm{d}\Omega = 0$$

and integrating by parts

$$\overline{\left(\frac{\partial^2 U}{\partial x^2}\right)^*} = \mathbf{M}^{-1}\left(\int_\Omega \mathbf{N}^\mathrm{T}\frac{\partial^2 U^h}{\partial x^2}\right) = -\mathbf{M}^{-1}\left(\int_\Omega \frac{\partial \mathbf{N}^\mathrm{T}}{\partial x}\frac{\partial \mathbf{N}}{\partial x}\mathrm{d}\Omega\right)\bar{\mathbf{U}}$$

where

$$\mathbf{M} = \int_\Omega \mathbf{N}^\mathrm{T}\mathbf{N}\,\mathrm{d}\Omega$$

which of course can be 'lumped'. This avoids the ambiguity mentioned below Eq. (14.27).

In the above, error estimations and control measures have been presented for a scalar variable U. A problem remains of the manner in which these should be extended to the vector quantity $\mathbf{U}$ occurring in the Euler and Navier–Stokes equations. Here various alternatives are open, e.g.:

1. The control is applied to each component of $\mathbf{U}$ in the way indicated above.
2. A key variable is chosen from the set $\mathbf{U}$.
3. Finally, a representative scalar composite variable is taken, such as, for example, $\mathbf{U}^T\mathbf{U}$.

None of these alternatives is perfect. The first, probably the most logical, is too costly. The last suffers from all kinds of scaling difficulties depending on the dimensionality of each component. In practice the choice of pressure, density or the Mach number for control has proved quite effective.

14.7.3 *The h-refinement process and mesh enrichment.* Once an approximate solution has been achieved on a given mesh, the local errors can be evaluated and new element sizes (and elongation directions if used) can be determined for each element. For some purposes it is convenient again to transfer such values to the nodes so that they can be interpolated continuously. The procedure here is of course identical to that of smoothing the derivatives discussed in the previous section.

To achieve the desired accuracy various procedures can be used. The most obvious is the process of *mesh enrichment* in which the existing mesh is locally subdivided into smaller elements still retaining the 'old' mesh in the configuration. Figure 14.6(*a*) shows how triangles can be readily subdivided in this way. With such enrichment an obvious connectivity difficulty appears. This concerns the manner in which the subdivided elements are connected to ones not so refined. A simple process is illustrated showing element halving in the manner of Fig. 14.6(*b*). Here of course it is fairly obvious that this process, first described in reference 26, can only be applied in a gradual manner to achieve the predicted subdivisions. However, element elongation is not possible with such mesh enrichment.

Despite such drawbacks the procedure is very effective in localizing (or capturing) shocks, as we illustrate in Figs 14.7 and 14.8.

In the first of these, i.e. Fig. 14.7, the theoretical solution is simply one of a line discontinuity shock of which a jump of all the components of $\mathbf{U}$ occurs. The original analysis carried out on a fairly uniform mesh shows a very considerable 'blurring' of the shock. In Fig. 14.7 we also show the refinement being carried out at two stages and we see how the shock is progressively reduced in width.

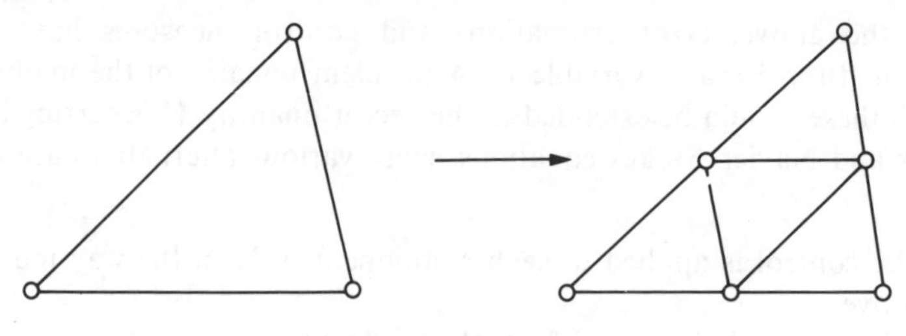

(*a*) Triangle subdivision

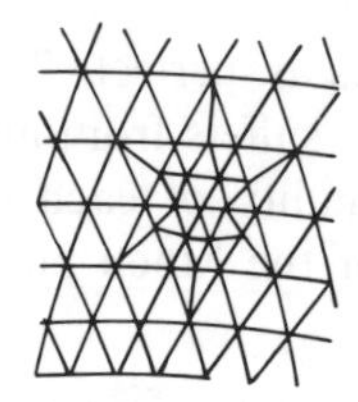

(*b*) Restoration of connectivity

Fig. 14.6 Mesh enrichment

Similar though less simply defined shock occurs and is resolved with an improvement of accuracy in Fig. 14.8, where a 'double wedge' airfoil is analysed.

In both the above examples, the mesh enrichment preserved the original, nearly equilateral, element form with no elongation possible.

Whenever a sharp discontinuity is present, local refinement will proceed indefinitely as curvatures increase without limit. Precisely the same difficulty indeed arises in mesh refinement near singularities for elliptic problems[40] if local refinement is the only guide. In such problems, however, the limits are generally set by the overall energy norm error consideration and the refinement ceases automatically. In the present case, the limit of refinement needs to be set and we generally achieve this by specifying the *smallest element size* in the mesh.

The *h* refinement of the type proposed can of course be applied in a similar manner to quadrilaterals. Here clever use of data storage allows the necessary refinement to be achieved in a few steps by ensuring proper transitions.[41]

14.7.4 *The h-refinement process and mesh regeneration. Two- and three-dimensional examples in the steady state.* Many difficulties mentioned above can be resolved by *automatic generation of meshes of a specified density*. Such automatic generation has been the subject of much research

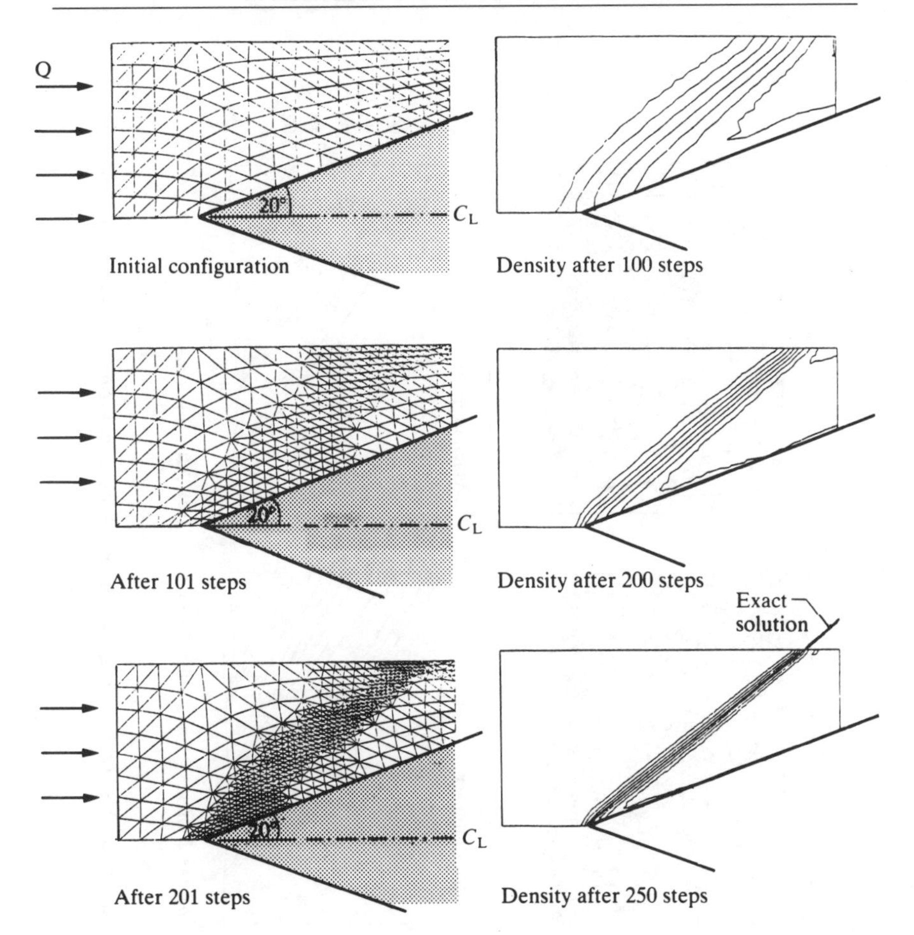

Fig. 14.7 Supersonic, Mach 3, flow past a wedge. Exact solution forms a stationary shock. Successive mesh enrichment and density contours

in many applications of finite element analysis. Again a few selected reports on such work are included in references 42 to 46, which present a mere sample of the research in progress. Probably the closest achievement of a prescribed element size and directionality can be obtained for triangles and tetrahedra. Here the procedures developed by Peraire *et al.*[32,39] are most direct and efficient, allowing element stretching in prescribed directions (though of course the amount of such stretching is sometimes restricted by practical considerations).

We refer the reader for details of such mesh generation to the original publications. In the examples that follow we shall exclusively use this type of mesh generation.

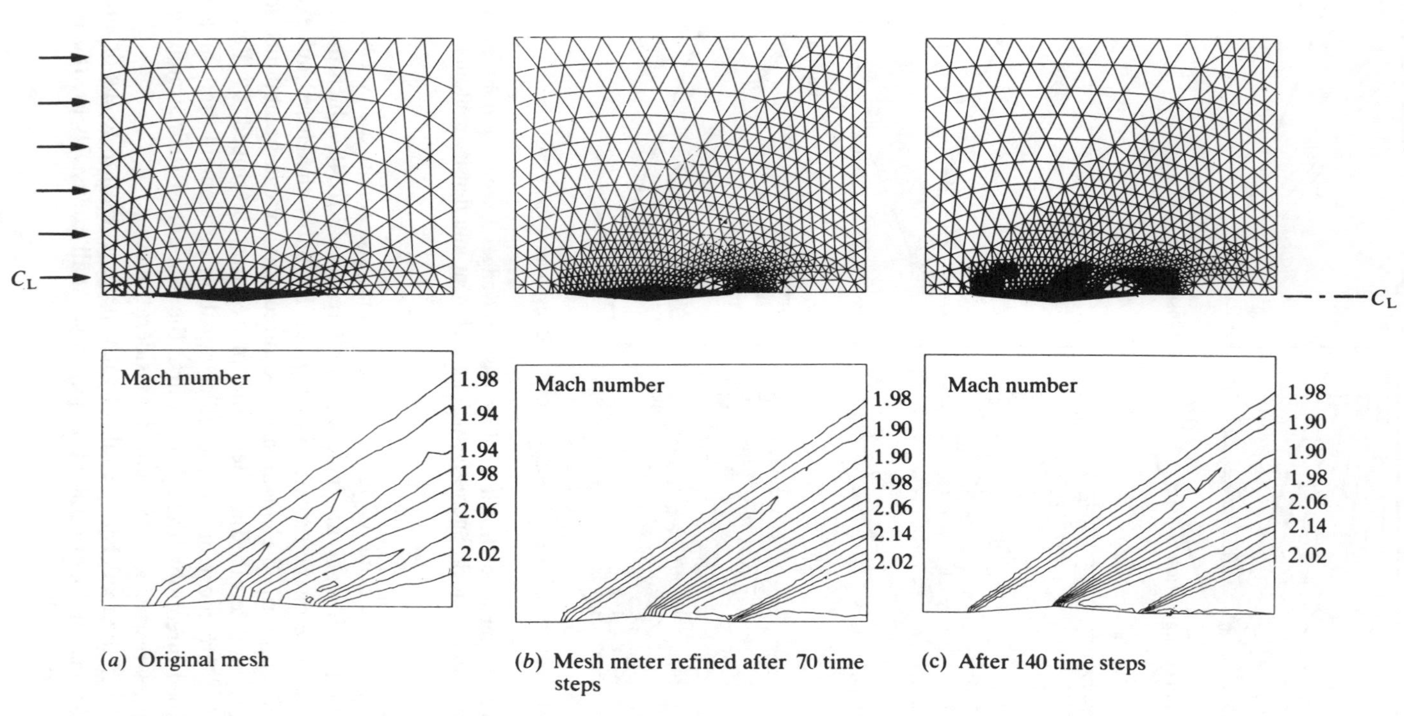

(*a*) Original mesh (*b*) Mesh meter refined after 70 time steps (c) After 140 time steps

Fig. 14.8 Supersonic, Mach 2, flow past a double wedge 'wing'. Mesh enrichment (contours of Mach mumber)

We have mentioned previously the advantages that can be gained with the use of structured meshes based on quadrilateral subdivisions. These are clearly desirable for parallel computation, though, of course, this is also efficient in the unstructured case. Further, storage requirements are generally larger for unstructured meshes. However, at the present stage of development such mesh generators are limited in scope and it seems that use of such structured meshes will at best only be able to approximate to the prescribed size distribution. It seems that the reduction of the total number of degrees of freedom which can be achieved with efficient adaptivity more than compensates the disadvantages of unstructured meshes.

In Fig. 14.9 we show a simple example[32] of the shock wave reflection from a solid wall. Here only a typical 'cut-out' is analysed with appropriate inlet and outlet conditions imposed. The elongation of the mesh along the discontinuity is clearly shown. The solution was remeshed after the iterations nearly reached a steady state.

In Fig. 14.10 a somewhat more complex example of hypersonic flow around a blunt obstacle is shown. Here it is of interest to note that:

1. A detached shock forms in front of the body.
2. A very coarse mesh suffices in front of such a shock where simple free stream flow continues and the mesh is made 'finite' by a maximum element size prescription.
3. For the same minimum element size a reduction of degrees of freedom is achieved by refinement which shows much improved accuracy.

Figure 14.11 shows a yet more sophisticated example in which an impinging shock interacts with a bow shock. Extremely fine mesh distribution was used here to compare results with experiments,[47] which were reproduced with high precision.

Automatic, adaptive, mesh generation provides the means for three-dimensional analysis where the needs for economy are most stringent. Here the number of nodes required for a reasonable analysis rises very rapidly, even with such meshing, and generally the use of structured meshes is prohibitive. Very efficient vectorized programs and the use of the largest computers is of course necessary. Plate 2 (see frontispiece) and Figs 14.12 and 14.13 illustrate some such three-dimensional solutions of Euler equations in specific applications for steady-state flows.[39] The large size of the ensuing problems should be noted.

The use of unstructured meshes in the examples shown has provided the means of analysing the aerodynamics of complete aircraft. Here again adaptive refinement of the kind previously discussed for two dimensions can be applied as shown in Fig. 14.12, though the computational costs now are substantial. Much research is therefore continuing and attempts to combine structured and unstructured meshes for maximum economy are reported.[48,49]

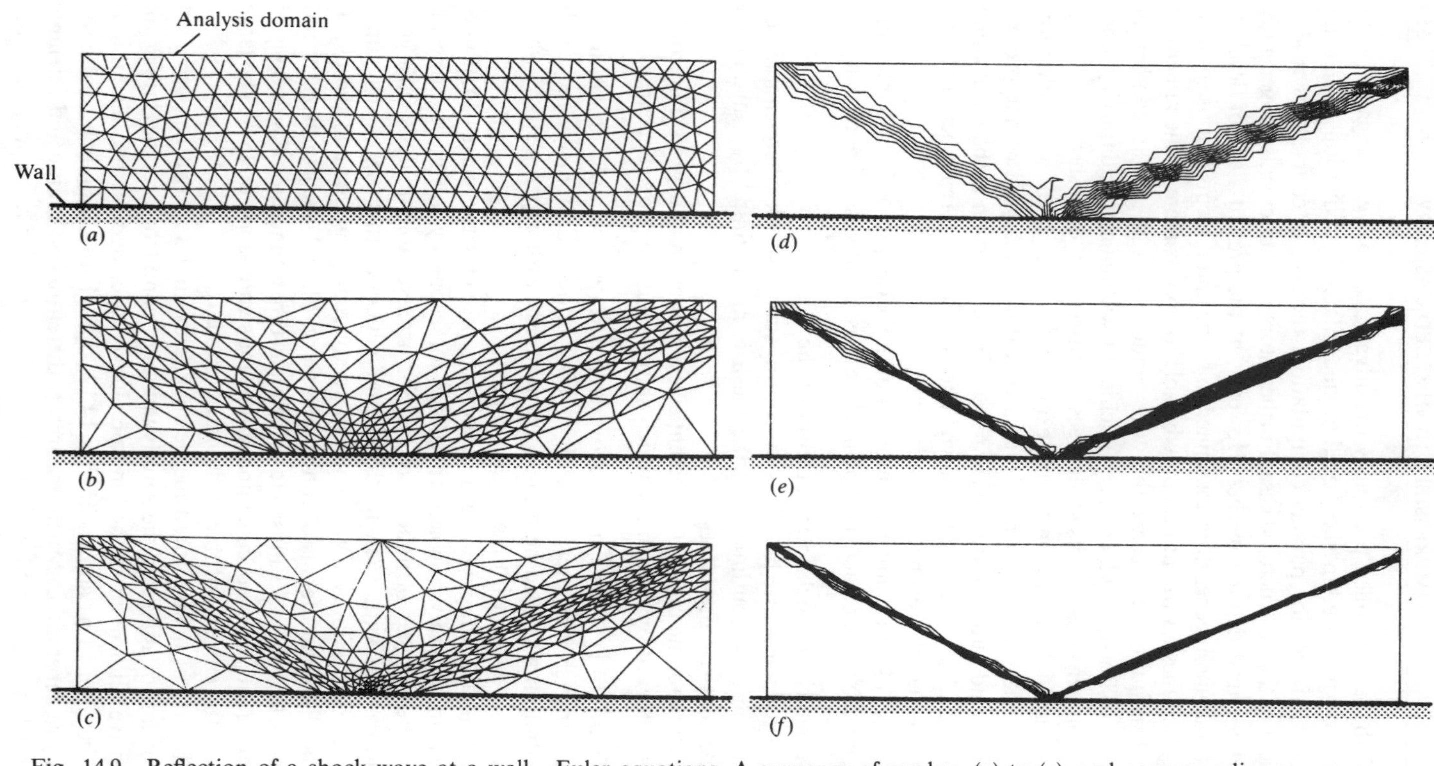

Fig. 14.9 Reflection of a shock wave at a wall—Euler equations. A sequence of meshes, (*a*) to (*c*), and corresponding pressure contours, (*d*) to (*f*)

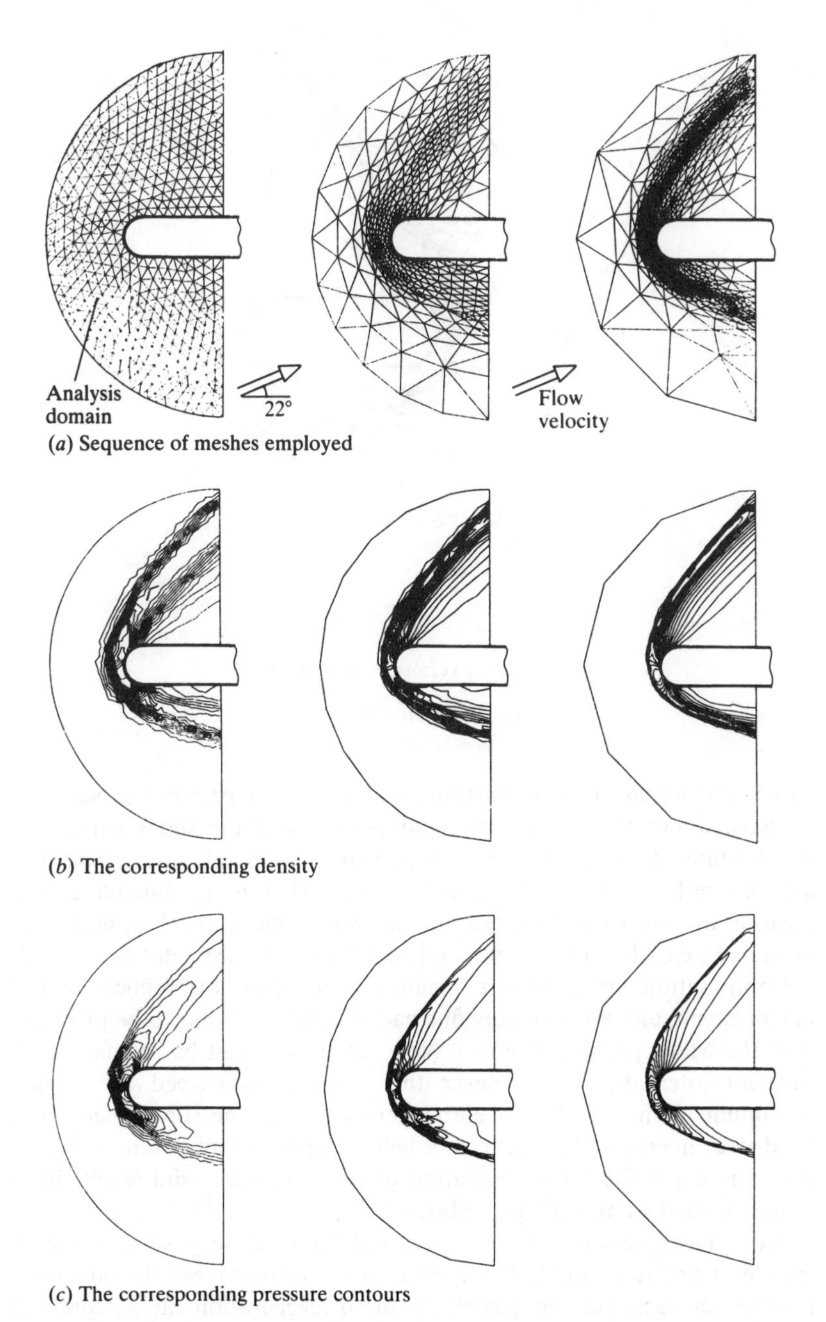

(*a*) Sequence of meshes employed

(*b*) The corresponding density

(*c*) The corresponding pressure contours

Fig. 14.10 Hypersonic flow past a blunt body at high angle of attack

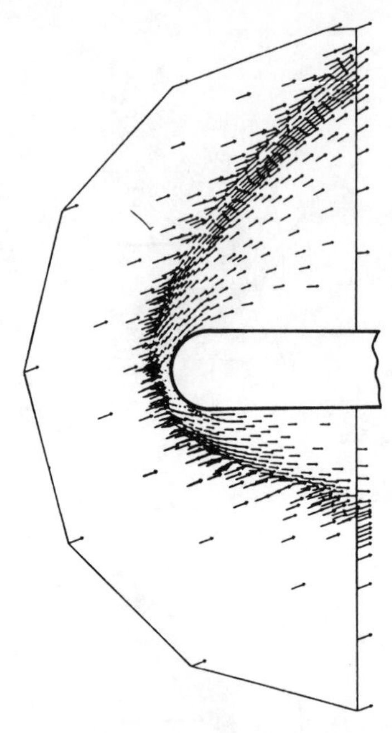

(*d*) Velocity vectors on final mesh

Fig. 14.10 (continued)

14.7.5 *The h-refinement in transient problems.* In the previous subsections we have shown the extraordinary improvement of the shock capture in steady-state situations. However, problems of transients are of considerable interest and here the process is somewhat more difficult as the position in which shocks occur changes with time. In such applications both mesh enrichment and mesh regeneration have been effectively used.

Computation with mesh enrichment now involves both refinement and derefinement, and a reasonably fine background mesh has to be provided over the whole problem domain. Now subdivision can be carried out at frequent intervals, i.e. whenever the local errors exceed a specified maximum. Removal of such enrichment can be done simultaneously if local error in enriched elements falls below a specified minimum, avoiding the more expensive mesh-generation phase. Very successful results have been obtained using such procedures.[50-53]

Mesh regeneration can, however, still be used very successfully in transients and is particularly advantageous if during these the boundary position changes. Here in general the mesh regeneration can be confined to the subregion where action is concentrated. Figure 14.14 shows such

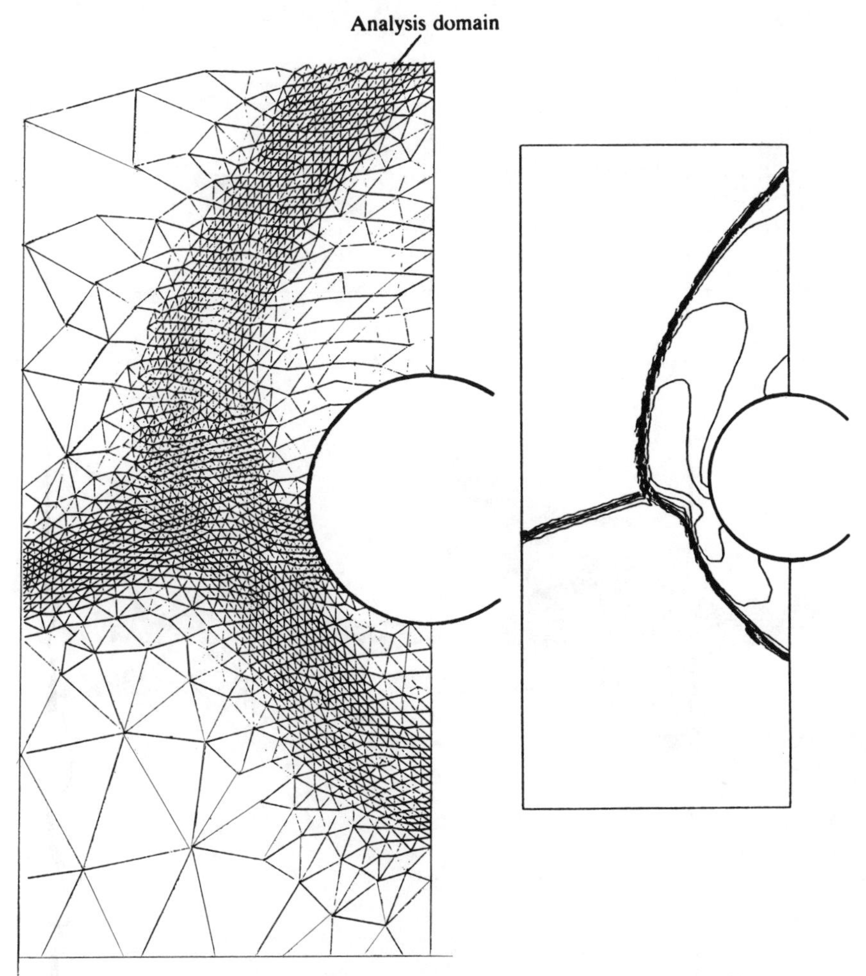

Fig. 14.11 An interaction of an impinging and bow shock wave. Fine mesh and pressure contours (water experiment not shown)

a transient computation for a problem in which an explosive failure of a pressure vessel is modelled with a lid motion.[54]

Similar computation is shown in Fig. 14.15, attempting to show some features of shuttle–rocket separation.

Adaptive computation in transients presents, of course, some extrapolation problems as the refinement in an explicit time-marching scheme 'anticipates' the flow configuration. Further, the refinement, being costly, cannot be carried out at every time step. Details of approximation techniques can be found in the references quoted.

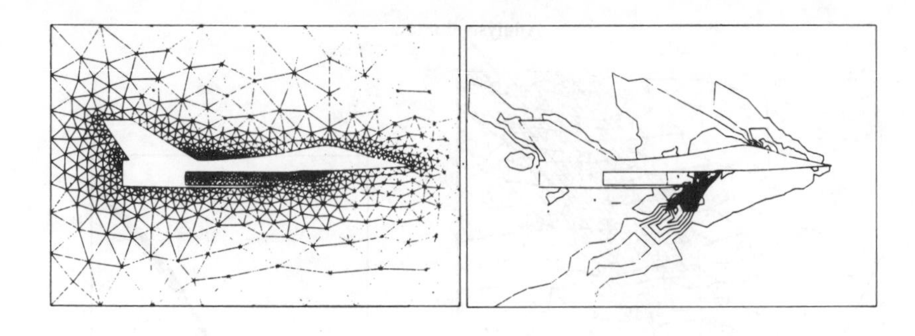

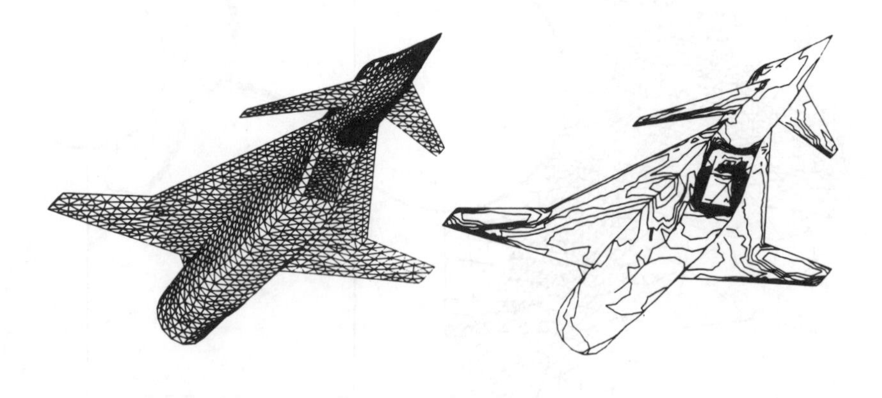

(*a*) **Original mesh and pressure contours**

Fig. 14.12 Adaptive three-dimensional solution of compressible inviscid flow around a high-speed (Mach 2) aircraft.[39] Details similar to Plate 2 frontispiece, 70 000 and 125 000 elements

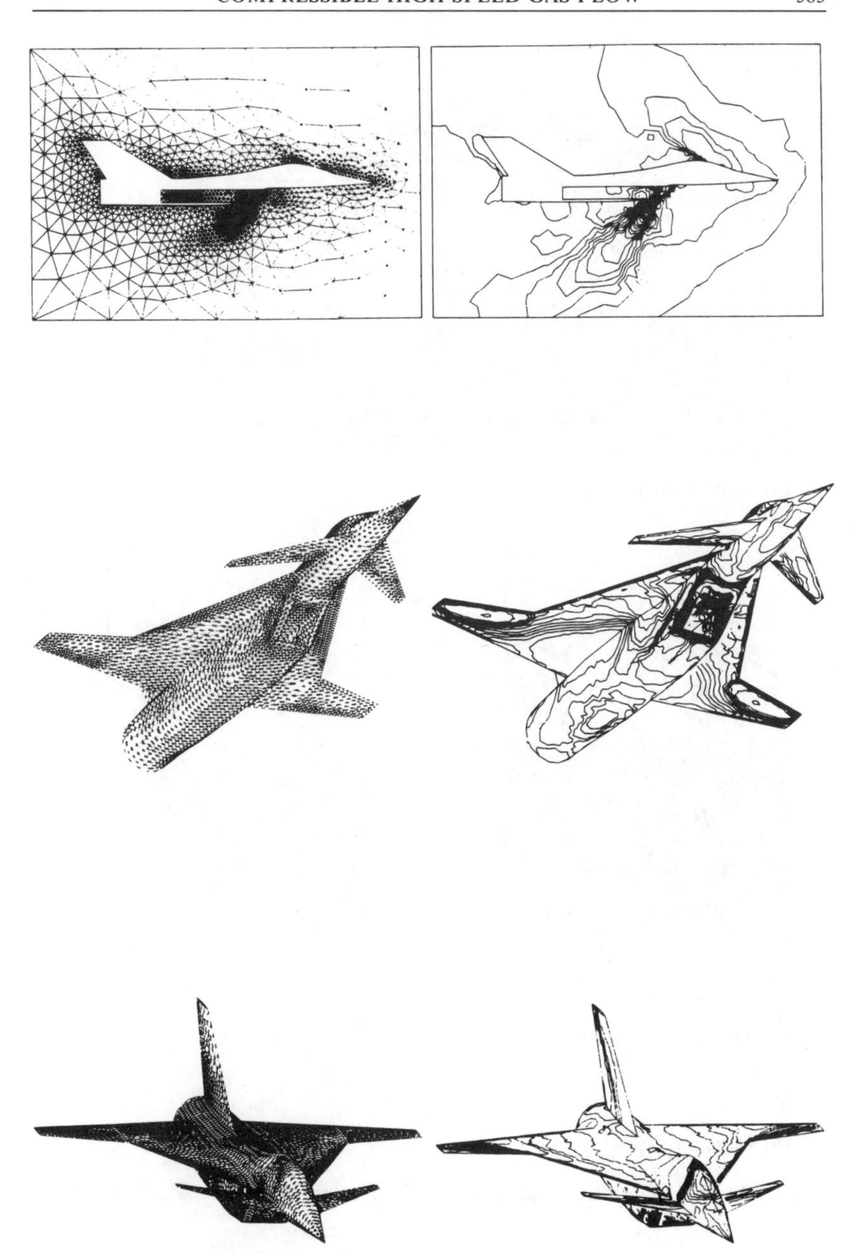

(*b*) Adaptively refined mesh and pressure contours

Fig. 14.12 (*continued*)

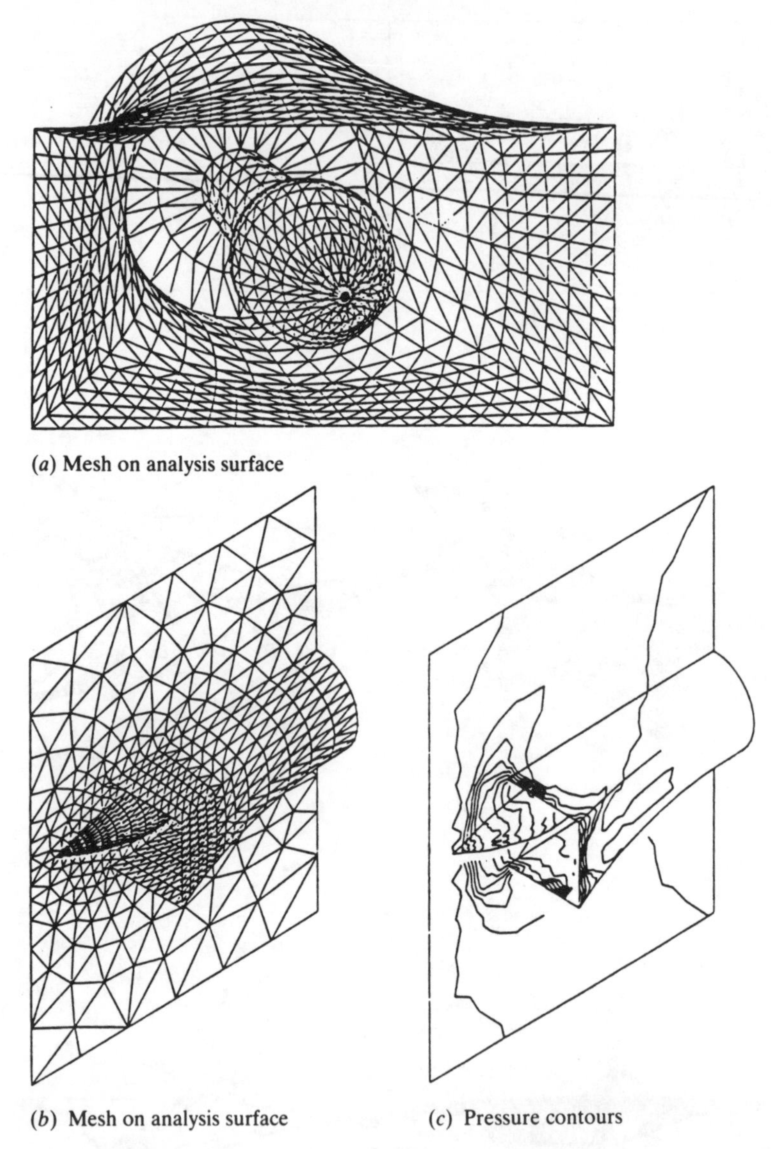

(*a*) Mesh on analysis surface

(*b*) Mesh on analysis surface

(*c*) Pressure contours

Fig. 14.13 Three-dimensional analysis of an engine intake at Mach 2 (14 000 elements)[39]

14.8 The Navier–Stokes problem—boundary layer refinement

The basic solution algorithm in its explicit form is of course applicable with equal ease to the Euler or Navier–Stokes equations. Indeed, the viscous term adds very little to the cost of computation. However, as is

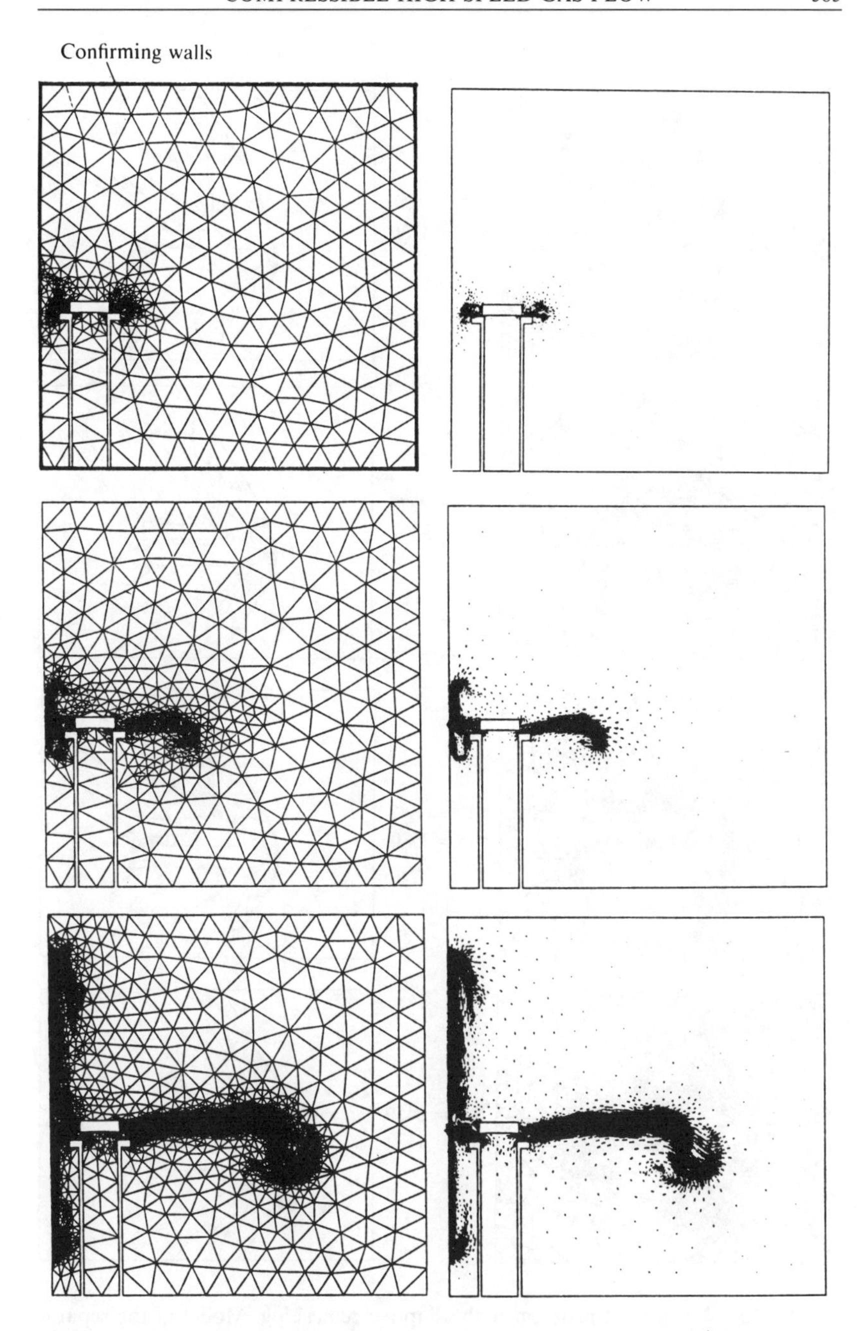

Fig. 14.14 A transient problem with adaptive remeshing. Simulation of a sudden failure of a pressure vessel cap. Velocity patterns and progression of refinement shown. Initial mesh 518 nodes[54]

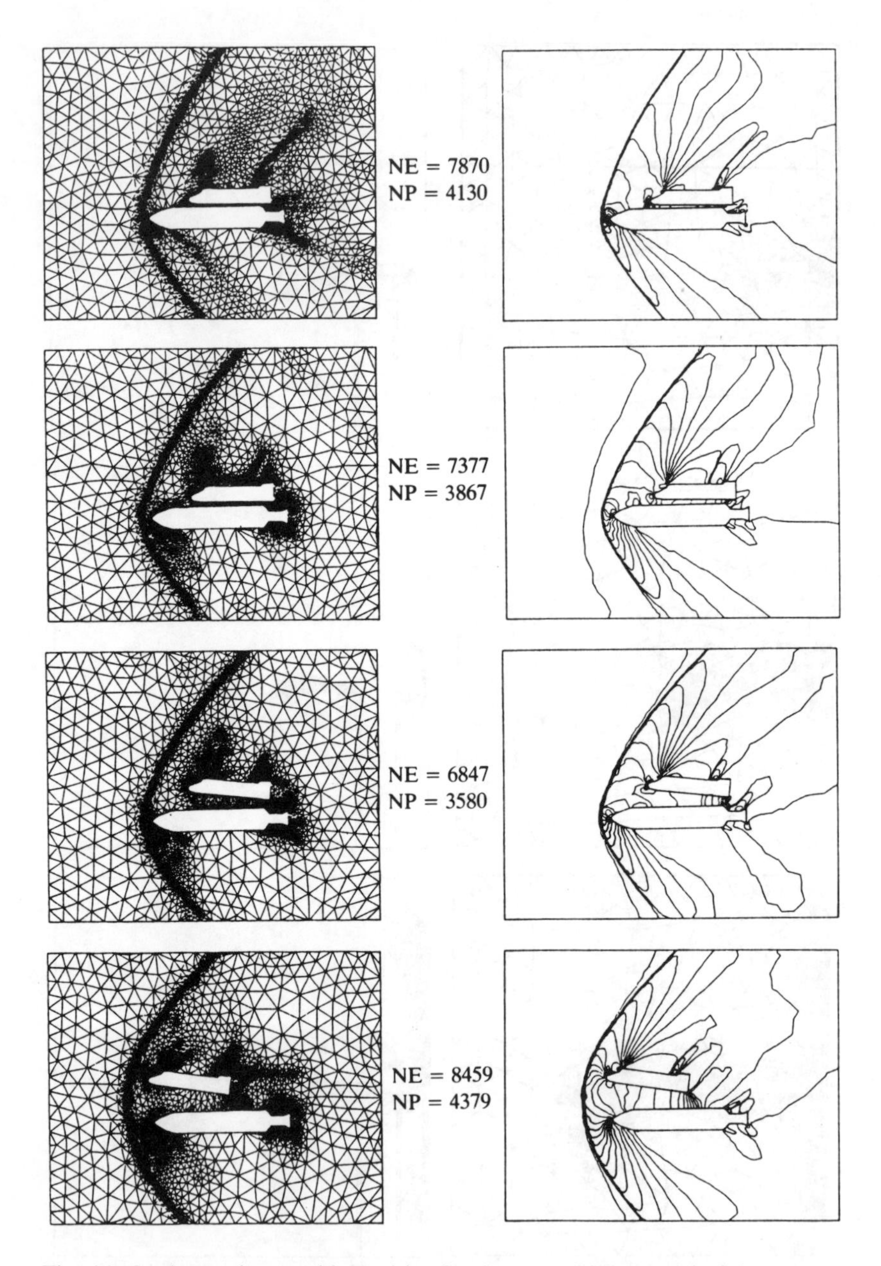

Fig. 14.15 A transient problem with adaptive remeshing. Model of the separation of shuttle and rocket. Mach 2, angle of attack $-4°$, initial mesh 4130 nodes[54]

well known, the 'no-slip' boundary condition will concentrate very rapid velocity gradients (and hence heat development) in a boundary layer attached to the surface of the body, and extremely fine meshes will now have to be used in this region, even if the flow is laminar and a steady-state solution is possible. (Of course, if turbulence occurs full steady-state solutions will not generally be possible and any resolution of the problem will need some additional modelling of the turbulent shears. However, even here steady solution states of average flow can exist if turbulence is small in scale.)

In Fig. 14.16 we show an early analysis which illustrates some of the features of such problems.[55] Here a 'bump' is introduced into a fully developed boundary layer in a stream of supersonic flow. This may represent an obstacle on a space shuttle. The data corresponding to the

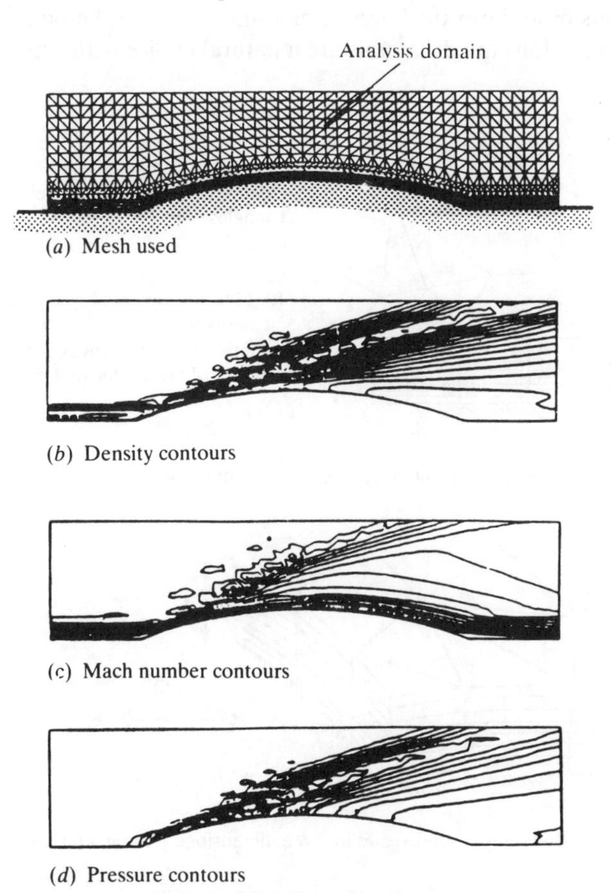

(*a*) Mesh used

(*b*) Density contours

(*c*) Mach number contours

(*d*) Pressure contours

Fig. 14.16 A surface 'bump' in a fully developed layer in hypersonic flow. Depth of bump $\sim 2 \times$ boundary layer thickness, Mach 7

boundary layer and free stream conditions at the inlet edge are introduced and a structured form of the mesh is used together with the explicit solver.

The results show clearly the formation of a shock (pressure contours) and the boundary layer (Mach number) where a fine mesh was used *a priori*.

Automatic mesh refinement for problems with boundary layers is difficult. In these, ideally, very elongated single elements will suffice as the main velocity gradients and curvatures occur in the directions normal to the surface. For problems involving such layers it is convenient to use a structured element form in a region encompassing the boundary layers and to use the previously described adaptive mesh outside, as shown in Fig. 14.17(*a*). Such a near 'body mesh' is conveniently divided into layers of elements of uniform thickness, decreasing in size to the boundary, and in two dimensions quadrilaterals are a natural choice with typically 10 to

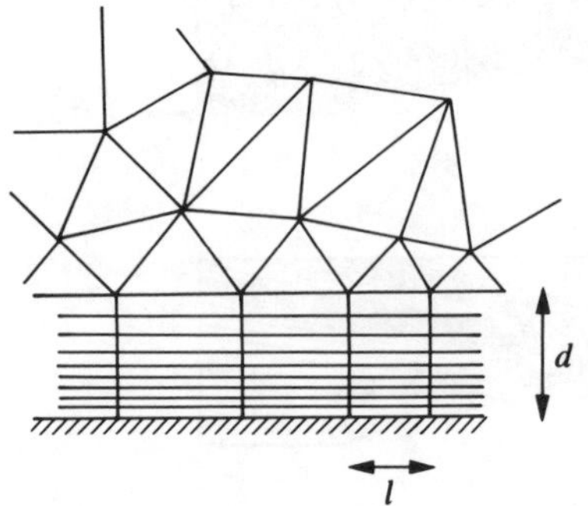

(*a*) A two-dimensional sublayer of structured quadrilaterals

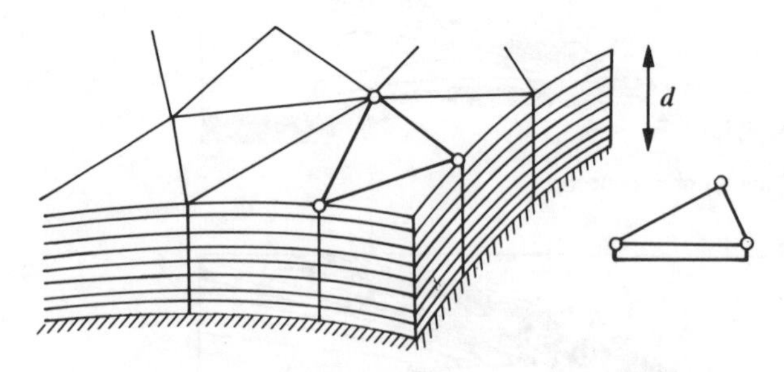

(*b*) A three-dimensional sublayer of prismatic elements

Fig. 14.17 Refinement in the boundary layer

15 elements in the thickness of the 'body layer'. In three dimensions where we desire to match an unstructural tetrahedral mesh, triangular prism layer elements are used as shown in Fig. 14.17(*b*).

The thickness of the body layer has to be decided upon by the user who can invoke previous experience here. All that is important is that

(*a*) the errors in each element be less than in the outer layers and
(*b*) it should be reasonably small to allow the main features of the boundary layer to be resolved.

In the application of such layers it was found[11,56,57] that a purely explicit algorithm is slow in converging to steady-state problems, even with the use of the time-step adjustment of Sec. 14.5. For this reason a locally implicit algorithm was developed using an alternating directed implicit (ADI) procedure. The examples we shall show have been solved with this process.

At this state it is worth while mentioning another possibility which at the time of writing has not yet been applied to the type of problem under discussion. In this an approximation is made to the governing equations in the 'body layer' or near the wall zone.[58,59] In this approximation all the derivatives in the equations corresponding to the tangential direction are omitted and the velocity in the normal direction is put as zero. Further, the pressure is assumed invariant in the normal direction to the surface. This approximation results in a 'one-dimensional' type of element coupling to the main flow field approximation achieved in the usual manner and the solution here appears economical.

14.9 Some examples of viscous compressible flow computation

In Figures 14.18 to 14.20, several computational results are shown in which full viscous effects are included. In all of these, structured element layers were used in the near body layer, and after the first solution an adaptive mesh was obtained outside of this. The qualitative improvement of result is readily observable.

Experimental results available[60] for the wave reflection and interaction with a boundary layer are used for comparison of results in Fig. 14.18. Noting that the negative skin friction results could not be measured, the agreement is quite impressive over most of the domain.

14.10 Compressible and incompressible behaviour

It is of interest to examine in more detail the basic governing equations (14.1) of compressible flow and to contrast these with the incompressible flow equations of Chapter 13 [viz. Eqs (13.10) and (13.11)].

Writing these now in the indicial notation we note that the mass

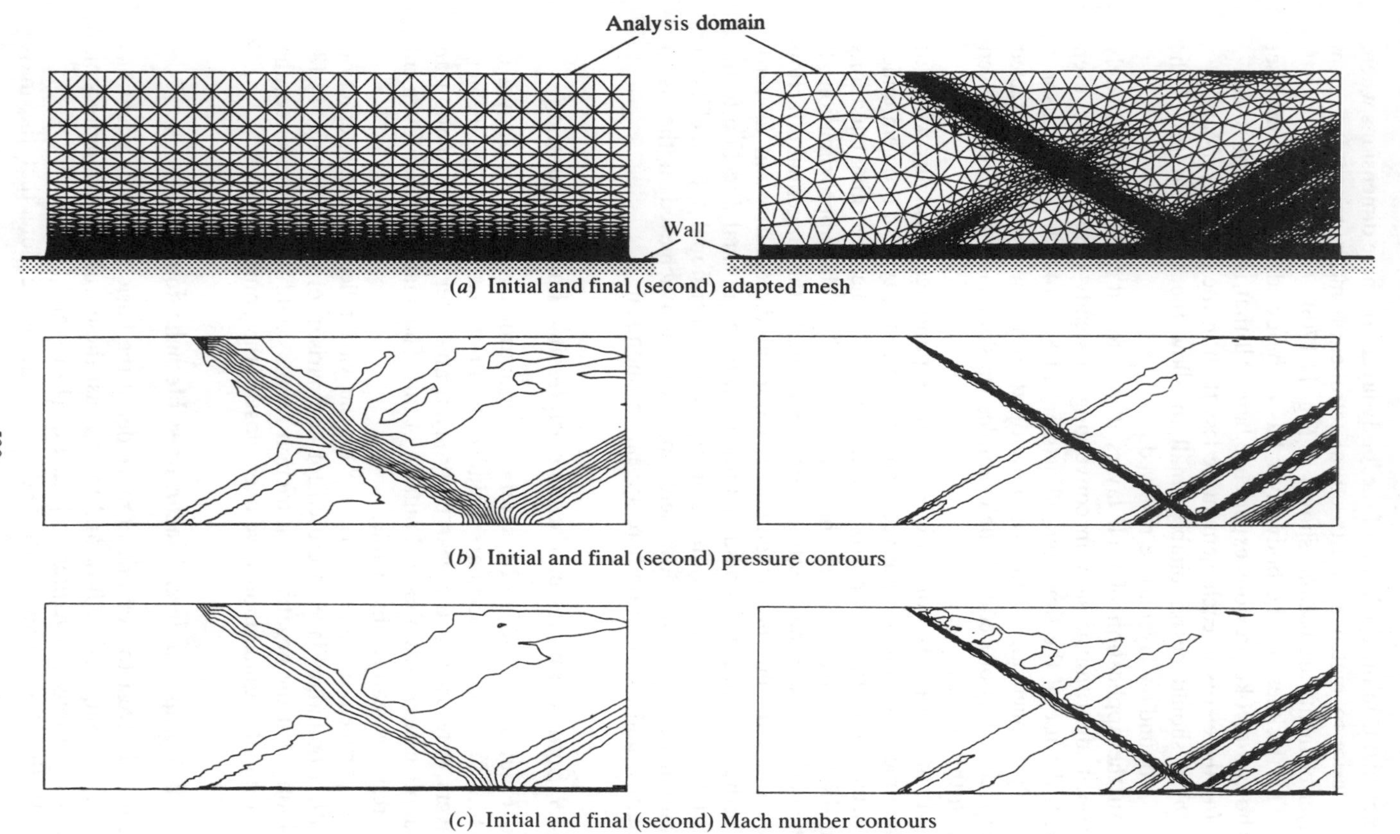

(*a*) Initial and final (second) adapted mesh

(*b*) Initial and final (second) pressure contours

(*c*) Initial and final (second) Mach number contours

Fig. 14.18 Shock–boundary layer interaction[11]

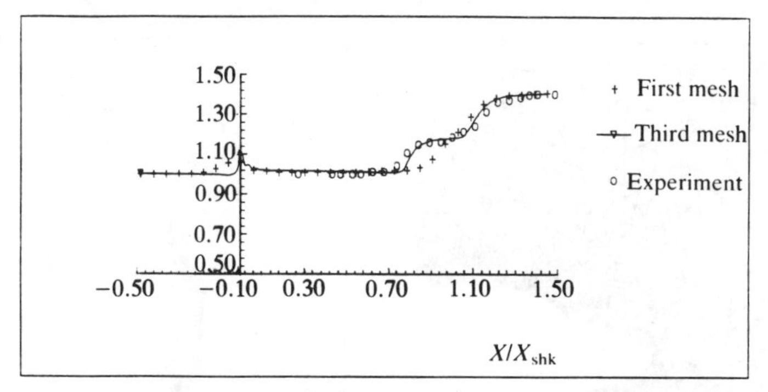

Surface pressure

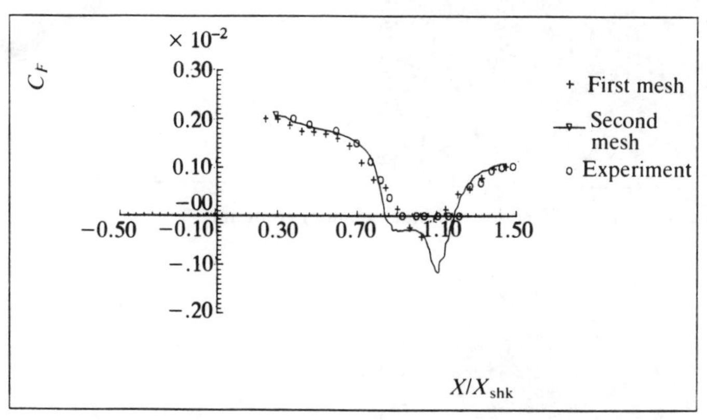

Skin friction

(*d*) Surface pressure and skin friction

Fig. 14.18 (*continued*)

conservation equation is

$$\frac{\partial \rho}{\partial t} + \left(\frac{\partial}{\partial x_i}\right)(\rho u_i) = 0 \tag{14.29}$$

and that the momentum conservation becomes

$$\left(\frac{\partial}{\partial t}\right)(\rho u_i) + \left(\frac{\partial}{\partial x_j}\right)(\rho u_i u_j) + \left(\frac{\partial}{\partial x_i}\right)p - \left(\frac{\partial}{\partial x_j}\right)\tau_{ij} + \rho f_i = 0 \tag{14.30}$$

Now differentiating the first we have

$$\frac{\partial \rho}{\partial t} + \frac{\rho \partial u_i}{\partial x_i} + \boxed{u_i \frac{\partial \rho}{\partial x_i}} = 0 \tag{14.31}$$

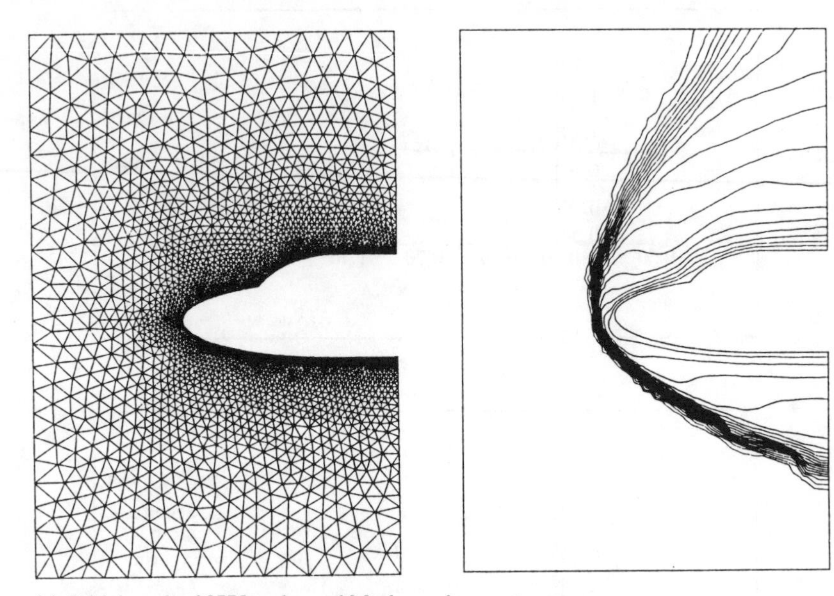

(*a*) Initial mesh of 2775 nodes and Mach number contours

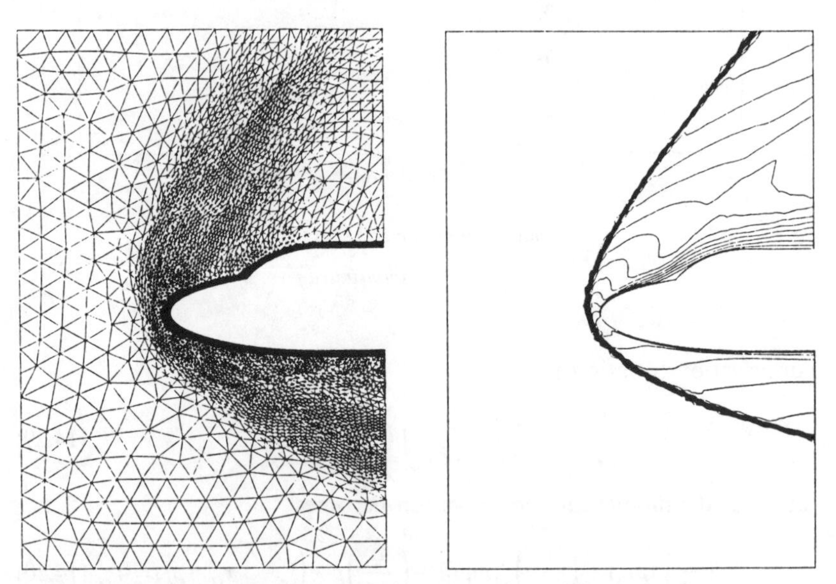

(*b*) First refined mesh of 2655 nodes on triangles and 15 × 200 structured nodes on sublayer and Mach number contours

Fig. 14.19 Viscous flow past a two-dimensional simulation of the forebody of a shuttle at Mach 2

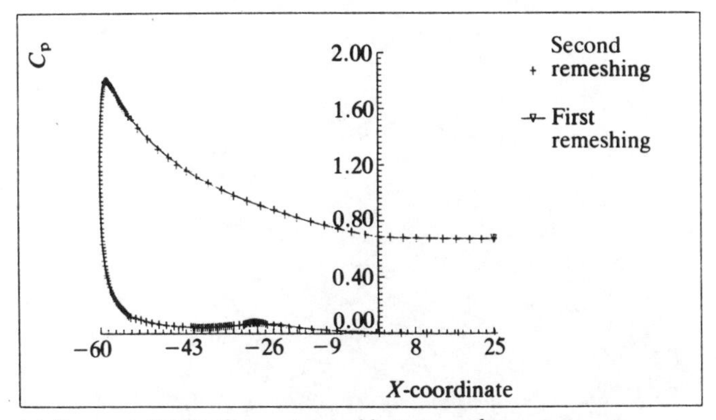

(c) Pressure distribution — top and bottom surfaces

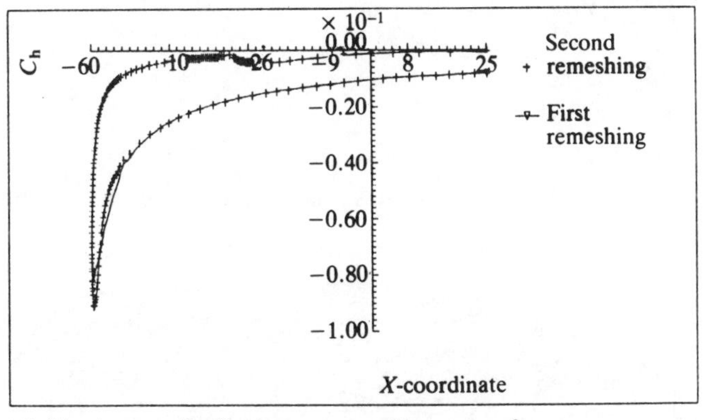

(d) Heating rate distribution — top and bottom surfaces

Fig. 14.19 *(continued)*

and on substraction of (14.29) from (14.30) we obtain

$$\rho\left(\frac{\partial u_i}{\partial t} + u_j\frac{\partial u_i}{\partial x_j}\right) + \frac{\partial p}{\partial x_i} - \frac{\partial \tau_{ij}}{\partial x_j} + \rho f_i = 0 \qquad (14.32)$$

Further, if we assume isothermal behaviour such that C is constant we have

$$\rho = \frac{p}{C^2}$$

and the equations become identical in steady state to those of incompressible flow *providing that the outlined term of Eq. (14.31) is negligible.* (This will occur if the sound velocity C is large or the flow velocities are small.)

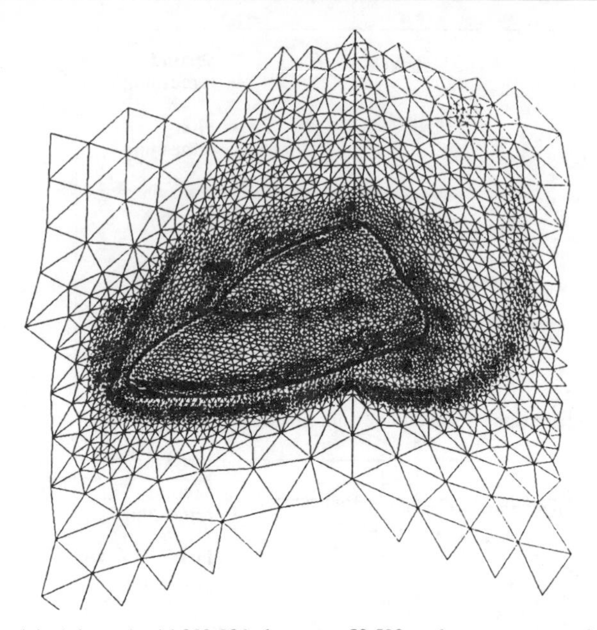

(*a*) Adapted grid 289 806 elements, 50 592 nodes, unstructured

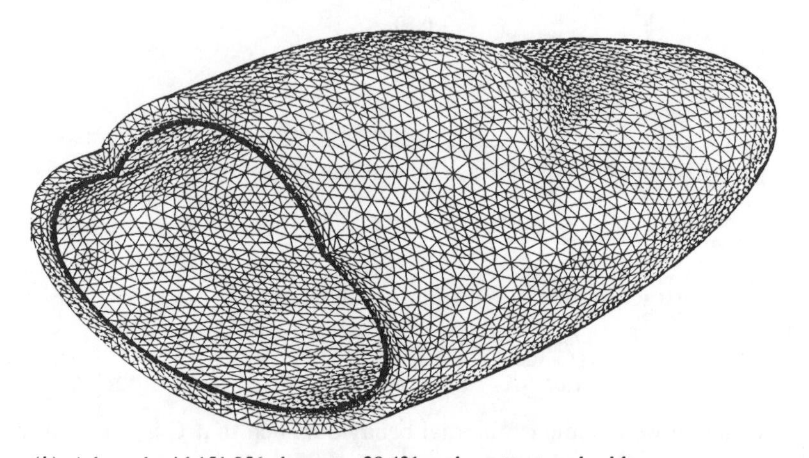

(*b*) Adapted grid 151 956 elements, 28 431 nodes, structured sublayer

Fig. 14.20 Viscous flow at Mach 8.15 around the forebody of a simulated shuttle. (Courtesy of Computation Dynamics Research Ltd, Swansea)

Immediately several conclusions can be made:

1. At small Mach numbers the flow patterns will correspond to those of incompressible flow—and in the absence of viscosity simple potential solutions are applicable.

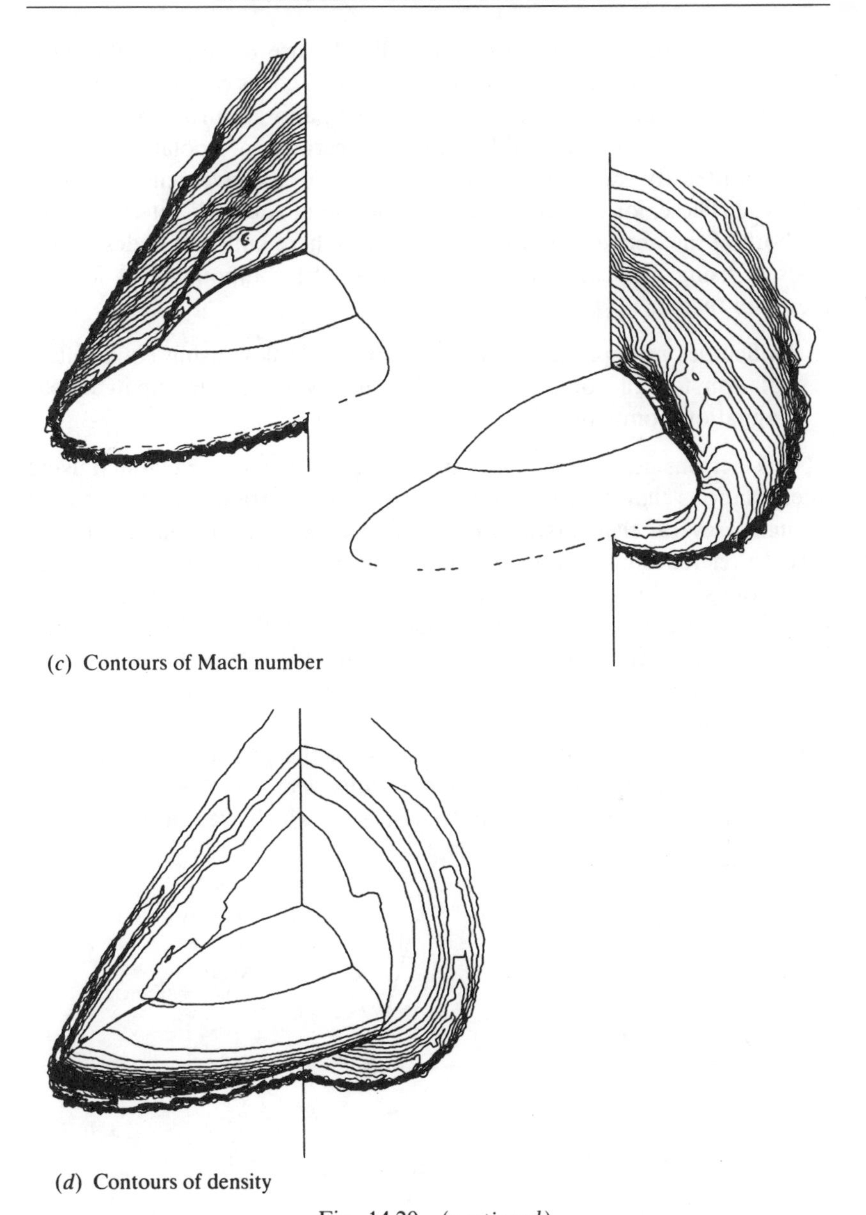

(*c*) Contours of Mach number

(*d*) Contours of density

Fig. 14.20 (*continued*)

2. At all Mach numbers of 'free' flow, near stagnation points of blunt bodies where the velocities are low, flow will be nearly incompressible in pattern.

The above means that, as suggested in Chapter 13, incompressible solutions can be frequently arrived at by the use of compressible flow codes. This occasionally is an advantage as such codes can deal with large problems using explicit procedures. However, another problem is posed. As we know in incompressible flow only certain interpolations of the pressure (density) and velocity are permissible if convergence is to be achieved. These, for instance, preclude the use of equal C_0 interpolations for both of the above equations, and we have here built our codes on the basis of using a linear interpolation for all the components. It follows therefore that:

3. Difficulties in approximation to compressible flow using equal interpolation for all components of **U** can always be anticipated near stagnation points of flow.

This fact was noted recently by Bristeau *et al.*[61-63] who suggested using interpolations that pass the incompressibility constraints. In the case of triangles, several elements illustrated in Fig. 14.21 are of course satisfactory. In reference 62 a dramatic improvement in results near the stagnation points by using such elements was illustrated, when a simple Galerkin approximation was used. Here, of course, the results using equal interpolation were totally unstable, but quite good performance was obtained without using any of the 'upwinding' techniques such as are implicit in the Taylor–Galerkin process.

It is of interest to observe that in most of the problems shown here, even though stagnation exists, good results were obtained in all cases using a certain amount of artificial viscosity, which was initially intro-

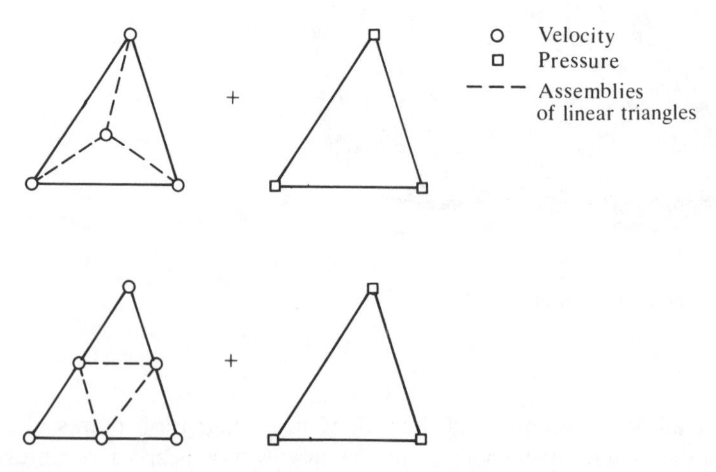

Fig. 14.21 Two triangular elements with C_0 interpolation of velocity and pressure which satisfy stability conditions

duced specifically to deal with shocks. Current research is being pursued to introduce such viscosity in a more rational manner to the mass conservation equation only to deal with the above incompressibility problem. Here it can be shown that with an operator split similar to the one used in Sec. 13.8.2 appropriate diffusion is automatically introduced and the problems outlined above avoided.[64]

14.11 Concluding remarks

The reader will have observed the dramatic progress in the solution of compressible flow problems by the finite element method that has occurred since 1984 when characteristic Galerkin procedures were first introduced. In the main we have concentrated here on a description of our methodology (using Taylor–Galerkin two-step algorithms and adaptivity based on error measures). This methodology proved most promising and problems previously insolvable using structured finite difference grids now yielded satisfactory solutions. However, many alternatives are possible and much yet remains to be done in making the computation more economical. Certainly in aeroengineering the growth of demand to utilize computational possibilities has been faster than that of the growth of computer hardware, and this provides a vital incentive for research.

References

1. C. Hirsch, *Numerical Computation of Internal and External Flows*, Vol. I, Wiley, Chichester, 1988.
2. L. Demkowicz, J. T. Oden and W. Rachowicz, 'A new finite element method for solving compressible Navier–Stokes equations based on an operator splitting method and h-p adaptivity', *Comp. Meth. Appl. Mech. Eng.*, **84**, 275–326, 1990.
3. R. Löhner, K. Morgan and O. C. Zienkiewicz, 'The solution of non-linear hyperbolic equation systems by the finite element method', *Int. J. Num. Meth. Fluids*, **4**, 1043–63, 1984.
4. R. Löhner, K. Morgan and O. C. Zienkiewicz, 'Domain splitting for an explicit hyperbolic solver', *Comp. Meth. Appl. Mech. Eng.*, **45**, 313–29, 1984.
5. O. C. Zienkiewicz, R. Löhner, K. Morgan and J. Peraire, 'High speed compressible flow and other advection dominated problems of fluid mechanics', in *Finite Elements in Fluids* (eds R. H. Gallagher, G. F. Carey, J. T. Oden and O. C. Zienkiewicz), Vol. 6, chap. 2, pp. 41–88, Wiley, Chichester, 1985.
6. R. Löhner, K. Morgan and O. C. Zienkiewicz, 'An adaptive finite element procedure for compressible high speed flows', *Comp. Meth. Appl. Mech. Eng.*, **51**, 441–65, 1985.
7. A. Lapidus, 'A detached shock calculation by second-order finite differences', *J. Comp. Phys.*, **2**, 154–77, 1967.
8. F. Angrant, V. Billey, A. Dervieux, C. Puoletty and B. Stoufflet,

'2-D and 3-D Euler flow calculations with a second order accurate Galerkin finite element method', in *18th Fluids Dynamics, Plasma Dynamics and Lasers Conf.*, 16–18 July, Cincinatti, Ohio, AIAA paper 85-1706, 1985.
9. R. LÖHNER, K. MORGAN, J. PERAIRE, O. C. ZIENKIEWICZ and L. KONG, 'Finite element methods for compressible flow', in *ICFD Conf. on Numerical Methods in Fluid Dynamics* (ed. K. W. Morton and M.J. Baines), Vol. II, pp. 27–53, Clarendon Press, Oxford, 1986.
10 R. W. MACCORMACK and B. S. BALDWIN, 'A numerical method for solving the Navier–Stokes equations with application to shock boundary interaction', AIAA paper 75-1, 1975.
11. O. HASSAN, 'Finite element computations of high speed viscous compressible flows', Ph.D. thesis, University of Wales, Swansea, 1990.
12. S. T. ZALESAK, 'Fully multidimensional flux-corrected transport algorithms for fluids', *J. Comp. Phys.*, **31**, 335–62, 1979.
13. J. P. BORIS and D. L. BOOK, 'Flux-corrected transport, I, SHASTA—a fluid transport algorithm that works', *J. Comp. Phys.*, **11**, 38–69, 1973.
14. G. ERLEBACHER, 'Solution adaptive triangular meshes with application to the simulation of plasma equilibrium', Ph.D. thesis, Columbia University, 1984.
15. R. LÖHNER, K. MORGAN, J. PERAIRE and M. VAHDATI, 'Finite element, flux corrected transport (FEM–FCT) for the Euler and Navier–Stokes equations', *Int. J. Num. Meth. Eng.*, **7**, 1093–109, 1987.
16. P. HANSBO and C. JOHNSON, 'Adaptive streamline diffusion methods for compressible flow using conservation variables', *Comp. Meth. Appl. Mech. Eng.* (to appear).
17. T. J. R. HUGHES and M. MALETT, 'A new finite element formulation for fluid dynamics, IV: A discontinuity capturing operator for multidimensional advective–diffusive systems', *Comp. Meth. Appl. Mech. Eng.*, **58**, 329–36, 1986.
18. G. SOD, 'A survey of several finite difference methods for systems of non-linear hyperbolic conservation laws', *J. Comp. Phys.*, **27**, 1–31, 1978.
19. T. E. TEZDUYAR and T. J. R. HUGHES,'Development of time accurate finite element techniques for first order hyperbolic systems with particular emphasis on Euler equation', Stanford University paper, 1983.
20. P. WOODWARD and P. COLELLA, 'The numerical simulation of two dimensional flow with strong shocks', *J. Comp. Phys.*, **54**, 115–73, 1984.
21. P. A. GNOFFO, 'A finite volume, adaptive grid algorithm applied to planetary entry flow fields', *JAIAA*, **29**, 1249–54, 1983.
22. R. LÖHNER, K. MORGAN and O. C. ZIENKIEWICZ, 'Adaptive grid refinement for the Euler and compressible Navier–Stokes equation', in *Proc. Int. Conf. on Accuracy Estimates and Adaptive Refinement in Finite Element Computations*, Lisbon, 1984.
23. M. J. BERGER and J. OLIGER, 'Adaptive mesh refinement for hyperbolic partial differential equations', *J. Comp. Phys.*, **53**, 484–512, 1984.
24. G. F. CAREY and H. DINH, 'Grading functions and mesh redistribution', *SIAM J. Num. Anal.*, **22**, 1028–40, 1985.
25. R. LÖHNER, K. MORGAN, J. PERAIRE and O. C. ZIENKIEWICZ, 'Finite element methods for high speed flows', AIAA paper 85-1531-CP, 1985.
26. O. C. ZIENKIEWICZ, K. MORGAN, J. PERAIRE, M. VAHDATI and R. LÖHNER, 'Finite elements for compressible gas flow and similar systems', in *7th Int. Conf. in Computational Methods in Applied Sciences and Engineering*, Versailles, December 1985.
27. B. PALMEIRO, V. BILLEY, A. DERVIAUX and J. PERIAUX, 'Self adaptive mesh

refinement and finite element methods for solving Euler equations', in *Proc. ICFD Conf. on Numerical Methods for Fluid Dynamics* (eds K. Morton and M. J. Baines), Clarendon Press, Oxford, 1986.

28. J. T. ODEN, P. DEVLOO and T. STROUBOULIS, 'Adaptive finite element methods for the analysis of inviscid compressible flow', *Comp. Meth. Appl. Mech. Eng.*, **59**, 327–62, 1986.
29. O. C. ZIENKIEWICZ, K. MORGAN, J. PERAIRE and J. Z. ZHU, 'Some expanding horizons for computational mechanics; error estimates, mesh generation and hyperbolic problems', in *Computational Mechanics–Advances and Trends* (ed. A. K. Noor), AMD 75, American Society of Mechanical Engineers, 1986.
30. J. F. DANNENHOFER and J. R. BARON, 'Robust grid adaptation for coupled trans-sonic flows', AIAA paper 86-0495, 1986.
31. R. LÖHNER, K. MORGAN and O. C. ZIENKIEWICZ, 'Adaptive grid refinement for the Euler and compressible Navier–Stokes equations', in *Accuracy Estimates and Adaptive Refinements in Finite Element Computations* (eds I. Babuska, O. C. Zienkiewicz, J. Gago and E. R. de A. Oliveira), chap. 15, pp. 281–98, Wiley, Chichester, 1986.
32. J. PERAIRE, M. VAHDATI, K. MORGAN and O. C. ZIENKIEWICZ, 'Adaptive remeshing for compressible flow computations', *J. Comp. Phys.*, **72**, 449–66, 1987.
33. J. T. ODEN, T. STROUBOULIS and P. DEVLOO, 'Adaptive finite element methods for high speed compressible flows', *Int. J. Num. Meth. Eng.*, **7**, 1211–28, 1987.
34. J. H. S. LEE, J. PERAIRE and O. C. ZIENKIEWICZ, 'The characteristic–Galerkin method for advection dominated problems—an assessment', *Comp. Meth. Appl. Mech. Eng.*, **61**, 359–69, 1987.
35. J. PERAIRE, K. MORGAN, J. PEIRO and O. C. ZIENKIEWICZ, 'An adaptive finite element method for high speed flows', in *AIAA 25th Aerospace Sciences Meeting*, Reno, Nevada, AIAA paper 87-0558, 1987.
36. O. C. ZIENKIEWICZ, J. Z. ZHU, Y. C. LIU, K. MORGAN and J. PERAIRE, 'Error estimates and adaptivity; from elasticity to high speed compressible flow', in *The Mathematics of Finite Elements and Application* (*MAFELAP 87*) (ed. J. R. Whiteman), pp. 483–512, Academic Press, London, 1988.
37. L. FORMAGGIA, J. PERAIRE and K. MORGAN, 'Simulation of state separation using the finite element method', *Appl. Math. Modelling*, **12**, 175–81, 1988.
38. O. C. ZIENKIEWICZ, K. MORGAN, J. PERAIRE, J. PEIRO and L. FORMAGGIA, 'Finite elements in fluid mechanics. Compressible flow, shallow water equations and transport', in *ASME Conf. on Recent Development in Fluid Dynamics*, AMD 95, American Society of Mechanical Engineers, December 1988.
39. J. PERAIRE, J. PEIRO, L. FORMAGGIA, K. MORGAN and O. C. ZIENKIEWICZ, 'Finite element Euler computations in 3-dimensions', *Int. J. Num. Meth. Eng.*, **26**, 2135–59, 1989. (See also same title: *AIAA 26th Aerospace Sciences Meeting*, Reno, AIAA paper 87-0032, 1988.)
40. O. C. ZIENKIEWICZ and J. Z. ZHU, 'A simple error estimator and adaptive procedure for practical engineering analysis', *Int. J. Num. Meth. Eng.*, **24**, 337–57, 1987.
41. J. T. ODEN and L. DEMKOWICZ, 'Advance in adaptive improvements: a survey of adaptive methods in computational fluid mechanics', in *State of the Art Survey in Computational Fluid Mechanics* (eds A. K. Noor and J. T. Oden), American Society of Mechanical Engineers, 1988.

42. S. H. Lo, 'A new mesh generation scheme for arbitrary planar domains', *Int. J. Num. Mech. Eng.*, **21**, 1403–26, 1985.
43. D. N. Shenton and Z. J. Cendes, 'Three dimensional finite element mesh generation using Delaunay tesselation', *IEEE Trans. on Magnetism*, **MAG-21**, 2535–8, 1985.
44. M. Tenemura, T. Ogawa and N. Ogitta, 'A new algorithm for three dimensional Voronoi tesselation', *J. Comp. Phys.*, **51**, 191–207, 1983.
45. M. A. Yerry and M. S. Shepherd, 'Automatic three dimensional mesh generation by a modified OCTREE technique', *Int. J. Num. Mech. Eng.*, **20**, 1965–90, 1984.
46. W. J. Schroeder and M. S. Shepherd, 'A combined Octree/Delaunay method for fully automatic 3-D mesh generation', *Int. J. Num. Meth. Eng.*, **29**, 37–56, 1990.
47. J. R. Stewart, R. R. Thareja, A. R. Wieting and K. Morgan, 'Application of finite elements and remeshing techniques to shock interference on a cylindrical leading edge', Reno, Nevada, AIAA paper 88-0368, 1988.
48. A. Jameson, T. J. Baker and N.P. Weatherill, 'Calculation of inviscid transonic flow over a complete aircraft', *AIAA 24th Aerospace Sci. Meeting*, Reno, Nevada, AIAA paper 86-0103, 1986.
49. N. P. Weatherill, 'Mixed structured–unstructured meshes for aerodynamic flow simulation', *Aeronaut. J.*, **94**, 111–23, 1990.
50. R. R. Thareja, J. R. Stewart, O. Hassan, K. Morgan and J. Peraire, 'A point implicit unstructured grid solver for the Euler and Navier–Stokes equation', *Int. J. Num. Meth. Fluids*, **9**, 405–25, 1989.
51. R. Löhner, 'Adaptive remeshing for transient problems with moving bodies', in *National Fluid Dynamics Congress*, Ohio, AIAA paper 88-3736, 1988.
52. R. Löhner, 'The efficient simulation of strongly unsteady flows by the finite element method', in *25th Aerospace Sci. Meeting*, Reno, Nevada, AIAA paper 87-0555, 1987.
53. R. Löhner, 'Adaptive remeshing for transient problems', *Comp. Meth. Appl. Mech. Eng.*, **75**, 195–214, 1989.
54. E. J. Probert, 'Finite element methods for transient compressible flow', Ph.D. thesis, University of Wales, Swansea, 1989.
55. J. Peraire, 'A finite element method for convection dominated flows', Ph.D. thesis, University of Wales, Swansea, 1986.
56. O. Hassan, K. Morgan and J. Peraire, 'An implicit–explicit scheme for compressible viscous high speed flows', *Comp. Meth. Appl. Mech. Eng.*, **76**, 245–58, 1989.
57. P. Rosland and B. Stoufflet, 'Simulation of compressible viscous flows by an implicit method on unstructured meshes', in *Proc. 7th Int. Conf. on Finite Element Methods in Flow Problems*, pp. 345–50, University of Alabama Press, Huntsville, 1989.
58. J. Y. Xia, C. Taylor and J. O. Medwell, 'Finite element modelling of the near wall zone of confined turbulent flows', *Eng. Comp.*, **6**, 127–32, 1989.
59. C. E. Thomas, K. Morgan and C. Taylor, 'A finite element analysis of flow over a backward facing step', *Comp. Fluids*, **9**, 265–78, 1981.
60. R. J. Hakkinen, I. Greberm, L. Trilling and S. S. Abarbanel, 'The interaction of an oblique shock with a laminar boundary layer', memo 2-28-59W, NASA, 1959.
61. M. O. Bristeau, R. Glowinski, B. Mantel, J. Periaux and G. Rogé, 'Self adaptive finite element method for 3-D, compressible, Navier–Stokes flow

simulation in aerospace engineering', in *Proc. 11th Conf. on Numerical Methods in Fluid Dynamics*, Williamsburg, Pa., 1988.
62. M. O. BRISTEAU, R. GLOWINSKI, L. DUTTO, J. PÉRIAUX and G. ROGÉ, 'Compressible viscous flow calculations using compatible finite element approximations', *Int. J. Num. Meth. Fluids*, **11**, 719–49, 1990.
63. M. O. BRISTEAU, M. MALLET, J. PERIAUX and G. ROGÉ, 'Developments of finite element methods for compressible Navier–Stokes flow simulations in aerospace design', Reno, Nevada, AIAA paper 90-0403, 1990.
64. O. C. ZIENKIEWICZ, 'Explicit (or semiexplicit) general algorithm for compressible and incompressible flows with equal finite element interpolation', Report 90.5, Chalmers University of Technology, Göteborg, Sweden, 1990.

15

Shallow water equations and waves

15.1 Introduction

The flow of water in shallow layers such as occur in coastal estuaries, oceans, rivers, etc., is of obvious practical importance. The prediction of tidal currents and elevations is vital for navigation and for determination of pollutant dispersal which, unfortunately, is still frequently deposited there. The transport of sediments associated with such flows is yet another field of interest.

In the free surface flow in relatively thin layers the horizontal velocities are of primary importance and the problem can be reasonably approximated in two dimensions. Here we find that the resulting equations, which include in addition to the horizontal velocities the free surface elevation, can once again be written in the conservation form [viz. Eqs (12.1) and (14.1)]:

$$\frac{\partial \mathbf{U}}{\partial t}+\frac{\partial \mathbf{F}_i}{\partial x_i}+\frac{\partial \mathbf{G}_i}{\partial x_i}+\mathbf{Q}=0 \qquad \text{for } i=1,2 \tag{15.1}$$

Indeed, the detailed form of these equations bears a striking similarity to those of compressible gas flow—despite the fact that now a purely incompressible fluid (water) is considered. It follows therefore that:

1. The methods developed in the previous chapter are in general applicable.
2. The type of phenomena (e.g. shocks, etc.) which we have encountered in Chapter 14 will again occur.

It will of course be found that practical interest focuses on different aspects. The objective of this, short, chapter is therefore to introduce the basis of the equation derivation and to illustrate the numerical approximation techniques by a series of examples.

The approximations made in the formulation of the flow in shallow water bodies are similar in essence to those describing the flow of air in the earth's environment and hence are widely used in meteorology. Here the vital subject of weather prediction involves their solution daily and a very large amount of computation. The interested reader will find much of the background in standard texts dealing with the subject, e.g. references 1 and 2.

A particular area of interest occurs in the linearized version of the shallow water equations which, in periodic response, are similar to those describing acoustic phenomena (viz. Chapter 11). In the second part of this chapter we shall therefore discuss some of these periodic phenomena involved in the action and forces due to waves.[3]

15.2 The basis of shallow water equations

In Chapter 13 we have introduced the essential Navier–Stokes equations and presented their incompressible, isothermal, form in Eqs (13.25a) and (13.25b) which we repeat below assuming full incompressibility. We have now the equations of mass conservation:

$$\frac{\partial u_i}{\partial x_i}=0 \tag{15.2a}$$

and momentum conservation:

$$\frac{\partial u_j}{\partial t}+\frac{\partial}{\partial x_i}(u_i u_j)+\frac{1}{\rho}\frac{\partial p}{\partial x_j}-\frac{1}{\rho}\frac{\partial}{\partial x_i}\tau_{ij}-f_j=0 \tag{15.2b}$$

with i, j being 1,2,3.

In the case of shallow water flow which we illustrate in Fig. 15.1 and where the direction x_3 is vertical, the vertical velocity u_3 is small and the corresponding accelerations negligible. The momentum equation in the vertical direction can therefore be reduced to

$$\frac{1}{\rho}\frac{\partial p}{\partial x_3}+g=0 \tag{15.3}$$

where $f_3=-g$, is the gravity acceleration. After integration this yields

$$p=\rho g(\eta-x_3)+p_a \tag{15.4}$$

as, when $x_3=\eta$, the pressure is atmospheric (p_a) (which can on occasion not be constant over the body of the water and thus influence its motion).

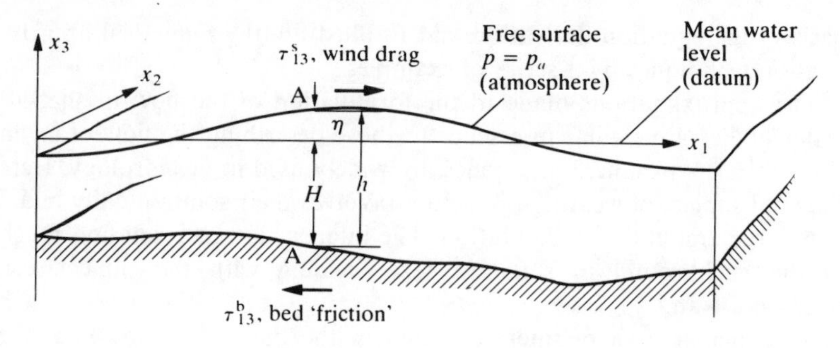

(*a*) Coordinates

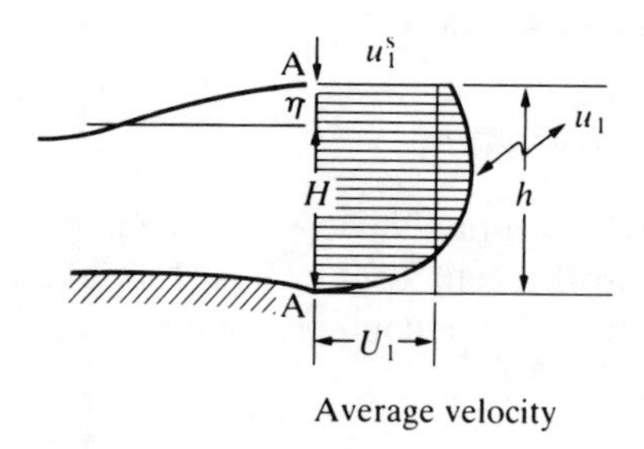

(*b*) Velocity distribution

Fig. 15.1 The shallow water problem. Notation

On the free surface the vertical velocity u_3 can of course be related to the total time derivative of the surface elevation, i.e.

$$u^s_3 = \frac{D\eta}{dt} \equiv \frac{\partial \eta}{\partial t} + u^s_1 \frac{\partial \eta}{\partial x_1} + u^s_2 \frac{\partial \eta}{\partial x_2} \tag{15.5a}$$

Similarly, at the bottom,

$$u^b_3 = \frac{DH}{dt} \equiv u^b_1 \frac{\partial \eta}{\partial x_1} + u^b_2 \frac{\partial \eta}{\partial x_2} \tag{15.5b}$$

as the total depth H, naturally, does not vary with time. Further, if we assume that for viscous flow no slip occurs then

$$u^b_1 = u^b_2 = 0 \tag{15.6}$$

and also by continuity

$$u^b_3 = 0$$

Now a further approximation will be made. In this the governing equations will be integrated with the depth coordinate x_3 and *depth-averaged governing equations* derived. We shall start with the continuity equation (15.2a) and integrate this in the x_3 direction, writing

$$\int_{-H}^{\eta} \frac{\partial u_3}{\partial x_3}\,\mathrm{d}x_3 + \int_{-H}^{\eta} \frac{\partial u_1}{\partial x_1}\,\mathrm{d}x_3 + \int_{-H}^{\eta} \frac{\partial u_2}{\partial x_2}\,\mathrm{d}x_3 = 0 \tag{15.7}$$

As the velocities u_1 and u_2 are unknown and are not uniform, as shown in Fig. 15.1(*b*), it is convenient at this stage to introduce the notion of average velocities defined so that

$$\int_{-H}^{\eta} u_i\,\mathrm{d}x_3 = U_i(H+\eta) \equiv U_i h \tag{15.8}$$

with $i=1,2$. We shall now recall the Leibnitz rule of integrals stating that for any function $F(r,s)$ we can write

$$\int_{-a}^{b} \frac{\partial}{\partial s}F(r,s)\,\mathrm{d}r \equiv \frac{\partial}{\partial s}\int_{-a}^{b} F(r,s)\,\mathrm{d}r - F(b,s)\,\frac{\partial b}{\partial s} + F(a,s)\,\frac{\partial a}{\partial s} \tag{15.9}$$

With the above we can rewrite the last two terms of Eq. (15.7) and introducing Eq. (15.6) we obtain

$$\int_{-H}^{\eta} \frac{\partial u_i}{\partial x_i}\,\mathrm{d}x_3 = \frac{\partial}{\partial x_i}(U_i h) - u_i^{\mathrm{s}}\,\frac{\partial \eta}{\partial x_i} \tag{15.10}$$

with $i=1,2$. The first term of Eq. (15.7) is, by simple integration, given as

$$\int_{-H}^{\eta} \frac{\partial u_3}{\partial x_3}\,\mathrm{d}x_3 = u_3^{\mathrm{s}} \tag{15.11}$$

which, on using (15.5a), becomes

$$\int_{-H}^{\eta} \frac{\partial u_3}{\partial x_3}\,\mathrm{d}x_3 = \frac{\partial \eta}{\partial t} + u_i^{\mathrm{s}}\,\frac{\partial \eta}{\partial x_i} \tag{15.12}$$

Addition of Eqs (15.10) and (15.12) gives the depth-averaged continuity conservation finally as

$$\frac{\partial \eta}{\partial t} + \frac{\partial (hU_i)}{\partial x_i} \equiv \frac{\partial h}{\partial t} + \frac{\partial (hU_i)}{\partial x_i} = 0 \tag{15.13}$$

Now we shall perform similar depth integration on the momentum equations in the horizontal directions. We have thus

$$\int_{-H}^{\eta} \left[\frac{\partial u_i}{\partial t} + \frac{\partial}{\partial x_j}(u_i\,u_j) + \frac{1}{\rho}\frac{\partial p}{\partial x_i} - \frac{1}{\rho}\frac{\partial \tau_{ij}}{\partial x_i} - f_i \right] \mathrm{d}x_3 = 0 \tag{15.14}$$

with $i=1,2$.

Proceeding as before we shall find after some algebraic manipulation that a conservative form of depth-averaged equations becomes

$$\frac{\partial(hU_i)}{\partial t}+\frac{\partial}{\partial x_j}\left[hU_iU_j+\delta_{ij}\frac{1}{2}g(h^2-H^2)-\frac{1}{\rho}\int_{-H}^{\eta}\tau_{ij}\,\mathrm{d}x_3\right]-$$

$$-\frac{1}{\rho}(\tau_{3i}^{\mathrm{s}}-\tau_{3i}^{\mathrm{b}})-hf_i-g(h-H)\frac{\partial H}{\partial x_i}+\frac{h}{\rho}\frac{\partial p_a}{\partial x_i}=0 \qquad (15.15)$$

In the above the shear stresses on the surface can be prescribed externally, given, say, the wind drag. The bottom shear is frequently expressed by suitable hydraulic resistance formulae, e.g. the Chézy expression, giving

$$\tau_{3i}^{\mathrm{b}}=\frac{\rho g|U|U_i}{Ch^2} \qquad (15.16)$$

where

$$|U|=\sqrt{U_i\,U_i}$$

and C is the Chézy coefficient.

In Eq. (15.15) f_i stands for the Coriolis accelerations, important in large-scale problems and defined as

$$f_1=\hat{f}U_2 \qquad f_2=-\hat{f}U_1 \qquad (15.17)$$

where $\hat{f}$ is the Coriolis parameter.

The τ_{ij} stresses require the definition of a viscosity coefficient, μ_H, generally of the averaged turbulent kind, and we have

$$\tau_{ij}=\mu_H\left(\frac{\partial u_i}{\partial x_j}+\frac{\partial u_j}{\partial x_i}-\frac{2}{3}\frac{\partial u_i}{\partial x_i}\right) \qquad (15.18)$$

Approximating in terms of average velocities the remaining integral of Eq. (15.15) can be written as

$$\frac{1}{\rho}\int_{-H}^{\eta}\tau_{ij}\,\mathrm{d}x_3\approx\frac{1}{\rho}\mu_H h\left(\frac{\partial U_i}{\partial x_j}+\frac{\partial U_j}{\partial x_i}-\frac{2}{3}\frac{\partial U_i}{\partial x_2}\right)=\frac{h}{\rho}\bar{\tau}_{ij} \qquad (15.19)$$

Equations (15.13) and (15.15) cast the shallow water problem in the general form of Eq. (15.1), where the appropriate vectors are defined below.

Thus, with $i=1,2$,

$$\mathbf{U}=\begin{Bmatrix} h \\ hU_1 \\ hU_2 \end{Bmatrix} \qquad (15.20a)$$

$$\mathbf{F}_i = \begin{Bmatrix} hU_i \\ hU_1U_i + \delta_{1i}\frac{1}{2}g(h^2 - H^2) \\ hU_2U_i + \delta_{2i}\frac{1}{2}g(h^2 - H^2) \end{Bmatrix} \tag{15.20b}$$

$$\mathbf{G}_i = \begin{Bmatrix} 0 \\ -(h/\rho)\bar{\tau}_{1i} \\ -(h/\rho)\tau_{2i} \end{Bmatrix} \tag{15.20c}$$

in which the relation (15.19) is used to give the internal average $\bar{\tau}$ in terms of the average velocity gradients and

$$\mathbf{Q} = \begin{Bmatrix} 0 \\ -h\hat{f}U_2 - g(h-H)\dfrac{\partial H}{\partial x_1} + \dfrac{h}{\rho}\dfrac{\partial p_a}{\partial x_1} - \dfrac{1}{\rho}\tau_{31}^s + \dfrac{gU_1|U|}{Ch^2} \\ h\hat{f}U_1 - g(h-H)\dfrac{\partial H}{\partial x_2} + \dfrac{h}{\rho}\dfrac{\partial p_a}{\partial x_2} - \dfrac{1}{\rho}\tau_{32}^s + \dfrac{gU_2|U|}{Ch^2} \end{Bmatrix} \tag{15.20d}$$

The above, conservative, form of shallow water equations was first presented in references 4 and 5 and is generally applicable. However, many variants of the general shallow water equations exist in the literature, introducing various approximations.

The following sections of this chapter will be divided into two parts. In the first, Part I, we shall discuss time-stepping solutions of the full set of the above equations in transient situations and in corresponding steady-state applications. Here non-linear behaviour will of course be included but for simplicity some terms will be dropped. In particular, we shall omit in most of the examples the consideration of viscous stresses such as $\bar{\tau}_{2i}, \bar{\tau}_{1i}$, whose influence is small as compared with the bottom drag stresses. This will, incidentally, help in the solution, as second-order derivatives now disappear and boundary layers can be eliminated.

The second, Part II, will deal with the linearized form of Eqs (15.13) and (15.15). Here we see immediately that on omission of all non-linear terms, bottom drag, etc., and approximating $h \sim H$, we can write these equations as

$$\frac{\partial h}{\partial t} + \frac{\partial}{\partial x_i}(HU_i) = 0 \tag{15.21a}$$

$$\frac{\partial (HU_i)}{\partial t} + gH\frac{\partial}{\partial x_i}(h - H) = 0 \tag{15.21b}$$

Noting that

$$\eta = h - H \qquad \text{and} \qquad \frac{\partial h}{\partial t} = \frac{\partial \eta}{\partial t}$$

the above becomes

$$\frac{\partial \eta}{\partial t} + \frac{\partial}{\partial x_i}(HU_i) = 0 \tag{15.22a}$$

$$\frac{\partial (HU_i)}{\partial t} + gH\frac{\partial \eta}{\partial x_i} = 0 \tag{15.22b}$$

Elimination of HU_i yields immediately

$$\frac{\partial^2 \eta}{\partial t^2} - \frac{\partial}{\partial x_i}\left(gH\frac{\partial \eta}{\partial x_i}\right) = 0 \tag{15.23}$$

or the standard Helmholtz wave equation. For this, of course, many special solutions are possible by utilizing the procedures of Chapters 9 to 11 and some attention will be given to these.

The shallow water equations derived in this section consider only the depth-averaged flows and hence cannot reproduce certain phenomena that occur in nature and in which some velocity variation with depth has to be allowed for. In many such problems the basic assumption of a vertically hydrostatic pressure distribution is still valid and a form of shallow water behaviour can be assumed.

The extension of the formulation can be achieved by an *a priori* division of the flow into strata in each of which different velocities occur. The final set of discretized equations consists then of several, coupled, two-dimensional approximations. Alternatively, the same effect can be introduced by using several different velocity 'trial functions' for the vertical distribution, as was suggested by Zienkiewicz and Heinrich.[6] Such generalizations are useful but outside the scope of the present text.

PART I:
DISCRETIZATION AND SOLUTION OF THE FULL SHALLOW WATER EQUATIONS

15.3 Numerical approximation

Both finite difference and finite element procedures have for many years been used widely in solving the shallow water equations. The latter approximation has been applied relatively recently and Kawahara[7] surveys the early applications to coastal and oceanographic engineering. In most of these the standard procedures of spatial discretization followed by suitable time-stepping schemes are adopted.[8-15] In meteorology the first application of the finite element method dates back to 1972, as reported in the survey given in reference 16, and the range of applications has been increasing steadily.[17-30] Once again the procedures used follow similar lines to those used for coastal and oceanographic problems and in both either fully implicit or semi-implicit methods have been used for the time domain. Similarly, for steady-state solution problems direct methods of gaussian elimination have generally been adopted.

As the equations are convection dominated and of a form very similar to those of compressible flow, we suggest that the procedure identical to that adopted in the previous chapter be used. This has already proved very effective and its application to shallow water explored.[4,5] Indeed, the assessment of this method shows a very favourable comparision with alternatives.[31] There is therefore no need to elaborate on this as the two-step explicit Taylor–Galerkin scheme has been fully described. Indeed, at this stage the reader may well observe that with the exception of certain source terms, the isothermal compressible flow equations can be transformed into shallow water equations with the variables being changed as follows:

$$\begin{array}{llll} \rho & \text{(density)} \longrightarrow & h & \text{(depth)} \\ u_i & \text{(velocity)} \longrightarrow & U_i & \text{(mean velocity)} \\ p & \text{(pressure)} \longrightarrow & \frac{1}{2}g(h^2-H^2) & \end{array}$$

Thus the writing of a separate solution program is now superfluous providing that the above changes are introduced.

In the examples that follow we shall illustrate several problems solved by the above procedures. However, we should remark that on occasion the time-step limitations set by the explicit method, which in general require that

$$\Delta t_{\text{ctit}} \leqslant \frac{h}{c} \tag{15.24}$$

where $c=\sqrt{gh}$ is the gravity wave velocity, may be too restrictive. For such situations various forms of the implicit solution procedures indicated in Chapter 12 may be required for use. As an alternative the operator split technique described in Chapter 13, Sec. 13.8.2, could well be adopted with some modifications.

15.4 Examples of application

15.4.1 *Transient one-dimensional problems—a performance assessment.* In this section we present some relatively simple examples in the one-space dimension to illustrate the applicability of the two-step algorithm.

The first, illustrated in Fig. 15.2, shows the progress of a solitary wave[32] onto a shelving beach. This frequently studied situation[33,34] shows well the progressive steepening of the wave often obscured by schemes that are very dissipative.

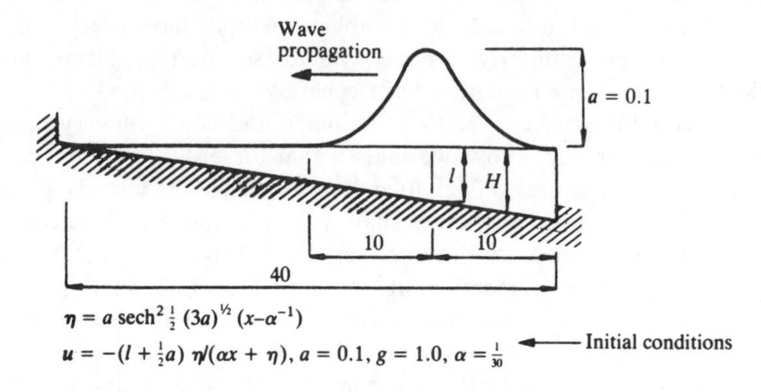

$\eta = a \operatorname{sech}^2 \frac{1}{2} (3a)^{1/2} (x - \alpha^{-1})$

$u = -(l + \frac{1}{2}a)\, \eta/(\alpha x + \eta),\ a = 0.1,\ g = 1.0,\ \alpha = \frac{1}{30}$ ← Initial conditions

(*a*) Problem statement

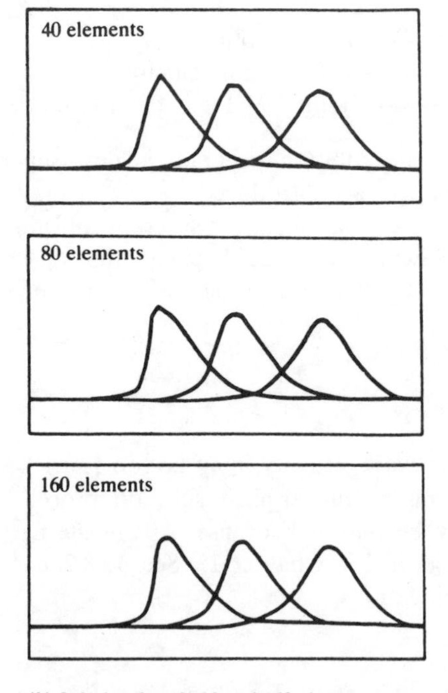

(*b*) Solution for 40, 80 and 160 elements at various times

Fig. 15.2 Shoaling of a wave

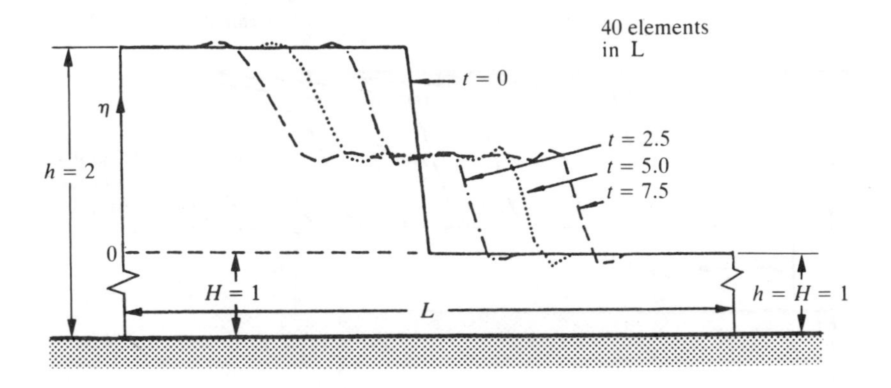

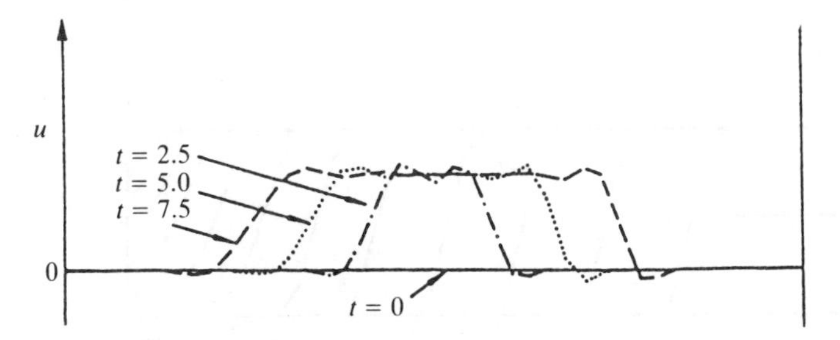

Fig. 15.3 Propagation of waves due to a dam break ($C_{\text{Lap}}=0$). 40 elements in analysis domain. $C=\sqrt{gH}=1$, $\Delta t=0.25$

The second example, of Fig. 15.3, illustrates the so-called 'dam break' problem diagrammatically. Here a dam separating two stationary water levels is suddenly removed and the almost vertical waves progress into the two domains. This problem, somewhat similar to those of a shock tube in compressible flow, has been solved quite successfully even without artificial diffusivity.

The final example of this section, Fig. 15.4, shows the formation of an idealized 'bore' or a steep wave progressing into a channel carrying water at a uniform speed caused by a gradual increase of the downstream water level. Despite the fact that the flow speed is 'subcritical' (i.e. velocity $<\sqrt{gh}$), a progressively steepening, travelling shock clearly develops.

15.4.2 *Two-dimensional periodic tidal motions.* The extension of the computation into two space dimensions follows the same pattern as that described in compressible formulations. Again linear triangles are used to interpolate the values of h, hU_1 and hU_2. The main difference in the solutions is that of emphasis. In the shallow water problem shocks either

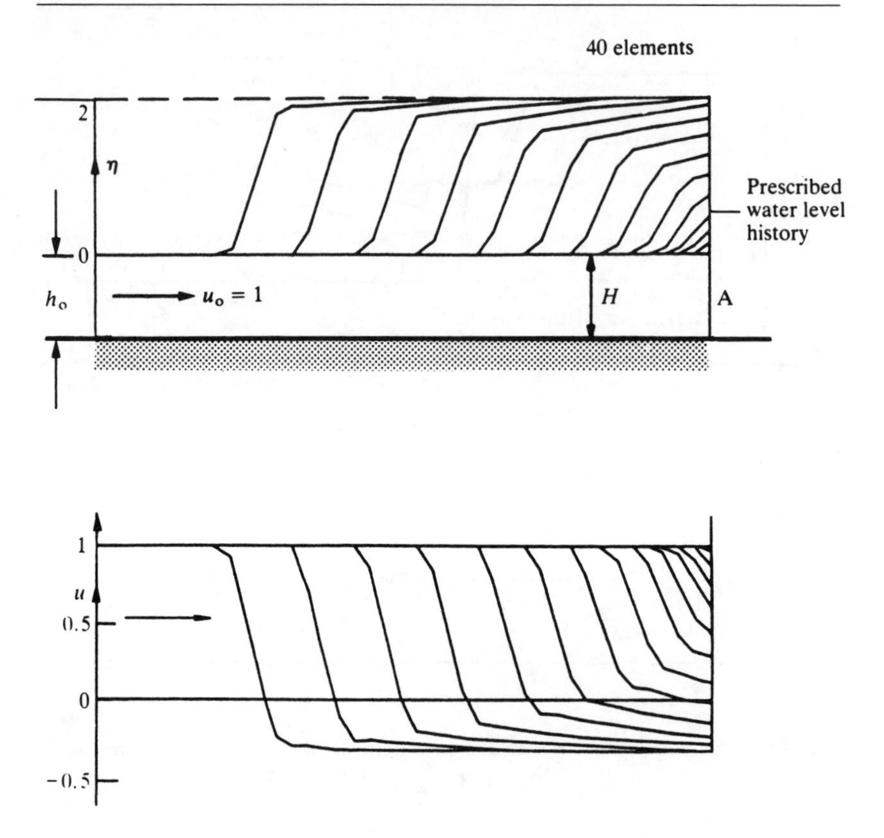

Fig. 15.4 A 'bore' created in a stream due to water level rise downstream (A). Level at A, $\eta = 1 - \cos \pi t/30 (0 \leqslant t \leqslant 30)$, $2 (30 \leqslant t)$. Levels and velocities at intervals of 5 time units, $\Delta t = 0.5$

do not develop or are sufficiently dissipated by the existence of bed friction so that the need for artificial viscosity and local refinement is not generally present. For this reason we have not introduced here the error measures and adaptivity—finding that meshes sufficiently fine to describe the geometry also usually prove sufficiently accurate.

The first example of Fig. 15.5 is presented merely as a test problem. Here the frictional resistance is linearized and an exact solution known for a periodic response[35] is used for comparison. This periodic response is obtained numerically by performing some five cycles with the input boundary conditions. Although the problem is essentially one dimensional, a two-dimensional uniform mesh was used and the agreement with analytical results is found to be quite remarkable.

In the second example we enter the domain of more realistic applications.[4,5] Here the 'test bed' is provided by the Bristol Channel and the

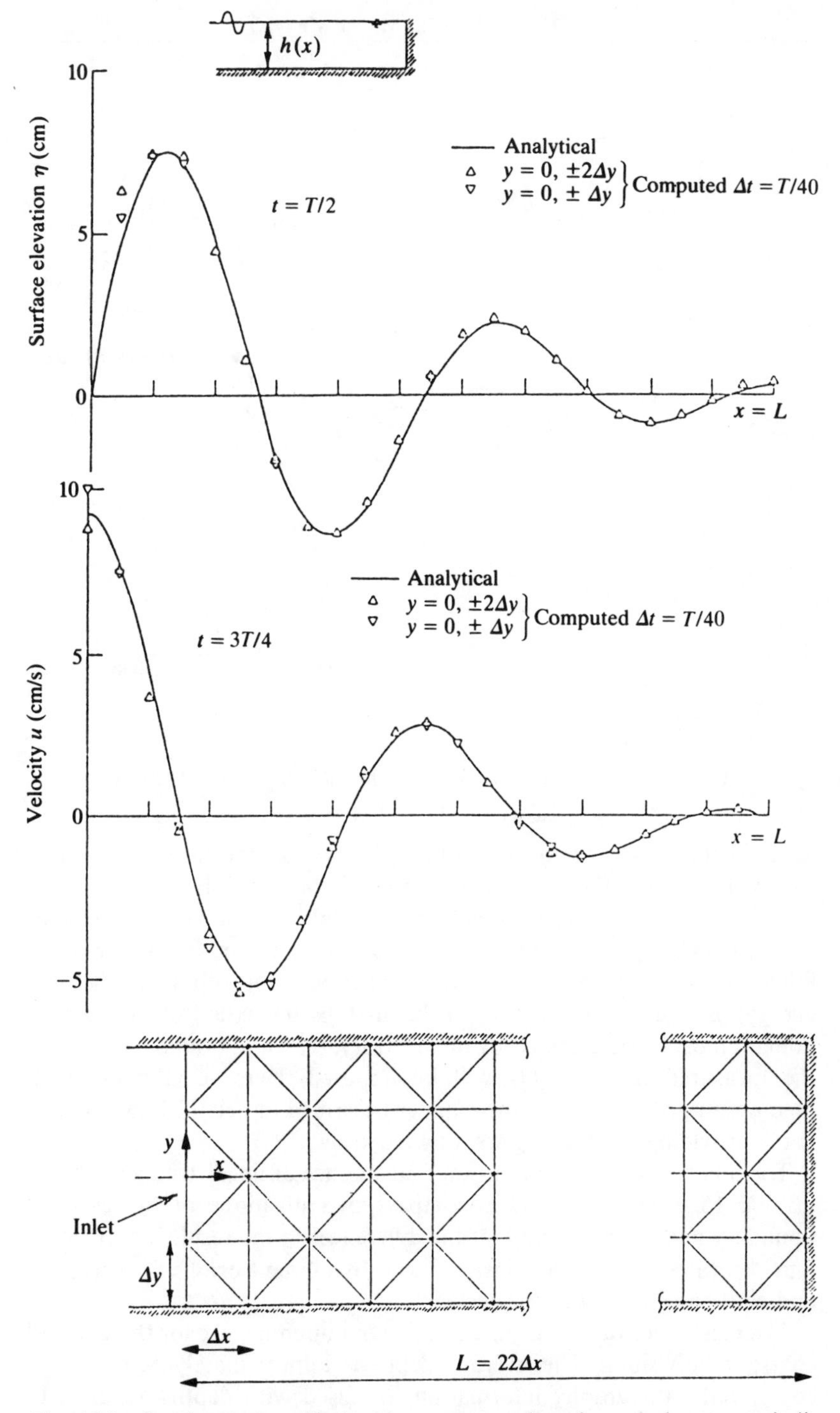

Fig. 15.5 Steady-state oscillation in a rectangular channel due to periodic forcing of surface elevation at inlet. Linear frictional dissipation[31]

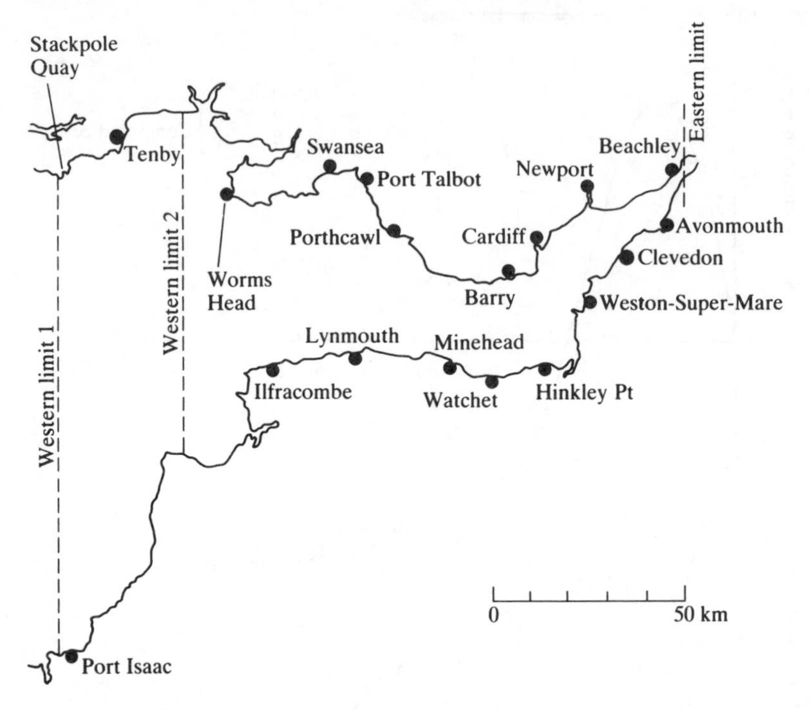

Fig. 15.6 Location map. Bristol Channel and Severn Estuary

Severn Estuary, known for some of the highest tidal motions in the world. Figure 15.6 shows the location and the scale of the problem.

The objective is here to determine tidal elevations and currents currently existing (as a possible preliminary to a subsequent study of the influence of a barrage which some day may be built to harness the tidal energy). Before commencement of the analysis the extent of the analysis domain must be determined by an arbitrary, seaward, boundary. On this the measured tidal heights will be imposed. Here of course use of ingoing/outgoing waves in the manner described in Sec. 12.12.3 could be made providing the ingoing waves were known.

The analysis was carried out on four meshes of linear triangles shown in Fig. 15.7. These meshes encompass two positions of the external boundary and it was found that the differences in the results obtained by four separate analyses were insignificant. In all, the Coriolis accelerations and eddy viscosities were neglected.

The mesh sizes ranged from 2 to 5 km in minimum size for the fine and coarse subdivisions. The average depth is approximately 50 m but of course full bathygraphy information was used with depths assigned to each nodal point.

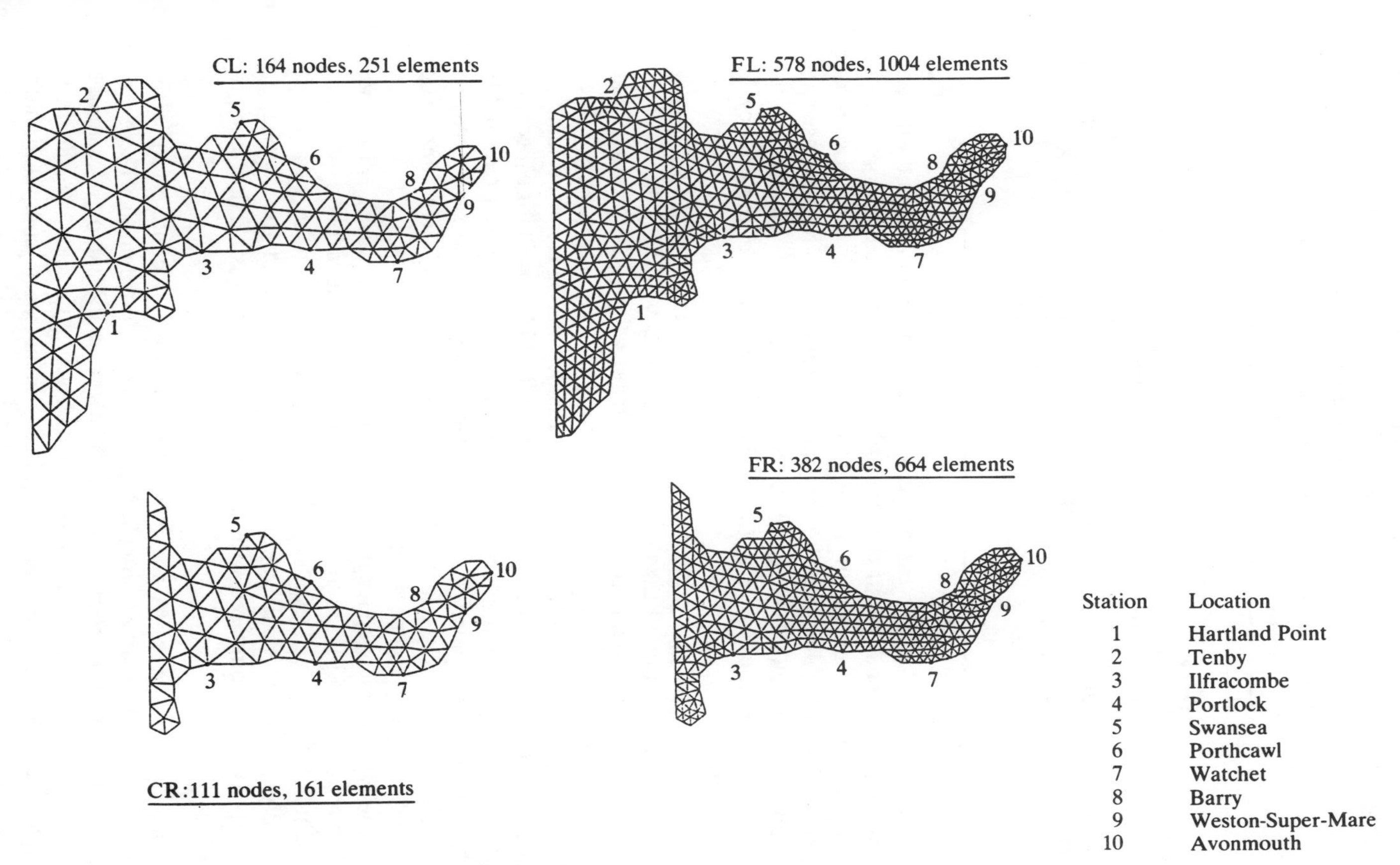

Fig. 15.7 Finite element meshes. Bristol Channel and Severn Estuary

TABLE 15.1
BRISTOL CHANNEL AND SEVERN ESTUARY—OBSERVED RESULTS AND FEM COMPUTATION (FL MESH) OF TIDAL HALF-AMPLITUDE ($m \times 10^2$)

Location	Observed	FEM
Tenby	262	260 (−1%)
Swansea	315	305 (−3%)
Port Talbot	316	316 (0%)
Porthcawl	317	327 (+3%)
Barry	382	394 (+3%)
Cardiff	409	411 (0%)
Newport	413	420 (+2%)
Ilfracombe	308	288 (−6%)
Minehead	358	362 (+1%)

The 'resonance' of the estuary is such that mean tidal input with an amplitude of some 4.5 m at the seaward side is augmented to some 9.5 m at the eastern extremity. In Table 15.1 we show some comparisons of predicted tidal heights with those actually observed at various measuring stations. The results show quite remarkable agreement. Finally, in Fig. 15.8, the distribution of velocities at different moments of the tidal cycle is indicated.

In the analysis just presented we have assumed a purely sinusoidal input and have omitted the details of the River Severn. It is, however, of some interest to extend the analysis domain as it is well known that during spring tides a quite spectalular 'bore' or a solitary wave is formed there. Thus in Fig. 15.9 we show an extended mesh. Here the analysis clearly indicates that the originally sinusoidal tidal oscillation becomes progressively distorted as the river is entered and represents reasonably the observed phenomena of the very rapid tidal rise and slow decay at the river mouth. Of course the detail of the actual wave cannot be modelled on this coarse representation.

The typical time-step length in the above calculation was $\Delta t = 1$ minute (2 min coarse mesh), requiring a large umber of time steps to represent the full tidal cycle. Despite this, the computation can be carried out rapidly and on a small computer as the total size of the problem is relatively small compared with that typical of compressible flow computation.

15.4.3 *Tsunami waves.* A problem of some considerable interest in earthquake zones is that of so-called tidal waves or tsunamis. These are caused by sudden movements in the earth's crust and can on occasion be extremely destructive. The analysis of such waves presents no difficulties in the general procedure demonstrated and indeed is computationally

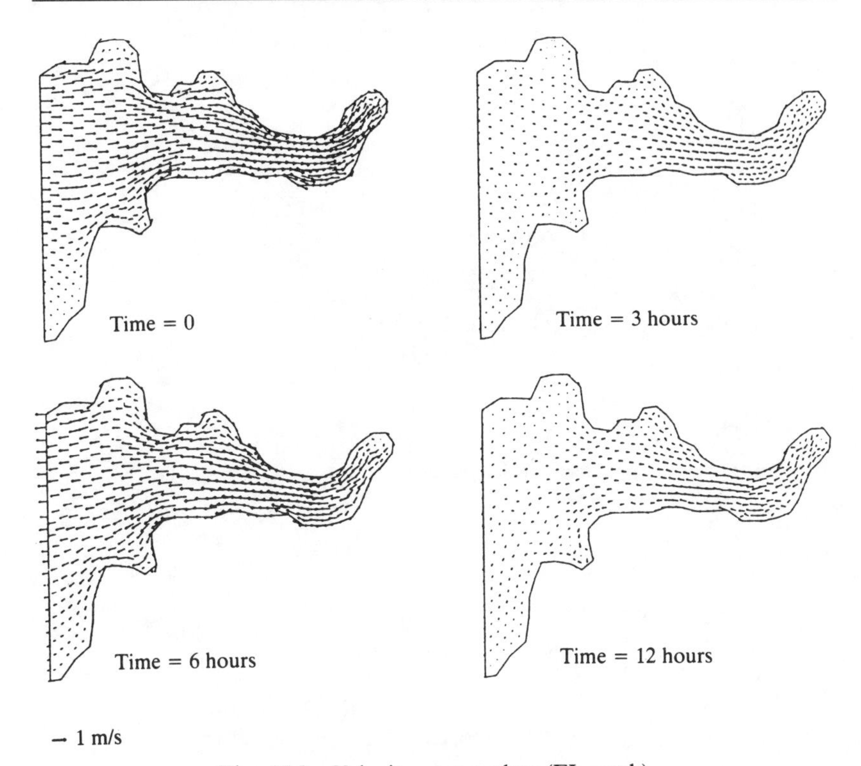

Fig. 15.8 Velocity vector plots (FL mesh)

cheaper as only relatively short periods of time need be considered. To illustrate a typical possible tsunami wave we have created one in the Severn Estuary just analysed (to save problems of mesh generation, etc., for another more likely configuration).

Here the tsunami is forced by an instantaneous raising of an element situated near the centre of the estuary by some 6 m and the previously designed mesh was used (FL). The progress of the wave is illustrated in Fig. 15.10. The tsunami wave was superposed on the tide at its highest level—though of course the tidal motion was allowed for.

One particular point only needs to be mentioned in this calculation. This is the boundary condition on the seaward, arbitrary, limit. Here the Riemann decomposition of the type discussed earlier (page 494) has to be made if tidal motion is to be incorporated and note taken of the fact that the tsunami forms only an outgoing wave. This, in the absence of tides, results simply in application of the free boundary condition there.

The clean way in which the tsunami is seen to leave the domain in Fig. 15.10 testifies to the effectiveness of this process.

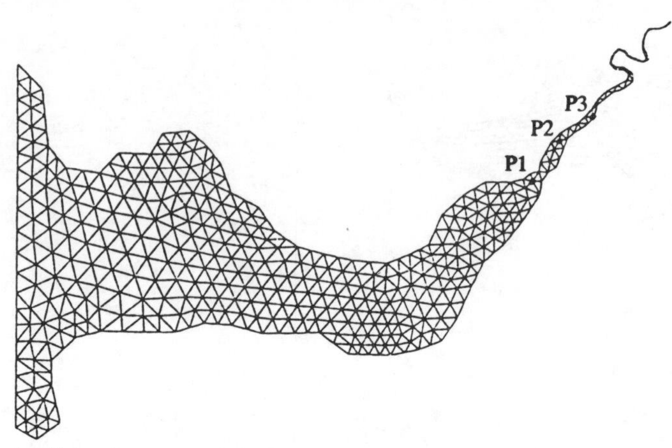

(*a*) Finite element mesh including river

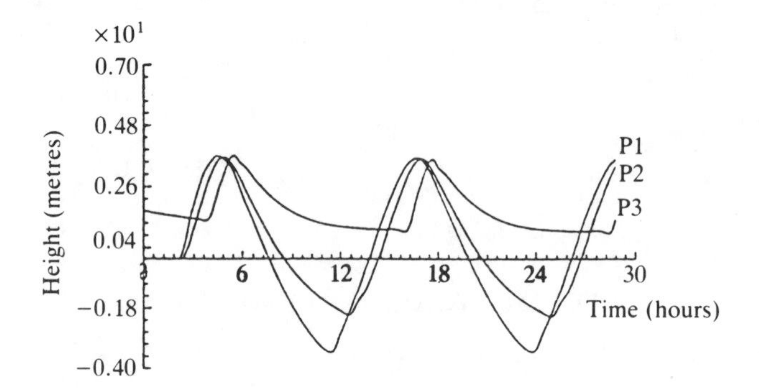

(*b*) Surface elevation—time curves at various points in river

Fig. 15.9 Severn bore

15.4.4 *Steady-state solutions.* On occasion steady-state currents such as may be caused by persistent wind motion or other influences have to be considered. Here once again the transient form of explicit computation proves very effective and convergence generally is more rapid than in compressible flow as the bed friction plays a greater role.

The interested reader will find many such steady-state solutions in the literature. In Fig. 15.11 we show a typical example. Here the currents are induced by the breaking of waves which occurs when these reach small depths creating so-called radiation stresses.[6,29,36] Obviously as a preliminary the wave patterns have to be computed using procedures to be given later. The 'forces' due to breaking are the cause of longshore currents and rip currents in general. The figure illustrates this effect on a harbour.

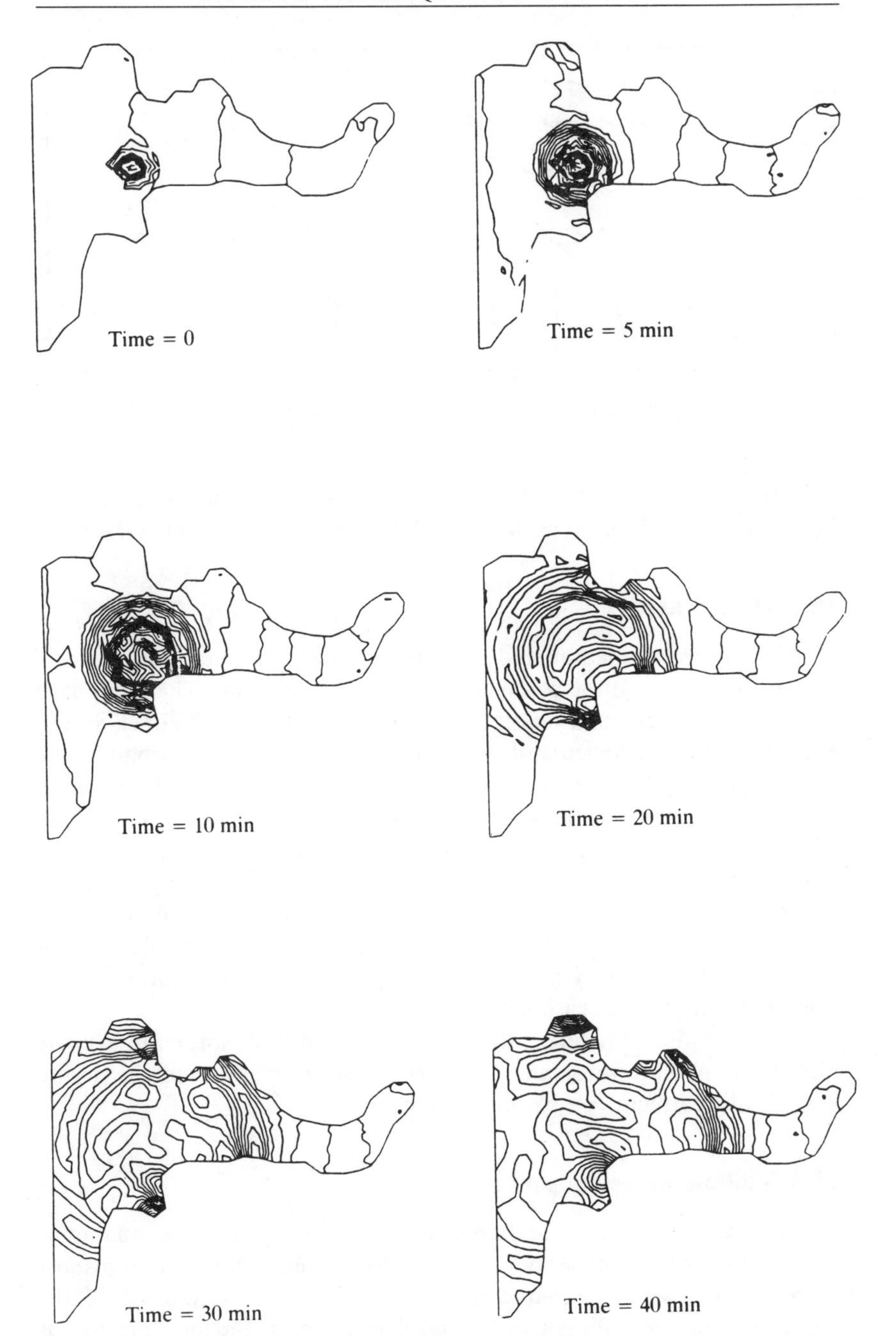

Fig. 15.10 Severn tsunami. Generation during high tide. Water height contours (times after generation)

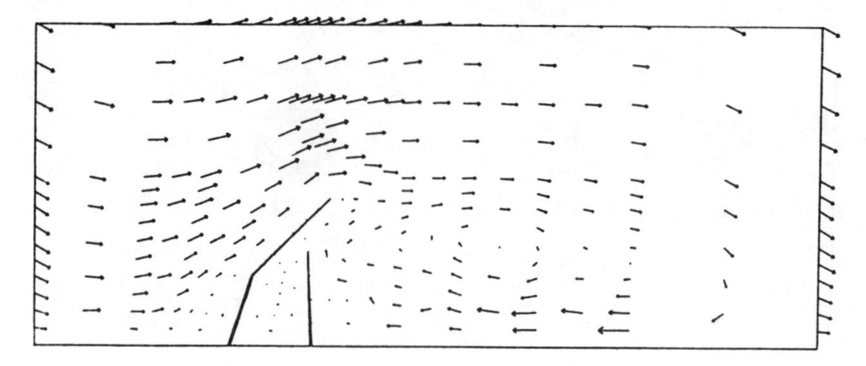

Fig. 15.11 Wave-induced steady-state flow past a harbour[29]

It is of interest to remark that in the problem discussed, the side boundaries have been 'repeated' to model an infinite harbour series.[36]

15.5 Drying areas

A special problem encountered in transient, tidal, computations is that of boundary change due to changes of water elevation. This has been ignored in the calculation presented for the Bristol Channel–Severn Estuary as the movements of the boundary are reasonably small in the scale analysed. However, in that example these may be of the order of 1 km and in tidal motions near Mont St Michel, France, can reach 12 km. Clearly on some occasions such movements need to be considered in the analysis and many different procedures for dealing with the problem have been suggested. In Fig. 15.12 we show the simplest of these which is effective if the total movement can be confined to one element size. Here the boundary nodes are repositioned along the normal direction as required by elevation changes $\Delta\eta$.

If the variations are larger than those that can be absorbed in a single element then, with automatic mesh regeneration confined to the drying zones, full treatment can be undertaken.

15.6 Shallow water transport

Shallow water currents are frequently the carrier for some quantities which may disperse or decay in the process. Typical here is the transport of hot water when discharge from power stations occurs, or of the sediment load or pollutants. The mechanism of sediment transport is quite complex[37] but in principle follows similar rules to that of the other equations. In all cases it is possible to write *depth-averaged transport*

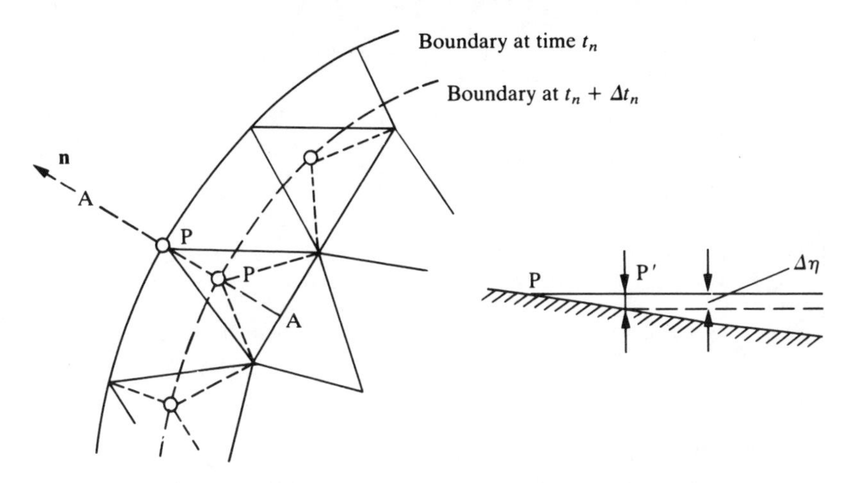

Fig. 15.12 Adjustment of boundary due to tidal height variation

equations in which the average velocities U_i have been determined independently.

A typical averaged equation can be written—using temperature (T) as the transported quantity—as

$$\frac{\partial(hT)}{\partial t}+\frac{\partial(hU_iT)}{\partial x_i}-\frac{\partial}{\partial x_i}\left(Hk\frac{\partial T}{\partial x_i}\right)+R(T-T_a)=0 \qquad \text{for } i=1,2 \quad (15.25)$$

where h and U_i are the previously defined and computed quantities, k is an appropriate diffusion coefficient, R a radiation constant and T_a the ambient air temperature.

Little has to be said about the method of solving such equations. Identical discretization solution procedures to those adopted in solving the flow problem can of course be used. Indeed, in the explicit form the calculation could be carried simultaneously for flow and dispersion. In general, however, we do not recommend such simultaneous calculation as:

1. The equations are not fully coupled and the flow equations can be solved independently.
2. The time increments which can be used in the computation of Eq. (15.25) can be much larger for stability, with

$$\Delta t_{\text{crit}}<\frac{h}{|U|}$$

depending on the transport, rather than the wave, velocities.

In Fig. 15.13 we show by way of an example the dispersion of a

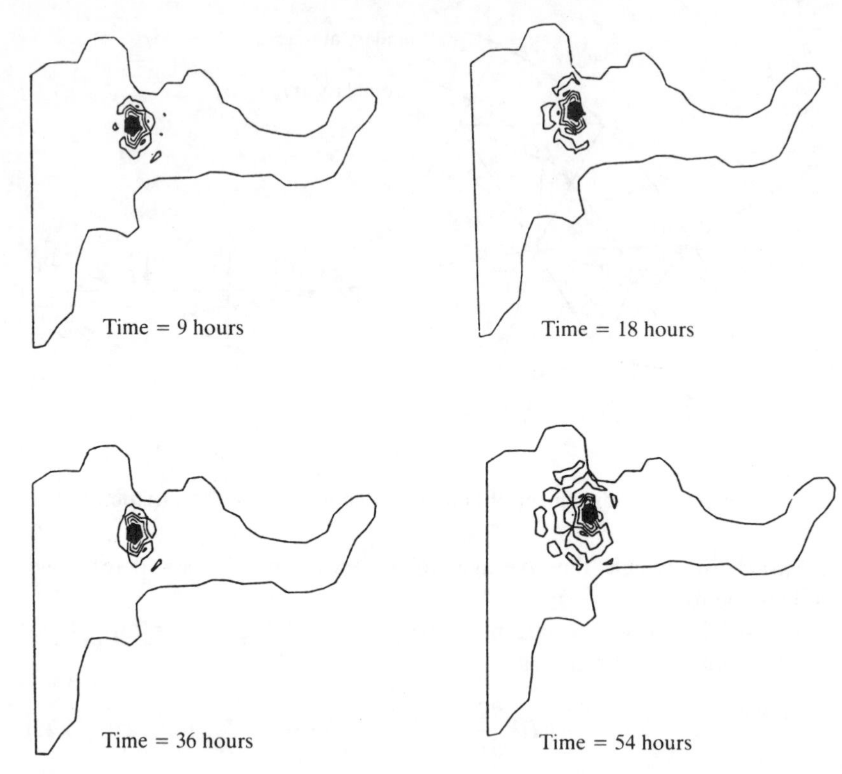

Fig. 15.13 Heat convection and diffusion in tidal currents. Temperature contours at several times after discharge of hot fluid

continuous hot water discharge in an area of the Severn Estuary. Here we note not only the convection movement but also the diffusion of the temperature contours.

PART II:
LINEARIZED SHALLOW WATER EQUATIONS AND WAVES†

15.7 Introduction and equations

We start from the wave equation (15.23) which was developed from the equations of momentum balance and mass conservation, and in which the following effects have been omitted: viscosity (real and eddy), Coriolis

† Contributed by Peter Bettess, Department of Marine Technology, University of Newcastle upon Tyne.

acceleration, Chézy bed friction and convective acceleration. In addition, the wave elevation, η, is small in comparison with the water depth, H. If the problem is periodic, we can write the wave elevation η quite generally as

$$\eta(x, y, t) = \bar{\eta}(x, y) \exp(\mathrm{i}\omega t) \tag{15.26}$$

where ω is the angular frequency and $\bar{\eta}$ may be complex. Equation (15.23) now becomes

$$\nabla(H\nabla\bar{\eta}) + \frac{\omega^2}{g}\bar{\eta} = 0 \qquad \text{or} \qquad \frac{\partial}{\partial x_i}\left(H\frac{\partial\bar{\eta}}{\partial x_i}\right) + \frac{\omega^2}{g}\bar{\eta} = 0 \tag{15.27}$$

and, for constant depth H,

$$\nabla^2\bar{\eta} + k^2\bar{\eta} = 0 \qquad \text{or} \qquad \frac{\partial^2\bar{\eta}}{\partial x^2} + k^2\bar{\eta} = 0 \tag{15.28}$$

where the wave number $k = \omega/\sqrt{gH}$. The wave speed $c = \omega/k$. Equation (15.28) is the Helmholtz equation, which models very many wave problems. This is only one form of the equation of surface waves, for which there is a very extensive literature.[38-41] From now on all problems will be taken to be periodic, and the overbar on η will be dropped.

Equation (15.27) is true when the wavelength $\lambda = 2\pi/k$ is large relative to the depth H. If this last condition is not true, then a more complicated analysis[38,39] shows that for constant depth of water the velocities and the velocity potential vary vertically as $\cosh kz$. Berkhoff[42] derived the intermediate depth water wave equation, for varying depth, using asymptotic expansions, which is a variant on Eq. (15.27). A full derivation of the equation is given by Zienkiewicz and Bettess[43] and it must be noted that now the assumption of uniform average velocities which we used in deriving the shallow water equations is dropped. The full equation can now be written as

$$\frac{\partial}{\partial x_i}\left(cc_g\frac{\partial\eta}{\partial x_i}\right) + \omega^2\frac{c_g}{c}\eta = 0 \qquad \text{or} \qquad \nabla(cc_g\nabla\eta) + \omega^2\frac{c_g}{c}\eta = 0 \tag{15.29}$$

where the group velocity $c_g = nc$,

$$n = \frac{1}{2}\left(1 + \frac{2kH}{\sinh 2kH}\right) \tag{15.30}$$

and the *dispersion relation*

$$\omega^2 = gk \tanh kH \tag{15.31}$$

links angular frequency ω and water depth H to wave number k. (This includes the shallow water equation given above, $k = \omega/\sqrt{gH}$, and the deep water equation, $k = \omega^2/g$, as special cases.) This relation is called the dispersion relation, because waves of different wavelengths, on water of intermediate depth, travel at different speeds, so that an arbitrary disturbance is resolved into components with different speeds, which disperse with time. For deep water Eq. (15.28) becomes

$$\nabla^2 \eta + \frac{\omega^4}{g^2}\eta = 0 \qquad \text{or} \qquad \frac{\partial^2 \eta}{\partial x_i^2} + \frac{\omega^4}{g^2} = 0 \tag{15.32}$$

15.8 Waves in closed basins—finite element models

We now consider a closed basin of any shape and varying depth. In plan, it can be divided into two-dimensional elements of any of the types discussed in Volume 1. The wave elevation η at any point (ξ, η) within the element can be expressed in terms of nodal values, using the element shape function $\mathbf{N}$; thus

$$\eta \approx \hat{\eta} = \mathbf{N}\bar{\boldsymbol{\eta}} \tag{15.33}$$

Next Eq. (15.23) is weighted with the shape function and integrated by parts in the usual way, to give

$$\int_\Omega \left(\frac{\partial \mathbf{N}}{\partial x_i} H \frac{\partial \mathbf{N}}{\partial x_i} - \mathbf{N}^{\mathrm{T}} \frac{\omega^2}{g} \mathbf{N} \right) \mathrm{d}\Omega \bar{\boldsymbol{\eta}} = 0 \tag{15.34}$$

The integral is taken over all the elements of the domain, and $\boldsymbol{\eta}_i$ represents all the nodal values of η. Clearly the integrals can be performed piecewise on all the elements used. The natural boundary condition which arises is $\partial \eta / \partial n = 0$, where n is the normal to the boundary. This corresponds to zero flow normal to the boundary, from the equation of momentum balance (15.15). Physically this corresponds to a vertical, perfectly reflecting wall. Equation (15.34) can be recast in the familiar form

$$(\mathbf{K} - \omega^2 \mathbf{M})\bar{\boldsymbol{\eta}} = 0 \tag{15.35}$$

where

$$\mathbf{M} = \sum_{\text{allelements}} \int \mathbf{N}^{\mathrm{T}} \frac{1}{g} \mathbf{N} \, \mathrm{d}\Omega_{\text{element}}$$

$$\mathbf{K} = \sum_{\text{allelements}} \int \mathbf{B}^{\mathrm{T}} \mathbf{D} \mathbf{B} \, \mathrm{d}\Omega_{\text{element}} \qquad \mathbf{D} = \begin{bmatrix} h & 0 \\ 0 & h \end{bmatrix}$$

It is thus an *eigenvalue* problem as discussed in Chapter 9. The **K** and **M** matrices are analogous to structure stiffness and mass matrices. The *eigenvalues* will give the natural frequencies of oscillation of the water in the basin and the *eigenvectors* give the mode shapes of the water surface. Such an analysis was first carried out using finite elements by Taylor *et al.*[44] and the results are shown in Fig. 9.10 of Chapter 9. There are of course analytical solutions for harbours of regular shape and constant depth.[38] The reader should find it easy to modify the standard element routine given in Chapter 15 of Volume 1 to generate the wave equation 'stiffness' and 'mass' matrices. In all wave models it is necessary to have a sufficiently fine element mesh, with about eight linear or four quadratic finite elements per wavelength. The model described above will give good results for harbour and basin resonance problems. Its main defects are the following:

1. Inaccuracy when the wave height becomes large. The equations are no longer valid when η becomes large, and for very large η the waves will break, which introduces energy loss.
2. Lack of modelling of bed friction. This will be discussed below.
3. Lack of modelling of separation at reentrant corners. At reentrant corners there is a singularity in the velocity of the form $1/\sqrt{r}$. The velocities become large, and physically the viscous effects, neglected above, become important. They cause retardation, flow separation and eddies. This effect can only be modelled in an approximate way. (The same phenomena cause roll damping in ships and other vessels.)

All of the above effects introduce energy loss into the system, which thus becomes a damped oscillation and is no longer an eigenvalue problem. Instead the response of the system can be determined for a given excitation frequency, as discussed in Chapter 9.

15.8.1 *Bed friction and other effects.* The Chézy bed friction term is non-linear and if it is included in its original form it makes the equations very difficult to solve. The usual procedure is to assume that its main effect is to damp the system, by absorbing energy, and to introduce a linear term, which in one period absorbs the same amount of energy as the Chézy term. The linearized bed friction version of Eq. (15.27) is

$$\frac{\partial}{\partial x_i}\left(H\frac{\partial \eta}{\partial x_i}\right)=-\frac{\omega^2}{g}\eta+\mathrm{i}\omega M\eta \quad \text{or} \quad \nabla(H\nabla\eta)=-\frac{\omega^2}{g}\eta+\mathrm{i}\omega M\eta \qquad (15.36)$$

where M is a linearized bed friction coefficient, given by $M = 8u_{max}/3\pi C^2 H$, C is the Chézy constant and u_{max} is the maximum velocity at the bed at that point. In general the results for η will now be complex and iteration has to be used, since M depends upon the unknown u_{max}. From the finite element point of view, there is no longer any point in separating the 'stiffness' and 'mass' matrices. Instead, Eq. (15.36) is weighted using the element shape function and the entire *complex* element matrix is formed. The matrix right-hand side arises from whatever exciting forces are present. The reentrant corner effect and wave-absorbing walls and permeable breakwaters can also be modelled in a similar way, as both of these introduce a damping effect, due to viscous dissipation. The method is explained in reference 45, where an example showing flow through a perforated wall in an offsore structure is solved.

15.9 Waves in unbounded domains (exterior surface wave problems)

Problems in this category include the diffraction and refraction of waves close to fixed and floating structures and the determination of wave forces and wave response for offshore structures and vessels and the determination of wave patterns adjacent to coastlines, open harbours and breakwaters. In the interior or finite part of the domain, finite elements, exactly as described in Sec. 15.8, can be used. However, special procedures must be adopted for the part of the domain extending to infinity. This is even more important than in the static problems described in Chapter 8 of Volume 1 because an incorrectly modelled boundary causes wave reflections that can give large errors in the interior region. Computationally the main difficulty is that the problem has no outer boundary. This necessitates the use of a *radiation* condition. Such a condition was introduced in Chapter 11 as Eq. (11.15) for the case of a one-dimensional wave, or a normally incident plane wave in two or more dimensions.

Until recently no complete version of the radiation condition was available in more than one dimension for transient problems. However, recent work by Bayliss *et al.*[46,47] has developed a suitable radiation condition in the form of an infinite series of operators. The starting point is the representation of the outgoing wave in the form of an infinite series. Each term in the series is then annihilated by using a boundary operator. The sequence of boundary operators thus constitutes the radiation condition. In addition there is a classical form of the boundary condition for periodic problems, given by Sommerfield.

A summary of all available radiation conditions is given in Table 15.2. The periodic boundary conditions are sometimes seen with different signs, because the time dependence can be introduced by factoring with $\exp(-i\omega t)$ or $\exp(i\omega t)$, which leads to a sign change on the terms involving time derivatives.

15.9.1 *Background to wave problems.* The simplest type of exterior or unbounded wave problem is that of some exciting device which sends out waves that do not return. A simple example would be a floating object in the middle of the sea, which is pushed up and down and which generates outgoing waves. The transient example is a stone thrown into

TABLE 15.2

RADIATION CONDITIONS FOR EXTERIOR WAVE PROBLEMS

Dimensions		
1	2	3
General boundary conditions		
Transient		
$\frac{\partial\eta}{\partial x}+\frac{1}{c}\frac{\partial\eta}{\partial t}=0$	$B_m\eta=0, m\to\infty$ $B_m=\prod_{j=1}^{m}\left(\frac{\partial}{\partial r}+\frac{\partial}{\partial t}+\frac{2j-\frac{3}{2}}{r}\right)$	$B_m\eta=0, m\to\infty$ $B_m=\prod_{j=1}^{m}\left(\frac{\partial}{\partial r}+\frac{\partial}{\partial t}+\frac{2j-1}{r}\right)$
Periodic		
$\frac{\partial\eta}{\partial x}+ik\eta=0$	$\lim_{r\to\infty}\sqrt{r}\left(\frac{\partial\eta}{\partial r}+ik\eta\right)=0$ or $B_m\eta=0, m\to\infty$ $B_m=\prod_{j=1}^{m}\left(\frac{\partial}{\partial r}+ik+\frac{2j-\frac{3}{2}}{r}\right)$	$\lim_{r\to\infty} r\left(\frac{\partial\eta}{\partial r}+ik\eta\right)=0$ or $B_m\eta=0, m\to\infty$ $B_m=\prod_{j=1}^{m}\left(\frac{\partial}{\partial r}+ik+\frac{2j-1}{r}\right)$
Symmetric boundary conditions		
Transient		
$\frac{\partial\eta}{\partial x}+\frac{1}{c}\frac{\partial\eta}{\partial t}=0$	$\frac{\partial\eta}{\partial r}+\frac{\eta}{2r}+\frac{1}{c}\frac{\partial\eta}{\partial t}=0$ Axisymmetric	$\frac{\partial\eta}{\partial r}+\frac{\eta}{r}+\frac{1}{c}\frac{\partial\eta}{\partial t}=0$ Spherically symmetric
Periodic		
$\frac{\partial\eta}{\partial x}+ik\eta=0$	$\frac{\partial\eta}{\partial r}+\left(\frac{1}{2r}+ik\right)\eta=0$ Axisymmetric	$\frac{\partial\eta}{\partial r}+\left(\frac{1}{r}+ik\right)\eta=0$ Spherically symmetric

the middle of a pond. This is usually termed the *radiation* problem, and we only need to deal with the outgoing wave. The next type of exterior wave problem is where we have a known incoming wave that encounters an object, is modified and then again radiates away to infinity. The example here would be of an incident wave hitting a large offshore structure and then radiating away, or in the transient case, an underwater shock wave hitting a submarine and radiating away. This case is known as the *diffraction* problem, and is more complicated, in as much as we have to deal with both incident and radiated waves. Even when both waves are linear, this can lead to complications. Finally, both of the above cases can be complicated by wave *refraction*, where the wave speeds change because of changes in the medium, in this case changes in water depth. Usually these phenomena lead to changes in the wave direction. Waves can also *reflect* from boundaries, both physical and computational.

15.9.2 *Wave diffraction.* We now consider the problem of an incident wave diffracted by an object. The problem consists of an object in some medium, which diffracts the incident waves. We divide the medium as shown in Fig. 15.14 into two regions, with boundaries $\Gamma_A, \Gamma_B, \Gamma_C$ and Γ_D. These boundaries have the following meanings: Γ_A is the boundary of the body which is diffracting the waves; Γ_B is the boundary between the two computational domains, that in which the total wave elevation (or other field variable) is used and that in which the elevation of the radiated wave is used; Γ_C is the outer boundary of the computational model; and Γ_D is the boundary at infinity. Some of these boundaries may be merged.

A variational treatment will be used, as described in Chapter 9 of Volume 1. A weighted residual treatment is also possible. The elevation of the total wave, η_T, is split into those for incident and radiated waves, η_I and η_R. Hence

$$\eta_T = \eta_I + \eta_R \tag{15.37}$$

The incident wave elevation, η_I, is assumed to be known. For the surface

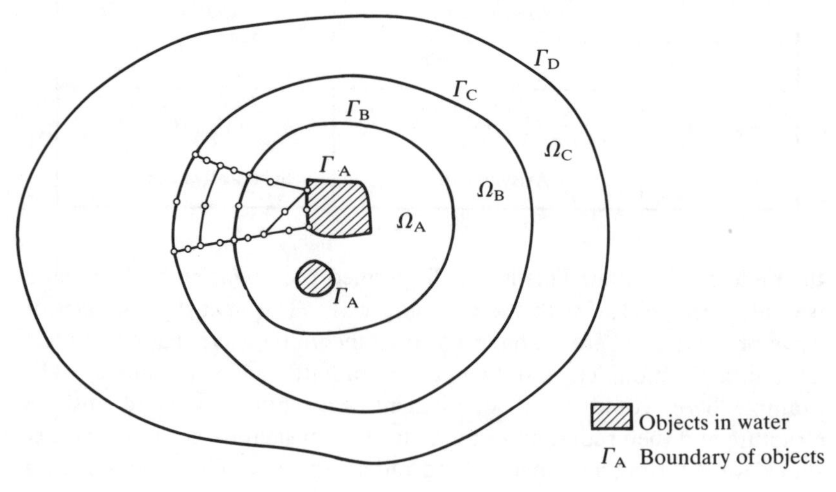

Fig. 15.14 General wave domains

wave problem, the functional for the exterior can be written

$$\Pi = \iint_{\Omega_B} \frac{1}{2}\left[H(\nabla\eta)^2 - \frac{\omega^2}{g}\eta^2 \right] \mathrm{d}x\,\mathrm{d}y \tag{15.38}$$

where making Π stationary with respect to variations in η corresponds to satisfying the shallow water wave equation (15.23) with the natural boundary condition $\partial\eta/\partial n = 0$, or zero velocity normal to the boundary. The functional is rewritten in terms of the incident and radiated elevations, and then Green's theorem in the plane (Appendix 6 of Volume 1) is applied on the domain exterior to Γ_B. For details, see reference 45. The final functional for the exterior is

$$\Pi = \iint_{\Omega_B} \frac{1}{2}\left[H(\nabla\eta_R)^2 - \frac{\omega^2}{g}\eta_R^2 \right] \mathrm{d}x\,\mathrm{d}y + \oint_{\Gamma_B} H\left(\frac{\partial\eta_I}{\partial x}\eta_R\,\mathrm{d}y - \frac{\partial\eta_I}{\partial y}\eta_R\,\mathrm{d}x \right) \tag{15.39}$$

The influence of the incident wave is thus to generate a 'forcing term' on the boundary Γ_B.

15.9.3 *Incident waves and domain integrals and nodal values.* It is possible to choose any known solution of the wave solution as the incident wave. Usually this is a plane monochromatic wave, for which the elevation is given by

$$\eta_I = a_0 \exp(ikr)\cos(\theta - \gamma) \tag{15.40}$$

where γ is the angle that the incident waves make to the positive x axis, r and θ are the polar coordinates and a_0 is the incident wave amplitude. On the boundary Γ_B, we have two types of variables, the total elevation η_T on the interior, and η_R, the radiation elevation, on the exterior. Clearly the nodal values of η in the finite element model must be unique, and on this boundary, as well as the line integral of Eq. (15.39), we must transform the nodal values, either to η_T or to η_R. This can be done simply by enforcing the change of variable, which leads to a contribution to the 'right-hand side' or 'forcing' term. The element matrix and 'r.h.s.' are written, for an element in the exterior, with nodal values of the radiated wave elevation η_R,

$$\mathbf{k}\eta_R = \mathbf{f}$$

and are partitioned into *set 1*, the nodes to be transformed into total elevation $\boldsymbol{\eta}_T$, and *set 2*, the nodes to stay as $\boldsymbol{\eta}_R$. Clearly $\boldsymbol{\eta}_T = \boldsymbol{\eta}_I + \boldsymbol{\eta}_R$. Then

$$\begin{aligned} \mathbf{k}_{11}\boldsymbol{\eta}_R^1 + \mathbf{k}_{12}\boldsymbol{\eta}_R^2 &= \mathbf{f}^1 \\ \mathbf{k}_{21}\boldsymbol{\eta}_R^1 + \mathbf{k}_{22}\boldsymbol{\eta}_R^2 &= \mathbf{f}^2 \end{aligned} \tag{15.41}$$

On substituting for $\boldsymbol{\eta}_R$, Eqs (15.41) become

$$\begin{aligned} \mathbf{k}_{11}\boldsymbol{\eta}_R^1+\mathbf{k}_{12}\boldsymbol{\eta}_T^2&=\mathbf{f}^1+\mathbf{k}_{12}\boldsymbol{\eta}_I \\ \mathbf{k}_{21}\boldsymbol{\eta}_R^1+\mathbf{k}_{22}\boldsymbol{\eta}_T^2&=\mathbf{f}^2+\mathbf{k}_{22}\boldsymbol{\eta}_I \end{aligned} \tag{15.42}$$

Thus there is just a simple modification of the element 'r.h.s.'.

There are several methods of dealing with exterior problems using finite elements in combination with other methods. These include:

(*a*) boundary dampers, both plane and cylindrical,
(*b*) linking to exterior solutions and
(*c*) infinite elements.

15.10 Boundary dampers

As we saw in Chapter 11, we can simply apply the plane damper at the boundary of the mesh. This was first done in fluid problems by Zienkiewicz and Newton.[48] However, the more sophisticated dampers proposed by Bayliss *et al.* can be used at little extra computational cost and a big increase in accuracy. The dampers are developed from the series given in Table 15.2. Full details are given in reference 49. For the case of two-dimensional waves the line integral which should be applied on the circular boundary of radius r is

$$A=\int_r\left[\frac{\alpha}{2}\eta^2+\frac{\beta}{2}\left(\frac{\partial\eta}{\partial s}\right)^2\right]\mathrm{d}s \tag{15.43}$$

where ds is an element of distance along the boundary and

$$\alpha=\frac{3/4r^2-2k^2+3\mathrm{i}k/r}{2/r+2\mathrm{i}k} \quad \text{and} \quad \beta=\frac{1}{2/r+2\mathrm{i}k} \tag{15.44}$$

For the plane damper, $\beta=0$ and $\alpha=\mathrm{i}k$. For the cylindrical damper $\beta=0$ and $\alpha=\mathrm{i}k-1/2r$. The corresponding expressions for three-dimensional waves are different. Non-circular boundaries can be handled but the expressions become *much* more complicated. Some results are given by Bando *et al.*[49] Figure 15.15 shows the waves diffracted by a cylinder problem for which there is a solution, due to MacCamy and Fuchs. The higher-order dampers are clearly a big improvement over the plane and cylindrical dampers, for little or no extra computational cost.

15.11 Lining to exterior solutions

The methodology for linking finite element to exterior solutions was proposed by Zienkiewicz *et al.*[50,51] and this is also discussed in Volume

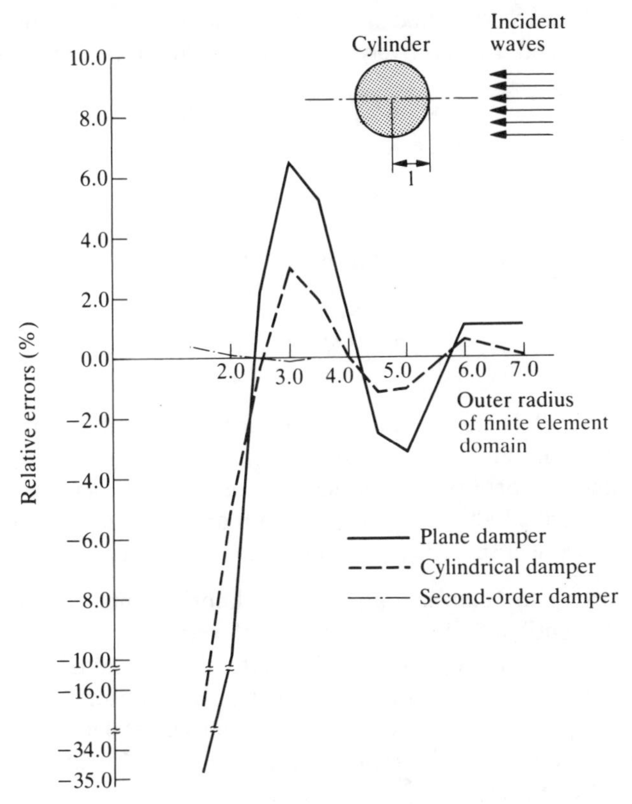

Fig. 15.15 Comparison of relative errors for various outer radii ($k = 1.0$). Relative error $= (|\eta_n^A| - |n_a^A|)/|\eta_a^A|$ [49]

1, particularly in Chapter 13. The exterior solution can take any form, and those chiefly used are (*a*) exterior series solutions and (*b*) exterior boundary integrals, although others are possible. The two main innovators in these cases were Berkhoff,[42,52] for coupling to boundary integrals, and Chen and Mei,[53,54] for coupling to exterior series solutions. Although the methods proposed are quite different, it is useful to cast them in the same general form. More details of this procedure are given in reference 45. Basically the energy functional given in Eqs (15.38) and (15.39) is again used. If the functions used in the exterior automatically satisfy the wave equation, then the contribution on the boundary reduces to a line integral of the form

$$\Pi = \frac{1}{2}\int_\Gamma \eta \frac{\partial \eta}{\partial n}\, \mathrm{d}\Gamma \tag{15.45}$$

It can be shown[45,50,51] that if the free parameters in the interior and exterior are **b** and **a** respectively, the coupled equations can be written

$$\begin{bmatrix} \mathbf{K} & \bar{\mathbf{K}}^{\mathrm{T}} \\ \bar{\mathbf{K}} & \hat{\mathbf{K}} \end{bmatrix} \begin{Bmatrix} \mathbf{a} \\ \mathbf{b} \end{Bmatrix} + \begin{Bmatrix} \mathbf{f} \\ \mathbf{0} \end{Bmatrix} = \begin{Bmatrix} \mathbf{0} \\ \mathbf{0} \end{Bmatrix} \tag{15.46}$$

where

$$\hat{K}_{ji} = \frac{1}{2}\int_{\Gamma} [(PN_j)^{\mathrm{T}} N_i + N_j^{\mathrm{T}}(PN_i)]\,\mathrm{d}\Gamma \tag{15.47}$$

$$\bar{K}_{ji} = \int_{\Gamma} (PN_j)^{\mathrm{T}} \bar{N}_i\,\mathrm{d}\Gamma \tag{15.48}$$

In the above P is an operator giving the normal derivative, that is $P \equiv \partial/\partial n$, $\bar{\mathbf{N}}$ is the finite element shape function, $\mathbf{N}$ is the exterior shape function and $\mathbf{K}$ corresponds to the normal finite element matrix. The approach described above can be used with any suitable form of exterior solution, as we will see. All the nodes on the boundary become coupled.

15.11.1 *Linking to boundary integrals.* Berkhoff adopted the simple expedient of identifying the nodal values of velocity potential obtained using the boundary integral with the finite element nodal values. This leads to a rather clumsy set of equations, part symmetrical, real and banded, and part unsymmetrical, complex and dense. The direct boundary integral method for the Helmholtz equation in the exterior leads to a matrix set of equations[50,51]

$$\mathbf{A}\boldsymbol{\eta} = \mathbf{B}\frac{\partial \boldsymbol{\eta}}{\partial n} \tag{15.49}$$

(The indirect boundary integral method can also be used.) The values of η and $\partial\eta/\partial n$ on the boundary are next expressed in terms of shape functions, so that

$$\eta \approx \hat{\eta} = \mathbf{N}\boldsymbol{\eta} \qquad \text{and} \qquad \frac{\partial \eta}{\partial n} \approx \frac{\partial \hat{\eta}}{\partial n} = \mathbf{M}\begin{Bmatrix} \dfrac{\partial \boldsymbol{\eta}}{\partial n} \end{Bmatrix} \tag{15.50}$$

$\mathbf{N}$ and $\mathbf{M}$ are equivalent to $\mathbf{N}$ in the previous section. Using this relation, the integral for the outer domain can be written as

$$\Pi = \frac{1}{2}\int_{\Gamma} \frac{\partial \boldsymbol{\eta}}{\partial n} \mathbf{M}^{\mathrm{T}} \mathbf{N} \boldsymbol{\eta}\,\mathrm{d}\Gamma \tag{15.51}$$

where Γ is the boundary between the finite elements and the boundary integrals. The normal derivatives can now be eliminated, using the relation (15.48), and $\boldsymbol{\eta}$ can be identified with the finite element nodal values to give

$$\Pi = \tfrac{1}{2}\mathbf{b}^{\mathrm{T}}(\mathbf{B}^{-1}\mathbf{A})^{\mathrm{T}}\int_{\Gamma}\mathbf{M}^{\mathrm{T}}\mathbf{N}\,\mathrm{d}\Gamma\mathbf{b} \tag{15.52}$$

Variations of this functional with respect to **b** can be set to zero to give

$$\frac{\partial \Pi}{\partial \mathbf{b}} = \frac{1}{2}\left\{(\mathbf{B}^{-1}\mathbf{A})\int_{\Gamma}\mathbf{M}^{\mathrm{T}}\mathbf{N}\,\mathrm{d}\Gamma + \left[(\mathbf{B}^{-1}\mathbf{A})\int_{\Gamma}\mathbf{M}^{\mathrm{T}}\mathbf{N}\,\mathrm{d}\Gamma\right]^{\mathrm{T}}\right\}\mathbf{b} = \check{\mathbf{K}}\mathbf{b} \tag{15.53}$$

where $\check{\mathbf{K}}$ is a 'stiffness' matrix for the exterior region. It is symmetrical and can be created and assembled like any other element matrix. The integrations involved must be carried out with care, as they involve singularities. Results obtained for the problem of an island on a parabolic shoal are shown in Fig. 15.16. They are compared with the analytical solution.

15.11.2 *Linking to series solutions.* Chen and Mei[53] took the series solution for waves in the exterior and worked out explicit expressions for the exterior and coupling matrices, $\bar{\mathbf{K}}$ and $\hat{\mathbf{K}}$, for piecewise linear shape functions, $\bar{\mathbf{N}}$, in the finite elements. The series used in the exterior consists of Hankel and trigonometric functions which automatically satisfy the governing equation (the Helmholtz equation) and the radiation condition:

$$\eta = \sum_{j=0}^{m} H_j(kr)(\alpha_j \cos j\theta + \beta_j \sin j\theta) \tag{15.54}$$

The method described above leads to the following matrices:

$$\bar{\mathbf{K}}^T = \frac{-knL_c}{2}\begin{bmatrix} 2H'_0 \cdots H'_n(\cos n\theta_p + \cos n\theta_1), & H'_n(\sin n\theta_p + \sin n\theta_1), & \ldots \\ 2H'_0 \cdots H'_n(\cos n\theta_1 + \cos n\theta_2), & H'_n(\sin n\theta_1 + \sin n\theta_2), & \ldots \\ 2H'_0 \cdots H'_n(\cos n\theta_2 + \cos n\theta_3), & H'_n(\sin n\theta_2 + \sin n\theta_3), & \ldots \\ \cdots & \cdots & \cdots \\ 2H'_0 \cdots H'_n(\cos n\theta_{p-1} + \cos n\theta_p), & H'_n(\sin n\theta_{p-1} + \sin n\theta_p), & \ldots \end{bmatrix} \tag{15.55}$$

$$\hat{\mathbf{K}} = \pi rkh\{\mathrm{diag}|2H_0H'_0, H_1H'_1, H_1H'_1, \ldots, H_mH'_m, H_mH'_m|\} \tag{15.56}$$

where m = number of terms in the Hankel function series

r = radius of the boundary

L_c = distance between the equidistant nodes on Γ

k = wave number

p = number of nodes

H_n and H'_n = Hankel functions and derivatives

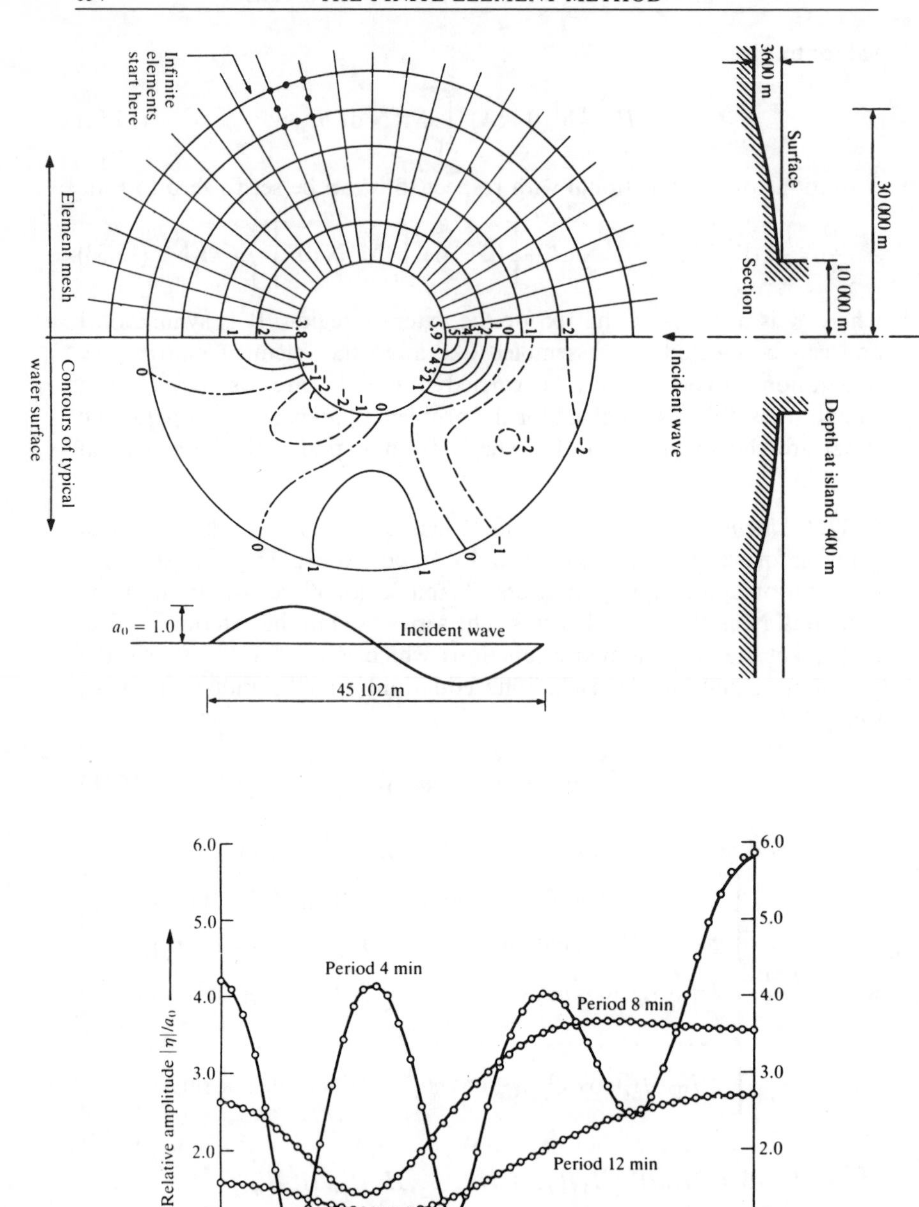

Fig. 15.16 Wave diffraction by a parabolic shoal and cylinder

Other authors have worked out the explicit forms of the above matrices for linear shape functions, and it is possible to work them out for any type of shape function, using, if necessary, numerical integration. It will be noticed that the matrix $\hat{\mathbf{K}}$ is diagonal. This is because the boundary Γ_B is circular and the Hankel functions are orthogonal. If a non-circular domain is used, $\hat{\mathbf{K}}$ will become dense. Chen and Mei[54] applied the method very successfully to a range of problems, most notably that of resonance effects in an artificial offshore harbour, the results for which are shown in Chapter 9, Fig. 9.12. The method was also utilized by Houston,[55] who applied it to a number of real problems, including resonance in Long Beach harbour, shown in Fig. 15.17.

15.12 Infinite elements

The methods described in Chapter 8 of Volume 1 can be developed to include periodic effects. This was first done by Bettess and Zienkiewicz, using so-called 'decay function' procedures, and they were very effective.[56] Figure 15.18 shows results obtained using these elements for diffraction by a breakwater.[57] An unusual feature of this problem is the singularity of the form $1/\sqrt{r}$ in the velocity at the tip of the breakwater. This was dealt with using 'quarter-point' singularity elements. Details of the method are given in Chapter 8 of Volume 1. Excellent agreement with the analytical solution (in terms of Fresnel integrals and originally derived in optics) was obtained. The problem is to some extent artificial, since in the real situation the high-velocity gradients cause high viscous forces, flow retardation, separation and the formation of eddies at the tip. Comparison of results with those of Chen and Mei for the artificial island problem are also shown in Fig. 9.12. Later 'mapped' infinite elements were developed for wave problems, and as these appear to be the most accurate to date, they will be described here.

15.12.1 *Mapped periodic infinite elements.* We briefly rehearse the theory developed in Chapter 8 of Volume 1 for static infinite elements, as it is used again in the periodic elements. Details are given in references 58 to 60. Consider first the geometry of the one-dimensional problem. As before the element extends from point x_1 through x_2 to x_3, which is at infinity. This element is to be mapped onto the finite domain $-1<\xi<1$. This is shown in Fig. 15.19. A suitable mapping expression is

$$x=\tilde{N}_0(\xi)x_0+\tilde{N}_2(\xi)x_2 \tag{15.57}$$

where

$$\tilde{N}_0(\xi)=\frac{-\xi}{1-\xi} \qquad \tilde{N}_2(\xi)=1+\frac{\xi}{1-\xi} \tag{15.58}$$

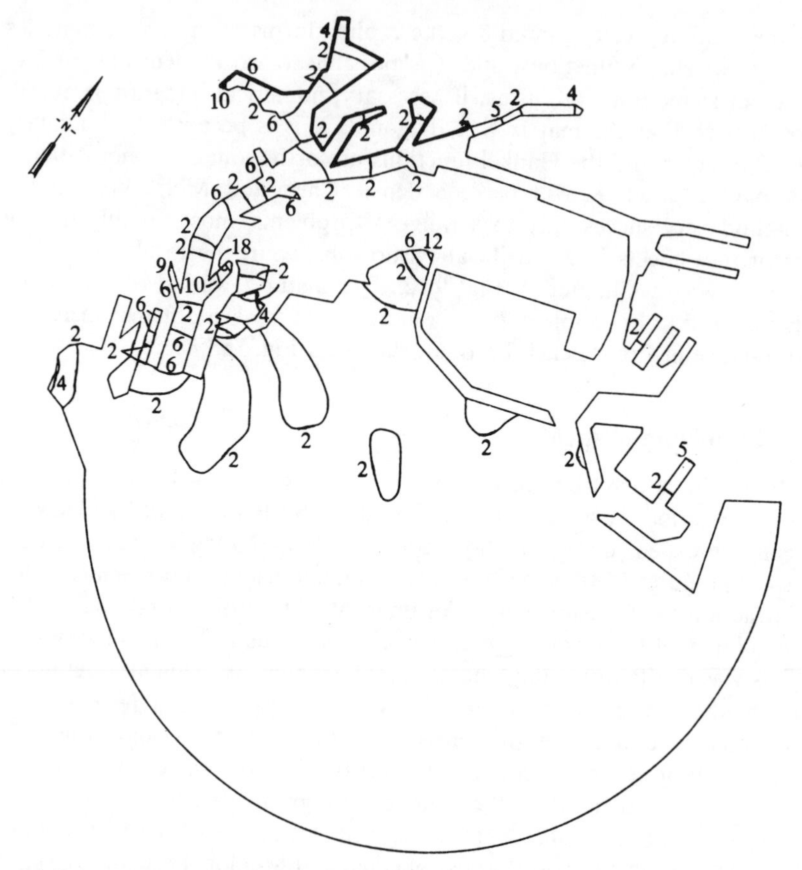

(*a*) Contours of wave height amplification, grid 3, 232-s wave period

Fig. 15.17 Finite element mesh and wave height amplification for Long Beach Harbour[55]

At

$$\begin{aligned}
\xi = 1: &\qquad x = \frac{\xi}{1-\xi}(x_2 - x_0) + x_2 = x_3 = \infty \\
\xi = 0: &\qquad x = x_2 \\
\xi = -1: &\qquad x = \frac{x_0 + x_2}{2} = x_1
\end{aligned} \tag{15.59}$$

The point at $\xi = -1$ is to correspond to the point x_1, which is now defined to be midway between x_0 and x_2. As we have seen previously, a

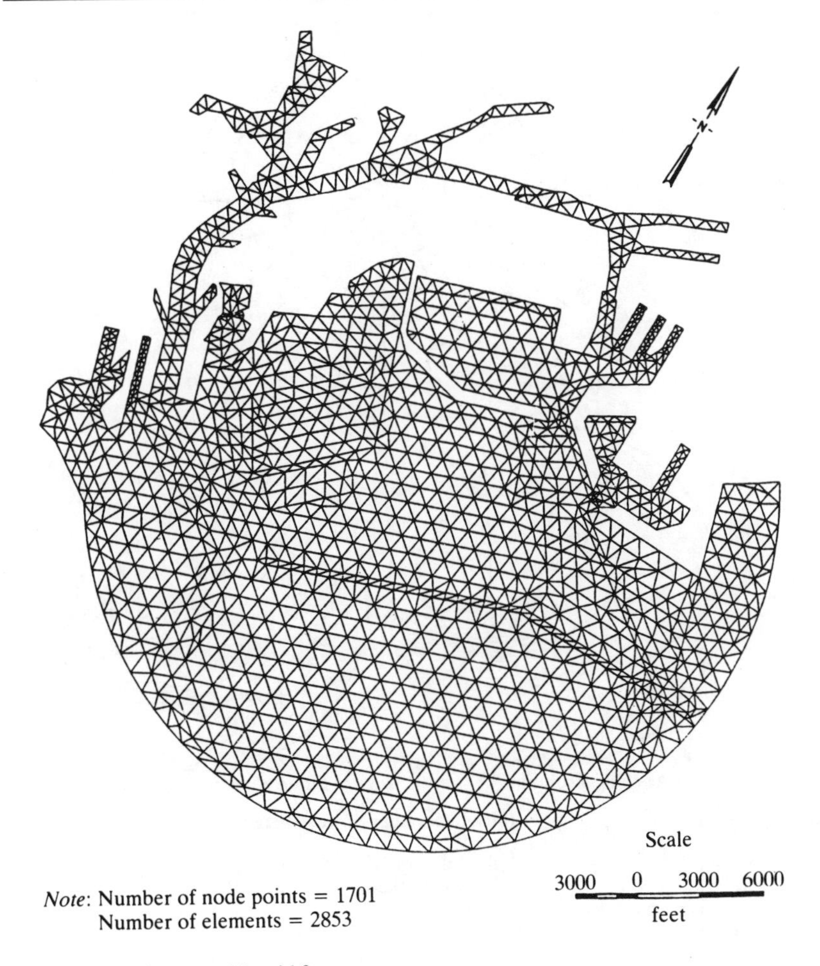

Note: Number of node points = 1701
Number of elements = 2853

(*b*) Finite element grid, grid 3

Fig. 15.17 (*continued*)

polynomial

$$P=\alpha_0+\alpha_1\xi+\alpha_2\xi^2+\alpha_3\xi^3+\cdots \tag{15.60}$$

in the finite domain, maps into a polynomial in inverse powers of r, in the infinite domain:

$$P=\beta_0+\frac{\beta_1}{r}+\frac{\beta_2}{r^2}+\frac{\beta_3}{r^3}+\cdots \tag{15.61}$$

where β_i can be determined from the α's and a. If the polynomial is required to decay to zero at infinity then $\beta_0=0$.

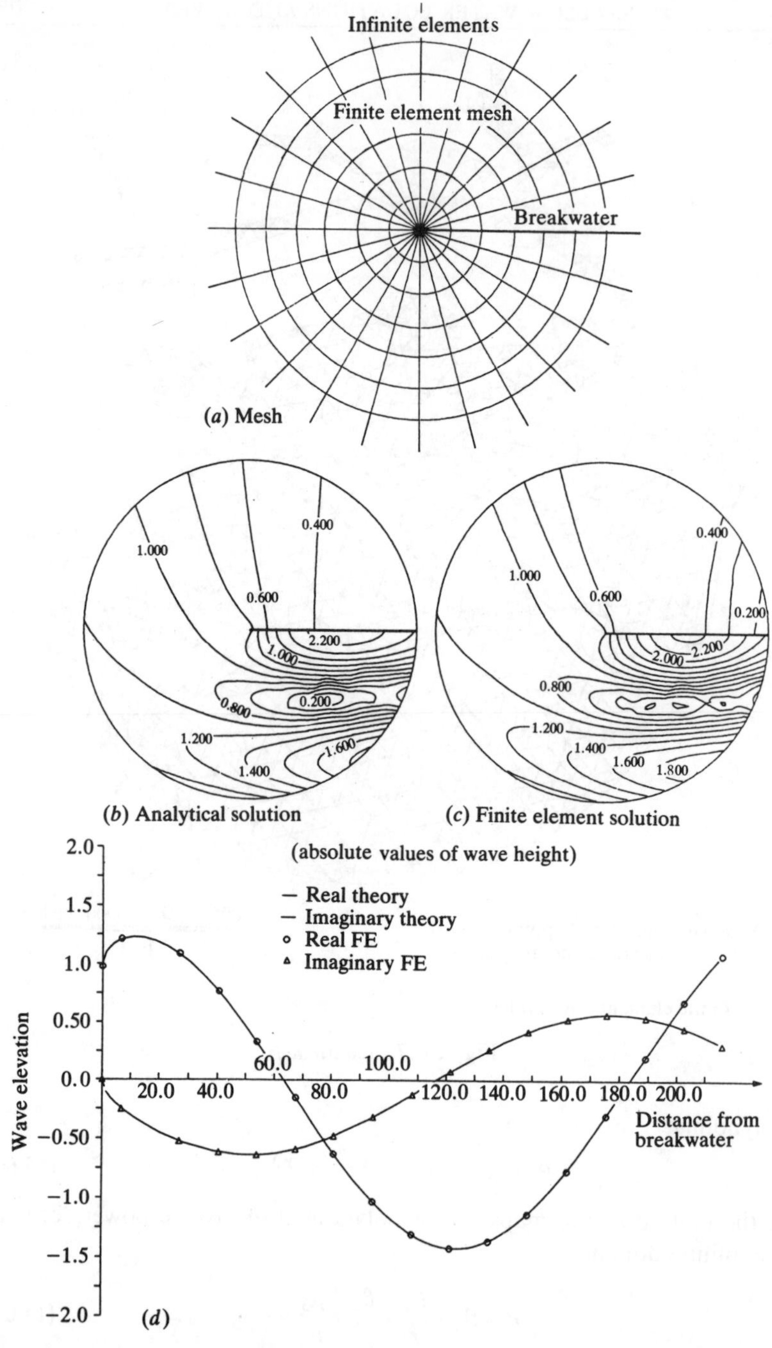

Fig. 15.18 Semi-infinite breakwater—heights of diffracted waves. The wave heights on a line perpendicular to breakwater through tip: angle of incidence 90° to breakwater; wavelength = 242.610 780 m; wave period = 20 s; mid-side nodes at quarter points

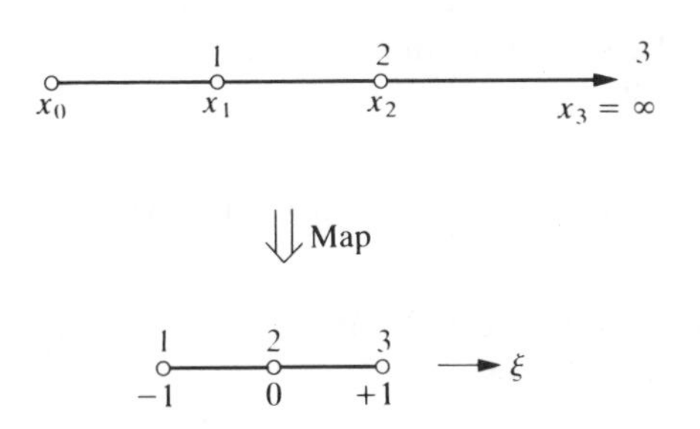

Fig. 15.19 One-dimensional mapping used in mapped infinite element formulation[58]

Many exterior wave problems have solutions in which the wave amplitude decays like $1/r$ and an advantage of this mapping is that such a decay can be accommodated. In some cases, however, the amplitude decays approximately as $1/\sqrt{r}$, and this case responds to a slightly different treatment, as we shall see. In general, accuracy can be increased by adding extra terms to the series (15.60). The point x_0 which has not been defined until now is seen to be the pole of the expansion of P.

15.12.2 *Introducing the wave component.* In two-dimensional exterior domains the solution to the Helmholtz equation can be described by a series of combined Hankel and trigonometric functions, the simplest solution to the Helmholtz equation being $H_0(kr)$. For large r the zeroth-order Hankel function oscillates roughly like $\cos(kr)+\mathrm{i}\,\sin(kr)$ while decaying in magnitude as $r^{-1/2}$. We will use a series of terms $1/r$, $1/r^2$, etc., generated by the mapping, multiplied by $r^{1/2}$ and the periodic component $\exp(\mathrm{i}kr)$ in order to model the $r^{-1/2}$ decay.

The shape function is therefore the original shape function described earlier, multiplied by these additional terms, giving

$$N^*(\xi,\eta)=M(\xi,\eta)r^{1/2}\exp(\mathrm{i}kr) \tag{15.62}$$

The mapping in Eq. (15.57) can also be written as

$$r=\frac{A}{1-\xi} \tag{15.63}$$

The shape function in Eq. (15.62) will now be

$$N^*(\xi,\eta)=M(\xi,\eta)\left(\frac{A}{1-\xi}\right)^{1/2}\exp\left(\frac{\mathrm{i}kA}{1-\xi}\right) \tag{15.64}$$

An important point to be mentioned is that if the shape functions are to be continuous between finite and infinite elements, the absolute value should be unity and the phase must be made zero at $\xi=-1$ (the boundary with the standard finite elements). At $\xi = -1$, the value of the shape function $N^*(\xi, \eta)$ is

$$N^*(-1, \eta) = M(-1, \eta)\left(\frac{A}{2}\right)^{1/2} \exp\left(\frac{ikA}{2}\right) \tag{15.65}$$

To ensure continuity, a factor $(2/A)^{1/2}\exp(-ikA/2)$ is introduced so that the shape function becomes

$$N^*(\xi, \eta) = M(\xi, \eta)\left(\frac{2}{A}\right)^{1/2}\left(\frac{A}{1-\xi}\right)^{1/2} \exp\left(\frac{-ikA}{2}\right)\exp\left(\frac{ikA}{1-\xi}\right) \tag{15.66}$$

The derivatives of the shape function can easily be found and used in Eq. (15.38), with integration over the infinite domain of the element.

15.12.3 *Integration procedure.* It turns out that in order to evaluate the element matrix some quite difficult integrals have to be performed. After integration by parts it is necessary to integrate terms of the form

$$\int_{B/2}^{\infty} \frac{\exp(iu)}{u} du \tag{15.67}$$

There is no analytical solution to Eq. (15.67) and special procedures had to be devised. Full details, including a listing of routines to form integration abscissas and weights, are given in reference 61. The routines use the Nag subroutine library routines S13ACF and S13ADF. The results obtained using the elements were extremely accurate. These routines can be used as a package, to set up the weights and abscissas, once and for all, at the start of each analysis. Figure 15.20 shows comparisons with results obtained using high-order dampers and exponential decay infinite elements for the MacCamy and Fuchs problem.

15.12.4 *Wave envelope infinite elements.* Astley has recently introduced a new type of infinite element, in which the weighting function is the *complex conjugate* of the shape function.[62,63] The great simplification that this introduces is that the oscillatory function $\exp(ikr)$ cancels after being multiplied by $\exp(-ikr)$ and the remaining terms are all polynomials, which can be integrated using standard techniques, like Gauss–Legendre integration (see Chapter 8 of Volume 1). This appears to be a big advance.

Figure 15.21 shows an example from acoustics, that of acoustical pressure in a hyperbolic duct. Good resuts were obtained despite using a coarse mesh. Unfortunately the method has not yet been fully tested for surface waves and it is still possible that there will be unforeseen

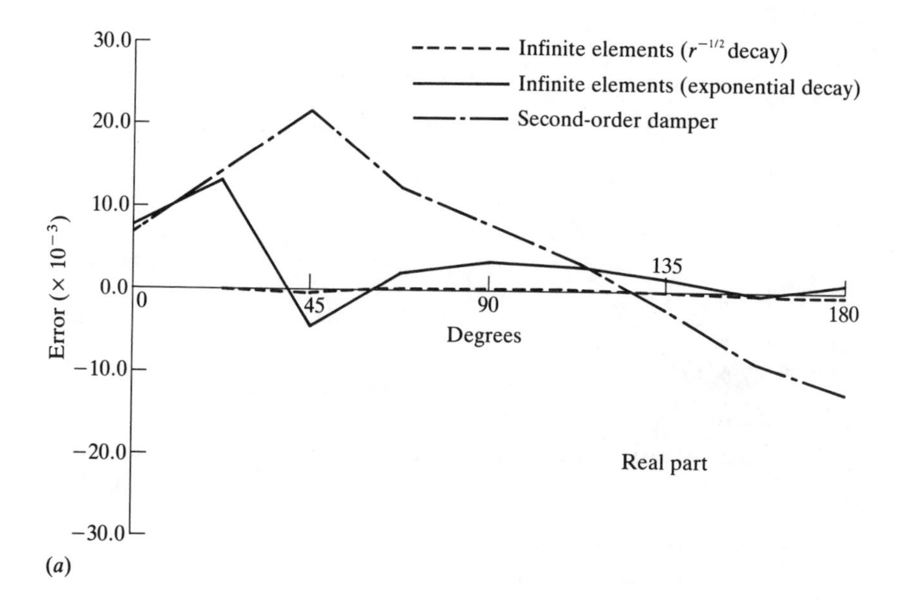

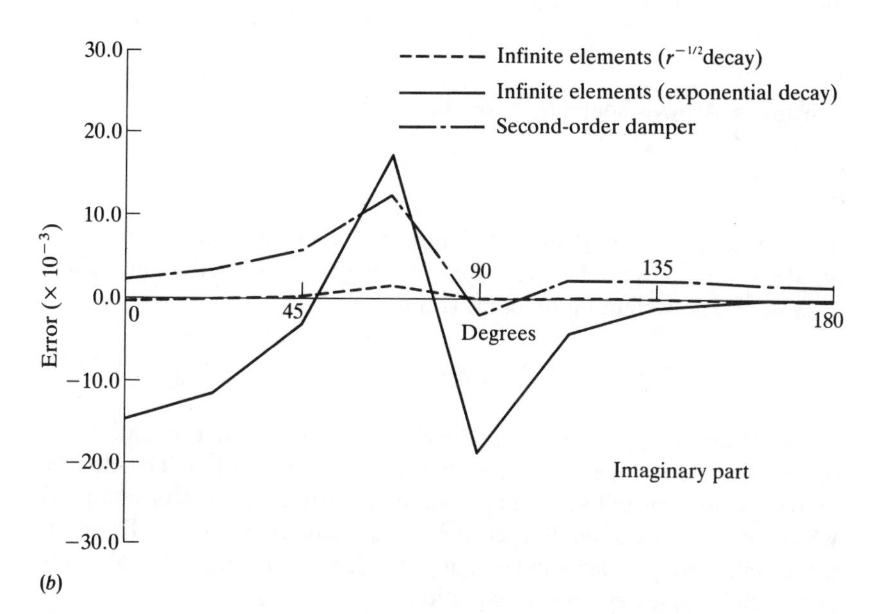

Fig. 15.20 Errors in the (*a*) real and (*b*) imaginary parts of the solution of a cylinder in shallow water produced using infinite elements with $r^{-1/2}$ decay and exponential decay and second-order damper elements

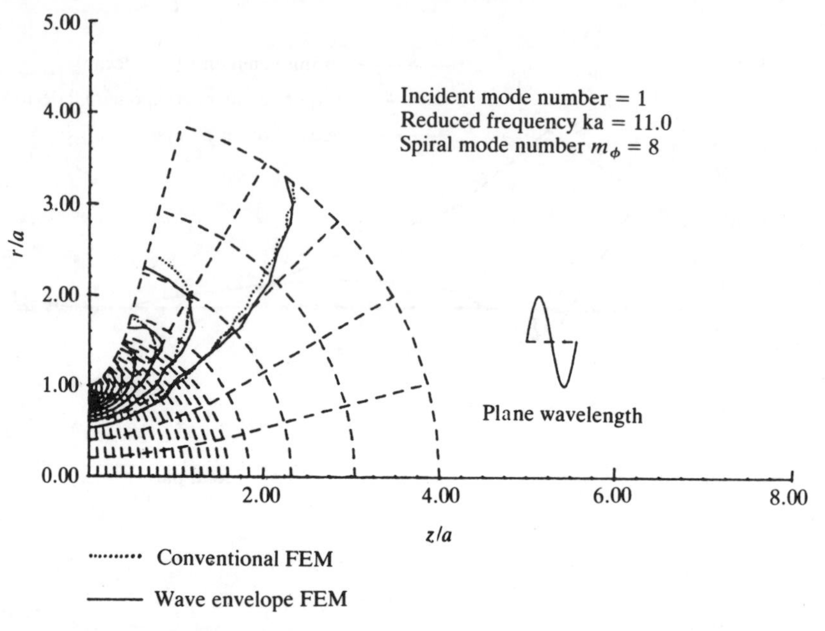

Fig. 15.21 Computed acoustical pressure contours for a hyperbolic duct ($\Theta_0 = 70°$, $ka = 11$, $m_\phi = 8$). Conventional and wave envelope solutions

difficulties. Astley's shape function was of the form

$$N_i(r, \theta)\frac{r_i}{r}e^{-ik(r-r_i)} \tag{15.68}$$

where N_i is the standard shape function. These shape functions incorporate decay and a wave-like variation, as suggested by Bettess and Zienkiewicz.[56] The weighting function is thus

$$N_i(r, \theta)\frac{r_i}{r}e^{+ik(r-r_i)} \tag{15.69}$$

Bettess[64] shows that for a one-dimensional model wave equation the infinite wave envelope element recovers the *exact* solution. The element matrix is now hermitian rather than symmetric (though still complex), which necessitates a small alteration to the equation solver. (There are not usually any problems in changing standard profile or front solvers to deal with complex systems of equations.)

15.13 Three-dimensional effects

As has already been described, when the water is deep in comparison with the wavelength, the shallow water theory is no longer adequate. For constant or slowly varying depth, Berkhoff's theory is valid. Also the geometry of the problem may necessitate another approach. The flow in the body of water is completely determined by the conservation of mass, which in the case of incompressible flow reduces to Laplace's equation. The free surface condition is zero pressure on the free surface. On using Bernoulli's equation and the kinematic condition, the free surface condition can be expressed, in terms of the velocity potential ϕ, as

$$\frac{\partial^2 \phi}{\partial t^2} + g\frac{\partial \phi}{\partial z} + 2\nabla\phi\nabla\frac{\partial \phi}{\partial t} + \frac{1}{2}\nabla\phi\nabla(\nabla\phi\nabla\phi) = 0 \tag{15.70}$$

where the velocities $u_i = \partial\phi/\partial x_i$. This condition is applied on the free surface, whose position is unknown *a priori.* If only linear terms are retained, Eq. (15.70) becomes, for transient and periodic problems,

$$\frac{\partial^2 \phi}{\partial t^2} + g\frac{\partial \phi}{\partial z} = 0 \qquad \text{or} \qquad \frac{\partial \phi}{\partial z} = \frac{\omega^2}{g}\phi \tag{15.71}$$

which is known as the *Cauchy–Poisson* free surface condition. It was derived in Chapter 11 as Eq. (11.11). Three-dimensional finite elements can be used to solve such problems. The actual three-dimensional element is very simple, being a potential element of the type described in Chapter 10 of Volume 1. The natural boundary condition is $\partial\phi/\partial n = 0$, where n is the outward normal, so to apply the free surface condition it is only necessary to add a surface integral to generate the $\omega^2\phi/g$ term from the Cauchy–Poisson condition [see Eq. (11.11)]. Two-dimensional elements in the far field can be linked to three-dimensional elements in the near field around the object of interest. Such models will predict velocity potentials, pressures throughout the fluid and wave elevations. They can also be used to predict fluid structure interaction. All the necessary equations are given in Chapter 11. More details of fluid structure interaction of this type are given by Zienkiewicz and Bettess.[65] Essentially the fluid equations must be solved for incident waves and for motion of the floating body in each of its degrees of freedom (usually six). The resulting fluid forces, masses, stiffnesses and damping are used in the equations of motion of the structure to determine its response. Figure 15.22 shows some results obtained by Hara *et al.*,[66] using the WAVE program for a floating breakwater. They obtained good agreement between the infinite elements and the methods of Sec. 15.11.

15.14 Large-amplitude waves

There is no complete wave theory that deals with the case when η is not small in comparison with the other dimensions of the problem. Various

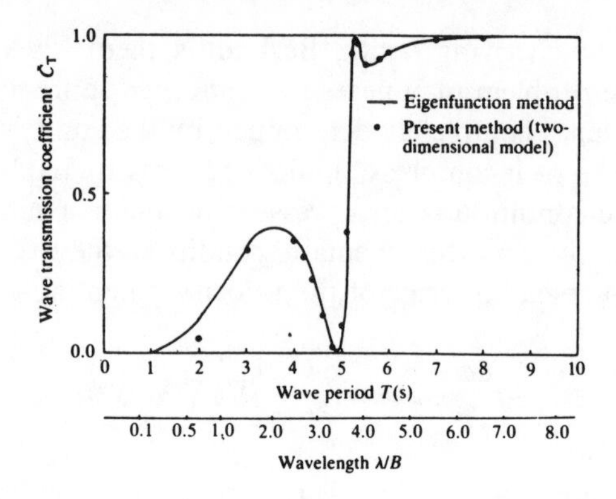

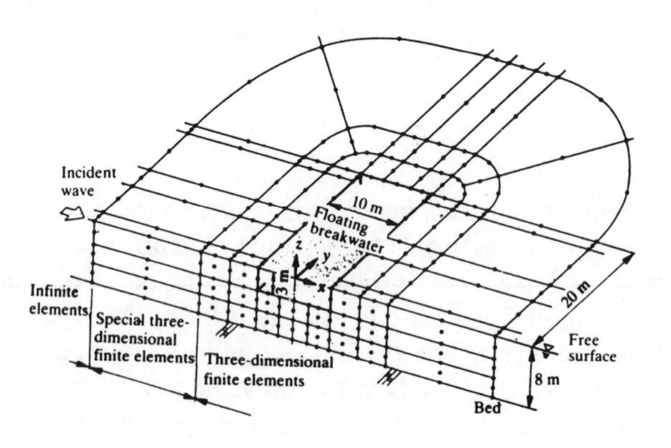

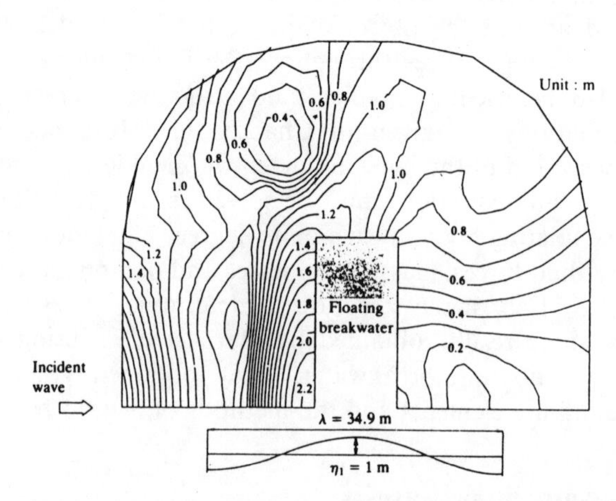

Fig. 15.22 Element mesh, contours of wave elevation and wave transmission coefficients for floating breakwater

special theories are invoked for different circumstances. We consider two of these, namely large wave elevations in shallow water and large wave elevations in intermediate to deep water.

15.14.1 *Cnoidal and solitary waves.* The equations modelled in Part I can deal with large amplitude waves on shallow water. These are called *cnoidal waves* when periodic and *solitary waves* when the period is infinite. For more details see references 38 to 41. The finite element methodology of Part I can be used to model the propagation of such waves. It is also possible to reduce the equations of momentum balance and mass conservation to corresponding wave equations in one variable, of which there are several different forms. One famous equation is the Korteweg–de Vries equation, which in physical variables is

$$\frac{\partial \eta}{\partial t}+\sqrt{gH}\left(1+\frac{3\eta}{2h}\right)\frac{\partial \eta}{\partial x}+\frac{h^2}{6}\sqrt{gH}\,\frac{\partial^3 \eta}{\partial x^3}=0 \tag{15.72}$$

This equation has been given a great deal of attention by mathematicians. It can be solved directly using finite element methods, and a general introduction to this field is given by Mitchell and Schoombie.[67]

15.15.2 *Stokes waves.* When the water is deep, a different asymptotic expansion can be used in which the velocity potential ϕ and the surface elevation η are expanded in terms of a small parameter, ε, which can be identified with the slope of the water surface. When these expressions are substituted into the free surface condition and terms with the same order in ε are collected, a series of free surface conditions is obtained. The equations were solved by Stokes initially, and then by other workers, to very high orders, to give solutions for large amplitude progressive waves on deep water. There is an extensive literature on these solutions, and they are used in the offshore industry for calculating loads on offshore structures. In recent years, attempts have been made to model the second-order wave diffraction problem, using finite elements and similar techniques. The first-order diffraction problem is as described in Sec. 15.9. In the second-order problem, the free surface condition now involves the first-order potential:

First order:

$$\frac{\partial \phi^{(1)}}{\partial z}-\frac{\omega^2}{g}\phi^{(1)}=0 \tag{15.73}$$

Second order:

$$\frac{\partial \phi^{(2)}}{\partial z}-\frac{\omega^2}{g}\phi^{(2)}=\alpha_D^{(2)} \tag{15.74}$$

$$\alpha_D^{(2)}=\alpha_{DI}^{(2)}+\alpha_{DD}^{(2)} \qquad \text{and} \qquad \nu=\frac{\omega^2}{g} \tag{15.75}$$

$$\alpha_{DI}^{(2)} = -\mathrm{i}\frac{\omega}{2g}\phi_D^{(1)}\left(\frac{\partial^2\phi_I^{(1)}}{\partial z^2} - \nu\frac{\partial\phi_I^{(1)}}{\partial z}\right) - \mathrm{i}\frac{\omega}{2g}\phi_I^{(1)}\left(\frac{\partial^2\phi_D^{(1)}}{\partial z^2} - \nu\frac{\partial\phi_D^{(1)}}{\partial z}\right) + \mathrm{i}\frac{2\omega}{g}\nabla\phi_I^{(1)}\nabla\phi_D^{(1)} \tag{15.76}$$

$$\alpha_{DD}^{(2)} = -\mathrm{i}\frac{\omega}{2g}\phi_D^{(1)}\left(\frac{\partial^2\phi_D^{(1)}}{\partial z^2} - \nu\frac{\partial\phi_D^{(1)}}{\partial z}\right) + \mathrm{i}\frac{\omega}{g}(\nabla\phi_D^{(1)})^2 \tag{15.77}$$

The second-order boundary condition can be thought of as identical to the first-order problem, but with a specified pressure applied over the entire free surface, of value α. Now there is no *a priori* reason why such a pressure distribution should give rise to outgoing waves as in the first-order problem, and so the usual radiation condition is not applicable. The usual procedure is to split the second-order wave into two parts, one the 'locked' wave in phase with the first-order wave and the other the 'free' wave, which is like the first-order wave but at twice the frequency, and with an appropriate wave number obtained from the dispersion relation. For further details of the theory, see Clark *et al.*[68] Figure 15.23 shows results for the second-order wave elevation around a

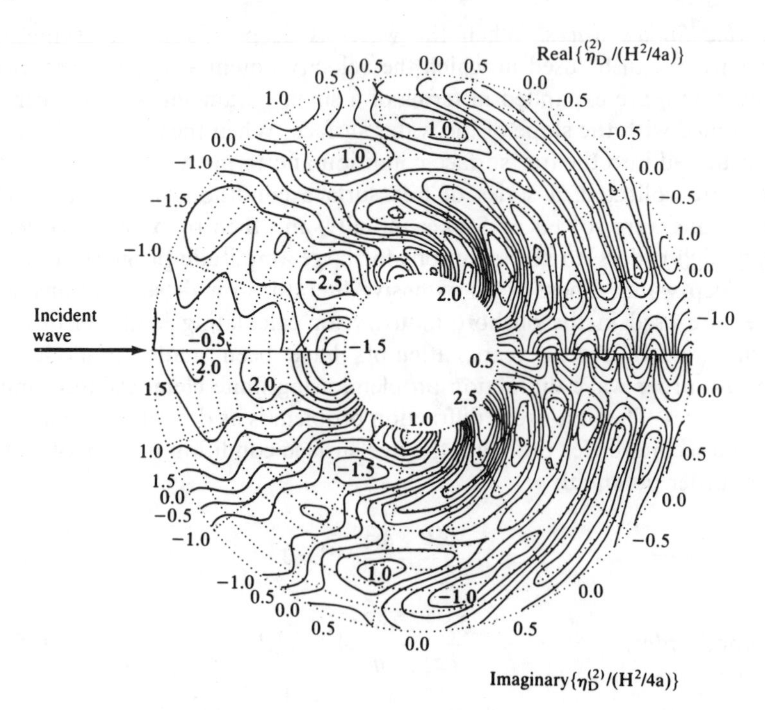

Fig. 15.23 Second-order wave elevations around cylinder—real and imaginary parts

circular cylinder, obtained by Clark *et al.* Although not shown, good agreement has been obtained with predictions made by boundary integrals. Preliminary results for wave forces only have also been produced by Lau *et al.*[69] A much finer finite element mesh is needed to resolve the details of the waves at second order. The second-order wave forces can be very significant for realistic values of the wave parameters (those encountered in the North Sea, for example). The first-order problem is solved first and the first-order potential is used to generate the forcing terms in Eqa. (15.75) to (15.77). These values have to be very accurate. In principle the method could be extended to third and higher orders, but in practice the difficulties multiply, and in particular the dispersion relation changes and the waves become unstable.[41]

References

1. M. B. Abbott, *Computational Hydraulics; Elements of the Theory of Free Surface Flows*, Pitman, London, 1979.
2. G. J. Haltiner and R. T. Williams, *Numerical Prediction and Dynamic Meteorology*, Wiley, New York, 1980.
3. O. C. Zienkiewicz, R. W. Lewis and K. G. Stagg (eds), *Numerical Methods in Offshore Engineering*, Wiley, Chichester, 1978.
4. J. Peraire, 'A finite element method for convection dominated flows', Ph.D. thesis, University of Wales, Swansea, 1986.
5 J. Peraire, O. C. Zienkiewicz and K. Morgan, 'Shallow water problems: a general explicit formulation', *Int. J. Num. Meth. Eng.*, **22**, 547–74, 1986.
6. O. C. Zienkiewicz and J. C. Heinrich, 'A unified treatment of steady state shallow water and two dimensional Navier–Stokes equations. Finite element penalty function approach', *Comp. Meth. Appl. Mech. Eng.*, **17/18**, 673–89, 1979.
7. M. Kawahara, 'On finite-element methods in shallow-water long-wave flow analysis', in *Computational Methods in Nonlinear Mechanics* (ed. J. T. Oden), pp. 261–87, North-Holland, Amsterdam, 1980.
8. J. J. Connor and C. A. Brebbia, *Finite-Element Techniques for Fluid Flow*, Newnes-Butterworth, London and Boston, 1976.
9. J. J. O'Brien and H. E. Hulburt, 'A numerical model of coastal upwelling', *J. Phys. Oceanogr.*, **2**, 14–26, 1972.
10. M. Crepon, M. C. Richez and M. Chartier, 'Effects of coastline geometry on upwellings', *J. Phys. Oceanogr.*, **14**, 365–82, 1984.
11. M. G. G. Foreman, 'An analysis of two-step time-discretisations in the solution of the linearized shallow-water equations', *J. Comp. Phys.*, **51**, 454–83, 1983.
12. W. R. Gray and D. R. Lynch, 'Finite-element simulation of shallow-water equations with moving boundaries', in *Proc. 2nd Conf. on Finite-Elements in Water Resources* (eds C. A. Brebbia *et al.*), pp. 2.23–2.42, 1978.
13. T. D. Malone and J. T. Kuo, 'Semi-implicit finite-element methods applied to the solution of the shallow-water equations', *J. Geophys. Res.*, **86**, 4029–40, 1981.
14. G. J. Fix, 'Finite-element models for ocean-circulation problems', *SIAM J. Appl. Math.*, **29**, 371–87, 1975.

15. C. TAYLOR and J. DAVIS, 'Tidal and long-wave propagation, a finite-element approach', *Computers and Fluids*, **3**, 125–48, 1975.
16. M. J. P. CULLEN, 'A simple finite element method for meteorological problems', *J. Inst. Math. Appl.*, **11**, 15–31, 1973.
17. H. H. WANG, P. HALPERN, J. DOUGLAS, Jr and I. DUPONT, 'Numerical solutions of the one-dimensional primitive equations using Galerkin approximations with localised basis functions', *Mon. Weekly Rev.*, **100**, 738–46, 1972.
18. I. M. NAVON, 'Finite-element simulation of the shallow-water equations model on a limited area domain', *Appl. Math. Modelling*, **3**, 337–48, 1979.
19. M. J. P. CULLEN, 'The finite element method', in *Numerical Methods Used in Atmosphere Models*, Vol. 2, chap. 5, pp. 330–8, WMO/GARP Publication Series 17, World Meteorological Organisation, Geneva, Switzerland, 1979.
20. M. J. P. CULLEN and C. D. HALL, 'Forecasting and general circulation results from finite-element models', *Q. J. Roy. Met. Soc.*, **102**, 571–92, 1979.
21. D. E. HINSMAN, R. T. WILLIAMS and E. WOODWARD, 'Recent advances in the Galerkin finite-element method as applied to the meteorological equations on variable resolution grids', in *Finite-Element Flow-Analysis* (ed. T. Kawai), University of Tokyo Press, Tokyo, 1982.
22. I. M. NAVON, 'A Numerov-Galerkin technique applied to a finite-element shallow-water equations model with enforced conservation of integral invariants and selective lumping', *J. Comp. Phys.*, **52**, 313–39, 1983.
23. I. M. NAVON and R. DE VILLIERS, 'GUSTAF, a quasi-Newton nonlinear ADI FORTRAN IV program for solving the shallow-water equations with augmented Lagrangians', *Computers and Geosci.*, **12**, 151–73, 1986.
24. A. N. STANIFORTH, 'A review of the application of the finite-element method to meteorological flows', in *Finite-Element Flow-Analysis* (ed. T Kawai), pp. 835–42, University of Tokyo Press, Tokyo, 1982.
25. A. N. STANIFORTH, 'The application of the finite element methods to meteorological simulations—a review', *Int. J. Num. Meth. Fluids*, **4**, 1–22, 1984.
26. R. T. WILLIAMS and O. C. ZIENKIEWICZ, 'Improved finite element forms for shallow-water wave equations', *Int. J. Num. Meth. Fluids*, **1**, 91–7, 1981.
27. M. G. G. FOREMAN, 'A two-dimensional dispersion analysis of selected methods for solving the linearised shallow-water equations', *J. Comp. Phys.*, **56**, 287–323, 1984.
28. I. M. NAVON, 'FEUDX: a two-stage, high-accuracy, finite-element FORTRAN program for solving shallow-water equations', *Computers and Geosci.*, **13**, 225–85, 1987.
29. P. BETTESS, C. A. FLEMING, J. C. HEINRICH, O. C. ZIENKIEWICZ and D. I. AUSTIN, 'A numerical model of longshore patterns due to a surf zone barrier', in *16th Coastal Engineering Conf.*, Hamburg, West Germany, October 1978.
30. M. KAWAHARA, N. TAKEUCHI and T. YOSHIDA, 'Two step explicit finite element method for tsunami wave propagation analysis', *Int. J. Num. Meth. Eng.*, **12**, 331–51, 1978.
31. J. H. W. LEE, J. PERAIRE and O. C. ZIENKIEWICZ, 'The characteristic Galerkin method for advection dominated problems—an assessment', *Comp. Meth. Appl. Mech. Eng.*, **61**, 359–69, 1987.
32. R. LÖHNER, K. MORGAN and O. C. ZIENKIEWICZ, 'The solution of non-linear hyperbolic equation systems by the finite element method', *Int. J. Num. Meth. Fluids*, **4**, 1043–63, 1984.
33. S. NAKAZAWA, D. W. KELLY, O. C. ZIENKIEWICZ, I. CHRISTIE and M.

KAWAHARA, 'An analysis of explicit finite element approximations for the shallow water wave equations', in *Proc. 3rd Int. Conf. on Finite Elements in Flow Problems*, Banff, Vol. 2, pp. 1–7, 1980.
34. M. KAWAHARA, H. HIRANO, K. TSUBOTA and K. INAGAKI, 'Selective lumping finite element method for shallow water flow', *Int. J. Num. Meth. Fluids*, **2**, 89–112, 1982.
35. D. R. LYNCH and W. G. GRAY, 'Analytic solutions for computer flow model testing', *Trans. ASCE, J. Hydr. Div.*, **104**(10), 1409–28, 1978.
36. D. I. AUSTIN and P. BETTESS, 'Longshore boundary conditions for numerical wave model', *Int. J. Num. Mech. Fluids*, **2**, 263–76, 1982.
37. C. K. ZIEGLER and W. LICK, 'The transport of fine grained sediments in shallow water', *Environ. Geol. Water Sci.*, **11**, 123–32, 1988.
38. H. LAMB, *Hydrodynamics*, 6th ed., Cambridge University Press, 1932.
39. G. B. WHITHAM, *Linear and Nonlinear Waves*, Wiley, New York, 1974.
40. C. C. MEI, *The Applied Dynamics of Ocean Surface Waves*, Wiley, New York, 1983.
41. M. J. LIGHTHILL, *Waves in Fluids*, Cambridge Univesity Press, 1978.
42. J. C. W. BERKHOFF, 'Linear wave propagation problems and the finite element method', in *Finite Elements in Fluids* (eds R. H. Gallagher *et al.*), Vol. 1, pp. 251–80, Wiley, Chichester, 1975.
43. O. C. ZIENKIEWICZ and P. BETTESS, 'Infinite elements in the study of fluid structure interaction problems', in *Proc. 2nd Int. Symp. on Computational Methods of Applied Science*, Versailles, 1975; also published in *Lecture Notes in Physics* (eds J. Ehlers *et al.*), Vol. 58, Springer-Verlag, Berlin, 1976.
44. C. TAYLOR, B. S. PATIL and O. C. ZIENKIEWICZ, 'Harbour oscillation: a numerical treatment for undamped natural modes', *Proc. Inst. Civ. Eng.*, **43**, 141–56, 1969.
45. O. C. ZIENKIEWICZ, P. BETTESS and D. W. KELLY, 'The finite element method for determining fluid loading on rigid structures: two- and three-dimensional formulations', in *Numerical Methods in Offshore Engineering* (eds O. C. Zienkiewicz, R. W. Lewis and K. G. Stagg), chap. 4, Wiley, Chichester, 1978.
46. A. BAYLISS and E. TURKEL, 'Radiation boundary conditions for wave-like equations', ICASE report 79–26, 1979.
47. A. BAYLISS, M. GUNZBERGER and E. TURKEL, 'Boundary conditions for the numerical solution of elliptic equations in exterior regions', ICASE report 80–1, 1980.
48. O. C. ZIENKIEWICZ and R. E. NEWTON, 'Coupled vibrations of a structure submerged in a compressible fluid', in *Proc. Int. Symp. on Finite Element Techniques*, Stuttgart, 1–15 May 1969.
49. K. BANDO, P. BETTESS and C. EMSON, 'The effectiveness of dampers for the analysis of exterior wave diffraction by cylinders and ellipsoids', *Int. J. Num. Meth. Fluids*, **4**, 599–617, 1984.
50. O. C. ZIENKIEWICZ, D. W. KELLY and P. BETTESS, 'The coupling of the finite element method and boundary solution procedures', *Int. J. Num. Meth. Eng.*, **11**, 355–75, 1977.
51. O. C. ZIENKIEWICZ, D. W. KELLY and P. BETTESS, 'Marriage à la mode—the best of both worlds (finite elements and boundary integrals)', in *Energy Methods in Finite Element Analysis* (eds R. Glowinski, E. Rodin and O. C. Zienkiewicz), chap. 5, pp. 81–106, Wiley, Chichester, 1980.
52. J. C. W. BERKHOFF, 'Computation of combined refraction–diffraction', in

Proc. 13th Int. Conf. on Coastal Engineering, Vancouver, 10–14 July 1972.

53. H. S. CHEN and C. C. MEI, 'Oscillations and wave forces in an offshore harbor', report 190, Parsons Laboratory, Massachusetts Institute of Technology, 1974.
54. H. S. CHEN and C. C. MEI, 'Oscillations and wave forces in a man-made harbor in the open sea', in *Proc. 10th. Symp. on Naval Hydrodynamics*, pp. 573–94, Office of Naval Research, 1974.
55. J. R. HOUSTON, 'Long Beach Harbor: numerical analysis of harbor oscillations', report 1, miscellaneous papers H-76-20, U.S. Army Engineering Waterways Experimental Station, Vicksburg, Miss., 1976.
56. P. BETTESS and O. C. ZIENKIEWICZ, 'Diffraction and refraction of surface waves using finite and infinite elements', *Int. J. Num. Meth. Eng.*, **11**, 1271–90, 1977.
57. P. BETTESS, S. C. LIANG and J. A. BETTESS, 'Diffraction of waves by semi-infinite breakwater using finite and infinite elements', *Int. J. Num. Meth. Fluids*, **4**, 813–32, 1984.
58. O. C. ZIENKIEWICZ, C. EMSON and P. BETTESS, 'A novel boundary infinite element', *Int. J Num. Meth. Eng.*, **19**, 393–404, 1983.
59. O. C. ZIENKIEWICZ, P. BETTESS, T. C. CHIAM and C. EMSON, *Numerical Methods of Unbounded Field Problems and a New Infinite Element Formulation*, AMD 46, pp. 115–48, American Society of Mechanical Engineers, New York, 1981.
60. P. BETTESS, C. EMSON and T. C. CHIAM, 'A new mapped infinite element for exterior wave problems, in *Numerical Methods in Coupled Systems* (eds R. W. Lewis, P. Bettess and E. Hinton), chap. 17, Wiley, Chichester, 1984.
61. O. C. ZIENKIEWICZ, K. BANDO, P. BETTESS, C. EMSON and T. C. CHIAM, 'Mapped infinite elements for exterior wave problems', *Int. J. Num. Meth. Eng.*, **21**, 1229–51, 1985.
62. R. J. ASTLEY, 'Wave envelope and infinite elements for acoustical radiation', *Int. J. Num. Meth. Fluids*, **3**, 507–26, 1983.
63. R. J. ASTLEY and W. EVERSMAN, 'A note on the utility of a wave envelope approach in finite element duct transmission studies', *J. Sound Vibr.*, **76**, 595–601, 1981.
64. P. BETTESS, 'A simple wave envelope element example', *Comm. Appl. Num. Meth.*, **3**, 77–80, 1987.
65. O. C. ZIENKIEWICZ and P. BETTESS, 'Fluid–structure interaction and wave forces. An introduction to numerical treatment', *Int. J. Num. Meth. Eng.*, **13**, 1–16, 1978.
66. H. HARA, K. KANEHIRO, H. ASHIDA, T. SUGAWARA and T. YOSHIMURA, 'Numerical simulation system for wave diffraction and response of offshore structures', Technical Bulletin TB 83–07, Mitsui Engineering and Shipbuilding Co., October 1983.
67. A. R. MITCHELL and S. W. SCHOOMBIE, 'Finite element studies of solitons', in *Numerical Methods in Coupled Systems* (eds R. W. Lewis, P. Bettess and E. Hinton), chap. 16, pp. 465–88, Wiley, Chichester, 1984.
68. P. J. CLARK, P. BETTESS, G. E. HEARN and M. J. DOWNIE, 'The application of finite element analysis to the solution of Stokes wave diffraction problems', *Int. J. Num. Meth. Fluids* (to appear).
69. S. L. LAU, K. K. WONG and Z. TAM, 'Nonlinear wave loads on large body by time–space finite element method', *Proc. Int. Conf. on Computers in Engineering*, Vol. 3, pp. 331–7, American Society of Mechanical Engineers, New York, 1987.

16

Computer procedures for finite element analysis

16.1 Introduction

In this chapter we extend the capabilites of the program presented in Volume 1 to include features to solve both non-linear and transient problems that are modelled by a finite element process. The material included in this chapter should be considered as supplementary to information contained in Chapter 15 of Volume 1.[1] Accordingly, throughout this chapter reference will be made to appropriate information in the first volume. It is suggested, however, that the reader review the material there prior to embarking on a study of this chapter.

The program included in this volume is intended for use by students and scholars who are undertaking a study of the finite element method and wish to implement new elements or solution strategies. The program is called PCFEAP to emphasize the fact that it may be used on computers that are as small as a personal computer. With very few exceptions, the entire program is written using standard Fortran 77; hence it may be implemented on personal computers, more powerful workstations or mainframe computers. Since the publication of Volume 1, the program has been tested using versions of the Microsoft Fortran Compiler[2] (Versions 3.31 and 5.0) and on workstations operating under the UNIX operating system.† The exception to standard Fortran is the use of INTEGER*2 storage of variables. Also, the other requirements to use the program in different environments are to furnish an appropriate timing routine and, if graphical output is desired, to implement a graphics interface. The routine needed to provide a timing routine for some environments is described in Sec. 16.10.2.

† Microsoft is a registered trademark of Microsoft Corporation and UNIX is a registered trademark of AT&T Corporation.

The current chapter is divided into several sections which describe different aspects of the program. Section 16.2 summarizes the additional program features and the macro solution commands that may be used to solve finite element problems. In Sec. 16.3 we present sample input data to describe the finite element mesh of some typical problems. These problems serve as test examples which are considered further in a later section. Some general solution strategies and the related macro commands for using the program to solve non-linear problems are presented in Sec. 16.4. Non-linear problems are often difficult to solve and time consuming in computer resources. In many applications the complete analysis may not be performed during one execution of the program; hence, techniques to stop the program at key points in the analysis for a later restart to continue the solution are presented in Sec. 16.5. Section 16.6 discusses the solution of general linear and non-linear transient problems using PCFEAP. The program includes capabilities to solve both first (diffusion-type)-order and second (vibration/wave-type)-order ordinary differential equations in time. In Sec. 16.7, some applications of an eigensystem to finite element problems are summarized together with the required macro solution statements. The program PCFEAP contains a simultaneous vector iteration algorithm (subspace method) to extract the eigenpairs nearest to a specified shift of the *tangent* matrix. Hence, the eigensystem may be used with either linear or non-linear problems. This section completes the description of new and extended solution options that have been added to the program.

As noted above, new elements have been included in the program PCFEAP and these are identified and briefly described in Sec. 16.8. This section also includes tables to describe the necessary user data for specifying the MATE data in the mesh inputs. Section 16.9 describes the solution steps for some typical problems that can be solved using PCFEAP and also presents numerical data that may be used to verify proper installation of the program on the reader's own computer. Finally, Sec. 16.10 includes installation information, listings and support information for the major changes to PCFEAP. A complete listing for the program is not included and it is necessary to consult Chapter 15 of Volume 1 to obtain some routines.

The program contained in this chapter has been developed and used in an educational and research environment over a period of nearly fifteen years. The concept of the macro solution algorithm has permitted several users, who are considering problems that differ widely in scope and concept, to use the system at the same time without the need for different versions of the program. Unique features for each user may be provided as new macro commands. The ability to treat problems whose coefficient matrix may be either symmetric or unsymmetric often proves useful for testing the performance of algorithms that advocate substitution of a symmetrized tangent matrix in place of an unsymmetric matrix resulting from a consistent linearization process. Also, the element interface is quite straightforward and, once understood, permits users to rapidly test new types of finite elements. Indeed, the present program has had a number of contributions from colleagues who should be acknowledged. Professor Juan C. Simo of Stanford University has helped in the development of the return-map plasticity element contained in this chapter and Dr Werner Wagner of the Technical University of Hannover furnished an early version of the axisymmetric shell element which was adapted for inclusion in this chapter. Dr Peter Wriggers, from the Technical University of Darmstadt, over a period of years has contributed to the development of the system with respect to solution algorithms for non-linear problems. Finally, the assistance of all our students over the years must be acknowledged, especially that of Mr Panayiotis Papadopoulos of the University of California at Berkeley, who carefully reviewed this chapter.

We believe that the program in this book provides a very powerful solution system to assist the interested reader in perfoming finite element analyses. The program PCFEAP is by no means a complete software system which can be used to solve any finite element problem. Some suggestions for extending the system are contained in this chapter, especially at the end of Sec. 16.9. While the program has been tested on several sample problems, it is likely that errors (blunders) still exist within the program modules. The authors need to be informed about these errors so that future printings may be corrected. We also welcome the reader's comments and suggestions concerning possible future improvements.

16.2 User instructions—description of additional program features

The user instructions given in Chapter 15 of Volume 1 also apply to the program contained in this volume. The next section presents a set of example problems which may be used to review applications of the program. These examples illustrate many of the features in generating a finite element model for problems that may be solved using the program. Some extensions to the capability of the program have been made and general information on using the extensions is contained in this section.

The macro solution program has been extended to permit simulation of a broader class of linear and non-linear applications. The principal additions relate to the solution of non-linear problems and the present program adds the ability to consider applications that have unsymmetric tangent 'stiffness' matrices. In addition, the program introduces a line search algorithm which may be invoked to permit convergence of Newton-type algorithms with rather large solution increments. The program also extends the algorithms to solve transient problems by including the SS11 and SS22 algorithms discussed in Chapter 10. Finally, the program has an eigensolution algorithm based upon subspace iteration.[3,4] A description to the user information required to invoke the added macro commands for these features is contained in Table 16.1.

16.3 Example problem descriptions

In this section we consider the specification of the mesh input data for several types of problems which are analysed later (see Sec. 16.9). These applications illustrate many of the features contained in the program to describe the mesh for a finite element simulation. The specification of a problem must begin with a FEAP record, which also may include

TABLE 16.1
LIST OF MACRO PROGRAMMING COMMANDS

Columns					
1–4	16–19	31–45	46–60	61–75	Description
BETA	SS11	V1			Set solution algorithm to SS11, V1 is value of θ
BETA	SS22	V1	V2		Set solution algorithm to SS22, V1, V2 are values of θ_1,θ_2
BETA	GN22	V1	V2		Set solution algorithm to GN22, V1, V2 are values of β,γ
BETA		V1	V2		Same as BETA GN22
EIGV		N1			Output N1 eigenpairs (after SUBS)
IDEN					Set the mass matrix to the identity
PLOT	EIGV	V1			Plot last eigenvector output, scale by V1 (after EIGV)
SOLV	LINE	V1			Solve for new displacements (after FORM). If LINE present, compute solution with line search, V1 controls initiation (see Sec. 16.4)
SUBS	PRIN	N1	N2		Perform eigenpair extraction for N1 values with N2 extra vectors. If PRIN print subspace arrays (after TANG and MASS or IDEN)
TANG	LINE	N1	V2	V3	Compute and factor symmetric tangent matrix (ISW = 3). See Note
UTAN	LINE	N1	V2	V3	Compute and factor unsymmetric tangent matrix (ISW = 3). See Note

Note: if N1 is non-zero a residual, solution and update to the displacements is performed. If V2 is non-zero, the tangent matrix is modified by subtracting the V2 times the mass matrix before computing the triangular factors. The mass matrix may be set to an identity matrix using the macro command IDEN. If LINE is specified as part of the TANG or UTAN command a linear line search is performed whenever the energy ratio between two successive iterations exceeds the value of V3 (default is 0.8).

additional alphanumerical information to describe the problem. Immediately following this record is a description of the problem size and other mesh parameters, and is given by

Data	*Description*
NUMNP	Number of nodal points
NUMEL	Number of elements
NUMMAT	Number of material property sets
NDM	Spatial dimension of mesh
NDF	Degrees of freedom/node (maximum)
NEN	Nodes/element (maximum)

Subsequent to these two records additional data describing the specific problem to be solved are specified. The information for each problem is different but must include data to specify the location of nodes (COOR or BLOC), the element connection lists (ELEM or BLOC) and the material parameters (MATE). Other data is optional, but normally includes the BOUNdary condition restraints and the FORCed loading conditions (non-zero nodal loads and/or displacements). The examples described below indicate use of specific features in PCFEAP.

16.3.1 *Straight beam.* We consider first a simply supported beam example which is to be analysed using two-node elements. We assume the beam is 10 units long and is directed horizontally along the x axis. The beam is modelled using 20 elements and thus there are 21 nodal points. The left boundary conditions are to model a pin joint and the right a roller condition. Loading is given by a single force at the centre, directed vertically downward with a magnitude of 5.0 units. The beam is to have two sets of material properties: one for the left half and one for the right. As shown, the properties are set to the same values, which permits an assessment of symmetry in results computed in Sec. 16.9. If desired, these may be changed to test other ratios of the parameters. A mesh for this problem is given by the statements:†

```
FEAP**SIMPLY SUPPORTED BEAM MODEL
21,20,2,2,3,2

COOR
1,1,0.0,0.0
11,1,5.0,0.0
21,0,10.0,0.0
```

† The alphanumeric data required by the program PCFEAP is shown in upper case letters, whereas optional information is shown in lower case. It should be noted, however, that the data may actually be entered in either upper or lower case, since the command parser recognizes both forms in an ASCII environment.

```
ELEM
1,1,1,2,1
11,2,11,12,1
BOUN
1,,1,1,0
21,,0,1,0
FORC
11,,0.0,-5.0,0.0
MATE
1,1
BEAM
100.,0.,1.0,0.,0.,1.
MATE
2,1
BEAM
100.,0.,1.0,0.,0.,1.
END
INTEractive
STOP
```

Descriptions for the specific data used in generating the mesh are given in Sec. 15.3 in Volume 1. The degrees of freedom for the beam/shell element are ordered as: u, horizontal displacement (x or r direction); v, vertical displacement (y or z direction); and θ, rotation about the normal to the xy (or rz) plane. Thus, the control record following the FEAP start/title record specifies NDF as 3. The interpretation of the material data is postponed until Sec. 16.9 where capabilities of each element are discussed. The data set indicates that an interactive solution is to be performed; a batch-type execution may be used by replacing the INTE statement by MACR followed by the specific solution steps to be performed. The steps for solution are selected from those included in Table 15.16 in Volume 1 or in Table 16.1.

16.3.2 *Spherical cap/circular ring.* The second example is a spherical cap (or circular ring) which is described in Fig. 14.7. The finite element model consists of two-node line elements. We assume that the cap covers a 35 degree sector in the first quadrant of the rz (or xy) plane. Symmetry boundaries are assumed at the y axis and the other boundary is assumed to be fixed. The radius of the cap is taken as 90 inches and the thickness is 3 inches. Loading is applied as a uniform external pressure with a magnitude of 1 lb/in^2 and Poisson's ratio is taken as $\frac{1}{6}$. The cap is modelled by 14 elements and the mesh data is given by the statements:

```
FEAP**SPHERICAL CAP MODEL
15,14,1,2,3,2

COOR
1,1,90.0,55.0
15,0,90.0,90.0

POLAr
1,15,1

ELEM
1,1,1,2,1

BOUN
1,,1,1,1
15,,1,0,1

MATE
1,1
SHELL
100,0.166667,3.0,1.0,0.,1.

END
INTEractive
STOP
```

The mesh described above first specifies the coordinates in the polar system (r, θ) and then uses the POLA statement to convert to cartesian coordinates. The symmetry conditions at the y axis of a cap are zero rotation and zero horizontal displacement. For a circular ring change SHELL to BEAM in the MATE data.

16.3.3 *Solid sphere—symmetrical loading.* We next consider a problem which in reality is one dimensional, but is used to test the performance of a two-dimensional element formulated in cylindrical coordinates. The mesh models a solid sphere with an outer radius of 1.0 unit. The region to be discretized lies in the rz plane between the r axis and the 45 degree line between the r and z axes. The example to be solved in Sec. 16.9 replicates a spontaneous ignition process as described in Sec. 10.6.4. Thus, the problem is governed by a differential equation with a scalar-dependent variable (temperature) and the finite element discretization has only one unknown at each node (NDF = 1). The boundary conditions are specified values of the dependent variable at the outer radius and zero flux on the lateral edges. The discretization of the region is peformed using nine-node isoparametric elements.

```
FEAP**THICK SPHERE – SYMMETRICAL LOADING
39,7,1,2,1,9
```

```
ELEM
1,1
7,1,31,33,35,39,32,34,38,36,37

BLOCk
6,4,6,1,1,1,,9
1,1.0.,0.
2,1.0.,45.
3,0.25.,45.
4,0.25.,0.

COOR
36,1,0.125,0.
38,0,0.125,45.
39,0,0.,0.

POLAr
1,38,1

BOUN
1,1,-1
5,0,1

FORCe
1,1,290.
5,0,290.

MATE
1,3
0.2,1.,1.,0.,0.,2

END
INTEractive
STOP
```

Again, it is convenient to generate the mesh in polar coordinates within the *rz* plane and then to transform the nodal coordinate values to a cartesian representation using the POLAr command. It should be noted that node 39 is at the origin and needs no transformation. The features of an isoparametric master block are used to generate a regular mesh of nine-node isoparametric elements. It should be noted that nodes on the master block must be numbered as shown in Fig. 15.25 in Volume 1. In particular, the nodes must be in sequence in an anticlockwise order around the master element. Before using the BLOCk command, a single element (element 7) is first specified (together with temporary data for the first six elements). Also, the coordinates for the interior wedge are generated using the COOR command. Some dependency on order exists in the generation: it is necessary to generate all the elements when using

ELEM; thus this must be done first. BLOCk and COOR can generate any part of the data and, thus, may appear at any location in the data. This mesh also uses the generation features for both boundary restraints (BOUN) and forced non-zero conditions (FORC) to minimize the amount of input data needed. In Sec. 16.9 an analysis of the spontaneous ignition problem, as described in Chapter 10, is performed with this mesh. The data shown above has been set up to compute the initial conditions for this problem. Thus, the data shown will compute a homogeneous, non-zero, set of temperatures at 290 K.

16.3.4 *Extension of a sheet with a circular hole.* In this problem we consider uniform extension of a strip of finite width which contains a circular hole symmetrically placed. Due to symmetry, only one quadrant of the mesh is modelled. The model is constructed using the BLOCk command to generate a group of elements. By careful selection of the numbering system for elements and nodes, the block of elements are joined together to form the total mesh. Furthermore, the NOPRint command is inserted to minimize the information written onto the output file.

```
FEAP**TENSION STRIP WITH A CIRCULAR HOLE
160,133,1,2,2,4
NOPRint

BLOCk 1
5,7,7,1,1,1
1,5.,0.
2,10.,0.
3,10.,8.
4,3.53553,3.53552
8,4.61940,1.91342

BLOCk 2
5,7,7,57,50,1,5
1,3.53553,3.53552
2,10.,8.
3,0.,12.
4,0.,5.
8,1.91342,4.61940

BLOCk 3
4,5,7,64,99,1,7
1,10.,8.
2,10.,18.
3,0.,18.
4,0.,12.
```

```
FORC
69,13,0.,1.
160,,0.,1.

BOUN
1,1,0,-1
8,0,0,-1
69,13,0,-1
160,-1,-1,1
148,0,1,0

MATE
1,2
0,.1
7000.,.2,24.3,0.,0.

END
INTEractive
STOP
```

A plot of the mesh generated by the above statements is shown in Fig. 16.1. The block structure used to construct the mesh together with the

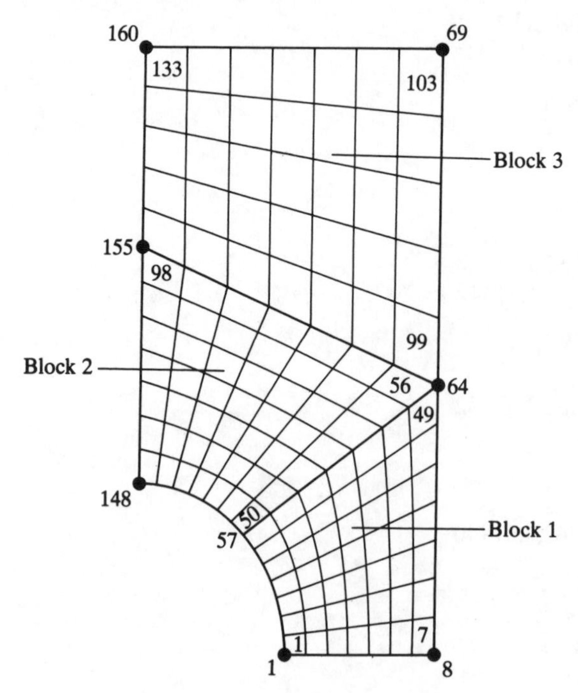

Fig. 16.1 Mesh for tension strip

numbers for the first node and element generated by each block are indicated in this figure. This example illustrates the use of multiple BLOCk commands to generate a mesh for the strip problem. The input statements use general characteristics of the BLOCk command to generate a consistent discretization. Users should especially note how the nodal increment feature of the BLOCk generation had been used in blocks 2 and 3, and how curved sides on blocks have been used in generating this mesh. Some care also has been used in selecting directions and in ordering of the blocks to minimize the front width (or profile) of the stiffness matrix. The program PCFEAP is currently dimensioned for a personal computer environment and with these values this problem is too large to solve using the variable band solver (file PASOLV.FOR); however, the problem can be solved using the frontal solver. Steps needed to perform an elastic–plastic analysis, using this mesh, are presented in Sec. 16.9.

The above examples illustrate many of the features available in the program presented in this book. Other mesh generation software may also be used provided the necessary input instructions for PCFEAP can be produced as output from the generation program.

16.4 Solution of non-linear problems

The solution of non-linear problems using the program contained in this volume is designed for a Newton or modified Newton-type algorithm.[5] In addition, the solution for transient non-linear problems may be achieved by combining a Newton-type algorithm with the transient integration methods described below.

We first consider a non-linear problem described by (see Chapter 7)

$$\boldsymbol{\psi}(\mathbf{a}) = \mathbf{P}(\mathbf{a}) - \mathbf{f} \tag{16.1}$$

where $\mathbf{f}$ is a vector of applied loads and $\mathbf{P}$ is the non-linear *internal* force vector which is indicated as a function of the nodal parameters $\mathbf{a}$. The vector $\boldsymbol{\psi}$ is known as the residual of the problem and a solution is defined as any set of nodal displacements, $\mathbf{a}$, for which the residual is zero. In general, there may be more than one set of displacements that define a solution and it is the responsibility of the user to ensure that a proper solution is obtained. This may be achieved by starting from a state that satisfies physical arguments for a solution and then applying small increments to the loading vector, $\mathbf{f}$. By taking small enough steps, a solution path may usually be traced. Thus, for any step, our objective is to find a set of values for the components of $\mathbf{a}$ such that

$$\boldsymbol{\psi}(\mathbf{a}) = \mathbf{0} \tag{16.2}$$

We assume some initial vector exists (initially in the program this vector is zero), from which we will seek a solution, and denote this as $\mathbf{a}^{(0)}$. Next we compute a set of iterates such that

$$\mathbf{a}^{(i+1)}=\mathbf{a}^{(i)}+\eta\delta\mathbf{a}^{(i)} \tag{16.3}$$

The scalar parameter η is introduced to control possible divergence during early stages of the iteration process and is often called *step size control.* A common algorithm to determine η is a *line search* defined by[6]

$$g=\min_{\eta\in 0,1} |G(\eta)| \tag{16.4}$$

where

$$G(\eta)=\delta\mathbf{a}^{(i)}\cdot\boldsymbol{\psi}(\mathbf{a}^{(i)}+\eta\delta\mathbf{a}^{(i)}) \tag{16.5}$$

An approximate solution to the line search is often advocated.[6]

It remains to deduce the vector $\delta\mathbf{a}^{(i)}$ for a given state $\mathbf{a}^{(i)}$. Newton's method is one algorithm that can be used to obtain incremental iterates. In this procedure we expand the residual $\boldsymbol{\psi}$ about the current state $\mathbf{a}^{(i)}$ in terms of the increment $\delta\mathbf{a}^{(i)}$ and set the linear part equal to zero. Accordingly,

$$\boldsymbol{\psi}(\mathbf{a}^{(i)})+\left.\frac{\partial\mathbf{P}}{\partial\mathbf{a}}\right|_{\mathbf{a}^{(i)}}\cdot\delta\mathbf{a}^{(i)}=\mathbf{0} \tag{16.6}$$

We define the tangent (or jacobian) matrix as

$$\mathbf{K}_T^{(i)}=\left.\frac{\partial\mathbf{P}}{\partial\mathbf{a}}\right|_{\mathbf{a}^{(i)}} \tag{16.7}$$

Thus, we obtain an increment

$$\delta\mathbf{a}^{(i)}=-(\mathbf{K}_T^{(i)})^{-1}\boldsymbol{\psi}(\mathbf{a}^{(i)}) \tag{16.8}$$

This step requires the solution of a set of simultaneous linear algebraic equations. We note that for a linear differential equation the finite element internal force vector may be written as

$$\mathbf{P}(\mathbf{a})=\mathbf{K}\mathbf{a} \tag{16.9}$$

where $\mathbf{K}$ is a *constant* matrix. Thus, (16.9) generates a constant tangent matrix and the process defined by (16.3) and (16.6) converges in one iteration provided a unit value of η is used.

For Newton's method the residual $\boldsymbol{\psi}$ must have a norm which becomes smaller for a sufficiently small $\delta\mathbf{a}^{(i)}$; accordingly a Newton step may be projected onto G as shown in Fig. 16.2. Generally, Newton's method is convergent if

$$G^{(i)}(1)<\alpha G^{(i)}(0), \qquad 0<\alpha<1 \tag{16.10}$$

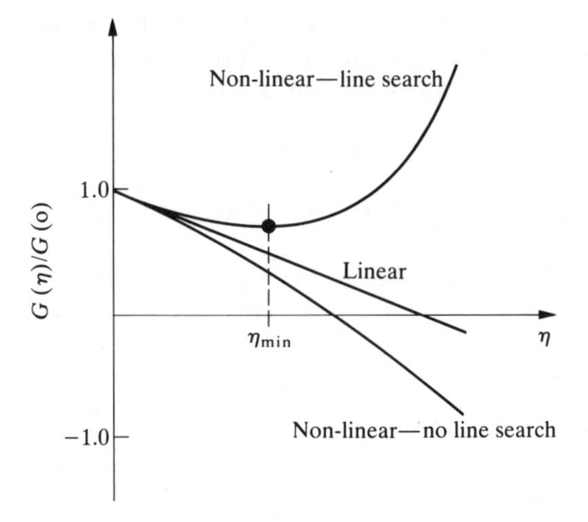

Fig. 16.2 Energy behaviour—line search use

for all iterations; however, convergence may not occur if this condition is not obtained.†

A Newton solution algorithm may be constructed using the macro programming language included in the program described in this chapter. A solution for a single loading step with a maximum of 10 iterations is given by‡

```
LOOP,newton,10
TANG
FORM
SOLV
NEXT,newton
```

or

```
LOOP,newton,10
TANG,,1
NEXT,newton
```

The second form is preferred since this will ensure that $\mathbf{K}_T^{(i)}$ and $\boldsymbol{\psi}^{(i)}$ are computed simultaneously for each element, whereas the original algo-

† For some problems (such as those defined by non-linear theories of beams, plates and shells) early iterations may produce shifts between one mode of behaviour (bending) and another (membrane) which can cause very large changes in $|\boldsymbol{\psi}|$. Later iterations, however, generally follow (16.10).

‡ Recall that information shown in upper case letters must be given; text indicated by lower case letters is optional. Finally, information given in italics must have numerical values assigned to define a proper macro statement.

rithm computes the two separately. If convergence occurs before 10 iterations are performed, the process will transfer to the macro statement following the NEXT statement. Convergence is based upon

$$G^{(i)}(0) < tol \cdot G^{(0)}(0)$$

where *tol* is specified by a TOL macro statement (a default value of 10^{-12} is set).

A line search may be added to the algorithm by modifying the macro commands to

```
LOOP,newton,10
TANG
FORM
SOLV,LINE,0.6
NEXT,newton
```

or

```
LOOP,newton,10
TANG,LINE,1,,0.6
NEXT,newton
```

Line search requires repeated computations of $\boldsymbol{\psi}(\mathbf{a}+\eta\delta\mathbf{a})$ which may increase solution times somewhat. Some assessment of need should be made before proceeding with large numbers of solution steps.

A modified Newton method may also be performed by removing the tangent computation from the loop. Accordingly,

```
TANG
LOOP,newton,10
FORM
SOLV,LINE,0.6
NEXT,newton
```

would compute only one tangent matrix and its associated triangular factors. The FORM command computes only $\boldsymbol{\psi}$ and SOLV solves the equations using previously computed triangular factors of the tangent matrix. Algorithms between full Newton and modified Newton may also be constructed. For example,

```
LOOP,,2
TANG
LOOP,newton,5
FORM
SOLV,LINE,0.6
NEXT,newton
NEXT
```

In this algorithm it should be noticed that convegence in the first five iterations would transfer to the outer NEXT statement and a second TANG would be computed followed by a single iteration in the inner loop before the entire algorithm is completed.

16.5 Restart option

The program described here permits the user to save a solution and subsequently use the data to continue the analysis from the point the data was saved. This is called a *restart* option. To use this feature, the file names given at initiation of the program must be appropriately specified.

Names for two files may be specified and are denoted on the screen as the RESTART (READ) and the RESTART (WRITE) files. The READ file is the name of a file that contains the data from a previous analysis. The screen will also indicate EXIST or NEW for the file. A NEW label indicates that no file with the specified name exists. The WRITE file is the name of a file to which data will be written as part of the current session. Care should be taken to avoid giving the same file name to the READ and WRITE files (unless neither exists) as a write to a existing file will destroy the original information. A WRITE file will be created for all *batch* executions or for *interactive* executions that terminate with an EXIT command. Interactive executions that terminate using QUIT will not save a restart WRITE file.

Once the file names are appropriately set, a restart may be performed for non-linear (or linear) problems that do not have inertia-type terms (see below for this class of problems) by entering the macro command

RESTart

The analysis proceeds from the point in the solution process where the restart file was written. Temporary file information is not retained by a restart. The temporary files include: triangular factors from a TANG command and eigenpairs from a SUBS command. Solution features for the recovered state may be displayed to the screen, placed in the output file, or plotted. Also the analysis may be continued to a new state by specification of new loadings, computation using new TANG statements, etc. It is recommended that analyses which require a large amount of computation be split into parts using the restart feature. This will ensure that computed results are retained and need not be determined again if an error or divergence occurs.

16.6 Solution of transient linear and non-linear problems

The solution of transient problems defined by Eqs (10.1) to (10.3) may be performed using the program described in this chapter. The program includes options to solve transient finite element problems which generate first- and second-order ordinary differential equations using the SS11, SS22 and CN22 algorithms described in Chapter 10. In implementing the algorithms a restriction applies with respect to specification of the parameters and it is not possible to perform any of the *explicit* algorithms described in Chapter 10. We consider each algorithm separately below.

16.6.1 *Solution of first-order problems using SS11.* Consider the linear problem described by

$$\mathbf{C}\dot{\mathbf{a}}+\mathbf{K}\mathbf{a}+\mathbf{f}=\mathbf{0} \tag{16.11}$$

If we introduce the SS11 algorithm, we have, at each time t_{n+1}, the discrete problem given by (10.41) as

$$\boldsymbol{\psi}(\boldsymbol{\alpha}_{n+1})=\mathbf{C}\boldsymbol{\alpha}_{n+1}+\mathbf{K}[\tilde{\mathbf{a}}_{n+1}+\theta\Delta t\boldsymbol{\alpha}_{n+1}]+\bar{\mathbf{f}}_{n+1}=\mathbf{0} \tag{16.12}$$

with

$$\tilde{\mathbf{a}}_{n+1}=\mathbf{a}_n \tag{16.13}$$

and from (10.40)

$$\mathbf{a}_{n+1}=\mathbf{a}_n+\Delta t\boldsymbol{\alpha}_{n+1} \tag{16.14}$$

We may also consider a non-linear, one-step extension to this problem, expressed by

$$\boldsymbol{\psi}(\boldsymbol{\alpha}_{n+1})=\mathbf{C}\boldsymbol{\alpha}_{n+1}+\mathbf{P}(\tilde{\mathbf{a}}_{n+1}+\theta\Delta t\boldsymbol{\alpha}_{n+1})+\bar{\mathbf{f}}_{n+1}=\mathbf{0} \tag{16.15}$$

where $\mathbf{P}$ is again the vector of non-linear internal forces. The solutions to either the linear or the non-linear problem may be expressed as

$$(\mathbf{C}+\theta\Delta t\mathbf{K}_T^{(i)})\delta\boldsymbol{\alpha}_{n+1}^{(i)}=-\boldsymbol{\psi}(\boldsymbol{\alpha}_{n+1}^{(i)}) \tag{16.16}$$

with

$$\boldsymbol{\alpha}_{n+1}^{(i+1)}=\boldsymbol{\alpha}_{n+1}^{(i)}+\eta\delta\boldsymbol{\alpha}_{n+1}^{(i)} \tag{16.17}$$

where η is the step size as described above for non-linear problems and for linear problems η is always taken as unity. For linear problems $\mathbf{K}_T^{(i)}\equiv\mathbf{K}$ while for non-linear problems

$$\mathbf{K}_T^{(i)}=\left.\frac{\partial\mathbf{P}}{\partial\mathbf{a}}\right|_{\mathbf{a}^{(i)}} \tag{16.18}$$

Finally, converged values are expressed without the (i) superscript.

The solution of the transient problem is achieved by satisfying the following steps:

1. Specify θ.
2. Specify Δt.

3. Compute **C**.
4. Specify the time, t_{n+1}, the number of time steps and set $i=0$.
5. For each time, t_{n+1}:
   (*a*) Compute $\psi(\boldsymbol{\alpha}_{n+1}^{(i)})$.
   (*b*) Compute $\mathbf{C}+\theta\Delta t\mathbf{K}_T^{(i)}$.
   (*c*) Solve for $\delta\boldsymbol{\alpha}_{n+1}^{(i)}$.
6. Check convergence for non-linear problems:
   (*a*) If satisfied terminate iteration.
   (*b*) If not satisfied set $i=i+1$ and repeat step 5.
7. Output solution information if needed.
8. Check time limit:
   (*a*) If $n \geqslant$ maximum number, stop, else.
   (*b*) If $n <$ maximum number, go to step 4.

For a typical problem these steps may be specified by the set of macro statements:

```
1.      BETA,SS11,0.5
2.      DT,,0.1
3.      MASS
4.      LOOP,time,20
        TIME
5.      LOOP,newton,10
   (a)  TANG
   (b)  FORM
   (c)  SOLV
6.      NEXT,newton
7.      DISP,ALL
8.      NEXT,time
```

As programmed, the MASS command formulates only a diagonal **C** matrix using the options discussed in Appendix 8 of Volume 1. The above algorithm works for both linear and non-linear problems. For linear problems the residual should be a numerical zero at the second iteration (if not there is a programming error!) and for efficiency purposes the macro commands

```
LOOP,newton,10
```

and

```
NEXT,newton
```

may be removed. Also, for all steps in which the time step is the same, a single tangent command may be used—in the above this can be accomplished by placing the TANG statement immediately *after* the MASS statement since step 5(*b*) needs **C** to compute the tangent matrix.

Any of the options for solving a non-linear problem may be used (e.g. modified Newton's method) by following the descriptions given above in the non-linear section. In particular, again for efficiency, one should use

```
LOOP,newton,10
TANG,,1
NEXT,newton
```

for a full Newton solution step.

Specification of time-dependent loading may be given for:

(*a*) proportional loading with a fixed spatial distribution of the nodal load vector and/or

(*b*) general time-varying loading.

For proportional loading

$$\bar{\mathbf{f}}_{n+1} = p(\bar{t}_{n+1})\mathbf{f}_0 \tag{16.19}$$

where, in the program,

$$\bar{t}_{n+1} = t_n + \theta\Delta t \tag{16.20}$$

The value of $\mathbf{f}_0$ is specified either as nodal forces (during description of the mesh using the FORC option) or is computed in each element as an element loading. For example, the heat source Q in (10.116) would generate element loads at node i as

$$f_i = \int_\Omega N_i Q \, d\Omega \tag{16.21}$$

The value of the proportional factor can be specified in the program as

$$p(t) = A_i + A_2 t + A_3 \sin^L A_4(t - t_{\min}), \qquad t_{\min} \leqslant t < t_{\max} \tag{16.22}$$

Details for input of the parameters are given in Sec. 15.4.2 in Volume 1. For proportional loading the macro program given above is modified by adding a step 0: specify the proportional loading function, $p(t)$. The macro command for this step for a single function of the type given above is

```
PROP,,1
```

Each specification of a new time will cause the program to recompute $p(\bar{t}_{n+1})$. The value of the proportional loading is passed to each element module as a REAL number which is the first entry in the ELDATA common statement (and is labelled DM in the routines included below) and may be used to multiply element loads to obtain the correct loading at each time.

General loading can be achieved only by reentering the mesh generation module and respecifying the nodal values. Accordingly, for this option a MESH macro statement must be inserted in the time-step loop.

For example, modify step 4 above to

```
LOOP,time,20
TIME
MESH
```

For each time step it is then necessary to specify the new nodal values for $\bar{\mathbf{f}}_{n+1}$. If the value at a node previously set to a non-zero condition becomes zero, the value *must be specified.* This may be achieved by specifying the node number only. For example,

```
FORCe
12
26,,5.0

END
```

would set all the components of the *force* at node 12 to zero and the first component of the force at node 26 to 5.0 units.

In batch execution the number of FORC–END paired statements must exactly match (or exceed) the number of time steps. In interactive execution a MESH > prompt will appear on the screen and the user must provide the necessary FORC data from the keyboard (terminating input with a blank line and an END statement). With a proper termination the prompt will once again indicate interactive inputs for macro statements.

While both proportional and general loads may be combined, extreme caution must be exercised as all nodal values will be multiplied by the current value of $p(t)$.

The value of the time increment may be changed by repeating step 2 and then performing steps 4 to 8 for the new increment. If a large number of different size time increments is involved this will be inefficient, and the program has an option to specify a new value for Δt as data. The macro program would be modified by replacing step 4 with

```
LOOP,time,20
DATA,DT
TIME
```

For each time step the DT macro statement is supplied as data. For interactive computations the user inputs

```
DT,,dt
```

where dt is the size of the time step to be used. In batch executions these commands follow the END statement which terminates the macro program. If other data is input (e.g. the FORC–END paired data), then the data statements must be in the order requested by the macro program statements.

A DATA macro program instruction may also be used to set a solution tolerance. The form would be

DATA,TOL

and the user would input a command

TOL,,1.E-10

to set the solution convergence tolerance to 10^{-10}.

16.6.2 *Restart option.* For transient problems it is recommended that periodic termination of the solution process be made to save a restart file. The program may be reentered using appropriate names for the desired restart read file (see the section on restarts above) and then giving the macro commands

BETA,SS11,0.5
MASS
RESTart

Failure to specify the BETA and MASS macro statements will usually lead to an error as the program will not read all the data for a transient solution as the RESTart information. An exception is when the data on the restart file was generated using an algorithm that did not contain the BETA macro statement. In this case a restart can be made and the algorithm changed to a transient algorithm by giving the commands as

RESTart
BETA,SS11,0.5
MASS
etc.

After a transient algorithm is initiated, return to a non-transient algorithm can only be achieved by using the restart mode. When a restart has been performed the time t is set to the value that existed when the restart file was written. The proportional loading tables must be reset by an appropriate macro command. Furthermore, the loading will be set to the values in the current mesh input file. The user may now specify any new macro commands and continue with the analysis.

16.6.3 *Solution of second-order systems.* For simplicity the damping matrix, **C**, in (10.1) is not included in any of the elements contained in this chapter. Steps to add such effects are summarized, however, at the end of this section. Thus, we consider the differential equation

$$\mathbf{M}\ddot{\mathbf{a}} + \mathbf{K}\mathbf{a} + \mathbf{f} = \mathbf{0} \tag{16.23}$$

for linear problems and

$$\mathbf{M\ddot{a}} + \mathbf{P(a)} + \mathbf{f} = \mathbf{0} \tag{16.24}$$

for non-linear problems.

Solution using the SS22 algorithm. If we introduce the SS22 algorithm described in Chapter 10, we have at each time t_{n+1} the discrete problem given by (10.45) (specialized to reflect $\mathbf{C}$ as zero)

$$\boldsymbol{\Psi}(\boldsymbol{\alpha}_{n+1}) = \mathbf{M}\boldsymbol{\alpha}_{n+1} + \mathbf{K}(\tilde{\mathbf{a}}_{n+1} + \tfrac{1}{2}\theta_2 \Delta t^2 \boldsymbol{\alpha}_{n+1}) + \bar{\mathbf{f}}_{n+1} = \mathbf{0} \tag{16.25}$$

with

$$\tilde{\mathbf{a}}_{n+1} = \mathbf{a}_n + \theta_1 \Delta t \dot{\mathbf{a}}_n \tag{16.26}$$

$$\tilde{\dot{\mathbf{a}}}_{n+1} = \dot{\mathbf{a}}_n \tag{16.27}$$

and

$$\mathbf{a}_{n+1} = \mathbf{a}_n + \Delta t \dot{\mathbf{a}}_n + \tfrac{1}{2}\Delta t^2 \boldsymbol{\alpha}_{n+1} \tag{16.28}$$

$$\dot{\mathbf{a}}_{n+1} = \dot{\mathbf{a}}_n + \Delta t \boldsymbol{\alpha}_{n+1} \tag{16.29}$$

The program also considers a one-step approximation to (16.25) for non-linear problems which is expressed as

$$\boldsymbol{\Psi}(\boldsymbol{\alpha}_{n+1}) = \mathbf{M}\boldsymbol{\alpha}_{n+1} + \mathbf{P}(\tilde{\mathbf{a}}_{n+1} + \tfrac{1}{2}\theta_2 \Delta t^2 \boldsymbol{\alpha}_{n+1}) + \bar{\mathbf{f}}_{n+1} = \mathbf{0} \tag{16.30}$$

The solution to either the linear or the non-linear problem may be expressed as

$$(\mathbf{M} + \tfrac{1}{2}\theta_2 \Delta t^2 \mathbf{K}_T^{(i)}) \delta\boldsymbol{\alpha}_{n+1}^{(i)} = -\boldsymbol{\psi}(\boldsymbol{\alpha}_{n+1}^{(i)}) \tag{16.31}$$

with

$$\boldsymbol{\alpha}_{n+1}^{(i+1)} = \boldsymbol{\alpha}_{n+1}^{(i)} + \eta \delta\boldsymbol{\alpha}_{n+1}^{(i)} \tag{16.32}$$

and at this stage the solution process is identical to that for SS11 except that now MASS forms a diagonal mass matrix $\mathbf{M}$ and the solution procedure for SS22 is invoked using

BETA,SS22,0.5,0.5

where the numerical values denote those used for θ_1 and θ_2 respectively.

Solution using the GN22 algorithm. The GN22 algorithm may also be selected and from (10.55) and (10.56) we have for a linear problem

$$\boldsymbol{\psi}(\ddot{\mathbf{a}}_{n+1}) = \mathbf{M}\ddot{\mathbf{a}}_{n+1} + \mathbf{K}(\bar{\mathbf{a}}_{n+1} + \beta \Delta t^2 \ddot{\mathbf{a}}_{n+1}) + \mathbf{f}_{n+1} = \mathbf{0} \tag{16.33}$$

or for a non-linear problem

$$\boldsymbol{\psi}(\ddot{\mathbf{a}}_{n+1}) = \mathbf{M}\ddot{\mathbf{a}}_{n+1} + \mathbf{P}(\bar{\mathbf{a}}_{n+1} + \beta \Delta t^2 \ddot{\mathbf{a}}_{n+1}) + \mathbf{f}_{n+1} = \mathbf{0} \tag{16.34}$$

where

$$\bar{\mathbf{a}}_{n+1}=\mathbf{a}_n+\Delta t\dot{\mathbf{a}}_n+\tfrac{1}{2}\Delta t^2\ddot{\mathbf{a}}_n \tag{16.35}$$

$$\dot{\bar{\mathbf{a}}}_{n+1}=\dot{\mathbf{a}}_n+\Delta t\ddot{\mathbf{a}}_n \tag{16.36}$$

$$\ddot{\bar{\mathbf{a}}}_{n+1}=\ddot{\mathbf{a}}_n \tag{16.37}$$

and

$$\mathbf{a}_{n+1}=\bar{\mathbf{a}}_{n+1}+\beta\Delta t^2(\ddot{\mathbf{a}}_{n+1}-\ddot{\mathbf{a}}_n) \tag{16.38}$$

$$\dot{\mathbf{a}}_{n+1}=\dot{\bar{\mathbf{a}}}_{n+1}+\gamma\Delta t(\ddot{\mathbf{a}}_{n+1}-\ddot{\mathbf{a}}_n) \tag{16.39}$$

In the above the standard Newmark parameters are introduced; however, they are related simply to the values given in Chapter 10 by the expressions

$$\gamma=\beta_1 \tag{16.40}$$

and

$$\beta=2\beta_2 \tag{16.41}$$

A solution to the GN22 problems may be expressed as

$$(\mathbf{M}+\beta\Delta t^2\mathbf{K}_T^{(i)})\delta\ddot{\mathbf{a}}_{n+1}^{(i)}=-\boldsymbol{\psi}(\ddot{\mathbf{a}}_{n+1}^{(i)}) \tag{16.42}$$

with

$$\ddot{\mathbf{a}}_{n+1}^{(i+1)}=\ddot{\mathbf{a}}_{n+1}^{(i)}+\eta\delta\ddot{\mathbf{a}}_{n+1}^{(i)} \tag{16.43}$$

Again the algorithm for solution is identical to the steps for SS11 or SS22 except that now it is initiated using the macro statement

BETA,GN22,0.25,0.5

where the two numerical values are for β and γ respectively (default values are $\beta=0.25$ and $\gamma=0.5$). Alternatively, the command

BETA

may be used for the GN22 algorithm with the default values.

As noted in Chapter 10, the GN22 algorithm requires the solution of (16.33) or (16.34) to define the initial acceleration. An exception is when the solution would yield a zero value for $\mathbf{a}_0$. To determine the initial solution, all the loading values must be assigned and then the macro statement

FORM,ACCEleration

may be used to compute the correct value of $\ddot{\mathbf{a}}_0$.

Adding damping effects. A damping matrix may be added by including in each element routine the appropriate terms. Thus, when computing a residual (see Chapter 15 of Volume1), with ISW = 3 or ISW = 6, the term

$$\mathbf{C}\dot{\mathbf{a}}_{n+1} \tag{16.44}$$

must be added to the equilibrium equation in each element. The value of $\dot{\mathbf{a}}_{n+1}$ localized for each element and adjusted for each algorithm is passed as part of the UL array (see Chapter 15 of Volume 1 for variable name descriptions). The UL array may be assumed to be dimensioned as

UL(NDF,NEN,IT)

where NDF is the number of unknowns at each node, NEN is the maximum number of nodes on any element and IT denotes the quantities as indicated in Table 16.2. Using these values and a definition for $\mathbf{C}$, the appropriate terms may be computed and added to the element residual vector $\mathbf{P}$. Similarly, the appropriate tangent stiffness term may be added to the element array $\mathbf{S}$ for ISW = 3.

16.7 Eigensolutions

The solution of a general linear eigenproblem is a useful feature included in the program continued in this chapter. The program can compute a set of the smallest real eigenvalues (in absolute value) and their associated eigenvectors for the problem

$$\mathbf{K}_T\mathbf{V} = \mathbf{M}\mathbf{V}\mathbf{\Lambda} \tag{16.45}$$

In the above, $\mathbf{K}_T$ is any symmetric tangent matrix that has been computed using a TANG macro statement; $\mathbf{M}$ is a diagonal mass or identity matrix computed using a MASS or IDEN macro statement respectively; the columns of $\mathbf{V}$ are the set of eigenvectors to be computed; and $\mathbf{\Lambda}$ is a diagonal matrix that contains the set of eigenvalues to be computed. For second-order equations the eigenvalues λ are the frequencies squared, ω^2. Accordingly, the program will compute and report the

TABLE 16.2
VALUES IN UL ARRAY FOR TRANSIENT ALGORITHMS

IT value	Algorithm		
	S11	SS22	GN22
1	$\tilde{\mathbf{a}} + \theta\Delta t\boldsymbol{\alpha}^{(i)}$	$\tilde{\mathbf{a}} + \frac{1}{2}\theta_2\Delta t^2\boldsymbol{\alpha}^{(i)}$	$\mathbf{a}^{(i)}$
2	$\theta\Delta t\boldsymbol{\alpha}^{(i)}$	$\frac{1}{2}\theta_2\Delta t^2\boldsymbol{\alpha}^{(i)}$	$\mathbf{a}^{(i)} - \mathbf{a}^{(0)}$
3	$\theta\Delta t\delta\boldsymbol{\alpha}^{(i)}$	$\frac{1}{2}\theta_2\Delta t^2\delta\boldsymbol{\alpha}^{(i)}$	$\mathbf{a}^{(i)} - \mathbf{a}^{(i-1)}$
4	—	$\dot{\tilde{\mathbf{a}}} + \theta_1\Delta t\boldsymbol{\alpha}^{(i)}$	$\dot{\mathbf{a}}^{(i)}$

square root of λ. Since negative values of λ can occur the square root of the absolute values is computed. For negative λ the reported values are in fact pure imaginary numbers.

The tangent matrix often has zero eigenvalues and, for this case, the algorithm used requires the problem to be transformed to

$$(\mathbf{K}_T - \alpha\mathbf{M})\mathbf{V} = \mathbf{M}\mathbf{V}\mathbf{\Lambda}_\alpha \tag{16.46}$$

where α is a parameter called the *shift*, which must be selected to make the coefficient matrix on the left side of (16.46) non-singular. $\mathbf{\Lambda}_\alpha$ are the eigenvalues of the shift which are related to the desired values by

$$\mathbf{\Lambda} = \mathbf{\Lambda}_\alpha + \alpha\mathbf{I} \tag{16.47}$$

The shift may also be used to compute the eigenpairs nearest to some specified value. The components of $\mathbf{\Lambda}$ are output as part of the eigenproblem solutions. In addition, the vectors may be output as numerical values or given as plots on the screen.

The program uses a subspace algorithm[3,4] to compute a small general eigenproblem defined as

$$\mathbf{K}^*\mathbf{x} = \mathbf{M}^*\mathbf{x}\boldsymbol{\lambda} \tag{16.48}$$

where

$$\mathbf{V} = \mathbf{Q}\mathbf{x} \tag{16.49}$$

$$\mathbf{K}^* = \mathbf{Q}^{\mathrm{T}}\mathbf{M}^{\mathrm{T}}(\mathbf{K}_{\mathrm{T}} - \alpha\mathbf{M})^{-1}\mathbf{M}\mathbf{Q} \tag{16.50}$$

$$\mathbf{M}^* = \mathbf{Q}^{\mathrm{T}}\mathbf{M}\mathbf{Q} \tag{16.51}$$

Accordingly, after the projection, the $\boldsymbol{\lambda}$ are reciprocals of $\mathbf{\Lambda}_\alpha$ (that is $\mathbf{\Lambda}_\alpha^{-1}$). An eigensolution of the small problem may be used to generate a sequence of iterates for $\mathbf{Q}$ which converge to the solution for the original problem (e.g. see reference 4). The solution of the projected small general problem is effected here using a transformation to a standard linear eigenproblem combined with a *QL* algorithm.[7,8]

The transformation is performed by computing the Choleski factors of $\mathbf{M}^*$ to define the standard linear eigenproblem

$$\mathbf{H}\mathbf{y} = \mathbf{y}\boldsymbol{\lambda} \tag{16.52}$$

where

$$\mathbf{M}^* = \mathbf{L}\mathbf{L}^{\mathrm{T}} \tag{16.53}$$

$$\mathbf{y} = \mathbf{L}^{\mathrm{T}}\mathbf{x} \tag{16.54}$$

and

$$\mathbf{H} = \mathbf{L}^{-1}\mathbf{K}^*\mathbf{L}^{-\mathrm{T}} \tag{16.55}$$

In the implementation described here scaling is introduced, which causes

$\mathbf{M}^*$ to converge to an identity matrix; hence the above transformation is numerically stable. Furthermore, use of a standard eigenproblem solution permits calculation of both positive and negative eigenvalues. The subspace algorithm implemented provides a means to compute a few eigenpairs for problems with many degrees of freedom or all of the eigenpairs of small problem. A subspace algorithm is based upon a power method to compute the dominant eigenvalues. Thus, the effectiveness of the solution strategy depends on the ratio of the absolute value of the largest eigenvalue sought in the subspace to that of the first eigenvalue not contained in the subspace. This ratio may be reduced by adding additional vectors to the subspace; that is if p pairs are sought, the subspace is taken as q vectors so that

$$\left| \frac{\lambda_p}{\lambda_{q+1}} \right| < 1 \tag{16.56}$$

Of course, the magnitude of this ratio is unknown before the problem is solved and some analysis is necessary to estimate its value. The program tracks the magnitude of the shifted reciprocal eigenvalues Λ and computes the change in values between successive iterations. If the subspace is too small, convergence will be extremely slow due to (16.56) having a ratio near unity. It may be desirable to increase the subspace size to speed the convergence. In some problems, characteristics of the eigenvalue magnitudes may be available to assist in the process. It should be especially noted that when p is specified as the total number of degrees of freedom in the problem (or q becomes this value), then λ_{q+1} is infinitely larger and the ratio given in (16.56) is zero. In this case subspace iteration converges in a single iteration, a fact that is noted by the program to limit the iterations to 1. Accordingly, it is usually more efficient to compute all the eigenpairs if q is very near the number of degree of freedoms.

Use of the subspace algorithm requires the steps:

1. Compute **M**.
2. Compute the tangent matrix $\mathbf{K}_T$ and apply the α shift if necessary.
3. Compute the eigenpairs.
4. Output the results.

The macro commands to achieve this algorithm are:

1. MASS (or IDEN)
2. TANG,,,α
3. SUBS,⟨PRINt⟩,p,q
4. EIGV,,n
   PLOT, EIGV,f

Note the specification of the shift value α as part of the TANG macro statement. The TANG macro computes both the matrix and its triangular factors; consequently, the mass must be available before using this command. The value for q is optional and, when omitted, is computed by the program as

$$q = \min(\text{NM}, \text{NEQ}, 2^*\text{p}, \text{p}+8)$$

where NM is the number of non-zero terms in the diagonal mass matrix and NEQ is the number of degrees of freedom in the problem (i.e. those not restrained by boundary constraints specified using BOUN).

The plot of eigenvectors may need to be increased or decreased by a factor f to permit proper viewing. Each eigenvector may be viewed but must first be output using the EIGV print instruction.

The eigenproblem for individual elements is a common procedure used to evaluate performance. It is necessary to describe a mesh without restraints to the degrees of freedom (i.e. BOUN should not be used in the mesh data). The element normally has zero eigenvalues; hence, a shift must be used for the analysis. Failure to use a shift will result in errors in the triangular factors of $\mathbf{K}_T$ (the program will detect a near singularity and output a warning) and generally all or a large number of the eigenpairs will collapse onto the singular subspace. If the shift is specified very close to an eigenvalue these types of errors may also occur. The user should monitor the outputs during the TANG and the SUBS commands to detect poor performance. If all or a large number of the eigenvalues are extremely near the shift, a second shift should be tried. In general, a shift should be picked about halfway between reported values. The program also counts the number of eigenvalues that are less than the shift. This may be used to ensure that the shift is not too large to determine the desired eigenvalues. Recall that only the p values nearest in absolute value to the shift are determined.

When properly used, the subspace method can produce accurate and reliable values for the eigenpairs of a finite element system problem. The method may be used to compute the vibration modes of structural systems and is implemented so that it may be used for both linear and non-linear finite element models. For non-linear problems this permits the dependence on frequency with load to be assessed. Thus, a dynamic buckling load where a frequency goes to zero may be computed. For non-linear models the static buckling loads for a problem may also be determined by solving the eigenproblem

$$\mathbf{K}_T\mathbf{V} = \mathbf{I}\mathbf{V}\mathbf{\Lambda} \tag{16.57}$$

for a set of loads and tracking the approach to zero of the smallest eigenvalue. A buckling load corresponds to a zero eigenvalue in (16.57).

As the buckling load is approached, a shift may be necessary to maintain high accuracy; however, since a collapsed subspace is the desired solution, this is usually unnecessary.

16.8 Description of elements

The program extensions described in this volume include four elements that are capable of analyzing linear and non-linear problems. Each of the elements has a consistently linearized tangent stiffness matrix which, when used with a Newton method of solution, achieves a quadratic rate of asymptotic convergence, as described in Chapter 7 and Sec. 16.4. In addition, the elements are capable of utilizing the time integration algorithms discussed in Sec. 16.6. The four elements included are:

ELMT01	An axisymmetric shell element which may also be specialized to a rectangular cross-section beam model
ELMT02	A plane strain/axisymmetric deformation element with elastic–plastic constitutive behaviour
ELMT03	A Laplace equation element with reactive loading for plane and axisymmetric models
ELMT04	A general truss element with elastic–plastic constitutive behaviour

Each of these elements may be used to illustrate some of the theory presented in the preceding chapters. A brief description of each element and the input data needed to specify the material and other parameters are contained below. The parameters NDM, NDF and NEN denote the required problem space dimension, number of degrees of freedom at each node and number of nodes for each element respectively. The control information (following the FEAP record) should contain the maximum of these for all elements included in the problem. In the next section the elements are used to solve the problems presented in Sec. 16.3.

16.8.1 *ELMT01: axisymmetric shell (beam) element.* This element is based on the theory presented in Sec. 4.7. An extension to permit limited non-linear geometric behaviour has been included by adding a term to the meridian strain given in (4.34) to obtain

$$\varepsilon_s = \frac{du}{ds} + \frac{1}{2}\left(\frac{dw}{ds}\right)^2$$

The material is assumed to be linear elastic. The above extension was originally proposed by Wagner (e.g. see references 9 and 10). The element, despite its simplicity, performs very well over a wide range of ap-

TABLE 16.3
MATERIAL PARAMETER SPECIFICATION FOR ELMT01

Property Record 1.) FORMAT—A5

Column	Description	Variable
1 to 5	BEAM or SHELL (alphanumeric)	typ

Property Record 2.) FORMAT—6F10.0

Column	Description	Variable
1 to 10	E, Young's modulus of elasticity	d(1)
11 to 20	ν, Poisson's ratio	d(2)
21 to 30	t, thickness	d(3)
31 to 40	p, uniform pressure on each element	d(4)
41 to 50	Linear flag for strain ε_s $0=$linear, $1=$non-linear	d(5)
51 to 60	ρ, mass density per unit volume	d(6)

plications. This element requires the problem parameters to be set as

$$\text{NDM}=2 \qquad \text{NDF}=3 \qquad \text{and} \qquad \text{NEN}=2$$

The data to specify the material parametes is given in Table 16.3. The listing of the subprograms for the beam/axisymmetric shell element is given in file PCELM1.FOR in Sec. 16.10.7(*a*).

16.8.2 *ELMT02*: *plane strain/axisymmetric elastic–plastic material model.* This element is identical to the element presented in Volume 1 (also as ELMT02 in file PCELM2.FOR) except that the material model has been replaced by a Prandl–Reuss elastic–plastic relationship with added isotropic and kinematic hardening. The stress and strain tensors are split into deviatoric and mean parts. The pressure–volume change constitutive equation is assumed to be linear elastic. The deviatoric part is elastic–plastic and yield is expressed as a von Mises function. The equations are integrated using the closest point, radial return algorithm discussed in Chapter 7. The resulting discretized equations for the deviatoric part are given by

$$\mathbf{s}_n = 2G(\mathbf{e}_n - \mathbf{e}_n^{\mathrm{p}})$$

where $\mathbf{s}_n$ is the deviatoric stress, $\mathbf{e}_n$ is the deviatoric strain, $\mathbf{e}_n^{\mathrm{p}}$ is a scalar effective plastic strain, all at time t_n, and G is the shear modulus of elasticity. The plastic strain is deviatoric and is recovered from a backward Euler integration scheme applied to the incremental equations defined in Chapter 7. For the constitutive model included in the program, the result is

$$\mathbf{e}_n^{\mathrm{p}} = \mathbf{e}_{n-1}^{\mathrm{p}} + \Delta\lambda_n \left.\frac{\partial F}{\partial \mathbf{s}}\right|_n$$

where F is the yield function and $\Delta\lambda_n$ is a discrete plastic consistency parameter. The yield behaviour of the material is expressed as a von Mises function where

$$F(\mathbf{s}_n, \boldsymbol{\alpha}_n, e^{\mathrm{p}}) = |\mathbf{s}_n - \boldsymbol{\alpha}_n| - R_n(e^{\mathrm{p}}) \leqslant 0$$

where
$$R_n(e^{\mathrm{p}}) = (\tfrac{2}{3})^{1/2}(Y_0 + H_i e_n^{\mathrm{p}})$$

where Y_0 defines the initial uniaxial yield stress in simple tension/compression, e_n^{p} is the effective plastic strain used to define isotropic hardening, H_i is a linear isotropic hardening parameter and $\boldsymbol{\alpha}_n$ is the 'back stress' introduced to give kinematic hardening. Finally,

$$|\mathbf{s} - \boldsymbol{\alpha}| = [(\mathbf{s} - \boldsymbol{\alpha})^{\mathrm{T}}(\mathbf{s} - \boldsymbol{\alpha})]^{1/2}$$

The effective plastic strain is computed incrementally from

$$e_n^{\mathrm{p}} = e_{n-1}^{\mathrm{p}} + (\tfrac{2}{3})^{1/2}\Delta\lambda_n$$

and the back stress is computed from

$$\boldsymbol{\alpha}_n = \boldsymbol{\alpha}_{n-1} + \tfrac{2}{3} H_k(\mathbf{e}_n^{\mathrm{p}} - \mathbf{e}_{n-1}^{\mathrm{p}})$$

where H_k is a kinematic hardening parameter.

The return map algorithm results from solving the above set of equations for a given strain state subject to

$$\mathbf{s}_n^{\mathrm{TR}} = 2G(\mathbf{e}_n - \mathbf{e}_{n-1}^{\mathrm{p}})$$

and
$$\Delta\lambda_n^{\mathrm{TR}} = 0$$

The quantities with the superscript TR are called trial values. If the trial values produce a state inside the yield function, the step is *elastic* and these values define the correct current solution state; however, if the trial values violate the yield condition, a *plastic* solution is computed using the trial values as initial conditions to solve the discrete constitutive equations. The solution process for the above problem is very simple and gives a linear problem to compute $\Delta\lambda_n$ from

$$|\mathbf{s}_n^{\mathrm{TR}} - \boldsymbol{\alpha}_{n-1}| - R_{n-1} = 2[G + \tfrac{1}{3}(H_i + H_k)]\Delta\lambda_n$$

Once $\Delta\lambda_n$ is known, the remaining part of the solution is easily evaluated from the above expressions. Additional details for constructing the solution to this problem and the consistent tangent matrix for computing the next solution step may be found in Chapter 7 and in reference 11.

The element module ELMT02 is restricted to a four-node isoparametric formulation using a B-bar treatment to avoid 'locking' in the nearly incompressible range, a condition that occurs for large plastic strains (compared to the elastic ones) as well as for a large Poisson ratio. See

TABLE 16.4
MATERIAL PARAMETER SPECIFICATION FOR ELMT02

Property Record 1.) FORMAT—I10, F10.0

Column	*Description*	*Variable*
1 to 10	Problem type (0 = plane, 1 = axisym)	it
11 to 20	ρ, mass density per unit volume	d(4)

Property Record 2.) FORMAT—5F10.0

Column	*Description*	*Variable*
1 to 10	E, Young's modulus of elasticity	ee
11 to 20	ν, Poisson's ratio	xnu
21 to 30	Y_0, initial yield stress	d(11)
31 to 40	H_i, isotropic hardening modulus	d(12)
51 to 50	H_k, kinematic hardening modulus	d(13)

Secs 12.5.2 and 15.8.6(*b*) in Volume 1 for additional information on the formulation. This element requires the problem parameters to be set as

$$\text{NDM} = 2 \qquad \text{NDF} = 3 \qquad \text{and} \qquad \text{NEN} = 2$$

and the material parameters are set as specified in Table 16.4. The listing for the material model subprograms is given in file PCELM2.FOR in Sec. 16.10.7(*b*).

16.8.3 *ELMT03*: *plane/axisymmetric Laplace equation.* This element is a three- to nine-node isoparametric element which may be used to model Laplace equation problems; for example the transient heat conduction described in Chapter 10. The general differential equations which may be considered using this element take one of the forms:

$$\nabla(K\nabla T) + Q(T) = 0 \tag{16.58}$$

for steady-state applications,

$$\nabla(K\nabla T) + Q(T) = \rho c \frac{\partial T}{\partial t} \tag{16.59}$$

for transient diffusion processes or

$$\nabla(K\nabla T) + Q(T) = \rho c \frac{\partial^2 T}{\partial t^2} \tag{16.60}$$

for wave propagation or vibration processes. The difference between (16.59) and (16.60) depends only on the time integration procedure used to solve the transient problem. Specifying the integration procedure SS11 implies a solution of Eq. (16.59), whereas specification of an SS22 or a GN22 method solves Eq. (16.60). In the equations, the parameters are as defined in Sec. 10.6.4. In particular a heat-generation term dependent upon temperature, T, as given by

$$Q = \delta\, e^{rT^n}$$

is included in the program. This representation of Q permits the analysis of steady-state ($\rho c = 0$) and transient problems for a fairly wide range of heat-generation types. Specification of r as zero gives a constant heat generation of magnitude δ. Specification of n as -1 gives the transient heat-conduction equation for a reactive solid obeying zero-order kinetics, whereas n as unity gives a heat generation of the type considered in a Frank–Kamenetskii approximation.[12] For the n equal to -1 case, the form of the heat source is modified slightly for more stable numerical treatment of the spontaneous combustion problem. The form used is[12]

$$Q = \bar{\delta}\, e^{r(1 - T_a/T)}$$

where T_a is a specified ambient temperature and $\bar{\delta}$ is a modified coefficient. For positive or zero n the first form of specifying the Q is used and the value of T_a is ignored.

The parameters for ELMT03 are

$$\text{NDM} = 2 \qquad \text{NDF} = 1 \qquad \text{and} \qquad \text{NEL} = 4 \text{ to } 9$$

and the material description is input as indicated in Table 16.5. The listing for the element subprograms is given in file PCELM3.FOR in Sec. 16.10.7(*c*).

16.8.4 *ELMT04: elastic–plastic truss model.* This element is nearly identical to the element presented in Volume 1 (also as ELMT04 in file PCELM4.FOR), except that the material model has been replaced by a simple one-dimensional elastic–plastic relationship with added isotropic and kinematic hardening. The equations are integrated using the closest point, radial return algorithm discussed in Chapter 7. The element has two nodes and uses linear interpolation in each element. Thus, at yield the entire element becomes plastic. For static applications with no hardening the system may become singular when a sufficient number of

TABLE 16.5
MATERIAL PARAMETERS SPECIFICATION FOR ELMT03

Column	*Description*	*Variable*
11 to 20	c, specific heat	d(2)
21 to 30	ρ, mass density	d(3)
31 to 40	δ, heat-generation factor	d(4)
41 to 50	r, heat-generation factor	d(5)
51 to 60	T_a, ambient temperature	d(6)
61 to 70	n, temperature power	d(7)
71 to 80	Geometry: 1 = plane, 2 = axisym	kat

elements has a plastic behaviour; consequently, it is recommended that for this class of problems some hardening be used. See Sec. 15.8.6(*d*) in Volume 1 for additional information on the element. Due to the number of changes required in the element routine the complete listing for this element is repeated in this volume. A slightly different type of coding has been employed which may be compared to the style used in Volume 1.

The data required for this element is

$$\text{NDM} = 1, 2 \text{ or } 3 \qquad \text{NDF} = \text{NDM} \qquad \text{and} \qquad \text{NEN} = 2$$

and material parameters are specified according to Table 16.6. The listing for the material model subprograms is given in file PCELM4.FOR in Sec. 16.10.7(*d*).

The above elements provide significant capability to the program within the limited space available in this chapter. By following the steps involved in coding each of the elements, users should be able to add elements to meet their specific needs. The requirements for interfacing an element with PCFEAP are contained in Sec. 15.5 in Volume 1. In addition to the information contained there, it is necessary to describe how any element variable needed to advance a non-linear problem from one time step to the next is to be stored.

16.8.5 *History variables.* PCFEAP provides capabilities to retain a set of variables for each element type as *history* variables. The number of variables for each element must be specified and returned to the program. This is accomplished during the input of the material parameters (i.e. when ISW = 1 in each element module) by including a common block in the module as

```
COMMON/HDATA/NH1,NH2
```

and setting NH1 to the number of parameters used by each element. The program will automatically assign storage within the memory and disk of the computer, and perform all the necessary input and output to make

TABLE 16.6

MATERIAL PARAMETERS SPECIFICATION FOR ELMT04

Property Record 1.) FORMAT—6F10.0

Column	*Description*	*Variable*
1 to 10	E, Young's modulus of elasticity	d(1)
11 to 20	A, area	d(2)
21 to 30	ρ, mass density	d(3)
31 to 40	Y, initial yield stress	d(4)
41 to 50	H_i, isotropic hardening modulus	d(5)
51 to 60	H_k, kinematic hardening modulus	d(6)

information available to each element as needed. The common block HDATA is also used to return a value in NH1 for each element to describe where in the blank common the history data may be retrieved and stored. Thus, by including a blank common in the element module as

```
COMMON H(1)
```

the first word of the history variables for the current element may be recovered at H(NH1) with other parameters immediately following up to the number set during ISW = 1. Information is returned to the program in the same locations. In general, the user may think of this information as being supplied for the time t_n and returned as values for time t_{n+1}. The program will store the values onto the disk when a TIME macro is given. Thus, a provision for a one-step integration of rate-type element equations is included automatically. Multistep-type algorithms may be accommodated by retaining data for each of the required levels as part of the history data and reassigning the locations during the return of data in the H array.

16.9 Solution of example problems

In this section some analyses using the program PCFEAP are presented. Results are generally presented in tables and reproduce the actual numerical results obtained from the program. A brief description of the macro solution program is also included to assist users in becoming familiar with the options available in the program.

The first step in using PCFEAP to solve a finite element problem is to create a file with the mesh data. It is useful to specify the disk file with a name beginning with I (e.g. IBEAM for a beam data file). The program then automatically provides default names for the output and restart files by stripping the I and adding an O, R or S (e.g. OBEAM for output, RBEAM for restart reads and SBEAM for restart saves). Once the data is available in the input file the program can be initiated by entering the program name. In what follows, it is assumed that the executable program is named PCFEAP; thus entering this name initiates an execution of the program. During the first execution of the program the user receives prompts for data needed to install the program. This information is saved in a file named FEAP.NAM; thus, if any of the installation parameters need to be changed, it is necessary to first delete this file and then rerun the program. Defaults for the input parameters are suggested and may be accepted by pressing the ENTER or ⟨CR⟩ key. The program always requests names for the input/output and restart files. During the first execution, a name must be provided for the input data file. Default names for the output and restart are provided and accepted by an

ENTER or ⟨CR⟩ or may be replaced by specifying the name of a file to be used. In subsequent executions of the program, the installation parameters and the names of the files from the last execution are used as the default values. Once the information is specified, the user may accept the names assigned by entering Y, repeat the file specification by entering N or stop execution by entering S (either upper or lower case letters may be given for all commands). Once a Y is entered, the program proceeds to input the data contained in the input file until either an INTEractive execution command or a STOP is encountered. If the file contains a MACRo execution command, then solution is performed in a batch-type mode and no user interactions are required. On the other hand, if an interactive execution mode is requested, the user must provide all the solution steps from the input keyboard. The inputs are given whenever the screen contains the line

Time=09:45:33 Macro 1>_

where the number following the 'Time=' is the clock or elapsed time for the computer and the _ indicates the computer cursor. Macro solution commands may then be entered as described in Table 15.16 of Volume 1 or Table 16.1 in this chapter. For example, entering the command

TANG,,1

performs a full solution step. Some examples of problem solutions using the data given in Sec. 16.3 are contained below.

16.9.1 *Analysis of a straight beam.* The straight beam model described by the input data in Sec. 16.3.1 is analysed for both static and dynamic loading. Firstly, a static analysis for the central point loading (a force on node 11 of the model) is performed. This analysis may be performed by specifying the macro command

TANG,,1

which causes the program to compute a tangent stiffness, a residual and a solution to the set of nodal equations (see Table 16.1). Once the command is entered, the program proceeds to perform these steps, and, when completed, outputs to the screen values of the residual norm, the energy (work) for the step and some values of the maximum and minimum in the diagonal matrix of the triangular decomposition step. Users should note these values. Some concern should be raised if the ratio of the diagonals for the decomposition step is near the precision level of the machine (which for double precision calculations is about 10^{16}). Values near the precision level indicate possible errors in the modelling. Factors that enter into this ratio are: boundary condition restraints (a rigid body mode or mechanism may exist); large ratios in material properties; large differences in length of beam elements, size of elements or bad aspect ratio of elements. Since the current problem is linear, a second specification of the TANG command should not change the solution. Thus, for well-behaved linear problems, the residual norm should always go to a *numerical* zero if a second iteration is performed. The energy norm in this iteration should also be zero. For poorly conditioned problems some non-zero residual and energy may occur.

Once the linear static solution is completed, the results may be displayed. A display of all results can be performed by giving the two macro commands:

```
DISP,ALL
STREss,ALL
```

Results appear on the screen after each command is entered. These results are also written into the output file and saved on the disk. The results may be reviewed after the analysis by displaying the output file. Should a user wish to enter both commands before the output appears on the screen a LOOP–NEXT pair may be used, i.e. give the commands

```
LOOP
DISP,ALL
STREss,ALL
NEXT
```

The results for the static analysis with a central loading on the beam are included in Tables 16.7 and 16.8. Results for nodes 13 to 21 are not shown

TABLE 16.7
STATIC BEAM DISPLACEMENTS
VERTICAL POINT LOAD AT CENTRE

Node	2 displ	3 displ
1	0.000000e+00	−3.750000e+00
2	−1.895625e+00	−3.712500e+00
3	−3.753750e+00	−3.600000e+00
4	−5.536875e+00	−3.412500e+00
5	−7.207500e+00	−3.150000e+00
6	−8.728125e+00	−2.812500e+00
7	−1.006125e+01	−2.400000e+00
8	−1.116938e+01	−1.912500e+00
9	−1.201500e+01	−1.350000e+00
10	−1.256063e+01	−7.125000e−01
11	−1.276875e+01	1.750162e−15
12	−1.256063e+01	7.125000e−01

TABLE 16.8
STATIC BEAM SHEARS AND MOMENTS
VERTICAL POINT LOAD AT CENTRE

Elmt	**MS**	**QS**
1	6.250e−01	−2.500e+00
2	1.875e+00	−2.500e+00
3	3.125e+00	−2.500e+00
4	4.375e+00	−2.500e+00
5	5.625e+00	−2.500e+00
6	6.875e+00	−2.500e+00
7	8.125e+00	−2.500e+00
8	9.375e+00	−2.500e+00
9	1.063e+01	−2.500e+00
10	1.188e+01	−2.500e+00
11	1.188e+01	2.500e+00

in Table 16.7; however, they are the same as those for nodes 9 to 1 respectively, except for the sign of the rotation (3 displ). Similarly, in Table 16.8 the moments and shears in elements 12 to 20 are not given but may be obtained from the results in elements 9 to 1 respectively by changing the sign of the shear.

After completing the static analysis, a dynamic analysis may also be performed. For example, the transient analysis for a sudden removal of the load may be considered (use a MESH command and FORC–END to remove the load). The program may be used to estimate the required time-step size for the analysis based upon the response of the modes associated with the lowest frequencies of the problem. Accordingly, a modal analysis is performed to obtain the lowest three values. The added commands (in addition to the TANG command already given) to perform this step are

```
MASS
SUBS,,3
```

The '3' on the SUBS command instructs the program to compute the three lowest modes (note that no shift has been specified on the TANG command, hence the three modes nearest to zero are the three lowest modes). The program includes an additional three vectors to speed the convergence in the subspace iteration algorithm used in PCFEAP. The results obtained for the eigenvalue analysis are shown in Table 16.9. The eigenvalues computed are the squares of the frequencies; hence, to assist users the square roots are also computed to give (for this problem) the actual frequencies. Also indicated are the accuracies of the values for the iteration process. It should be noted that the three values requested have been obtained to the accuracy of the program default tolerance (10^{-12}),

TABLE 16.9
FREQUENCY PROPERTIES FOR STRAIGHT BEAM

	Solution for eigenvalues, iteration 9	
0.79132391d−01	0.11784609d+01	0.24661330d+01
0.53609721d+01	0.14876488d+02	0.22185429d+02
	Square root of eigenvalues	
0.281130480d+00	0.10855694d+01	0.15703926d+01
0.23153773d+01	0.38570050d+01	0.47101410d+01
	Current residuals	
0.17570920d−16	0.81772151d−16	0.54724929d−13
0.35527975d−09	0.30864848d−04	0.37500592d−02

whereas the extra vectors have values that are not converged to this accuracy. All values and their associated 'vectors' are available for outputs and plots; however, use of the extra vectors should be limited as, due to lack of convergence, they are not necessarily natural modes of the problem.

The frequencies may be used to compute a time step. For example, if a solution is to be performed with about 16 time steps for each period of the third mode, T_3, the Δt value must satisfy the relationship

$$\omega_3 T_3 = \omega_3 16 \Delta t = 2\pi$$

Using this relationship and ω_3 from Table 16.9 gives a value of Δt equal to about 0.25 time units. We can now proceed to perform the transient analysis by giving the following macro commands:

```
DT,,0.25
BETA
FORM,ACCE
LOOP,,50
TIME
TANG,,1
DISP,,11
NEXT
```

Again recall that in the interactive mode PCFEAP performs each command (except for LOOP–NEXT pairs) immediately after it is entered. Some additional information may be given after each command. For example, the BETA command specifies a Newmark integration for the equations of motion and the program informs the user that the default values of $\beta = 0.25$ and $\gamma = 0.5$ will be used. Should other values or another integration type (e.g. SS22) be desired, the BETA command should be entered again. Note the use of the FORM,ACCE command to compute the initial accelerations for the Newmark method. This command is not required for the SS22 algorithm and will not change the transient solution if specified. Use of the LOOP–NEXT pair indicated performs 50 time steps and saves only the displacement of node 11 in the output file

TABLE 16.10
BEAM VERTICAL DISPLACEMENT VERSUS TIME
GN22 SOLUTION—$\Delta t = 0.25$

t	$v(5.,t)$	t	$v(5.,t)$	t	$v(5.,t)$
0.00	−1.27687e+01	0.25	−1.26535e+01	0.50	−1.24567e+01
0.75	−1.22291e+01	1.00	−1.19352e+01	1.25	−1.16296e+01
1.50	−1.12774e+01	1.75	−1.09136e+01	2.00	−1.05486e+01
2.25	−1.01621e+01	2.50	−9.73598e+00	2.75	−9.17223e+00
3.00	−8.47580e+00	3.25	−7.71089e+00	3.50	−6.93037e+00
3.75	−6.12728e+00	4.00	−5.25853e+00	4.25	−4.40904e+00
4.50	−3.62202e+00	4.75	−2.86062e+00	5.00	−2.11626e+00
5.25	−1.34278e+00	5.50	−4.84249e−01	5.75	4.51830e−01
6.00	1.37438e+00	6.25	2.28409e+00	6.50	3.24920e+00
6.75	4.19304e+00	7.00	5.05339e+00	7.25	5.81280e+00
7.50	6.47234e+00	7.75	7.11163e+00	8.00	7.74784e+00
8.25	8.38005e+00	8.50	9.01887e+00	8.75	9.62533e+00
9.00	1.02058e+01	9.25	1.08077e+01	9.50	1.13639e+01
9.75	1.17441e+01	10.00	1.19935e+01	10.25	1.21800e+01
10.50	1.22851e+01	10.75	1.23522e+01	11.00	1.23812e+01
11.25	1.23651e+01	11.50	1.23369e+01	11.75	1.23061e+01
12.00	1.22754e+01	12.25	1.21429e+01	12.50	1.18475e+01

TABLE 16.11
BEAM VERTICAL DISPLACEMENT VERSUS TIME
SS22 SOLUTION—$\Delta t = 0.25$

t	$v(5.,t)$	t	$v(5.,t)$	t	$v(5.,t)$
0.00	−1.27687e+01	0.25	−1.27111e+01	0.50	−1.25551e+01
0.75	−1.23429e+01	1.00	−1.20821e+01	1.25	−1.178243+01
1.50	−1.14535e+01	1.75	−1.10955e+01	2.00	−1.07311e+01
2.25	−1.03553e+01	2.50	−9.94907e+00	2.75	−9.45411e+00
3.00	−8.82402e+00	3.25	−8.09335e+00	3.50	−7.32063e+00
3.75	−6.52883e+00	4.00	−5.69291e+00	4.25	−4.83378e+00
4.50	−4.01553e+00	4.75	−3.24132e+00	5.00	−248844e+00
5.25	−1.72952e+00	5.50	−9.13515e−01	5.75	−1.62090e−02
6.00	9.13107e−01	6.25	1.82924e+00	6.50	2.76665e+00
6.75	3.72112e+00	7.00	4.62322e+00	7.25	5.43309e+00
7.50	6.14257e+00	7.75	6.79199e+00	8.00	7.42974e+00
8.25	8.06395e+00	8.50	8.69946e+00	8.75	9.32210e+00
9.00	9.91557e+00	9.25	1.05068e+01	9.50	1.10858e+01
9.75	1.15540e+01	10.00	1.18688e+01	10.25	1.20867e+01
10.50	1.22326e+01	10.75	1.23187e+01	11.00	1.23667e+01
11.25	1.23732e+01	11.50	1.23510e+01	11.75	1.23215e+01
12.00	1.22908e+01	12.25	1.22091e+01	12.50	1.19952e+01

(it will probably scroll past the screen too quickly to be read; however, use of toggle for screen output creates a pause to permit reading the information). Tables 16.10 and 16.11 present the results obtained at node 11 for GN22 (Newmark) and SS22 time integration methods.

The above results may be used to verify proper installation of the program on the user's computer. Solutions may also be performed using larger and smaller time increments to verify overall solution accuracy using a finite element model. Indeed, many other solution possibilities may be performed using the ELMT01 for the BEAM/SHELL finite element model. Since the element is non-linear, the program may be used to compute the buckling of Euler beams. This may be performed by including in the element model the non-linear terms (set *lin* to 1 to include these non-linear geometric terms), changing to pure axial loading, i.e. set the axial force on node 21 to -1.0 (compresseive), and removing the vertical force applied to node 11. Then by systematically increasing the axial loading the first buckling load may be found. One algorithm to perform this is as follows:

```
PROP,,1
IDEN
LOOP,load,20
DATA,DT
TIME
LOOP,newton,10
TANG,,1
NEXT,newton
SUBS,,1
NEXT,load
```

This algorithm solves the non-linear eigenproblem given by

$$\mathbf{K}_T(\mathbf{u})\mathbf{V} = \mathbf{V}\boldsymbol{\Lambda}$$

where $\mathbf{u}$ is the solution of the axially loaded bar (without lateral loading the Newton solution step converges in two iterations) and $\mathbf{K}_T$ is the tangent matrix for this solution. At a buckling load the value of $\boldsymbol{\Lambda}$ goes to zero. By using the DATA,DT command the program prompts the user to specify a value for the time step; input is a command

DT,,Δt

with Δt specified at the value desired. Using a ramp loading (the default) for the PROPortional loading function generates solutions for successive values of the load which is computed from

$$\mathbf{f}(t) = p(t)\mathbf{f}_0$$

Both positive and negative time increments may be specified once a load is bounded (the eigenvalue is positive below the critical load and negative above). Other algorithms may be used. However, the current program does not provide storage for a non-diagonal global matrix in the general linear eigenproblem. This is one feature that students (and other users) are suggested to add on their own.

16.9.2 *Analysis of a spherical cap.* The data for a finite element model for a spherical cap is presented in Sec. 16.3.2. This problem is analysed for static loading to indicate the results that may be achieved for an axisymmetric shell problem using ELMT01. The solution steps are merely

```
TANG,,1
DISP,ALL
STRE,ALL
```

and results obtained are included in Tables 16.12 and 16.13. The solution for this problem and results using other elements are included in Fig. 14.7 of this volume. The answers using ELMT01 agree very well with the results in this figure.

16.9.3 *Analysis for spontaneous ignition.* The formulation included as ELMT03 in the program permits the solution of problems that are modelled by the differential equation (16.59) described in Sec. 16.8.3. This differential equation together with the heat source term Q available in ELMT03 permits the analysis of the spontaneous ignition problem for a sphere as described in Sec. 10.6.3 and shown in Fig. 10.17. An analysis of

TABLE 16.12
SPHERICAL CAP: u AND v NODAL DISPLACEMENTS

1	0.000000e+00	0.000000e+00
2	−2.180519e−01	−1.181020e+00
3	−1.027568e+00	−3.461537e+00
4	−1.930318e+00	−6.149003e+00
5	−2.659180e+00	−8.774480e+00
6	−3.107350e+00	−1.106299e+01
7	−3.266646e+00	−1.289060e+01
8	−3.180168e+00	−1.423935e+01
9	−2.909851e+00	−1.515696e+01
10	−2.516794e+00	−1.572456e+01
11	−2.051297e+00	−1.603338e+01
12	−1.549500e+00	−1.616963e+01
13	−1.033944e+00	−1.620599e+01
14	−5.161404e−01	−1.619811e+01
15	0.000000e+00	−1.618896e+01

TABLE 16.13
SPHERICAL CAP: ELEMENT STRESS VALUES

Elmt	**NS**	**NPHI**	**MS**	**MPHI**	**QS**
1	−3.743e+01	−6.892e+00	2.740e+01	3.687e+00	5.053e+00
2	−3.927e+01	−1.055e+01	1.180e+01	−4.206e−01	3.470e+00
3	−4.114e+01	−1.711e+01	1.367e+00	−3.007e+00	2.111e+00
4	−4.285e+01	−2.444e+01	−4.771e+00	−4.346e+00	1.056e+00
5	−4.429e+01	−3.125e+01	−7.677e+00	−4.755e+00	3.099e−01
6	−4.541e+01	−3.689e+01	−8.368e+00	−4.533e+00	−1.645e−01
7	−4.624e+01	−4.114e+01	−7.701e+00	−3.935e+00	−4.232e−01
8	−4.679e+01	−4.408e+01	−6.333e+00	−3.159e+00	−5.253e−01
9	−4.713e+01	−4.592e+01	−4.726e+00	−2.347e+00	−5.235e−01
10	−4.730e+01	−4.692e+01	−3.172e+00	−1.587e+00	−4.604e−01
11	−4.737e+01	−4.735e+01	−1.835e+00	−9.316e−01	−3.660e−01
12	−4.737e+01	−4.745e+01	−7.852e−01	−3.999e−01	−2.604e−01
13	−4.734e+01	−4.740e+01	−2.723e−02	1.684e−02	−1.538e−01
14	−4.730e+01	−4.733e+01	5.937e−01	3.119e−01	−5.037e−02

the problem for the spontaneous ignition case shown in Fig. 10.17(*b*) (that is $\delta = 16$) may be performed by first computing the steady-state solution using the data given in Sec. 16.3.3. Thus, by using the macro command TANG,,1 the steady-state solution, which is a homogeneous state of 290 K, is recovered. The boundary conditions may now be changed to the value of 500 K and the material properties changed to specify the heat-generation term by using the command

```
MESH
```

followed by (note that the screen should now have a 'Mesh>' prompt instead of the one for the macro command entry)

```
FORC
1,1,500.
5,,500.

MATE
1,3
0.2,1.,1.,−80.,20.,500.,−1,2

END
```

The first set of FORC statements change the boundary conditions, while the MATE statements reset the material parameters to include the heat-generation term (if this had been included in the original set of properties, the wrong steady-state initial conditions would result due to the presence of the heat-generation terms). The material parameters correspond to the values

$$K = 0.20$$
$$c = 1.00$$
$$\rho = 1.00$$
$$\bar{\delta} = -80.0$$
$$r = 20.0$$
$$T_a = 500.0$$

Prompts appear on the screen at each step to remind the user that data is to be input. After each set of statements is completed, the program outputs the new data input to the screen and the output file. The END command returns the program to macro execution. At this time the heat capacity matrix and the specification of the parameters for the transient solution should be given. The solution reported in this section was computed using the SS11 method with θ equal to 0.50 and 20 time steps with Δt equal to 0.01 followed by 30 time steps with a Δt of 0.04 units. The macro commands to be entered to perform these steps may be specified as

```
MASS
BETA,SS11,0.5
DT,,.01
LOOP,time,20
TIME
LOOP,newton,10
TANG,,1
NEXT,newton
DISP,,4,39,5
NEXT,time
DT,,.04
LOOP,time,30
TIME
LOOP,newton,10
TANG,,1
NEXT,newton
DISP,,4,39,5
NEXT,time
```

The problem is non-linear due to the effects of the heat-generation term and, thus, a Newton solution is performed for each time step. The DISP print describes the location of the temperatures tabulated at five levels of time and shown in Table 16.14. These values agree well with the ones reported in Fig. 10.17. Solutions beyond the time 1.40 depend quite strongly on the particular choice of time increments used and an

TABLE 16.14

SPONTANEOUS IGNITION TEMPERATURE STATES VERSUS TIME

	Time values				
Radius	0.0	0.13	0.60	1.00	1.40
1.000	290.0000	500.0000	500.0000	500.0000	500.0000
0.875	290.0000	427.0221	484.0392	497.8513	508.9534
0.750	290.0000	364.1205	462.9881	490.9011	513.8165
0.625	290.0000	322.0308	440.2810	480.9882	514.2681
0.500	290.0000	300.4395	418.8623	470.3554	511.7056
0.375	290.0000	292.9916	400.3157	460.5371	507.4656
0.250	290.0000	290.2667	386.2729	452.8100	503.3086
0.000	290.0000	289.8447	372.3680	444.9784	498.6067

automatic stepping procedure (not available in PCFEAP) was used to produce the results shown in Fig. 10.17. The solution is unstable once the temperature exceeds about 600 K due to the nature of the exponential function used to describe Q.

16.9.4 *Analysis of an elastic–plastic tension strip.* The analysis of the finite width strip with a circular hole subjected to axial deformation is performed for an elastic–plastic material without hardening. The analysis uses ELMT02 with material properties specified as

$$E = 7000$$

$$\nu = 0.2$$

$$Y_0 = 24.3$$

The axial loading is applied by restraining boundary conditions at the top and bottom of the mesh to indicate specified displacement values in the FORCed inputs. The unit values specified at the top of the mesh provide the axial loading and control is described using a proportional loading function equal to t together with specified values of the time increment. Accordingly, the first loading increment is applied with a time increment set to 0.04 using the commands

```
PROP,,1
DT,,0.04
TIME
```

A Newton iteration step is then performed using the commands

```
LOOP,,10
TANG,,1
NEXT
```

The solution step produces some yielding at the narrow section adjacent

TABLE 16.15
RESIDUAL NORMS FOR TENSION STRIP

Item number	Loading time 0.04	0.05	0.06	0.07	0.08
1	7.3165e+02	1.8291e+02	1.8291e+02	1.8291e+02	1.8291e+02
2	5.7335e+00	3.3490e+00	6.9462e+00	2.5496e+00	1.4204e+00
3	1.0605e+00	6.7595e−01	3.1487e+00	1.8722e−01	2.2917e−01
4	1.0554e−01	7.0917e−03	1.1493e−01	8.6060e−05	5.1782e−04
5	1.3849e−04	2.0422e−06	3.3520e−05	1.2271e−10	6.1778e−09
6	1.6046e−09	2.0755e−13	1.5476e−1		

TABLE 16.16
ENERGY NORMS FOR TENSION STRIP

Item number	Loading time 0.04	0.05	0.06	0.07	0.08
1	7.5010e+01	4.7314e+00	4.7936e+00	4.9280e+00	4.9524e+00
2	1.6565e−02	6.8191e−03	3.4727e−02	3.2724e−03	1.0295e−03
3	7.9394e−04	1.3852e−04	1.5827e−03	5.0961e−06	1.1134e−05
4	2.1674e−06	2.2567e−08	1.7065e−06	1.8806e−12	6.8850e−11
5	1.0337e−11	1.7323e−15	2.7846e−13	3.8667e−24	1.8542e−20
6	1.2317e−21	1.8038e−29	7.2369e−26		

to the circular hole. The values of the residual and energy norms produced for this step are given in Tables 16.15 and 16.16 respectively. Several additional steps are also performed for a smaller time increment of 0.01 and the results for the first four are also included in the tables. The yield patterns produced by the solution are similar to the ones included in Chapter 7.

The resultant load on the tension strip is plotted as a function of time in Fig. 16.3. It is evident that a limit load is achieved in five loading

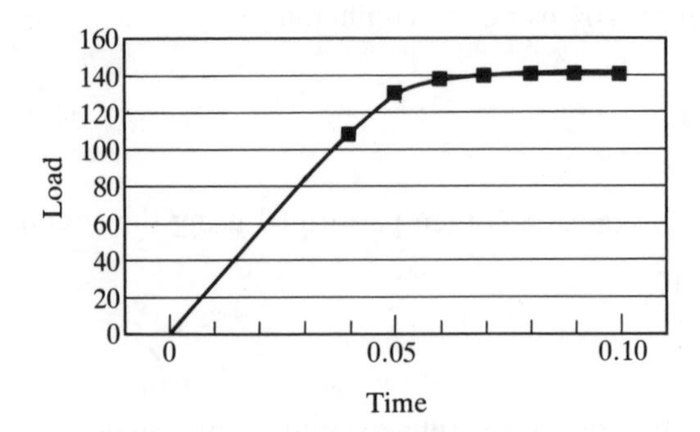

Fig. 16.3 Tension strip: load versus time

increments. Noting the quadratic asymptotic rate of convergence for the last three iterations of each loading step illustrates the performance of a properly implemented Newton algorithm.

The above analyses have indicated some of the types of problems that may be considered using the finite element program PCFEAP included in this chapter. The examples have illustrated the application of the program to the solution of *static*, *quasi-static* (time-dependent problems in which the differential equation has no rate effects) and *transient* problems. Both linear and non-linear problems have been addressed. The program includes a subspace algorithm to extract the smallest eigenpairs of both standard and general linear eigenproblems. In the algorithm included in this chapter only diagonalized matrices are permitted for the 'mass' term in the general linear eigenproblem. The reader has undoubtedly observed other features that may be added to the program to improve its utility. For example, a very useful feature is an ability to abandon a solution step, by restoring the states that existed at the beginning of a 'time' increment. This requires that the time integration algorithm be solved in an inverse or backward direction. However, this is easily achieved for one-step methods such as the SSij and GN22 (Newmark) algorithms used here. Other improvements include automatic time-step control, such as proposed in reference 12 to solve the spontaneous combustion problem, and continuation methods to solve problems that possess limit points (often called arc length methods, e.g. see Chapter 7). Finally, we have noted during the development of the plate elements discussed in Chapter 2 that it is convenient to describe the unknowns associated with elements in a hierarchical process; namely to associate the element degrees of freedom with nodes, edges and (for three-dimensional problems) faces. This improvement has not been included here due to space limitations and, again, is left as a user improvement.

In the final section of this chapter, listings for the additions and changes to the program are included.

16.10 Installation information and listings for program modules

This section contains the information for installing the program on computer systems as well as listings for the changes and additions to the program.

16.10.1 *Installation information.* The program may be installed on a variety of different computer systems, ranging in size from small personal computers to the largest machines available. The program is written for a *scalar computing mode* and, thus, will not achieve optimal performance on large vector or array-type computers. The basic installation steps include:

1. Selection of a timing routine [see Sec. 16.10.3(*b*) below].
2. Selection of a plot interface (see Sec. 16.10.6 below).
3. Selection of a solution option: variable-band solution uses Files PASOLV.FOR and PARSOL.FOR; frontal uses Files PFSOLV.FOR and PFRSOL.FOR.
4. Compilation and linking of the above routines with the other files named in Table 16.17.

When used on a personal computer, all integer variables are declared to be 2 byte words to keep the executable program as small as possible. For the dimensions currently specified for large arrays, no problems result from this, with the exception of the variable ITREC in the labelled common block named TEMFL2. This variable is used to count bytes for direct input/output to the disk and is now forced to be type INTEGER*4. The original specification does not cause difficulties with the Microsoft Fortran Version 3.31 compiler, but does lead to errors for large problems when using Version 5.0. Accordingly, all files that contain this array should be modified to

```
      integer*4 itrec
      common /temfl2/ itrec(4),nw1,nw2
```

The parameters NW1 and NW2 may remain as type INTEGER*2.

TABLE 16.17
PCFEAP FILES FOR SOURCE PROGRAM

File name	*Description*	*Section*
PCFEAP.FOR	Main program, file assignment, installation	15.8.2(*a*)
PCDEPT.FOR	Installation or compiler-dependent routines	15.8.2(*b*), 16.10.3(*b*)
PCMESH.FOR	Mesh definition routines	15.8.2(*c*)
PCMAC1.FOR	Macro program routines—part 1	15.8.3(*a*), 16.10.4(*a*)
PCMAC2.FOR	Macro program routines—part 2	15.8.3(*b*), 16.10.4(*b*)
PCMAC3.FOR	Macro program routines—part 3	15.8.3(*c*)
PCEIGN.FOR	Subspace eigenpair computation routines	16.10.4(*d*)
PASOLV.FOR	Variable-band solution routines	15.8.4(*a*), 16.10.5(*a*)
PFSOLV.FOR	Frontal solution routines	15.8.4(*b*)
PARSOL.FOR	Variable-band resolution	16.10.5(*c*)
PFRSOL.FOR	Frontal resolution	16.10.5(*d*)
PCPLOT.FOR	Plot routines	15.8.5, 16.10.56(*a*)
PCFOR5.FOR	Graphics interface (Microsoft 5.0)	16.10.6(*b*)
PCTEKT.FOR	Graphics interface (Tektronix 4012)	16.10.6(*c*)
PCELM1.FOR	Element routine for non-linear beam/axisymmetric elastic shell	16.10.7(*a*)
PCELM2.FOR	Element routine for **B**-bar model—elastoplastic material	16.10.7(*b*)
PCELM3.FOR	Element routine for Laplace equation—reaction-type source term	16.10.7(*c*)
PCELM4.FOR	Element routine for general truss model—elastoplastic material	16.10.7(*d*)

16.10.2 *Listing of finite element computer program.* We include here and in the following sections the extensions to the finite element computer program originally given in Chapter 15 of Volume 1. The routines to be used are indicated in Table 16.17. Label 15.*m.n* refers to Chapter 15, Sec. *m.n.* in Volume 1 whereas a label 16.*m.n.* is for Sec. *m.n.* in this chapter.

16.10.3 *Control and data input modules.* The control and data input subprograms are contained in the files PCFEAP.FOR, PCDEPT.FOR and PCMESH.FOR. Except for the timing routine, no substantive changes have been made to the routines and listings are as given in Volume 1, Chapter 15, pp. 497–515. For completeness, tables listing the subroutines are given below. The small number of changes in these routines is indicated in the notes following each table and the timing routine for the various options is given in Sec. 16.10.3(*b*).

(*a*) File PCFEAP.FOR contains the following subprograms:

Name	*Type*	*Description*	*Chapter*
PCFEAP	PROGRAM	Main program	15
PDEFIL	SUBROUTINE	Deletes scratch disk files	15
FILNAM	SUBROUTINE	Input file names for I/O	15
PCONTR	SUBROUTINE	Control problem solution	15
PLTSTR	SUBROUTINE	Compute nodal stress projection	15
PRTHED	SUBROUTINE	Print head to output file	15

Note: in SUBROUTINE PCONTR (Volume 1, p. 502) line PCO 41 should be changed to read:

```
call psetm(nn, 5*nen*ndf,        ipd,tfl)
```

(*b*) File PCDEPT.FOR contains the following subprograms (note that routines in this file may need to be changed for various computer systems):

Name	*Type*	*Description*	*Chapter*
TIME	INTERFACE	Time interface	15
PCTIME	SUBROUTINE	Set elapsed time for outputs	15, 16
PCOMP	FUNCTION	Compare ASCII character data	15
PINTIO	SUBROUTINE	Input data from screen or file	15
PCLEAR	SUBROUTINE	DOS clear screen	15

The timing routine for the Microsoft Fortran Version 5.0 or UNIX compilers is given by:

```
      subroutine pctime(xtime)                                    pct  1
      character*10 xtime                                          pct  2
c                                                                 pct  3
c.... Time routine for Microsoft Version 5.0                       pct  4
c     call gettim (ihr,imin,isec,ihth)                            pct  5
c     write(xtime,2000) ihr,imin,isec                             pct  6
c2000 format(i3,':',i2,':',i2)                                    pct  7
c                                                                 pct  8
c.... Time routine for UNIX: Elapsed cpu time                      pct  9
      real tarry(2)                                               pct 10
      call etime(tarry)                                           pct 11
      write(xtime,2001) tarry(1)                                  pct 12
 2001 format(f9.2)                                                pct 13
      end                                                         pct 14
```

(*c*) File PCMESH.FOR contains the following subprograms:

Name	*Type*	*Description*	*Chapter*
PMESH	SUBROUTINE	Control for mesh data inputs	15
PMATIN	SUBROUTINE	Input material property sets	15
BLKGEN	SUBROUTINE	Input block of nodes/elements	15
GENVEC	SUBROUTINE	Generate real nodal vector/array	15
PBCIN	SUBROUTINE	Input boundary restraint codes	15
PELIN	SUBROUTINE	Input element connection lists	15
POLAR	SUBROUTINE	Convert polar coordinates to cartesian	15
SBLK	SUBROUTINE	Generate nodes/elements for BLKGEN	15
SETHIS	SUBROUTINE	Write initial history data to disk	15

Note: in SUBROUTINE SETHIS (Volume 1, p. 514) the following should be added after line SET 5:

```
integer*4 itfil
```

16.10.4 *Macro solution and output modules.* The solution and output for each problem is controlled by the subprograms contained in the files PCMAC1.FOR, PCMAC2.FOR, PCMAC3.FOR and PCEIGN.FOR. Subprograms where substantial changes have been made are included in this volume. The tables indicate Chapter 16 for these routines; subprograms where no changes have been made are indicated as appearing in Chapter 15 of Volume 1. Also included are new routines which have been added to the program and are marked with an (*) following their name in the tables.

(*a*) File PCMAC1.FOR contains the following subprograms:

Name	*Type*	*Description*	*Chapter*
PMACR	SUBROUTINE	Controls macro solution process	16
PINITC	SUBROUTINE	Set initial values to parameters	15
PMACIO	SUBROUTINE	Input and compile statements	15
PMACR1	SUBROUTINE	Control FE macro command execution	16
PMACR2	SUBROUTINE	Control other macro command execution	16
FORMFE	SUBROUTINE	Set up calls for FE array computation	15
INACCL(*)	SUBROUTINE	Compute initial acceleration	16
PHSTIO	SUBROUTINE	I/O of data to disk	15
SERCHL(*)	SUBROUTINE	Line search algorithm	16
GAMMA(*)	FUNCTION	Function evaluation for line search	16

```
      subroutine pmacr (ul,xl,tl,ld,p,s,ie,d,id,x,ix,f,t,jd,f0,b,          pma  1
     1  dr,ndf,ndm,nen1,nst,prt)                                           pma  2
c.... macro instruction subprograms                                        pma  3
      logical fl,prt,hfl,hout                                              pma  4
      integer*2 ld(*),ie(*),id(*),ix(*),jd(*),jct,lvs,lve,im               pma  5
      integer*4 itrec,iz                                                   pma  6
      real    ct(3,100),xl(*),tl(*),x(*),f(*),f0(*),t(*)                   pma  7
      real*8 d(*),ul(*),p(*),s(*),b(*),dr(*),dm,engy,rnmax,prop            pma  8
      character*4 wd(24),lct,tary*10,tfile*12                              pma  9
      common dm(1),rm(1),im(1)                                             pma 10
      common /cdata/ numnp,numel,nummat,nen,neq                            pma 11
      common /fdata/ fl(11)                                                pma 12
      common /hdatb/ nhi,nhf,ihbuff,irec,jrec,nrec,hfl,hout                pma 13
      common /iofild/ iodr,iodw,ipd,ipr,ipi                                pma 14
      common /iofile/ ior,iow                                              pma 15
      common /ldata/ l,lv,lvs(9),lve(9),jct(100)                           pma 16
      common /ldatb/ lct(100)                                              pma 17
      common /ndata/ nv,nw,nl                                              pma 18
      common /rdata/ engy,rnmax,tol,shift                                  pma 19
      common /tbeta/ beta,gamm,theta,nop,nt                                pma 20
      common /tdata/ ttim,dt,c1,c2,c3,c4,c5,c6                             pma 21
```

```
      common /temfl1/ tfile(6)                                          pma 22
      common /temfl2/ itrec(4),nw1,nw2                                  pma 23
      common /prlod/ prop,a(6,10),iexp(10),ik(10),npld                  pma 24
      data wd/'stre','utan','tang','form','mass','reac','chec','disp',  pma 25
     1        'solv','mesh','rest',                                     pma 26
     2        'tol ','dt  ','loop','next','prop','data','time','beta',  pma 27
     2        'newf','subs','eigv','iden',                              pma 28
     p        'plot'/                                                   pma 29
c.... nmi = no. macro commands in 'pmacri'; nlp = loop number           pma 30
      data nm1,nm2/11,12/,iz/0/                                         pma 31
      nlp = nm1 + 3                                                     pma 32
c.... set initial values of parameters                                  pma 33
      call pinitc(engy,rnmax,shift,tol,dt,prop,ttim,npld)               pma 34
      nw1  = nm1                                                        pma 35
      nw2  = nm2 + nw1                                                  pma 36
      nneq = ndf*numnp                                                  pma 37
      call psetm (nt,nneq*2,ipr,fl(1))                                  pma 38
      call pconsr(rm(nt),nneq*2,0.0)                                    pma 39
c.... input the macro commands                                          pma 40
100   call pmacio (jct,lct,ct,wd,nw2+1,nlp,ll)                          pma 41
      if(ll.le.0) go to 300                                             pma 42
c.... execute macro instruction program                                 pma 43
      lv = 0                                                            pma 44
      l = 1                                                             pma 45
200   j = jct(l)                                                        pma 46
      i = l - 1                                                         pma 47
      call pctime (tary)                                                pma 48
      if(l.ne.1.and.l.ne.ll) then                                       pma 49
       write(iow,2001) i,wd(j),lct(l),(ct(k,l),k=1,3),tary              pma 50
       if(ior.lt.0) write(*,2001) i,wd(j),lct(l),(ct(k,l),k=1,3),tary   pma 51
      endif                                                             pma 52
      if(j.le.nw1) call pmacr1(id,ie,ix,ld,d,s,x,f,f0,t,jd,b,dr,        pma 53
     1                   lct,ct,ndf,ndm,nen1,nst,nneq,prt,j)            pma 54
      if(j.ge.nw1+1.and.j.le.nw2)                                       pma 55
     1           call pmacr2(id,f,f0,jd,b,dr,lct,ct,ndf,nneq,j-nw1)     pma 56
c.... plot macro call                                                   pma 57
      if(j.eq.nw2+1) then                                               pma 58
        call pplotf(x,ix,b,lct(l),ct(1,l),ndf,ndm,nen1,nneq)            pma 59
      endif                                                             pma 60
      l = l + 1                                                         pma 61
      if(l.le.ll) go to 200                                             pma 62
      if (ior.lt.0) go to 100                                           pma 63
300   call pctime(tary)                                                 pma 64
      write(iow,2000) tary                                              pma 65
      if(ior.lt.0) write(*,2000) tary                                   pma 66
      if(.not.fl(4)) close(4,status='delete')                           pma 67
c.... save restart information                                          pma 68
      if(ll.lt.-1.or.fl(7)) close(3,status='delete')                    pma 69
      if(ll.lt.-1.or.fl(7)) return                                      pma 70
      open (7,file=tfile(6),form='unformatted',status='new')            pma 71
      rewind 7                                                          pma 72
      write(7) numnp,numel,nummat,ndm,ndf,nhi,nhf,nrec                  pma 73
      write(7) ttim,(b(i),i=1,3*nneq)                                   pma 74
      write(7) (dm(i),i=nt,nt+nneq)                                     pma 75
      if(fl(9)) write(7) (dm(i),i=nv,nv+4*neq)                          pma 76
      if(nrec.gt.0) then                                                pma 77
        do 400 j = 1,nrec                                               pma 78
          call phstio(3,j,dm(nhi),nhf-nhi+1,1,tfile(2),itrec(2))        pma 79
          call phstio(7,j,dm(nhi),nhf-nhi+1,22,tfile(6),iz)             pma 80
400     continue                                                        pma 81
```

```
        call pdefil(tfile,2,2)                                        pma 82
      endif                                                           pma 83
      close(7)                                                        pma 84
c.... formats                                                         pma 85
2000  format(' *End of macro execution*',15x,'time=',a10)             pma 86
2001  format(' *Macro ',i3,' *',2(a4,1x),                             pma 87
     1   'V1=',g10.3,' V2=',g10.3,' V3=',g10.3/40x,'time=',a10)       pma 88
      end                                                             pma 89
c
      subroutine pmacr1(id,ie,ix,ld,d,s,x,f,f0,t,jd,b,dr,             pma  1
     1                  lct,ct,ndf,ndm,nen1,nst,nneq,prt,j)           pma  2
c.... macro instruction subprograms                                   pma  3
      logical fa,tr,fl,pcomp,prt,sflg,tflg,hfl,hout                   pma  4
      character lct(*)*4,tfile*12,tt*10                               pma  5
      integer*2 id(*),ie(*),ix(*),jd(*),ld(*),lvs,lve,jct,            pma  6
     1          ia,im                                                 pma  7
      integer*4 itrec,iz                                              pma  8
      real   ct(3,*),x(*),f(*),f0(*),t(*)                             pma  9
      real*8 b(*),d(*),dr(*),s(*),dm,aa,engy,rnorm,                   pma 10
     1  rnmax,dot,prop,eerror,elproj,ecproj,efem,enerr,ebar           pma 11
      common /adata/ aa(1),ap(15998)                                  pma 12
      common /cdata/ numnp,numel,nummat,nen,neq                       pma 13
      common /errind/ eerror,elproj,ecproj,efem,enerr,ebar            pma 14
      common /fdata/ fl(11)                                           pma 15
      common /frdata/ maxf                                            pma 16
      common /hdatb/ nhi,nhf,ihbuff,irec,jrec,nrec,hfl,hout           pma 17
      common /iofile/ ior,iow                                         pma 18
      common /iofild/ iodr,iodw,ipd,ipr,ipi                           pma 19
      common /ldata/ l,lv,lvs(9),lve(9),jct(100)                      pma 20
      common /mdat2/ n11a,n11b,n11c,ia(2,11)                          pma 21
      common /ndata/ nv,nw,nl                                         pma 22
      common /prlod/ prop,a(6,10),iexp(10),ik(10),npld                pma 23
      common /rdata/ engy,rnmax,tol,shift                             pma 24
      common /tbeta/ beta,gamm,theta,nop,nt                           pma 25
      common /tdata/ ttim,dt,c1,c2,c3,c4,c5,c6                        pma 26
      common /temfl1/ tfile(6)                                        pma 27
      common /temfl2/ itrec(4),nw1,nw2                                pma 28
      common dm(1),rm(1),im(1)                                        pma 29
      data zero/0.0/,fa,tr/.false.,.true./,iz/0/                      pma 30
c.... transfer to correct process                                     pma 31
      n1 = 1                                                          pma 32
      n3 = 1                                                          pma 33
      go to (1,2,2,3,4,5,6,5,2,9,10), j                               pma 34
c.... print stress values                                             pma 35
1     n1 = ct(1,l)                                                    pma 36
      n2 = ct(2,l)                                                    pma 37
      n3 = ct(3,l)                                                    pma 38
      n3 = max(n3,1)                                                  pma 39
      n4 = numnp - 1                                                  pma 40
      n0 = -2                                                         pma 41
      if (pcomp(lct(l),'node')) then                                  pma 42
        n1 = max(1,min(numnp,n1))                                     pma 43
        n2 = max(n1,min(numnp,n2))                                    pma 44
        enerr = 0.0                                                   pma 45
        if(.not.fl(11)) then                                          pma 46
          call pconsr(ap(n0),8*numnp,0.0)                             pma 47
          call formfe(b,dr,fa,fa,fa,fa,8,1,numel,1)                   pma 48
          call pltstr(aa,ap(n4),numnp)                                pma 49
        endif                                                         pma 50
        call prtstr(aa,ap(n4),numnp,n1,n2,n3)                         pma 51
```

```
        fl(11) = tr                                                    pma 52
      elseif (pcomp(lct(l),'erro')) then                               pma 53
        n1 = max(n1,1)                                                 pma 54
        n2 = 8*numnp                                                   pma 55
        call pconsr(ap(n2),n2,0.0)                                     pma 56
        enerr = 0.0                                                    pma 57
        do 110 i = 1,n1                                                pma 58
          call pconsr(ap(n0),n2,0.0)                                   pma 59
          call formfe(b,dr,fa,fa,fa,fa,8,1,numel,1)                    pma 60
          call pltstr(aa,ap(n4),numnp)                                 pma 61
          call addvec(ap(n2),ap(n4),n2-numnp)                          pma 62
110     continue                                                       pma 63
        fl(11) = tr                                                    pma 64
        eerror = 0.0                                                   pma 65
        eproj  = 0.0                                                   pma 66
        efem   = 0.0                                                   pma 67
        ebar   = 0.05*sqrt(enerr/numel)                                pma 68
        ietyp  = 1                                                     pma 69
        call formfe(b,dr,fa,fa,fa,fa,7,1,numel,1)                      pma 70
        call prterr                                                    pma 71
      else                                                             pma 72
        if(pcomp(lct(l),'all ')) then                                  pma 73
          n2 = numel                                                   pma 74
        else                                                           pma 75
          n1 = max(1,min(numel,n1))                                    pma 76
          n2 = max(n1,min(numel,n2))                                   pma 77
        endif                                                          pma 78
        call formfe(b,dr,fa,fa,fa,fa,4,n1,n2,n3)                       pma 79
      endif                                                            pma 80
      return                                                           pma 81
c.... solution step                                                    pma 82
2     shft = c1                                                        pma 83
      sflg = fl(9)                                                     pma 84
      if(j.eq.9) then                                                  pma 85
        if(.not.fl(8)) return                                          pma 86
        fl(7) = .false.                                                pma 87
        tflg  = .false.                                                pma 88
c.... form tangent stiffness                                           pma 89
      else                                                             pma 90
c....  unsymmetric tangent (must use profile solver)                   pma 91
        if(j.eq.2) fl(6) = .true.                                      pma 92
c....  symmetric tangent                                               pma 93
        if(j.eq.3) fl(6) = .false.                                     pma 94
        if(ct(1,l).ne.zero) then                                       pma 95
          fl(8) = tr                                                   pma 96
          fl(7) = fa                                                   pma 97
          call raxpb(f, f0, prop, nneq, rm(nt+nneq))                   pma 98
          call pload(id,rm(nt),dr,nneq,dm(nl),dm(nw))                  pma 99
        endif                                                          pma100
        shift= 0.                                                      pma101
        tflg = tr                                                      pma102
        if(.not.fl(9).and.ct(2,l).ne.zero) then                        pma103
          if(fl(2)) then                                               pma104
            sflg = tr                                                  pma105
            shift= ct(2,l)                                             pma106
            shft = -shift                                              pma107
            if(ior.lt.0) write(*,2006) shift                           pma108
            write(iow,2006) shift                                      pma109
          else                                                         pma110
            if(ior.lt.0) write(*,2007)                                 pma111
```

```
            write(iow,2007)                                            pma112
            if(ior.gt.0) stop                                          pma113
            return                                                     pma114
          endif                                                        pma115
        endif                                                          pma116
      endif                                                            pma117
c.... call the solve routine to assemble and solve the tangent matrix  pma118
      na = maxf + 1                                                    pma119
      nal= (maxf*(maxf+1))/2 + na                                      pma120
      call psolve(b,aa(na),aa,dr,aa(nal),dm(nl),s,ld,jd,im(n11c),nst,1, pma121
     1            tflg,fl(8),sflg,shft,4,rnorm,engy,1)                 pma122
      call pctime(tt)                                                  pma123
      if(fl(8)) then                                                   pma124
        fl(8) = fa                                                     pma125
        if(tflg) write(iow,2001) rnorm,tt                              pma126
        if(tflg .and. ior.lt.0) write(*,2001) rnorm,tt                 pma127
        if (rnmax.eq.0.0d0) then                                       pma128
          rnmax = abs(engy)                                            pma129
          if(ct(3,l).le.0.0) ct(3,l) = 0.6                             pma130
          enold = rnmax*0.9999                                         pma131
        endif                                                          pma132
        write(iow,2004) rnmax,engy,tol                                 pma133
        if(ior.lt.0) write(*,2004) rnmax,engy,tol                      pma134
        if(abs(engy).le.tol*rnmax) then                                pma135
          ct(1,lve(lv)) = ct(1,lvs(lv))                                pma136
          l = lve(lv) - 1                                              pma137
        elseif(pcomp(lct(l),'line')) then                              pma138
c.... line search                                                      pma139
          if(abs(engy).gt.ct(3,l)*enold) then                          pma140
            ml1 = 1 + nneq                                             pma141
            call serchl(rm(nt),id,engy,aa,b,dr,ct(3,l),aa(ml1),nneq)   pma142
          endif                                                        pma143
        endif                                                          pma144
        call update(id,rm(nt),b,dm(nv),dr,nneq,fl(9),2)                pma145
      else                                                             pma146
        write(iow,2002) tt                                             pma147
        if(ior.lt.0) write(*,2002) tt                                  pma148
      endif                                                            pma149
      return                                                           pma150
c.... form out of balance force for time step/iteration                pma151
3     if(fl(8)) return                                                 pma152
      call raxpb(f, f0, prop, nneq, rm(nt+nneq))                       pma153
      call pload(id,rm(nt),dr,nneq,dm(nl),dm(nw))                      pma154
      call formfe(b,dr,fa,tr,fa,fa,6,1,numel,1)                        pma155
      if(pcomp(lct(l),'acce').and.ttim.eq.0.0 .and. nop.eq.3) then     pma156
        call inaccl(id,dr,dm(nl),dm(nw),nneq)                          pma157
      endif                                                            pma158
      rnorm = sqrt(dot(dr,dr,neq))                                     pma159
      write(iow,2003) rnorm                                            pma160
      if(ior.lt.0) write(*,2003) rnorm                                 pma161
      fl(8) = tr                                                       pma162
      return                                                           pma163
c.... form a lumped mass approximation                                 pma164
4     if(fl(5)) call psetm(nl,neq,ipd,fl(5))                           pma165
      call pconsd(dm(nl),neq,0.0d0)                                    pma166
      fl(2) = tr                                                       pma167
      call formfe(b,dm(nl),fa,tr,fa,fa,5,1,numel,1)                    pma168
      return                                                           pma169
c.... compute reactions and print                                      pma170
5     if(pcomp(lct(l),'all ')) then                                    pma171
```

```
          n2 = numnp                                                     pma172
        else                                                             pma173
          n1 = ct(1,l)                                                   pma174
          n2 = ct(2,l)                                                   pma175
          n3 = ct(3,l)                                                   pma176
          n1 = max(1,min(numnp,n1))                                      pma177
          n2 = max(n1,min(numnp,n2))                                     pma178
          n3 = max(1,n3)                                                 pma179
        endif                                                            pma180
        if(j.eq.6) then                                                  pma181
          call pconsd(dr,nneq,0.0d0)                                     pma182
          call formfe(b,dr,fa,tr,fa,tr,6,1,numel,1)                      pma183
          call prtrea(dr,ndf,numnp,n1,n2,n3)                             pma184
        else                                                             pma185
          if(nop.le.2) then                                              pma186
            call prtdis(x,dm(nw+nneq),ttim,prop,ndm,ndf,n1,n2,n3)        pma187
          else                                                           pma188
            call prtdis(x,b,ttim,prop,ndm,ndf,n1,n2,n3)                  pma189
          endif                                                          pma190
        endif                                                            pma191
        return                                                           pma192
c.... check mesh for input errors                                        pma193
6       call formfe(b,dr,fa,fa,fa,fa,2,1,numel,1)                        pma194
        return                                                           pma195
c.... modify mesh data (cannot change profile of stiffness/mass)         pma196
9       i = -1                                                           pma197
        call pmesh(ld,ie,d,id,x,ix,f,t,ndf,ndm,nen1,i,prt)               pma198
        if (i.gt.0) go to 400                                            pma199
        return                                                           pma200
c.... restart previously run problem                                     pma201
10      open (7,file=tfile(5),form='unformatted',status='old')           pma202
        read(7) nnpo,nnlo,nnmo,ndmo,ndfo,nhio,nhfo,nrco                  pma203
        if((nnpo.eq.numnp).and.(nnlo.eq.numel).and.(nnmo.eq.nummat)      pma204
     1   .and.(ndmo.eq.ndm).and.(ndfo.eq.ndf).and.(nrco.eq.nrec)         pma205
     2   .and.(nhfo-nhio.eq.nhf-nhi) ) then                              pma206
           read(7) ttim,(b(i),i=1,3*nneq)                                pma207
           read(7) (dm(i),i=nt,nt+nneq)                                  pma208
           if(fl(9)) read(7) (dm(i),i=nv,nv+4*neq)                       pma209
           if(nrec.gt.0) then                                            pma210
             do 101 j = 1,nrec                                           pma211
               call phstio(7,j,dm(nhi),nhf-nhi+1,11,tfile(5),iz)         pma212
               call phstio(3,j,dm(nhi),nhf-nhi+1,2,tfile(2),itrec(2))    pma213
101          continue                                                    pma214
           endif                                                         pma215
           close(7)                                                      pma216
        else                                                             pma217
           if(ior.gt.0) write(iow,3001)                                  pma218
           if(ior.lt.0) write(  *,3001)                                  pma219
        endif                                                            pma220
        return                                                           pma221
c.... error diagnostics                                                  pma222
400     write(iow,3002)                                                  pma223
        if(ior.gt.0) stop                                                pma224
        write(  *,3002)                                                  pma225
c.... formats                                                            pma226
2001    format('    | R(i) | =',1pe15.7,12x,'time=',a10)                 pma227
2002    format(40x,'time=',a10)                                          pma228
2003    format('    | R(i) | = ',1pe15.7)                                pma229
2004    format('   Energy Convergence Test'/'    E(1)=',1pe21.14,        pma230
     1  ', E(i)=',1pe21.14,', Tol.=',1pe11.4)                            pma231
```

```
2006  format('   Shift of',1pe11.4,' applied with mass')               pma232
2007  format('   Shift requested but no mass matrix exists.')          pma233
3001  format(' **ERROR** File incompatible with current problem.')     pma234
3002  format(' **ERROR** Attempt to change profile during mesh.')      pma235
      end                                                               pma236
c
      subroutine pmacr2(id,f,f0,jd,b,dr,lct,ct,ndf,nneq,j)             pma   1
c.... macro instruction subprograms                                     pma   2
      logical fl,pcomp,hfl,hout,sfl                                     pma   3
      integer*2 id(*),jd(*),lvs,lve,jct,im                              pma   4
      real      f0(*),f(*),ct(3,*),xtl                                  pma   5
      real*8    b(*),dr(*),uu,dm,engy,rnmax,prop                        pma   6
      character*4 lct(*),ctl(2),tfile*12,yyy*80                         pma   7
      common /cdata/ numnp,numel,nummat,nen,neq                         pma   8
      common /fdata/ fl(11)                                             pma   9
      common /hdatb/ nhi,nhf,ihbuff,irec,jrec,nrec,hfl,hout             pma  10
      common /iofile/ ior,iow                                           pma  11
      common /iofild/ iodr,iodw,ipd,ipr,ipi                             pma  12
      common /ldata/ l,lv,lvs(9),lve(9),jct(100)                        pma  13
      common /ndata/ nv,nw,nl                                           pma  14
      common /prlod/ prop,a(6,10),iexp(10),ik(10),npld                  pma  15
      common /psize/ maxm,ne                                            pma  16
      common /rdata/ engy,rnmax,tol,shift                               pma  17
      common /tbeta/ beta,gamm,theta,nop,nt                             pma  18
      common /tdata/ ttim,dt,c1,c2,c3,c4,c5,c6                          pma  19
      common /temfl1/ tfile(6)                                          pma  20
      common /udata/ uu(4000)                                           pma  21
      common dm(1),rm(1),im(1)                                          pma  22
c.... transfer to correct process                                       pma  23
      go to (1,2,3,4,5,6,7,8,9,10,11,12), j                             pma  24
c.... set solution tolerance                                            pma  25
1     tol = ct(1,l)                                                     pma  26
      return                                                            pma  27
c.... set time increment                                                pma  28
2     dt = ct(1,l)                                                      pma  29
      if(fl(9)) call setci(ior)                                         pma  30
      return                                                            pma  31
c.... set loop start indicators                                         pma  32
3     lv = lv + 1                                                       pma  33
      lvs(lv) = l                                                       pma  34
      lve(lv) = ct(2,l)                                                 pma  35
      ct(1,lve(lv)) = 1.                                                pma  36
      return                                                            pma  37
c.... loop terminator control                                           pma  38
4     n = ct(2,l)                                                       pma  39
      ct(1,l) = ct(1,l) + 1.0                                           pma  40
      if(ct(1,l).gt.ct(1,n)) lv = lv - 1                                pma  41
      if(ct(1,l).le.ct(1,n)) l = n                                      pma  42
      return                                                            pma  43
c.... input proportional load table                                     pma  44
5     npld = ct(1,l)                                                    pma  45
      npld = max(1,min(npld,10))                                        pma  46
      prop = propld (ttim,npld)                                         pma  47
      return                                                            pma  48
c.... data command                                                      pma  49
6     if(ior.lt.0) write(*,3000) lct(l)                                 pma  50
      call pintio(yyy,10)                                               pma  51
      read(yyy,1000,err=61) (ctl(i),i=1,2),xtl                          pma  52
      if(.not.pcomp(lct(l),ctl(1))) go to 402                           pma  53
      if(pcomp(ctl(1),'tol ')) tol = xtl                                pma  54
```

```
      if(pcomp(ctl(1),'dt  ')) dt  = xtl                              pma 55
      return                                                          pma 56
61    call perror('PMACR2',yyy)                                       pma 57
      go to 6                                                         pma 58
c.... increment time - initialize force / solution vectors            pma 59
7     ttim = ttim + dt                                                pma 60
      do 71 i = 0,nneq-1                                              pma 61
        rm(nt+i) = rm(nt+nneq+i)                                      pma 62
71    continue                                                        pma 63
      if(npld.gt.0) prop = propld(ttim,0)                             pma 64
      write(iow,2002) ttim,prop                                       pma 65
      if(ior.lt.0) write(  *,2002) ttim,prop                          pma 66
      engy = 0.0                                                      pma 67
      rnmax = 0.0                                                     pma 68
c.... update history on the disk                                      pma 69
      if(.not.hfl) then                                               pma 70
        hout = .true.                                                 pma 71
        call formfe(b,dr,.false.,.false.,.false.,.false.,6,1,numel,1) pma 72
        hout = .false.                                                pma 73
      endif                                                           pma 74
c.... update dynamic vectors for time step                             pma 75
      if(fl(9)) then                                                  pma 76
        call setci(ior)                                               pma 77
        call update(id,rm(nt),b,dm(nv),dr,nneq,fl(9),1)               pma 78
      endif                                                           pma 79
c.... zero displacement increment for next time step                   pma 80
      call pconsd(b(nneq+1),nneq+nneq,0.0d0)                          pma 81
      fl(10) = .true.                                                 pma 82
      return                                                          pma 83
c.... input integration parameters and initialize vectors              pma 84
8     call param(lct(l),ct(1,l))                                      pma 85
      if(fl(9)) return                                                pma 86
      call psetm(nv,nneq*4,ipd,fl(9))                                 pma 87
      call pconsd(dm(nv),nneq*4,0.0d0)                                pma 88
      nw = nv + nneq                                                  pma 89
c.... set initial condition for transient solution                      pma 90
      n1 = nw + nneq - 1                                              pma 91
      do 81 i = 1,nneq                                                pma 92
        dm(n1+i) = b(i)                                               pma 93
81    continue                                                        pma 94
      fl(9) = .true.                                                  pma 95
      return                                                          pma 96
c.... update the current force vector f0                                pma 97
9     call raxpb(f, f0, prop, nneq, f0)                               pma 98
      return                                                          pma 99
c.... subspace eigencomputations                                       pma100
10    mf = ct(1,l)                                                    pma101
      n2 = ct(2,l)                                                    pma102
      mf = min(neq,max(1,mf))                                         pma103
      mq = min(mf+mf,mf+8,neq)                                        pma104
      if(n2.gt.0) mq = min(mf+n2,neq)                                 pma105
      if(fl(2)) call numass(dm(nl),neq,mq)                            pma106
      if(mq.lt.mf) write(iow,2001) mq                                 pma107
      if(mq.lt.mf.and.ior.lt.0) write(  *,2001) mq                    pma108
      mf = min(mf,mq)                                                 pma109
      if(mf.le.0) return                                              pma110
      sfl = pcomp(lct(l),'prin')                                      pma111
c.... establish addresses for eigen solutions                           pma112
      md =  1                                                         pma113
      mv = md + mq                                                    pma114
```

```
      mg = mv + mq*neq                                                 pma115
      mh = mg + mq*(mq+1)/2                                            pma116
      mdp= mh + mq*(mq+1)/2                                            pma117
      mdt= mdp+ mq                                                     pma118
      mp = mdt+ mq                                                     pma119
      nlm= max(mp+mq*mq,neq*mq+neq)                                    pma120
      if(nlm.gt.maxm/ipd) then                                         pma121
        if(ior.lt.0) then                                              pma122
          write( *,3001) nlm,maxm/ipd                                  pma123
          return                                                       pma124
        else                                                           pma125
          write(iow,3001) nlm,maxm/ipd                                 pma126
          stop                                                         pma127
        endif                                                          pma128
      endif                                                            pma129
      nlm = ne/ipd + 1                                                 pma130
      open(7,file=tfile(3),form='unformatted',status='new')            pma131
      write(7) (dm(i),i=1,nlm)                                         pma132
      close (7)                                                        pma133
      nlm = nl - 1                                                     pma134
      n1  = neq*ipd - ipd - ipr                                        pma135
      do 101 i = 1,neq                                                 pma136
        dm(i)     = dm(i+nlm)                                          pma137
        im(n1+i) = jd(i)                                               pma138
101   continue                                                         pma139
      nlm = neq + (neq + ipd -1)/ipd + 1                               pma140
      call subsp(dm(1),uu(mv),uu(mg),uu(mh),uu(md),uu(mdp),            pma141
     1 uu(mdt),uu(mp),dm(nlm),mf,mq,neq,shift,tol,sfl,25)              pma142
c.... restore the solution to continue macro executions                 pma143
      nlm = ne/ipd + 1                                                 pma144
      open(7,file=tfile(3),form='unformatted',status='old')            pma145
      rewind 7                                                         pma146
      read(7) (dm(i),i=1,nlm)                                          pma147
      close(7,status='delete')                                         pma148
      return                                                           pma149
c.... print eigenvectors                                                pma150
11    if(mf.gt.0) then                                                 pma151
        n1    = ct(1,l)                                                pma152
        n1    = max(1,min(mq,n1))                                      pma153
        n2    = mv + neq*(n1-1) -1                                     pma154
c.... expand and move eigenvector for prints                            pma155
        call pconsd(dm,nneq,0.0d0)                                     pma156
        do 111 i = 1,nneq                                              pma157
          n = id(i)                                                    pma158
          if(n.ne.0) dm(i) = uu(n2+n)                                  pma159
111     continue                                                       pma160
        call prtdis(x,dm,ttim,uu(md+n1-1),ndm,ndf,1,numnp,1)           pma161
      else                                                             pma162
        write(iow,3002)                                                pma163
        if(ior.lt.0) write(*,3002)                                     pma164
      endif                                                            pma165
      return                                                           pma166
c.... set identity matrix for mass                                      pma167
12    if(fl(5)) call psetm(nl,neq,ipd,fl(5))                           pma168
      fl(2) = .true.                                                   pma169
      call pconsd(dm(nl),neq,1.d0)                                     pma170
      return                                                           pma171
c.... error diagnostics                                                 pma172
402   if(ior.gt.0) write(iow,3003)                                     pma173
      if(ior.gt.0)  stop                                               pma174
```

```
      write(*,3003)                                                    pma175
      go to 6                                                          pma176
c.... formats                                                          pma177
1000  format(a4,6x,a4,6x,f15.0)                                        pma178
2001  format(' **WARNING** Subspace set to',i4,' number of mass terms.')pma179
2002  format('   Computing solution for time',1pe12.5/                  pma180
     1       '   Proportional load value is ',1pe12.5)                 pma181
3000  format(' Input ',a4,' macro >',$)                                pma182
3001  format(' **ERROR** Subspace memory too large'/                   pma183
     1 5x,'Need =',i7,' : Available =',i7)                             pma184
3002  format(' **ERROR** Need eigensolution.')                         pma185
3003  format(' **ERROR** Macro label mismatch on data command.')       pma186
      end                                                              pma187
c
      subroutine inaccl(id,dr,xm,a,nneq)                               ina   1
      integer*2 id(nneq)                                               ina   2
      real*8  dr(*),xm(*),a(nneq)                                      ina   3
      do 100 n = 1,nneq                                                ina   4
        j = id(n)                                                      ina   5
        if(j.gt.0) a(n) = dr(j)/xm(j)                                  ina   6
100   continue                                                         ina   7
      end                                                              ina   8
c
      subroutine serchl(fn,id,g0,rsd,u,d,stol,t,nneq)                  ser   1
c                                                                      ser   2
c.... linear line search for nonlinear problems                        ser   3
c                                                                      ser   4
      integer*2 id(*)                                                  ser   5
      real      fn(nneq,2)                                             ser   6
      real*8    rsd(*),u(*),d(*),t(*),gamma,g,g0,ga,gb,sa,sb           ser   7
      common /cdata/  numnp,numel,nummat,nen,neq                       ser   8
      common /iofile/ ior,iow                                          ser   9
c.... compute step size for line search in direction d                 ser  10
      linmax = 10                                                      ser  11
      sb     = 0.0                                                     ser  12
      sa     = 1.0                                                     ser  13
      s      = 1.0                                                     ser  14
      g      = gamma(fn,id,u,rsd,d,t,s,neq,nneq)                       ser  15
c.... find bracket on zero                                             ser  16
      if(g*g0.gt.0.0d0) then                                           ser  17
        write(iow,3000)                                                ser  18
        if(ior.lt.0) write(*,3000)                                     ser  19
      else                                                             ser  20
        j    = 0                                                       ser  21
        gb   = g0                                                      ser  22
        ga   = g                                                       ser  23
        sb   = 0.0d0                                                   ser  24
        sa   = 1.0d0                                                   ser  25
10      j    = j + 1                                                   ser  26
        step = sa - ga*(sa-sb)/(ga-gb)                                 ser  27
        g    = gamma(fn,id,u,rsd,d,t,step,neq,nneq)                    ser  28
        gb   = 0.5d0*gb                                                ser  29
        if (g*ga.lt.0.0d0) then                                        ser  30
          sb = sa                                                      ser  31
          gb = ga                                                      ser  32
        endif                                                          ser  33
        sa = step                                                      ser  34
        ga = g                                                         ser  35
        write(iow,3001) j,step,g                                       ser  36
        if(ior.lt.0) write(  *,3001) j,step,g                          ser  37
        if (j.lt.linmax) then                                          ser  38
```

```
          if(abs(g)     .gt. stol*abs(g0)      ) go to 10                   ser 39
          if(abs(sb-sa) .gt. stol*0.5d0*(sa+sb)) go to 10                   ser 40
        endif                                                               ser 41
        do 20 j = 1, neq                                                    ser 42
          d(j) = step*d(j)                                                  ser 43
20      continue                                                            ser 44
      endif                                                                 ser 45
3000  format(4x,'No line search, end points both positive.')                ser 46
3001  format(4x,'Iter =',i2,' Step Size =',e12.5,' Energy =',e12.5)         ser 47
      end                                                                   ser 48
c
      double precision function gamma(fn,id,u,dr,du,t,s,nqe,nneq)           gam  1
      logical fa,tr,fl                                                      gam  2
      integer*2 id(*),m                                                     gam  3
      real   fn(nneq,2)                                                     gam  4
      real*8 u(*),dr(*),du(*),t(*),dm,db,dot,prop                           gam  5
      common /cdata/ numnp,numel,nummat,nen,neq                             gam  6
      common /fdata/ fl(11)                                                 gam  7
      common /mdata/ nn,n0,n1,n2,n3,n4,n5,n6,n7,n8,n9,n10,n11,n12,n13       gam  8
      common /ndata/ nv,nw,nl                                               gam  9
      common /prlod/ prop,a(6,10),iexp(10),ik(10),npld                      gam 10
      common /tbeta/ beta,gamm,theta,nop,nt                                 gam 11
      common /tdata/ ttim,dt,c1,c2,c3,c4,c5,c6                              gam 12
      common dm(1),rm(1),m(1)                                               gam 13
      data fa,tr/.false.,.true./                                            gam 14
c.... get a search displacement                                             gam 15
      nneq2 = nneq + nneq                                                   gam 16
      ctem  = theta                                                         gam 17
      theta = s*theta                                                       gam 18
      do 100 n = 1,nneq                                                     gam 19
        j = id(n)                                                           gam 20
        if(j.gt.0) then                                                     gam 21
          db          = s*du(j)                                             gam 22
          t(n)        = u(n)        + db                                    gam 23
          t(n+nneq )  = u(n+nneq)  + db                                     gam 24
          t(n+nneq2)  = db                                                  gam 25
        else                                                                gam 26
          db = theta*fn(n,2) + (1.0-theta)*fn(n,1)                          gam 27
          t(n+nneq2)  = db - u(n)                                           gam 28
          t(n+nneq )  = u(n+nneq)  - u(n) + db                              gam 29
          t(n)        = db                                                  gam 30
        endif                                                               gam 31
100   continue                                                              gam 32
c.... compute a residual                                                    gam 33
      call pload(id,fn,dr,nneq,dm(nl),dm(nw))                               gam 34
      call formfe(t,dr,fa,tr,fa,fa,6,1,numel,1)                             gam 35
      theta = ctem                                                          gam 36
c.... update the residual for lumped mass inertial effects                  gam 37
      if(fl(9)) then                                                        gam 38
        ctem = s*c1                                                         gam 39
        do 110 n = 1,nneq                                                   gam 40
          j = id(n)                                                         gam 41
          if(j.gt.0) then                                                   gam 42
            dr(j) = dr(j) - dm(nl+j-1)*(dm(nw+n-1) + ctem*du(j))            gam 43
          endif                                                             gam 44
110     continue                                                            gam 45
      endif                                                                 gam 46
c.... compute the value of gamma                                            gam 47
      gamma = dot (du,dr,nqe)                                               gam 48
      end                                                                   gam 49
```

(*b*) File PCMAC2.FOR contains the following subprograms:

Name	*Type*	*Description*	*Chapter*
ADDVEC	SUBROUTINE	Accumulate a vector	15
JUST	SUBROUTINE	Justify data for parser	15
MODIFY	SUBROUTINE	Modify vector for specified boundary conditions	15
PANGL	SUBROUTINE	Set angles for sloping DOF	15
PFORM	SUBROUTINE	Form FE arrays	16
PLOAD	SUBROUTINE	Set up load/inertia vector	16
PROPLD	FUNCTION	Input or set proportional load value	15
PRTDIS	SUBROUTINE	Output nodal displacements	15
PRTERR	SUBROUTINE	Output error estimates	15
PRTREA	SUBROUTINE	Output nodal reactions and sums	15
PRTSTR	SUBROUTINE	Output nodal stresses	15
PTRANS	SUBROUTINE	Transform arrays for sloping boundary conditions	15
RAXPB	SUBROUTINE	Scalar times real vector plus real vector	16
PARAM	SUBROUTINE	Input parameters for transient algorithms	16
SETCI	SUBROUTINE	Set transient integration parameters	16
UPDATE	SUBROUTINE	Update solution vectors	16

```
      subroutine pform(ul,xl,tl,ld,p,s,ie,d,id,x,ix,f,t,idl,             pfo  1
     1                 f0,u,b,ndf,ndm,nen1,nst)                           pfo  2
c.... compute element arrays and assemble global arrays                  pfo  3
      logical   afl,bfl,cfl,dfl,efl,hfl,hout,fl,pfr                      pfo  4
      character*12 tfile                                                  pfo  5
      integer*2 ld(ndf,1),ie(9,1),id(ndf,1),ix(nen1,1),idl(1),ia,im      pfo  6
      integer*4 itrec                                                     pfo  7
      real      xl(ndm,1),x(ndm,1),f(ndf,1),f0(ndf,1),tl(1),t(1)        pfo  8
      real*8    d(18,1),p(1),s(nst,1),b(1),ul(ndf,1),u(ndf,1),dm,       pfo  9
     1          dun,un,prop                                               pfo 10
      common /cdata/ numnp,numel,nummat,nen,neq                           pfo 11
      common /eldata/ dq,n,ma,mct,iel,nel                                 pfo 12
      common /fdata/ fl(11),pfr                                           pfo 13
      common /mdat2/ n11a,n11b,n11c,ia(2,11)                              pfo 14
      common /ndata/ nv,nw,nl                                             pfo 15
      common /hdata/ nh1,nh2                                              pfo 16
      common /hdatb/ nhi,nhf,ihbuff,irec,jrec,nrec,hfl,hout               pfo 17
      common /prlod/ prop,ap(6,10),iexp(10),ik(10),npld                   pfo 18
      common /tbeta/ beta,gamm,theta,nop,nt                               pfo 19
      common /temfl1/ tfile(6)                                            pfo 20
      common /temfl2/ itrec(4),nw1,nw2                                    pfo 21
      common /xdata/ isw,nn1,nn2,nn3,afl,bfl,cfl,dfl                      pfo 22
      common dm(1),rm(1),im(1)                                            pfo 23
```

```
c.... set up local arrays before calling element library           pfo 24
      iel = 0                                                        pfo 25
      tm  = 1. - theta                                               pfo 26
      efl = .false.                                                  pfo 27
      if(.not.dfl.and.isw.eq.6) efl = .true.                         pfo 28
      if(bfl.and.isw.eq.3)      efl = .true.                         pfo 29
      if(isw.ne.3.or.nn1.eq.1)  irec = 0                             pfo 30
      ne2 = nen + nen                                                pfo 31
      ne3 = nen + ne2                                                pfo 32
      ne4 = nen + ne3                                                pfo 33
      nt0 = nt  - 1                                                  pfo 34
      nt1 = nt0 + ndf*numnp                                          pfo 35
      numnp2 = numnp + numnp                                         pfo 36
      do 110 nu = 1,numel                                            pfo 37
       n = idl(nu)                                                   pfo 38
       if( (n.ge.nn1 .and. n.le.nn2) .and. (mod(n-nn1,nn3).eq.0) ) then pfo 39
c.... set up history terms                                           pfo 40
         ma  = ix(nen1,n)                                            pfo 41
         nh1 = ix(nen+1,n)                                           pfo 42
         nh2 = nh1                                                   pfo 43
         if(.not.hfl) then                                           pfo 44
           jrec= ix(nen+2,n)                                         pfo 45
           if(jrec.ne.irec) then                                     pfo 46
             if(hout .and. irec.ne.0) then                           pfo 47
               call phstio(3,irec,dm(nhi),nhf-nhi+1,2,tfile(2),itrec(2))pfo 48
             endif                                                   pfo 49
             call phstio(3,jrec,dm(nhi),nhf-nhi+1,1,tfile(2),itrec(2))  pfo 50
             irec = jrec                                             pfo 51
           endif                                                     pfo 52
         endif                                                       pfo 53
         call pconsd(ul,5*nen*ndf,0.0d0)                             pfo 54
         call pconsr(xl,nen*ndm,0.0)                                 pfo 55
         call pconsr(tl,nen,0.0)                                     pfo 56
         call pconsr(rm(n11a),nen,0.0)                               pfo 57
         call pconsi(ld,nst,0)                                       pfo 58
         un = 0.0                                                    pfo 59
         dun= 0.0                                                    pfo 60
         call pangl(ix(1,n),nen,rm(n11a),rm(n11b),nrot)              pfo 61
         do 108 i = 1,nen                                            pfo 62
           ixi= ix(i,n)                                              pfo 63
           ii = abs(ixi)                                             pfo 64
           if(ii.gt.0) then                                          pfo 65
             iid = ii*ndf - ndf                                      pfo 66
             nel = i                                                 pfo 67
             tl(i) = t(ii)                                           pfo 68
             do 106 j = 1,ndm                                        pfo 69
               xl(j,i) = x(j,ii)                                     pfo 70
106          continue                                                pfo 71
             do 107 j = 1,ndf                                        pfo 72
               jj = ie(j,ma)                                         pfo 73
               if(jj.le.0) go to 107                                 pfo 74
               ijj     = iid + jj                                    pfo 75
               k       = id(jj,ii)                                   pfo 76
               ul(j,i) = u(jj,ii)                                    pfo 77
               ul(j,i+nen) = u(jj,ii+numnp)                          pfo 78
               ul(j,i+ne2) = u(jj,ii+numnp2)                         pfo 79
               if(fl(9)) ul(j,i+ne3) = dm(nv+ijj)                    pfo 80
               if(k.le.0) then                                       pfo 81
                 ul(j,i+ne4) = theta*rm(nt1+ijj)+tm*rm(nt0+ijj)-u(jj,ii)pfo 82
                 dun = max(dun,abs(ul(j,i+ne4)))                     pfo 83
```

```
              endif                                                        pfo 84
              un = max(un,abs(ul(j,i)))                                    pfo 85
              if(dfl) then                                                 pfo 86
                ld(j,i) = ijj                                              pfo 87
              else                                                         pfo 88
                ld(j,i) = sign(k,ixi)                                      pfo 89
              endif                                                        pfo 90
107         continue                                                       pfo 91
          endif                                                            pfo 92
108     continue                                                           pfo 93
c.... form element array                                                   pfo 94
        dq  = prop                                                         pfo 95
        if(ie(7,ma).ne.iel) mct = 0                                        pfo 96
        iel = ie(7,ma)                                                     pfo 97
        isx = isw                                                          pfo 98
        if(efl .and. dun.gt.0.0000001d0*un .and. .not.afl) isx = 3         pfo 99
        if(nrot.gt.0)                                                      pfo100
     1  call ptrans(ia(1,iel),rm(n11a),ul,p,s,nel,nen,ndf,nst,1)           pfo101
        call elmlib(d(1,ma),ul,xl,ix(1,n),tl,s,p,ndf,ndm,nst,iel,isx)     pfo102
        if(nrot.gt.0)                                                      pfo103
     1  call ptrans(ia(1,iel),rm(n11a),ul,p,s,nel,nen,ndf,nst,2)           pfo104
c.... modify for non-zero displacement boundary conditions                  pfo105
        if(efl) call modify(p,s,ul(1,ne4+1),nst)                           pfo106
c.... assemble a vector if needed                                          pfo107
        if(bfl) then                                                       pfo108
          do 109 i = 1,nst                                                 pfo109
            j = abs(ld(i,1))                                               pfo110
            if(j.ne.0) b(j) = b(j) + p(i)                                  pfo111
109       continue                                                         pfo112
        endif                                                              pfo113
      endif                                                                pfo114
110   continue                                                             pfo115
c.... put the last history state on the disk                                pfo116
      if(hout) call phstio(3,jrec,dm(nhi),nhf-nhi+1,2,tfile(2),itrec(2))pfo117
      end                                                                  pfo118
c
      subroutine pload(id,fn,b,nn,xm,ac)                                   plo  1
c.... form load vector in compact form                                     plo  2
      logical fl,pfr                                                       plo  3
      integer*2 id(*)                                                      plo  4
      real fn(nn,2)                                                        plo  5
      real*8 b(*),xm(*),ac(*)                                              plo  6
      common /fdata/ fl(11),pfr                                            plo  7
      common /tbeta/ beta,gamm,theta,nop,nt                                plo  8
      fl(11) = .false.                                                     plo  9
      call pconsd(b,nn,0.0d0)                                              plo 10
      do 100 n = 1,nn                                                      plo 11
        j = id(n)                                                          plo 12
        if(j.gt.0) then                                                    plo 13
          b(j) = theta*fn(n,2) + (1. - theta)*fn(n,1) + b(j)               plo 14
          if(fl(9)) b(j) =  b(j) - xm(j)*ac(n)                             plo 15
        endif                                                              plo 16
100   continue                                                             plo 17
      end                                                                  plo 18

      subroutine raxpb (a,b,x,n,c)                                         rax  1
      real   a(1),b(1),c(1)                                                rax  2
      real*8 x                                                             rax  3
c... vector times scalar added to second vector                            rax  4
      do 10 k=1,n                                                          rax  5
```

```
        c(k) = a(k)*x +b(k)                                              rax  6
   10 continue                                                           rax  7
      end                                                                rax  8
c
      subroutine param(ctl,ct)                                           par  1
c.... set appropriate time integration parameters                        par  2
      logical      pcomp                                                 par  3
      character*4 ctl                                                    par  4
      real         ct(3)                                                 par  5
      common /iofile/ ior,iow                                            par  6
      common /tbeta/ beta,gamm,theta,nop,nt                              par  7
c.... set the integration parameters                                     par  8
      beta = ct(1)                                                       par  9
      gamm = ct(2)                                                       par 10
c.... ss11 algorithm                                                     par 11
      if(pcomp(ctl,'ss11')) then                                         par 12
        nop = 1                                                          par 13
        if(beta.le.0.0) beta = 0.50                                      par 14
        write(iow,2001) beta                                             par 15
        if(ior.lt.0) write(*,2001) beta                                  par 16
        theta = beta                                                     par 17
c.... ss22 algorithm                                                     par 18
      elseif(pcomp(ctl,'ss22')) then                                     par 19
        nop = 2                                                          par 20
        if(beta.le.0.0) beta = 0.50                                      par 21
        if(gamm.le.0.0) gamm = 0.50                                      par 22
        write(iow,2002) beta,gamm                                        par 23
        theta = gamm                                                     par 24
        if(ior.lt.0) write(*,2002) beta,gamm                             par 25
c.... newmark algorithm                                                  par 26
      else                                                               par 27
        nop = 3                                                          par 28
        if(beta.le.0.0) beta = 0.25                                      par 29
        if(gamm.le.0.0) gamm = 0.50                                      par 30
        write(iow,2003) beta,gamm                                        par 31
        if(ior.lt.0) write(*,2003) beta,gamm                             par 32
        theta = beta                                                     par 33
      endif                                                              par 34
2001  format(' SS-11 Method Parameter: theta = ',f9.4)                   par 35
2002  format(' SS-22 Method Parameters'/                                 par 36
     1        ' theta-1 = ',f9.4,' ;  theta-2 = ',f9.4)                  par 37
2003  format(' Newmark Method Parameters'/                               par 38
     1        ' beta = ',f9.4,' ;  gamma = ',f9.4)                       par 39
      end                                                                par 40
c
      subroutine setci(ior)                                              set  1
c.... compute integration constants 'c1' to 'c5' for current 'dt'        set  2
      common /tbeta/ beta,gamm,theta,nop,nt                              set  3
      common /tdata/ ttim,dt,c1,c2,c3,c4,c5,c6                           set  4
      if(dt.le.0.0) then                                                 set  5
        write(*,2000)                                                    set  6
        if(ior.gt.0) stop                                                set  7
        return                                                           set  8
      endif                                                              set  9
c.... compute integration constants 'c1' to 'c5' for current 'dt'        set 10
      if(nop.eq.1) then                                                  set 11
        c1 = 1.d0/(beta*dt)                                              set 12
        c2 = c1                                                          set 13
        c3 = 1.d0/beta                                                   set 14
        c6 = beta                                                        set 15
```

```
      elseif(nop.eq.2) then                                              set 16
        c5 = 2.d0/(gamm*dt)                                              set 17
        c1 = c5/dt                                                       set 18
        c2 = c5*beta                                                     set 19
        c3 = dt*beta                                                     set 20
        c4 = 1.d0/gamm                                                   set 21
        c6 = gamm                                                        set 22
      else                                                               set 23
        c1 = 1.d0/(beta*dt*dt)                                           set 24
        c2 = gamm/(dt*beta)                                              set 25
        c3 = 1.d0 - 1.d0/(beta+beta)                                     set 26
        c4 = 1.d0 - gamm/beta                                            set 27
        c5 = (1.d0 - gamm/(beta+beta))*dt                                set 28
        c6 = dt*c1                                                       set 29
      endif                                                              set 30
2000  format(' **ERROR** Input DT as nonzero number.')                   set 31
      end                                                                set 32
c
      subroutine update(id,fn,u,v,du,nneq,fdyn,isw)                      upd  1
c.... update the displacements (and velocities and accelerations)        upd  2
      logical fdyn                                                       upd  3
      integer*2 id(*)                                                    upd  4
      real      fn(nneq,2)                                               upd  5
      real*8    u(nneq,*),v(nneq,*),du(*),ur1,ur2,dot                    upd  6
      common /iofile/ ior,iow                                            upd  7
      common /tbeta/ beta,gamm,theta,nop,nt                              upd  8
      common /tdata/ ttim,dt,c1,c2,c3,c4,c5,c6                           upd  9
c.... update solution vectors to begin a step                            upd 10
      if(isw.eq.1) then                                                  upd 11
        ur1 = sqrt(dot(v(1,1),v(1,1),nneq))                              upd 12
        ur2 = sqrt(dot(v(1,2),v(1,2),nneq))                              upd 13
        write(iow,2000) ur1,ur2                                          upd 14
        if(ior.lt.0) write(*,2000) ur1,ur2                               upd 15
        do 100 n = 1,nneq                                                upd 16
c....    ss11 algorithm u=u-bar; v1=v-bar; v2=a=abar; v3=u;              upd 17
          if(nop.eq.1) then                                              upd 18
            u(n,1) = v(n,3)                                              upd 19
            v(n,1) = 0.0d0                                               upd 20
            v(n,2) = 0.0d0                                               upd 21
c....    ss22 algorithm u=u-bar; v1=v-bar; v2=a=abar; v3=u; v4=v         upd 22
          elseif(nop.eq.2) then                                          upd 23
            u(n,1) = v(n,3) + c3*v(n,4)                                  upd 24
            v(n,1) = v(n,4)                                              upd 25
            v(n,2) = 0.0d0                                               upd 26
            v(n,3) = v(n,3) + dt*v(n,4)                                  upd 27
c....    newmark algorithm                                               upd 28
          else                                                           upd 29
            ur2    = - c6*v(n,1) + c3*v(n,2)                             upd 30
            v(n,1) =   c4*v(n,1) + c5*v(n,2)                             upd 31
            v(n,2) =   ur2                                               upd 32
          endif                                                          upd 33
100     continue                                                         upd 34
      else                                                               upd 35
c.... update displacement and its increments within the time step        upd 36
        ur2 = 1. - theta                                                 upd 37
        do 200 n = 1,nneq                                                upd 38
          j = id(n)                                                      upd 39
          if (j.gt.0) then                                               upd 40
c.... for the active degrees-of-freedom compute values from solution     upd 41
            u(n,1) = du(j) + u(n,1)                                      upd 42
```

```
          u(n,2) = du(j) + u(n,2)                                    upd 43
          u(n,3) = du(j)                                             upd 44
        else                                                         upd 45
c.... for the fixed degrees-of-freedom compute values from forced inputsupd 46
          ur1    = theta*fn(n,2) + ur2*fn(n,1)                       upd 47
          u(n,3) = ur1    - u(n,1)                                   upd 48
          u(n,2) = u(n,2) + u(n,3)                                   upd 49
          u(n,1) = ur1                                               upd 50
        endif                                                        upd 51
200     continue                                                     upd 52
c.... for time dependent solutions update the rate terms              upd 53
       if(fdyn) then                                                 upd 54
         do 210 n = 1,nneq                                           upd 55
           v(n,1) = v(n,1) + c2*u(n,3)                               upd 56
           v(n,2) = v(n,2) + c1*u(n,3)                               upd 57
c....  ss11 algorithm                                                 upd 58
           if(nop.eq.1) then                                         upd 59
             v(n,3) = v(n,3) + c3*u(n,3)                             upd 60
c....  ss22 algorithm                                                 upd 61
           elseif(nop.eq.2) then                                     upd 62
             v(n,3) = v(n,3) + c4*u(n,3)                             upd 63
             v(n,4) = v(n,4) + c5*u(n,3)                             upd 64
           endif                                                     upd 65
210      continue                                                    upd 66
       endif                                                         upd 67
      endif                                                          upd 68
2000  format('   N o r m s   f o r   D y n a m i c s'/               upd 69
     1   10x,'Velocity:',e13.5,' Acceleration:',e13.5)               upd 70
      end                                                            upd 71
```

(*c*) File PCMAC3.FOR contains the following subprograms:

Name	*Type*	*Description*	*Chapter*
ACHECK	SUBROUTINE	Parser of alphanumerical input	15
CKISOP	SUBROUTINE	Check isoparametric element errors	15
DOT	FUNCTION	Vector dot product	15
ELMLIB	SUBROUTINE	Element library	15
PCONSD	SUBROUTINE	Set REAL*8 array to constant	15
PSCONSI	SUBROUTINE	Set INTEGER array to constant	15
PCONSR	SUBROUTINE	Set REAL array to constant	15
PDISK	SUBROUTINE	Add disk character to filename	15
PEND	SUBROUTINE	Output error on end-of-file read	15
PERROR	SUBROUTINE	Output on input data error	15
PGAUSS	SUBROUTINE	Two-dimensional Gauss points/weights	15
PHELP	SUBROUTINE	Help information	15
PSETM	SUBROUTINE	Array pointer assignments	15
PSTRES	SUBROUTINE	Two-dimensional principal stresses	15
SAXPB	SUBROUTINE	Scalar times vector plus vector	15
SHAP2	SUBROUTINE	Two-dimensional biquadratic shape functions	15
SHAPE	SUBROUTINE	Two-dimensional bilinear shape functions	15

(*d*) File PCEIGN.FOR, which is completely new for this volume, contains the following subprograms:

Name	*Type*	*Description*	*Chapter*
SUBSP	SUBROUTINE	Subspace main algorithm	16
CHLBAC	SUBROUTINE	Choleski backsubstitution	16
CHLFWD	SUBROUTINE	Choleski forward solution	16
CHLFAC	SUBROUTINE	Choleski decomposition	16
COLBAC	SUBROUTINE	Backsubstitution routine	16
EISQL	SUBROUTINE	Solve standard eigenproblem	16
GEIG	SUBROUTINE	Solve general eigenproblem	16
NUMASS	SUBROUTINE	Count number of masses	16
SCALEV	SUBROUTINE	Scale vector to unit length	16
SPROJA	SUBROUTINE	Project **K** matrix	16
SPROJB	SUBROUTINE	Project **M** matrix	16
WPROJM	SUBROUTINE	Output projected matrices	16

```
      subroutine subsp(b,v,g,h,d,dp,dtol,p,z,nf,nev,neq,                 sub  1
     1                  shift,tol,prt,its)                               sub  2
c                                                                        sub  3
c.... subspace iteration to extract the lowest nf eigenpairs             sub  4
c                                                                        sub  5
      real*8 b(1),v(neq,1),g(1),h(1),d(1),dp(1),                         sub  6
     1       dtol(1),p(nev,1),z(neq,1),dm,tolmx,told                     sub  7
      logical conv,prt                                                   sub  8
      common /iofile/ ior,iow                                            sub  9
      data ipd/4/                                                        sub 10
c                                                                        sub 11
c.... compute the initial iteration vectors                              sub 12
c                                                                        sub 13
      call pconsd(v,nev*neq,0.0d0)                                       sub 14
      nmas = 0                                                           sub 15
      do 100 n = 1,neq                                                   sub 16
c.... count the number of nonzero masses                                 sub 17
        if(b(n).ne.0.0d0) nmas = nmas + 1                                sub 18
100   continue                                                           sub 19
      nmas = nmas/nev                                                    sub 20
      i = 0                                                              sub 21
      j = 1                                                              sub 22
      do 110 n = 1,neq                                                   sub 23
        dm = b(n)                                                        sub 24
        if(dm.ne.0.0d0) then                                             sub 25
          v(n,j) = dm                                                    sub 26
          i = i + 1                                                      sub 27
          if(mod(i,nmas).eq.0) j = j + 1                                 sub 28
          j = min(j,nev)                                                 sub 29
        endif                                                            sub 30
110   continue                                                           sub 31
      do 120 i = 1,nev                                                   sub 32
        dp(i)   = 0.0d0                                                  sub 33
        dtol(i) = 1.0d0                                                  sub 34
        call scalev(v(1,i),neq)                                          sub 35
120   continue                                                           sub 36
c                                                                        sub 37
```

```
c.... compute the new vectors and project 'a' onto 'g'                  sub 38
      told  = tol                                                       sub 39
      conv  = .false.                                                   sub 40
      itlim = its                                                       sub 41
      if(nev.eq.nf) itlim = 1                                           sub 42
      do 300 it = 1,itlim                                               sub 43
        itt = it                                                        sub 44
c                                                                       sub 45
c.... project the 'b' matrix to form 'h' and compute 'z' vectors        sub 46
        call sprojb(b,v,z,h,neq,nev)                                    sub 47
c                                                                       sub 48
c.... project the 'a' matrix to form 'g'                                sub 49
        call sproja(v,z,g,neq,nev,ipd)                                  sub 50
c                                                                       sub 51
c.... solve the reduced eigenproblem 'g*p = h*p*d'                      sub 52
        call geig(g,h,d,p,v,nev,prt)                                    sub 53
c                                                                       sub 54
c.... check for convergence                                             sub 55
        tolmx = 0.0d0                                                   sub 56
        do 200 n = 1,nev                                                sub 57
          if(d(n).ne.0.0d0) dtol(n) = abs((d(n)-dp(n))/d(n))            sub 58
          dp(n) = d(n)                                                  sub 59
          if(n.le.nf) tolmx = max(tolmx,dtol(n))                        sub 60
200     continue                                                        sub 61
        if(prt) then                                                    sub 62
          write(iow,2000) it,(d(n),n=1,nev)                             sub 63
          if(ior.lt.0) write(*,2000) it,(d(n),n=1,nev)                  sub 64
          if(itlim.gt.1) write(iow,2003) (dtol(n),n=1,nev)              sub 65
          if(ior.lt.0.and.itlim.gt.1)                                   sub 66
     1      write(*,2003) (dtol(n),n=1,nev)                             sub 67
        else                                                            sub 68
          if(ior.lt.0) write(*,2004) it,tolmx                           sub 69
        endif                                                           sub 70
c..... tolerance check                                                  sub 71
        do 210 n = 1,nf                                                 sub 72
          if(dtol(n).gt.told) go to 220                                 sub 73
210     continue                                                        sub 74
        conv = .true.                                                   sub 75
c                                                                       sub 76
c... divide eigenvectors by eigenvalue to prevent overflows             sub 77
220     do 235 i = 1,nev                                                sub 78
          dm = d(i)                                                     sub 79
          if(p(i,i).lt.-0.00001d0) dm = -dm                             sub 80
          do 230 j = 1,nev                                              sub 81
            p(j,i) = p(j,i)/dm                                          sub 82
230       continue                                                      sub 83
235     continue                                                        sub 84
c                                                                       sub 85
c.... compute the new iteration vector 'u' from 'z'                     sub 86
        do 250 i = 1,neq                                                sub 87
        do 250 j = 1,nev                                                sub 88
          v(i,j) = 0.0d0                                                sub 89
          do 240 k = 1,nev                                              sub 90
            v(i,j) = v(i,j) + z(i,k)*p(k,j)                             sub 91
240       continue                                                      sub 92
250     continue                                                        sub 93
        if(conv) go to 305                                              sub 94
300   continue                                                          sub 95
c                                                                       sub 96
c.... scale the vectors to have maximum element of 1.0                  sub 97
```

```
305   do 310 n = 1,nev                                                     sub 98
        d(n)  = 1.0/d(n) + shift                                           sub 99
        dp(n) = sqrt(abs(d(n)))                                            sub100
        call scalev(v(1,n),neq)                                            sub101
310   continue                                                             sub102
      write(iow,2001) itt,(d(n),n=1,nev)                                   sub103
      write(iow,2002)     (dp(n),n=1,nev)                                  sub104
      if(itt.gt.1) write(iow,2003) (dtol(n),n=1,nev)                       sub105
      if(ior.lt.0) then                                                    sub106
        write(*,2001) itt,(d(n),n=1,nev)                                   sub107
        write(*,2002)     (dp(n),n=1,nev)                                  sub108
        if(itt.gt.1) write(*,2003) (dtol(n),n=1,nev)                       sub109
      endif                                                                sub110
2000  format(/5x,'Current reciprocal shifted eigenvalues, iteration',      sub111
     1       i4/(4d20.8))                                                  sub112
2001  format(/5x,'Solution  for  eigenvalues, iteration',i4/(4d20.8))      sub113
2002  format( 5x,'Square root of eigenvalues'/(4d20.8))                    sub114
2003  format( 5x,'Current residuals'/(4d20.8))                             sub115
2004  format( 5x,'End of iteration',i3,' Max tol =',1p1e11.4)              sub116
      end                                                                  sub117
c
      subroutine chlbac(u,s,nn)                                            chl  1
      real*8 u(1),s(nn,nn)                                                 chl  2
c                                                                          chl  3
c.... compute eigenvalues of general linear problem by backsubstitution   chl  4
      j  = nn                                                              chl  5
      jd = nn*(nn+1)/2                                                     chl  6
      do 100 i = 1,nn                                                      chl  7
        s(nn,i) = s(nn,i)/u(jd)                                            chl  8
100   continue                                                             chl  9
200   jd = jd - j                                                          chl 10
      j  = j - 1                                                           chl 11
      if(j.le.0) return                                                    chl 12
      do 300 i = 1,nn                                                      chl 13
        call colbac(u(jd+1),s(1,i),u(jd),j)                                chl 14
300   continue                                                             chl 15
      go to 200                                                            chl 16
      end                                                                  chl 17
c
      subroutine chlfwd(u,g,s,nn)                                          chl  1
c                                                                          chl  2
c.... use the choleski factors to project onto a standard eigenproblem     chl  3
      real*8 u(1),g(1),s(nn,nn),dot                                        chl  4
      s(1,1) = g(1)/u(1)                                                   chl  5
      if(nn.eq.1) go to 300                                                chl  6
      id = 1                                                               chl  7
      do 200 i = 2,nn                                                      chl  8
        s(1,i) = g(id+1)/u(1)                                              chl  9
        im     = i - 1                                                     chl 10
        jd     = 0                                                         chl 11
        do 100 j = 1,im                                                    chl 12
          s(i,j)              = (g(id+j) -dot(u(id+1),s(1,j),im))/u(id+i) chl 13
          if(j.gt.1) s(j,i)   = (g(id+j) -dot(u(jd+1),s(1,i),j-1))/u(jd+j)chl 14
          jd                  = jd + j                                     chl 15
100     continue                                                           chl 16
        id     = id + i                                                    chl 17
        s(i,i) = (g(id) - dot(u(id-im),s(1,i),im))/u(id)                   chl 18
200   continue                                                             chl 19
c                                                                          chl 20
c.... complete projection                                                  chl 21
```

```
300   g(1) = s(1,1)/u(1)                                                    chl 22
      if(nn.eq.1) return                                                    chl 23
      jd = 2                                                                chl 24
      do 500 j = 2,nn                                                       chl 25
        g(jd) = s(j,1)/u(1)                                                 chl 26
        id    = 2                                                           chl 27
        do 400 i = 2,j                                                      chl 28
          im       = i - 1                                                  chl 29
          g(jd+im) = (s(j,i) - dot(u(id),g(jd),im))/u(id+im)                chl 30
          id       = id + i                                                 chl 31
400     continue                                                            chl 32
        jd = jd + j                                                         chl 33
500   continue                                                              chl 34
      end                                                                   chl 35
c
      subroutine chlfac(a,nn)                                               chl  1
c                                                                           chl  2
c.... choleski factorization of a symmetric, positive definite matrix       chl  3
      real*8 a(1),dot                                                       chl  4
      a(1) = sqrt(a(1))                                                     chl  5
      if(nn.eq.1) return                                                    chl  6
      jd = 1                                                                chl  7
      do 200 j = 2,nn                                                       chl  8
        jm = j - 1                                                          chl  9
        id = 0                                                              chl 10
        do 100 i = 1,jm                                                     chl 11
          if(i-1.gt.0) a(jd+i) = a(jd+i) - dot(a(id+1),a(jd+1),i-1)         chl 12
          id      = id + i                                                  chl 13
          a(jd+i) = a(jd+i)/a(id)                                           chl 14
100     continue                                                            chl 15
        a(jd+j) = sqrt(a(jd+j) - dot(a(jd+1),a(jd+1),jm))                   chl 16
        jd      = jd + j                                                    chl 17
200   continue                                                              chl 18
      end                                                                   chl 19
c
      subroutine colbac(u,s,d,jj)                                           col  1
c                                                                           col  2
c.... backsubstitution macro                                                col  3
      real*8 u(1),s(1),d,dd                                                 col  4
      dd = s(jj+1)                                                          col  5
      do 100 j = 1,jj                                                       col  6
        s(j) = s(j) - dd*u(j)                                               col  7
100   continue                                                              col  8
      s(jj) = s(jj)/d                                                       col  9
      end                                                                   col 10
c
      subroutine eisql(a,d,e,z,n,ierr)                                      eis  1
      implicit real*8 (a-h,o-z)                                             eis  2
      real*8 a(1),d(1),e(1),z(n,n),machep                                   eis  3
      data machep/0.222044605d-15/                                          eis  4
c                                                                           eis  5
c.... eispac ql algorithm: adapted from 'tred2' and 'tql2'                  eis  6
      n2 = 0                                                                eis  7
      do 100 i = 1,n                                                        eis  8
      do 100 j = 1,i                                                        eis  9
        n2      = n2 + 1                                                    eis 10
        z(i,j) = a(n2)                                                      eis 11
100   continue                                                              eis 12
      if(n.eq.1) go to 320                                                  eis 13
      n2 = n + 2                                                            eis 14
```

```
      do 300 ii = 2,n                                                eis 15
        i     = n2 - ii                                              eis 16
        l     = i - 1                                                eis 17
        h     = 0.0d0                                                eis 18
        scale = 0.0d0                                                eis 19
        if(l.lt.2) go to 130                                         eis 20
        do 120 k = 1,l                                               eis 21
          scale = scale + abs(z(i,k))                                eis 22
120     continue                                                     eis 23
        if(scale.ne.0.0d0) go to 140                                 eis 24
130     e(i) = z(i,l)                                                eis 25
        go to 290                                                    eis 26
140     do 150 k = 1,l                                               eis 27
          z(i,k) = z(i,k)/scale                                      eis 28
          h      = h + z(i,k)*z(i,k)                                 eis 29
150     continue                                                     eis 30
        f      = z(i,l)                                              eis 31
        g      = -sign(sqrt(h),f)                                    eis 32
        e(i)   = scale*g                                             eis 33
        h      = h - f*g                                             eis 34
        z(i,l) = f - g                                               eis 35
        f      = 0.0d0                                               eis 36
        do 240 j = 1,l                                               eis 37
          z(j,i) = z(i,j)/h                                          eis 38
          g      = 0.0d0                                             eis 39
          do 180 k = 1,j                                             eis 40
            g = g + z(j,k)*z(i,k)                                    eis 41
180       continue                                                   eis 42
          jp1 = j + 1                                                eis 43
          if(l.lt.jp1) go to 220                                     eis 44
          do 200 k = jp1,l                                           eis 45
            g = g + z(k,j)*z(i,k)                                    eis 46
200       continue                                                   eis 47
220       e(j) = g/h                                                 eis 48
          f    = f + e(j)*z(i,j)                                     eis 49
240     continue                                                     eis 50
        hh = f/(h+h)                                                 eis 51
        do 270 j = 1,l                                               eis 52
          f    = z(i,j)                                              eis 53
          g    = e(j) - hh*f                                         eis 54
          e(j) = g                                                   eis 55
          do 260 k = 1,j                                             eis 56
            z(j,k) = z(j,k) - f*e(k) - g*z(i,k)                      eis 57
260       continue                                                   eis 58
270     continue                                                     eis 59
290     d(i) = h                                                     eis 60
300   continue                                                       eis 61
c.... set transformation array for ql                                eis 62
320   d(1)   = z(1,1)                                                eis 63
      z(1,1) = 1.0d0                                                 eis 64
      e(1)   = 0.0d0                                                 eis 65
      ierr   = 0                                                     eis 66
      if(n.eq.1) go to 950                                           eis 67
      do 500 i = 2,n                                                 eis 68
        l = i - 1                                                    eis 69
        if(d(i).eq.0.0d0) go to 380                                  eis 70
        do 370 j = 1,l                                               eis 71
          g = 0.0d0                                                  eis 72
          do 340 k = 1,l                                             eis 73
            g = g + z(i,k)*z(k,j)                                    eis 74
```

```
340         continue                                          eis 75
            do 360 k = 1,l                                    eis 76
              z(k,j) = z(k,j) - g*z(k,i)                      eis 77
360         continue                                          eis 78
370       continue                                            eis 79
380       d(i)   = z(i,i)                                     eis 80
          z(i,i) = 1.0d0                                      eis 81
          do 400 j = 1,l                                      eis 82
            z(i,j) = 0.0d0                                    eis 83
            z(j,i) = 0.0d0                                    eis 84
400       continue                                            eis 85
500     continue                                              eis 86
c.... begin 'ql' algorithm on tridagonal matrix               eis 87
        do 600 i = 2,n                                        eis 88
          e(i-1) = e(i)                                       eis 89
600     continue                                              eis 90
        f    = 0.0d0                                          eis 91
        b    = 0.0d0                                          eis 92
        e(n) = 0.0d0                                          eis 93
        do 840 l = 1,n                                        eis 94
          j = 0                                               eis 95
          h = machep*(abs(d(l)) + abs(e(l)))                  eis 96
          if(b.lt.h) b = h                                    eis 97
          do 710 m = l,n                                      eis 98
            if(abs(e(m)).le.b) go to 720                      eis 99
710       continue                                            eis100
720       if(m.eq.l) go to 820                                eis101
730       if(j.eq.30) go to 1000                              eis102
          j    = j + 1                                        eis103
          l1   = l + 1                                        eis104
          g    = d(l)                                         eis105
          p    = (d(l1)-g)/(e(l)+e(l))                        eis106
          r    = sqrt(p*p+1.0d0)                              eis107
          d(l) = e(l)/(p+sign(r,p))                           eis108
          h    = g - d(l)                                     eis109
          do 740 i = l1,n                                     eis110
            d(i) = d(i) - h                                   eis111
740       continue                                            eis112
          f   = f + h                                         eis113
          p   = d(m)                                          eis114
          c   = 1.0d0                                         eis115
          s   = 0.0d0                                         eis116
          mml = m - l                                         eis117
          do 800 ii = 1,mml                                   eis118
            i = m - ii                                        eis119
            g = c*e(i)                                        eis120
            h = c*p                                           eis121
            if(abs(p).ge.abs(e(i))) then                      eis122
              c      = e(i)/p                                 eis123
              r      = sqrt(c*c+1.0d0)                        eis124
              e(i+1) = s*p*r                                  eis125
              s      = c/r                                    eis126
              c      = 1.0d0/r                                eis127
            else                                              eis128
              c      = p/e(i)                                 eis129
              r      = sqrt(c*c+1.0d0)                        eis130
              e(i+1) = s*e(i)*r                               eis131
              s      = 1.0d0/r                                eis132
              c      = c*s                                    eis133
            endif                                             eis134
```

```
            p       = c*d(i) - s*g                                 eis135
            d(i+1) = h + s*(c*g + s*d(i))                          eis136
            do 780 k = 1,n                                         eis137
              h          = z(k,i+1)                                eis138
              z(k,i+1) = s*z(k,i) + c*h                            eis139
              z(k,i  ) = c*z(k,i) - s*h                            eis140
780         continue                                               eis141
800       continue                                                 eis142
          e(l) = s*p                                               eis143
          d(l) = c*p                                               eis144
          if(abs(e(l)).gt.b) go to 730                             eis145
820       d(l) = d(l) + f                                          eis146
840     continue                                                   eis147
        do 900 ii = 2,n                                            eis148
          i = ii - 1                                               eis149
          k = i                                                    eis150
          p = d(i)                                                 eis151
          do 860 j = ii,n                                          eis152
          if(abs(d(j)).gt.abs(p)) then                             eis153
            k = j                                                  eis154
            p = d(j)                                               eis155
          endif                                                    eis156
860       continue                                                 eis157
          if(k.ne.i) then                                          eis158
            d(k) = d(i)                                            eis159
            d(i) = p                                               eis160
            do 880 j = 1,n                                         eis161
              p      = z(j,i)                                      eis162
              z(j,i) = z(j,k)                                      eis163
              z(j,k) = p                                           eis164
880         continue                                               eis165
          endif                                                    eis166
900     continue                                                   eis167
950     return                                                     eis168
1000    ierr = l                                                   eis169
        end                                                        eis170
c
        subroutine geig(g,h,d,p,t,nev,prt)                         gei  1
c                                                                  gei  2
c.... solve the general eigenproblem 'g*p = h*p*d'                 gei  3
        logical prt                                                gei  4
        real*8 g(1),h(1),d(1),p(nev,1),t(1)                        gei  5
c                                                                  gei  6
c.... compute the choleski factors of 'h'                          gei  7
        if(prt) call wprojm(g,nev,1)                               gei  8
        if(prt) call wprojm(h,nev,2)                               gei  9
        call chlfac(h,nev)                                         gei 10
c                                                                  gei 11
c.... compute the standard eigenvalue problem matrix 'c'           gei 12
        call chlfwd(h,g,p,nev)                                     gei 13
c                                                                  gei 14
c.... perform the eignfunction decomposition of 'c'                gei 15
        call eisql(g,d,t,p,nev,ir)                                 gei 16
c                                                                  gei 17
c.... compute the vectors of the original problem                  gei 18
        call chlbac(h,p,nev)                                       gei 19
        end                                                        gei 20
c
        subroutine numass(b,neq,mq)                                num  1
        real*8 b(1)                                                num  2
```

```
      common /iofile/ ior,iow                                          num  3
      nn = 0                                                           num  4
      do 10 n = 1,neq                                                  num  5
        if(b(n).ne.0.0d0) nn = nn + 1                                  num  6
10    continue                                                         num  7
      if(nn.lt.mq) write(iow,2000) nn                                  num  8
      if(ior.lt.0.and.nn.lt.mq) write(*,2000) nn                       num  9
      mq = min(mq,nn)                                                  num 10
2000  format(1x,'Subspace reduced to',i4,' by number of nonzero lumped',num 11
     1       ' mass terms')                                            num 12
      end                                                              num 13
c
      subroutine scalev(v,nn)                                          sca  1
c                                                                      sca  2
c.... scale a vector to have maximum element of 1.0                    sca  3
      real*8 v(1),vmax                                                 sca  4
      vmax = abs(v(1))                                                 sca  5
      do 100 n = 1,nn                                                  sca  6
        vmax = max(vmax,abs(v(n)))                                     sca  7
100   continue                                                         sca  8
      do 110 n = 1,nn                                                  sca  9
        v(n) = v(n)/vmax                                               sca 10
110   continue                                                         sca 11
      end                                                              sca 12
c
      subroutine sproja(v,z,g,neq,nev,ipd)                             spr  1
c                                                                      spr  2
c.... compute the subspace projection of 'aa' to form 'g'              spr  3
      real*8 v(neq,1),z(neq,1),g(1),aa,dot,engy,dimx,dimn              spr  4
      common /adata/  aa(8000)                                         spr  5
      common /frdata/ maxf                                             spr  6
      common /nfrta/  dimx,dimn,nvb,npl                                spr  7
c.... forward reduce the eigenvector estimates                         spr  8
      ma = maxf*nev + 1                                                spr  9
c.... copy vectors 'v' into 'z'                                        spr 10
      do 100 i = 1,nev                                                 spr 11
      do 100 j = 1,neq                                                 spr 12
        z(j,i) = v(j,i)                                                spr 13
100   continue                                                         spr 14
c.... solve the equations                                              spr 15
      call rsolve(aa,z,aa(ma),ipd,maxf,nvb,neq,nev,engy,1)             spr 16
c.... compute the projection of the stiffness                          spr 17
      k = 0                                                            spr 18
      do 200 j = 1,nev                                                 spr 19
      do 200 i = 1,j                                                   spr 20
        k    = k + 1                                                   spr 21
        g(k) = dot(v(1,i),z(1,j),neq)                                  spr 22
200   continue                                                         spr 23
      end                                                              spr 24
c
      subroutine sprojb(b,v,t,h,neq,nev)                               spr  1
c                                                                      spr  2
c.... compute the subspace projection of 'b' to form 'h'               spr  3
      real*8 b(1),v(neq,1),t(1),h(1),dot                               spr  4
c.... compute 'z' and the 'b' projection to form 'h'                   spr  5
      do 130 j = 1,nev                                                 spr  6
c.... compute 'z'for a lumped mass                                     spr  7
        do 100 i = 1,neq                                               spr  8
          t(i) = v(i,j)*b(i)                                           spr  9
100     continue                                                       spr 10
```

```
c.... project the'z' and 'v' vectors to form 'h'                 spr 11
        k = j*(j+1)/2                                            spr 12
        do 110 i = j,nev                                         spr 13
          h(k) = dot(t,v(1,i),neq)                               spr 14
          k    = k + i                                           spr 15
110     continue                                                 spr 16
        do 120 i = 1,neq                                         spr 17
          v(i,j) = t(i)                                          spr 18
120     continue                                                 spr 19
130   continue                                                   spr 20
      end                                                        spr 21
c
      subroutine wprojm(a,nn,ia)                                 wpr  1
      character*1 ah(2)                                          wpr  2
      real*8 a(1)                                                wpr  3
      common /iofile/ ior,iow                                    wpr  4
      data ah(1),ah(2) /'g','h'/                                 wpr  5
      write(iow,2000) ah(ia)                                     wpr  6
      if(ior.lt.0) write(*,2000) ah(ia)                          wpr  7
      i = 1                                                      wpr  8
      do 100 n = 1,nn                                            wpr  9
        j = i + n - 1                                            wpr 10
        write(iow,2001) (a(k),k=i,j)                             wpr 11
        if(ior.lt.0) write(*,2001) (a(k),k=i,j)                  wpr 12
        i = i + n                                                wpr 13
100   continue                                                   wpr 14
2000  format(1x,'Matrix ',a1)                                    wpr 15
2001  format(1p8d10.2)                                           wpr 16
      end                                                        wpr 17
```

16.10.5 *Equation solution modules.* The two equation solution options of (*a*) variable-band and (*b*) frontal method are included in this volume. The options for the variable-band solver have been enhanced to permit solution of problems for which the coefficient (tangent) matrix is unsymmetric. No changes have been made to the basic frontal solution system; consequently, it may only be used to solve problems that have a symmetric coefficient matrix. Both solution systems have been enhanced to include a resolution option for use in the eigenproblem program included in PCEIGN.FOR. Thus, when selecting a solution option it is necessary to load the solver (PASOLV.FOR or PFSOLV.FOR) and the resolution program (PARSOL.FOR or PFRSOL.FOR respectively).

(*a*) File PASOLV.FOR contains the following subprograms:

Name	*Type*	*Description*	*Chapter*
PSOLVE	SUBROUTINE	Controls equation solution	16
DASBLY	SUBROUTINE	Assemble variable-band arrays	15
DASOL	SUBROUTINE	Solve equations	15
DATEST	SUBROUTINE	Test equations	15
DATRI	SUBROUTINE	Triangular decomposition	15
DREDU	SUBROUTINE	Compute reduced diagonal	15
PROFIL	SUBROUTINE	Compute equation numbers and profile	15

```
      subroutine psolve(u,a,b,dr,m,xm,s,ld,ig,idl,nst,nrs,afac,solv,     pso  1
     1                  dyn,cl,ipd,rnorm,aengy,ifl)                      pso  2
c...  active column assembly and solution of equations                  pso  3
      logical afac,afl,solv,dyn,fl,fa                                    pso  4
      character*12 tfile                                                 pso  5
      real*8 u(1),a(1),b(1),dr(1),xm(1),s(nst,1),aengy,rnorm,dot         pso  6
      integer*2 m(1),ld(1),ig(1),idl(1)                                  pso  7
      integer*4 itrec                                                    pso  8
      common /cdata/  numnp,numel,nummat,nen,neq                         pso  9
      common /fdata/  fl(11)                                             pso 10
      common /iofile/ ior,iow                                            pso 11
      common /temfl1/ tfile(6)                                           pso 12
      common /temfl2/ itrec(4),nw1,nw2                                   pso 13
      data fa/.false./                                                   pso 14
c.... form and assemble the matrix                                       pso 15
      if(afac) then                                                      pso 16
        if(fl(6)) then                                                   pso 17
          nep = neq + ig(neq)                                            pso 18
          afl = .true.                                                   pso 19
        else                                                             pso 20
          nep = neq                                                      pso 21
          afl = .false.                                                  pso 22
        endif                                                            pso 23
        if(fl(3).or.fl(4)) then                                          pso 24
          ibuf = ig(neq)+neq                                             pso 25
          if(fl(3)) then                                                 pso 26
            ibuf  = ibuf + ig(neq)                                       pso 27
            fl(3) = fa                                                   pso 28
```

```
          endif                                                           pso 29
          if(ibuf.gt.8000) stop 'profile too large'                       pso 30
          itrec(1) = ibuf*8                                               pso 31
          open (4,file=tfile(1),status='new',access='direct',             pso 32
     1           form='unformatted',recl=itrec(1))                        pso 33
          close(4)                                                        pso 34
          fl(4) = fa                                                      pso 35
        endif                                                             pso 36
        call pconsd(a,ibuf,0.0d0)                                         pso 37
c.... modify tangent form lumped mass effects                              pso 38
        if(dyn) then                                                      pso 39
          do 310 n = 1,neq                                                pso 40
            a(n)  =  c1*m(n)                                              pso 41
310       continue                                                        pso 42
        endif                                                             pso 43
        do 320 n = 1,numel                                                pso 44
c...  compute and assemble element arrays                                  pso 45
          ne = n                                                          pso 46
          call formfe(u,dr,.true.,solv,fa,fa,3,ne,ne,1)                   pso 47
          if(ior.lt.0 .and. mod(n,20).eq.0) write(*,2000) n               pso 48
          call dasbly(s,s,ld,ig,nst,afl,afac,fa,dr,a(nep+1),a(neq+1),a) pso 49
320     continue                                                          pso 50
        rnorm = sqrt(dot(dr,dr,neq))                                      pso 51
        call datri(a(nep+1),a(neq+1),a,ig,neq,afl)                        pso 52
        call phstio(4,1,a,ibuf,2,tfile(1),itrec(1))                       pso 53
      endif                                                               pso 54
      if(solv) then                                                       pso 55
        if(.not.afac) call phstio(4,1,a,ibuf,1,tfile(1),itrec(1))         pso 56
        do 330 n = 1,nrs                                                  pso 57
          ne = (n-1)*neq + 1                                              pso 58
          call dasol(a(nep+1),a(neq+1),a,dr(ne),ig,neq, aengy)            pso 59
330     continue                                                          pso 60
      endif                                                               pso 61
2000  format(5x,'**',i4,' Elements completed.')                           pso 62
      end                                                                 pso 63
```

(*b*) File PFSOLV.FOR contains the following subprograms:

Name	*Type*	*Description*	*Chapter*
PSOLVE	SUBROUTINE	Frontal equation solution	15
PBUFF	SUBROUTINE	Disk I/O of front buffer array	15
PFRTAS	SUBROUTINE	Assemble front arrays	15
PRFTBK	SUBROUTINE	Backsubstitution solution	15
PFRTFW	SUBROUTINE	Forward solution	15
PREFRT	SUBROUTINE	Compute frontal order	15
PROFIL	SUBROUTINE	Compute equation numbers and front size	15
PFRTB	SUBROUTINE	Backsubstitution macro	15
PFRTD	SUBROUTINE	Triangular decomposition	15
PFRTF	SUBROUTINE	Forward solution macro	15

Note: it may be necessary to place the last three subroutines in a separate file to avoid compiler warnings.

(*c*) File PARSOL.FOR contains the following subprogram:

Name	*Type*	*Description*	*Chapter*
RSOLVE	SUBROUTINE	Controls equation re-solution for variable-band option	16

```
      subroutine rsolve(b,dr,a,ipd,maxf,nv,neq,nev,engy,ifl)           rso  1
c.... resolution for profile solution                                  rso  2
      real*8     a(1),b(1),dr(neq,1),engy,dm                           rso  3
      real       rm                                                    rso  4
      integer*2 im                                                     rso  5
      common dm(1),rm(1),im(1)                                         rso  6
c                                                                      rso  7
      n12 = neq*ipd - ipd - 1                                          rso  8
      do 100 ne = 1,nev                                                rso  9
        call dasol(a(neq+1),a(neq+1),a,dr(1,ne),im(n12),neq, engy)     rso 10
100   continue                                                         rso 11
c                                                                      rso 12
      end                                                              rso 13
```

(*d*) File PFRSOL.FOR contains the following subprogram:

Name	*Type*	*Description*	*Chapter*
RSOLVE	SUBROUTINE	Controls equation re-solution for frontal option	16

```
      subroutine rsolve(b,dr,m,ipd,maxf,nv,neq,nev,aengy,ifl)              rso  1
c.... resolution for frontal solution                                      rso  2
      real*8 b(maxf,1),dr(neq,1),aengy                                     rso  3
      integer*2 m(1)                                                       rso  4
      integer*4 itrec                                                      rso  5
      common /temfl2/ itrec(4),nw1,nw2                                     rso  6
      ibuf = (itrec(1) - 4)/2                                              rso  7
      call pfrtfw(b,dr,m,ipd,ibuf,maxf,nv,neq,nev,ifl)                     rso  8
      call pfrtbk(b,dr,m,ipd,ibuf,maxf,nv,neq,nev,aengy,ifl)               rso  9
c                                                                          rso 10
      end                                                                  rso 11
```

16.10.6 *Plot module.* The plot interface for installation on personal computers may be achieved using the Graphical Development Toolkit (GDT),[13] as described in Volume 1, or using the interface for the Microsoft Fortran Version 5.0 graphics capability, as given in the listings in 16.10.6(*b*). Alternatively, the program may be installed on any computer that supports a Tektronix 4012 compatible interface using the routines given in 16.10.6(*c*). With any of the interfaces, plots of the undeformed and the deformed mesh may be obtained for two-dimensional problems. In addition plots of the eigenvectors for this class of problem may be displayed. The deformed configuration is constructed by adding the values for the first degrees of freedom to the corresponding coordinates. Scaling of the deformation pattern may be necessary to enhance the plot characteristics.

(*a*) File PCPLOT.FOR contains the following subprograms:

Name	*Type*	*Description*	*Chapter*
PPLOTF	SUBROUTINE	Controls plot sequence	16
DPLOT	SUBROUTINE	Draw lines to screen	15
PDEVCL	SUBROUTINE	Close plot, return to macro	15
PDEVOP	SUBROUTINE	Open plot, draw border	15
PLOTL	SUBROUTINE	Scale plot point to screen	16
FRAME	SUBROUTINE	Determine scale factors	16
PDEFM	SUBROUTINE	Compute deformed plot coordinates	15
PLINE	SUBROUTINE	Draw mesh or outline	15

```
      subroutine pplotf(x,ix,b,lci,ct,ndf,ndm,nen1,nneq)                ppl   1
c.... plot control subroutine for feap                                   ppl   2
      logical pcomp,oflg                                                ppl   3
      character lci*4                                                   ppl   4
      integer*2 ixy,devnam,status,vslcol,il,ix(1),coli                  ppl   5
      real   x(ndm,1),ct(2)                                             ppl   6
      real*8 dm,b(1)                                                    ppl   7
      common dm(1)                                                      ppl   8
      common /adata/ dr(1),il(31998)                                    ppl   9
      common /cdata/ numnp,numel,nummat,nen,neq                         ppl  10
      common /pdata2/ ixy(4),devnam                                     ppl  11
c.... open kernel system and plot mesh or outline of parts              ppl  12
      call pdevop(devnam)                                               ppl  13
      call frame(x,ndm,numnp)                                           ppl  14
c.... plot mesh or outline of parts                                     ppl  15
      oflg = .not.pcomp(lci,'outl')                                     ppl  16
      if(ct(1).ne.0.0.or.pcomp(lci,'eigv')) then                        ppl  17
        ic = 2                                                          ppl  18
        if(ct(1).eq.0.0) ct(1) = 1.0                                    ppl  19
      else                                                              ppl  20
        ic = 1                                                          ppl  21
      endif                                                             ppl  22
      c  = 0.0                                                          ppl  23
      n1 = 2*ndm*numnp                                                  ppl  24
```

```
      do 100 i = ic,1,-1                                              ppl 25
        coli    = 8 - 2*i                                             ppl 26
        status = vslcol(devnam,coli)                                  ppl 27
        if(pcomp(lci,'eigv')) then                                    ppl 28
          call pdefm(x,dm,c,ndm,ndf,numnp, dr)                        ppl 29
        else                                                          ppl 30
          call pdefm(x, b,c,ndm,ndf,numnp, dr)                        ppl 31
        endif                                                         ppl 32
        call pline(dr,ix,il(nl),numnp,numel,ndm,nenl,nen,ct(2),oflg)  ppl 33
        c   = ct(1)                                                   ppl 34
100   continue                                                        ppl 35
c.... close plot                                                      ppl 36
      call pdevcl(devnam)                                             ppl 37
      end                                                             ppl 38
c
      subroutine plotl(x1,x2,x3,ipen)                                 plo  1
c.... line drawing command                                            plo  2
      common /pdata1/ scale,dx(2),sx(2)                               plo  3
c.... compute the normal coordinates                                  plo  4
      s1 = max(0.0,min(1.45,scale*(x1 + x1 - sx(1)) + 0.725))         plo  5
      s2 = max(0.0,min(1.00,scale*(x2 + x2 - sx(2)) + 0.500))         plo  6
      call dplot(s1,s2,ipen)                                          plo  7
      end                                                             plo  8
c
      subroutine frame(x,ndm,numnp)                                   fra  1
c.... compute scaling for plot area                                   fra  2
      logical iflg                                                    fra  3
      real      x(ndm,1),xmn(2),xmx(2),xmin(3),xmax(3)                fra  4
      common /pdata1/ scale,dx(2),sx(2)                               fra  5
c.... determine window coordinates                                    fra  6
      if(ndm.eq.1) then                                               fra  7
        dx(2) = 0.                                                    fra  8
        sx(2) = 0.0                                                   fra  9
      endif                                                           fra 10
      ii = min(ndm,3)                                                 fra 11
      ij = min(ndm,2)                                                 fra 12
c.... find the minimum and maximum coordinate of input nodes          fra 13
      iflg = .true.                                                   fra 14
      do 104 n = 1,numnp                                              fra 15
        if(x(1,n).ne. -999.) then                                     fra 16
          if(iflg) then                                               fra 17
            do 100 i = 1,ii                                           fra 18
              xmin(i) = x(i,n)                                        fra 19
              xmax(i) = x(i,n)                                        fra 20
100         continue                                                  fra 21
            iflg = .false.                                            fra 22
          else                                                        fra 23
            do 102 i = 1,ii                                           fra 24
              xmin(i) = min(xmin(i),x(i,n))                           fra 25
              xmax(i) = max(xmax(i),x(i,n))                           fra 26
102         continue                                                  fra 27
          endif                                                       fra 28
        endif                                                         fra 29
104   continue                                                        fra 30
      scale  = max(xmax(1)-xmin(1),xmax(2)-xmin(2))                   fra 31
c.... plot region determination                                       fra 32
      do 110 i = 1,ij                                                 fra 33
        xmn(i) = min(xmin(i),xmax(i))                                 fra 34
        xmx(i) = max(xmin(i),xmax(i))                                 fra 35
        dx(i) = xmx(i) - xmn(i)                                       fra 36
```

```
      sx(i) = xmx(i) + xmn(i)                                    fra 37
110   continue                                                   fra 38
c.... rescale window                                             fra 39
      if(dx(1).gt.1.45*dx(2)) then                               fra 40
        xmn(2) = (sx(2) - dx(1))/2.0                             fra 41
        xmx(2) = (sx(2) + dx(1))/2.0                             fra 42
        fact   = 0.58                                            fra 43
      else                                                       fra 44
        xmn(1) = (sx(1) - dx(2))/2.0                             fra 45
        xmx(1) = (sx(1) + dx(2))/2.0                             fra 46
        fact   = 0.40                                            fra 47
      endif                                                      fra 48
      do 112 i = 1,ij                                            fra 49
        xmin(i) = max(xmin(i),xmn(i)) - scale/100.               fra 50
        xmax(i) = min(xmax(i),xmx(i)) + scale/100.               fra 51
112   continue                                                   fra 52
c.... reset values for deformed plotting                         fra 53
      scale  = max(xmax(1)-xmin(1),xmax(2)-xmin(2))              fra 54
      do 114 i = 1,ij                                            fra 55
        xcen = xmax(i)+xmin(i)                                   fra 56
        xmax(i) = (xcen + 1.1*scale)/2.                          fra 57
        xmin(i) = (xcen - 1.1*scale)/2.                          fra 58
114   continue                                                   fra 59
      scale = fact/scale                                         fra 60
      end                                                        fra 61
```

(*b*) File PCFOR5.FOR may be used to interface the program with the graphics features contained in the Microsoft Fortran Version 5.0 library. The routines are written using the GDT names to simplify the process. The following file contains the subprograms:

Name	*Type*	*Description*	*Chapter*
VOPNWK	FUNCTION	Open for plot	16
VCLRWK	FUNCTION	Clear plot screen	16
VCLSWK	FUNCTION	Close plot screen	16
VRQSTR	FUNCTION	Input character	16
VSLCOL	FUNCTION	Set line colour	16
VPLINE	FUNCTION	Draw line	16
VENCUR	FUNCTION	Return to text mode	16

```
      include 'fgraph.fi'
c
      integer*2 function vopnwk(wkin,devnam,wkout)                    vop  1
      implicit integer*2 (a-z)                                        vop  2
      include 'fgraph.fd'                                             vop  3
      logical lfil                                                    vop  4
      integer*2 wkin(*),wkout(*)                                      vop  5
      record /videoconfig/ myscreen                                   vop  6
      common /instl2/ jfill,jplot,lfil                                vop  7
      common /vgraph/ idxl,idyl                                       vop  8
c.....open the workstation, home cursor, set up scaling               vop  9
      call getvideoconfig(myscreen)                                   vop 10
      status = setvideomode( $MAXRESMODE )                            vop 11
      call getvideoconfig( myscreen )                                 vop 12
      ixln   = myscreen.numxpixels - 1                                vop 13
      iyln   = myscreen.numypixels - 1                                vop 14
      idxl   = 32640.0/(ixln+1) + 0.5                                 vop 15
      idyl   = 22480.0/(iyln+1) + 0.5                                 vop 16
      jfill  = 2                                                      vop 17
      if(myscreen.numcolors .le. 4) jfill = 1                         vop 18
      status = displaycursor ( $GCURSOROFF )                          vop 19
      vopnwk = 0                                                      vop 20
      end                                                             vop 21
c
      integer*2 function vclrwk(devnam)                               vcl  1
      implicit integer*2 (a-z)                                        vcl  2
      include 'fgraph.fd'                                             vcl  3
c.....clear the workstation and home cursor                           vcl  4
      call clearscreen( $GCLEARSCREEN )                               vcl  5
      vclrwk = displaycursor( $GCURSOROFF )                           vcl  6
      end                                                             vcl  7
c
      integer*2 function vclswk(devnam)                               vcl  1
      implicit integer*2 (a-z)                                        vcl  2
      include 'fgraph.fd'                                             vcl  3
      record /rccoord / s                                             vcl  4
c.... home cursor - text mode                                         vcl  5
      call settextposition( 1 , 1 , s )                               vcl  6
      vclswk = displaycursor( $GCURSORON )                            vcl  7
      vclswk = setvideomode ( $TEXTC80 )                              vcl  8
      vclswk = settextcursor( #0607 )                                 vcl  9
```

```
      vclswk = settextcolor(7)                                           vcl 10
      end                                                                vcl 11
c
      integer*2 function vrqstr(devnam,len,echoh,ixy,cstr)               vrq  1
c.... input a single character                                            vrq  2
      character*1 cstr                                                   vrq  3
      read(*,1000) cstr                                                  vrq  4
1000  format(a1)                                                         vrq  5
      vrqstr = 0                                                         vrq  6
      end                                                                vrq  7
c
      integer*2 function vslcol(devnam,it)                               vsl  1
      implicit integer*2 (a-z)                                           vsl  2
      include 'fgraph.fd'                                                vsl  3
      logical lfil                                                       vsl  4
      integer*2 ipal(7)                                                  vsl  5
      common /instl2/ jfill,jplot,lfil                                   vsl  6
c.... set line color                                                      vsl  7
      data ipal/ 15, 4, 2, 1, 14, 3, 5/                                  vsl  8
      if(it.gt.0 .and. it.le.7) then                                     vsl  9
        icll = ipal(it)                                                  vsl 10
        if(jfill.lt.2) icll = 1                                          vsl 11
      else                                                               vsl 12
        icll = 0                                                         vsl 13
      endif                                                              vsl 14
      vslcol = setcolor( icll )                                          vsl 15
      vslcol = displaycursor( $GCURSOROFF )                              vsl 16
      end                                                                vsl 17
c
      integer*2 function vpline(devnam,npt,ixy)                          vpl  1
      implicit integer*2 (a-z)                                           vpl  2
      include 'fgraph.fd'                                                vpl  3
      integer*2 ixy(2,*)                                                 vpl  4
      record /xycoord/ xy                                                vpl  5
      common /vgraph/ idxl,idyl                                          vpl  6
c.... draw line                                                           vpl  7
      call moveto( ixy(1,1)/idxl , (22200 - ixy (2,1))/idyl , xy )       vpl  8
      do 100 n = 2,npt                                                   vpl  9
        vpline =  lineto(ixy(1,n)/idxl ,(22200 - ixy(2,n))/idyl )        vpl 10
100   continue                                                           vpl 11
      vpline = displaycursor( $GCURSOROFF )                              vpl 12
      end                                                                vpl 13
c
      integer*2 function vencur(devnam)                                  ven  1
      implicit integer*2 (a-z)                                           ven  2
c.... display cursor                                                      ven  3
      vencur = displaycursor( $GCURSOROFF )                              ven  4
      end                                                                ven  5
```

(*c*) File PCTEKT.FOR may be used to interface the program to any system that supports a Tektronix 4012 graphics interface. Most engineering workstations and National Center for Supercomputer Applications Telnet software support an emulation. The following file contains the subprograms (the DPLOT subroutine replaces the one in the file PCPLOT above for this interface):

Name	*Type*	*Description*	*Chapter*
VOPNWK	FUNCTION	Open for plot	16
VCLRWK	FUNCTION	Clear plot screen	16
VSLCOL	FUNCTION	Set line type	16
VRQSTR	FUNCTION	Input character	16
VENCUR	FUNCTION	Return to text mode	16
VCLSWK	FUNCTION	Close plot screen	16
DPLOT	SUBROUTINE	Draw lines to screen	16
BRK4	SUBROUTINE	Tektronix 4012 code generator	16

```
      integer*2 function vopnwk(wkin,devnam,wkout)                vop   1
c.... open tektronix 4012 device                                  vop   2
      vopnwk = 0                                                  vop   3
      end                                                         vop   4
c
      integer*2 function vclrwk(devnam)                           vcl   1
c.... clear tektronix 4012 device                                 vcl   2
      write(*,1000) char(27),char(12)                             vcl   3
1000  format(2a1)                                                 vcl   4
      end                                                         vcl   5
c
      integer*2 function vslcol(devnam,icol)                      vsl   1
c.... set line type                                               vsl   2
      iln = 0                                                     vsl   3
      if(icol.eq.4) iln = 3                                       vsl   4
      write(*,1000) char(27),char(iln+96)                         vsl   5
1000  format(2a1)                                                 vsl   6
      end                                                         vsl   7
c
      integer*2 function vrqstr(devnam,i1,i2,ixy,xx)              vrq   1
c.... input character to quit plot                                vrq   2
      character*1 xx                                              vrq   3
      read(*,1000) xx                                             vrq   4
1000  format(a1)                                                  vrq   5
      end                                                         vrq   6
c
      integer*2 function vencur(devnam)                           ven   1
c.... enter text mode with cursor                                 ven   2
      write(   *,1000) char(31),char(27),char(50),char(27),       ven   3
     .                 char(91),char(50),char(74)                 ven   4
1000  format(7a1)                                                 ven   5
      end                                                         ven   6
c
      integer*2 function vclswk(devnam)                           vcl   1
c.... close tektronix 4012 device                                 vcl   2
      vclswk = 0                                                  vcl   3
```

```
      end                                                          vcl   4
c
      subroutine dplot(x,y,ipen)                                   dpl   1
      character str4*4,stro*4                                      dpl   2
      real      x,y                                                dpl   3
      common /tektl/ stro                                          dpl   4
c.... tektronix 4012 device or emulator                             dpl   5
c     pen command motions: ipen = 1, move to position x,y          dpl   6
c                          ipen = 2, drawline from to x,y          dpl   7
      jxl = x*700                                                  dpl   8
      jyl = y*770                                                  dpl   9
      call brk4(jxl,jyl,str4)                                      dpl  10
      if(ipen.eq.2) then                                           dpl  11
        write(*,1000) char(13),char(29),stro,str4                  dpl  12
      endif                                                        dpl  13
      stro = str4                                                  dpl  14
c.... format statements                                            dpl  15
1000  format(2a1,2a4)                                              dpl  16
      end                                                          dpl  17
c
      subroutine brk4(jxl,jyl, str)                                brk   1
      character*1 str(4)                                           brk   2
c.... convert integers for tektronix 4012                           brk   3
      str(1) = char(32 + jyl/32)                                   brk   4
      str(2) = char(96 + mod(jyl,32))                              brk   5
      str(3) = char(32 + jxl/32)                                   brk   6
      str(4) = char(64 + mod(jxl,32))                              brk   7
      end                                                          brk   8
```

16.10.7 *Element modules.* Volume 1 contains the listings for four element modules that may be used with the enhanced program included in this volume. An additional four elements, which extend the capabilities of the program into the transient non-linear range, are included in this volume. The class of problems addressed by these elements and the associated data input formats have been described in Secs 16.8 and 16.9. The listing for each element is included below.

(*a*) File PCELM1.FOR contains the element module for a linear elastic beam or axisymmetric shell model. Non-linear geometric behaviour is included by using strain measures with quadratic displacement gradient terms.

Name	*Type*	*Description*	*Chapter*
ELMT01	SUBROUTINE	Beam/shell element routine	16
BMAT01	SUBROUTINE	Compute **B** matrix	16
MATL01	SUBROUTINE	Input material properties	16
MODL01	SUBROUTINE	Compute stress, tangent arrays	16
TRAN01	SUBROUTINE	Transform to global coordinates	16

```
      subroutine elmt01(d,ul,xl,ix,tl,s,p,ndf,ndm,nst,isw)             elm  1
      implicit real*8(a-h,o-z)                                         elm  2
c                                                                      elm  3
c.....nonlinear beam/axisymmetric shell                                elm  4
c                                                                      elm  5
      character*4 head,yyy*80                                          elm  6
      logical     bs                                                   elm  7
      integer*2   ix(1)                                                elm  8
      real        xl(ndm,1),tl(1),dm                                   elm  9
      real*8      d(7),s(nst,1),shp(2),dd(5,5),btd(5,3),               elm 10
     1            bm(5,3),sig(5),ul(ndf,1),vl(3,2),p(nst),dot          elm 11
      common /bdata/  head(20)                                         elm 12
      common /cdata/  numnp,numel,nummat,nen,neq                       elm 13
      common /eldata/ dm,n,ma,mct,iel,nel                              elm 14
      common /iofile/ ior,iow                                          e!m 15
      common /ydata/  yyy                                              elm 16
c.... input material properties                                        elm 17
      if(isw.eq.1) then                                                elm 18
        call matl01(d)                                                 elm 19
        call pconsd(dd,25,0.0d0)                                       elm 20
        call pconsd(bm,15,0.0d0)                                       elm 21
      else                                                             elm 22
c.... length, angle, radius, jacobian                                  elm 23
        sn = xl(2,2)-xl(2,1)                                           elm 24
        cs = xl(1,2)-xl(1,1)                                           elm 25
        sl = sqrt(cs*cs + sn*sn)                                       elm 26
        bs = d(13).eq.0.0d0                                            elm 27
        if(isw.eq.2) then                                              elm 28
c.... check mesh for error                                             elm 29
          if(sl.le.0.0d0) then                                         elm 30
            write(iow,3001) n                                          elm 31
            if(ior.lt.0) write(*,3001) n                               elm 32
```

```
          endif                                                    elm 33
        elseif(isw.eq.5) then                                      elm 34
c.... form a lumped mass matrix; length, volume*density             elm 35
          dv    = d(11)*sl/8.0                                     elm 36
          if(bs) then                                              elm 37
            p(1)     = 4.0*dv                                      elm 38
            p(ndf+1) = p(1)                                        elm 39
          else                                                     elm 40
            p(1)     = dv*(3.*xl(1,1) +    xl(1,2))                elm 41
            p(ndf+1) = dv*(   xl(1,1) + 3.*xl(1,2))                elm 42
          endif                                                    elm 43
          p(2)     = p(1)                                          elm 44
          p(3)     = p(1)*d(12)                                    elm 45
          p(ndf+2) = p(ndf+1)                                      elm 46
          p(ndf+3) = p(ndf+1)*d(12)                                elm 47
        else                                                       elm 48
c.... form stiffness/residual                                       elm 49
          sn   = sn/sl                                             elm 50
          cs   = cs/sl                                             elm 51
          if(bs) then                                              elm 52
            r    = 1.0                                             elm 53
            recr = 0.0                                             elm 54
          else                                                     elm 55
            r    = 0.5*(xl(1,1) + xl(1,2))                         elm 56
            recr = 1.0/r                                           elm 57
          endif                                                    elm 58
          dv   = sl*r                                              elm 59
c.... shape function derivatives                                    elm 60
          shp(2) =  1.0/sl                                         elm 61
          shp(1) = -shp(2)                                         elm 62
c.... local  displacements                                          elm 63
          do 300 k = 1,2                                           elm 64
            vl(1,k) = cs*ul(1,k) + sn*ul(2,k)                      elm 65
            vl(2,k) =-sn*ul(1,k) + cs*ul(2,k)                      elm 66
            vl(3,k) = ul(3,k)                                      elm 67
300       continue                                                 elm 68
c.... derivative w,s                                                elm 69
          lin = d(5)                                               elm 70
          if(lin.eq.0) then                                        elm 71
            wks = 0.0                                              elm 72
          else                                                     elm 73
            wks = (vl(2,2) - vl(2,1))/sl                           elm 74
          endif                                                    elm 75
c.... stresses, strains, d-matrix                                   elm 76
          call modl01(sig,vl,dd,d,sn,cs,sl,recr,wks)               elm 77
          if(mod(isw,3).eq.0) then                                 elm 78
            i    = ndf + 2                                         elm 79
c.... Load vector in local coordinates (reference system)           elm 80
            f = d(4)*sl/8.0*dm                                     elm 81
            if(bs) then                                            elm 82
              p(2) = f*4.0                                         elm 83
              p(i) = p(2)                                          elm 84
            else                                                   elm 85
              p(2) = f*(3.*xl(1,1) +    xl(1,2))                   elm 86
              p(i) = f*(   xl(1,1) + 3.*xl(1,2))                   elm 87
            endif                                                  elm 88
c.... K-sigma tangent matrix                                        elm 89
            if(lin.ne.0) then                                      elm 90
              s(2,2) =   r*sig(1)/sl                               elm 91
              s(i,i) =   s(2,2)                                    elm 92
```

```
            s(2,i) = - s(2,2)                                                elm 93
          endif                                                              elm 94
c.... multiply stress and moduli by jacobian                                 elm 95
          do 302 k = 1,5                                                     elm 96
            sig(k) = sig(k)*dv                                               elm 97
            do 301 j = 1,5                                                   elm 98
              dd(j,k) = dd(j,k)*dv                                           elm 99
301         continue                                                         elm100
302       continue                                                           elm101
          i1=0                                                               elm102
          do 309 ii = 1,2                                                    elm103
c.... residual G = P - Bt*S                                                  elm104
            call bmat01(bm,shp(ii),sn,cs,recr,wks)                           elm105
            do 303 i = 1,3                                                   elm106
              p(i1+i) = p(i1+i) - dot(bm(1,i),sig,5)                         elm107
303         continue                                                         elm108
c.... tangent stiffness matrix                                               elm109
            if(isw.eq.3) then                                                elm110
              do  305 i = 1,3                                                elm111
                do 304 k = 1,5                                               elm112
                  btd(k,i) = dot(bm(1,i),dd(1,k),5)                          elm113
304             continue                                                     elm114
305           continue                                                       elm115
              j1 = i1                                                        elm116
              do 308 jj = ii,2                                               elm117
                call bmat01(bm,shp(jj),sn,cs,recr,wks)                       elm118
                do 307  i = 1,3                                              elm119
                  do 306  j = 1,3                                            elm120
                    s(i1+i,j1+j)=s(i1+i,j1+j)+dot(btd(1,i),bm(1,j),5)        elm121
306               continue                                                   elm122
307             continue                                                     elm123
                j1 = j1 + ndf                                                elm124
308           continue                                                       elm125
            endif                                                            elm126
            i1 = i1 + ndf                                                    elm127
309       continue                                                           elm128
c.... lower part of stiffness matrix and transformation to global frame      elm129
          if(isw.eq.3) then                                                  elm130
            do 310 i = 1,3                                                   elm131
            do 310 j = 1,3                                                   elm132
              s(i+ndf,j) = s(j,i+ndf)                                        elm133
310         continue                                                         elm134
            call tran01(s,cs,sn,nst,ndf,1)                                   elm135
          endif                                                              elm136
          call tran01(p,cs,sn,nst,ndf,2)                                     elm137
c.... output stresses (N, M, Q)                                              elm138
        elseif(isw.eq.4) then                                                elm139
          mct = mct - 1                                                      elm140
          if(mct.le.0) then                                                  elm141
             write(iow,2001) head                                            elm142
             if(ior.lt.0) write(*,2001) head                                 elm143
             mct = 50                                                        elm144
          endif                                                              elm145
          r = 0.5*(xl(1,1) + xl(1,2))                                        elm146
          z = 0.5*(xl(2,1) + xl(2,2))                                        elm147
          write(iow,2002) n,ma,r,z,sig                                       elm148
          if(ior.lt.0) write(*,2002) n,ma,r,z,sig                            elm149
        endif                                                                elm150
      endif                                                                  elm151
    endif                                                                    elm152
```

```
c.... format statements                                                  elm153
2001  format(1x,20a4//2x,'E L E M E N T   S T R E S S E S'//             elm154
     1 '  El  Mat  1-Coor  2-Coor    **NS**     **NPHI**',               elm155
     2 '     **MS**     **MPHI**     **QS**'/1x)                         elm156
2002  format(i5,i4,0p2f8.3,5(1x,1p1e10.3))                               elm157
3001  format(' *ERROR* Element',i5,' has zero length.')                  elm158
      end                                                                elm159
c
      subroutine bmat01(bm,shp,sn,cs,recr,wks)                           bma  1
      implicit real*8(a-h,o-z)                                           bma  2
      real*8 bm(5,3)                                                     bma  3
c                                                                        bma  4
c.... nonlinear B-matrix for axisymmetric shells                         bma  5
      bm(1,1) = shp                                                      bma  6
      bm(1,2) = shp*wks                                                  bma  7
      bm(2,1) = 0.5*cs*recr                                              bma  8
      bm(2,2) =-0.5*sn*recr                                              bma  9
      bm(3,3) = shp                                                      bma 10
      bm(4,1) = bm(2,1)*sn*recr                                          bma 11
      bm(4,2) = bm(2,2)*sn*recr                                          bma 12
      bm(4,3) = bm(2,1)                                                  bma 13
      bm(5,2) = shp                                                      bma 14
      bm(5,3) =-0.5                                                      bma 15
c                                                                        bma 16
      end                                                                bma 17
c
      subroutine matl01(d)                                               mat  1
      logical pcomp                                                      mat  2
      character yyy*80,typ*5,type*5                                      mat  3
      real*8    d(6)                                                     mat  4
      common /iofile/ ior,iow                                            mat  5
      common /ydata/  yyy                                                mat  6
      data   type/'beam '/                                               mat  7
c                                                                        mat  8
 1    if(ior.lt.0) then                                                  mat  9
        write(*,3000)                                                    mat 10
        read(*,1000) typ                                                 mat 11
      else                                                               mat 12
        read(ior,1000) typ                                               mat 13
      endif                                                              mat 14
      call pintio(yyy,10)                                                mat 15
      read(yyy,1001,err=2) d                                             mat 16
      if(pcomp(typ,type)) then                                           mat 17
        d(2)  = 0.0                                                      mat 18
        d(13) = 0.0                                                      mat 19
      else                                                               mat 20
        d(13) = 1.0                                                      mat 21
      endif                                                              mat 22
      write(iow,2000) typ,d                                              mat 23
      if(ior.lt.0) write(*,2000) typ,d                                   mat 24
c.... set beam/shell in-plane and bending stiffness values               mat 25
      d(7)  = d(1)*d(3)/(1.0 - d(2)*d(2))                                mat 26
      d(8)  = d(2)*d(7)                                                  mat 27
      d(9)  = d(7)*d(3)*d(3)/12.                                         mat 28
      d(10) = d(2)*d(9)                                                  mat 29
      d(11) = d(6)*d(3)                                                  mat 30
      d(12) = d(3)*d(3)/12.                                              mat 31
      return                                                             mat 32
 2    call perror('PCEL01',yyy)                                          mat 33
      go to 1                                                            mat 34
```

```
c                                                                    mat 35
1000  format(a5)                                                     mat 36
1001  format(7f10.0)                                                 mat 37
2000  format(5x,'Rectangular Beam/Axisymmetric Shell Model'/         mat 38
     x 5x,'Type: ',a5/5x,'Elastic Modulus',g15.4/                    mat 39
     1 5x,'Poisson Ratio',g17.4/5x,'Thickness  ',g19.4/              mat 40
     2 5x,'Normal Load   ',g17.4/                                    mat 41
     3 5x,'Linear(0=l 1=nl)',f9.1/5x,'Mass Density',g18.4/1x)        mat 42
3000  format(' Input: 1. Type (beam or shell)'/                      mat 43
     1        '          2. E nu h press lin rho'/3x,'>',$)          mat 44
      end                                                            mat 45
c
      subroutine modl01(sig,vl,dd,d,sn,cs,sl,recr,wks)               mod  1
      implicit real*8(a-h,o-z)                                       mod  2
      real*8 sig(5),eps(5),vl(3,2),dd(5,5),d(1)                      mod  3
c                                                                    mod  4
c.... elasticity matrix for axisymmetric shells                      mod  5
      dd(1,1) = d(7)                                                 mod  6
      dd(2,2) = d(7)                                                 mod  7
      dd(1,2) = d(8)                                                 mod  8
      dd(2,1) = d(8)                                                 mod  9
      dd(3,3) = d(9)                                                 mod 10
      dd(4,4) = d(9)                                                 mod 11
      dd(3,4) = d(10)                                                mod 12
      dd(4,3) = d(10)                                                mod 13
      dd(5,5) = 5.*d(7)*(1.0 - d(2))/12.0                            mod 14
c...  strains                                                        mod 15
      v1      = 0.5*((vl(1,1) + vl(1,2))*cs - (vl(2,1) + vl(2,2))*sn) mod 16
      eps(1) = (vl(1,2) - vl(1,1))/sl + 0.5*wks*wks                  mod 17
      eps(2) =  v1*recr                                              mod 18
      eps(3) = (vl(3,2) - vl(3,1))/sl                                mod 19
      eps(4) = (v1*sn*recr +0.5*(vl(3,1) + vl(3,2))*cs)*recr         mod 20
      eps(5) = (vl(2,2) - vl(2,1))/sl - 0.5*(vl(3,1) + vl(3,2))      mod 21
c.... stresses                                                       mod 22
      sig(1) = dd(1,1)*eps(1) + dd(1,2)*eps(2)                       mod 23
      sig(2) = dd(2,1)*eps(1) + dd(2,2)*eps(2)                       mod 24
      sig(3) = dd(3,3)*eps(3) + dd(3,4)*eps(4)                       mod 25
      sig(4) = dd(4,3)*eps(3) + dd(4,4)*eps(4)                       mod 26
      sig(5) = dd(5,5)*eps(5)                                        mod 27
c                                                                    mod 28
      end                                                            mod 29
c
      subroutine tran01(s,cs,sn,nst,ndf,itype)                       tra  1
      implicit real*8 (a-h,o-z)                                      tra  2
      real*8 s(nst,1)                                                tra  3
c                                                                    tra  4
c.... itype: 1  Transform matrix s(nst,nst)                          tra  5
c            2  Transform vector s(nst,1)                            tra  6
c                                                                    tra  7
      if(itype.eq.1) then                                            tra  8
        do 12 i = 1,nst,ndf                                          tra  9
          do 11 j = 1,nst                                            tra 10
            t        = s(j,i)*cs - s(j,i+1)*sn                       tra 11
            s(j,i+1) = s(j,i)*sn + s(j,i+1)*cs                       tra 12
            s(j,i  ) = t                                             tra 13
11        continue                                                   tra 14
12      continue                                                     tra 15
        nn = nst                                                     tra 16
      else                                                           tra 17
        nn = 1                                                       tra 18
```

```
      endif                                                  tra 19
c                                                            tra 20
      do 14 i = 1,nst,ndf                                    tra 21
        do 13 j = 1,nn                                       tra 22
          t         = s(i,j)*cs - s(i+1,j)*sn                tra 23
          s(i+1,j) = s(i,j)*sn + s(i+1,j)*cs                 tra 24
          s(i  ,j) = t                                       tra 25
13      continue                                             tra 26
14    continue                                               tra 27
c                                                            tra 28
      end                                                    tra 29
```

(*b*) File PCELM2.FOR contains an element module for two-dimensional analysis of plane strain and axisymmetric behaviour of solids modelled by elastoplastic constitutive behaviour. Only the material model subprograms are included, while the remainder may be found in Chapter 15 of Volume 1.

Name	*Type*	*Description*	*Chapter*
ELMT02	SUBROUTINE	**B**-bar element routine	15
GVC2	SUBROUTINE	Compute volume integrals of shape functions	15
BMAT02	SUBROUTINE	Compute strain displacement matrix	15
STRN02	SUBROUTINE	Compute strains at point	15
STCN02	SUBROUTINE	Compute nodal stress integrals	15
MATL02	SUBROUTINE	Input elastoplastic properties	16
MODL02	SUBROUTINE	Scale stress, tangent moduli	16
ELPL02	SUBROUTINE	Compute stress, tangent moduli	16

```
      subroutine matl02(d,it,ib)                                        mat  1
      implicit real*8 (a-h,o-z)                                         mat  2
      real*8 d(15)                                                      mat  3
      character*24 wa(2),yyy*80                                         mat  4
c.... parameter specification for FEAP materials                        mat  5
      common /hdata/ nh1,nh2                                            mat  6
      common /iofile/ ior,iow                                           mat  7
      common /ydata/ yyy                                                mat  8
c.... output parameters                                                 mat  9
c          d(*)  -  constants for constitutive equation.                mat 10
c          it    -  geometry type (0=plane, 1=axisymmetric)             mat 11
c          ib    -  incompressibility treatment (ib = 0 b-bar,          mat 12
c                     ib = 1 normal b matrix).                          mat 13
c          nh    -  number of history variables needed for each         mat 14
c                     stress point (i.e., at gauss points in element).  mat 15
c                                                                       mat 16
      data wa/' P l a n e   S t r a i n',' A x i s y m m e t r i c'/   mat 17
c                                                                       mat 18
c....  input the material parameters                                    mat 19
  1   if(ior.lt.0) write(*,3000)                                        mat 20
      call pintio(yyy,10)                                               mat 21
      read(yyy,1000,err=101) it,d(4)                                    mat 22
      it  = max(0,min(1,it))                                            mat 23
      ib  = 0                                                           mat 24
      nh  = 9                                                           mat 25
      nh1 = 36                                                          mat 26
      if(ior.lt.0) write(*,3001)                                        mat 27
      call pintio(yyy,10)                                               mat 28
      read(yyy,1001,err=101) ee,xnu,d(11),d(12),d(13)                   mat 29
      d(1)     = ee/(1. - 2.*xnu)/3.0d0                                 mat 30
      d(2)     = ee/(1.+xnu)/2.                                         mat 31
      write(iow,2000) wa(it+1),ee,xnu,d(4),d(11),d(12),d(13)            mat 32
      if(ior.lt.0) write(*,2000) wa(it+1),ee,xnu,d(4),d(11),d(12),d(13) mat 33
      return                                                            mat 34
```

```
101    call perror('PCELM2',yyy)                                          mat 35
       go to 1                                                            mat 36
c....  formats                                                            mat 37
1000   format(i10,f10.0)                                                  mat 38
1001   format(8f10.0)                                                     mat 39
2000   format(2x,'E l a s t i c / P l a s t i c   M a t e r i a l'/      mat 40
     +        2x,'v o n   M i s e s   Y i e l d - ',a24/                  mat 41
     +  10x,'Youngs Modulus (E)',e17.5/                                   mat 42
     +  10x,'Poisson Ratio (nu)',e17.5/                                   mat 43
     +  10x,'Mass Density (rho)',e17.5/                                   mat 44
     +  10x,'Yield Stress       ',e17.5/                                  mat 45
     +   8x,'Linear hardening moduli'/                                    mat 46
     +  10x,'Isotropic Hardening',e16.5/                                  mat 47
     +  10x,'Kinematic Hardening',e16.5/)                                 mat 48
3000   format(' Input: it, rho'/                                          mat 49
     +  4x,'it = 0: Plane'/4x,'it = 1: Axisymmetric'/' >',$)              mat 50
3001   format(' Input: E, nu, Y, H-isotropic, H-kinematic'/3x,'>',$)      mat 51
       end                                                                mat 52
c
       subroutine modl02(d,ul,eps,sig,xsj,ndf,ib)                         mod  1
       implicit real*8 (a-h,o-z)                                          mod  2
       real*8 d(*),ul(ndf,*),eps(4),sig(5)                                mod  3
       common /elcom2/ g(2,4),ad(4,4)                                     mod  4
       common /hdata/  nh1,nh2                                            mod  5
       common h(1)                                                        mod  6
c....  input parameters                                                   mod  7
c           d(*)     -  material parameters                               mod  8
c           eps(4)   -  current strains at point                          mod  9
c           h(*)     -  history terms at point                            mod 10
c           nh       -  number of history terms                           mod 11
c....  ouput parameters                                                   mod 12
c           ad(4,4) -  current material tangent moduli                    mod 13
c           sig(4)  -  stresses at point.                                 mod 14
c           sig(5)  -  yield at point.                                    mod 15
c                                                                         mod 16
c....   compute material moduli and stresses                              mod 17
       call pconsd(ad,16,0.0d0)                                           mod 18
       call elpl02(d,eps,h(nh2).h(nh2+4),h(nh2+8),sig,ul,ndf,ib)          mod 19
       nh2 = nh2 + 9                                                      mod 20
c                                                                         mod 21
c....   multiply by jacobian                                              mod 22
       do 110 i = 1,4                                                     mod 23
         sig(i) = sig(i)*xsj                                              mod 24
         do 100 j = 1,4                                                   mod 25
            ad(i,j) = ad(i,j)*xsj                                         mod 26
100      continue                                                         mod 27
110    continue                                                           mod 28
c                                                                         mod 29
c....  reorder stresses for stress divergence calculations and prints     mod 30
       temp   = sig(4)                                                    mod 31
       sig(4) = sig(3)                                                    mod 32
       sig(3) = sig(2)                                                    mod 33
       sig(2) = temp                                                      mod 34
       end                                                                mod 35
c
       subroutine elpl02(d,eps,epln,alph,ep,sig,ul,ndf,ib)                elp  1
       implicit real*8 (a-h,o-z)                                          elp  2
       real   dm                                                          elp  3
       real*8 d(*),ul(ndf,*),eps(4),sig(5),alph(4),epln(4),psi(4),en(4)   elp  4
       common /eldata/ dm,n,ma,mct,iel,nel                                elp  5
```

```
      common /elcom2/ g(2,4),ad(4,4)                                  elp  6
c                                                                     elp  7
c.... elasto-plastic model with isotropic / kinematic hardening       elp  8
c                                                                     elp  9
c.... input parameters - for t-n                                      elp 10
c        d       - array of material constants                        elp 11
c        eps     - strains (at t-n+1)                                 elp 12
c        sig     - stresses                                           elp 13
c        alph    - back stress                                        elp 14
c        ep      - effective plastic strain                           elp 15
c.... output parameters - at t-n+1                                    elp 16
c        ep      - effective plastic strain                           elp 17
c        sig     - stress tensor                                      elp 18
c        alph    - back stress tensor                                 elp 19
c        ad      - "tangent" matrix                                   elp 20
c                                                                     elp 21
c.... set parameters                                                  elp 22
      data tt/.816496580927726D+00/                                   elp 23
      oneg = d(2)                                                     elp 24
      twog = d(2) + d(2)                                              elp 25
      bulk = d(1)                                                     elp 26
      elam = bulk - twog/3.                                           elp 27
c                                                                     elp 28
c.... compute the trial deviator stress                               elp 29
      treps  = (eps(1) + eps(2) + eps(3))/3.0d0                       elp 30
      do 100 i = 1,3                                                  elp 31
        eps(i) = eps(i) - treps                                       elp 32
100   continue                                                        elp 33
      do 110 i = 1,4                                                  elp 34
        sig(i) = twog*(eps(i) - epln(i))                              elp 35
        psi(i) =       sig(i) - alph(i)                               elp 36
110   continue                                                        elp 37
c                                                                     elp 38
c.... set up elastic tangent                                          elp 39
      do 130 i = 1,3                                                  elp 40
        do 120 j = 1,3                                                elp 41
          ad(i,j) = elam                                              elp 42
120     continue                                                      elp 43
        ad(i,i) = elam + twog                                         elp 44
130   continue                                                        elp 45
      ad(4,4) = oneg                                                  elp 46
c                                                                     elp 47
c.... compute the yield state - J2d                                   elp 48
      if(d(11).gt.0.0d0) then                                         elp 49
        radius = tt*(d(11) + d(12)*ep)                                elp 50
        psitr  = sqrt(psi(1)**2+psi(2)**2+psi(3)**2+2.d0*psi(4)**2)   elp 51
        sig(5) = psitr/tt/d(11)                                       elp 52
c                                                                     elp 53
c.... compute plasticity solution state                               elp 54
        if (psitr.gt.radius) then                                     elp 55
          beta   = 1.d0/(1.d0 + (d(12) + d(13))/(3.d0*oneg))          elp 56
          gamn   = beta*(psitr - radius)/twog                         elp 57
          ep     = ep + tt*gamn                                       elp 58
c                                                                     elp 59
c.... gam3 ensures stress is slightly outside yield surface.          elp 60
          gam1   = gamn*twog                                          elp 61
          gam2   = (d(13)+d(13))*gamn/3.d0                            elp 62
          gam3   = (1.d0 - 1.d-8)*gamn                                elp 63
          sig(5) = (psitr - gam1 - gam2)/tt/d(11)                     elp 64
          do 140 i = 1,4                                              elp 65
```

```
          en(i)    = psi(i)/psitr                                     elp 66
          sig(i)   =  sig(i)  - gam1*en(i)                            elp 67
          alph(i) =  alph(i) + gam2*en(i)                             elp 68
          epln(i) =  epln(i) + gam3*en(i)                             elp 69
140     continue                                                      elp 70
c                                                                     elp 71
c.... plastic modification for tangent                                elp 72
        gam1 = twog*gam1/psitr                                        elp 73
        gam2 = gam1/3.0d0                                             elp 74
        do 160 i = 1,3                                                elp 75
          do 150 j = 1,3                                              elp 76
            ad(i,j) = ad(i,j) + gam2                                  elp 77
 150      continue                                                    elp 78
          ad(i,i) = ad(i,i) - gam1                                    elp 79
 160    continue                                                      elp 80
        ad(4,4) = ad(4,4) - 0.5*gam1                                  elp 81
        gam1 = gam1 - twog*beta                                       elp 82
        do 180 i = 1,4                                                elp 83
          gam2    = gam1*en(i)                                        elp 84
          do 170 j = 1,4                                              elp 85
            ad(i,j) = ad(i,j) + gam2*en(j)                            elp 86
 170      continue                                                    elp 87
 180    continue                                                      elp 88
      endif                                                           elp 89
    endif                                                             elp 90
c                                                                     elp 91
c.... compute trace of strain and add pressure term                   elp 92
      if(ib.eq.0) then                                                elp 93
        treps = 0.0d0                                                 elp 94
        do 190 i = 1,4                                                elp 95
          treps = treps + g(1,i)*ul(1,i) + g(2,i)*ul(2,i)             elp 96
 190    continue                                                      elp 97
      else                                                            elp 98
        treps = treps*3.d0                                            elp 99
      endif                                                           elp100
      press = bulk*treps                                              elp101
      do 200 i = 1,3                                                  elp102
        sig(i) = sig(i) + press                                       elp103
 200  continue                                                        elp104
c                                                                     elp105
      end                                                             elp106
```

(*c*) File PCELM3.FOR contains an element module to solve problems characterized by a Laplace equation in two dimensions. Reactive type loadings are included.

Name	*Type*	*Description*	*Chapter*
ELMT03	SUBROUTINE	Laplace equation element with reactive term	16
COORD	FUNCTION	Compute coordinates	16
FLO03	SUBROUTINE	Compute flux terms	16
STCN03	SUBROUTINE	Compute nodal flux	16

```
      subroutine elmt03(d,ul,xl,ix,tl,s,p,ndf,ndm,nst,isw)                elm  1
      implicit real*8 (a-h,o-z)                                           elm  2
c                                                                         elm  3
c.... two dimensional laplace equation with a reaction term              elm  4
c                                                                         elm  5
      character*4 head,wlab(2)*12,yyy*80                                  elm  6
      integer*2   ix(1)                                                   elm  7
      real        xl(ndm,*),tl(*),dm,aa                                   elm  8
      real*8      d(*),ul(ndf,*),s(nst,nst),p(nst),shp(3,9),              elm  9
     1            sg(9),tg(9),wg(9),coord                                 elm 10
      common /adata/  aa(16000)                                           elm 11
      common /bdata/  head(20)                                            elm 12
      common /cdata/  numnp,numel,nummat,nen,neq                          elm 13
      common /eldata/ dm,n,ma,mct,iel,nel                                 elm 14
      common /iofile/ ior,iow                                             elm 15
      common /ydata / yyy                                                 elm 16
      data wlab/'  p l a n e ','axisymmetric'/                            elm 17
c.... compute the quadrature points                                      elm 18
      if(isw.gt.2) then                                                   elm 19
        l = max(1,min(nel/2,3))                                           elm 20
        call pgauss(l,lint,sg,tg,wg)                                      elm 21
      endif                                                               elm 22
c.... input material properties                                          elm 23
      if(isw.eq.1) then                                                   elm 24
 1      if(ior.lt.0) write(*,3000)                                        elm 25
        call pintio(yyy,10)                                               elm 26
        read(yyy,1000,err=110) (d(i),i=1,6),nn,kat                        elm 27
        if(kat.ne.2) kat=1                                                elm 28
        nn = max(-1,min(1,nn))                                            elm 29
        write(iow,2000) (d(i),i=1,6),nn,wlab(kat)                         elm 30
        if(ior.lt.0) write(*,2000) (d(i),i=1,6),nn,wlab(kat)              elm 31
        d(7) = nn                                                         elm 32
        d(8) = kat                                                        elm 33
        d(9) = d(2)*d(3)                                                  elm 34
        return                                                            elm 35
110     call perror('PCELM3',yyy)                                         elm 36
        go to 1                                                           elm 37
c.... compute conductivity matrix and residual                           elm 38
      elseif(mod(isw,3).eq.0) then                                        elm 39
        nn  = d(7)                                                        elm 40
        kat = d(8)                                                        elm 41
        do 330 l=1,lint                                                   elm 42
          call shape(sg(l),tg(l),xl,shp,xsj,ndm,nel,ix,.false.)           elm 43
          xsj = xsj*wg(l)                                                 elm 44
          if(kat.eq.2) xsj = xsj*coord(xl,shp,ndm,nel)                    elm 45
```

```
          call flo03(1.0d0,shp,ul,q1,q2,qm,uu,ndf,nel)                 elm 46
          if(nn.eq.0) then                                             elm 47
            qq = dm*d(4)                                               elm 48
            dq = 0.0                                                   elm 49
          elseif(nn.eq.1) then                                         elm 50
            qq = dm*d(4)*exp(d(5)*uu)                                  elm 51
            dq = d(5)                                                  elm 52
          elseif(uu.ne.0.0d0) then                                     elm 53
            qq = dm*d(4)*exp(d(5)-d(5)*d(6)/uu)                        elm 54
            dq = d(5)*d(6)/uu**2                                       elm 55
          else                                                         elm 56
            write(*,*) ' ** ELMT03 ERROR ** T = 0.0: stop'             elm 57
            stop                                                       elm 58
          endif                                                        elm 59
          j1 = 1                                                       elm 60
          do 320 j=1,nel                                               elm 61
            a1    = d(1)*shp(1,j)*xsj                                  elm 62
            a2    = d(1)*shp(2,j)*xsj                                  elm 63
            a3    = qq*shp(3,j)*xsj                                    elm 64
c....   compute residual                                               elm 65
            p(j1) = p(j1) + a1*q1 + a2*q2 - a3                         elm 66
            a3    = a3*dq                                              elm 67
c....   compute tangent                                                elm 68
            if(isw.eq.3) then                                          elm 69
              i1  = 1                                                  elm 70
              do 310 i=1,nel                                           elm 71
                s(i1,j1) = s(i1,j1)+a1*shp(1,i)+a2*shp(2,i)+a3*shp(3,i) elm 72
                i1 = i1 + ndf                                          elm 73
310           continue                                                 elm 74
            endif                                                      elm 75
            j1 = j1 + ndf                                              elm 76
320       continue                                                     elm 77
330     continue                                                       elm 78
c.... compute the flows in each element                                 elm 79
      elseif(isw.eq.4) then                                            elm 80
        call shape(0.0d0,0.0d0,xl,shp,xsj,ndm,nel,ix,.false.)          elm 81
        call flo03(d,shp,ul,q1,q2,qm,uu,ndf,nel)                       elm 82
        rr  = coord(xl(1,1),shp,ndm,nel)                               elm 83
        zz  = coord(xl(2,1),shp,ndm,nel)                               elm 84
        mct = mct - 1                                                  elm 85
        if(mct.lt.0) then                                              elm 86
          mct = 50                                                     elm 87
          write(iow,2001) head                                         elm 88
          if(ior.lt.0) write(*,2001) head                              elm 89
        endif                                                          elm 90
        write(iow,2002) n,ma,rr,zz,q1,q2,qm,uu                         elm 91
        if(ior.lt.0) write(*,2002) n,ma,rr,zz,q1,q2,qm,uu              elm 92
c.... compute heat capacity (mass) matrix                               elm 93
      elseif(isw.eq.5) then                                            elm 94
        do 520 l=1,lint                                                elm 95
          call shape(sg(l),tg(l),xl,shp,xsj,ndm,nel,ix,.true.)         elm 96
          xsj = xsj*wg(l)                                              elm 97
          if(kat.eq.2) xsj = xsj*coord(xl,shp,ndm,nel)                 elm 98
          j1 = 1                                                       elm 99
          do 510 j=1,nel                                               elm100
            p(j1) = p(j1) + d(9)*shp(3,j)*xsj                          elm101
            j1 = j1 + ndf                                              elm102
510       continue                                                     elm103
520     continue                                                       elm104
c.... compute the nodal flow values                                     elm105
```

```
      elseif(isw.eq.8) then                                                elm106
        call stcn03(ix,d,xl,ul,shp,aa,aa(numnp+1),ndf,ndm,nel,             elm107
     1              numnp,sg,tg,wg,lint)                                   elm108
      endif                                                                elm109
c.... formats                                                               elm110
1000  format(6f10.0,2i10)                                                  elm111
2000  format(3x,'Laplace Element with Reaction Loading'//                  elm112
     1 4x,'Conductivity  ',e12.4/4x,'Specific Heat ',e12.4/                elm113
     x 4x,'Density       ',e12.4/4x,'Load Amplitude',e12.4/                elm114
     2 4x,'Reaction Exp. ',e12.4/4x,'Ambient Temp. ',e12.4/                elm115
     3 4x,'n - (Temp**n) ',i5/   4x,a12,' analysis')                       elm116
2001  format(1x,20a4//'  L a p l a c e   E q u a t i o n'//                elm117
     1 ' elem  mat    1-coord    2-coord    1-flow     2-flow',            elm118
     2 '    max flow     U-value'/)                                        elm119
2002  format(2i5,0p2f11.3,1p4e11.3)                                        elm120
3000  format(' Input:K, c, rho, Q, r, Ta, nn, geom(1=plane,2=axisym)'      elm121
     1       /3x,'>',$)                                                    elm122
      end                                                                  elm123
c
      double precision function coord(xl,shp,ndm,nel)                     coo  1
      real   xl(ndm,*)                                                     coo  2
      real*8 shp(3,*)                                                      coo  3
c                                                                          coo  4
      coord = 0.d0                                                         coo  5
      do 100 i = 1,nel                                                     coo  6
        coord = coord + shp(3,i)*xl(1,i)                                   coo  7
100   continue                                                             coo  8
      end                                                                  coo  9
c
      subroutine flo03(d,shp,ul,q1,q2,qm,uu,ndf,nel)                       flo  1
      implicit real*8 (a-h,o-z)                                            flo  2
      real*8   shp(3,*),ul(ndf,*)                                          flo  3
c.... compute flows at current point                                       flo  4
      q1 = 0.0d0                                                           flo  5
      q2 = 0.0d0                                                           flo  6
      uu = 0.0d0                                                           flo  7
      do 100 i = 1,nel                                                     flo  8
        q1 = q1 - d*shp(1,i)*ul(1,i)                                       flo  9
        q2 = q2 - d*shp(2,i)*ul(1,i)                                       flo 10
        uu = uu +   shp(3,i)*ul(1,i)                                       flo 11
100   continue                                                             flo 12
      qm =    sqrt(q1*q1 + q2*q2)                                          flo 13
      end                                                                  flo 14
c
      subroutine stcn03(ix,d,xl,ul,shp,dt,st,ndf,ndm,nel,numnp,            stc  1
     1                  sg,tg,wg,lint)                                     stc  2
      implicit real*8 (a-h,o-z)                                            stc  3
c....  project values to nodes                                             stc  4
      integer*2 ix(*)                                                      stc  5
      real      dt(numnp),st(numnp,*),xl(ndm,*)                            stc  6
      real*8    d(*),ul(ndf,*),sg(*),tg(*),wg(*),shp(3,*)                  stc  7
      do 110 l = 1,lint                                                    stc  8
        call shape(sg(l),tg(l),xl,shp,xsj,ndm,nel,ix,.false.)              stc  9
        call flo03(d,shp,ul,q1,q2,qm,uu,ndf,nel)                           stc 10
        do 100 ii = 1,nel                                                  stc 11
          ll = abs(ix(ii))                                                 stc 12
          if(ll.gt.0) then                                                 stc 13
            xsji     = xsj*shp(3,ii)*wg(l)                                 stc 14
            dt(ll)   = dt(ll)    + xsji                                    stc 15
            st(ll,1) = st(ll,1) + q1*xsji                                  stc 16
```

```
         st(ll,3) = st(ll,3) + q2*xsji                                    stc 17
         st(ll,4) = st(ll,4) + qm*xsji                                    stc 18
         st(ll,5) = st(ll,5) + ul(1,ii)*xsji                              stc 19
       endif                                                              stc 20
100    continue                                                           stc 21
110  continue                                                             stc 22
     end                                                                  stc 23
```

(*d*) File PCELM4.FOR contains an element module for a one-, two- or three-dimensional truss with an elastoplastic constitutive behaviour.

Name	*Type*	*Description*	*Chapter*
ELMT04	SUBROUTINE	General truss element	16
MODL04	SUBROUTINE	Elastoplastic routine	16

```
      subroutine elmt04(d,u,x,ix,t,s,p,ndf,ndm,nst,isw)                 elm  1
c.... any dimensional truss element routine                             elm  2
      implicit real*8 (a-h,o-z)                                         elm  3
      real   x(ndm,1),dm                                                elm  4
      real*8 d(1),u(ndf,1),s(nst,1),p(1),db(3),dx(3),xx(3)              elm  5
      character*4 yyy*80                                                elm  6
      common /cdata/ numnp,numel,nummat,nen,neq                         elm  7
      common /eldata/ dm,n,ma,mct,iel,nel                               elm  8
      common /hdata/  nh1,nh2                                           elm  9
      common /iofile/ ior,iow                                           elm 10
      go to (1,2,3,3,3,3,2,2),isw                                       elm 11
c.... input material properties                                         elm 12
1     if(ior.lt.0) write(*,3000)                                        elm 13
      call pintio(yyy,10)                                               elm 14
      read(yyy,1000,err=110) (d(i),i=1,6)                               elm 15
      write(iow,2000) (d(i),i=1,6)                                      elm 16
      d( 7) = d(1)*d(2)                                                 elm 17
      d( 8) = d(4)*d(2)                                                 elm 18
      d( 9) = d(5)*d(2)                                                 elm 19
      d(10) = d(6)*d(2)                                                 elm 20
      d(11) = d(3)*d(2)                                                 elm 21
      call pconsr(xx,3,0.0)                                             elm 22
      write(iow,2000) (d(i),i=1,6)                                      elm 23
      if(ior.lt.0) write(*,2000) (d(i),i=1,6)                           elm 24
      nh1 = 3                                                           elm 25
      return                                                            elm 26
110   call perror('PCELM4',yyy)                                         elm 27
      go to 1                                                           elm 28
2     return                                                            elm 29
c.... compute element arrays                                            elm 30
3     xl  = 0.0                                                         elm 31
      eps = 0.0                                                         elm 32
      do 30 i = 1,ndm                                                   elm 33
        dx(i) = x(i,2) - x(i,1)                                         elm 34
        xl = xl + dx(i)**2                                              elm 35
        eps = eps + dx(i)*(u(i,2)-u(i,1))                               elm 36
        xx(i) = (x(i,2) + x(i,1))/2.                                    elm 37
30    continue                                                          elm 38
      eps = eps/xl                                                      elm 39
      if(mod(isw,3).eq.0) then                                          elm 40
        call modl04(d,eps, sig,ad)                                      elm 41
c.... form a residual                                                   elm 42
        sig = sig/sqrt(xl)                                              elm 43
        do 31 i = 1,ndf                                                 elm 44
          p(i) = dx(i)*sig                                              elm 45
          p(i+ndf) = -p(i)                                              elm 46
31      continue                                                        elm 47
c.... compute tangent stiffness                                         elm 48
        if(isw.eq.3) then                                               elm 49
          xl = xl*sqrt(xl)                                              elm 50
          do 32 i = 1,ndm                                               elm 51
```

```
          db(i) = ad*dx(i)                                         elm 52
          dx(i) = dx(i)/xl                                         elm 53
32      continue                                                   elm 54
        i1 = 0                                                     elm 55
        do 36 ii = 1,2                                             elm 56
          j1 = i1                                                  elm 57
          do 35 jj = ii,2                                          elm 58
            do 33 i = 1,ndm                                        elm 59
            do 33 j = 1,ndm                                        elm 60
              s(i+i1,j+j1) = db(i)*dx(j)                           elm 61
33          continue                                               elm 62
            j1 = j1 + ndf                                          elm 63
            do 34 j = 1,ndm                                        elm 64
              dx(j) = -dx(j)                                       elm 65
34          continue                                               elm 66
35        continue                                                 elm 67
          i1 = i1 + ndf                                            elm 68
36      continue                                                   elm 69
        do 37 i = 1,ndm                                            elm 70
        do 37 j = 1,ndm                                            elm 71
          s(i+ndf,j) = s(j,i+ndf)                                  elm 72
37      continue                                                   elm 73
      endif                                                        elm 74
c.... output stress and strain in element                          elm 75
     elseif(isw.eq.4) then                                         elm 76
       call mod104(d,eps, sig,ad)                                  elm 77
       mct = mct - 1                                               elm 78
       if(mct.le.0) then                                           elm 79
         call prthed(iow)                                          elm 80
         write(iow,2001)                                           elm 81
         if(ior.lt.0) write(*,2001)                                elm 82
         mct = 50                                                  elm 83
       endif                                                       elm 84
       write(iow,2002) n,ma,xx,sig,eps                             elm 85
       if(ior.lt.0) write(*,2002) n,ma,xx,sig,eps                  elm 86
c.... compute element lumped mass matrix                            elm 87
     elseif(isw.eq.5) then                                         elm 88
       sm = d(11)*sqrt(xl)/2.0                                     elm 89
       do 52 i = 1,ndm                                             elm 90
         p(i    ) = sm                                             elm 91
         p(i+ndf) = sm                                             elm 92
52     continue                                                    elm 93
     endif                                                         elm 94
c.... formats                                                       elm 95
1000 format(8f10.0)                                                elm 96
2000 format(5x,'t r u s s    e l e m e n t'//10x,'Modulus  =',e12.5/   elm 97
    1  10x,'Area     =',e12.5/10x,'Density  =',e12.5/              elm 98
    1  10x,'Yield    =',e12.5/10x,'Iso. Hard=',e12.5/              elm 99
    +  10x,'Kin. Hard=',e12.5)                                     elm100
2001 format(5x,'t r u s s    e l e m e n t'//' elem mate',         elm101
    1 4x,'1-coord',4x,'2-coord',4x,'3-coord',5x,'force',7x,'strain')  elm102
2002 format(2i5,3f11.4,2e13.5)                                     elm103
3000 format(' Input: E, A, rho, Y, H-iso, H-Kin'/3x,'>',$)         elm104
     end                                                           elm105
c
     subroutine mod104(d,eps, sig,ad)                              mod  1
     real*8 d(10), eps, sig, ad, yld, hi, sum, gam, h              mod  2
     common /hdata/  nh1,nh2                                       mod  3
     common h(1)                                                   mod  4
c.... trial stress                                                  mod  5
```

```
      ad  = d(7)                                                        mod  6
      sig = d(7)*(eps - h(nh1))                                         mod  7
      if(d(8).gt.0.0d0) then                                            mod  8
c.... compute plastic corrections                                       mod  9
        yld = d(8) + d(9)*h(nh1+2)                                      mod 10
        phi = abs(sig - h(nh1+1))                                       mod 11
        if(phi.gt.yld) then                                             mod 12
          sum        = d(7) + d(9) + d(10)                              mod 13
          gam        = (phi-yld)/sum                                    mod 14
          sig        = sig - d(7)*gam                                   mod 15
          ad         = ad -  d(7)**2/sum                                mod 16
c.... update the history terms                                          mod 17
          sum        = gam*(sig - h(nh1+1))/phi                         mod 18
          h(nh1  )   = h(nh1  ) + gam*sum                               mod 19
          h(nh1+1)   = h(nh1+1) + gam*sum*d(10)                         mod 20
          h(nh1+2)   = h(nh1+2) + gam                                   mod 21
        endif                                                           mod 22
      endif                                                             mod 23
      end                                                               mod 24
```

References

1. O. C. ZIENKIEWICZ and R. L. TAYLOR, *The Finite Element Method*, 4th ed., Vol. 1., McGraw-Hill, London, 1989.
2. *Microsoft Fortran Reference*, Version 5.0, Microsoft Corporation, 1989.
3. J.H. WILKINSON and C. REINSCH, *Linear Algebra. Handbook for Automatic Computation*, Vol. II, Springer-Verlag, Berlin, 1981.
4. K.-J. BATHE and E. L. WILSON, *Numerical Methods in Finite Element Analysis*, Prentice-Hall, Englewood Cliffs, N.J., 1976.
5. L. COLLATZ, *Functional Analysis and Numerical Mathematics*, Academic Press, New York, 1966.
6. H. MATTHIES and G. STRANG, 'The solution of nonlinear finite element equations', *Int. J. Num. Meth. Eng.*, **14**, 1613–26, 1979.
7. B. N. PARLETT, *The Symmetric Eigenvalue Problem*, Prentice-Hall, Englewood Cliffs, N.J., 1980.
8. *EISPACK–The Eigensystem Solution Package*, Argonne National Laboratories, Ill.
9. E. STEIN, W. WAGNER and P. WRIGGERS, 'Postbuckling analysis of stability problems with nonlinear contact constraints', in *Finite Element Methods for Nonlinear Problems* (eds. K.-J. Bathe, P. Bergan and W. Wunderlich), Springer-Verlag, Berlin, 1986.
10. P. WRIGGERS, W. WAGNER and E. STEIN, 'Algorithms for non-linear contact constraints with application to stability problems of rod and shells', *Comp. Mech.*, **2**, 215–30, 1987.
11. J. C. SIMO and R. L. TAYLOR, 'Consistent tangent operators for rate independent elasto-plasticity', *Comp. Meth. Appl. Mech. Eng.*, **48**, 101–18, 1985.
12. C. A. ANDERSON and O. C. ZIENKIEWICZ, 'Spontaneous ignition: finite element solutions for steady and transient conditions', *Trans. ASME, J. Heat Transfer*, 398–404, 1974.
13. *Graphics Development Toolkit*, IBM Personal Computer Software, Boca Raton, Fla., 1984.

Author Index

Numbers in bold type refer to the list of references at the end of each chapter

Subject Index

Words in capitals, such as BOUN, usually refer to the finite element macro programming language, Chapter 16.